Non-conventional Yeasts: from Basic Research to Application

Andriy Sibirny
Editor

Non-conventional Yeasts: from Basic Research to Application

Springer

Editor
Andriy Sibirny
Department of Molecular Genetics
and Biotechnology, Institute of Cell Biology
NAS of Ukraine
Lviv, Ukraine

Department of Microbiology and Biotechnology
University of Rzeszow
Rzeszow, Poland

ISBN 978-3-030-21109-7 ISBN 978-3-030-21110-3 (eBook)
https://doi.org/10.1007/978-3-030-21110-3

This Springer imprint is published by the registered company Springer Nature Switzerland AG.
The registered company address is: Gewerbestrasse 11, 6330 Cham, Switzerland

Contents

Chapter 1
Utilization of α-Glucosidic Disaccharides by *Ogataea* (*Hansenula*) *polymorpha*: Genes, Proteins, and Regulation

Tiina Alamäe, Katrin Viigand, and Kristina Põšnograjeva

Abstract Utilization of α-glucosidic sugars such as maltose, maltotriose, isomaltose and sucrose has been extensively studied in a conventional yeast *Saccharomyces cerevisiae* as many important processes such as baking, brewing, and bioethanol production rely on fermentation of these sugars. In 1998, a non-conventional yeast *Ogataea* (formerly *Hansenula*) *polymorpha* was reported to grow on α-glucosidic disaccharides maltose and sucrose using intracellular α-glucosidase for their hydrolysis. Later on, the list of α-glucosidic sugars assimilated by *O. polymorpha* and hydrolyzed by its α-glucosidase was extended by adding maltotriose, isomaltose, palatinose, maltulose, and some others. In this chapter, we review the data on genetics, genomics, transport, and intracellular hydrolysis of α-glucosidic sugars in *O. polymorpha*. We also address evolution of yeast α-glucosidases and regulation of α-glucosidase and permease genes. Relevant data on other yeasts, mostly on *S. cerevisiae*, are used for comparison.

Keywords Maltose · Methylotrophic yeast · Sugar transport · Gene cluster · Genome mining

1.1 Introduction

Yeasts prefer sugars over other carbon sources and thrive in sugar-rich environment. In addition to glucose and fructose, many yeasts assimilate α-glucosidic oligosaccharides. α-Glucosidic oligosaccharides, maltose, isomaltose, and maltotriose (Fig. 1.1), emerge from degradation of starch and glycogen by amylases (Janecek 2009). Many plants (such as sugarcane and sugar beet) and berries contain a big amount of sucrose, a disaccharide of glucose and fructose. Sucrose is

T. Alamäe (✉) · K. Viigand · K. Põšnograjeva
Department of Genetics, Institute of Molecular and Cell Biology, University of Tartu, Tartu, Estonia
e-mail: tiina@alamae.eu

A. Sibirny (ed.), *Non-conventional Yeasts: from Basic Research to Application*,
https://doi.org/10.1007/978-3-030-21110-3_1

also synthesized by cyanobacteria and proteobacteria (Lunn 2002). Importantly, sucrose can be converted to isomers by enzymes of many organisms including plants, yeasts, filamentous fungi, bacteria, and even insects as shown in (Lee et al. 2011) and references therein. So, turanose, palatinose, and maltulose (Fig. 1.1) present in honey are isomerization products of sucrose. Palatinose is currently enzymatically produced from sucrose at a large scale and advertised as a novel healthy sugar with low glycemic index and no cariogenic effect (Sawale et al. 2017). Melezitose (Fig. 1.1) that is also present in honey is a main constituent of aphid honeydew (Daudé et al. 2012). Fig. 1.1 shows α-glucosidic sugars assimilated by a methylotrophic yeast *O. polymorpha* (Viigand 2018).

A question arises: Why do yeast communities have members that assimilate both methanol and disaccharides? The answer may lie in the natural habitat of these yeasts. For instance, the species of *Ogataea* have been isolated from spoiled orange juice, rotting plants, plant leaf surfaces and exudates, soil samples, and insect gut (Morais et al. 2004; Limtong et al. 2008; Naumov et al. 2017). These habitats certainly contain α-glucosidic sugars but may also contain methanol. In the soil, methanol is produced as a result of pectin and lignin degradation (Dorokhov et al. 2015). Furthermore, even living plant leaves emit methanol due to the recycling of pectin and lignin of the cell wall (Keppler et al. 2006). Thus, it is not surprising that yeast species metabolizing both disaccharides and methanol are found in nature.

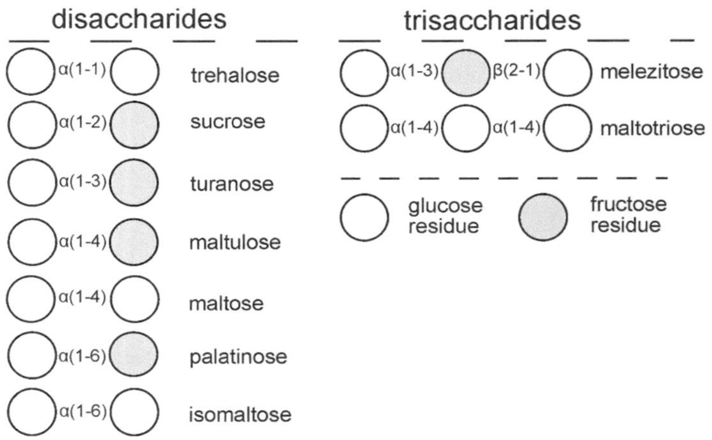

Fig. 1.1 α-Glucosidic di- and trisaccharides and the linkages between the sugar residues. (Viigand 2018)

1.2 *MAL* Clusters in Yeast Genomes

1.2.1 Saccharomyces cerevisiae

Metabolism of maltose has been thoroughly studied in *Saccharomyces* yeasts as they are commonly used in brewing – a process largely based on fermentation of maltose and maltotriose. Indeed, the beer wort produced by grain starch hydrolysis contains 50–60% of maltose and 15–20% maltotriose. During brewing, these oligosaccharides are transported into the yeast cell, hydrolyzed to glucose by maltases and isomaltases, and fermented to ethanol and CO_2 (Stewart 2016). In *S. cerevisiae*, the genes for maltose metabolism are genomically clustered comprising five *MAL* clusters, *MAL1*, *MAL2*, *MAL3*, *MAL4*, and *MAL6*, located near the telomeres of chromosomes VII, III, II, XI, and VIII (Brown et al. 2010; Charron et al. 1986; Needleman 1991; Vanoni et al. 1989). Each *MAL* cluster in *S. cerevisiae* consists of three genes (Fig. 1.2): *MALx1* (maltose permease gene), *MALx2* (maltase gene), and *MALx3* (*MAL*-activator gene) (Charron et al. 1986; Needleman 1991; Chang et al. 1988; Dubin et al. 1985). The “x” refers to the number of the cluster.

It should be noted that genomic clustering of functionally related genes is quite exceptional in eukaryotes. In addition to *MAL* clusters, gene clusters for utilization of galactose, allantoin, and nitrate are known in yeasts and filamentous fungi (Ávila et al. 2002; Kunze et al. 2014; Slot and Rokas 2010; Wong and Wolfe 2005).

1.2.2 Ogataea polymorpha

The maltase structural gene (*MAL1*) of *O. polymorpha* CBS 4732 was isolated from a genomic library of this strain in 2001 (Liiv et al. 2001). An open reading frame of 1695 bp encoding a 564 aa protein having 58% identity with *Candida albicans* maltase was characterized in the library clone p51 (Liiv et al. 2001). A scheme of the genomic insert of the p51 plasmid is shown in Fig.1.3.

Inspection of *O. polymorpha* CBS 4732 genomic library clones sequenced in the Génolevures project (Blandin et al. 2000; Feldmann 2000) identified several clones

Fig. 1.2 Composition of the *S. cerevisiae MAL3* cluster. *MAL31*, α-glucoside permease gene; *MAL32*, maltase gene; *MAL33*, *MAL*-activator gene. The lower panel depicts **position** of the *MAL* cluster (a dark vertical stripe) in chromosome 2 of *S. cerevisiae* S288C. Subtelomeric regions (50 kbp from the chromosome end) are shown in gray. (Viigand 2018)

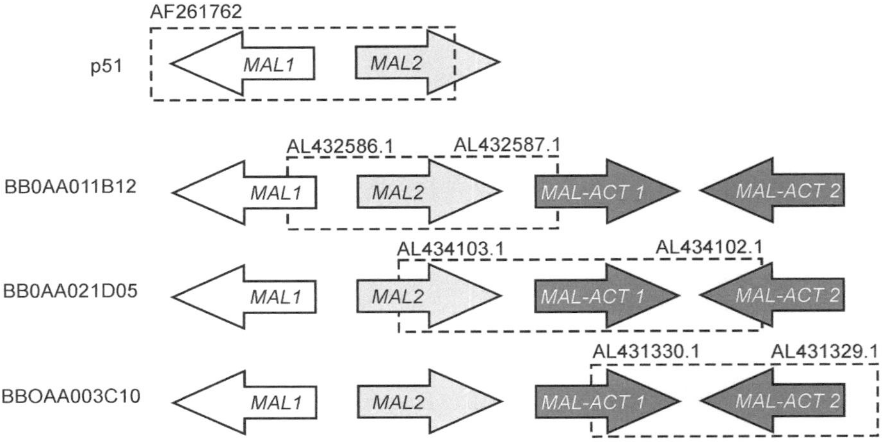

Fig. 1.3 Composition of the *O. polymorpha MAL* locus. Genomic insert in the Génolevures library clones is shown within the frame. Respective GenBank numbers of sequences belonging to *MAL* genes are shown above the frames. *MAL1* gene encodes a maltase (α-glucosidase), *MAL2* gene encodes an α-glucoside permease, and *MAL-ACT* genes encode putative *MAL*-activators. (Modified from Viigand 2018)

(BB0AA021D05, BB0AA011B12, and BB0AA003C10) (see Fig. 1.3) that contained fragments of *MAL* genes (Viigand 2018; Viigand et al. 2005).

Further sequencing of the Génolevures clone inserts revealed the composition of *O. polymorpha MAL* locus (Fig. 1.3). Similar to *S. cerevisiae*, a bidirectional promoter region was identified between the *MAL1* and *MAL2* genes of *O. polymorpha*. Two hypothetical transcriptional activator genes, *MAL*-activator 1 (*MAL-ACT 1*) and *MAL*-activator 2 (*MAL-ACT 2*), were detected next to the *MAL2* permease gene. Full-length sequence of the *MAL* locus of *O. polymorpha* CBS 4732 is accessible in the GenBank under number MH252366.

As the *MAL* loci of *S. cerevisiae* are subtelomeric, chromosomal location of the *O. polymorpha MAL* locus was inspected. The first *O. polymorpha* genome – of strain CBS 4732 – was sequenced in 2003 (Ramezani-Rad et al. 2003), but the sequence is not yet public. Therefore, the genome of *O. polymorpha* strain NCYC 495 leu1.1 (Grigoriev et al. 2014) was assayed instead. Comparison of genomic sequences of *MAL* clusters from *O. polymorpha* strains NCYC 495 leu1.1 and CBS 4732 revealed their identity, and the *MAL* cluster in *O. polymorpha* NCYC 495 leu1.1 was proven as not subtelomeric (Viigand 2018; Viigand et al. 2018).

Gene disruption experiments confirmed that *MAL1* and *MAL2* genes are, respectively, responsible for the hydrolysis and transport of α-glucosidic sugars in *O. polymorpha* (Viigand and Alamäe 2007; Viigand et al. 2016). Although hypothetical α-glucoside permease and α-glucosidase genes were recently detected outside the *O. polymorpha MAL* cluster, they were considered unnecessary for assimilation of α-glucosidic sugars (Viigand et al. 2018).

We conclude that *O. polymorpha* has a single *MAL* locus that consists of four genes coding for maltase MAL1 (syn. α-glucosidase, maltase-isomaltase), α-glucoside permease MAL2, and two putative MAL-activators. The functionality and role of *MAL*-activators have still to be clarified. However, there is indirect support of functionality for at least *MAL*-activator 1 gene. Namely, the promoter of this gene was regulated by carbon sources in the same way as the promoters of *MAL1* and *MAL2* genes (Viigand and Alamäe 2007). A four-gene non-telomeric *MAL* cluster similar to that of *O. polymorpha* was also detected in *O. parapolymorpha* which was formerly named *O. polymorpha* DL-1 (Viigand et al. 2018; Ravin et al. 2013).

1.2.3 Other Non-conventional Yeasts

Besides *O. polymorpha*, *MAL* clusters have been revealed in many yeasts such as *Lodderomyces elongisporus*, *Torulaspora delbrueckii*, *Meyerozyma guilliermondii*, *Cyberlindnera fabianii*, *Debaryomyces hansenii*, *Lipomyces starkeyi* (Viigand et al. 2018), *Scheffersomyces stipitis* (Jeffries and Van Vleet 2009), *Kluyveromyces lactis* (Fairhead and Dujon 2006; Leifso et al. 2007), and a filamentous fungus *Aspergillus oryzae* (Hasegawa et al. 2010). A recent genome mining study (Viigand et al. 2018) discovered the highest number of *MAL* clusters in *L. starkeyi* (Fig. 1.4), whereas no *MAL* clusters were found in *Blastobotrys adeninivorans* and *Schizosaccharomyces pombe*. Figure 1.4 illustrates the composition and number of *MAL* clusters in the genomes of non-conventional yeasts.

1.3 Transport of α-Glucosidic Sugars by *O. polymorpha*

Permeases responsible for maltose uptake have been experimentally characterized in *Saccharomyces* strains (Cheng and Michels 1989; Cheng and Michels 1991), *Torulaspora delbrueckii* (Alves-Araújo et al. 2004), *Candida utilis* (Peinado et al. 1987), *Schizosaccharomyces pombe* (Reinders and Ward 2001), and *O. polymorpha* (Viigand and Alamäe 2007).

The putative maltose permease gene *MAL2* from the *MAL* cluster of *O. polymorpha* was cloned in 2005. The MAL2 protein (582 amino acids) deduced from the respective gene had 51 and 57% identity to respective hypothetical maltose permeases of *Candida albicans* and *Debaryomyces hansenii* (Viigand et al. 2005). Inspection of genomes of a set of non-conventional yeasts thereafter revealed closer homologues of the MAL2: potential α-glucoside permeases of *O. parapolymorpha* (87% sequence identity to MAL2) and *Cyberlindnera fabianii* (67% identity to MAL2). Currently, the AGT1 protein of *S. cerevisiae* is the closest counterpart (37% sequence identity) of *O. polymorpha* MAL2 among experimentally studied permeases (Viigand et al. 2018). The *MAL2* disruption mutant of *O. polymorpha* lost the

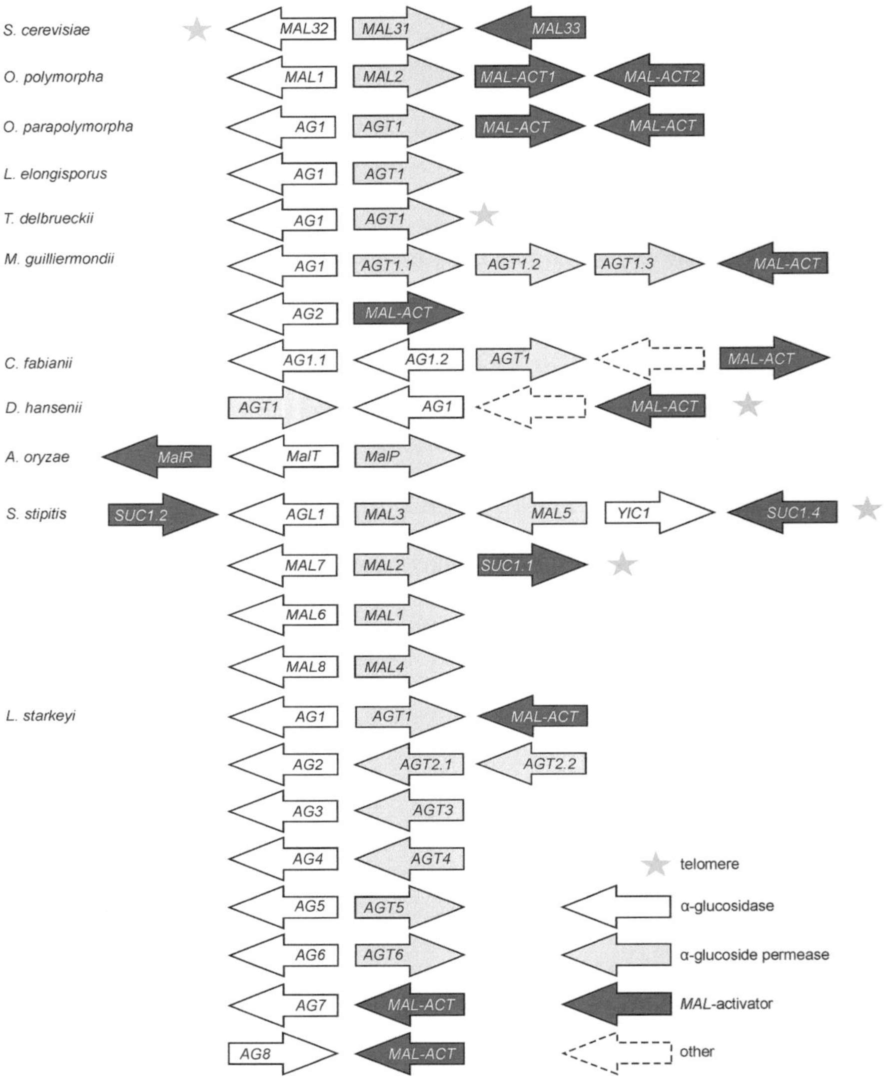

Fig. 1.4 *MAL* loci of some non-conventional yeasts (Viigand et al. 2018). Genes (potentially) encoding α-glucosidases (AG), α-glucoside transporters (AGT), and *MAL*-activators (*MAL*-ACT) are shown with different shading. Accession numbers and annotation data of the *AG* and *AGT* genes are given in Tables S1 and S2 of (Viigand et al. 2018)

ability to grow on maltose and sucrose, and complementation of the mutant with the *MAL2* gene on a plasmid restored the growth on these disaccharides (Viigand and Alamäe 2007). The MAL2 of *O. polymorpha* was also functional in *S. cerevisiae* (Viigand et al. 2005).

Similar to α-glucoside transporters of other yeasts (Reinders and Ward 2001; Stambuk et al. 2000; Stambuk and de Araujo 2001), the *O. polymorpha* MAL2 functioned as an energy-dependent proton symport (Viigand and Alamäe 2007). In agreement with that, the transport by MAL2 was highly sensitive to protonophores and energy uncouplers and had the highest activity at pH 5.0 (Viigand and Alamäe 2007).

Most α-glucoside transporters of *S. cerevisiae* (e.g., MAL61 and MAL31) have a narrow substrate range – they transport just maltose and turanose. In contrast to these transporters, the AGT1 permease of *S. cerevisiae* is highly promiscuous – it transports sucrose, maltose, turanose, PNPG (*p*-nitrophenyl-α-D-glucopyranoside), trehalose, α-methylglucoside, maltotriose, isomaltose, palatinose, and melezitose (Stambuk and de Araujo 2001; Han et al. 1995; Hollatz and Stambuk 2001; Day et al. 2002a). Notably, PNPG can be used as a chromogenic substrate to measure the activity of this type of permeases (Viigand and Alamäe 2007; Hollatz and Stambuk 2001). A recent study (Kulikova-Borovikova et al. 2018) shows that PNPG hydrolysis by the cells can also be used to evaluate permeabilization status of cell membrane in the case of PNPG-transporting yeasts (Kulikova-Borovikova et al. 2018).

In the first study (Viigand and Alamäe 2007), the MAL2 permease was reported to transport maltose, sucrose, turanose, trehalose, maltotriose, and PNPG. Further assay (Viigand et al. 2016) revealed that the MAL2 permease is also required for the growth of *O. polymorpha* on maltulose, melezitose, palatinose, and isomaltooligosaccharides. Because of the relaxed substrate specificity, the *O. polymorpha* MAL2 resembles the AGT1 of *S. cerevisiae* (Stambuk and de Araujo 2001; Han et al. 1995; Hollatz and Stambuk 2001) and should be considered an α-glucoside transporter. Even though the MAL2 is indispensable for the growth of *O. polymorpha* on trehalose (an α-1,1-linked disaccharide of glucose), the α-glucosidase (MAL1) protein does not hydrolyze this sugar (Liiv et al. 2001). It was shown later that for the hydrolysis of trehalose, *O. polymorpha* uses a specific enzyme – trehalase (Ishchuk et al. 2009).

The MAL2 transports PNPG with high affinity (K_m 0.51 mM) (Viigand and Alamäe 2007). The K_m value of the AGT1 permease of *S. cerevisiae* for PNPG is ~3 mM (Han et al. 1995). The MALx1 transporters from *S. cerevisiae* do not transport PNPG (Hollatz and Stambuk 2001). Inhibition of PNPG transport by various α-glucosidic substrates was used to evaluate affinity of the MAL2 permease of *O. polymorpha* for these substrates (Viigand and Alamäe 2007). Sucrose, maltose, trehalose, maltotriose, turanose, and α-methylglucoside (α-MG, also methyl-α-D-glucopyranoside) competitively inhibited the PNPG transport by MAL2 in *O. polymorpha* (respective K_i values were between 0.23 and 1.47 mM). However, though α-MG probably binds the MAL2 permease of *O. polymorpha*, this yeast does not grow on this synthetic sugar (Viigand et al. 2016). The *S. cerevisiae* AGT1 has lower affinity for its substrates than the MAL2 permease of *O. polymorpha*: K_m of the AGT1 for maltose is 5.1–17.8 mM (Stambuk and de Araujo 2001; Day et al. 2002b), K_m for sucrose is ~ 8 mM (Stambuk et al. 2000), and K_m for maltotriose is 4–18.1 mM (Stambuk et al. 2000; Day et al. 2002a). As the MAL2 permease of

O. polymorpha has a considerably high affinity for α-glucosidic sugars, concentrations of these substrates are expected to be low in the habitat of *O. polymorpha*. Interestingly, the AGT1 protein of *S. cerevisiae* was also capable of glucose transport if overexpressed in a hxt1–17 gal2 deletion strain (Wieczorke et al. 1999).

The affinity of the MAL2 permease for maltose and sucrose is much higher than that of the maltase for these sugars (Viigand 2018; Liiv et al. 2001; Viigand and Alamäe 2007; Viigand et al. 2016), suggesting that these substrates must be concentrated into the cell to enable their efficient hydrolysis. The presence of intracellular substrate (e.g., maltose) is crucial for the induction of the *MAL* genes as reported for *O. polymorpha* (Viigand and Alamäe 2007) and *S. cerevisiae* (Wang et al. 2002). So, if a *MAL2* disruptant was incubated in the medium containing maltose, no maltase induction was recorded in the cells (Viigand and Alamäe 2007).

The AGT1 permease of *S. cerevisiae* was recently studied by Trichez et al. using mutagenesis (Trichez et al. 2018). Given that charged residues are usually responsible for sugar and proton binding by H^+-symporters, Trichez et al. selected four conserved charged residues in the predicted transmembrane α-helices 1 (Glu120, Asp123), 2 (Glu167), and 11 (Arg504) of the AGT1 permease for mutagenesis. Mutation of Glu120 and Arg504 to alanine most severely reduced the active transport of maltotriose and other α-glucosidic sugars by the yeast. In the case of Asp123 and Glu167, a double mutation (Asp123Gly/Glu167Ala) was required to reduce the transport (Trichez et al. 2018). Figure 1.5a shows alignment of *O. polymorpha* MAL2 transporter with AGT1 and MAL31 of *S. cerevisiae* in regions predicted to encode transmembrane domains (TMDs) 1, 2, and 11. Conserved charged residues proven crucial for active transport by the AGT1 are highlighted in the alignment, and predicted TMDs of MAL2 are shown overlined. Fig. 1.5b shows predicted structure of the *O. polymorpha* MAL2 permease. The charged residues of TMDs 1, 2, and 11 that are inferred to participate in active transport of α-glucosidic sugars are shown on dark background.

1.4 The MAL1 of *O. polymorpha* Among Yeast α-Glucosidases

The MAL1 protein of *O. polymorpha* was first characterized in 2001 (Liiv et al. 2001). This subchapter characterizes not only catalytic properties of the MAL1 protein but also its phylogenetic position among α-glucosidases of yeasts.

A

```
MAL2_Op   KTFPKAACWSIVLSTAIIMEGYDTTLLNSLYSMQSFA 104
AGT1_Sc   LKYPKAALWSILVSTTLVMEGYDTALLSALYALPVFQ 137      TMD1 (aa 73-94)
MAL31_Sc  KTYPKAAAWSLLVSTTLIQEGYDTAILGAFYALPVFQ 130
Cc          .:**** **:::**::: *****::*.::*::  *

MAL2_Op   QYQVPAKWQTSLSMSTYVGEIVGLYIAGLVAEKWGYRR 153
AGT1_Sc   SYEITSQWQIGLNMCVLCGEMIGLQITTYMVEFMGNRY 185     TMD2 (aa 125-145)
MAL31_Sc  DYEISVSWQIGLCLCYMAGEIVGLQMTGPSVDYMGNRY 179
Cc         .*::  .** .* :.  **::** ::   .:  * *

MAL2_Op    VALARNWYNLSQIPLSIVTPYMLNPTAWNWKAKA 500
AGT1_Sc    IVLARICYNLMAVINAILTPYMLNVSDWNWGAKT 531    TMD11 (aa 468-489)
MAL31_Sc   IILARNAYNVIQVVVTVLIMYQLNSEKWNWGAKS 525
Cc         : ***  **:  :  :::  * **   *** **
```

B

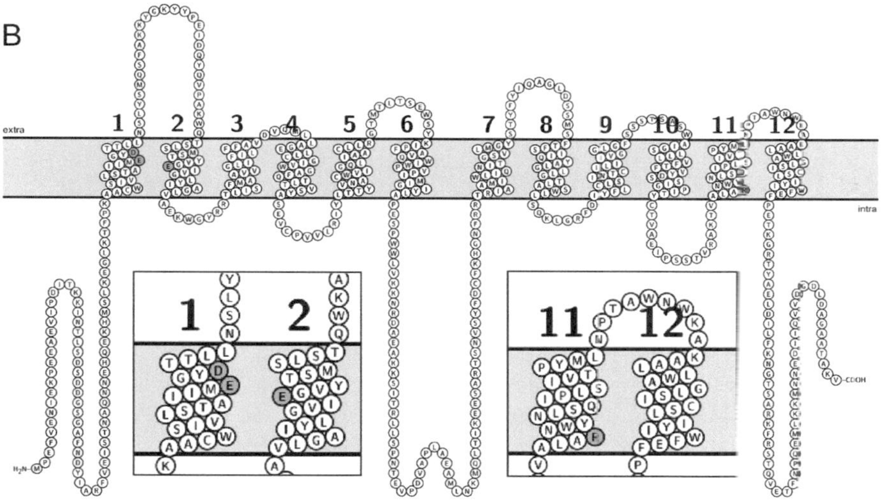

Fig. 1.5 (**a**) Extract of ClustalW alignment of the *O. polymorpha* MAL2 with AGT1 and MAL31 of *S. cerevisiae*. Sequences comprising the transmembrane domains (TMDs) 1, 2, and 11 are shown overlined. (**b**) Predicted TMDs of the MAL2 permease of *O. polymorpha*. The insert highlights charged residues of TMD1 (Glu87 and Asp90), TMD2 (Glu135), and TMD11 (Arg471) of *O. polymorpha* MAL2 assumed crucial for active transport of α-glucosidic sugars. TMDs were predicted through the CCTOP website (Dobson et al. 2015). The MAL2 permease was visualized using the Protter program (Omasits et al. 2014)

1.4.1 α-Glucosidases in CAZy

Yeast α-glucosidases belong to glycoside hydrolases (EC 3.2.1.–). Glycoside hydrolases (GHs) hydrolyze the glycosidic bond between two or more carbohydrates or between a carbohydrate and a non-carbohydrate moiety (Lombard et al. 2014). For example, the α-glucosidase MAL1 of *O. polymorpha* also hydrolyzes an α-glycosidic bond in α-MG and PNPG liberating methanol and *p*-nitrophenol from respective substrates (Liiv et al. 2001). Hydrolysis of a chromogenic substrate PNPG is widely used to quantitate catalytic activity of α-glucosidases (Liiv et al. 2001; Zimmermann et al. 1977). CAZy, a protein sequence-based database of Carbohydrate-Active enZymes, classifies maltases and isomaltases to family GH13 (Lombard et al. 2014). This family also includes α-amylases, pullulanases, neopullulanases, isoamylases, trehalose synthases, trehalose-6-phosphate hydrolases, and many others. The similarity of the amino acid sequences within the GH13 family proteins is low, yet they all share highly conserved regions and a catalytic machinery – three acidic catalytic residues located in conserved regions (Yamamoto et al. 2010). In the MAL1 protein of *O. polymorpha*, Asp199 was predicted as catalytic nucleophile, Glu257 as acid-base catalyst, and Asp338 as transition state stabilizer. The importance of Asp199 for the catalysis was experimentally proven: the Asp199Ala (D199A) mutant of MAL1 was catalytically inactive (Viigand et al. 2016).

1.4.2 Maltases and Isomaltases of Saccharomyces cerevisiae

S. cerevisiae has two types of α-glucosidases for the hydrolysis of α-glucosidic sugars: maltases (EC 3.2.1.20) and isomaltases (EC 3.2.1.10). Isomaltases have also been referred to as oligo-1,6-glucosidases, sucrase-isomaltases, and α-methylglucosidases. Maltases (e.g., MAL12 and MAL62) degrade maltose and so-called "maltose-like" sugars, maltotriose, sucrose, turanose, and maltulose, but are not able to degrade isomaltose and "isomaltose-like" sugars such as palatinose and α-MG (Viigand 2018; Needleman et al. 1978; Krakenaĭte and Glemzha 1983; Voordeckers et al. 2012). For composition and linkages in these sugars, see Fig. 1.1. For isomaltose degradation, *Saccharomyces* has isomaltases IMA1 to IMA5 (Naumoff and Naumov 2010; Teste et al. 2010). In addition to isomaltose, palatinose, and α-methylglucoside, isomaltases of *S. cerevisiae* also hydrolyze PNPG and sucrose (Teste et al. 2010; Deng et al. 2014).

1.4.3 Substrate Specificity of the O. polymorpha *MAL1*

Enzymatic assay of crude extract from *E. coli* cells expressing the MAL1 protein revealed maltose, sucrose, and α-methylglucoside as substrates for the enzyme, whereas trehalose, melibiose, and cellobiose were not hydrolyzed, indicating that the enzyme acts on α-1,4 (as in maltose) and α-1,2 (as in sucrose) glycosidic linkages (Liiv et al. 2001). In (Viigand et al. 2016), the substrate specificity of heterologously expressed and purified MAL1 protein was addressed, and additional substrates were revealed for the enzyme. It was shown that the MAL1 can hydrolyze the following maltose-like substrates with affinities decreasing in the order: maltulose, maltotriose, sucrose, turanose, maltose, and melezitose. From

Table 1.1 Substrate specificity of the *O. polymorpha* MAL1 (Viigand et al. 2016)

Substrate	Kinetic parameters		
	K_m (mM) ±SD	k_{cat} (1/min) ±SD	k_{cat}/K_m (mM/min)
Maltose-like substrates			
PNPG	0.52 ± 0.05	8476.4 ± 297.8	16300.7
Maltose	51.8 ± 3.7	10395.3 ± 297.8	200.7
Sucrose	25.1 ± 2.4	10984.2 ± 416.9	437.6
Turanose	34.0 ± 2.6	11864.3 ± 370.5	348.9
Maltotriose	20.2 ± 1.7	13597.9 ± 377.2	673.2
Maltulose	15.8 ± 0.8	2038.0 ± 33.1	129.0
Melezitose	115.0 ± 8.2	3870.9 ± 145.6	33.7
Isomaltose-like substrates			
α-Methylglucoside	35.5 ± 3.3	1806.4 ± 66.2	50.9
Isomaltose	27.3 ± 2.3	1191.1 ± 72.9	43.6
Palatinose	6.8 ± 0.4	1495.4 ± 26.5	219.9

SD standard deviation, *PNPG* *p*-nitrophenyl-α-D-glucopyranoside

isomaltose-like substrates, palatinose was the most suitable substrate, followed by isomaltose and α-MG. PNPG was the best substrate for MAL1 (Table 1.1).

Interestingly, the MAL1 protein could also hydrolyze fructooligosaccharides 1-kestose and 6-kestose (Viigand et al. 2016).This property has not been shown before for α-glucosidases. In the case of hydrolysis of trisaccharides melezitose and panose (see Fig. 1.1), the linkages hydrolyzed first by the enzyme were α-1,3 and α-1,6, respectively (Viigand 2018; Viigand et al. 2016). Though in early publications, the *O. polymorpha* MAL1 has been defined as a maltase, according to its substrate specificity, it rather belongs to maltase-isomaltases.

Phylogenetic analysis of yeast and fungal α-glucosidases revealed clustering of *O. polymorpha* MAL1 with α-glucosidases of *O. parapolymorpha*, *Cyberlindnera fabianii*, *Scheffersomyces stipitis*, *Lodderomyces elongisporus*, and *Meyerozyma guilliermondii* (Viigand et al. 2018). Of all these enzymes, only the *S. stipitis* MAL7, MAL8, and MAL9 (Viigand et al. 2018) and an α-glucosidase of *L. elongisporus* (Voordeckers et al. 2012) have been experimentally studied – these enzymes were reported to hydrolyze both maltose-like and isomaltose-like substrates. α-Glucosidases have been also studied in a phylogenetically old yeast *Schizosaccharomyces pombe*. The intracellular MAL1 protein of *S. pombe* hydrolyzed not only PNPG, maltose, and sucrose but also polymeric α-glucans dextrin and soluble starch (Chi et al. 2008). In addition to MAL1, *S. pombe* has an extracellular maltase AGL1, which was shown strictly specific for maltose – it did not hydrolyze other maltose-like sugars such as maltotriose and turanose (Jansen et al. 2006).

The malt extract and isomaltooligosaccharides were also substrates for *O. polymorpha* MAL1: the DP4 oligosaccharide was the longest substrate for the enzyme (Viigand et al. 2016). At the same time, the extracellular α-glucosidase of *S. pombe* used maltooligosaccharides with size up to DP7 – maltoheptaose (Okuyama et al.

2005). Some bacterial α-glucosidases also hydrolyze longer oligosaccharides (up to maltoheptaose) and in some cases also polysaccharides starch and dextrin (Egeter and Brückner 1995; Schönert et al. 1998, 1999; Cihan et al. 2011).

Similar to *S. cerevisiae* maltase (Kim et al. 1999), the MAL1 of *O. polymorpha* was very strongly inhibited by glucose and a diabetes drug acarbose. Fructose, a hydrolysis product of sucrose, turanose, maltulose, and palatinose (see Fig. 1.1), exerted much lower inhibiting power (Viigand et al. 2016). However, as in living yeast cells, glucose released from di- and trisaccharides is metabolized further; the in vivo inhibitory effect of glucose is probably lower than that recorded in vitro.

From yeast α-glucosidases, a three-dimensional structure has been resolved only for the *S. cerevisiae* isomaltase 1 (IMA1; PDB ID 3AJ7 and 3A4A). Structures of the IMA1 variants in complex with sugar ligands (maltose, isomaltose) have uncovered the amino acids bordering the active site (Y158, V216, G217, S218, L219, M278, Q279, D307, E411) with Val216 being crucial for selective binding of the substrate (Yamamoto et al. 2010, 2011; Deng et al. 2014). These nine positions vary between the maltases, isomaltases, and maltase-isomaltases (Voordeckers et al. 2012), and amino acids residing at these positions have been used as a signature in phylogenetic and substrate specificity analyses of α-glucosidases (Viigand et al. 2016, 2018; Voordeckers et al. 2012).

Table 1.2 illustrates signature amino acid patterns of α-glucosidases with known substrate specificity. α-Glucosidases hydrolyzing the α-1,6-glycosidic linkage have a Val residue at position equivalent to Val216 of IMA1 – a residue next to catalytic nucleophile Asp215. The corresponding residue of α-glucosidases that are able to hydrolyze maltose-like substrates is either Thr (Yamamoto et al. 2010) or in some cases Ala (Viigand et al. 2018; Tsujimoto et al. 2007). Respective residues are shown in bold in Table 1.2.

If Val216 of the *S. cerevisiae* IMA1 was replaced with a Thr; the enzyme gained the ability to hydrolyze maltose (Yamamoto et al. 2004). The *O. polymorpha* MAL1 has a Thr at a respective position (Table 1.2). Substitution of Thr200 with a Val drastically reduced the hydrolysis of maltose-like substrates by the *O. polymorpha* MAL1, whereas hydrolysis of isomaltose-like sugars was not affected (Fig. 1.6). Thus, it was concluded that Thr200 is required for efficient hydrolysis of maltose and maltose-like sugars. Indeed, the mutant Thr200Val became similar to isomaltases (Fig. 1.6).

1.4.4 The MAL1 of **O. polymorpha** *Is Ancient-Like*

Voordeckers et al. (Voordeckers et al. 2012) raised a hypothesis suggesting that modern maltases and isomaltases as those present in *S. cerevisiae* have evolved from a common promiscuous ancestor. The protein sequence of the ancestral protein (ancMalS) was predicted in silico, resurrected by heterologous synthesis in *E. coli*, and studied for enzymatic properties. Though the ancMalS was primarily

Table 1.2 Signature amino acids of α-glucosidases with known substrate specificity (Viigand et al. 2018)

α-Glucosidase	Signature amino acids (numbering as in *Sc* IMA1)									Substrate specificity
	158	216	217	218	219	278	279	307	411	
ancMALS	F	**T**	A	G	L	V	G	D	E	M-I
Le α-glucosidase	H	**T**	A	G	M	V	G	D	N	M-I
Op MAL1	F	**T**	A	G	L	V	G	D	N	M-I
Ss MAL7	F	**T**	A	G	L	V	G	T	N	M-I
Ss MAL8	Y	**T**	A	G	L	V	G	E	N	M-I
Ss MAL9	Y	**T**	A	G	M	V	G	E	N	M-I
Sc MAL32	F	**T**	A	G	L	V	A	E	D	M
Sp Mal1	Y	**A**	I	N	M	M	P	D	E	M
Ao MalT	I	**T**	V	N	M	L	P	D	D	M
Bs α-1,4-glucosidase	I	**A**	I	S	H	A	N	G	A	M
Sc IMA1	Y	**V**	G	S	L	M	Q	D	E	I
Sc IMA2	Y	**V**	G	S	L	M	Q	D	E	I
Sc IMA3/4	Y	**V**	G	S	L	M	R	D	E	I
Sc IMA5	F	**V**	G	S	M	V	G	S	E	I
An AgdC	F	**V**	I	N	F	M	P	D	D	I
Fo Foagl1	F	**V**	I	N	F	M	P	D	D	I
Bt oligo-1,6-glucosidase	V	**V**	I	N	M	T	P	D	E	I

M-I maltase-isomaltase, M maltase, I isomaltase. The amino acid at position corresponding to Val216 of *S. cerevisiae* IMA1 is shown in bold. AncMALS a resurrected hypothetical ancestor protein of *S. cerevisiae* maltases/isomaltases (Voordeckers et al. 2012), *Le Lodderomyces elongisporus, Sc S. cerevisiae, Ss Scheffersomyces stipitis, Sp Schizosaccharomyces pombe, Bs Bacillus stearothermophilus, Bt Bacillus thermoglucosidasius, Ao Aspergillus oryzae, Fo Fusarium oxysporum, An Aspergillus niger*

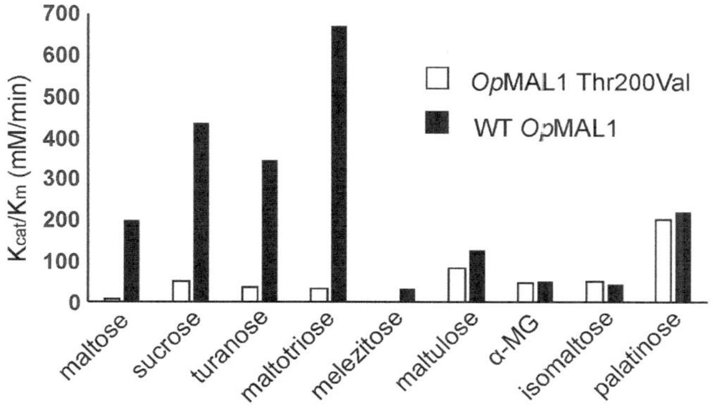

Fig. 1.6 According to catalytic efficiency k_{cat}/K_m (mM/min), a Thr200Val mutant of the *O. polymorpha* MAL1 protein is similar to an isomaltase. Compiled using data from Table 2 of (Viigand et al. 2016)

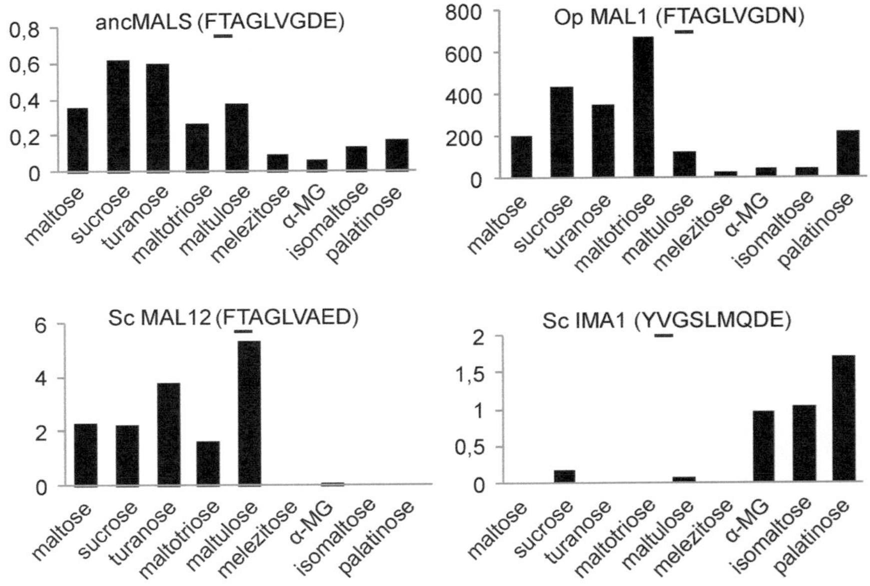

Fig. 1.7 Catalytic efficiencies, k_{cat}/K_m (mM/min), of the *O. polymorpha* (*Op*) maltase-isomaltase MAL1, ancestral maltase ancMALS, maltase MAL12 of *S. cerevisiae* (*Sc*), and isomaltase IMA1 of *Sc*. The signature amino acids of the enzymes (see Table 1.2) are presented in parentheses with the amino acid equivalent to Val216 of IMA1 shown underlined (Viigand et al. 2016)

active on maltose-like substrates, it also had a minor activity on isomaltose-like sugars (Voordeckers et al. 2012). The present-day α-glucosidases of *S. cerevisiae* preferentially hydrolyze either isomaltose-like sugars (IMA1, IMA2, and IMA5) or maltose-like sugars (MAL12, MAL32, MAL62) (Voordeckers et al. 2012); see Fig. 1.7 for MAL12 and IMA1).

Intriguingly, according to substrate specificity and the signature amino acids, the *O. polymorpha* MAL1 is highly similar to ancMALS (Table 1.2 and Fig. 1.7). Even though it has been claimed that both maltase and isomaltase activities cannot be fully optimized in a single enzyme (Voordeckers et al. 2012), the MAL1 properties confirm that it is still possible (Fig. 1.7).

When studying evolutionary origin of α-glucosidases, Gabriško (Gabriško 2013) noted close relatedness of fungal and bacterial enzymes suggesting bacterial ancestry of fungal α-glucosidases. In agreement with this suggestion, the maltase Mal1 of an "ancient" yeast *S. pombe* and a bacterial maltase (from *B. stearothermophilus*) both have an Ala and Ile at positions corresponding to respective amino acids Val216 and Gly217 of the *S. cerevisiae* IMA1 protein. The *S. cerevisiae* maltases have Thr and Ala at respective positions (Table 1.3). Interestingly, the *MAL1* gene of *O. polymorpha* has also a property of a bacterial gene – its promoter region is perfectly recognized in a bacterium *Escherichia coli* as it possesses two pairs of sigma 70-like sequences (Alamäe et al. 2003).

1.5 Regulation of the *MAL* Genes by Carbon Sources in *O. polymorpha*

1.5.1 General Features

Maltase synthesis in *O. polymorpha* is regulated by carbon sources in the growth medium: maltose and sucrose act as strong inducers, glucose as a repressor, and glycerol and ethanol allow moderate derepression (Alamäe and Liiv 1998). If *O. polymorpha* is grown on maltose or sucrose in the presence of a high concentration (2%) of glucose or fructose, induction of maltase is prevented. Interestingly, maltase activity is slightly repressed even during growth on maltose or sucrose if these substrates are provided at a high (2%) concentration. Therefore, glucose and fructose arising from the hydrolysis of maltose and sucrose in the cell most probably have some negative effect on the expression of *MAL* genes (Suppi et al. 2013).

Reporter gene assay of the bidirectional *MAL1-MAL2* promoter showed co-regulated expression in both directions, repression by glucose and induction by maltose, whereas the basal expression was higher in the direction of the permease gene (Viigand et al. 2005). It seems reasonable, because the permease activity is first required to provide intracellular maltose that is needed for induction of the *MAL* genes (Viigand and Alamäe 2007). Induction of the *O. polymorpha MAL1-MAL2* promoter by maltose (and sucrose) was stronger in the direction of maltase gene (Viigand and Alamäe 2007), and induced strength of the *MAL1* promoter was shown to constitute up to 70% of that of the *MOX* promoter (Alamäe et al. 2003). This knowledge can be used in biotechnological applications. The *MAL1* promoter has already been successfully used to overexpress and purify a biotechnologically relevant levansucrase protein from *E. coli* (Visnapuu et al. 2008).

1.5.2 Phosphorylated and Unphosphorylated Hexoses as Regulatory Signals

S. cerevisiae has three hexose kinases: hexokinase PI (HXK1) and PII (HXK2) that phosphorylate both fructose and glucose and a glucose-specific glucokinase (GLK1) (Zimmermann and Entian 1997). One of the hexokinases, the HXK2 of *S. cerevisiae*, has evolved to play a key role in establishing glucose repression (Gancedo 1998; Vega et al. 2016). Therefore, *S. cerevisiae* mutants defective in HXK2 lack glucose repression of many enzymes, including the maltase (Moreno and Herrero 2002; Zimmermann and Scheel 1977).

O. polymorpha has two hexose kinases: a hexokinase phosphorylating both glucose and fructose and a glucose-specific glucokinase (Kramarenko et al. 2000; Laht et al. 2002). Quite unexpectedly, hexokinase-negative mutants from chemical mutagenesis retained glucose repression of maltase synthesis losing only fructose repression (Kramarenko et al. 2000). Further assay of respective gene disruption mutants

(Suppi et al. 2013) confirmed that in contrast to *S. cerevisiae*, hexokinase has no specific role in establishing glucose repression in *O. polymorpha* – instead the ability of the cell to phosphorylate hexoses was required for downregulation of sugar-repressed genes, including the *MAL1*. In this report, glucose-6-phosphate was proposed as a signaling metabolite to trigger sugar repression (Suppi et al. 2013).

Highly intriguing results regarding the regulation of *MAL* genes were obtained through the study of double kinase-negative mutants of *O. polymorpha* (Suppi et al. 2013). As these mutants have no enzymes for glucose and fructose phosphorylation, they cannot grow on sugars (glucose, fructose, sucrose, maltose, etc.), but they grow perfectly on glycerol, methanol, and ethanol. Though not growing on sugars, these mutants are capable of sugar transport (Suppi et al. 2013). Importantly, when double kinase-negative mutants were cultivated on glycerol in the presence of glucose or fructose, a very high maltase activity was recorded in the cells (Suppi et al. 2013). Thus, sugars considered to be repressors of the *MAL* genes promoted their activation in these mutants. Based on these results, a hypothesis was raised according to which presence of disaccharides in the environment is sensed by *O. polymorpha* inside the cell through a transient increase of "free" glucose and fructose (Fig. 1.8). This signal is captured by a mechanism that is yet unknown and results in initial activation of the *MAL* genes (Viigand 2018; Suppi et al. 2013).

Indeed, *O. polymorpha* growing on non-sugar substrates (e.g., glycerol) is prepared to transport and hydrolyze maltose and sucrose – basal levels of both activities are detected in the cells (Viigand et al. 2005; Viigand and Alamäe 2007). In agreement with that, the maltase and maltose permease genes were, respectively, 89- and and 181-fold derepressed after the shift of *O. polymorpha* from glucose to methanol medium (van Zutphen et al. 2010). As the MAL1 protein has a very low affinity for disaccharides (Liiv et al. 2001; Viigand et al. 2016), its substrates should be concentrated in the cell by energy-dependent transport (Viigand and Alamäe 2007) to enable their efficient intracellular hydrolysis. Therefore, the hydrolysis reaction of disaccharides should produce a significant amount of free glucose (and fructose) inside the cell. As glucose- and fructose-phosphorylating activity in

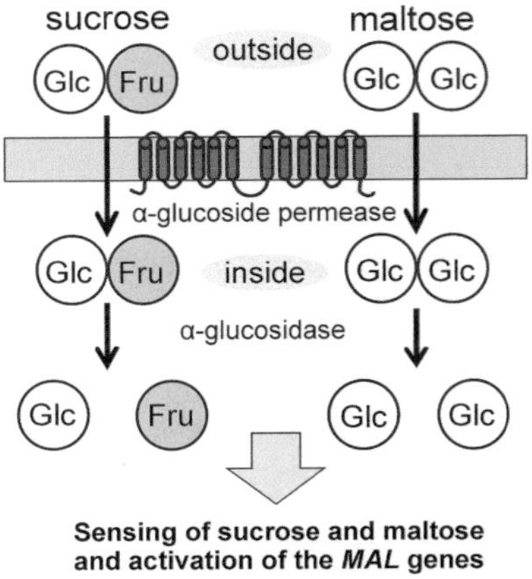

Fig. 1.8 Activation of the *MAL* genes in *O. polymorpha* by unphosphorylated glucose and fructose accumulating in the cell

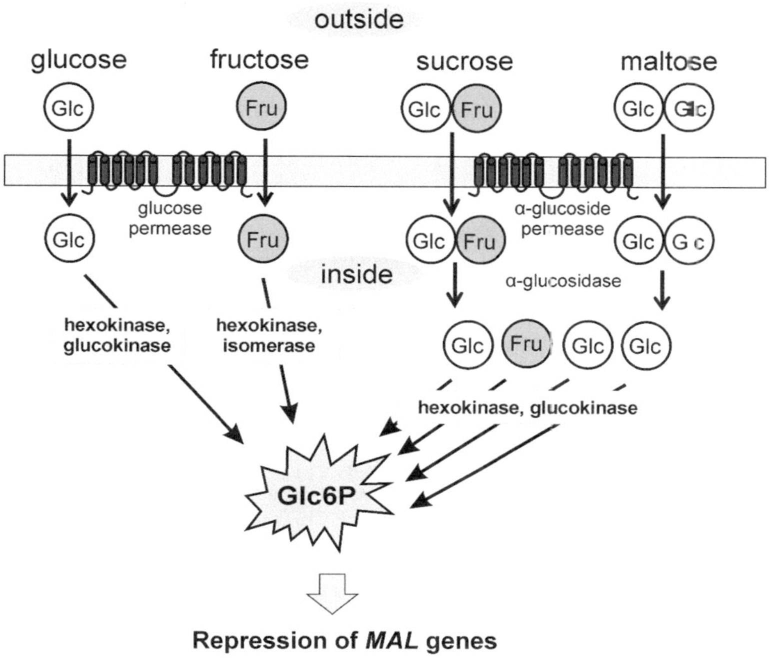

Fig. 1.9 Glucose-6-phosphate as a proposed repressing metabolite in the expression of *MAL* genes in *O. polymorpha*

O. polymorpha cells growing on gluconeogenic carbon sources is low (Kramarenko et al. 2000; Parpinello et al. 1998); at least part of glucose and fructose should stay unphosphorylated and trigger initial activation of the *MAL* genes.

Following internalization and hydrolysis of the disaccharides, the hydrolysis products are phosphorylated and channeled to glycolysis. As accumulation of phosphorylated derivatives of glucose and fructose causes repression of the *MAL* genes, a fine balance should exist between intracellular concentrations of free and phosphorylated species of monosaccharides, glucose, and fructose (Fig. 1.9). We assume that glucose-6-phosphate (Glc6P) signals for both glucose and fructose repression of *MAL* genes (Fig. 1.9).

Metabolic balance is also very important for *S. cerevisiae* growing on maltose. Therefore, if *S. cerevisiae* cannot manage intracellular metabolism of glucose resulting from maltose hydrolysis, excess of intracellular glucose will be exported by hexose transporters to prevent toxic effects exerted by high glucose concentration (Jansen et al. 2002).

Acknowledgments This book chapter is based on experimental work supported by grants from the Estonian Research Council (ETF 3923, ETF 5676, ETF 7528; ETF 9072 and PUT1050).

References

Alamäe T, Liiv L (1998) Glucose repression of maltase and methanol-oxidizing enzymes in the methylotrophic yeast *Hansenula polymorpha*: Isolation and study of regulatory mutants. Folia Microbiol (Praha) 43:443–452. https://doi.org/10.1007/BF02820789

Alamäe T, Pärn P, Viigand K, Karp H (2003) Regulation of the *Hansenula polymorpha* maltase gene promoter in *H. polymorpha* and *Saccharomyces cerevisiae*. FEMS Yeast Res 4:165–173. https://doi.org/10.1016/S1567-1356(03)00142-9

Alves-Araújo C, Hernandez-Lopez MJ, Sousa MJ et al (2004) Cloning and characterization of the MAL11 gene encoding a high-affinity maltose transporter from *Torulaspora delbrueckii*. FEMS Yeast Res 4:467–476. https://doi.org/10.1016/S1567-1356(03)00208-3

Ávila J, González C, Brito N et al (2002) A second Zn(II)2Cys6 transcriptional factor encoded by the *YNA2* gene is indispensable for the transcriptional activation of the genes involved in nitrate assimilation in the yeast *Hansenula polymorpha*. Yeast 19:537–544. https://doi.org/10.1002/yea.847

Blandin G, Llorente B, Malpertuy A et al (2000) Genomic exploration of the hemiascomycetous yeasts: 13. *Pichia angusta*. FEBS Lett 487:76–81. https://doi.org/10.1016/S0014-5793(00)02284-5

Brown CA, Murray AW, Verstrepen KJ (2010) Rapid expansion and functional divergence of subtelomeric gene families in yeasts. Curr Biol CB 20:895–903. https://doi.org/10.1016/j.cub.2010.04.027

Chang YS, Dubin RA, Perkins E et al (1988) *MAL63* codes for a positive regulator of maltose fermentation in *Saccharomyces cerevisiae*. Curr Genet 14:201–209. https://doi.org/10.1007/BF00376740

Charron MJ, Dubin RA, Michels CA (1986) Structural and functional analysis of the *MAL1* locus of *Saccharomyces cerevisiae*. Mol Cell Biol 6:3891–3899

Cheng Q, Michels CA (1989) The maltose permease encoded by the *MAL61* gene of *Saccharomyces cerevisiae* exhibits both sequence and structural homology to other sugar transporters. Genetics 123:477–484

Cheng Q, Michels CA (1991) *MAL11* and *MAL61* encode the inducible high-affinity maltose transporter of *Saccharomyces cerevisiae*. J Bacteriol 173:1817–1820

Chi Z, Ni X, Yao S (2008) Cloning and overexpression of a maltase gene from *Schizosaccharomyces pombe* in *Escherichia coli* and characterization of the recombinant maltase. Mycol Res 112:983–989. https://doi.org/10.1016/j.mycres.2008.01.024

Cihan A, Ozcan B, Tekin N, Cokmus C (2011) Characterization of a thermostable α-glucosidase from *Geobacillus thermodenitrificans* F84a, pp 945–955

Daudé D, Remaud-Siméon M, André I (2012) Sucrose analogs: an attractive (bio)source for glycodiversification. Nat Prod Rep 29:945–960. https://doi.org/10.1039/c2np20054f

Day RE, Rogers PJ, Dawes IW, Higgins VJ (2002a) Molecular analysis of maltotriose transport and utilization by *Saccharomyces cerevisiae*. Appl Environ Microbiol 68:5326–5335. https://doi.org/10.1128/AEM.68.11.5326-5335.2002

Day RE, Higgins VJ, Rogers PJ, Dawes IW (2002b) Characterization of the putative maltose transporters encoded by *YDL247w* and *YJR160c*. Yeast 19:1015–1027. https://doi.org/10.1002/yea.894

Deng X, Petitjean M, Teste M-A et al (2014) Similarities and differences in the biochemical and enzymological properties of the four isomaltases from *Saccharomyces cerevisiae*. FEBS Open Bio 4:200–212. https://doi.org/10.1016/j.fob.2014.02.004

Dobson L, Reményi I, Tusnády GE (2015) CCTOP: a Consensus Constrained TOPology prediction web server. Nucleic Acids Res 43:W408–W412. https://doi.org/10.1093/nar/gkv451

Dorokhov YL, Shindyapina AV, Sheshukova EV, Komarova TV (2015) Metabolic methanol: molecular pathways and physiological roles. Physiol Rev 95:603–644. https://doi.org/10.1152/physrev.00034.2014

Dubin RA, Needleman RB, Gossett D, Michels CA (1985) Identification of the structural gene encoding maltase within the *MAL6* locus of *Saccharomyces carlsbergensis*. J Bacteriol 164:605–610

Egeter O, Brückner R (1995) Characterization of a genetic locus essential for maltose-maltotriose utilization in *Staphylococcus xylosus*. J Bacteriol 177:2408–2415

Fairhead C, Dujon B (2006) Structure of *Kluyveromyces lactis* subtelomeres: duplications and gene content. FEMS Yeast Res 6:428–441. https://doi.org/10.1111/j.1567-1364.2006.00033.x

Feldmann H (2000) Génolevures – a novel approach to "evolutionary genomics". FEBS Lett 487:1–2

Gabriško M (2013) Evolutionary history of eukaryotic α-glucosidases from the α-amylase family. J Mol Evol 76:129–145. https://doi.org/10.1007/s00239-013-9545-4

Gancedo JM (1998) Yeast carbon catabolite repression. Microbiol Mol Biol Rev MMBR 62:334–361

Grigoriev IV, Nikitin R, Haridas S et al (2014) MycoCosm portal: gearing up for 1000 fungal genomes. Nucleic Acids Res 42:D699–D704. https://doi.org/10.1093/nar/gkt1183

Han EK, Cotty F, Sottas C et al (1995) Characterization of *AGT1* encoding a general α-glucoside transporter from *Saccharomyces*. Mol Microbiol 17:1093–1107

Hasegawa S, Takizawa M, Suyama H et al (2010) Characterization and expression analysis of a maltose-utilizing (*MAL*) cluster in *Aspergillus oryzae*. Fungal Genet Biol 47:1–9. https://doi.org/10.1016/j.fgb.2009.10.005

Hollatz C, Stambuk BU (2001) Colorimetric determination of active α-glucoside transport in *Saccharomyces cerevisiae*. J Microbiol Methods 46:253–259

Ishchuk OP, Voronovsky AY, Abbas CA, Sibirny AA (2009) Construction of *Hansenula polymorpha* strains with improved thermotolerance. Biotechnol Bioeng 104:911–919. https://doi.org/10.1002/bit.22457

Janecek S (2009) Amylolytic enzymes-focus on the alpha-amylases from archaea and plants. Nova Biotechnol 9

Jansen MLA, De Winde JH, Pronk JT (2002) Hxt-carrier-mediated glucose efflux upon exposure of *Saccharomyces cerevisiae* to excess maltose. Appl Environ Microbiol 68:4259–4265. https://doi.org/10.1128/AEM.68.9.4259-4265.2002

Jansen MLA, Krook DJJ, De Graaf K et al (2006) Physiological characterization and fed-batch production of an extracellular maltase of *Schizosaccharomyces pombe* CBS 356. FEMS Yeast Res 6:888–901. https://doi.org/10.1111/j.1567-1364.2006.00091.x

Jeffries TW, Van Vleet JRH (2009) *Pichia stipitis* genomics, transcriptomics, and gene clusters. Fems Yeast Res 9:793–807. https://doi.org/10.1111/j.1567-1364.2009.00525.x

Keppler F, Hamilton JTG, Braß M, Röckmann T (2006) Methane emissions from terrestrial plants under aerobic conditions. Nature 439:187–191

Kim M-J, Lee S-B, Lee H-S et al (1999) Comparative study of the inhibition of α-glucosidase, α-amylase, and cyclomaltodextrin glucanosyltransferase by acarbose, isoacarbose, and acarviosine–glucose. Arch Biochem Biophys 371:277–283. https://doi.org/10.1006/abbi.1999.1423

Krakenaïte RP, Glemzha AA (1983) Some properties of two forms of alpha-glucosidase from *Saccharomyces cerevisiae*-II. Biokhimiia Mosc Russ 48:62–68

Kramarenko T, Karp H, Järviste A, Alamäe T (2000) Sugar repression in the methylotrophic yeast *Hansenula polymorpha* studied by using hexokinase-negative, glucokinase-negative and double kinase-negative mutants. Folia Microbiol (Praha) 45:521–529. https://doi.org/10.1007/BF02818721

Kulikova-Borovikova D, Lisi S, Dauss E et al (2018) Activity of the α-glucoside transporter Agt1 in *Saccharomyces cerevisiae* cells during dehydration-rehydration events. Fungal Biol 122:613–620. https://doi.org/10.1016/j.funbio.2018.03.006

Kunze G, Gaillardin C, Czernicka M et al (2014) The complete genome of *Blastobotrys* (*Arxula*) *adeninivorans* LS3 - a yeast of biotechnological interest. Biotechnol Biofuels 7:66. https://doi.org/10.1186/1754-6834-7-66

Laht S, Karp H, Kotka P et al (2002) Cloning and characterization of glucokinase from a methylotrophic yeast *Hansenula polymorpha*: different effects on glucose repression in *H. polymorpha* and *Saccharomyces cerevisiae*. Gene 296:195–203

Lee G-Y, Jung J-H, Seo D-H et al (2011) Isomaltulose production via yeast surface display of sucrose isomerase from *Enterobacter* sp. FMB-1 on *Saccharomyces cerevisiae*. Bioresour Technol 102:9179–9184. https://doi.org/10.1016/j.biortech.2011.06.081

Leifso KR, Williams D, Hintz WE (2007) Heterologous expression of cyan and yellow fluorescent proteins from the *Kluyveromyces lactis KlMAL21–KlMAL22* bi-directional promoter. Biotechnol Lett 29:1233–1241. https://doi.org/10.1007/s10529-007-9381-y

Liiv L, Pärn P, Alamäe T (2001) Cloning of maltase gene from a methylotrophic yeast, *Hansenula polymorpha*. Gene 265:77–85. https://doi.org/10.1016/S0378-1119(01)00359-6

Limtong S, Srisuk N, Yongmanitchai W et al (2008) *Ogataea chonburiensis* sp. nov. and *Ogataea nakhonphanomensis* sp. nov., thermotolerant, methylotrophic yeast species isolated in Thailand, and transfer of *Pichia siamensis* and *Pichia thermomethanolica* to the genus *Ogataea*. Int J Syst Evol Microbiol 58:302–307. https://doi.org/10.1099/ijs.0.65380-0

Lombard V, Golaconda Ramulu H, Drula E et al (2014) The carbohydrate-active enzymes database (CAZy) in 2013. Nucleic Acids Res 42:D490–D495. https://doi.org/10.1093/nar/gkt1178

Lunn JE (2002) Evolution of sucrose synthesis. Plant Physiol 128:1490–1500. https://doi.org/10.1104/pp.010898

Morais PB, Teixeira LCRS, Bowles JM et al (2004) *Ogataea falcaomoraisii* sp. nov., a sporogenous methylotrophic yeast from tree exudates. FEMS Yeast Res 5:81–85. https://doi.org/10.1016/j.femsyr.2004.05.006

Moreno F, Herrero P (2002) The hexokinase 2-dependent glucose signal transduction pathway of *Saccharomyces cerevisiae*. FEMS Microbiol Rev. 26:83–90

Naumoff DG, Naumov GI (2010) Discovery of a novel family of α-glucosidase *IMA* genes in yeast *Saccharomyces cerevisiae*. Dokl Biochem Biophys 432:114–116. https://doi.org/10.1134/S1607672910030051

Naumov GI, Naumova ES, Lee C-F (2017) *Ogataea haglerorum* sp. nov., a novel member of the species complex, *Ogataea* (*Hansenula*) *polymorpha*. Int J Syst Evol Microbiol 67:2465–2469. https://doi.org/10.1099/ijsem.0.002012

Needleman R (1991) Control of maltase synthesis in yeast. Mol Microbiol 5:2079–2084

Needleman RB, Federoff HJ, Eccleshall TR et al (1978) Purification and characterization of an alpha-glucosidase from *Saccharomyces carlsbergensis*. Biochemistry 17:4657–4661

Okuyama M, Tanimoto Y, Ito T et al (2005) Purification and characterization of the hyperglycosylated extracellular α-glucosidase from *Schizosaccharomyces pombe*. Enzyme Microb Technol 37:472–480. https://doi.org/10.1016/j.enzmictec.2004.06.018

Omasits U, Ahrens CH, Müller S, Wollscheid B (2014) Protter: interactive protein feature visualization and integration with experimental proteomic data. Bioinformatics 30:884–886. https://doi.org/10.1093/bioinformatics/btt607

Parpinello G, Berardi E, Strabbioli R (1998) A regulatory mutant of *Hansenula polymorpha* exhibiting methanol utilization metabolism and peroxisome proliferation in glucose. J Bacteriol 180:2958–2967

Peinado JM, Barbero A, van UN (1987) Repression and inactivation by glucose of the maltose transport system of *Candida utilis*. Appl Microbiol Biotechnol 26:154–157. https://doi.org/10.1007/BF00253901

Ramezani-Rad M, Hollenberg CP, Lauber J et al (2003) The *Hansenula polymorpha* (strain CBS4732) genome sequencing and analysis. FEMS Yeast Res 4:207–215. https://doi.org/10.1016/S1567-1356(03)00125-9

Ravin NV, Eldarov MA, Kadnikov VV et al (2013) Genome sequence and analysis of methylotrophic yeast *Hansenula polymorpha* DL1. BMC Genomics 14:837

Reinders A, Ward JM (2001) Functional characterization of the α-glucoside transporter Sut1p from *Schizosaccharomyces pombe*, the first fungal homologue of plant sucrose transporters. Mol Microbiol 39:445–455. https://doi.org/10.1046/j.1365-2958.2001.02237.x

Sawale PD, Shendurse AM, Mohan MS, Patil GR (2017) Isomaltulose (palatinose) – an emerging carbohydrate. Food Biosci 18:46–52. https://doi.org/10.1016/j.fbio.2017.04.003

Schönert S, Buder T, Dahl MK (1998) Identification and enzymatic characterization of the maltose-inducible α-glucosidase MalL (sucrase-isomaltase-maltase) of *Bacillus subtilis*. J Bacteriol 180:2574–2578

Schönert S, Buder T, Dahl MK (1999) Properties of maltose-inducible α-glucosidase MalL (sucrase-isomaltase-maltase) in *Bacillus subtilis*: evidence for its contribution to maltodextrin utilization. Res Microbiol 150:167–177

Slot JC, Rokas A (2010) Multiple *GAL* pathway gene clusters evolved independently and by different mechanisms in fungi. Proc Natl Acad Sci 107:10136–10,141

Stambuk BU, de Araujo PS (2001) Kinetics of active α-glucoside transport in *Saccharomyces cerevisiae*. FEMS Yeast Res 1:73–78

Stambuk BU, Batista AS, De Araujo PS (2000) Kinetics of active sucrose transport in *Saccharomyces cerevisiae*. J Biosci Bioeng 89:212–214

Stewart GG (2016) *Saccharomyces* species in the production of beer. Beverages 2. https://doi.org/10.3390/beverages2040034

Suppi S, Michelson T, Viigand K, Alamäe T (2013) Repression vs. activation of *MOX, FMD, MPP1* and *MAL1* promoters by sugars in *Hansenula polymorpha*: the outcome depends on cell's ability to phosphorylate sugar. FEMS Yeast Res 13:219–232. https://doi.org/10.1111/1567-1364.12023

Teste M-A, Francois JM, Parrou J-L (2010) Characterization of a new multigene family encoding isomaltases in the yeast *Saccharomyces cerevisiae*, the *IMA* family. J Biol Chem 285:26815–26,824. https://doi.org/10.1074/jbc.M110.145946

Trichez D, Knychala MM, Figueiredo CM et al (2018) Key amino acid residues of the AGT1 permease required for maltotriose consumption and fermentation by *Saccharomyces cerevisiae*. J Appl Microbiol.. https://doi.org/10.1111/jam.14161

Tsujimoto Y, Tanaka H, Takemura R et al (2007) Molecular determinants of substrate recognition in thermostable α-glucosidases belonging to glycoside hydrolase family 13. J Biochem (Tokyo) 142:87–93. https://doi.org/10.1093/jb/mvm110

van Zutphen T, Baerends RJ, Susanna KA et al (2010) Adaptation of *Hansenula polymorpha* to methanol: a transcriptome analysis. BMC Genomics 11(1). https://doi.org/10.1186/1471-2164-11-1

Vanoni M, Sollitti P, Goldenthal M, Marmur J (1989) Structure and regulation of the multigene family controlling maltose fermentation in budding yeast. Prog Nucleic Acid Res Mol Biol 37:281–322

Vega M, Riera A, Fernández-Cid A et al (2016) Hexokinase 2 is an intracellular glucose sensor of yeast cells that maintains the structure and activity of Mig1 protein repressor complex. J Biol Chem 291:7267–7285. https://doi.org/10.1074/jbc.M115.711408

Viigand K (2018) Utilization of α-glucosidic sugars by *Ogataea* (*Hansenula*) *polymorpha*. Dissertation, University of Tartu. http://hdl.handle.net/10062/61743

Viigand K, Alamäe T (2007) Further study of the *Hansenula polymorpha MAL* locus: characterization of the α-glucoside permease encoded by the *HpMAL2* gene. FEMS Yeast Res 7:1134–1144. https://doi.org/10.1111/j.1567-1364.2007.00257.x

Viigand K, Tammus K, Alamäe T (2005) Clustering of *MAL* genes in *Hansenula polymorpha*: cloning of the maltose permease gene and expression from the divergent intergenic region between the maltose permease and maltase genes. FEMS Yeast Res 5:1019–1028. https://doi.org/10.1016/j.femsyr.2005.06.003

Viigand K, Visnapuu T, Mardo K et al (2016) Maltase protein of *Ogataea* (*Hansenula*) *polymorpha* is a counterpart to the resurrected ancestor protein ancMALS of yeast maltases and isomaltases. Yeast 33:415–432. https://doi.org/10.1002/yea.3157

Viigand K, Põšnograjeva K, Visnapuu T, Alamäe T (2018) Genome mining of non-conventional yeasts: search and analysis of *MAL* clusters and proteins. Genes 9 354. https://doi.org/10.3390/genes9070354

Visnapuu T, Mäe A, Alamäe T (2008) *Hansenula polymorpha* maltase gene promoter with sigma 70-like elements is feasible for *Escherichia coli*-based biotechnological applications: Expression of three genomic levansucrase genes of *Pseudomonas syringae* pv. *tomato*. Process Biochem 43:414–422. https://doi.org/10.1016/j.procbio.2008.01.002

Voordeckers K, Brown CA, Vanneste K et al (2012) Reconstruction of ancestral metabolic enzymes reveals molecular mechanisms underlying evolutionary innovation through gene duplication. PLoS Biol 10:e1001446. https://doi.org/10.1371/journal.pbio.1001446

Wang X, Bali M, Medintz I, Michels CA (2002) Intracellular maltose is sufficient to induce *MAL* gene expression in *Saccharomyces cerevisiae*. Eukaryot Cell 1:696–703. https://doi.org/10.1128/EC.1.5.696-703.2002

Wieczorke R, Krampe S, Weierstall T et al (1999) Concurrent knock-out of at least 20 transporter genes is required to block uptake of hexoses in *Saccharomyces cerevisiae*. FEBS Lett 464:123–128

Wong S, Wolfe KH (2005) Birth of a metabolic gene cluster in yeast by adaptive gene relocation. Nat Genet 37:777–782

Yamamoto K, Nakayama A, Yamamoto Y, Tabata S (2004) Val216 decides the substrate specificity of α-glucosidase in *Saccharomyces cerevisiae*: Substrate specificity of α-glucosidase. Eur J Biochem 271:3414–3420. https://doi.org/10.1111/j.1432-1033.2004.04276.x

Yamamoto K, Miyake H, Kusunoki M, Osaki S (2010) Crystal structures of isomaltase from *Saccharomyces cerevisiae* and in complex with its competitive inhibitor maltose: Crystal structure of isomaltase. FEBS J 277:4205–4214. https://doi.org/10.1111/j.1742-4658.2010.07810.x

Yamamoto K, Miyake H, Kusunoki M, Osaki S (2011) Steric hindrance by 2 amino acid residues determines the substrate specificity of isomaltase from *Saccharomyces cerevisiae*. J Biosci Bioeng 112:545–550. https://doi.org/10.1016/j.jbiosc.2011.08.016

Zimmermann FK, Entian K-D (1997) Yeast Sugar Metabolism. CRC Press, Boca Raton

Zimmermann FK, Scheel I (1977) Mutants of *Saccharomyces cerevisiae* resistant to carbon catabolite repression. Mol Gen Genet MGG 154:75–82

Zimmermann FK, Kaufmann I, Rasenberger H, Haubetamann P (1977) Genetics of carbon catabolite repression in *Saccharomycess cerevisiae*: genes involved in the derepression process. Mol Gen Genet MGG 151:95–103

Chapter 2
Biotechnological Application of Non-conventional Yeasts for Xylose Valorization

Jessica C. Bergmann, Débora Trichez, Wilson Galvão de Morais Junior, Talita Gabriela Salles Ramos, Thályta Fraga Pacheco, Clara Vida G. C. Carneiro, Victor Mendes Honorato, Luana Assis Serra, and João Ricardo M. Almeida

Abstract Demands for clean and sustainable processes and products that are environmentally friendly are challenging biotechnologists to develop new strategies to produce fuels and chemicals. As the petroleum demands rise together with the concern of climatic and environmental changes, there is an increasing interest for renewable energy. Sugars present in the lignocellulosic biomass can be used as raw material in biotechnological processes employing yeasts as catalysts. Several known yeasts such as *Saccharomyces cerevisiae* assimilate glucose but lack the efficiency to consume xylose. Due to industrial interest, there has been an increasing effort to discover and construct new xylose-assimilating yeast strains. In this sense, due to the diversity and metabolic potential, several non-conventional yeasts species were isolated, identified, and physiologically and genetically characterized in the last years. The current review sought to summarize the main characteristics as well as the biotechnological applications of non-conventional yeasts for xylose utilization. First, it will present and discuss the data about non-conventional yeasts that naturally and efficiently assimilate xylose as *Scheffersomyces*, *Meyerozyma*, *Candida*, *Spathaspora*, and *Kluyveromyces*. Then the yeasts *Komagataella*, *Yarrowia*, and *Ogataea* that do not assimilate xylose or poorly assimilate xylose justifying genetic manipulation to increase xylose utilization will also be presented. In each case, basic information about yeast taxonomy, morphology, and physiology will be presented, and the clearest biotechnological application will be introduced.

Keywords Xylose · Biorefinery · Yeast · Biofuels · Renewable chemicals

J. C. Bergmann · D. Trichez · W. G. de Morais Junior · T. G. S. Ramos · T. F. Pacheco
Clara Vida G. C. Carneiro · V. M. Honorato · L. A. Serra · J. R. M. Almeida (✉)
Embrapa Agroenergia, Brasília-DF, Brazil
e-mail: joao.almeida@embrapa.br

A. Sibirny (ed.), *Non-conventional Yeasts: from Basic Research to Application*,
https://doi.org/10.1007/978-3-030-21110-3_2

2.1 Introduction

2.1.1 Lignocellulosic Biomass

The energetic source most used nowadays is petroleum, which is employed as fuel and raw material for the production of chemicals with several industrial applications (Sun et al. 2013; Gallo and Trapp 2017). Since petroleum stocks are decreasing, prices are volatile and its burn is contributing to global pollution, development of sustainable and environmentally friendly processes to produce fuels and chemicals is required. Lignocellulosic biomass is a widely available resource that can be employed in such processes (Gallo and Trapp 2017).

Biomass is the most abundant renewable feedstock resource available on the planet. Briefly, lignocellulosic biomass (vegetal cell wall) is composed of the sugar polymers cellulose and hemicellulose and the macromolecule lignin (Chandel et al. 2011). Cellulose, the most abundant polymer in biomass (Table 2.1), is composed of a homo-polysaccharide D-glucose (C6-hexose) linked by B-1,4-glucosidic bonds organized in complex microfibriles (Jørgensen et al. 2007). Hemicellulose is a heteropolysaccharide polymer composed of pentose and hexose sugars (C5 and C6, respectively). The nature of the main chain is formed by D-xylose or D-galactose, depending on the biomass source, which is ramified with other sugars, like L-arabinose, D-glucuronic acid, D-glucose, D-mannose, and other molecules, such as acids (van Wyk 2001). Lignin is a complex macromolecule composed of phenylpropane units that protect cellulose-hemicellulose giving integrity to the lignocellulose structure (Canilha et al. 2010; Kim 2018).

Several biotechnological processes for the production of fuels and other chemicals from biomass have been developed and evaluated from laboratorial to industrial scale (Garrote et al. 2002). A common requirement in the processes is the hydrolysis of lignocellulose to release the monosaccharides present in the cellulose and

Table 2.1 Compositions of lignocellulosic biomass (% dry basis). Biomass presented has a high xylan content in hemicellulose fraction

Biomass	Cellulose	Hemicellulose	Lignin	References
Barley hull	33.6	37.2	19.3	Kim et al. (2008)
Corn fiber	14.3	16.8	8.4	Mosier (2005)
Corn pericarp	22.5	23.7	4.7	Kim et al. (2017)
Corn stover	37.0	22.7	18.6	Kim et al. (2016)
Wheat straw	30.2	21.0	17.0	Ballesteros et al. (2006)
Rice straw	31.1	22.3	13.3	Chen et al. (2011)
Rye straw	30.9	21.5	22.1	García-Cubero et al. (2009)
Switch grass	39.5	20.3	17.8	Li et al. (2010)
Sugarcane bagasse	43.1	31.1	11.4	Martín et al. (2007)
Sunflower stalks	33.8	20.2	17.3	Ruiz et al. (2008)

Based on the review: Kim (2018)

hemicellulose fractions, which finally can be converted by the microorganism of choice to the desired product (Ji et al. 2017).

The composition and concentration of sugars in the lignocellulosic hydrolysate will vary according to biomass and pretreatment/hydrolysis conditions employed (Almeida et al. 2011). Glucose and xylose are, respectively, the most abundant hexose and pentose sugars in nature and consequently the most available in biomass hydrolysates (Table 2.1). Glucose is promptly metabolized and converted into a variety of fuels and chemicals by microorganisms (Elshahed 2010), whereas xylose consumption is limited to a reduced number of species (Jeffries 1983). Thus, xylose valorization processes have gained much attention lately.

2.1.2 *Xylose Application Overview*

The usage of xylose started to gain attention in the latest years of the 1920s when the National Bureau of Standards in the USA began to evaluate the possibility to aggregate value to crop wastes (Schreiber et al. 1930). After the discovery that some bacteria produced acetic and lactic acids from xylose fermentation, methods to recover xylose from different biomasses started to be investigated (Schreiber et al. 1930). Nowadays, D-xylose can be extracted from several different crops (Table 2.1) (Satish and Murthy 2010). Many efforts are being performed to recover as much xylose (and other sugars) as possible from the biomass with minimal generation of degradation products (Christopher 2012). The main process to obtain xylose is through a pretreatment process that hydrolyzes the hemicellulose with sulfuric acid (Zhang et al. 2014; Gallo and Trapp 2017).

Xylose can be applied in different industrial sectors, including food and medicine (Gunah 2011). It can be dehydrated by acid catalysis to produce furfural (Bozell and Petersen 2010), and it also can be reduced using nickel catalysts under high temperature and pressure to produce xylitol (Akinterinwa et al. 2008). Biologically, it can be converted to organic acids, lipids, alcohols, and other chemicals by pentose-assimilating microorganisms. For example it can be fermented to ethanol (Webb and Lee 1990; Bajwa et al. 2011), reduced to xylitol and arabitol (Gírio et al. 2000; Faria et al. 2014; Martins et al. 2018), or oxidized to xylonic and xylaric acid (Gallo and Trapp 2017) depending on the microorganism employed. With the increasing interest to produce chemicals from biomass through biotechnological processes, new xylose-assimilating microorganisms are being identified and/or constructed by bioprospecting and metabolic engineering strategies (Jeffries 2006; Mans et al. 2018).

2.1.3 Microbial Utilization of Xylose

Bacteria, eukaryotes (yeasts and fungus), and archaea are capable of using xylose as a carbon source (Sampaio et al. 2003; Johnsen et al. 2009; Kręgiel et al. 2017; Kim and Woo 2018), however, through specific pentose assimilation pathways (Martins et al. 2018). Most of the bacteria and some anaerobic fungi such as *Piromyces* (Harhangi et al. 2003) and *Orpinomyces* (Madhavan et al. 2009) use an isomerization pathway to assimilate xylose. In this pathway, a xylose isomerase (XI, EC 5.3.1.5) interconverts xylose to xylulose, which is then phosphorylated by the xylulokinase enzyme (XK, EC 2.7.1.17) to xylulose-5-phosphate, to enter the pentose phosphate pathway (PPP) (Kwak and Jin 2017). The archaea employ an oxidative pathway where xylose is oxidized to α-ketoglutarate (tricarboxylic acid cycle intermediate) by D-xylose dehydrogenase, xylonate dehydratase, 2-keto-3-deoxyxylonate dehydratase, and α-ketoglutarate semialdehyde dehydrogenase (Johnsen et al. 2009)

Yeast employs an oxidoreductase pathway (Fig. 2.1). In this pathway, xylose is reduced to xylitol by an NAD(P)H-dependent xylose reductase (XR, EC 1.1.1.30).

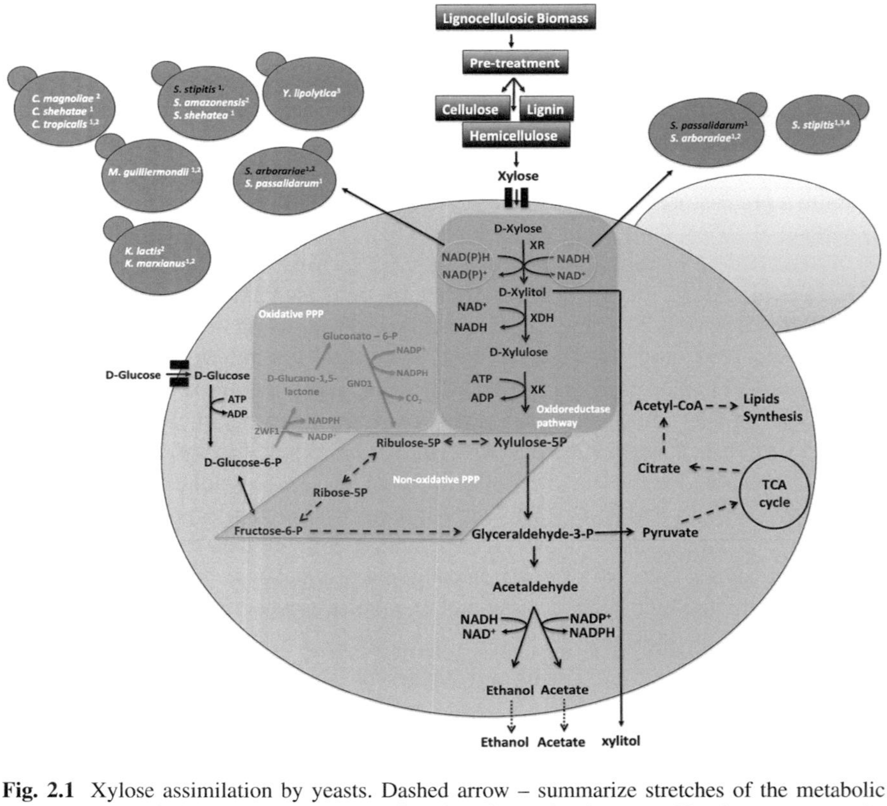

Fig. 2.1 Xylose assimilation by yeasts. Dashed arrow – summarize stretches of the metabolic pathway; names in black represent the preferred co-factor for the yeast. Numbers represent the product converted naturally by the yeasts using xylose as carbon source: [1]Ethanol producer; [2]Xylitol producer; [3]Lipids producer

Then, an NAD+-dependent xylitol dehydrogenase (XDH, EC 1.1.1.9) oxidates xylitol to D-xylulose, which in turn is phosphorylated to D-xylulose-5P by the enzyme xylulose kinase (XK, EC 2.7.1.17). D-xylulose-5P can be metabolized by the pentose phosphate pathway (PPP) and glycolysis (Wang et al. 1980).

2.1.3.1 Xylose Conversion by Non-conventional Yeast (NCY)

Fermentation technology has continued to evolve in producing chemicals with industrial applications, and microbial physiology has become a fast-growing and increasingly impressive branch of science (Palsson et al. 1981). *Saccharomyces cerevisiae*, the yeast most employed in the industry nowadays, can convert efficiently glucose to ethanol and has been described as a producer of a large range of chemical compounds as biofuels (ethanol, farnese, isobutanol), pharmaceutical drugs, and carboxylic acids (Abbott et al. 2009). Its robustness makes it tolerant to contamination and inhospitable environment (Jeffries 2006). These phenotypes, among others, have led to the widespread study of *S. cerevisiae* and its development as a model eukaryotic host for chemical biosynthesis (Löbs et al. 2017). Although *S. cerevisiae* presents some limitations regarding metabolic profile, it is incapable of fermenting five-carbon sugars, such as xylose (Hou et al. 2017). Thus, several *S. cerevisiae* strains able to convert xylose to different chemicals have been constructed through genetically engineering strategies (Kim et al. 2013; Moysés et al. 2016). This choice is usually made because of the range of strains and genetic engineering tools available for this yeast. However, xylose metabolism by recombinant strains rarely reaches rates comparable to the glucose, due to different metabolic constraints (reviewed at Moysés et al. 2016).

On the other hand, several yeast species naturally capable of using xylose as a sole carbon source have been isolated and identified, and they present unique characteristics of industrial interest. Species-specific characteristics have been identified among such yeasts, some showing more fermentative metabolism and producing ethanol than others. For instance, *Scheffersomyces stipitis* and *Spathaspora passalidarum* are capable of fermenting xylose to ethanol with high yields and productivity (Veras et al. 2017), whereas *Yarrowia lipolytica* showed a respiratory metabolism, and it is capable of synthesizing and accumulate high levels of intracellular lipids (Beopoulos et al. 2009). In addition to the low number of xylose-consuming yeast strains reported among almost 1000 yeast taxa, there is still few information about physiology and genetics of many species (Nguyen et al. 2006; Kurtzman 2011b).

Due to the incomplete understanding of genetics, metabolism, and cellular physiology and availability of genome-editing tools, when compared with *S. cerevisiae* (Löbs et al. 2017), those yeasts have been named as non-conventional yeasts. The term "non-conventional" yeast is being used by microbiologists and biotechnologists to define yeasts that do not belong to *Saccharomyces cerevisiae* and *Schizosaccharomyces pombe* (Sreekrishna and Kropp 1996; Kręgiel et al. 2017). Different metabolic pathways, variability of fermentative profile, and growth physiology are some of the advantages of using non-conventional yeasts over *S. cerevisiae* (Kręgiel et al. 2017; Rebello et al. 2018)

As the molecular information is becoming available, the divergence of class, genus, and species is being made every year, and reclassification of non-conventional yeast species became common. Non-conventional yeasts present a huge, yet barely exploited, diversity that is an excellent opportunity to achieve some new strategies in biotechnologist research. Many of these non-conventional yeast species exhibit industrially relevant traits such as the ability to utilize complex substrates as nutrients and extreme tolerance against stress and fermentation inhibitors (Mukherjee et al. 2017). Such yeasts may be employed directly in biotechnological processes or used to identify traits of industrial interest for transferring these properties (genes that confers the specific advantage) to other yeasts (Mukherjee et al. 2017). Indeed, physiological, molecular, and genetic characterization and development of toolkits for non-conventional yeasts is readily increasing (Yaguchi et al. 2018), and the potential applications of non-conventional yeast species have been rising considerably (Radecka et al. 2015).

The current review sought to summarize the main characteristics as well as the biotechnological applications of non-conventional yeasts for xylose utilization. First, it will present and discuss the data about the most known non-conventional yeasts focusing on yeasts that naturally assimilate xylose in an efficient way such as *Scheffersomyces*, *Meyerozyma*, *Candida*, *Spathaspora*, and *Kluyveromyces*. Then, it will present data about the yeasts that do not assimilate xylose as *Komagataella* and poorly assimilate xylose as *Yarrowia* and *Ogatae*, which are being genetically modified with the purpose of generating and increasing the xylose incorporation. In each case, basic information about yeast taxonomy, morphology, and physiology will be presented as well as the clearest biotechnological application.

2.2 Yeast Descriptions and Biotechnological Applications

2.2.1 Scheffersomyces

Originally, some of the species assigned to *Scheffersomyces* belonged to *Pichia* genus because of its morphology, phenotype, and absence of nitrate assimilation. In an attempt to revise the polyphyletic genus *Pichia*, Kurtzman and Suzuki (2010) performed a comparative analysis of D1/D2 LSU and SSU rRNA gene sequences, resulting in a strong support for a change of genus of some yeasts within the *Pichia* clade, which included *P. stipitis*, *P. segobiensis*, *P. spartinae*, and some species of *Candida* (Kurtzman et al. 2010). Therefore, the ascosporic species of *Pichia* clade were transferred to the new genus *Scheffersomyces* and renamed as *Scheffersomyces stipitis*, *S. segobiensis*, and *S. spartinae* (Kurtzman et al. 2010).

Additional phylogenetic analysis suggests that the common ancestor of the *Scheffersomyces* genus probably was able to assimilate and ferment D-xylose and cellobiose. Currently, all the yeasts placed in the *Scheffersomyces* genus are either a D-xylose or a cellobiose fermenter; the only exception is the *S. spartinae* species, which does not ferment xylose or cellobiose (Urbina and Blackwell 2012). Thus, a subdivision of the *Scheffersomyces* genus into three smaller subclades was suggested to gather all the yeasts with the different characteristics (Table 2.2). Based on the D-xylose fermentation, and lack of bootstrap support in the D1/D2-SSU tree (52%), the placement of the *S. spartinae* in the *Scheffersomyces* genus is uncertain (Kurtzman et al. 2010).

Scheffersomyces yeasts have some traits that are interesting for several biotechnological applications. *S. stipitis* is one of the most efficient microorganism for xylose and lignocellulose fermentation, while *S. amazonensis* and *S. stambukii* have been highlighted as good xylitol producers (Jeffries et al. 2007; Lopes et al. 2018). *S. stipitis* is also capable of metabolizing several sugars found in lignocellulosic biomass, such as glucose, mannose, galactose, and cellobiose (Agbogbo and Coward-Kelly 2008; Jeffries and Van Vleet 2009). It is a predominantly haploid, homothallic, and hemiascomycetous yeast. During vegetative growth, budded cells are spherical to elongate and pseudohyphae are formed. Asci produce one or two hat-shaped ascospores (Jeffries et al. 2007; Kurtzman and Suzuki 2010). Phylogenetic analyses using the LSU D1/D2 sequences placed *S. stipitis* next to the *Spathaspora* and *Lodderomyces* clade (Fig. 2.2).

Many studies have pointed the ethanol production from xylose by *S. stipitis* cells. It was found that batch fermentations of *S. stipitis* produce high ethanol amounts, reaching yields of 0.40 to 0.48 g/g xylose (Agbogbo and Coward-Kelly 2008; Veras et al. 2017), values that are close to the theoretical maximum yield. *S. stipitis* shows lower cell yields and higher ethanol yields when cultivated in xylose compared to that in glucose as sole carbon source and under low aeration condition (Su et al. 2015). The same behavior was also observed for the xylose-fermenting yeast *S. passalidarum* (Su et al. 2015). To explain this observation, Su et al. (2015) suggested that the net ATP yields are substantially lower when yeast is growing on xylose; consequently, cells need to consume more carbon to make the same amount of biomass. It is important to note that the energy expended per mole of sugar should be

Table 2.2 Example of *Scheffersomyces* species included in the three subclades

Main physiological characteristic of each subclade		
Xylose-fermenting yeast	Cellobiose-fermenting yeast	No xylose assimilation yeast
Scheffersomyces stipitis, S. lignosus, S. illinoinensis, S. insectosa, S. quercinus, S. segobiensis, S. shehatae, S. virginianus	*S. amazonensis, S. coipomoensis, S. ergatensis, S. lignicola, S. queiroziae*	*S. spartinae, S. gosingicus*[a]

[a]A cellobiose-fermenting yeast

The yeasts are mainly isolated from the gut of beetle and rotting wood. Only *S. spartinae* is known to be found exclusively in water environments

Based on data from Kurtzman and Suzuki (2010), Urbina and Blackwell (2012), and Lopes et al. (2018)

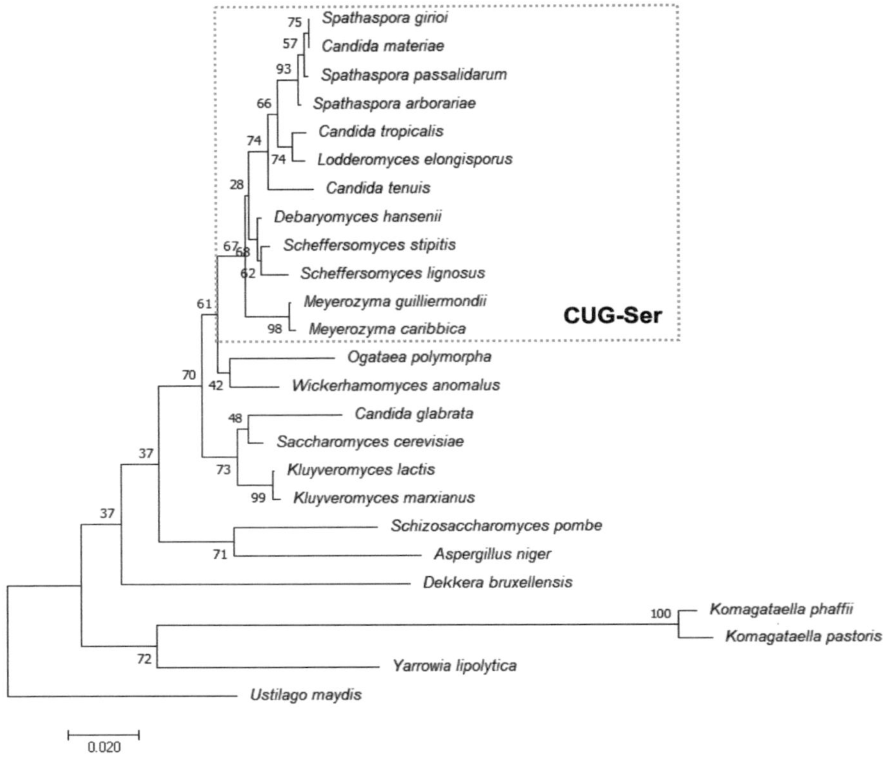

Fig. 2.2 Phylogenetic tree based on LSU-D1/D2 sequence from different microorganism. Sequences were withdrawn from Yeast IP database and NCBI. Alignment was performed using Clustal W program. Tree was reconstructed using neighbor-joining analysis using 25 sequences. It was used default settings and 1000 bootstrap replicates. For a more detailed phylogeny analysis, see Riley et al. (2016)

similar for a C5 or C6 sugar considering sugar uptake and xylulose phosphorylation but the C moles of carbon would be 17% less in each step.

Apart from the great potential of *S. stipitis* for 2G (second generation) bioethanol production, some critical points should be addressed to achieve the employment of this yeast at industrial scale, including improved conversion rates, tolerance to hydrolysate inhibitors, tolerance to high concentrations of ethanol, and necessity of controlled aeration conditions. *S. stipitis* has evolved in low-sugar environment prevailing in the beetle midgut; consequently, its glycolytic flux, ethanol tolerance, and productivity from glucose and xylose are lower than that obtained from *S. cerevisiae* when cultivated on glucose (Jeffries and Van Vleet 2009). In addition, glucose represses genes involved in xylose assimilation as *XYL1*, and *XYL2*, *XYL3* (encoding XR, XDH, XK, respectively), and some genes from PPP (Jeffries et al. 2007; Jeffries and Van Vleet 2009). Indeed, acetic acid, furaldehydes, and phenolic compounds present in lignocellulosic hydrolysates can inhibit growth, viability, and fermentative performance by this yeast (Ma et al. 2017).

Several studies focus on solving some of these problems. Different pretreatment processes and detoxification steps have been suggested to alleviate the formation of toxic compounds or eliminate them before fermentation (Ferreira et al. 2011; Nakanishi et al. 2017; Artifon et al. 2018). Using cell recycling and decreasing temperature for each batch, *S. stipitis* showed increased xylose consumption and ethanol production during cultivation on a detoxified sugarcane bagasse hydrolysate, pretreated with NaOH/ anthraquinone (AQ), and then enzymatically hydrolyzed (Nakanishi et al. 2017). In turn, engineered *S. stipitis* strains with improved traits have been developed mainly by random mutagenesis approaches, adaptive evolution, protoplast fusion, and genome shuffling (Ma et al. 2017; Löbs et al. 2017). Through UV mutagenesis and evolutionary adaptation, Ma et al. (2017) isolated a *S. stipitis* mutant with improved fermentation performance and tolerance against ethanol and lignocellulosic derived inhibitors. Glucose and xylose consumption rates of the mutant were 4.18 g/L.h and 2.39 g/L.h, respectively, values approximately two times higher than those of the parental strain (2.09 g/L.h and 1.4 g/L.h, respectively) in batch fermentation using YP medium with 100 g/L sugar. Similar improvement was observed for ethanol productivity (mutant strain, 2.01 and 1.10 g/L.h; parental strain, 0.99 and 0.62 g/L.h on glucose and xylose, respectively). Besides that, employing membrane integrated continuous fermentation system and rice straw hydrolysate non-detoxified as feedstock, 43.2 g/L ethanol titer and 2.16 g/L/h ethanol productivity were achieved by the yeast (Ma et al. 2017). These values are one of the best results obtained so far (Table 2.3).

Rational metabolic engineering of non-conventional yeasts requires a better understanding of their physiology, biochemistry, and genetics (Jeffries and Van Vleet 2009). Moreover, due to a lack of efficient genome-editing tools adapted to *Scheffersomyces*, relatively few studies have attempted the rational modification of these yeasts (Löbs et al. 2017). In an example, an engineered *S. stipitis* strain, expressing a heterologous *LDH* gene coding for lactate dehydrogenase, produced 41 or 58 g/L lactate from approximately 100 g/L of glucose or xylose, respectively (Ilmen et al. 2007). A more elaborate pathway was introduced in *S. stipitis* for fumaric acid production from xylose, based on the overexpression of a heterologous reductive pathway from *Rhizopus oryzae*. Additionally, codon optimization of the gene sequences introduced deletion of native fumarase genes, and overexpression of heterologous C4-dicarboxylic acid transporter led to 4.67 g/L fumaric acid titer from 20 g/L xylose (Wei et al. 2015). Table 2.3 summarizes some results of synthetic medium or lignocellulosic hydrolysate fermentation by *S. stipitis* strains.

Another interesting biotechnological application of *S. stipitis* in the biorefinery context was in the production of a biosurfactant (Franco Marcelino et al. 2017). Hemicellulosic sugarcane bagasse hydrolysate was used as raw material for the production of a green glycolipidic biosurfactant by *S. stipitis*, which shows significant larvicidal proprieties against *Aedes aegypti*, a vector of neglected tropical diseases. In addition, the biosurfactant showed emulsifying property in hydrophobic

Table 2.3 Fermentation parameters and product formation by *Scheffersomyces* strains

Strain	Feedstock	Condition	Product	Yield (g/g)	Productivity (g/L.h)	Final concentration (g/L)	References
S. stipitis NRRL-Y7124	Medium: minimal Sugar: 40 g/L xylose	Batch: bioreactor; oxygen-limited condition; low initial cell density	Ethanol	0.45	n.d	16.48	Veras et al. (2017)
S. amazonensis UFMG-CM-Y493T	Medium: YP Sugar: 50 g/L xylose	Batch: flasks; low initial cell density; moderate oxygen-limited condition	Xylitol	0.554	0.732	26.40	Cadete et al. (2016b)
S. amazonensis UFMG-CM-Y493T	Medium: YP Sugar: 50 g/L xylose	Batch: flasks; low initial cell density; severe oxygen-limited condition	Xylitol	0.75	0.476	34.24	Cadete et al. (2016b)
S. amazonensis UFMG-CM-Y493T	Medium: rice hull hydrolysate detoxified Sugar: 20 g/L xylose, 4.6 g/L glucose	Batch: flasks; high initial cell density; severe oxygen-limited condition	Xylitol	1.04	0.053	6.30	Cadete et al. (2016b)
S. stambukii UFMG-CM-Y427T	Medium: YP Sugar: 50 g/L xylose	Batch: flasks; low initial cell density; moderate oxygen-limited condition	Xylitol	0.663	0.687	33.0	Lopes et al. (2018)
S. stipitis NRRL Y-7124	Medium: sugarcane bagasse hydrolysate pre-treated with NaOH/AQ; Sugar: 43 g/L glucose;15 g/L xylose	Batch: bioreactor-sequential fed batch cell recycle; high initial cell density	Ethanol	0.32	0.36	18.52	Nakanishi et al. (2017)

*S. stipitis MA301 *mutagenesis and adaptation*	Medium: rice straw hydrolysate non-detoxified, Sugar: 59.3 g/L glucose, 43.7 g/L xylose, w/o supplementation	Batch: bioreactor - continuous fermentation with membrane cell recycle system	Ethanol	0.43	2.16	43.2	Ma et al. (2017)
*S. stipitis ldhL *engineered strain*	Medium: YNB Sugar: 100 g/L xylose, buffered with $CaCO_3$	Batch: flasks; high initial cell density	Lactic acid	0.58	n.d	58	Ilmen et al. (2007)
*S. stipitis ldhL *engineered strain*	Medium: YNB Sugar: 94 g/L glucose, buffered with $CaCO_3$	Batch: flasks; high initial cell density	Lactic acid	0.44	n.d	41	Ilmen et al. (2007)
*S stipitis *engineered strain*	Medium: synthetic medium Sugar: 50 g/L xylose, with 10 g/L $CaCO_3$	Batch: flasks; low initial cell density	Fumaric acid	0.10	n.d	4.67	Wei et al. (2015)

YP yeast peptone, *YNB* yeast nitrogen base, *w/o* without, *n.d.* no data available, *g/L* grams per liter, *g/g* grams per gram, *g/L.h* grams per liter-hour

compounds that has potential for applications in bioremediation processes as well as cosmetic, agricultural, and food formulations.

To gain more information about the yeast *S. stipitis*, its genome was sequenced and analyzed. The genome was estimated in 15.4 Mb, accounting with 5841 predicted genes of which 72% have a single exon (Jeffries et al. 2007; Table 2.4). The species uses the alternative yeast nuclear codon that substitutes serine for leucine when CUG is specified, placing *S. stipitis* within the CUG-serine clade, together with other xylose-fermenting yeasts (Fig. 2.2) (Jeffries et al. 2007; Cadete and Rosa 2018). Kinases, helicases, transporters (sugar and major facilitator superfamily), and domains involved in transcriptional regulation are the most frequent domains characterized. Genes related to xylose assimilation, oxidative pentose phosphate pathway, glycolytic cycle, tricarboxylic acid cycle (TCA), and ethanol production were present in isoforms similar to those found in other yeasts (Jeffries et al. 2007).

In several yeasts, the redox imbalance concern to the first two steps of xylose metabolism, involving xylose reductase and xylitol dehydrogenase enzymes, has a significant impact on xylose fermentation and leads to higher xylitol production (Bruinenberg et al. 1983; de Albuquerque et al. 2014; Veras et al. 2017). However, *S. stipitis* presents various strategies for NAD^+ and NADPH regeneration (Jeffries et al. 2007), and besides that its xylose reductase enzyme, encoded by *XYL1* gene, can use both NADH and NADPH cofactors, although it shows higher affinity toward NADPH (Fig. 2.1) (Verduyn et al. 1985). All these factors may explain the efficient fermentation of xylose observed for *S. stipitis* yeasts.

Although *S. stipitis* is the most studied species from this clade, other *Scheffersomyces* yeasts also possess interesting biotechnological traits. Currently, around 20 yeast species belong to the *Scheffersomyces* genus (Table 2.2). Several of them have the ability to ferment D-xylose to ethanol, including *S. insectosa*, *S. lignosus*, *S. parashehatae*, *S. quercinus*, *S titanus*, *S. xylosifermentans*, and *S. shehatae*. However, all of them showed less efficient xylose fermentation capabilities than *S. stipitis* (du Preez and van der Walt 1983; Liu et al. 2016). In turn, other yeasts were reported as good xylitol producers. In a comparative analysis of two species of *Scheffersomyces*, *S. amazonensis* was able to produce 26.4 g/L xylitol, achieving a yield of 0.554 g/g and a volumetric productivity of 0.732 g/L/h, while *S. stambukii* produced 33.0 g/L xylitol, reaching a yield of 0.663 g/g and a volumetric productivity of 0.687 g/L/h (Cadete et al. 2016a; Lopes et al. 2018). These data reveal the great potential of these strains for the biotechnological production of xylitol (Table 2.3).

2.2.2 Meyerozyma

The genus *Meyerozyma* was proposed to englobe two closely related ascosporic species of yeasts belonging to the genus *Pichia* and, consequently, reduce the polyphyletic characteristic of the genus (Kurtzman et al. 2010). *Pichia guilliermondii* (anamorph *Candida guilliermondii*) and *Pichia caribbica* (anamorph *Candida*

Table 2.4 Description of genomes from different NCY

Yeast	Genome size (Mb)	Number of chromosomes	Gene	GC %	References
Genera: *Scheffersomyces*					
S. stipitis CBS 6064	15.40	8	5.841	41.1	Jeffries et al. (2007)
S. lignosus JCM 9837	16.59	n.d	n.d	41.2	RIKEN (n.d.)
S. shehatae NBRC 1983T	16.75	n.d	6.162	40.9	Okada et al. (2017)
S. stambukii UFMG-CM-Y427	13.24	n.d	5.384	44.7	Lopes et al. (2018)
Genera: *Meyerozyma*					
M. guilliermondii ATCC 6260	10.57	n.d	6.062	43.8	Butler et al. (2009)
M. caribbica MG20W	10.61	n.d	5.390	n.d	De Marco et al. (2018b)
Genera: *Candida*					
C. tropicalis	14.50	n.d	6258	33.1	Butler et al. (2009)
Genera: *Spathaspora*					
S. passalidarum NRRL Y-27907T	13.18	n.d	5.983	37.4	Wohlbach et al. (2011b)
S. arborariae UFMG-HM19.1AT	12.70	8	5.625	31.7	Lobo et al. (2014b)
S. boniae UFMG-CM-Y306T	12.30	n.d	6.046	53.9	Morais et al. (2017)
S. girioi UFMG-CM-Y302	14.81	n.d	7.026	35.0	Lopes et al. (2016)
S. gorwiae UFMG-CM-Y312	14.89	n.d	7.077	35.2	Lopes et al. (2016)
S. xylofermentans UFMG-HMD23.3	15.10	n.d	5.948	35.3	Lopes et al. (2017)
Spathaspora sp. JA1	14.5	n.d	5.268	34.6	Unpublished data
Genera: *Kluyveromyces*					
K. aestuarii NRRL YB-4510	10.05	n.d	n.d	38.3	GenBank GCA_003707555.1
K. dobzhanskii NRRL Y-1974	10.83	n.d	n.d	41.5	GenBank GCA_003705805.2
K. lactis NRRL Y-1140	10.72	6	5.085	38.7	GenBank GCA_000002515.1
K. marxianus DMKU3-1042	10.9	8	4.952	40.1	GenBank GCA_001417885.1
K. nonfermentans NRRL Y-27343	9.56	n.d.	n.d.	35.9	GenBank GCA_003670155.1
K. wickerhamii UCD 54-210	9.80	n.d.	n.d.	40.9	GenBank GCA_000179415.1
Genera: *Komagataella*					
K. pastoris DSMZ 70382	9.4	4	5274	41.34	Mattanovich et al. (2009)

(continued)

Table 2.4 (continued)

Yeast	Genome size (Mb)	Number of chromosomes	Gene	GC %	References
K. phaffii GS115	9.3	4	5313	41.1	De Schutter et al. (2009)
Genera: *Yarrowia*					
Y. lipolytica	20.5	6	510	49.0	Dujon et al. (2004)
Y. porcina	30.4	n.d	n.d	43.7	GenBank GCA_900519025.1
Y. divulgata	21.34	n.d	n.d	50.15	GenBank ID: 71639
Y. yakushimensis	18.83	n.d	n.d	48.30	GenBank GCA_900518995
Y. bubula	20.81	n.d	n.d	46.55	GenBank ID: 71638
Y. deformans	20.99	n.d	n.d	49.75	GenBank ID: 44230
Y. alimentaria	19.83	n.d	n.d	49.20	GenBank GCA_900518985.1
Genera: *Ogataea*					
O. minuta	8.97	7	5.251	47.9	Riley et al. (2016)

Mb megabase, *n.d.* no data available

fermentati) were reclassified to *Meyerozyma guilliermondii* and *Meyerozyma caribbica*, respectively (Wrent et al. 2016), because of its phylogenetic proximity and common production of Coenzyme Q-9 (Kurtzman et al. 2010).

Such classifications were proposed through analysis of the similarity between the DNA sequences of the domains D1/D2 LSU rRNA and the sequence SSU rRNA of the yeasts that have the common ability to produce coenzyme Q-9 (CoQ-9) (Yurkov et al. 2017). Those analyses provided enough data for allowing the redistribution of the CoQ-9-producing species in a distinct genus and also reuniting those species in a different group, named *M. guilliermondii* species complex (Kurtzman et al. 2010).

Like *Scheffersomyces* yeasts, the *Meyerozyma* genus belongs to the *Saccharomycotina* CTG clade and currently englobe two sexual species, *M. guilliermondii* and *M. caribbica* (Fig. 2.2) (Yurkov et al. 2017). The phylogenetic structure of the yeasts that belongs to the clade CTG has been studied for more than four decades and is still not fully understood, in part due to the morphological and metabolic similarity between the yeasts but is also due to the large range of synonyms for each species, what is related with the different names given to the same yeasts but in different sexual states (anamorph/teleomorph) (De Marco et al. 2018a). For example, the yeast *M. guilliermondii* is the sexual state of *C. guilliermondii* (Wickerham and Burton 1954), and a very few wild strains of *C. guilliermondii* are sexually reactive, what represents a teleomorph-anamorph pair (Kurtzman et al. 2010).

Most species belonging to the *Meyerozyma* genus have been found in insects or related environments like fruits and soil (Yurkov et al. 2017). The type yeast of the genus *M. guilliermondii* has been found for the first time in eggs of insects in the

USA, and the other component of the genus *M. caribbica* was found in sugarcane juice, in Cuba (Kurtzman et al. 2010). In addition to *M. guilliermondii* and *M. caribbica*, it was recently proposed the yeast *Meyerozyma amylolytica* A. M. Yurkov and Péter sp. Nov. (Yurkov et al. 2017) as a new species into the *Meyerozyma* genus (Yurkov et al. 2017). This species was found in Germany trees, and like other members of the genus, its origin is related to insects (Yurkov et al. 2017).

Some studies have shown that the yeast *M. guilliermondii* displays antimicrobial activity. This yeast has the ability to inhibit the growth of fungi (Coda et al. 2013), bacteria (Zhao et al. 2010), and even protozoa (Dantán-González et al. 2015). Thus it can be used as a biocontrol agent in the food and agriculture industries, acting on the preservation of fruits and vegetables after the harvest (Yan et al 2018). But even if *M. guilliermondii* is not reported as a pathogen in laboratory tests, the anamorph state of the *yeast M. guilliermondii* (*Candida guilliermondii*) is an opportunistic pathogen (Corte et al. 2015). Because of the proximity with *C. guilliermondii* (more than 98% of nuclear DNA compatibility) (Kurtzman et al. 2010), *M. guilliermondii* have a questionable GRAS (generally recognized as safe) status, confusing the health legislation in many countries (Corte et al. 2015).

M. guilliermondii is used in industrial production of vitamin B2 (riboflavin), due to its flavinogenic potential (Romi et al. 2014). Indeed, most yeasts from the *Candida* genus have the potential to overproduce riboflavin under iron-limitation conditions. Three genes that regulate the riboflavin production in *M. guilliermondii* (*P. guilliermondii*) were identified. Although the specific role of the genes in the cell metabolism is not fully understood, it was also shown that the regulation of iron and riboflavin production is closely related in this yeast, (Protchenko et al. 2011).

Strains of *Meyerozyma* genus have also been evaluated for biotechnological application for production of chemicals, due to their natural ability to assimilate and utilize pentose in its metabolism. When the subject is xylose fermentation, the strains of the anamorph pair *C. guilliermondii* FTI 20037 has received more attention, because it can produce xylitol, with higher yields and productivity than the most strains in its complex, as shown in Table 2.5 (Vaz de Arruda et al. 2017). Xylitol is an important sugar-alcohol with five carbons derived from xylose that can be used as a substitute for the currently used sugar (sucrose). It has many healthy properties, such as it prevents tooth decay and airway inflammations and can also be an intermediate in the production of other chemicals of interest, like ethylene glycol, an important lubricant (Venkateswar Rao et al. 2015).

The *Meyerozyma* species are also able to assimilate the xylose present in lignocellulosic hydrolysates, and even in the presence of inhibitors, they were capable of producing xylitol and/or ethanol (Table 2.5) (Martini et al. 2016; Vaz de Arruda et al. 2017).

M. caribbica is another member of *Meyerozyma* genus but is less studied. *M. caribbica* is also a hexose and pentose naturally ferment yeast, and it can be used as a biocontrol microorganism as well. In addition to that, this species has a nonpathogenic anamorph state, known as *Candida fermentati*, which may facilitate its biotechnological utilization in food and health industries (Romi et al. 2014). Phylogenetically *M. caribbica* is closely related with *M. guilliermondii* (Fig. 2.2),

Table 2.5 Fermentation parameters and product formation by *Meyerozyma* strains

Yeast	Feedstock	Condition	Product	Yield (g/g)	Productivity (g/L.h)	Final concentration (g/L)	References
M. guilliermondii FT 200037	Medium: sugarcane straw hemicellulosic hydrolysate Sugar: 50.6 g/L xylose; 9.75 g/L glucose	Batch: flasks	Xylitol	0.67	0.18	8.69	Hernández-Pérez et al. (2016)
M. guilliermondii FT 200037	Medium: sugarcane bagasse hydrolysate Sugar: 65.56 g/L xylose; 7.89 g/L glucose	Batch: bioreactors	Xylitol	0.67	0.29	20.80	Vaz de Arruda et al. (2017)
M. guilliermondii BL13	Medium: soy bean hull hydrolysate Sugar: 18.25 g/L xylose; 17.75 g/L glucose	Batch: flasks	Ethanol	0.41	1.4	16.8	da Cunha-Pereira et al. (2017)
M. guilliermondii CCT7783	Medium: sugarcane bagasse hydrolysate Sugar: 40 g/L xylose	Batch: flasks	Xylitol Ethanol	X: 0.19 E: 0.14	X: n.d E: 0.22	n.d	Martini et al. (2016)

X xylitol, *E* ethanol, *n.d.* no data available, *g/L* grams per liter, *g/g* grams per gram, *g/L.h* grams per liter-hour

and these species are often misidentified by the classic molecular differentiation method employing the sequence of the D1/D2 domains (Romi et al. 2014). Romi et al. (2014) proposed a new protocol to differentiate the species using ITS-RFLP combined with an in silico selection of restriction enzymes with further in vitro validations that proved work for *M. guilliermondii* and *M. caribbica* species differentiation (Romi et al. 2014).

The *Meyerozyma* species have been studied for a long time, but the biotechnological utilization is currently around the riboflavin production and prevention of mold decay in fruits and vegetables (Corte et al. 2015). Research from the past decade are focusing in the great natural potential of xylose assimilation and fermentation capability, shown by the *Meyerozyma* yeasts, for industrial uses of biomass in biorefineries (Hernández-Pérez et al. 2016).

2.2.3 Candida

The genus *Candida* belongs to the Ascomycota phylum, Hemiascomycetes class, and Saccharomycetales order and covers all imperfect ascomycetes species whose phylogenetic relationships have not been defined (Hommel and Ahnert 1999). *Candida* is a dimorphic fungus that can present as yeast globose, ellipsoidal, cylindroidal or elongate, occasionally ogival, triangular, or lunate (Kurtzman 2011b). In addition, this genus presents a great variety of CoQ types, with the species that were found to sporulate having teleomorphic counterparts in 11 different genera (Schauer and Hanschke 1999).

Candida is phylogenetically heterogeneous and covers more than 310 species. The identification of *Candida* species is complex and based on morphological, physiological, chemotaxonomic characteristics and in the sequence analysis of the D1/D2 domain of rDNA 26S (Odds 2010). *Candida* species show a wide range of physiological properties and are widespread in natural habitats with a high content of organic matter, with low or high temperature, and high osmolarity. The majority is mesophilic and grow well in temperatures between 25 and 30 °C (Spencer and Spencer 1997; Odds 2010). The yeasts in this genus are aerobic, not growing in an anaerobic condition, but some strains survive and reproduce under microaerophilic conditions (Lachance 2011).

Candida species are mostly found in nutrient-rich habitats such as soil, rotting vegetation (fruit peel and decaying fruit), plants (leaf surfaces, nectaries, and nectar of flowers, flower petals, and other flower parts), and insects that feed on plants (Hommel and Ahnert 1999). They are able to assimilate and ferment xylose (from wood degradation) and cellobiose, and also a smaller number of species oxidize aliphatic hydrocarbons (components of the cuticle plant), degrade starch, or use methanol as a possible metabolite of pectin catabolism (Schauer and Hanschke 1999). Naturally, most of them require individual vitamins produced mainly in plant materials.

Like other yeasts, *Candida* species are able to metabolize xylose through via oxido-reductive pathway (Fig. 2.1) (Chakravorty et al. 1962). Among the xylitol-producing yeasts, *Candida* strains have shown high XR-specific activity, associated with dependence on the NADPH cofactor, which favors the production of xylitol (Veras et al. 2017). Candida strain application on fermentative processes has gained attention due to their capacity to produce xylitol with high yields, up to 65 and 85% of theoretical maximum (Bier et al. 2007). Recently, different *Candida* species have been considered for the production of chemicals, mainly ethanol and xylitol, using different substrates, such as horticultural waste (Zhang et al. 2012), olive tree pruning (Mateo et al. 2015), rice straw (Swain and Krishnan 2015), corncob (Kumar et al. 2018), and sugarcane bagasse (Tizazu et al. 2018). The most common *Candida* species evaluated in biotechnological processes are discussed in the next topics.

2.2.3.1 *Candida magnoliae*

Candida magnoliae is a member of the of *Starmerella* clade, presents CoQ 9 (Yamada and Kondo 1972) and % mol GC of 60% (Nakase and Komagata 1971). Based on D1/D2 LSU rRNA gene sequences, *C. magnoliae* is a close relative to *C. sorbosivorans* and *C. geochares* (Rosa et al. 2003). They were collected and isolated from bees and their habitats such as flower of *Magnolia* sp. (Magnoliaceae), concentrated orange juice, and gut of a bee (Kurtzman 2011b). This species is interesting in biotechnology mainly in the production of substitute sweeteners such as erythritol from glucose with productivity of 0.54 g/L.h and 43% of conversion yield based on glucose (Yang et al. 1999), mannitol in the presence of fructose and glucose as carbon sources reaching a productivity of 1.03 g/L.h and 83% of conversion (Song et al. 2002), and xylitol from xylose-containing lignocellulosic hydrolysates with productivity of 0.21 g/L.h and 54.5% of conversion yield based on xylose (Wannawilai et al. 2017).

2.2.3.2 *Candida tropicalis*

Candida tropicalis, a member of the phylogenetic placement of *Lodderomyces-Spathaspora*, also presents CoQ 9 (Yamada and Kondo 1972) with % mol GC between 34.4 and 36.1% (Stenderup and Bak 1968; Nakase and Komagata 1971; Meyer et al. 1975). The first strains of this species were found in Jamaica from fruits of *Stenocereus hystrix* (Cactaceae) and *Royen cactus* flower (*Cephalocereus royenii*, Cactaceae) and as a contaminant of a clinical specimen of *Candida albicans* in Brazil (Kurtzman 2011b). *C. tropicalis* has a high degree of similarity with related species and exhibits variability among lineages, so their identification by growth characteristics is problematic. Despite this being one of the most frequently found clinical yeast species after *C. albicans* (Moran 2004), in recent years many kinds of research have been carried out using *C. tropicalis* in the fermentation of xylose to produce xylitol (Table 2.6).

Table 2.6 Fermentation parameters and product formation by *Candida* strains

Yeast	Feedstock	Condition	Product	Yield (g/g)	Productivity (g/L.h)	Final concentration (g/L)	References
C. shehatea D45-6	Medium: YP Sugar: 40 g/L xylose	Batch: bioreactor	Ethanol	0.41	0.71	34.3	Guan et al. (2013)
C. tropicalis	Medium: rice straw hydrolysate supplemented with yeast extract and beef extract Sugar: 50 g/L glucose; 20gL xylose; two-stage	Batch: flasks	Ethanol Xylitol	E: 0.50 glucose X:0.19 xylose	E: 1.43 X:0.06	E: 26.2 X: 3.6	Swain and Krishnan (2015)
C. tropicalis NBRC0618	Medium: olive tree pruning acid hydrolysate enriched with $MgSO_4$, KH_2PO_4 Sugar: 25 g/L xylose	Batch: flasks	Ethanol Xylitol	E: 0.38 X: 0.23	E: 0.0328* X: 0.017*	n.d	Mateo et al. (2015)
C. tropicalis	Medium: detoxified hemicellulosic liquor from sugarcane bagasse Sugar: 104.1 g/L xylose	Batch: flasks	Xylitol	0.46	0.27	32.0	Vallejos et al. (2016)
C. magnoliae TISTR 5663	Medium: lignocellulosic hydrolysates with furfural Sugar: 30 g/L xylose; 300 mg/L furfural	Batch: fed-batch flasks	Xylitol	n.d	0.21	15.4	Wannawilai et al. (2017)
C. shehatae ATCC 22984	Medium: rice straw hydrolysate Sugar: 40 g/L glucose; 20 g/L xylose	Batch: flasks (immobilized cells)	Ethanol	0.50 glucose	0.12	17.2	Yuvadetkun et al. (2018)
C. tropicalis MTCC 6192	Medium: detoxified corn cob hydrolysate pH 5.0 Sugar: 56 g/L xylose	Batch: bioreactor	Xylitol	0.85	0.41	29.6	Kumar et al. (2018)

(continued)

Table 2.6 (continued)

Yeast	Feedstock	Condition	Product	Yield (g/g)	Productivity (g/L.h)	Final concentration (g/L)	References
C. tropicalis MTCC 184	Medium: sugarcane bagasse hydrolysate treated with ultrasonic enhancement Sugar: 20.9 g/L xylose	Batch: flasks (immobilized cells)	Xylitol	0.66	0.80	11.9	Tizazu et al. (2018)

X xylitol, *E* ethanol, *n.d.* no data available, *g/L* grams per liter, *g/g* grams per gram, *g/L.h* grams per liter-hour, ∗*Kg/Kg.h* kilograms per kilograms-hour

C. tropicalis has been applied in different bioprocesses, especially for the production of ethanol or xylitol. It usually shows better productivity and yields when compared to other *Candida* strains. *C. shehatae* produced ethanol with a productivity of 0.12 g/L.h and yield of 0.50 g/g in a medium containing rice straw hydrolysate with 20 g/L of xylose and 40 g/L of glucose (Yuvadetkun et al. 2018). On the other hand, *C. tropicalis* produced ethanol in a medium containing rice straw hydrolysate presenting productivity 2 times and yields 12 times higher than *C. shehatae* (Table 2.6) (Swain and Krishnan 2015). The differences are significant even if a direct comparison cannot be made due to the variations in the experimental conditions.

In relation to the production of xylitol, *C. tropicalis* has been outstanding for achieving better productivity and yields, especially when applied in a medium containing lignocellulosic hydrolysates. C. *tropicalis* reached xylitol productivity and yields up to 0.80 g/L.h and 0.85 g/g in a medium containing lignocellulosic hydrolysates (Table 2.6). *C. tropicalis'* ability to produce xylitol efficiently has been associated with its high NADPH-XR-specific activity, which favors higher xylose consumption rate and xylitol secretion (Cadete et al. 2016a; unpublished data).

2.2.4 Spathaspora

The genus *Spathaspora* was described by Nguyen et al. (2006) to accommodate the teleomorphic species *Spathaspora passalidarum*, which form specific allantoid asci, with a single, elongated, and curve-ended ascospore, very distinct from any other known yeast. The first species was isolated from the gut of the wood-boring beetle *Odontotaenius disjunctus* (Coleoptera: Passalidae), collected in Louisiana (USA). The vegetative cells are predominantly globose, formed by budding, and septate hyphae and pseudohyphae are present. In sexual reproduction, an allantoid ascus is formed without conjugation and contains the single ascospore surrounded by a persistent membrane (Nguyen et al. 2006; Kurtzman 2011b). Supported by phylogenetic analyses of D1/D2 LSU, the genus also included other related taxa as the anamorphic species *Candida jeffriesii* and *Candida materiae* (Nguyen et al. 2006; Barbosa et al. 2009). The phylogeny in Fig. 2.2 shows the placement of the *Spathaspora* and *C. materiae* strains in the same clade.

Other isolates of *S. passalidarum* were collected later from rotting wood samples (Cadete et al. 2012; Ren et al. 2014), wood-boring beetles, and galleries (Souza et al. 2017) or soil (Rodrussamee et al. 2018). The higher number of isolates obtained from decaying wood and soil/gallery samples suggests that the species might be more associated with this environment rather than the gut of beetles. However, the hypothesis of a symbioses yeast-insect cannot be discarded until further and extensive ecological studies are carried out (Nguyen et al. 2006; Cadete and Rosa 2018).

The mainly metabolic trait of *S. passalidarum* is its high capacity to ferment xylose. Until now, only few yeast strains have been described as natural xylose fermenters, among them are *Scheffersomyces (Pichia) stipitis*, *Candida tenuis*, and *Pachysolen tannophilus* (Slininger et al. 1982; Suh et al. 2003; Wohlbach et al. 2011a). Together with *S. stipitis*, *S. passalidarum* has the highest native capacity for xylose fermentation (Jeffries and Van Vleet 2009; Veras et al. 2017; Cadete and Rosa 2018). Batch cultivation reaches ethanol yields up to 0.48 g/g xylose, showing higher specific ethanol productivity under oxygen-limited condition than aerobic condition (Table 2.7) (Cadete et al. 2016a, b; Veras et al. 2017). In addition, *S. passalidarum* strains have the advantageous ability to co-utilize different substrates like glucose, xylose, and cellobiose and to produce higher ethanol yields on xylose than on glucose (Long et al. 2012; Su et al. 2015).

The high ethanol yield and productivity of *S. passalidarum* make clear the great potential of this strain for 2G bioethanol processes. However, like other microorganisms, *S. passalidarum* are inhibited by toxic compounds present in the lignocellulosic hydrolysates (Almeida et al. 2011; Su et al. 2018). Therefore, research efforts have focused on surpassing the toxicity of hydrolysate to achieve the practical use of this strain in biotechnological processes. Previous studies have shown that *S. passalidarum* was able to co-ferment glucose, xylose, and cellobiose mixtures in synthetic medium (ethanol yield of 0.42g/g) or hardwood hydrolysates without acetic acid or furaldehydes (ethanol yield of 0.34 g/g). Nevertheless, a significant inhibition and delay in the fermentation were observed when another kind of hydrolysate (AFEX) was used, which contains approximately 1.5 g/L of acetic acid (Long et al. 2012). Changing pretreatment process, a sugarcane bagasse hydrolysate with 43 g/L of glucose and 15 g/L of xylose, with no acetic acid, furfural, or hydroxymethylfurfural was obtained (Nakanishi et al. 2017). Employing this hydrolysate, the maximum yield achieved for fed-batch fermentation by *S. passalidarum* with cell recycle was 0.46 g/g, with a productivity of 0.81 g/L/h.

Evolutionary strategies have been successfully applied to develop *S. passalidarum* strains with improved traits. An evolved strain obtained by Morales et al. (2017) showed increased tolerance to acetic acid, achieving an ethanol yield of 0.48 g/g in the presence of 4.5 g/l acetic acid or 0.36 g/g in a medium containing 100% of non-detoxified *Eucalyptus* autohydrolysate. In another study, a *S. passalidarum* strain (E11) was selected after several rounds of batch adaptation, cell mating, and high-throughput screening, with traits that confer resistance to toxins found in hydrolysates and improved xylose fermentation abilities (Su et al. 2018).

The second species of the genus *Spathaspora* was isolated from rooting wood collected in the Brazilian Atlantic Rainforest and Cerrado biomes, and it was named as *S. arborariae* (Cadete et al. 2009). The species present similar asci and ascospore to the type strain *S. passalidarum*. Differently from this one, the new species can assimilate L-sorbose and produce both ethanol and xylitol as major products from xylose (Cadete et al. 2009, 2016b). Recently, other ten species were identified as belonging to *Spathaspora* genus. Cadete et al. (2013) described the species *S. brasiliensis*, *S. roraimensis*, *S. suhii*, and *S. xylofermentans*, which were mainly xylitol producers, albeit some of them are also able to produce ethanol from xylose. Wang et al. (2016) proposed the new species *S. allomyrinae* for a yeast isolated from the gut of beetles in China. Lopes et al. (2016) reported three novel strains *S.*

Table 2.7 Fermentation parameters and product formation by *Spathaspora* strains

Strain	Feedstock	Condition	Product	Yield (g/g)	Productivity (g/L.h)	Final concentration (g/L)	References
S. passalidarum NRRL-Y-27907	Medium: minimal Sugar: 40 g/L xylose	Batch: bioreactor; oxygen-limited condition; low initial cell density	Ethanol	0.44	n.d.	16.36	Veras et al. (2017)
S. passalidarum NRRL-Y-27907	Medium: maple hemicellulosic hydrolyzate Sugar: 65 g/L xylose; 35 g/L glucose and nutrient suplementation	Batch: bioreactor; low initial cell density	Ethanol	0.34	0.6	38.00	Long et al. (2012)
S. passalidarum NRRL-Y-27907	Medium: sugarcane bagasse hydrolyzate pretreated with NaOH/AQ Sugar: 43 g/L of glucose; 15 g/L xylose and nutrient suplementation	Batch: bioreactor - sequential fed-batch cell recycle; high initial cell density	Ethanol	0.46	0.81	23.30	Nakanishi et al. (2017)
S. passalidarum strain (E11) *mutagenesis and adaptation	Medium: synthetic medium with 4.5 g/L acetic acid	Batch: flasks; microaerobic conditions	Ethanol	0.48	0.23	n.d.	Morales et al. (2017)
S. passalidarum strain (E11) *mutagenesis and adaptation	Medium: 100% of non-detoxified *Eucaliptus* autohydrolysate, Sugar: 80 g/L glucose; less than 20 g/L xylose	Batch: flasks; microaerobic conditions	Ethanol	0.36	0.55	n.d	Morales et al. (2017)
S. roraimanensis UFMG-CM-Y477T	Medium: YP Sugar: 40–50 g/L xylose	Batch: flasks; low initial cell density	Xylitol	0.56	n.d.	27.4	Cadete et al. (2016a)

(continued)

Table 2.7 (continued)

Strain	Feedstock	Condition	Product	Yield (g/g)	Productivity (g/L.h)	Final concentration (g/L)	References
S. roraimanensis UFMG-CM-Y477T	Medium: detoxified sugarcane hemicellulosic hydrolysate, Sugar: 50.2 g/L D-xylose, 5.3 g/L glucose, supplemented with yeast extract	Batch: flasks; low initial cell density	Xylitol	0.61	n.d.	17.1	Cadete et al. (2012)
S. arborariae UFMG-CM-Y352T	Medium: YP Sugar: 40–50 g/L xylose	Batch: flasks; low initial cell density	Ethanol Xylitol	E: 0.32 X: 0.18	n.d.	E: 16.0; X: 9.0	Cadete et al. (2016a)
Spathaspora. sp. JA1	Medium: mineral Sugar: 40 g/L xylose	Batch: bioreactor; low initial cell density	Xylitol	0.75	0.20	22.62	Unpublished data

X xylitol, *E* ethanol, *n.d.* no data available, *YP* Yeast peptone, *g/L* grams per liter, *g/g* grams per gram, *g/L.h* grams per liter-hour

girioi, *S. gorwiae*, and *S. hagerdaliae*, which are also isolated from rotting wood in Brazil. Finally, other two species were isolated from rotting wood samples in Brazil, *S. boniae* and *S. piracicabensis*, reported by Morais et al. (2017) and Varize et al. (2018).

Among the *Spathaspora* strains already characterized, *S. passalidarum*, *S. arborariae*, *S. gorwiae*, *S. hagerdaliae*, and *S. piracicabensis* are mainly ethanol producers from xylose, while other *Spathaspora* are mostly xylitol producers (Cadete et al. 2016b; Cadete and Rosa 2018; Varize et al. 2018). In addition to *Candida tropicalis* strains that are characterized as a great xylitol producers (de Albuquerque et al. 2014), *Spathaspora* species have shown an interesting potential for xylitol production, reaching yields in the range of 0.21–0.61 g/g xylose depending on fermentation medium and process conditions (Cadete et al. 2012, 2016b). Until now, the best xylitol production by *Spathaspora* strains on xylose fermentation medium was obtained for the *S. roraimanensis* UFMG-CM-Y477T (27.4 g/L xylitol, Y = 0.56), *S. xylofermentans* UFMG-CM-Y478T (24.4 g/L xylitol, Y= 0.51) (Cadete et al. 2012, 2016b), and for a not completely identified strain *Spathaspora* sp. JA1 (22.6 g/L xylitol, Y= 0.75) (Trichez et al. unpublished data). In the case of detoxified sugarcane hemicellulosic hydrolysate as feedstock, an uncharacterized *Spathaspora* UFMG-XMD-23.2 strain showed significant xylitol yield (17.1 g/L xylitol, Y = 0.61 g/g) (Cadete et al. 2012, 2016b).

Genomic information about native xylose-assimilating yeasts may contribute to understand xylose metabolism and be a source of genes to improve xylose fermentation. The genomic sequence of *S. passalidarum* was reported by Wohlbach et al. (2011a), and more recently other *Spathaspora* genomes have been deposited (Lobo et al. 2014a; Lopes et al. 2017; Morais et al. 2017; Varize et al. 2018) (Table 2.4).

Like other yeasts, the *Spathaspora* strains show the genes involved in the xylose oxido-reductase pathway *XYL1*, *XYL2*, and *XYL3* that encodes for a xylose reductase (XR), a xylitol dehydrogenase (XDH), and a xylulokinase (XK), respectively (Wohlbach et al. 2011a; Lobo et al. 2014a; Cadete et al. 2016b; Lopes et al. 2017; Morais et al. 2017). Genome sequencing revealed that differently from other *Spathaspora* spp., *S. passalidarum* contains two copies of xylose reductase-encoding gene. The second copy (*XYL1.2*) encodes an NADH-preferred XR greatly expressed in oxygen-limited conditions, while *XYL1.1* encodes a strictly NADPH-dependent enzyme (Wohlbach et al. 2011a; Cadete et al. 2016b; Cadete and Rosa 2018). Previous studies have suggested that the better balance between cofactor supply and demand through the XR and XDH leads to higher ethanol yields and efficient xylose fermentation by *S. passalidarum* cells (Hou 2012; Cadete et al. 2016b). In addition to *S. passalidarum*, apparently, *S. arborariae*, *S. boniae*, and *S. gorwiae* also have XR enzymes that use both cofactors NADH and NADPH, although showing a preference for the last one (Fig. 2.1) (Lopes et al. 2016; Cadete et al. 2016b; Morais et al. 2017). In turn, xylitol producers, as *S. brasiliensis*, *S. roraimanensis*, *S. suhii*, and *S. xylofermentans*, have genes encoding strictly NADPH-dependent XR enzymes (Cadete et al. 2016b). Thus, since all *Spathaspora* spp. characterized until now have shown XDH activities strictly NAD^+-dependent, the imbalance of cofactors, in relation to the first steps of xylose metabolism, apparently favors the high xylitol yields observed in these strains.

In summary, *Spathaspora* species are able to use xylose and lignocellulosic hydrolysates in the fermentation process. *S. passalidarum* is one of the best ethanol producers from xylose. The high yields and productivities show the great potential of this yeast for 2G ethanol production. In addition, the capacity of some *Spathaspora* to produce xylitol opens the possibility of the employment of these yeasts in production processes for other bio-products. Bioprospection of novel yeasts, data from genome sequencing, metabolic engineering approaches, and process optimization may contribute to the development of yeast strains or better process conditions to an effective industrial application of these yeasts.

2.2.5 Kluyveromyces

The *Kluyveromyes* genus, located in the Saccharomycetes class, was created in 1956 by van der Walt in order to fit the yeast *Kluyveromyces polysporus* (currently a member of the *Vanderwaltozyma* genus). In the following years, it was noted that the type yeast *K. polysporus* presented some resemblances features to species previously accepted in the *Saccharomyces* genus like ascus deliquescence, ascospores, and robust fermentation (van der Walt 1965; Lachance 2007). That resulted in the transfer of important yeasts like *S. fragilis*, *S. marxianus*, and *S. lactis* to the *Kluyveromyces* genus (van der Walt 1965; Fonseca et al. 2008). Since then, many changes in the number of accepted species followed as taxonomic studies were being made. These changes kept happening until advances in genetic sequencing and multigene analysis tools enabled more accurate studies, decreasing the number of species for the current six members, *K. aestuarii*, *K. dobzhanskii*, *K. lactis*, *K. marxianus*, *K. nonfermentans*, and *K. wickerhamii*, distributing the species previously attributed to *Kluyveromyces* in the genera *Kazachstania*, *Nakaseomyces*, *Tetrapisispora*, *Vanderwaltozyma*, and *Lachancea* (Kurtzman and Robnett 2003; Kurtzman 2011b). The current type species is *Kluyveromyces marxianus* (Lachance 2007).

As described by Lachance (2011), generally, *Kluyveromyces* cells can be ovoid, ellipsoid, cylindrical, or, elongate. Its members can reproduce by asexual means, with multilateral budding, where pseudohyphae may be formed but never true hyphae. Sexual reproduction also occurs with or without the production of ascus preceding conjugation (Lachance 2011). They are considered to be thermotolerant yeasts, with the maximum growing temperature ranging from 35 to 52°C. With the exception of *K. nonfermentans*, all species are capable of fermenting glucose (Lane and Morrissey 2010). Genomic analysis of the *Kluyveromyces* genus representative members shows that genome sizes, in general, vary from 9.5 to 11 Mb, having in mind that other information like chromosome number and protein count are more available for *K. marxianus* and *K. lactis* (Table 2.4).

Among the six species present in the genus, *K. marxianus* and *K. lactis* are the ones largely used in biotechnological studies and industrial processes. A rare trait present in both species is lactose assimilation, enabled by the presence of genes *LAC12*, which encodes a permease responsible for lactose uptake, and *LAC4*, encoding for β-galactosidase, the enzyme responsible for hydrolyzing lactose molecules to glucose and galactose. In addition, the recurrent isolation of both yeasts from dairy products, like cheese, yogurt, and kefir, along with many years of secure use, has granted them the GRAS and QPS status (generally regarded as safe and qualified presumption of safety) (Morrissey et al. 2015). Moreover, more details about *K. marxianus* and *K. lactis* are presented, and the possible application of these yeasts on the biotechnological valorization of xylose is discussed.

2.2.5.1 *Kluyveromyces marxianus*

Despite not having many studies involving matters like physiology and metabolism, the type species *Kluyveromyces marxianus* is largely applied in industries (Lane and Morrissey 2010). Some features like the assimilation of inulin, by the activity of the *INU1* gene; high thermotolerance (up to 52°C), one of the fastest growth rates among eukaryote organisms (Groeneveld et al. 2009); and its respiro-fermentative nature are differentials that make this yeast very attractive for use in industrial processes (Morrissey et al. 2015). For this reason, not only different applications for this species are being largely studied but also biotechnological tools are being developed in order to overcome bottlenecks related to this yeast and to establish it as a new synthetic biology platform (Cernak et al. 2018).

Industrial applications of *K. marxianus* are mainly related to protein expression, enzyme production, and ethanol production (Fonseca et al. 2008; Lane and Morrissey 2010). Having that in mind, among the most important commercial applications are the production of its native enzymes. One of them is inulinase, encoded by the previously mentioned *INU1* gene. Inulinase (β-2,1-D-fructan fructanohydrolase) targets the β-2,1 linkage of inulin, a polyfructan consisting of linear β-2,1-linked fructose, and hydrolyzes it into fructose. Thus, this enzyme can be used for the production of syrups with high fructose content. Another possible application of *K. marxianus* is the production of ethanol using whey permeate and Jerusalem artichokes as substrate (Yang et al. 2015). As mentioned before, tolerance to high temperatures is an interesting trait, once it allows this yeast to be utilized together with enzymes responsible for liberating monomers from the lignocellulosic feedstocks during simultaneous saccharification and fermentation (SSF). Another use of *K. marxianus* that has been recently reviewed and drawn attention to is the production of high alcohols, such as 2-phenylethanol (Morrissey et al. 2015). Characterized by having a strong rose scent, 2-PE can be used in cosmetics and ffragrance applications, and as a flavor additive in the food industry (Morrissey et al. 2015). The biological production, via the Ehrlich pathway, is an important alternative for the chemical route production, as the latter involves the use of benzene, which can be hazardous when applied in food and cosmetics.

The metabolism of xylose is present in *K. marxianus* once it possesses *XYL1*, *XYL2*, and *XKS1* genes. The transport of pentoses to the intracellular environment is also possible in this yeast, but its kinetic parameters depend on substrates and conditions applied. In 2003, Stambuk et al. demonstrated that not only the presence of glucose in the medium is responsible for inhibiting the uptake of xylose but also that under anaerobic conditions, the transport presents low affinity and high capacity, while aerobic conditions induce the production of the high-affinity xylose transporter system (Stambuk et al. 2003). Although the necessary genes for xylose conversion to ethanol are present, unbalanced redox reactions in the xylose reductase pathway (XR and XDH) usually result in low ethanol yield and formation of undesired co-products under low-oxygen conditions (Varela et al. 2017). Therefore, metabolic engineering strategies have been employed to optimize xylose conversion through XR-XDH or XI pathways. One of these studies has constructed the strain K. *marxianus* YZJ088 by the combination of XR from *N. crassa* and XDH from *S. stipitis*, achieving high productivity (2.49 g/L/h) and yield (0.38 g/g) of ethanol at high temperatures with xylose as the only carbon source (Zhang et al. 2015). It was also reported that the recombinant *K. marxianus* YZJ119 strain was able to produce 44.58 g/L of ethanol and 32.03 g/L of xylitol from detoxified pre-treated corncob lignocellulosic hydrolysate (Table 2.8) (Zhang et al. 2016).

Xylitol, a sugar alcohol considered a very promising bio-based chemical, is also one of the possible products originated from xylose utilization by *K. marxianus*. Many genetic engineering strategies had also been applied for this end. Among them, the overexpression of heterologous transporters allowed to generate *K. marxianus* YZJ074 strain, which produces 101.30 xylitol with a productivity of 2.81 g/L/h even at temperatures as high as 45°C (Table 2.8) (Zhang et al. 2014).

2.2.5.2 *Kluyveromyces lactis*

K. lactis is considered one of the most studied yeasts and a model organism among the "non-conventional" group, making it unavoidable to compare it with *S. cerevisiae*. A major difference between them is the preference for respiration in *K. lactis* over sugar fermentation in *S. cerevisiae* (Dias et al. 2012). The low-cost production of bovine chymosin by *K. lactis* is considered to be a milestone for the expression of high-eukaryote enzymes in microorganisms (van Ooyen et al. 2006). By 2006 more than 40 proteins were already produced by *K. lactis*, and this number increased to around 100 proteins in 2016 (van Ooyen et al. 2006; Spohner et al. 2016). The applications of this yeast and its products also have a wide range, like the production of β-galactosidase and chymosin in the food industry and the production of human interleukin 1-β and interferon-α for the pharmaceutical industry (Rodicio and Heinisch 2013; Spohner et al. 2016).

Not many studies were made with *K. lactis* to analyze its potential of using lignocellulosic biomass in bioprocesses. However, it was already observed that this yeast is capable of metabolizing xylose and obtaining products like d-xylonate and xylitol (Nygård et al. 2011). In this work, a xylose dehydrogenase gene from

Table 2.8 Fermentation parameters and product formation by *Kluyveromyces* strains

Strain	Feedstock	Condition	Product	Yield (g/g)	Productivity (g/L.h)	Final concentration (g/L)	References
K. marxianus YZJ088	Medium: YP Sugar: 128.46 g/L xylose	Batch: flasks; oxygen limited condition; high cell density	Ethanol Xylitol	E: 0.38 X: 0.09	E: 2.49	E: 44.95 X: 11.09	Zhang et al. (2015)
K. marxianus YZJ088	Medium: YP Sugar: 103.97 g/L xylose; 40.96 g/L glucose	Batch: flasks; oxygen limited condition; high cell density	Ethanol Xylitol	E: 0.38 X: 0.16	E: 2.14	E: 51.43 X: 15,04	Zhang et al. (2015)
K. marxianus YZJ074	Medium: YP Sugar: 101.41 g/L xylose; 37.79 g/L glycerol	Batch; bioreactor; 45°C	Xylitol	0.99	2.81	101.30	Zhang et al. (2015)
K. marxianus YZJ119	Medium: YP Sugar: 150 g/L corncob residue (56.91 g/L xylose, 12.91 g/L glucose, 6.76 g/L acetic acid)	Batch: flasks; simultaneous saccharification and co-fermentation	Ethanol Xylitol	n.d	E: 0.74 X:0.53	E: 44.58 X: 32.03	Zhang et al. (2016)

X xylitol, *E* ethanol, *n.d.* no data available, *YP* yeast peptone, *g/L* grams per liter, *g/g* grams per gram, *g/L.h* grams per liter hour

Trichoderma reesei was expressed, and the resulting strain produced xylonic acid with higher productivity and yield than a *S. cerevisiae* strain expressing the same gene (Nygård et al. 2011). This study can indicate not only possible advantages of this yeast over *S. cerevisiae*, such as better xylose transport and tolerance to acid accumulation, but also points out that these traits represent the potential of the application *K. lactis* in studies involving xylose and lignocellulosic biomass (Nygård et al. 2011).

Although there are not many studies regarding xylose utilization involving the remaining species of this genus, interesting findings involving them can be observed in recent years. As examples, we can cite *K. aestuarii* being depicted as a quality bioindicator (Araujo and Hagler 2011) and the application of a toxin produced by *K. wickerhamii* for controlling spoilage yeasts in winemaking (Comitini and Ciani 2011).

2.2.6 Komagataella

Komagataella belongs to Saccharomycetales order and Saccharomycetaceae family. The genus was proposed by Yamada et al (1994) when the partial sequences of 18S and 26S rRNAs subunits of methanol-assimilating yeasts were phylogenetically analyzed. The study exhibited that partial sequences of *Pichia pastoris* were different when compared to other strains. Therefore, it was proposed the new genus *Komagataella* for *P. pastoris* that was phylogenetically distinct from the other genera examined. The proposal was not initially accepted because just a few numbers of strains were used in the comparison. However, Kurtzman and Robnett (1998) analyzed the divergence of the D1/D2 LSU rRNA gene sequences among approximately 500 species of Ascomycetous yeasts. When they compared D1/D2 sequences, the divergences found in *P. pastoris* were confirmed. Subsequently, multigenic sequence analysis supported the phylogenetic difference position of *Pichia pastoris* (renamed to *Komagataella* pastoris). Thus, the genus became accepted and currently has seven species.

The majority of *Komagataella* species was isolated from habitats such as tree exudates and decomposing woods, sources capable of supplying methanol for the growth of these yeasts. The type species of *K. pastoris* was first isolated in 1919 from a chestnut tree in France by Guilliermond. Furthermore, most strains of *K. pastoris* and *K. pseudopastoris* were found in exudates and decomposing wood in Hungary. The species *K. phaffii*, *K. populi*, *K. ulmi*, *K. kurtzmanii*, and *K. mondaviorum* were isolated from exudates of different trees at the USA in the states of California, Arizona, and Illinois. The growth and isolation of these species were mostly done in YM agar medium at 25 °C, as described by Yarrow (1998) and Kurtzman (2011a).

The colonies of *Komagataella* sp. have spherical to ovoid shapes; the coloration can be white to cream and may have moderately lobate margins. In the asexual reproduction, haploid cells divide by multilateral budding and do not present pseudohyphae or true hyphae (Kurtzman 2011b). For sexual reproduction, the formation of ascospore is hat-shaped which may range from 1 to 4 and can be conjugated or not. Also, known species are homothallic. An important characteristic of the genus is the ability to grow at high cell densities and ferment glucose. Because those species are methylotrophic, they can use methanol as the only carbon source (Kurtzman 2011b). The nitrate cannot be used as the only source for nitrogen during growth. The seven species have very similar phenotypic and fermentative characteristics. Therefore, they cannot be differentiated with tests that are usually used for yeast taxonomy. In this way, multigenic sequence analysis, such as D1/D2 LSU rRNA, ITS, EF-1α, and other gene regions, are required for species identification of the genus (Kurtzman 2011b).

Komagataella species play a major rule in biotechnological applications. These yeasts provide an efficient host system for heterologous protein production, having both industrial and pharmaceutical importance. In addition, metabolic engineering has become an interesting application for these yeasts to produce different chemicals. On account of that, the genome sequencing of strains *K. pastoris DMSZ* and *K. phaffii* GS115 were performed (Mattanovich et al. 2009; De Schutter et al. 2009). The approximate size of the genome of both species is 9.4 Mb, containing 4 chromosomes and more than 5000 genes (Table 2.4). The information provided by genome data highlights the advantages of *Komagataella* sp. as a heterologous expression system, and it is essential for metabolic engineering studies.

Initially, the Phillips Petroleum Company used *Komagataella (Pichia) pastoris* as single-cell protein (SCP) because of its ability to assimilate methanol as a carbon source and ability to grow in high cell densities. The SCP was marketed for high protein animal additive. Posteriorly, the oil crisis in the 1970s increased the methanol price, and the SCP system was no longer favorable. In the following years, the genetic manipulation of *K. pastoris* strains and the development of protocols, vectors, and isolation of the AOX gene led to the creation of the heterologous expression system known today. Invitrogen Company developed the "*Pichia* Expression Kit," which is widely commercialized and used in laboratories for heterologous protein expression. The strains used in the expression kit were phenotypically identified as *K. (Pichia) pastoris*. However, multigene sequence analysis determined that strains used in the Invitrogen Expression Kit were from *K. phaffii* (Kurtzman 2009; Ahmad et al. 2014).

The success of the *Komagataella* sp. expression system is due to established genetic manipulation techniques, with recombinant proteins that can be expressed both extracellular and intracellular. In addition, the expressed proteins may undergo post-translational eukaryotic modifications such as glycosylation, disulfide bonds, and proteolytic processing. The vectors integrated into its genome facilitate the genetic stability of the recombinant elements even in continuous and large-scale fermentation processes. Furthermore, the species have the "generally recognized as safe" (GRAS), status, and for this reason, more than 500 biopharmaceutical proteins and a great number of other recombinant enzymes have been produced since 2009 (Yang and Zhang 2018). Many studies have been reporting the expression of recombinant proteins that can be useful in food, feed, detergent, clothing industries, and more (Table 2.9).

Table 2.9 Protein expression of different gene donor organism by *Komagatella*

Yeast	Gene donor organism	Feedstock	Condition	Product	Protein expression (g/L)	Activity	References
K. phaffi GS115	*Thermomyces lanuginosusThicoderma reese*	Medium: basal salt	Batch: bioreactor; fed-batch	Cellobiohydrolaseb-mannanaseXylanase	C: 5.984 M: 1.142 X: 1.2	C: 18.21 U/mL M: 109 U/mL X: 138 U/mL	Mellitzer et al. (2012)
K. phaffi X33	*Yarrowia lipolytica*	Medium: basal salt	Batch: bioreactor	Lipase	2.82	38500 U/mL	Zhou et al. (2015)
K. phaffi X33	Human	Medium: basal salt	Batch	Inter-leukin 25	0.1*	n.d	Liu et al. (2013)
K. phaffi GS115	Human	Medium: BMMY	Batch: flasks	Irisin hormone	0. 077*	n.d	Duan et al. (2015)
K. phaffi GS115	*Trichoderma koningii*	Medium: basal salt	Batch: bioreactor; fed-batch	Serine protease	3.2	5200 U/mg	Shu et al. (2016)
K. phaffi KM71	Aspergillus aculeatus ATCC16872	Medium: FM22	Batch: bioreactor; fed-batch	Polygalacturonase	n.d.	2408 U/mL	Abdulrachman et al. (2017)

X xylanase, *C* cellobiohydrolase, *M BMMY* buffered methanol complex media, b-mannanase; *n.d.* no data available, *g/L* grams per liter, *g/g* grams per gram, *U/mL* units per mililiter, *U/mg* Units per miligram, *numbers converted

There is some research showing the potential of some *Komagataella* strains to use xylose and glucose from lignocellulosic biomass to achieve the production of value-added products. In this context, *K. pastoris* DSM 70877 was cultivated on mixtures of glucose and xylose to produce chitin-glucan complex (CGC) (Araújo et al. 2017). This biopolymer is a component of the inner cell wall of yeast and has several applications in food, cosmetics, and pharmaceutical industries. When cultivated in the glucose/xylose mixture, the CGC and xylitol were produced. The xylose started to be assimilated after the glucose depletion, but no cell growth was detected. However, the production of xylitol reached a final concentration of 7.64 g/L. Therefore, this work presents great potential for co-production of bioproducts using the lignocellulosic biomass.

According to Li et al. (2015), *K. phaffii* GS115 strain is able to use the xylose at low rates for specific growth (0.0075 h^{-1}). In order to enhance the efficiency in xylose assimilation, the host strain GS115 was engineered with XI (xylose isomerase). After the introduction of XI pathway, an evolutionary engineering strategy was also applied, resulting in a strain (GS-XI^{SB50}) capable to assimilate xylose at higher rates (specific growth: 0.0193 h^{-1}). Also, this engineered strain was used for heterologous expression of β-mannanase enzyme produced in a medium containing xylose. Finally, this engineered strain is able to grow in xylose medium producing enzymes with industrial importance.

A few numbers of microorganisms have been engineered to produce xylonic acid by the overexpression of xylose dehydrogenases either from *Caulobacter crescentus* or *Trichoderma reesei*. *K. phaffii* is a promising yeast for production of xylonic acid because of its tolerance to low pHs and ability to grow at very high cell densities. The xylonic acid can be used as a precursor for other chemicals and has several industrial applications such as building blocks of polymers. Recently, new putative xylose dehydrogenase (XDH) genes from bacteria and fungi were identified by phylogenetic analysis. Then, three of those were chosen for genetic engineering of *K. phaffii* X33. Recombinant strains expressing each gene were evaluated by their ability to produce xylonic acid, and the best candidate genes were chosen for further analysis. Strains were able to produce up to 36.2 g/L of xylonic acid from 40 g/L xylose, which accounts for a yield of 0.95 g/g, under the best fermentative evaluated conditions (patent under subscription by the number BR102018001359-9_870180005782). Finally, strain's capability to produce xylonic acid on sugarcane biomass hydrolysate was demonstrated.

K. phaffii was also engineered for production of acid lactic. This acid has applications such as in food, pharmaceutical, and textile industries. Also, it is a monomer of a biodegradable plastic called poly-lactic acid (PCA). In that way, de Lima et al. (2016) genetically modified strains of *K. phaffii* (X-33 and GS115) that were capable to produce lactic acid using glycerol as the only carbon source at limited oxygen conditions. They used a bovine lactate dehydrogenase (LDH) and lactate transporter to make the production of this acid possible. The best strain (named GLS) obtained by the introduction of both genes reached a yield of 0.7 g/g that was really close to the maximum theoretical yield (1 g/g). However, the production of lactic acid from xylose using recombinant K. *phaffii* strains was not reported.

As a conclusion, *Komagataella* species can be explored in a range of biotechnological applications (Mattanovich et al. 2009). The main use for those species is the heterologous expression of recombinant proteins that have important uses in several industry sectors. Also, with the genome sequencing, new studies are showing potential as a platform of metabolic engineering and synthetic biology. In that way, new approaches and methodologies are being explored to achieve future developments of heterologous expression, metabolic engineering, and synthetic biology of these yeasts.

2.2.7 Yarrowia

Yarrowia (van der Walt and von Arx 1980), a fungal genus of the family Diposdascaceae, constructed from the previously ascosporic state of *Candida lipolytica*, is composed of 13 species. This genus was, prior DNA comparisons, considered monotypic, containing the single species *Yarrowia lipolytica* (Dujon et al. 2004; Kurtzman 2005; Kurtzman et al. 2010). Some descriptions of new *Yarrowia* species clade were recognized in the last years based on D1/D2 sequence comparisons (Nagy et al. 2013). For example, Groenewald and Smith (2013) have described two novel species, *Y. deformans* and *Y. yakushimensi*, Nagy et al. (2013) described *Y. divulgata*, and Crous et al. (2017) described another species called *Y. parophonii*. These new species are phenotypically indistinguishable from Y. *lipolytica*. Kurtzman (2005) and Crous et al. (2017) based on phylogenetic analysis of nucleotide sequences from domains D1/D2 26S rDNA have proposed new relocation in the genus *Yarrowia*, such as *Y. galli* (basionym *C. galli*) and *Y. oslonensis* (basionym *C. oslonensis*). The discovery of new species is very recent, so *Y. lipolytica* is still considered the most important yeast of the *Yarrowia* genus. The other species have not yet been widely studied, and absence of growth is recorded for xylose (Kurtzman 2005; Nagy et al. 2013, 2014; Crous et al. 2017). Therefore, only *Y. lipolytica* description will be covered in this section. But it is important to note that some yeast isolates identified as *Y. lipolytica* based on phenotypic characteristics and used in industries may really be a member of other *Yarrowia* clade yet incorrectly identified (Nagy 2015).

Yarrowia lipolytica is strictly an aerobic nonpathogenic ascomycetous yeast. This yeast shares some properties common to filamentous fungi being distantly associated with other yeasts (Kurtzman and Robnett 1998; Dujon et al. 2004; Gonçalves et al. 2014). It was originally isolated from lipid-rich materials as rancid butter and first identified, in the 1960s, as *Candida lipolytica* and reclassified as *Endomycopsis lipolytica*, *Saccharomycopsis lipolytica*, and finally *Yarrowia lipolytica* (Liu et al. 2015). Strains of this species are usually isolated as contaminant or spoilage organisms from lipid-rich or protein-rich environments such as dairy products, meat, poultry, and olive oil (Najjar et al. 2011; Nagy et al. 2013; Liu et al. 2015; Li and Alper 2016). *Y. lipolytica* has a haplodiplontic cycle. The most of natural isolated strains are haploid. A diploid strain has little occurrence, but both states are stable in laboratory conditions (Dujon et al. 2004; Liu et al. 2015). The process of genetic regulation that leads to dimorphism is still unknown (Liu et al. 2015).

Y. lipolytica is classified as oleaginous yeast and has the ability to assimilate different carbon sources that can be soluble in water, such as glucose, glycerol, alcohols, and acetate, or insoluble in water, such as fatty acids, triacylglycerols, and alkanes. Besides, it is tolerant to environmental stress like presence of salt, low temperatures, and acidic and alkaline pH (Gonçalves et al. 2014; Liu et al. 2015). These characteristics would allow the use of low-cost raw materials, such as wastewaters, industrial fat, glycerol co-product of biodiesel production, and molasses. *Y. lipolytica* is also exceptionally resistant to inhibitors associated with lignocellulosic biomass pretreatments (Ryu and Trinh 2017).

Thus, *Y. lipolytica* has multiple biotechnological and industrial applications such as the production of citric, isocitric, α-ketoglutaric and succinic acids, biomass to be used as single-cell protein, biosurfactants, and flavoring lactones to be used in the food industry (Groenewald and Smith 2013; Nagy et al. 2013). It can also secrete enzymes with functional applications, like lipases, acid or alkaline proteases, and RNase (Kurtzman et al. 2010; Najjar et al. 2011). Currently, one of the main interests in *Y. lipolytica* is its ability to produce and store lipids (Fig. 2.1) (up to 36% of its dry cell weight), which can be used for biodiesel production (Abghari and Chen 2014).

There have been inconsistent reports regarding xylose utilization by *Y. lipolytica* strains (Ledesma-Amaro et al. 2016; Spagnuolo et al. 2018). Some studies claiming native xylose metabolism demonstrated that xylose can be converted into lipids when cells of *Y. lipolytica* Po1g are transferred in the stationary phase into high xylose concentrations. Even in this condition, xylose was not a primary source for biomass production (Blazeck et al. 2014). In addition, Tsigie et al. (2011) showed that the maximum growth of *Y. lipolytica* Po1g (derivative of the wild-type strain W29 with a series of genetic modification) (Yeastern Biotech Co. Ltd.) in a medium containing xylose as only carbon source is 9.89 g/L with a short lag period and without adaptation time and have produced 6.68 g/L of lipids using sugarcane bagasse hydrolysate as a carbon source. On the other hand, the majority of studies suggests that *Y. lipolytica* is unable to grow with xylose as a primary carbon source without subjecting it to adaptation or starvation periods (Kurtzman 2011b; Stephanopoulos 2013; Zhao et al. 2015). These reports conflicted even among identical strains.

The xylose pathway in *Y. lipolytica* was not proven to be functional. Even though this yeast has putative genes coding for XR, XDH, and pentose transport genes, XK was the only enzyme with activity in this pathway (Ryu et al. 2016; Li and Alper 2016; Ryu and Trinh 2017). Then, in order to improve xylose catabolic phenotype in *Y. lipolytica*, pathway engineering is necessary. Overexpression of XR and XDH of *S. stipitis* in *Y. lipolytica* allowed moderate growth in xylose. When cultured in co-fermentation of 20 g/L glycerol and 80 g/L xylose, the strain produced 18 g/L of biomass, 7.64 g/L of lipids, and 9.13 g/L of citrate (Stephanopoulos 2013; Li and Alper 2016). Li and Alper (2016), introducing heterologous XR and XDH genes from *S. stipitis*, established a *Y. lipolytica* strain that is able to use xylose as sole carbon source and produce over 15 g/L of intracellular lipids, with a productivity of 0.19 g/L.h. Recently, Niehus et al. (2018) showed that overexpression of native XDH, XR, and XK genes allowed *Y. lipolytica* to grow on xylose as a sole carbon source and produced 16.5 g/L of lipids in high yield (3.44 g/g sugars) on a non-detoxified agave bagasse hydrolysate.

The activation of the native xylose pathway in *Y. lipolytica* ATCC MYA-2613 could be achieved by adaptation of the strain to grow on xylose as sole carbon source through serial culture transfers (Ryu et al. 2016). In the first transfer, the cell growth was very low and started only after 3 days. After 15 generations, *Y. lipolytica* grew in xylose with a specific growth rate of 0.10 h^{-1} and consumed 2.23 g/L xylose, producing 0.18 g/L xylitol, in 72 h. By transcriptomic analysis of the adapted strain, it was shown that *Y. lipolytica* has 16 putative xylose transporters and xylose-degrading metabolic enzymes (XR, XDH, and XK) required for xylose assimilation. In comparison with growth on glucose, *Y. lipolytica* reached a much lower OD, producing a high yield of xylitol, and did not completely consume xylose. This phenotype suggests that xylose consumption is not efficient because the XDH step was limiting. In addition, it was demonstrated that XDH is transcriptionally repressed by glucose. Ryu and Trinh (2017) have reported that overexpression of pentose transporters YALI0C04730p and YALI0B00396p and rate-limiting D-xylitol dehydrogenases (XDH) allowed activation of the dormant pentose metabolism of *Y. lipolytica* ATCC MYA-2613 and improved xylose assimilation approximately 50% in comparison with the parental strain. However, the improved strain grows on xylose ten times slower than on glucose.

The overexpression of XR and XDH of *S. stipitis* is necessary but not sufficient to permit *Y. lipolytica*'s growth on xylose. Thus, to improve oils and citric acid production from lignocellulosic materials by *Y. lipolytica*, an additional overexpression of the endogenous xylulokinase (XK) was necessary (Ledesma-Amaro et al. 2016). The recombinant strain was able to produce high titers of lipids, up to 20 g/L, on a medium containing xylose as sole carbon source and 50 g/L of lipids on xylose medium co-fed with glycerol using high xylose/nitrogen ratio. When lower xylose/nitrogen ratio and higher pH were used, citric acid was produced up to 80 g/L.

It can be observed that the mechanisms of many enzymes involved in different biochemical reactions to xylose assimilation remain unclear in *Y. lipolytica* (Liu et al. 2015; Spagnuolo et al. 2018). However, the fully sequenced genome, physiological capacity, and particular genetic advantages are what make *Y. lipolytica* a promising platform to produce added-value chemicals and biofuels (Table 2.10) (Dujon et al. 2004; Stephanopoulos 2013; Li and Alper 2016). Among the advantages of *Y. lipolytica*, the following can be mentioned: (1) protein is secreted primarily by the co-transcription pathway; (2) it has a high secretion capacity and low glycosylation modification; (3) it is a non-pathogenic yeast; (4) versatility of metabolites is produced; (5) it is robust in culture; (6) it is an obligate aerobe; (7) it is able to grownon a variety of substrates; (8) simple alterations in the fermentative process can modulate metabolite production (Gonçalves et al. 2014; Liu et al. 2015; Spagnuolo et al. 2018).

Table 2.10 Fermentation parameters and product formation by *Yarrowia* strains

Yeast	Feedstock	Condition	Product	Yield (g/g)	Productivity (g/L.h)	Final concentration (g/L)	References
Y. lipolytica Po1g	Medium: sugarcane bagassse hydrolysate Sugar: 13.6 g/L xylose; 4 g/L glucose	Batch: flasks	Lipid	0.58	0.073	6.68	Tsigie et al. (2011)
Y. lipolytica XR and XDH of S. stipitis overexpressed	Medium: synthetic medium with C/N ratio adjusted to 100 Sugar: 20 g/L glycerol; 80 g/L xylose	Batch: bioreactor	Biomass lipids citrate	B: 0,18 L: 0.07 C: 0.9	L: 0.033	B: 18 L: 7.64 C: 9.13	Stephanopoulos (2013)
Y. lipolytica XR and XDH of S. stipitis overexpressed	Medium: YNB Sugar: 160 g/L xylose	Batch: bioreactor	Lipids	0.12	0.19	15	Li and Alper (2016)
Y. lipolytica overexpressed native XDH, XR and XK	Medium: non-detoxified agave bagasse hydrolysate Sugar: 15.5 g/L glucose; 20.0 g/L xylose	Batch: flasks	Lipids	3.44	1.85	16.5	Niehus et al. (2018)
Y. lipolytica XR and XDH of S. stipitis overexpressed	Medium: YNB Sugar: 150 g/L xylose	Batch: bioreactor	Citric acid	0.53	0.91	80	Ledesma-Amaro et al. (2016)

XR xylose reductase, *XDH* xylose dehidrogenase, *XK* xylulokinase, *B* biomass, *L* lipids, *C* citrate, *C/N* Carbon/Nitrogen, *YNB* Yeast nitrogen base

2.2.8 Ogataea

First, the genus *Hansenula* included species with ascopores in pellicle shape, hat shape, or saturn shape. After analyzing the yeast with saturn form, it was transferred to the *Williopsis* genus (Yamada et al. 1994). Nuclear DNA analyses revealed that 75% of the *Hansenula minuta* DNA was related to *Pichia lindneri*, the *Pichia*-type species. Kurtzman et al. (2011b) concluded with these analyses that the nitrate assimilation ability was not enough to separate the genera *Hansenula* and *Pichia*. After the analyses, all ascopores species with hat shape and nitrate assimilation ability were classified as *Pichia* (previously classified as *Hansenula*) (Yamada et al. 1994).

The results of 18s RNA analyses of *P. membraefariens* and *P. anomala* showed that the two species were phylogenetically different. Therefore, was proposed the creation of three new genera: *Ogataea*, *Kuraishia*, and *Kakazawea* (Yamada et al. 1994). The genus *Ogataea* was proposed in 1994 by Yamada, which was composed of yeasts previously identified as *Pichia* and *Hansenula.* At the beginning, five species and two varieties were proposed to the *Ogataea* genus: *Ogataea clucozyma*, *Ogataea minuta minuta variatie*, *Ogataea minuta nonfermentans variatie*, *Ogataea philodendra*, *Ogataea polymorpha*, and *Ogataea henricie* (Yamada et al. 1994). In the last edition of the book *The Yeasts: A Taxonomic Study*, 31 species of *Ogataea* were accepted to the genus (Kurtzman 2011b).

The genus belongs to Saccharomycetables family. Colonies are butyroid to mucoid. The cells can be globose, ellipsoid, ovoid, or cylindrical. In the asexual reproduction, cell division is by multilateral budding on a narrow base, and budded cells are spherical to ellipsoidal. True hyphas are not formed, but pseudohyphae if formed consist of few elongated T cells (Kurtzman 2011a).

Among the several species of *Ogatae*, *O. methanolica*, *O. minuta*, *O. thermomethanolica*, and *O. polymorpha* are being studied and applied in biotechnology processes. The first three species are employed mainly as heterologous expression systems (Kuroda et al. 2008; Tsai and Huang 2008; Puseenam et al. 2018).

After the yeast *Kluveromyces marxianus*, *Ogataea polymorpha* is the second yeast thermotolerant that can grow in temperature of 50° C, an advantage of using simultaneous saccharification and fermentation (SSF). *O. polymorpha* can grow using glucose, cellobiose, and xylose as carbon source; the yeast can also produce ethanol using glycerol. The ethanol yield and productivity of *O. polymorpha* using xylose are very low. But it could be increased raising the fermentation temperature (from 30°C to 37°C) in a simultaneous saccharification and fermentation (SSF) process (Ryabova et al. 2003) and through genetic engineering techniques such a riboflavin-deficient mutant under suboptimal supply with flavins. The flavin limitation apparently makes the pyruvate be redistributed via a flavin-independent pathway to ethanol production (Ryabova et al. 2003).

In a different approach, ethanol yield and productivity of *O. polymorpha* were improved by engineering the enzyme xylose reductase (XR). Firstly, the native XR was engineered to reduce affinity toward NADPH. After genes coding for modified XR, native XDH and XK were overexpressed in strain CBS4732. Then, the strain CBS4732 was modified to be unable to utilize ethanol as a carbon source, and the gene PDC1, coding for pyruvate decarboxylase (PDC) was cloned and overexpressed. The strain resulting was named 2EthOH$^-$. Finally the strain 2EthOH$^-$ was able to produce up to 10 g of ethanol per liter at 45 °C using xylose as carbon source (Kurylenko et al. 2014).

2.3 Conclusions

Production of fuels and chemicals through environmentally friendly and cost-effective processes has been proposed and evaluated based on the utilization of sugars present in the lignocellulosic biomass as raw material and yeast as the catalyst. As yeasts promptly assimilate hexose sugars (glucose), much effort has been employed to obtain strains with high efficiency in pentose (xylose) assimilation and conversion. In this sense, due to the diversity and metabolic potential, several non-conventional yeasts species were isolated, identified, and physiologically and genetically characterized in the last years.

Bioprospecting of novel species and increasing the knowledge about non-conventional yeast genome and physiology opened new opportunities to explore the available biodiversity as well as identify new industrially relevant features. In addition, metabolic and evolutionary engineering strategies allowed further improvement and diversification of the bioconversion efficiency of xylose-containing hydrolysates to products of interest.

The non-conventional yeasts can be divided into two groups, the first with species naturally capable of efficiently metabolize xylose, belonging to *Scheffersomyces*, *Meyerozyma*, *Candida*, and *Spathaspora*. The second group, with species that have limited capacity to naturally metabolize xylose, but efficient strains have mainly been obtained mainly by genetic engineering strategies, belonging to *Komagataella*, *Ogatae*, and *Yarrowia*. Yeasts from *Komagataella*, *Ogatae*, and *Kluyveromyces* have been mostly used for the production of heterologous proteins of industrial application, in the pharmaceutical, food, feeding, detergents, and clothing industries. But recent developments allowed production of chemicals, such as xylonic acid, ethanol, and xylitol, with *K. phaffii*, *Ogatae*, and *Kluyvermomyces*. *S. passalidarum* and *S. stipitis* show great potential for production of ethanol 2G due to their high efficiency to ferment xylose, whereas other species as *Spathaspora*, *Scheffersomyces*, *Candida*, and *Meyerozyma* shows great potential for xylitol production. Finally, *Yarrowia* strains have been mostly employed for the production of lipids and organic acids.

In conclusion, robust strains for several applications have been obtained and evaluated in the most varied conditions. Even if this field has space to explore the improvement of strain's performance before industrial use, new products based on non-conventional yeast processes should be in the market in the coming years.

References

Abbott DA, Zelle RM, Pronk JT, van Maris AJA (2009) Metabolic engineering of *Saccharomyces cerevisiae* for production of carboxylic acids: current status and challenges. FEMS Yeast Res 9:1123–1136. https://doi.org/10.1111/j.1567-1364.2009.00537.x

Abdulrachman D, Thongkred P, Kocharin K, Nakpathom M, Somboon B, Narumol N, Champreda V, Eurwilaichitr L, Suwanto A, Nimchua T, Chantasingh D (2017) Heterologous expression of Aspergillus aculeatus endo-polygalacturonase in Pichia pastoris by high cell density fermentation and its application in textile scouring. BMC Biotechnol 17. https://doi.org/10.1186/s12896-017-0334-9

Abghari A, Chen S (2014) Yarrowia lipolytica as an oleaginous cell factory platform for production of fatty acid-based biofuel and bioproducts. Front Energy Res 2. https://doi.org/10.3389/fenrg.2014.00021

Agbogbo FK, Coward-Kelly G (2008) Cellulosic ethanol production using the naturally occurring xylose-fermenting yeast, Pichia stipitis. Biotechnol Lett 30:1515–1524. https://doi.org/10.1007/s10529-008-9728-z

Ahmad M, Hirz M, Pichler H, Schwab H (2014) Protein expression in Pichia pastoris: recent achievements and perspectives for heterologous protein production. Appl Microbiol Biotechnol 98:5301–5317. https://doi.org/10.1007/s00253-014-5732-5

Akinterinwa O, Khankal R, Cirino PC (2008) Metabolic engineering for bioproduction of sugar alcohols. Curr Opin Biotechnol 19:461–467. https://doi.org/10.1016/j.copbio.2008.08.002

Almeida JRM, Runquist D, Sànchez Nogué V, Lidén G, Gorwa-Grauslund MF (2011) Stress-related challenges in pentose fermentation to ethanol by the yeast Saccharomyces cerevisiae. Biotechnol J 6:286–299. https://doi.org/10.1002/biot.201000301

Araujo FV, Hagler AN (2011) Kluyveromyces aestuarii, a potential environmental quality indicator yeast for mangroves in the State of Rio de Janeiro, Brazil. Braz J Microbiol 42:954–958

Araújo D, Freitas F, Sevrin C, Grandfils C, Reis MAM (2017) Co-production of chitin-glucan complex and xylitol by Komagataella pastoris using glucose and xylose mixtures as carbon source. Carbohydr Polym 166:24–30. https://doi.org/10.1016/j.carbpol.2017.02.088

Artifon W, Bonatto C, Bordin ER, Bazoti SF, Dervanoski A, Alves SL, Treichel H (2018) Bioethanol Production From Hydrolyzed Lignocellulosic After Detoxification Via Adsorption With Activated Carbon and Dried Air Stripping. Front Bioeng Biotechnol 6. https://doi.org/10.3389/fbioe.2018.00107

Bajwa PK, Phaenark C, Grant N, Zhang X, Paice M, Martin VJJ, Trevors JT, Lee H (2011) Ethanol production from selected lignocellulosic hydrolysates by genome shuffled strains of Scheffersomyces stipitis. Bioresour Technol 102:9965–9969. https://doi.org/10.1016/j.biortech.2011.08.027

Ballesteros I, Negro MJ, Oliva JM, Cabañas A, Manzanares P, Ballesteros M (2006) Ethanol production from steam-explosion pretreated wheat straw. In: McMillan JD, Adney WS, Mielenz JR, Klasson KT (eds) Twenty-seventh symposium on biotechnology for fuels and chemicals. Humana Press, Totowa, pp 496–508

Barbosa AC, Cadete RM, Gomes FCO, Lachance M-A, Rosa CA (2009) Candida materiae sp. nov., a yeast species isolated from rotting wood in the Atlantic Rain Forest. Int J Syst Evol Microbiol 59:2104–2106. https://doi.org/10.1099/ijs.0.009175-0

Beopoulos A, Cescut J, Haddouche R, Uribelarrea J-L, Molina-Jouve C, Nicaud J-M (2009) Yarrowia lipolytica as a model for bio-oil production. Prog Lipid Res 48:375–387. https://doi.org/10.1016/j.plipres.2009.08.005

Bier MCJ, Maranho LT, Azevedo JAM, Junios LS d S (2007) Crescimento e consumo de Xilose de Candida guilliermondii na fermentação submersa utilizando-se bagaço de cana-de-açúcar. Evidência 7:119–130

Blazeck J, Hill A, Liu L, Knight R, Miller J, Pan A, Otoupal P, Alper HS (2014) Harnessing Yarrowia lipolytica lipogenesis to create a platform for lipid and biofuel production. Nat Commun 5. https://doi.org/10.1038/ncomms4131

Bozell JJ, Petersen GR (2010) Technology development for the production of biobased products from biorefinery carbohydrates—the US Department of Energy's "Top 10" revisited. Green Chem 12:539. https://doi.org/10.1039/b922014c

Bruinenberg PM, Bot PHM, Dijken JP, Scheffers WA (1983) The role of redox balances in the anaerobic fermentation of xylose by yeasts. Eur J Appl Microbiol Biotechnol 18:287–292. https://doi.org/10.1007/BF00500493

Butler G, Rasmussen MD, Lin MF, Santos MAS, Sakthikumar S, Munro CA, Rheinbay E, Grabherr M, Forche A, Reedy JL, Agrafioti I, Arnaud MB, Bates S, Brown AJP, Brunke S, Costanzo MC, Fitzpatrick DA, de Groot PWJ, Harris D, Hoyer LL, Hube B, Klis FM, Kodira C, Lennard N, Logue ME, Martin R, Neiman AM, Nikolaou E, Quail MA, Quinn J, Santos MC, Schmitzberger FF, Sherlock G, Shah P, Silverstein KAT, Skrzypek MS, Soll D, Staggs R, Stansfield I, Stumpf MPH, Sudbery PE, Srikantha T, Zeng Q, Berman J, Berriman M, Heitman J, Gow NAR, Lorenz MC, Birren BW, Kellis M, Cuomo CA (2009) Evolution of pathogenicity and sexual reproduction in eight Candida genomes. Nature 459:657–662. https://doi.org/10.1038/nature08064

Cadete RM, Rosa CA (2018) The yeasts of the genus *Spathaspora* : potential candidates for second-generation biofuel production: Spathaspora yeasts as candidates for 2G bioethanol production. Yeast 35:191–199. https://doi.org/10.1002/yea.3279

Cadete RM, Santos RO, Melo MA, Mouro A, DL GÃ§a, Stambuk BU, Gomes FCO, Lachance M-A, Rosa CA (2009) *Spathaspora arborariae* sp. nov., a d-xylose-fermenting yeast species isolated from rotting wood in Brazil. FEMS Yeast Res 9:1338–1342. https://doi.org/10.1111/j.1567-1364.2009.00582.x

Cadete RM, Melo MA, Dussán KJ, Rodrigues RCLB, Silva SS, Zilli JE, Vital MJS, Gomes FCO, Lachance M-A, Rosa CA (2012) Diversity and Physiological Characterization of D-Xylose-Fermenting Yeasts Isolated from the Brazilian Amazonian Forest. PLoS One 7:e43135. https://doi.org/10.1371/journal.pone.0043135

Cadete RM, Melo MA, Zilli JE, Vital MJS, Mouro A, Prompt AH, Gomes FCO, Stambuk BU, Lachance M-A, Rosa CA (2013) Spathaspora brasiliensis sp. nov., Spathaspora suhii sp. nov., Spathaspora roraimanensis sp. nov. and Spathaspora xylofermentans sp. nov., four novel d-xylose-fermenting yeast species from Brazilian Amazonian forest. Antonie Van Leeuwenhoek 103:421–431. https://doi.org/10.1007/s10482-012-9822-z

Cadete RM, de las Heras AM, Sandström AG, Ferreira C, Gírio F, Gorwa-Grauslund M-F, Rosa CA, Fonseca C (2016a) Exploring xylose metabolism in Spathaspora species: XYL1.2 from Spathaspora passalidarum as the key for efficient anaerobic xylose fermentation in metabolic engineered Saccharomyces cerevisiae. Biotechnol Biofuels 9. https://doi.org/10.1186/s13068-016-0570-6

Cadete RM, Melo-Cheab MA, Viana AL, Oliveira ES, Fonseca C, Rosa CA (2016b) The yeast Scheffersomyces amazonensis is an efficient xylitol producer. World J Microbiol Biotechnol 32. https://doi.org/10.1007/s11274-016-2166-5

Canilha L, Milagres AMF, Silva SS, Almeida e Silva JB, Felipe MGA, Rocha GJM, Ferraz A, Carvalho W (2010) Sacarificação da Biomassa Lignocelulósica através de pré-hidrólise ácida seguida por hidrólise enzimática: uma estratégia de "desconstrução" da fibra vegetal. Revista Analytica 44

Cernak P, Estrela R, Poddar S, Skerker JM, Cheng Y-F, Carlson AK, Chen B, Glynn VM, Furlan M, Ryan OW, Donnelly MK, Arkin AP, Taylor JW, Cate JHD (2018) Engineering *Kluyveromyces marxianus* as a robust synthetic biology platform host. mBio 9. https://doi.org/10.1128/mBio.01410-18

Chakravorty M, Veiga LA, Bacila M, Horecker BL (1962) Pentose Metabolism in Candida. II. The diphosphopyridine nucleotide-specific polyol dehydrogenase of candida utilis. J Biol Chem 237:1014–1020

Chandel AK, Singh OV, Rao LV, Chandrasekhar G, Narasu ML (2011) Bioconversion of novel substrate Saccharum spontaneum, a weedy material, into ethanol by Pichia stipitis NCIM3498. Bioresour Technol 102:1709–1714. https://doi.org/10.1016/j.biortech.2010.08.016

Chen W-H, Pen B-L, Yu C-T, Hwang W-S (2011) Pretreatment efficiency and structural characterization of rice straw by an integrated process of dilute-acid and steam explosion for bioethanol production. Bioresour Technol 102:2916–2924. https://doi.org/10.1016/j.biortech.2010.11.052
Christopher L (2012) Adding value prior to pulping: bioproducts from hemicellulose. In: Okia DCA (ed) Global perspectives on sustainable forest management. InTech
Coda R, Rizzello CG, Di Cagno R, Trani A, Cardinali G, Gobbetti M (2013) Antifungal activity of Meyerozyma guilliermondii: Identification of active compounds synthesized during dough fermentation and their effect on long-term storage of wheat bread. Food Microbiol 33:243–251. https://doi.org/10.1016/j.fm.2012.09.023
Comitini F, Ciani M (2011) Kluyveromyces wickerhamii killer toxin: purification and activity towards Brettanomyces/Dekkera yeasts in grape must: Kwkt killer toxin purification. FEMS Microbiol Lett 316:77–82. https://doi.org/10.1111/j.1574-6968.2010.02194.x
Corte L, di Cagno R, Groenewald M, Roscini L, Colabella C, Gobbetti M, Cardinali G (2015) Phenotypic and molecular diversity of Meyerozyma guilliermondii strains isolated from food and other environmental niches, hints for an incipient speciation. Food Microbiol 48:206–215. https://doi.org/10.1016/j.fm.2014.12.014
Crous PW, Wingfield MJ, Burgess TI, Carnegie AJ, Hardy GESJ, Smith D, Summerell BA, Cano-Lira JF, Guarro J, Houbraken J, Lombard L, Martín MP, Sandoval-Denis M, Alexandrova AV, Barnes CW, Baseia IG, Bezerra JDP, Guarnaccia V, May TW, Hernández-Restrepo M, Stchigel AM, Miller AN, Ordoñez ME, Abreu VP, Accioly T, Agnello C, Agustin Colmán A, Albuquerque CC, Alfredo DS, Alvarado P, Araújo-Magalhães GR, Arauzo S, Atkinson T, Barili A, Barreto RW, Bezerra JL, Cabral TS, Camello Rodríguez F, Cruz RHSF, Daniëls PP, da Silva BDB, de Almeida DAC, de Carvalho Júnior AA, Decock CA, Delgat L, Denman S, Dimitrov RA, Edwards J, Fedosova AG, Ferreira RJ, Firmino AL, Flores JA, García D, Gené J, Giraldo A, Góis JS, Gomes AAM, Gonçalves CM, Gouliamova DE, Groenewald M, Guéorguiev BV, Guevara-Suarez M, Gusmão LFP, Hosaka K, Hubka V, Huhndorf SM, Jadan M, Jurjević Ž, Kraak B, Kučera V, Kumar TKA, Kušan I, Lacerda SR, Lamlertthon S, Lisboa WS, Loizides M, Luangsa-ard JJ, Lysková P, Mac Cormack WP, Macedo DM, Machado AR, Malysheva EF, Marinho P, Matočec N, Meijer M, Mešić A, Mongkolsamrit S, Moreira KA, Morozova OV, Nair KU, Nakamura N, Noisripoom W, Olariaga I, Oliveira RJV, Paiva LM, Pawar P, Pereira OL, Peterson SW, Prieto M, Rodríguez-Andrade E, Rojo De Blas C, Roy M, Santos ES, Sharma R, Silva GA, Souza-Motta CM, Takeuchi-Kaneko Y, Tanaka C, Thakur A, Smith MT, Tkalčec Z, Valenzuela-Lopez N, van der Kleij P, Verbeken A, Viana MG, Wang XW, Groenewald JZ (2017) Fungal Planet description sheets: 625–715. Persoonia Mol Phylog Evolut Fungi. https://doi.org/10.3767/persoonia.2017.39.11
da Cunha-Pereira F, Hickert LR, Rech R, Dillon AP, Záchia Ayub MA (2017) Fermentation of hexoses and pentoses from hydrolyzed soybean Hull into ethanol and xylitol BY *Candida guilliermondii* BL 13. Braz J Chem Eng 34:927–936
Dantán-González E, Quiroz-Castañeda RE, Cobaxin-Cárdenas M, Valle-Hernández J, Gama-Martínez Y, Tinoco-Valencia JR, Serrano-Carreón L, Ortiz-Hernández L (2015) Impact of Meyerozyma guilliermondii isolated from chickens against Eimeria sp. protozoan, an in vitro analysis. BMC Vet Res 11:1–11. https://doi.org/10.1186/s12917-015-0589-0
de Albuquerque TL, da Silva IJ, de Macedo GR, Rocha MVP (2014) Biotechnological production of xylitol from lignocellulosic wastes: a review. Process Biochem 49:1779–1789. https://doi.org/10.1016/j.procbio.2014.07.010
de Lima PBA, Mulder KCL, Melo NTM, Carvalho LS, Menino GS, Mulinari E, de Castro VH, dos Reis TF, Goldman GH, Magalhães BS, Parachin NS (2016) Novel homologous lactate transporter improves l-lactic acid production from glycerol in recombinant strains of Pichia pastoris. Microb Cell Factories 15. https://doi.org/10.1186/s12934-016-0557-9
De Marco L, Epis S, Capone A, Martin E, Bozic J, Crotti E, Ricci I, Sassera D 2018a The genomes of four *Meyerozyma caribbica* isolates and novel insights into the *Meyerozyma guilliermondii* species complex. G3 (Bethesda) 8(3):755–759. https://doi.org/10.1534/g3.117.300316

De Marco L, Epis S, Capone A, Martin E, Bozic J, Crotti E, Ricci I, Sassera D (2018b) The genomes of four *Meyerozyma caribbica* isolates and novel insights into the *Meyerozyma guilliermondii* Species Complex. G3 (Bethesda) 8(3):755–759. https://doi.org/10.1534/g3.117.300316
De Schutter K, Lin Y-C, Tiels P, Van Hecke A, Glinka S, Weber-Lehmann J, Rouzé P, Van de Peer Y, Callewaert N (2009) Genome sequence of the recombinant protein production host Pichia pastoris. Nat Biotechnol 27:561–566. https://doi.org/10.1038/nbt.1544
de Souza G, Luana TCNV, Samila RPN, Maxwel AA (2017) Efficient production of second generation ethanol and xylitol by yeasts from Amazonian beetles (Coleoptera) and their galleries. Afr J Microbiol Res 11:814–824. https://doi.org/10.5897/AJMR2017.8522
Dias O, Gombert AK, Ferreira EC, Rocha I (2012) Genome-wide metabolic (re-) annotation of Kluyveromyces lactis. BMC Genomics 13:517. https://doi.org/10.1186/1471-2164-13-517
du Preez JC, van der Walt JP (1983) Fermentation of D-xylose to ethanol by a strain ofCandida shehatae. Biotechnol Lett 5:357–362. https://doi.org/10.1007/BF01141138
Duan H, Wang H, Ma B, Jiang P, Tu P, Ni Z, Li X, Li M, Ma X, Wang B, Wu R, Li M (2015) Codon optimization and expression of irisin in Pichia pastoris GS115. Int J Biol Macromol 79:21–26. https://doi.org/10.1016/j.ijbiomac.2015.04.030
Dujon B, Sherman D, Fischer G, Durrens P, Casaregola S, Lafontaine I, de Montigny J, Marck C, Neuvéglise C, Talla E, Goffard N, Frangeul L, Aigle M, Anthouard V, Babour A, Barbe V, Barnay S, Blanchin S, Beckerich J-M, Beyne E, Bleykasten C, Boisramé A, Boyer J, Cattolico L, Confanioleri F, de Daruvar A, Despons L, Fabre E, Fairhead C, Ferry-Dumazet H, Groppi A, Hantraye F, Hennequin C, Jauniaux N, Joyet P, Kachouri R, Kerrest A, Koszul R, Lemaire M, Lesur I, Ma L, Muller H, Nicaud J-M, Nikolski M, Oztas S, Ozier-Kalogeropoulos O, Pellenz S, Potier S, Richard G-F, Straub M-L, Suleau A, Swennen D, Tekaia F, Wésolowski-Louvel M, Westhof E, Wirth B, Zeniou-Meyer M, Zivanovic I, Bolotin-Fukuhara M, Thierry A, Bouchier C, Caudron B, Scarpelli C, Gaillardin C, Weissenbach J, Wincker P, Souciet J-L (2004) Genome evolution in yeasts. Nature 430:35–44. https://doi.org/10.1038/nature02579
Elshahed MS (2010) Microbiological aspects of biofuel production: Current status and future directions. J Adv Res 1:103–111. https://doi.org/10.1016/j.jare.2010.03.001
Faria NT, Santos MV, Fernandes P, Fonseca LL, Fonseca C, Ferreira FC (2014) Production of glycolipid biosurfactants, mannosylerythritol lipids, from pentoses and d-glucose/d-xylose mixtures by Pseudozyma yeast strains. Process Biochem 49:1790–1799. https://doi.org/10.1016/j.procbio.2014.08.004
Ferreira AD, Mussatto SI, Cadete RM, Rosa CA, Silva SS (2011) Ethanol production by a new pentose-fermenting yeast strain, Scheffersomyces stipitis UFMG-IMH 43.2, isolated from the Brazilian forest. Yeast 28:547–554. https://doi.org/10.1002/yea.1858
Fonseca GG, Heinzle E, Wittmann C, Gombert AK (2008) The yeast Kluyveromyces marxianus and its biotechnological potential. Appl Microbiol Biotechnol 79:339–354. https://doi.org/10.1007/s00253-008-1458-6
Franco Marcelino PR, da Silva VL, Rodrigues Philippini R, Von Zuben CJ, Contiero J, dos Santos JC, da Silva SS (2017) Biosurfactants produced by Scheffersomyces stipitis cultured in sugarcane bagasse hydrolysate as new green larvicides for the control of Aedes aegypti, a vector of neglected tropical diseases. PLoS One 12:e0187125. https://doi.org/10.1371/journal.pone.0187125
Gallo R, Trapp A (2017) The chemical conversion of biomass-derived saccharides: an overview. J Braz Chem Soc. https://doi.org/10.21577/0103-5053.20170009
García-Cubero MT, González-Benito G, Indacoechea I, Coca M, Bolado S (2009) Effect of ozonolysis pretreatment on enzymatic digestibility of wheat and rye straw. Bioresour Technol 100:1608–1613. https://doi.org/10.1016/j.biortech.2008.09.012
Garrote G, Dominguez H, Parajo JC (2002) Autohydrolysis of corncob: study of non-isothermal operation for xylooligasaccharide production. J Food Eng 52(3):211–218
Gírio F, Amaro C, Azinheira H, Pelica F, Amaral-Collaço M (2000) Polyols production during single and mixed substrate fermentations in Debaryomyces hansenii. Bioresour Technol 71:245–251. https://doi.org/10.1016/S0960-8524(99)00078-4

Gonçalves FAG, Colen G, Takahashi JA (2014) *Yarrowia lipolytica* and its multiple applications in the biotechnological industry. Sci World J 2014:1–14. https://doi.org/10.1155/2014/476207

Groeneveld P, Stouthamer AH, Westerhoff HV (2009) Super life – how and why 'cell selection' leads to the fastest-growing eukaryote: Control of highest eukaryotic growth rate. FEBS J 276:254–270. https://doi.org/10.1111/j.1742-4658.2008.06778.x

Groenewald M, Smith MT (2013) The teleomorph state of Candida deformans Langeron & Guerra and description of Yarrowia yakushimensis comb. nov. Antonie Van Leeuwenhoek 103:1023–1028. https://doi.org/10.1007/s10482-013-9882-8

Guan D, Li Y, Shiroma R, Ike M, Tokuyasu K (2013) Sequential incubation of Candida shehatae and ethanol-tolerant yeast cells for efficient ethanol production from a mixture of glucose, xylose and cellobiose. Bioresour Technol 132:419–422. https://doi.org/10.1016/j.biortech.2012.12.040

Gunah P (2011) Optimization of xylose production from sugarcane bagasse using response surface methodology (RSM). University of Malayia Pahang, Bachelor

Harhangi HR, Akhmanova AS, Emmens R, van der Drift C, de Laat WTAM, van Dijken JP, Jetten MSM, Pronk JT, Op den Camp HJM (2003) Xylose metabolism in the anaerobic fungus Piromyces sp. strain E2 follows the bacterial pathway. Arch Microbiol 180:134–141. https://doi.org/10.1007/s00203-003-0565-0

Hernández-Pérez AF, de Arruda PV, Felipe M das G de A (2016) Sugarcane straw as a feedstock for xylitol production by Candida guilliermondii FTI 20037. Braz J Microbiol 47:489–496. https://doi.org/10.1016/j.bjm.2016.01.019

Hommel RK, Ahnert P (1999) Candida. In: Encyclopedia of food microbiology, 1st edn. Elsevier Ltd, London

Hou X (2012) Anaerobic xylose fermentation by Spathaspora passalidarum. Appl Microbiol Biotechnol 94:205–214. https://doi.org/10.1007/s00253-011-3694-4

Hou J, Qiu C, Shen Y, Li H, Bao X (2017) Engineering of Saccharomyces cerevisiae for the efficient co-utilization of glucose and xylose. FEMS Yeast Res 17. https://doi.org/10.1093/femsyr/fox034

Ilmen M, Koivuranta K, Ruohonen L, Suominen P, Penttila M (2007) Efficient production of L-lactic acid from xylose by Pichia stipitis. Appl Environ Microbiol 73:117–123. https://doi.org/10.1128/AEM.01311-06

Jeffries TW (1983) Utilization of xylose by bacteria, yeasts, and fungi. In: Pentoses and lignin. Springer, Berlin/Heidelberg, pp 1–32

Jeffries TW (2006) Engineering yeasts for xylose metabolism. Curr Opin Biotechnol 17:320–326. https://doi.org/10.1016/j.copbio.2006.05.008

Jeffries TW, Van Vleet JRH (2009) *Pichia stipitis* genomics, transcriptomics, and gene clusters. FEMS Yeast Res 9:793–807. https://doi.org/10.1111/j.1567-1364.2009.00525.x

Jeffries TW, Grigoriev IV, Grimwood J, Laplaza JM, Aerts A, Salamov A, Schmutz J, Lindquist E, Dehal P, Shapiro H, Jin Y-S, Passoth V, Richardson PM (2007) Genome sequence of the lignocellulose-bioconverting and xylose-fermenting yeast Pichia stipitis. Nat Biotechnol 25:319–326. https://doi.org/10.1038/nbt1290

Ji X, Ma H, Tian Z, Lyu G, Fang G, Chen J, Saeed HAM (2017) Production of xylose from diluted sulfuric acid hydrolysis of wheat straw. Bioresources 12:7084–7095

Johnsen U, Dambeck M, Zaiss H, Fuhrer T, Soppa J, Sauer U, Schönheit P (2009) d-xylose degradation pathway in the halophilic Archaeon *Haloferax volcanii*. J Biol Chem 284:27290–27303. https://doi.org/10.1074/jbc.M109.003814

Jørgensen H, Kristensen JB, Felby C (2007) Enzymatic conversion of lignocellulose into fermentable sugars: challenges and opportunities. Biofuels Bioprod Biorefin 1:119–134. https://doi.org/10.1002/bbb.4

Kim D (2018) Physico-chemical conversion of lignocellulose: inhibitor effects and detoxification strategies: a mini review. Molecules 23:309. https://doi.org/10.3390/molecules23020309

Kim D, Woo HM (2018) Deciphering bacterial xylose metabolism and metabolic engineering of industrial microorganisms for use as efficient microbial cell factories. Appl Microbiol Biotechnol 102:9471–9480. https://doi.org/10.1007/s00253-018-9353-2

Kim TH, Taylor F, Hicks KB (2008) Bioethanol production from barley hull using SAA (soaking in aqueous ammonia) pretreatment. Bioresour Technol 99:5694–5702. https://doi.org/10.1016/j.biortech.2007.10.055

Kim SR, Park Y-C, Jin Y-S, Seo J-H (2013) Strain engineering of Saccharomyces cerevisiae for enhanced xylose metabolism. Biotechnol Adv 31:851–861. https://doi.org/10.1016/j.biotechadv.2013.03.004

Kim D, Ximenes EA, Nichols NN, Cao G, Frazer SE, Ladisch MR (2016) Maleic acid treatment of biologically detoxified corn stover liquor. Bioresour Technol 216:437–445. https://doi.org/10.1016/j.biortech.2016.05.086

Kim D, Orrego D, Ximenes EA, Ladisch MR (2017) Cellulose conversion of corn pericarp without pretreatment. Bioresour Technol 245:511–517. https://doi.org/10.1016/j.biortech.2017.08.156

Kręgiel D, Pawlikowska E, Antolak H (2017) Non-conventional yeasts in fermentation processes: potentialities and limitations. In: Lucas C, Pais C (eds) Old yeasts – new questions. InTech

Kumar V, Krishania M, Preet Sandhu P, Ahluwalia V, Gnansounou E, Sangwan RS (2018) Efficient detoxification of corn cob hydrolysate with ion-exchange resins for enhanced xylitol production by Candida tropicalis MTCC 6192. Bioresour Technol 251:416–419. https://doi.org/10.1016/j.biortech.2017.11.039

Kuroda K, Kobayashi K, Kitagawa Y, Nakagawa T, Tsumura H, Komeda T, Shinmi D, Mori E, Motoki K, Fuju K, Sakai T, Nonaka K, Suzuki T, Ichikawa K, Chiba Y, Jigami Y (2008) Efficient antibody production upon suppression of O mannosylation in the yeast Ogataea minuta. Appl Environ Microbiol 74:446–453. https://doi.org/10.1128/AEM.02106-07

Kurtzman CP (2005) New species and a new combination in the Hyphopichia and Yarrowia yeast clades. Antonie Van Leeuwenhoek 88:121–130. https://doi.org/10.1007/s10482-005-2495-0

Kurtzman CP (2009) Biotechnological strains of Komagataella (Pichia) pastoris are Komagataella phaffii as determined from multigene sequence analysis. J Ind Microbiol Biotechnol 36:1435–1438. https://doi.org/10.1007/s10295-009-0638-4

Kurtzman CP (2011a) Yarrowia van der Walt & von Arx (1980). In: The yeasts. Elsevier, Amsterdam, pp 927–929

Kurtzman CP (ed) (2011b) The yeasts: a taxonomic study 5 Elsevier, Amsterdam

Kurtzman CP, Robnett CJ (1998) Identification and phylogeny of ascomycetous yeasts from analysis of nuclear large subunit (26S) ribosomal DNA partial sequences. Antonie Van Leeuwenhoek 73:331

Kurtzman C, Robnett C (2003) Phylogenetic relationships among yeasts of the ? complex? determined from multigene sequence analyses. FEMS Yeast Res 3:417–432. https://doi.org/10.1016/S1567-1356(03)00012-6

Kurtzman CP, Suzuki M (2010) Phylogenetic analysis of ascomycete yeasts that form coenzyme Q-9 and the proposal of the new genera Babjeviella, Meyerozyma, Millerozyma, Priceomyces, and Scheffersomyces. Mycoscience 51:2–14. https://doi.org/10.1007/S10267-009-0011-5

Kurtzman M, Suzuki M, Kurtzman CP, Clade C, Clade C, U CY-T, Ay MY-T, Ay CY-T, Ay CY-T, Ay CY-T, U DY-T (2010) Meyerozyma Kurtzman & M. Suzuki (2010). Elsevier BV, London

Kurylenko OO, Ruchala J, Hryniv OB, Abbas CA, Dmytruk KV, Sibirny AA (2014) Metabolic engineering and classical selectionof the methylotrophic thermotolerant yeast Hansenulapolymorpha for improvement of high-temperature xylose alcoholicfermentation. Microb Cell Factories 13:122. https://doi.org/10.1186/s12934-014-0122-3

Kwak S, Jin Y-S (2017) Production of fuels and chemicals from xylose by engineered Saccharomyces cerevisiae: a review and perspective. Microb Cell Factories 16. https://doi.org/10.1186/s12934-017-0694-9

Lachance M-A (2007) Current status of *Kluyveromyces* systematics. FEMS Yeast Res 7:642–645. https://doi.org/10.1111/j.1567-1364.2006.00197.x

Lachance M-A (2011) Kluyveromyces van der Walt (1971). In: The yeasts. Elsevier, Amsterdam, pp 471–481

Lane MM, Morrissey JP (2010) Kluyveromyces marxianus: A yeast emerging from its sister's shadow. Fungal Biol Rev 24:17–26. https://doi.org/10.1016/j.fbr.2010.01.001

Ledesma-Amaro R, Lazar Z, Rakicka M, Guo Z, Fouchard F, Coq A-MC-L, Nicaud J-M (2016) Metabolic engineering of Yarrowia lipolytica to produce chemicals and fuels from xylose. Metab Eng 38:115–124. https://doi.org/10.1016/j.ymben.2016.07.001

Li H, Alper HS (2016) Enabling xylose utilization in *Yarrowia lipolytica* for lipid production. Biotechnol J 11:1230–1240. https://doi.org/10.1002/biot.201600210

Li C, Knierim B, Manisseri C, Arora R, Scheller HV, Auer M, Vogel KP, Simmons BA, Singh S (2010) Comparison of dilute acid and ionic liquid pretreatment of switchgrass: Biomass recalcitrance, delignification and enzymatic saccharification. Bioresour Technol 101:4900–4906. https://doi.org/10.1016/j.biortech.2009.10.066

Li P, Sun H, Chen Z, Li Y, Zhu T (2015) Construction of efficient xylose utilizing Pichia pastoris for industrial enzyme production. Microb Cell Factories 14. https://doi.org/10.1186/s12934-015-0206-8

Liu Y, Wu C, Wang J, Mo W, Yu M (2013) Codon optimization, expression, purification, and functional characterization of recombinant human IL-25 in Pichia pastoris. Appl Microbiol Biotechnol 97:10349–10358. https://doi.org/10.1007/s00253-013-5264-4

Liu H-H, Ji X-J, Huang H (2015) Biotechnological applications of Yarrowia lipolytica: Past, present and future. Biotechnol Adv 33:1522–1546. https://doi.org/10.1016/j.biotechadv.2015.07.010

Liu X-J, Cao W-N, Ren Y-C, Xu L-L, Yi Z-H, Liu Z, Hui F-L (2016) Taxonomy and physiological characterisation of Scheffersomyces titanus sp. nov., a new D-xylose-fermenting yeast species from China. Sci Rep 6. https://doi.org/10.1038/srep32181

Lobo FP, Goncalves DL, Alves SL, Gerber AL, de Vasconcelos ATR, Basso LC, Franco GR, Soares MA, Cadete RM, Rosa CA, Stambuk BU (2014a) Draft genome sequence of the D-xylose-fermenting yeast *Spathaspora arborariae* UFMG-HM19.1AT. Genome Announ 2. https://doi.org/10.1128/genomeA.01163-13

Lobo FP, Goncalves DL, Alves SL, Gerber AL, de Vasconcelos ATR, Basso LC, Franco GR, Soares MA, Cadete RM, Rosa CA, Stambuk BU (2014b) Draft genome sequence of the D-xylose-fermenting yeast *Spathaspora arborariae* UFMG-HM19.1AT. Genome Announ 2. https://doi.org/10.1128/genomeA.01163-13

Löbs A-K, Schwartz C, Wheeldon I (2017) Genome and metabolic engineering in non-conventional yeasts: Current advances and applications. Synthet Syst Biotechnol 2:198–207. https://doi.org/10.1016/j.synbio.2017.08.002

Long TM, Su Y-K, Headman J, Higbee A, Willis LB, Jeffries TW (2012) Cofermentation of glucose, xylose, and cellobiose by the beetle-associated yeast Spathaspora passalidarum. Appl Environ Microbiol 78:5492–5500. https://doi.org/10.1128/AEM.00374-12

Lopes MR, Morais CG, Kominek J, Cadete RM, Soares MA, Uetanabaro APT, Fonseca C, Lachance M-A, Hittinger CT, Rosa CA (2016) Genomic analysis and D-xylose fermentation of three novel *Spathaspora* species: *Spathaspora girioi* sp. nov., *Spathaspora hagerdaliae* f. a., sp. nov. and *Spathaspora gorwiae* f. a., sp. nov. FEMS Yeast Res 16:fow044. https://doi.org/10.1093/femsyr/fow044

Lopes DD, Cibulski SP, Mayer FQ, Siqueira FM, Rosa CA, Hector RE, Ayub MAZ (2017) Draft genome sequence of the d-xylose-fermenting yeast *Spathaspora xylofermentans* UFMG-HMD23.3. Genome Announ 5. https://doi.org/10.1128/genomeA.00815-17

Lopes MR, Batista TM, Franco GR, Ribeiro LR, Santos ARO, Furtado C, Moreira RG, Goes-Neto A, Vital MJS, Rosa LH, Lachance M-A, Rosa CA (2018) Scheffersomyces stambukii f.a., sp. nov., a d-xylose-fermenting species isolated from rotting wood. Int J Syst Evol Microbiol 68:2306–2312. https://doi.org/10.1099/ijsem.0.002834

Ma K, He M, You H, Pan L, Hu G, Cui Y, Maeda T (2017) Enhanced fuel ethanol production from rice straw hydrolysate by an inhibitor-tolerant mutant strain of Scheffersomyces stipitis. RSC Adv 7:31180–31188. https://doi.org/10.1039/C7RA04049K

Madhavan A, Tamalampudi S, Ushida K, Kanai D, Katahira S, Srivastava A, Fukuda H, Bisaria VS, Kondo A (2009) Xylose isomerase from polycentric fungus Orpinomyces: gene sequencing, cloning, and expression in Saccharomyces cerevisiae for bioconversion of xylose to ethanol. Appl Microbiol Biotechnol 82:1067–1078. https://doi.org/10.1007/s00253-008-1794-6

Mans R, Daran J-MG, Pronk JT (2018) Under pressure: evolutionary engineering of yeast strains for improved performance in fuels and chemicals production. Curr Opin Biotechnol 50:47–56. https://doi.org/10.1016/j.copbio.2017.10.011

Martín C, Klinke HB, Thomsen AB (2007) Wet oxidation as a pretreatment method for enhancing the enzymatic convertibility of sugarcane bagasse. Enzym Microb Technol 40:426–432. https://doi.org/10.1016/j.enzmictec.2006.07.015

Martini C, Tauk-Tornisielo SM, Codato CB, Bastos RG, Ceccato-Antonini SR (2016) A strain of Meyerozyma guilliermondii isolated from sugarcane juice is able to grow and ferment pentoses in synthetic and bagasse hydrolysate media. World J Microbiol Biotechnol 32. https://doi.org/10.1007/s11274-016-2036-1

Martins GM, Bocchini-Martins DA, Bezzerra-Bussoli C, Pagnocca FC, Boscolo M, Monteiro DA, da Silva R, Gomes E (2018) The isolation of pentose-assimilating yeasts and their xylose fermentation potential. Braz J Microbiol 49:162–168. https://doi.org/10.1016/j.bjm.2016.11.014

Mateo S, Puentes JG, Moya AJ, Sánchez S (2015) Ethanol and xylitol production by fermentation of acid hydrolysate from olive pruning with Candida tropicalis NBRC 0618. Bioresour Technol 190:1–6. https://doi.org/10.1016/j.biortech.2015.04.045

Mattanovich D, Graf A, Stadlmann J, Dragosits M, Redl A, Maurer M, Kleinheinz M, Sauer M, Altmann F, Gasser B (2009) Genome, secretome and glucose transport highlight unique features of the protein production host Pichia pastoris. Microb Cell Factories 8:29. https://doi.org/10.1186/1475-2859-8-29

Mellitzer A, Weis R, Glieder A, Flicker K (2012) Expression of lignocellulolytic enzymes in Pichia pastoris. Microb Cell Factories 11:61. https://doi.org/10.1186/1475-2859-11-61

Meyer SA, Anderson K, Brown RE, Smith MT, Yarrow D, Mitchell G, Ahearn DG (1975) Physiological and DNA characterization of Candida maltose, a hydrocarbon-utilizing yeast. Arch Microbiol 104:225–231

Morais CG, Batista TM, Kominek J, Borelli BM, Furtado C, Moreira RG, Franco GR, Rosa LH, Fonseca C, Hittinger CT, Lachance M-A, Rosa CA (2017) Spathaspora boniae sp. nov., a D-xylose-fermenting species in the Candida albicans/Lodderomyces clade. Int J Syst Evol Microbiol 67:3798–3805. https://doi.org/10.1099/ijsem.0.002186

Morales P, Gentina JC, Aroca G, Mussatto SI (2017) Development of an acetic acid tolerant Spathaspora passalidarum strain through evolutionary engineering with resistance to inhibitors compounds of autohydrolysate of Eucalyptus globulus. Ind Crop Prod 106 5–11. https://doi.org/10.1016/j.indcrop.2016.12.023

Moran G (2004) Comparative genomics using Candida albicans DNA microarrays reveals absence and divergence of virulence-associated genes in Candida dubliniensis. Microbiology 150:3363–3382. https://doi.org/10.1099/mic.0.27221-0

Morrissey JP, Etschmann MMW, Schrader J, de Billerbeck GM (2015) Cell factory applications of the yeast Kluyveromyces marxianus for the biotechnological production of natural flavour and fragrance molecules. Yeast 32:3–16

Mosier N (2005) Features of promising technologies for pretreatment of lignocellulosic biomass. Bioresour Technol 96:673–686. https://doi.org/10.1016/j.biortech.2004.06.025

Moysés D, Reis V, Almeida J, Moraes L, Torres F (2016) Xylose Fermentation by Saccharomyces cerevisiae: Challenges and Prospects. Int J Mol Sci 17:207. https://doi.org/10.3390/ijms17030207

Mukherjee V, Radecka D, Aerts G, Verstrepen KJ, Lievens B, Thevelein JM (2017) Phenotypic landscape of non-conventional yeast species for different stress tolerance traits desirable in bioethanol fermentation. Biotechnol Biofuels 10. https://doi.org/10.1186/s13068-017-0899-5

Nagy E (2015) Biodiversity of food spoilage Yarrowia group in different kinds of food. Dissertation, University of Budapest

Nagy E, Niss M, Dlauchy D, Arneborg N, Nielsen DS, Peter G (2013) Yarrowia divulgata f.a., sp. nov., a yeast species from animal-related and marine sources. Int J Syst Evol Microbiol 63:4818–4823. https://doi.org/10.1099/ijs.0.057208-0

Nagy E, Dlauchy D, Medeiros AO, Péter G, Rosa CA (2014) Yarrowia porcina sp. nov. and Yarrowia bubula f.a. sp. nov., two yeast species from meat and river sediment. Antonie Van Leeuwenhoek 105:697–707. https://doi.org/10.1007/s10482-014-0125-4

Najjar A, Robert S, Guérin C, Violet-Asther M, Carrière F (2011) Quantitative study of lipase secretion, extracellular lipolysis, and lipid storage in the yeast Yarrowia lipolytica grown in the presence of olive oil: analogies with lipolysis in humans. Appl Microbiol Biotechnol 89:1947–1962. https://doi.org/10.1007/s00253-010-2993-5

Nakanishi SC, Soares LB, Biazi LE, Nascimento VM, Costa AC, Rocha GJM, Ienczak JL (2017) Fermentation strategy for second generation ethanol production from sugarcane bagasse hydrolyzate by *Spathaspora passalidarum* and *Scheffersomyces stipitis*: Fermentation Strategy for Second Generation Ethanol Production. Biotechnol Bioeng 114:2211–2221. https://doi.org/10.1002/bit.26357

Nakase T, Komagata K (1971) Significance of DNA base composition in the classification of the yeast genus Candida. J Gen Appl Microbiol 17:259–179

Nguyen NH, Suh S-O, Marshall CJ, Blackwell M (2006) Morphological and ecological similarities: wood-boring beetles associated with novel xylose-fermenting yeasts, Spathaspora passalidarum gen. sp. nov. and Candida jeffriesii sp. nov. Mycol Res 110:1232–1241. https://doi.org/10.1016/j.mycres.2006.07.002

Niehus X, Crutz-Le Coq A-M, Sandoval G, Nicaud J-M, Ledesma-Amaro R (2018) Engineering Yarrowia lipolytica to enhance lipid production from lignocellulosic materials. Biotechnol Biofuels 11. https://doi.org/10.1186/s13068-018-1010-6

Nygård Y, Toivari MH, Penttilä M, Ruohonen L, Wiebe MG (2011) Bioconversion of d-xylose to d-xylonate with Kluyveromyces lactis. Metab Eng 13:383–391. https://doi.org/10.1016/j.ymben.2011.04.001

Odds FC (2010) Molecular phylogenetics and epidemiology of *Candida albicans*. Future Microbiol 5:67–79. https://doi.org/10.2217/fmb.09.113

Okada N, Tanimura A, Hirakawa H, Takashima M, Ogawa J, Shima J (2017) Draft Genome Sequences of the Xylose-Fermenting Yeast *Scheffersomyces shehatae* NBRC 1983 ^{T} and a Thermotolerant Isolate of *S. shehatae* ATY839 (JCM 18690). Genome Announ 5. https://doi.org/10.1128/genomeA.00347-17

Palsson BO, Fathi-Afshar S, Rudd DF, Lightfoot EN (1981) Biomass as a Source of Chemical Feedstocks: An Economic Evaluation. Science 213:513–517. https://doi.org/10.1126/science.213.4507.513

Protchenko O, Philpott CC, Sibirny AA (2011) Insertion mutagenesis. doi:https://doi.org/10.1111/j.1567-1364.2011.00720.x

Puseenam A, Kocharin K, Tanapongpipat S, Eurwilaichitr L, Ingsriswang S, Roongsawang N (2018) A novel sucrose-based expression system for heterologous proteins expression in thermotolerant methylotrophic yeast Ogataea thermomethanolica. FEMS Microbiol Lett 365. https://doi.org/10.1093/femsle/fny238

Radecka D, Mukherjee V, Mateo RQ, Stojiljkovic M, Foulquié-Moreno MR, Thevelein JM (2015) Looking beyond *Saccharomyces* : the potential of non-conventional yeast species for desirable traits in bioethanol fermentation. FEMS Yeast Res 15:fov053. https://doi.org/10.1093/femsyr/fov053

Rebello S, Abraham A, Madhavan A, Sindhu R, Binod P, Babu AK, Aneesh EM, Pandey A (2018) Non-conventional Yeast cell factories for sustainable bioprocesses. FEMS Microbiol Lett. https://doi.org/10.1093/femsle/fny222

Ren Y, Chen L, Niu Q, Hui F (2014) Description of Scheffersomyces henanensis sp. nov., a new D-xylose-fermenting yeast species isolated from rotten wood. PLoS One 9:e92315. https://doi.org/10.1371/journal.pone.0092315

Riley R, Haridas S, Wolfe KH, Lopes MR, Hittinger CT, Göker M, Salamov AA, Wisecaver JH, Long TM, Calvey CH, Aerts AL, Barry KW, Choi C, Clum A, Coughlan AY, Deshpande S, Douglass AP, Hanson SJ, Klenk H-P, LaButti KM, Lapidus A, Lindquist EA, Lipzen AM, Meier-Kolthoff JP, Ohm RA, Otillar RP, Pangilinan JL, Peng Y, Rokas A, Rosa CA, Scheuner C, Sibirny AA, Slot JC, Stielow JB, Sun H, Kurtzman CP, Blackwell M, Grigoriev IV, Jeffries TW (2016) Comparative genomics of biotechnologically important yeasts. Proc Natl Acad Sci 113:9882–9887. https://doi.org/10.1073/pnas.1603941113

Rodicio R, Heinisch JJ (2013) Yeast on the milky way: genetics, physiology and biotechnology of *Kluyveromyces lactis*: Kluyveromyces lactis. Yeast 30:165–177. https://doi.org/10.1002/yea.2954

Rodrussamee N, Sattayawat P, Yamada M (2018) Highly efficient conversion of xylose to ethanol without glucose repression by newly isolated thermotolerant Spathaspora passalidarum CMUWF1–2. BMC Microbiol 18. https://doi.org/10.1186/s12866-018-1218-4

Romi W, Keisam S, Ahmed G, Jeyaram K (2014) Reliable differentiation of Meyerozyma guilliermondii from Meyerozyma caribbica by internal transcribed spacer restriction fingerprinting. BMC Microbiol 14:52. https://doi.org/10.1186/1471-2180-14-52

Rosa C, Lachance M, Silva J, Teixeira A, Marini M, Antonini Y, Martins R (2003) Yeast communities associated with stingless bees. FEMS Yeast Res 4 271–275. https://doi.org/10.1016/S1567-1356(03)00173-9

Ruiz E, Cara C, Manzanares P, Ballesteros M, Castro E (2008) Evaluation of steam explosion pretreatment for enzymatic hydrolysis of sunflower stalks. Enzym Microb Technol 42:160–166. https://doi.org/10.1016/j.enzmictec.2007.09.002

Ryabova O, Chmil O, Sibirny A (2003) Xylose and cellobiose fermentation to ethanol by the thermotolerant methylotrophic yeast. FEMS Yeast Res 4:157–164. https://doi.org/10.1016/S1567-1356(03)00146-6

Ryu S, Trinh CT (2017) Understanding functional roles of native pentose-specific transporters for activating dormant pentose metabolism in *Yarrowia lipolytica*. Appl Environ Microbiol 84. https://doi.org/10.1128/AEM.02146-17

Ryu S, Hipp J, Trinh CT (2016) Activating and Elucidating Metabolism of Complex Sugars in Yarrowia lipolytica. Appl Environ Microbiol 82:1334–1345. https://doi.org/10.1128/AEM.03582-15

Sampaio FC, da Silveira WB, Chaves-Alves VM, Passos FML, Coelho JLC (2003) Screening of filamentous fungi for production of xylitol from D-xylose. Braz J Microbiol 34:321–324. https://doi.org/10.1590/S1517-83822003000400007

Satish T, Murthy NYS (2010) Optimisation of xylose production using xylanase. Intl J Sci 8:909–913

Schauer F, Hanschke R (1999) Taxonomy and ecology of the genus Candida. Mycosis 42:12–21

Schreiber WT, Geib NV, Wingfield B, Acree SF (1930) Semi-commercial production of xylose. Ind Eng Chem 22:497–501. https://doi.org/10.1021/ie50245a020

Shu M, Shen W, Yang S, Wang X, Wang F, Wang Y, Ma L (2016) High-level expression and characterization of a novel serine protease in Pichia pastoris by multi-copy integration. Enzym Microb Technol 92:56–66. https://doi.org/10.1016/j.enzmictec.2016.06.007

Slininger PJ, Bothast RJ, Van Cauwenberge JE, Kurtzman CP (1982) Conversion of D-xylose to ethanol by the yeastPachysolen tannophilus. Biotechnol Bioeng 24:371–384. https://doi.org/10.1002/bit.260240210

Song KH, Song JY, Hong SG, Baek H, Kim SY, Hyun HH (2002) Production of mannitol by a novel strain of Candida magnoliae. Biotechnol Lett 24:9–12

Spagnuolo M, Shabbir Hussain M, Gambill L, Blenner M (2018) Alternative substrate metabolism in Yarrowia lipolytica. Front Microbiol 9. https://doi.org/10.3389/fmicb.2018.01077

Spencer JFT, Spencer DM (eds) (1997) Yeasts in natural and artificial habitats. Springer, Berlin/Heidelberg

Spohner SC, Schaum V, Quitmann H, Czermak P (2016) Kluyveromyces lactis: an emerging tool in biotechnology. J Biotechnol 222:104–116. https://doi.org/10.1016/j.jbiotec.2016.02.023

Sreekrishna K, Kropp KE (1996) Pichia pastoris. In: Nonconventional yeasts in biotechnology. Springer, Berlin/Heidelberg, pp 203–253
Stambuk BU, Franden MA, Singh A, Zhang M (2003) D-xylose transport by Candida succiphila and Kluyveromyces marxianus. Appl Biochem Biotechnol 106:255–264. https://doi.org/10.1385/ABAB:106:1-3:255
Stenderup A, Bak AL (1968) Deoxyribonucleic acid base composition of some species within the genus Candida. J Gen Microbiol 52:231–236
Stephanopoulos G TM (2013) Engineered microbes and methods for microbial oil overproduction from cellulosic materials
Su Y-K, Willis LB, Jeffries TW (2015) Effects of aeration on growth, ethanol and polyol accumulation by *Spathaspora passalidarum* NRRL Y-27907 and *Scheffersomyces stipitis* NRRL Y-7124: Aeration Effects on Xylose Fermenting Yeasts. Biotechnol Bioeng 112:457–469. https://doi.org/10.1002/bit.25445
Su Y-K, Willis LB, Rehmann L, Smith DR, Jeffries TW (2018) Spathaspora passalidarum selected for resistance to AFEX hydrolysate shows decreased cell yield. FEMS Yeast Res 18. https://doi.org/10.1093/femsyr/foy011
Suh S-O, Marshall CJ, Mchugh JV, Blackwell M (2003) Wood ingestion by passalid beetles in the presence of xylose-fermenting gut yeasts: gut yeasts of passalid beetles. Mol Ecol 12:3137–3145. https://doi.org/10.1046/j.1365-294X.2003.01973.x
Sun J, Ding S-Y, Doran-Peterson J (2013) Chapter 1: biomass and its biorefinery: novel approaches from nature-inspired strategies and technology. In: Sun J, Ding S-Y, Peterson JD (eds) Energy and environment series. Royal Society of Chemistry, Cambridge, pp 1–13
Swain MR, Krishnan C (2015) Improved conversion of rice straw to ethanol and xylitol by combination of moderate temperature ammonia pretreatment and sequential fermentation using Candida tropicalis. Ind Crop Prod 77:1039–1046. https://doi.org/10.1016/j.indcrop.2015.10.013
Tizazu BZ, Roy K, Moholkar VS (2018) Ultrasonic enhancement of xylitol production from sugarcane bagasse using immobilized Candida tropicalis MTCC 184. Bioresour Technol 268:247–258. https://doi.org/10.1016/j.biortech.2018.07.141
Tsai CT, Huang C-T (2008) Overexpression of the Neocallimastix frontalis xylanase gene in the methylotrophic yeasts Pichia pastoris and Pichia methanolica. Enzym Microb Technol 42:459–465. https://doi.org/10.1016/j.enzmictec.2008.01.018
Tsigie YA, Wang C-Y, Truong C-T, Ju Y-H (2011) Lipid production from Yarrowia lipolytica Po1g grown in sugarcane bagasse hydrolysate. Bioresour Technol 102:9216–9222. https://doi.org/10.1016/j.biortech.2011.06.047
Urbina H, Blackwell M (2012) Multilocus phylogenetic study of the scheffersomyces yeast clade and characterization of the N-terminal region of xylose reductase gene. PLoS ONE 7. https://doi.org/10.1371/journal.pone.0039128
Vallejos ME, Chade M, Mereles EB, Bengoechea DI, Brizuela JG, Felissia FE, Area MC (2016) Strategies of detoxification and fermentation for biotechnological production of xylitol from sugarcane bagasse. Ind Crop Prod 91:161–169. https://doi.org/10.1016/j.indcrop.2016.07.007
van der Walt JP (1965) The emendation of the genus Kluyveromyces v. d. Walt. Antonie Van Leeuwenhoek 31:341–348
van der Walt JP, von Arx JA (1980) The yeast genus Yarrowia gen. nov. Antonie Van Leeuwenhoek 46:517–521. https://doi.org/10.1007/BF00394008
van Ooyen AJJ, Dekker P, Huang M, Olsthoorn MMA, Jacobs DI, Colussi PA, Taron CH (2006) Heterologous protein production in the yeast *Kluyveromyces lactis*. FEMS Yeast Res 6:381–392. https://doi.org/10.1111/j.1567-1364.2006.00049.x
van Wyk JP (2001) Biotechnology and the utilization of biowaste as a resource for bioproduct development. Trends Biotechnol 19:172–177
Varela JA, Gethins L, Stanton C, Ross P, Morrissey JP (2017) Applications of Kluyveromyces marxianus in biotechnology. In: Satyanarayana T, Kunze G (eds) Yeast diversity in human welfare. Springer, Singapore, pp 439–453

Varize CS, Cadete RM, Lopes LD, Christofoleti-Furlan RM, Lachance M-A, Rosa CA, Basso LC (2018) Spathaspora piracicabensis f. a., sp. nov., a d-xylose-fermenting yeast species isolated from rotting wood in Brazil. Antonie Van Leeuwenhoek 111:525–531. https://doi.org/10.1007/s10482-017-0974-8

Vaz de Arruda P, dos Santos JC, de Cássia Lacerda Brambilla Rodrigues R, da Silva DDV, Yamakawa CK, de Moraes Rocha GJ, Júnior JN, da Cruz Pradella JG, Vaz Rossell CE, das Graças de Almeida Felipe M (2017) Scale up of xylitol production from sugarcane bagasse hemicellulosic hydrolysate by Candida guilliermondii FTI 20037. J Ind Eng Chem 47:297–302. https://doi.org/10.1016/j.jiec.2016.11.046

Venkateswar Rao L, Goli JK, Gentela J, Koti S (2015) Bioconversion of lignocellulosic biomass to xylitol: an overview. Bioresour Technol 213:299–310. https://doi.org/10.1016/j.biortech.2016.04.092

Veras HCT, Parachin NS, Almeida JRM (2017) Comparative assessment of fermentative capacity of different xylose-consuming yeasts. Microb Cell Factories 16. https://doi.org/10.1186/s12934-017-0766-x

Verduyn C, Van Kleef R, Frank J, Schreuder H, Van Dijken JP, Scheffers WA (1985) Properties of the NAD(P)H-dependent xylose reductase from the xylose-fermenting yeast Pichia stipitis. Biochem J 226:669–677

Wang PY, Shopsis C, Schneider H (1980) Fermentation of a pentose by yeasts. Biochem Biophys Res Commun 94:248–254. https://doi.org/10.1016/S0006-291X(80)80213-0

Wang Y, Ren Y-C, Zhang Z-T, Ke T, Hui F-L (2016) Spathaspora allomyrinae sp. nov., a d-xylose-fermenting yeast species isolated from a scarabeid beetle Allomyrina dichotoma. Int J Syst Evol Microbiol 66:2008–2012. https://doi.org/10.1099/ijsem.0.000979

Wannawilai S, Chisti Y, Sirisansaneeyakul S (2017) A model of furfural-inhibited growth and xylitol production by Candida magnoliae TISTR 5663. Food Bioprod Process 105:129–140. https://doi.org/10.1016/j.fbp.2017.07.002

Webb SR, Lee H (1990) Regulation of d-xylose utilization by hexoses in pentose-fermenting yeasts. Biotechnol Adv 8:685–697. https://doi.org/10.1016/0734-9750(90)91991-O

Wei L, Liu J, Qi H, Wen J (2015) Engineering Scheffersomyces stipitis for fumaric acid production from xylose. Bioresour Technol 187:246–254. https://doi.org/10.1016/j.biortech.2015.03.122

Wickerham LJ, Burton KA (1954) A clarification of the relationship of Candida Guilliermondii to other yeasts by a study of their mating types. J Bacteriol 68(5):594–597

Wohlbach DJ, Kuo A, Sato TK, Potts KM, Salamov AA, LaButti KM, Sun H, Clum A, Pangilinan JL, Lindquist EA, Lucas S, Lapidus A, Jin M, Gunawan C, Balan V, Dale BE, Jeffries TW, Zinkel R, Barry KW, Grigoriev IV, Gasch AP (2011a) Comparative genomics of xylose-fermenting fungi for enhanced biofuel production. Proc Natl Acad Sci 108:13212–13217. https://doi.org/10.1073/pnas.1103039108

Wohlbach DJ, Kuo A, Sato TK, Potts KM, Salamov AA, LaButti KM, Sun H, Clum A, Pangilinan JL, Lindquist EA, Lucas S, Lapidus A, Jin M, Gunawan C, Balan V, Dale BE, Jeffries TW, Zinkel R, Barry KW, Grigoriev IV, Gasch AP (2011b) Comparative genomics of xylose-fermenting fungi for enhanced biofuel production. Proc Natl Acad Sci 108:13212–13217. https://doi.org/10.1073/pnas.1103039108

Wrent P, Rivas EM, Peinado JM, de Silóniz MI (2016) Development of an affordable typing method for Meyerozyma guilliermondii using microsatellite markers. Int J Food Microbiol 217:1–6. https://doi.org/10.1016/j.ijfoodmicro.2015.10.008

Yaguchi A, Spagnuolo M, Blenner M (2018) Engineering yeast for utilization of alternative feedstocks. Curr Opin Biotechnol 53:122–129. https://doi.org/10.1016/j.copbio.2017.12.003

Yamada Y, Kondo K (1972) Taxonomic significance of coenzyme Q system in yeasts and yeast-like fungi (1). Yeast 363:373

Yamada Y, Maeda K, Mikata K (1994) The Phylogenetic Relationships of the Hat-shaped Ascospore-forming, Nitrate-assimilating *Pichia* Species, Formerly Classified in the Genus *Hansenula* Sydow et Sydow, Based on the Partial Sequences of 18S and 26S Ribosomal

RNAs (Saccharomycetaceae): The Proposals of Three New Genera, *Ogataea* , *Kuraishia* , and *Nakazawaea*. Biosci Biotechnol Biochem 58:1245–1257. https://doi.org/10.1271/bbb.58.1245
Yan Y, Zhang X, Zheng X, Apaliya MT, Yang Q, Zhao L, Gu X, Zhang H (2018) Control of post-harvest blue mold decay in pears by Meyerozyma guilliermondii and it's effects on the protein expression profile of pears. Postharvest Biol Technol 136:124–131. https://doi.org/10.1016/j.postharvbio.2017.10.016
Yang Z, Zhang Z (2018) Engineering strategies for enhanced production of protein and bio-products in Pichia pastoris: a review. Biotechnol Adv 36:182–195. https://doi.org/10.1016/j.biotechadv.2017.11.002
Yang SW, Park JB, Han NS, Ryu YW, Seo JH (1999) Production of erythritol from glucose by an osmophilic mutant of Candida magnolia. Biotechnol Lett 21:887–890
Yang L, He QS, Corscadden K, Udenigwe CC (2015) The prospects of Jerusalem artichoke in functional food ingredients and bioenergy production. Biotechnol Rep 5:77–88. https://doi.org/10.1016/j.btre.2014.12.004
Yarrow D (1998) Methods for the isolation, maintenance and identification of yeasts. In: The yeasts. Elsevier, Amsterdam, pp 77–100
Yurkov AM, Dlauchy D, Péter G (2017) Meyerozyma amylolytica sp. nov. from temperate deciduous trees and the transfer of five candida species to the genus meyerozyma. Int J Syst Evol Microbiol 67:3977–3981. https://doi.org/10.1099/ijsem.0.002232
Yuvadetkun P, Reungsang A, Boonmee M (2018) Comparison between free cells and immobilized cells of Candida shehatae in ethanol production from rice straw hydrolysate using repeated batch cultivation. Renew Energy 115:634–640. https://doi.org/10.1016/j.renene.2017.08.033
Zhang J, Geng A, Yao C, Lu Y, Li Q (2012) Xylitol production from d-xylose and horticultural waste hemicellulosic hydrolysate by a new isolate of Candida athensensis SB18. Bioresour Technol 105:134–141. https://doi.org/10.1016/j.biortech.2011.11.119
Zhang H-J, Fan X-G, Qiu X-L, Zhang Q-X, Wang W-Y, Li S-X, Deng L-H, Koffas MAG, Wei D-S, Yuan Q-P (2014) A novel cleaning process for industrial production of xylose in pilot scale from corncob by using screw-steam-explosive extruder. Bioprocess Biosyst Eng 37:2425–2436. https://doi.org/10.1007/s00449-014-1219-0
Zhang J, Zhang B, Wang D, Gao X, Sun L, Hong J (2015) Rapid ethanol production at elevated temperatures by engineered thermotolerant Kluyveromyces marxianus via the NADP(H)-preferring xylose reductase-xylitol dehydrogenase pathway. Metab Eng 31:140–152. https://doi.org/10.1016/j.ymben.2015.07.008
Zhang B, Zhang J, Wang D, Han R, Ding R, Gao X, Sun L, Hong J (2016) Simultaneous fermentation of glucose and xylose at elevated temperatures co-produces ethanol and xylitol through overexpression of a xylose-specific transporter in engineered Kluyveromyces marxianus. Bioresour Technol 216:227–237. https://doi.org/10.1016/j.biortech.2016.05.068
Zhao J, Mou Y, Shan T, Li Y, Zhou L, Wang M, Wang J (2010) Antimicrobial metabolites from the endophytic fungus pichia guilliermondii Isolated from Paris polyphylla var. yunnanensis. Molecules 15:7961–7970. https://doi.org/10.3390/molecules15117961
Zhao C, Gu D, Nambou K, Wei L, Chen J, Imanaka T, Hua Q (2015) Metabolome analysis and pathway abundance profiling of Yarrowia lipolytica cultivated on different carbon sources. J Biotechnol 206:42–51. https://doi.org/10.1016/j.jbiotec.2015.04.005
Zhou W-J, Yang J-K, Mao L, Miao L-H (2015) Codon optimization, promoter and expression system selection that achieved high-level production of Yarrowia lipolytica lipase in Pichia pastoris. Enzym Microb Technol 71:66–72. https://doi.org/10.1016/j.enzmictec.2014.10.007

Chapter 3
Application of Non-*Saccharomyces* Yeasts in Wine Production

Santiago Benito, Javier Ruiz, Ignacio Belda, Florian Kiene, Beata Beisert, Eva Navascués, Domingo Marquina, Fernando Calderón, Antonio Santos, and Doris Rauhut

Abstract In the past, *Saccharomyces* spp. yeasts were almost the only option for use in modern winemaking due to their unparalleled ability to metabolize all grape juice sugar into ethanol. For that reason, until some years ago, all commercial dry yeasts were *Saccharomyces* spp. For several years, non-*Saccharomyces* were forgotten at industrial level, and even some of them were considered as spoilage microorganisms. Non-*Saccharomyces* only played a significant role in limited productions that perform spontaneous fermentations following organic polities. However, during the last decade, several researchers have proved numerous non-*Saccharomyces* to be able to improve wine quality and to solve some modern enology challenges. Some of the factors that can improve are acidity, aromatic complexity, glycerol content, ethanol reduction, mannoproteins, anthocyanins, and polysaccharide concentrations. They can also decrease the concentrations of unwanted compounds that affect food safety, such as ochratoxin A, ethyl carbamate, and biogenic amines. Due to all those scientific advances, the main manufacturers have just started to commercialize dry non-*Saccharomyces* such as *Torulaspora delbrueckii*, *Schizosaccharomyces pombe*, *Metschnikowia pulcherrima*, *Lachancea thermotolerans*, and *Pichia kluyveri*. Other non-*Saccharomyces* species with special enology abilities such as *Candida zemplinina*, *Kloeckera apiculata*, *Hanseniaspora vineae*,

S. Benito (✉) · I. Belda
Department of Chemistry and Food Technology. Escuela Técnica Superior de Ingeniería Agronómica, Alimentaria y de Biosistemas, Polytechnic University of Madrid, Madrid, Spain
e-mail: santiago.benito@upm.es; ignacio.belda@urjc.es

J. Ruiz · E. Navascués · D. Marquina · F. Calderón · A. Santos
Department of Genetics, Physiology and Microbiology (Microbiology Unit). Biology Faculty, Complutense University of Madrid, Madrid, Spain
e-mail: javiru02@ucm.es; eva.navascues@upm.es; dommarq@bio.ucm.es; fernando.calderon@upm.es; ansantos@ucm.es

F. Kiene · B. Beisert · D. Rauhut
Department of Microbiology and Biochemistry, Hochschule Geisenheim University, Geisenheim, Germany
e-mail: florian.kiene@hs-gm.de; beata.beisert@hs-gm.de; doris.rauhut@hs-gm.de

A. Sibirny (ed.), *Non-conventional Yeasts: from Basic Research to Application*,
https://doi.org/10.1007/978-3-030-21110-3_3

Hanseniaspora uvarum, *C. stellata*, *Kazachstania aerobia*, or *Schizosaccharomyces japonicus* could follow a similar progress. The aim of the chapter is to show which are the main abilities and advantages of these non-*Saccharomyces* in modern winemaking.

Keywords Non-*Saccharomyces* · Winemaking · *Torulaspora delbrueckii* · *Schizosaccharomyces pombe* · *Schizosaccharomyces japonicus* · *Metschnikowia pulcherrima* · *Lachancea thermotolerans* · *Pichia kluyveri* · *Pichia guilliermondii* · *Hanseniaspora* spp.

3.1 Introduction

Comprehensive studies gave attention to *Saccharomyces* and non-*Saccharomyces* yeasts, the main microorganisms involved in spontaneous wine fermentation (Fleet 1993; Fleet and Heard 1993; Lonvaud-Funel 1996). A wide variety of yeast genera and species belong to the indigenous yeasts occurring in grape musts. In spontaneous fermentations a sequential activity of the various non-*Saccharomyces* yeasts is coming up with abundance of apiculate yeasts until autochthonous *Saccharomyces cerevisiae* strains dominate (Varela et al. 2009; Jolly et al. 2014). Consequently, many yeast species and strains participate in the fermentation and the overall quality of the resulting wines (Padilla et al. 2016; Varela 2016). Indigenous yeasts originate from the vineyard, the grapes, the surfaces, and the equipment of the winery (Schütz and Gafner 1993; Fleet 2008). The composition of this yeast genera, species, and strain community differs every year and, therefore, a persistence of certain yeast strains in the winery environment cannot be expected (Rosini 1984; Lonvaud-Funel 1996; Pretorius 2000). In addition, most of the non-*Saccharomyces* yeasts have weak fermenting capacities. Hence, spontaneous fermentations have to be seen as an uncontrollable risk that can lead to sluggish or stuck fermentations and spoiled wines. Consequently, some of the non-*Saccharomyces* yeasts have been considered as spoilage microorganisms (Padilla et al. 2016; Varela 2016). The intensified use of pure dry yeast cultures of the species *S. cerevisiae* since the 1980s offered the winemakers the opportunity for controlled alcoholic fermentations and the production of predictable wines with established and consistent quality criteria (Varela 2016). In contrast it is discussed that wines fermented with pure mono *S. cerevisiae* cultures can lack the diversity and complexity of flavor caused by specific autochthonous yeasts (Lambrechts and Pretorius 2000; Romano et al. 2003; Padilla et al. 2016). On the other hand, intensive research demonstrated that certain non-conventional and selected strains can positively impact on wine quality (Ciani and Maccarelli 1998; Jolly et al. 2006; Fleet 2008; Jolly et al. 2014) due to their ability to enhance the aroma profile, to increase flavor diversity, and/or to contribute to other metabolites that influence the wine character and style (e.g., content of ethanol, acidity, glycerol, mannoproteins, and color stability) or affect wine safety (e.g., decrease of ochratoxin A, ethyl carbamate, and/or biogenic amines) (Quirós et al. 2014; Benito et al. 2015b; Ciani et al. 2016; Mylona et al. 2016). Furthermore, non-*Saccharomyces* yeasts are discussed to produce enzymes and metabolites of enological importance (Belda et al. 2016; Padilla et al. 2016).

The application of mixed and selected non-*Saccharomyces* yeasts with *S. cerevisiae* strains is intensively investigated within the last decade to combine the advantages and to improve wine quality. Thus, it is possible to intensify the opportunity of great specific and distinct wine types and styles, with a wide range of organoleptic characteristics and with lower risks of sluggish and stuck fermentations and spoilage (Romano et al. 2003; Ciani et al. 2006; Padilla et al. 2016).

The following chapter gives brief review on the latest research related to the use of non-*Saccharomyces* yeasts as single or mixed starter cultures for wine production and presents already existing applications.

3.2 Non-*Saccharomyces* Species

3.2.1 Torulaspora delbrueckii

Among the non-*Saccharomyces* species studied for their application to face the modern challenges of winemaking, *Torulaspora delbrueckii* is the best studied and more utilized yeast at industrial level (Benito 2018a; Benito et al. 2018). Unlike most of the non-*Saccharomyces* yeasts, the fermentative capacity of *T. delbrueckii* (Quirós et al. 2014) allows its implantation during the first phases of the alcoholic fermentation. Even though *T. delbrueckii* shows limited ethanol resistance compared with *S. cerevisiae*, it seems to be able to tolerate over 9% (v/v) (Bely et al. 2008). Because of that, this species can have an important role during the fermentation process, contributing positivity to several wine quality parameters. In addition, *T. delbrueckii* can properly complete the entire fermentation in beer or sparkling base wine production, due to the lower alcohol level of these beverages. In the case of wine fermentations, sequential fermentation with *S. cerevisiae*, with 2 to 5 days of delay, must be carried out to achieve the effects that *T. delbrueckii* might impact on wines. *T. delbrueckii* primarily has influence on the acetic acid content of wine. Fermentations in which *T. delbrueckii* is involved typically showed low levels of acetic acid (Bely et al. 2008). Acetic acid decreases between 0.13 and 0.27 g/L have been observed in *T. delbrueckii* sequential fermentation (Benito 2018a). Also, this yeast is able to lower the ethanol concentration of wines, over 1% (v/v) (Contreras et al. 2014). This impact is important in warmer viticultural regions (most affected by the climatic change) where musts reach higher concentrations of sugars and, therefore, higher concentration of ethanol after fermentation. Increases of glycerol content, which can have great influence on wine sensory properties, have been reported in a range of 0.1 to 1 g/L (González-Royo et al. 2015; Belda et al. 2017; Medina-Trujillo et al. 2017; Puertas et al. 2017).

T. delbrueckii is reported to consume between 20% (Belda et al. 2015) and 25% (Chen et al. 2018) of the initial malic acid concentration in must. Regarding succinic acid production, Puertas et al. (2017) reported an increase of 0.46 g/L, compared to pure *S. cerevisiae* fermentation. Higher amounts of mannoproteins and polysaccharides are also produced by *T. delbrueckii*, impacting the mouthfeel properties of wines (Belda et al. 2015). Reduction on acetaldehyde concentration in wine has been observed in most of research works using *T. delbrueckii*, in comparison to *S. cerevisiae* fermentations. Decrease from 20 to 40 mg/L has been reported (Belda et al. 2017; Puertas et al. 2017).

In relation to wine color quality, lower concentration of some anthocyanins was shown by *T. delbrueckii* and *S. cerevisiae* sequential fermentation than by single *S. cerevisiae* fermentation (Belda et al. 2015). Opposite results were observed regarding other anthocyanins (Bañuelos et al. 2016).

Aroma profiles of wines are also positively affected by *T. delbrueckii*, due to higher production of desirable volatile compounds, especially on aroma complexity and fruity character. Higher concentrations of fruity esters have been reported in sequential fermentation with *T. delbrueckii* (Renault et al. 2015; Belda et al. 2017), although a decrease in ester amount has also been observed (Azzolini et al. 2015; Puertas et al. 2017). As most of non-*Saccharomyces* yeasts, *T. delbrueckii* can reduce the concentration of higher alcohols in sequential fermentations (Milanovic et al. 2012; Belda et al. 2015). This is important because of the improvement of varietal compounds perception in final wine flavor. Nevertheless, increase of higher alcohol production has also been reported (Azzolini et al. 2015).

Concerning varietal compounds, significant increases of terpene compounds have been observed in *T. delbrueckii* sequential fermentation (Cus and Jenko, 2013; Whitener et al. 2017). Among non-*Saccharomyces* species, *T. delbrueckii* stands out because of its capability to release varietal thiols from their odorless precursors in musts. Renault et al. (2016) observed increase on 3-sulfanylhexan-1-ol (3SH) thiol release, but not on 4-methyl-4-sulfanylpentan-2-one (4MSP) release, in *T. delbrueckii* fermentations. Nevertheless, important increases on 4MSP productions have been reported with another *T. delbrueckii* strain (Belda et al. 2017).

In spite these impacts on wine quality of *T. delbrueckii*, high strain-dependent effects have been reported for *T. delbrueckii* (Azzolini et al. 2015; Escribano et al. 2018). Regarding commercialization of *T. delbrueckii* in wine industry, five strains are available for winemaking: PRELUDE™ (Chr. Hansen Holding A/S, Hoersholm, Denmark), BIODIVA™ (Lallemand Inc., Rexdale, Canada), ZYMAFLORE® ALPHA (LAFFORT®, Bordeaux, France), Viniferm NS-TD® (Agrovin S.A., AlcÁzar de San Juan, Spain) and PRIMAFLORA® VB BIO (Sud et Bio, Lattes, France). All of them are recommended to improve wine quality by decreasing volatile acidity and improving flavor complexity.

3.2.2 Schizosaccharomyces *spp.*

3.2.2.1 *Schizosaccharomyces pombe*

Schizosaccharomyces pombe is unique among the other non-*Saccharomyces* yeast because of its capacity to reach ethanol concentrations of about 15% (v/v) (depending on the strain), in regular wine fermentations. In addition, this yeast has the exclusive ability to deacidify wines through the conversion of malic acid into small amounts of ethanol and CO_2 (Benito et al. 2014). For that reason, the initial application of this species was focused in decreasing the acidity of wines from northern European cold wine regions, where the low grape maturity supposes high contents in malic acid, over 6 g/L. Under those circumstances, *S. pombe* metabolizes all the malic acid during the alcoholic fermentation generating increases in pH of about 0.5 units, producing smoother wines.

However, collateral effects such as high concentrations of volatile acidity, which occasioned strong vinegar character, are commonly reported when *S. pombe* ferments as a single inoculum. Most *S. pombe* strains tend to produce concentrations in acetic acid over 0.8 g/L, although some strains show moderate productions (Benito et al. 2014; Benito et al. 2016). This undesirable effect was initially minimized by using *S. pombe* strains in combination with selected *S. cerevisiae* strains (Benito et al. 2012). The main problem about selecting proper *S. pombe* strains is its low incidence in nature, lower than 0.5% in grapes and other fruits. Therefore, it is very difficult to isolate strains to accomplish a proper selection process. Nowadays the development of specific selective-differential growth media for *S. pombe* has solved that problem (Benito et al. 2018). During the last years, other industry processes, such as sparkling wine, plum wine, apple fermentation, or bilberry fermentation, which present higher concentration of malic acid than grapes, start to use *S. pombe* during their alcoholic fermentations in order to reduce malic acid content of their wines.

S. pombe has also been used in red wines (Benito et al. 2012) in totally different circumstances than the cool northern European viticultural regions. Warm viticulture areas from the south of Europe present contents of malic acid about 1 g/L in wines, with pH of about 3.9 and probable alcohol levels of about 15% (v/v). Under those circumstances, performing the microbiological stabilization process of malolactic fermentation by lactic acid bacteria is highly risky. The main risks are difficult alcoholic fermentation endings where lactic bacteria consume residual sugars generating high volatile acidity concentrations. In addition, the glycosidase activity and cell absorption of most lactic bacteria usually decrease the color of red wine in about 10 to 20%. If the malolactic fermentation is avoided after alcoholic fermentation, this process usually will take place in the bottle generating turbidity or deterioration of the top cork. Additionally, health problems for human beings such as the production of high concentrations of biogenic amines or ethyl carbamate usually take place when lactic bacteria perform in high pH. In this context, the use of

S. pombe, combined with *Lachancea thermotolerans*, can compensate the low acidity of warmer viticultural regions musts (Benito et al. 2015a).

Other beneficial ability of *S. pombe* is its production of high pyruvic acid concentration, five times higher than *S. cerevisiae*. This fact allows the production of high-color pigments such as Vitisin A, composed of pyruvic acid and anthocyanin (Benito et al. 2017). The production of moderate levels of acetaldehyde by some specific *S. pombe* strains, below its undesirable perception threshold, also increases the concentration of the stable color pigment Vitisin B.

The *S. pombe* cell structure is rich in α-galactomannose and β-glucans (Domizio et al. 2017; Domizio et al. 2018; Benito et al. 2019). This fact makes *S. pombe* able to release higher concentrations of polysaccharides (from 2.5 to 5 times higher than *S. cerevisiae*). Those polysaccharides and mannoprotein contents usually increase sensory parameters related to the wine structure and reduce the wine astringency.

Another interesting property of *S. pombe* is its urease enzymatic activity that allows removing all the urea from wine (Benito et al. 2015a). Urea is the main precursor of the carcinogenic compound ethyl carbamate. Additionally, the wines fermented by *S. pombe* do not contain nutrients that could be metabolized by lactic acid bacteria such as malic acid. Therefore, the wines are also stable against possible undesirable productions of biogenic amines such as histamine from lactic acid bacteria.

Some specific strains of *S. pombe* can remove gluconic acid from grape juice during the alcoholic fermentation (Peinado et al. 2009). This activity allows to avoid the possible negative effects produced by this compound when it is presented in high concentrations in rotten grapes.

Finally, *S. pombe*, other non-*Saccharomyces* yeasts, produces lower concentrations in higher alcohols, regarding *S. cerevisiae* (Benito et al. 2016), varying from 25 to 50% depending on the strain and the specific higher alcohol. That effect produces in some occasions fruity wines, with strong varietal character not masked by higher alcohols and without lactic notes from malolactic fermentation.

3.2.2.2 *Schizosaccharomyces japonicus*

Although most studies regarding *Schizosaccharomyces* genus focus on *S. pombe* applications in fermentation industries, another species from that genus named *Schizosaccharomyces japonicus* stands out because of its beneficial effects on winemaking (Domizio et al. 2018). *S. japonicus* also shows the ability to degrade malic acid (from 71 to 82%), and it also possesses a high fermentative power, up to 14% (v/v) of ethanol, being slower fermenter than *S. cerevisiae*. Some studies report that *S. japonicus* can produce higher concentrations of glycerol than *S. pombe*. The main undesired effect of the species is that they tend to produce high concentrations of acetic acid, about 0.76 g/L. That problem is nowadays palliated using immobilized cells in alginate beads that allow producing moderate final acetic acid concentrations of about 0.4 g/L. As *S. pombe*, *S. japonicus* produces low concentrations of higher alcohols. Nevertheless, some strains are reported to produce higher concentrations in 2-phenyl ethanol and isoamyl acetate than the *S. cerevisiae* controls. Finally, it has been reported higher concentrations of polysaccharides and the end of fermentation when *S. japonicus* is used, even higher than *S. pombe* in about 15%.

3.2.3 Metschnikowia pulcherrima

At the moment, there is one strain of *Metschnikowia pulcherrima* commercially available. The manufacturer advertises a competitive advantage based on its α-L-arabinofuranosidase activity. When the *M. pulcherrima* strain FLAVIA™ Mp346 is inoculated in grape juice, it increases the release of varietal compounds such as terpenes and volatile thiols (Lallemand Inc., Rexdale, Canada). However, co-fermentation with *S. cerevisiae* should be carried out to ensure a proper alcoholic fermentation ending. Other authors report that the compatibility between *M. pulcherrima* and *S. cerevisiae* must be considered before inoculation in order to avoid possible delays due to the inhibition effect of killer toxins production (Jolly et al. 2014).

Regarding basic fermentation parameters, sequential fermentations with *M. pulcherrima* and *S. cerevisiae* result in a slight increase of glycerol (0.2 g/L) and reductions of malic acid (0.2 g/L) and acetaldehyde (10 mg/L), while the final acetic acid levels do not increase (Ruiz et al. 2018). Although some strains are reported to be able to reduce the ethanol content during sequential fermentation in about 1% (v/v) (Contreras et al. 2014), most reported benefits about *M. pulcherrima* strains are related to the volatile aroma composition.

Most studies report *M. pulcherrima* as a lower producer of higher alcohols, which can mask varietal aromas such as terpenes or thiols. Ruiz et al. (2018) report a reduction of about 25% respect to the *S. cerevisiae* controls. Some authors mention *M. pulcherrima* as a higher producer of fruity total esters (Jolly et al. 2014), while other scientist did not observe evident differences, based on the *S. cerevisiae* strain (Ruiz et al. 2018). In other studies only the increases of specific esters such as ethyl octanoate, which are related to pleasant aromas like pine apple, were observed (Benito et al. 2015b).

Some studies report no influence on total terpenes or even decreases in their concentration (Benito et al. 2015b); however, great differences are reported in the case of thiols (Ruiz et al. 2018). The release of the polyfunctional thiol 4MSP is the most notorious advantage, observing an increase of seven times higher concentrations than the *S. cerevisiae* control (Ruiz et al. 2018). That effect notably enhances the varietal character of thiolic varieties such as Verdejo or Sauvignon blanc. Nevertheless, this effect is not observed by other researches (Sadoudi et al. 2012) because this impact is strain-dependent. Therefore, *M. pulcherrima* strains must be previously selected regarding this specific trait, mainly due to its cystathionine-β-lyase activity, although other unknown metabolic pathways could be involved.

3.2.4 Lachancea thermotolerans

Yeasts belonging to *Lachancea* genus stand out due to its potential application in biotechnological processes. Currently, this genus comprises 11 species, based on D1/D2 sequence analysis. These species can be isolated from multiple substrates, including both wild and human anthropogenic niches (Porter et al. 2019). *L. thermotolerans*, formerly classified as *Kluyveromyces thermotolerans, was selected as the type species. Among Lachancea* spp., *L. thermotolerans*, *L. fermentati*, and, to a lesser extent, *L. lanzarotensis* have been associated with grape must and wine environments, presenting biochemical traits with enological interest (Porter et al. 2019) and, therefore, have potential applicability on winemaking.

Hranilovic et al. 2017a, studied the population structure of *L. thermotolerans*, revealing clusters explained by multiple domestication events. They propose adaptation to different niches. Also, the phenotyping of the various strains revealed a concordance between these clusters and their fermentation capacity and their volatile profile obtained.

Despite *L. thermotolerans* is a moderate fermentative species, during wine production a co-fermentation with more vigorous yeasts as *S. cerevisiae* (Gobbi et al. 2013) or *Schizosaccharomyces pombe* (Benito et al. 2017) is carried out. *L. thermotolerans strains are* already commercialized for its use in winemaking, as a single strain (CONCERTO™) and in a blend with *T. delbrueckii* and *S. cerevisiae* strains (MELODY™, Chr. Hansen Holding A/S, Hoersholm, Denmark).

Due to its ability to produce L-lactic acid during alcoholic fermentation, an uncommon trait among yeasts, *L. thermotolerans* is the most frequently used species in wine industry for improving wine quality by acidification of grape musts. This fact is of great relevance in warmer viticultural regions due to the increase of sugars and the decrease of acidity in grapes caused by global warming. Lactic acid production by *L. thermotolerans* has been reported from 0.3 to 9.6 g/L and an increase of the pH value from 3.3 to 3.5 has been observed (Gobbi et al. 2013). In red wines from warm viticultural regions, the level of lactic acid produced by this yeast can be higher, even if there is a bacterial lactic acid production during malolactic fermentation (Benito 2018b).

Although lactic acid production is the main application of this yeast, the use of *L. thermotolerans* in winemaking can improve other wine quality parameters mentioned below, even in a strain-dependent manner.

L. thermotolerans can be used in the wine industry for the prevention of toxins produced during fermentation. Because of its role as a biocontrol agent against the growth of ochratoxigenic fungi, its inhibition impact on ochratoxin A accumulation has been reported (Ponsone et al. 2016). Some studies about the impact of *L. thermotolerans* on winemaking report malic acid degradation from 8% (Gobbi et al. 2013) to 25% (Kapsopoulou et al. 2005); however this ability is highly strain-dependent. Relating to acetic acid, low concentrations are obtained during fermentations with *L. thermotolerans*. Comitini et al. (2011) observed that different strains of *L. thermotolerans* have a 50% lower production of volatile acids than *S. cerevisiae* strains, ranging from 0.32 to 0.58 g/L.

Due to their low fermentation power, *L. thermotolerans* strains produce low final levels of ethanol, leaving residual sugars in the final wine. It has been reported that in sequential fermentation with *S. cerevisiae*, an ethanol decrease of 0.20% (v/v) (Benito et al. 2015b) or 0.40% (v/v) (Hranilovic et al. 2017b) can be achieved. In contrast, several studies observed an increase in the formation of glycerol, e.g., from 0.29 to 0.69 g/L in Gobbi et al. (2013) and Benito et al. (2015b). Decreases in acetaldehyde production have been shown in single *L. thermotolerans* fermentations regarding single *S. cerevisiae* fermentations (Ciani et al. 2006). All these reported *L. thermotolerans* effects require in most cases sequential fermentations with *S. cerevisiae* or *S. pombe* strains (Benito 2018b).

Several works have reported the influence of *L. thermotolerans* strains on wine aroma profile. Due to the reported glucosidase activity for some *L. thermotolerans* strains, an increase in the release of terpenes has been observed during single and sequential fermentations with this species (Benito et al. 2015a, b; Comitini et al. 2011). Against that a decrease in the production of higher alcohols, especially in 2-methyl-1-propanol, 3-methyl-1-butanol, and 2-methyl-1-butanol has been noticed during co-fermentation with *L. thermotolerans* and *S. cerevisiae*, in comparison to a single fermentation with *S. cerevisiae* (Gobbi et al. 2013; Benito et al. 2015b; Balikci et al. 2016; Escribano et al. 2018). Opposite results were also observed (Comitini et al. 2011; Chen et al. 2018), demonstrating again the variability between *L. thermotolerans* strains. Similar effects were perceived regarding ester production. Both increase (Benito et al. 2015b; Hranilovic et al. 2017b) and decrease (Escribano et al. 2018) on ester production during wine fermentation have been reported.

Other parameters related to wine quality can also be improved by using *L. thermotolerans*. Several studies have reported increase in color intensity due to the increment of anthocyanin concentration (Benito et al. 2015a; Hranilovic et al. 2017b; Chen et al. 2018). Polysaccharides and mannoproteins have a great influence on mouthfeel of wine. Some works reported increases on polysaccharide concentrations when *L. thermotolerans* participated in the fermentation, compared to pure cultures of *S. cerevisiae* (Gobbi et al. 2013; Domizio et al. 2014). This increasing effect has not been observed in mannoproteins with the use of *L. thermotolerans* in aging over lees (Benito 2018b).

Special mention deserves the combined use of *L. thermotolerans* and *S. pombe* in wine fermentation. *S. pombe* has been used to avoid the negative effects of malolactic fermentation by lactic acid bacteria, e.g., toxin production, loss of color, or increase in acetic acid (Domizio et al. 2017). Nevertheless, *S. pombe* use might reduce acidity of wine. For this reason, combination of *L. thermotolerans* and *S. pombe* is a good solution in grape must with low acidity. Thus, *L. thermotolerans* lactic acid production compensates for the loss of acidity produced by *S. pombe* due to its malic acid metabolism, without malolactic fermentation by lactic acid bacteria (Benito et al. 2015a).

In conclusion, *L. thermotolerans* strains have a great potential use for the winemaking industry in an improvement of wine parameters, during a combined use with higher fermentative yeasts like *S. cerevisiae* or *S. pombe*. Despite some undesirable traits have been reported in this non-*Saccharomyces* species, these disadvantages seem to be strain-dependent. Therefore, a good strain selection procedure allows to select suitable strains for the application in wine quality improvement.

3.2.5 Pichia *spp.*

3.2.5.1 *Pichia kluyveri*

There is one strain of *Pichia kluyveri* commercially available (FROOTZEN®, Chr. Hansen Holding A/S, Horsholm, Denmark). *P. kluyveri* is known to have superior capabilities of releasing 3SH and 3-sulfanylhexyl acetate (3SHA) from precursors found in grape must (Anfang et al. 2009). These compounds impart passion fruit and other tropical aromas in thiolic white grape varieties. Their growth is inhibited by 4 to 5% ethanol. In co-fermentations with *S. cerevisiae*, *P. kluyveri* showed to produce higher levels of specific esters such as 2-phenylethyl acetate and ethyl octanoate in about 23% and 10% more than the *S. cerevisiae* control (Benito et al. 2015b). The same study also observed an increase of about 20% in total terpenes. According to the sensorial analysis, the wines fermented by *P. kluyveri* showed higher Riesling typicity than the control.

3.2.6 Hanseniaspora *spp.*

Despite their medium/low fermentation capacity and the current lack of any commercial strains, *Hanseniaspora* genus yeasts stand out due to their emerging potential in the winemaking industry. This apiculate yeast genus, composed of ten species, presents an extremely high occurrence on grape-associated microflora. In addition, positive enological traits have been reported for these species that can contribute to the chemical composition of wines, in combination with *S. cerevisiae* in sequential fermentation (Martin et al. 2018). Some authors observed that *H. vineae* can contribute to the fruity aroma of wines by increasing the concentration of acetate esters and isoprenoids during wine fermentation, accompanied with a decrease of fatty acid and ethyl esters concentrations compared to *S. cerevisiae* strains (Martín et al. 2019). Giorello et al. (2018) also reported these effects on *H. vineae*, identifying gene duplication and absences compared to *S. cerevisiae* genome that explain this impact on wine flavor. With regard to the release of varietal compounds, *H. uvarum* and *H. vineae* showed β-glucosidase and β-xylosidase enzymatic activities that can potentially increase the production of terpenes during the fermentation (López et al. 2015). Mendes-Ferreira et al. (2001) also demonstrated the ability of a *H. uvarum* strain to release monoterpenols such as linalool, geraniol, or α-terpineol. Although these apiculate yeasts have been associated with the production of high volatile acidity, this trait seems to be strain-dependent (Martin et al. 2018). Also, *H. uvarum* and *H. guilliermondii* strains have been described as producers of high levels of heavy sulfur-containing aromatics (Moreira et al. 2005). Regarding its contribution to the color quality of red wines, several *Hanseniaspora* genus species have been shown to add to contribute to the polyphenolic composition of sequential fermentation wines (Lleixa et al. 2016). Finally, *Hanseniaspora* spp. yeasts not only can impact on wine parameters directly but indirectly by affecting the genetic expression profile of *S. cerevisiae*. Barbosa et al. (2015) demonstrate effect of *H. guilliermondii* on the transcriptomic response of *S. cerevisiae*, particularly on the expression of flavor-active compound-associated genes.

3.3 Conclusions

The increasing and already numerous research studies indicate the growing interest in the application of non-*Saccharomyces* yeasts as pure single or mixed cultures for controlled mixed fermentations with simultaneous or sequential inoculations to achieve more flavor complexity and stylistic features (Padilla et al. 2016; Varela et al. 2017). In addition, the mixed fermentations with selected non-*Saccharomyces* and *Saccharomyces* yeasts are more and more used to meet the challenges of the climatic change as lowering the alcohol content (González-Royo et al. 2015; Contreras et al. 2015; Morales et al. 2015) and the increase or decrease of acidity (Kapsopoulou et al. 2007; Benito et al. 2015). Further applications like wine color stabilization (Benito et al. 2017); low formation of hydrogen sulfide (H_2S), sulfites (SO_2), acetaldehyde, and other SO_2-bindung compounds; ability to reduce the copper content, biogenic amines, etc. will be of an increasing demand to optimize wine quality and safety (Comitini et al. 2017).

Continuative research is required to get more experience with the application of mixed yeast cultures and single non-*Saccharomyces* fermentations and to optimize the control over mixed culture fermentations. In particular, strain-dependent interactions and cell-to-cell contact in mixed cultures which might affect the entire metabolism and regulation processes have to be studied in detail (Nissen et al. 2003; Kemsawasd et al. 2015; Padilla et al. 2016) for an improved and adequate design of mixed yeast cultures for additional and new applications related to the requirements and demands of the wine industry.

References

Anfang N, Brajkovich M, Goddard MR (2009) Co-fermentation with *Pichia kluyveri* increases varietal thiol concentrations in Sauvignon Blanc. Aust J Grape Wine Res 15:1–8

Azzolini M, Tosi E, Lorenzini M, Finato F, Zapparoli G (2015) Contribution to thearoma of white wines by controlled *Torulaspora delbrueckii* cultures in association with *Saccharomyces cerevisiae*. World J Microbiol Biotechnol 31:277–293

Balikci EK, Tanguler H, Jolly NP, Erten H (2016) Influence of *Lachancea thermotolerans* on cv. Emir wine fermentation. Yeast 33:313–321

Bañuelos MA, Loira I, Escott C, Del Fresno JM, Morata A, Sanz PD, Otero L, JA S–L (2016) Grape processing by high hydrostatic pressure: effect on use of non–*Saccharomyces* in must fermentation. Food Bioprocess Technol 9:1769–1778

Barbosa C, Mendes-Faia A, Lage P, Mira NP, Mendes-Ferreira A (2015) Genomic expression program of *Saccharomyces cerevisiae* along a mixed-culture wine fermentation with *Hanseniaspora guillermondii*. Microb Cell Factories 14:124

Belda I, Navascués E, Marquina D, Santos A, Calderon F, Benito S (2015) Dynamic analysis of physiological properties of *Torulaspora delbrueckii* in wine fermentations and its incidence on wine quality. Appl Microbiol Biotechnol 99:1911–1922

Belda I, Ruiz J, Alastruey-Izquierdo A, Navascués E, Marquina D, Santos A (2016) Unraveling the enzymatic basis of wine "flavorome": a phylo-functional study of wine related yeast species. Front Microbiol 7:12

Belda I, Ruiz J, Beisert B, Navascués E, Marquina D, Calderón F, Rauhut D, Benito S, Santos A (2017) Influence of *Torulaspora delbrueckii* in varietal thiol (3–SH and 4–MSP) release in wine sequential fermentations. Int J Food Microbiol 257:183–191

Bely M, Stoeckle P, Masneuf–Pomarède I, Dubourdieu D (2008) Impact of mixed *Torulaspora delbrueckii–Saccharomyces cerevisiae* culture on high–sugar fermentation. Int J Food Microbiol 122:312–320

Benito S (2018a) The impact of *Torulaspora delbrueckii* yeast in winemaking. Appl Microbiol Biotechnol 102:3081–3094

Benito S (2018b) The impacts of *Lachancea thermotolerans* yeast strains on winemaking. App Microbiol and Biotechnol 6:1–6

Benito S (2019) The impacts of Schizosaccharomyces on winemaking. Appl Microbiol Biotechnol 103(11):4291–4312

Benito S, Palomero F, Morata A, Calderón F, Suárez-Lepe JA (2012) New applications for Schizosaccharomyces pombe in the alcoholic fermentation of red wines. Int J Food Sci Technol 47(10):2101–2108

Benito S, Palomero F, Calderón F, Palmero D, JA S–L (2014) Schizosaccharomyces. In: Batt CA, Tortorelo ML (eds) Encyclopedia of food microbiology, 2nd edn. V3. Elsevier, Amsterdam, pp 365–370

Benito Á, Calderón F, Palomero F, Benito S (2015a) Combine use of selected *Schizosaccharomyces pombe* and *Lachancea thermotolerans* yeast strains as an alternative to the traditional malolactic fermentation in red wine production. Molecules 20:9510–9523

Benito S, Hofmann T, Laier M, Lochbühler B, Schüttler A, Ebert K, Fritsch S, Röcker J, Rauhut D (2015b) Effect on quality and composition of Riesling wines fermented by sequential inoculation with non–*Saccharomyces* and *Saccharomyces cerevisiae*. Eur Food Res Technol 241:707–717

Benito Á, Calderón F, Benito S (2016) Combined use of S. pombe and L. thermotolerans in winemaking. Beneficial effects determined through the study of wines' analytical characteristics. Molecules 21(12):1744

Benito A, Calderon F, Benito S (2017) The combined use of *Schizosaccharomyces pombe* and *Lachancea thermotolerans*–effect on the anthocyanin wine composition. Molecules 22:739

Benito Á, Calderón F, Benito S (2018) *Schizosaccharomyces pombe* isolation protocol. In: Schizosaccharomyces pombe. Springer, New York, NY, pp 227–234

Benito Á, Calderón F, Benito S (2019) Mixed alcoholic fermentation of *Schizosaccharomyces pombe* and *Lachancea thermotolerans* and its influence on mannose–containing polysaccharides wine composition. AMB Express 9:17

Chen K, Escott C, Loira I, del Fresno JM, Morata A, Tesfaye W, Calderon F, JA S–L, Han S, Benito S (2018) Use of non–*Saccharomyces* yeasts and oenological tannin in red winemaking: influence on colour, aroma and sensorial properties of young wines. Food Microbiol 69:51–63

Ciani M, Maccarelli F (1998) Oenological properties of non-*Saccharomyces* yeasts associated with wine-making. World J Microbiol Biotechnol 14:199–203

Ciani M, Beco L, Comitini F (2006) Fermentation behaviour and metabolic interactions of multistarter wine yeast fermentations. Int J Food Microbiol 108:239–245

Ciani M, Morales P, Comitini F, Tronchoni J, Canonico L, Curiel JA, Oro L, Rodrigues AJ, Gonzalez R (2016) Non–conventional yeast species for lowering ethanol content of wines. Front Microbiol 7:642

Comitini F, Gobbi M, Domizio P, Romani C, Lencioni L, Mannazzu I, Ciani M (2011) Selected non–*Saccharomyces* wine yeasts in controlled multistarter fermentations with *Saccharomyces cerevisiae*. Food Microbiol 28:873–882

Comitini F, Capece A, Ciani M, Romano P (2017) New insights on the use of wine yeasts. Curr Opin Food Sci 13:44–49

Contreras A, Hidalgo C, Henschke PA, Chambers PJ, Curtin C, Varela C (2014) Evaluation of non–*Saccharomyces* yeasts for the reduction of alcohol content in wine. Appl Environ Microbiol 80:1670–1678

Contreras A, Hidalgo C, Schmidt S, Henschke PA, Curtin C, Varela C (2015) The application of non-*Saccharomyces* yeast in fermentations with limited aeration as a strategy for the production of wine with reduced alcohol content. Int J Food Microbiol 205:7–15

Cus F, Jenko M (2013) The influence of yeast strains on the composition and sensory quality of Gewürztraminer wine. Food Technol Biotechnol 51:547–553

Domizio P, Liu Y, Bisson LF, Barile D (2014) Use of non–*Saccharomyces* wine yeasts as novel sources of mannoproteins in wine. Food Microbiol 43:5–15

Domizio P, Liu Y, Bisson L, Barile D (2017) Cell wall polysaccharides released during the alcoholic fermentation by *Schizosaccharomyces pombe* and *S. japonicus*: quantification and characterization. Food Microbiol 61:136–149

Domizio P, Lencioni L, Calamai L, Portaro L, Bisson LF (2018) Evaluation of the Yeast *Schizosaccharomyces japonicus* for Use in Wine Production. Am J Enol Vitic 69:266–277

Escribano R, González–Arenzana L, Portu J, Garijo P, López-Alfaro I, López R, Santamaría P, Gutiérrez AR (2018) Aromatic compound production and fermentative behavior within different non–*Saccharomyces* species and clones. J Appl Microbiol 124:1521–1531

Fleet GH (1993) The microorganisms of winemaking-isolation numeration and identification. In: Fleet GH (ed) Wine Microbiology and Biotechnology, 1st edn. Harwood Academic Publishers, Chur, pp 1–25

Fleet GH (2008) Wine yeasts for the future. FEMS Yeast Res 8:979–995

Fleet GH, Heard GM (1993) Yeast-growth during winemaking. In Fleet GH (ed) Wine microbiology and biotechnology, 1st edn. Harwood Academic Publishers, Chur, pp 27–54

Giorello F, Valera MJ, Martín V, Parada A, Salzman V, Camesasca L, Fariña L, Boido E, Medina K, Dellacassa E, Berna L, Aguilar PS, Mas A, Gaggero C, Carrau F (2018) Genomic and Transcriptomic Basis of *Hanseniaspora vineae*'s Impact on Flavor Diversity and Wine Quality. Appl Environ Microbiol 85:e01959–e01918

Gobbi M, Comitini F, Domizio P, Romani C, Lencioni L, Mannazzu I, Ciani M (2013) *Lachancea thermotolerans* and *Saccharomyces cerevisiae* in simultaneous and sequential co–fermentation: a strategy to enhance acidity and improve the overall quality of wine. Food Microbiol 33:271–281

González-Royo E, Pascual O, Kontoudakis N, Esteruelas M, Esteve-Zarzoso B, Mas A, Canals JM, Zamora F (2015) Oenological consequences of sequential inoculation with non–*Saccharomyces* yeasts (*Torulaspora delbrueckii* or *Metschnikowia pulcherrima*) and *Saccharomyces cerevisiae* in base wine for sparkling wine production. Eur Food Res Technol 240:999–1012

Hranilovic A, Bely M, Masneuf-Pomarede I, Jiranek V, Albertin W (2017a) The evolution of is driven by geographical determination, anthropisation and flux between different ecosystems. PLoS One 12:e0184652

Hranilovic A, Li S, Boss PK, Bindon K, Ristic R, Grbin PR, Van der Westhuizen T, Jiranek V (2017b) Chemical and sensory profiling of Shiraz wines co–fermented with commercial non-*Saccharomyces* inocula. Aust J Grape Wine Res 24:166–180

Jolly NP, Augustyn OHP, Pretorius IS (2006) The role and use of non-Saccharomyces yeasts in wine production. S Afr J Enol Vitic 27:15–39

Jolly NP, Varela C, Pretorius IS (2014) Not your ordinary yeast: non–*Saccharomyces* yeasts in wine production uncovered. FEMS Yeast Res 14:215–237

Kapsopoulou K, Kapaklis A, Spyropoulos H (2005) Growth and fermentation characteristics of a strain of the wine yeast *Kluyveromyces thermotolerans* isolated in Greece. World J Microbiol Biotechnol 21:1599–1602

Kapsopoulou K, Mourtzini A, Anthoulas M, Nerantzis E (2007) Biological acidification during grape must fermentation using mixed cultures of *Kluyveromyces thermotolerans* and *Saccharomyces cerevisiae*. World J Microbiol Biotechnol 23:735–739

Kemsawasd V, Branco P, Almeida MG, Caldeira J, Albergaria H, Arneborg N (2015) Cell-to-cell contact and antimicrobial peptides play a combined role in the death of *Lachanchea thermotolerans* during mixed-culture alcoholic fermentation with Saccharomyces cerevisiae. FEMS Microbiol Let 14:1–8

Lambrechts MG, Pretorius IS (2000) Yeast and its importance to wine aroma–areview. S Afr J Enol Vitic 21:97–129

Lleixa J, Martin V, Portillo C, Carrau F, Beltran G, Mas A (2016) Comparison of the performances of *Hanseniaspora vineae* and *Saccharomyces cerevisiae* during winemaking. Front Microbiol 7:338

Lonvaud-Funel A (1996) Microorganisms of winemaking. Cerevisia 21:55–58

López MC, Mateo JJ, Maicas S (2015) Screening of β-glucosidase and β- xylosidase activities in four non-*Saccharomyces* yeast isolates. J Food Sci 80:1696–1704

Martin V, Valera MJ, Medina K, Boido E, Carrau F (2018) Oenological Impact of the *Hanseniaspora/Kloeckera* Yeast Genus on Wines—A Review. Fermentation 4:76

Martín V, Fariña L, Medina K, Boido E, Dellacassa E, Mas A, Carrau F (2019) Oenological attributes of the yeast *Hanseniaspora vineae* and its application for white and red winemaking. BIO Web Conf 12:02010

Medina-Trujillo L, González-Royo E, Sieczkowski N, Heras J, Canals JM, Zamora F (2017) Effect of sequential inoculation (*Torulaspora delbrueckii/Saccharomyces cerevisiae*) in the first fermentation on the foaming properties of sparkling wine. Eur Food Res Technol 243:681–688

Mendes-Ferreira A, Climaco MC, Mendes Faia A (2001) The role of non-*Saccharomyces* species in releasing glycosidic bound fraction of grape aroma components—A preliminary study. J Appl Microbiol 91:67–71

Milanovic V, Ciani M, Oro L, Comitini F (2012) *Starmerella bombicola* influences the metabolism of *Saccharomyces cerevisiae* at pyruvate decarboxylase and alcohol dehydrogenase level during mixed wine fermentation. Microb Cell Factories 11:18

Morales P, Rojas V, Quirós M, González R (2015) The impact of oxygen in the final alcohol content of wine fermented by a mixed starter culture. Appl Microbiol Biotechnol 99:3993–4003

Moreira N, Mendes F, Hogg T, Vasconcelos I (2005) Alcohols, esters and heavy sulfur compounds production by pure and mixed cultures of apiculate wine yeasts. Int J Food Microbiol 103:285–294

Mylona AE, Del Fresno JM, Palomero F, Loira I, Bañuelos MA, Morata A, Calderón F, Benito S, Suárez-Lepe JA (2016) Use of Schizosaccharomyces strains for wine fermentation—effect on the wine composition and food safety. Int J Food Microbiol 232:63–72

Nissen P, Nielsen D, Arneborg N (2003) Viable Saccharomyces cerevisiae cells at high concentrations cause early growth arrest of non-Saccharomyces yeasts in mixed cultures by a cell-cell contact-mediated mechanism. Yeast 20:331–341

Padilla B, Gil JV, Manzanares P (2016) Past and Future of Non-Saccharomyces Yeasts: From Spoilage Microorganisms to Biotechnological Tools for Improving Wine Aroma Complexity. Front in Microbiol 7:411

Peinado RA, Maestre O, Mauricio JC, Moreno JJ (2009) Use of a *Schizosaccharomyces pombe* mutant to reduce the content in gluconic acid of must obtained from rotten grapes. J Agric Food Chem 57:2368–2377

Ponsone ML, Nally MC, Chiotta ML, Combina M, Köhl J, Chulze SN (2016) Evaluation of the effectiveness of potential biocontrol yeasts against black sur rot and ochratoxin A occurring under greenhouse and field grape production conditions. Biol Control 103:78–85

Porter TM, Divol B, Setati ME (2019) *Lachancea* yeast species: Origin, biochemical characteristics and oenological significance. Food Res Int 119:378–389

Pretorius IS (2000) Tailoring wine yeast for the new millennium: novel approaches to the ancient art of winemaking. Yeast 16:675–729

Puertas B, Jiménez MJ, Cantos–Villar E, Cantoral JM, Rodríguez ME (2017) Use of *Torulaspora delbrueckii* and *Saccharomyces cerevisiae* in semi–industrial sequential inoculation to improve quality of palomino and chardonnay wines in warm climates. J Appl Microbiol 122:733–746

Quirós M, Rojas V, Gonzalez R, Morales P (2014) Selection of non–*Saccharomyces* yeast strains for reducing alcohol levels in wine by sugar respiration. Int J Food Microbiol 181:85–91

Renault P, Coulon J, de Revel G, Barbe J–C, Bely M (2015) Increase of fruity aroma during mixed *T. delbrueckii/S. cerevisiae* wine fermentation is linked to specific esters enhancement. Int J Food Microbiol 207:40–48

Renault P, Coulon J, Moine V, Thibon C, Bely M (2016) Enhanced 3– sulfanylhexan–1–ol production in sequential mixed fermentation with *Torulaspora delbrueckii/Saccharomyces cerevisiae* reveals a situation of synergistic interaction between two industrial strains. Front Microbiol 7:293

Romano P, Fiore C, Paraggio M, Caruso M, Capece A (2003) Function of yeast species and strains in wine flavour. Int J Food Microbiol 86:169–180

Rosini G (1984) Assessment of dominance of added yeast in wine fermentation ripening. Microb Ecol 8:83–89

Ruiz J, Belda I, Beisert B, Navascués E, Marquina D, Calderón F, Rauhut D, Santos A, Benito S (2018) Analytical impact of *Metschnikowia pulcherrima* in the volatile profile of Verdejo white wines. Appl Microbiol Biotechnol 102:8501–8509

Sadoudi M, Tourdot-Maréchal R, Rousseaux S, Steyer D, Gallardo-Chacón JJ, Ballester J, Vichi S, Guérin-Schneider R, Caixach J, Alexandre H (2012) Yeast-yeast interactions revealed by aromatic profile analysis of Sauvignon Blanc wine fermented by single or co–culture of non–*Saccharomyces* and *Saccharomyces yeasts* Food Microbiol 32:243–253

Schütz M, Gafner J (1993) Analysis of yeast diversity during spontaneous and induced alcoholic fermentations. J Appl Bacteriol 75:551–558

Varela C (2016) The impact of non-*Saccharomyces* yeasts in the production of alcoholic beverages. Appl Microbiol Biotechnol 100:9861–9874

Varela C, Siebert T, Cozzolino D, Rose L, McLean H, Henschke PA (2009) Discovering a chemical basis for differentiating wines made by fermentation with 'wild' indigenous and inoculated yeasts: role of yeast volatile compounds. Aust J Grape Wine Res 15:238–248

Varela C, Barker A, Tran T, Borneman A, Curtin C (2017) Sensory profile and volatile aroma composition of reduced alcohol Merlot wines fermented with *Metschnikowia pulcherrima* and *Saccharomyces uvarum*. Int J Food Microbiol 252:1–9

Whitener MEB, Stanstrup J, Carlin S, Divol B, Du Toit M, Vrhovsek U (2017) Effect of non–*Saccharomyces* yeasts on the volatile chemical profile of Shiraz wine. Aust J Grape Wine Res 23:179–192

Chapter 4
Citric Acid Production by *Yarrowia lipolytica*

Erdem Carsanba, Seraphim Papanikolaou, Patrick Fickers, Bilal Agirman, and Huseyin Erten

Abstract Recently, with an increasing annual demand of more than two million tons, citric acid (CA) has become the main additive and functional component in the food, pharmaceutical, and chemical industries. This rising demand has induced research to search for alternative and cheap ways to fulfil CA requirements in industry. Lately, *Yarrowia lipolytica* has been considered as a promising microorganism in the production of CA since it has many advantages over moulds: mainly high productivity, easier cultivation (convenient for continuous process), and the capability to use a wide range of agricultural or industrial by-products and wastes as cheap carbon sources. CA production by this yeast depends on certain factors such as medium composition (type and concentrations of carbon, nitrogen, and trace elements) and the type of strain used (wild, mutant, or genetically engineered), as well as cultivation conditions (pH, temperature, dissolved oxygen, etc.). This review principally details recent studies concerning CA production by *Y. lipolytica* with an emphasis on techniques to increase productivity and yield to meet the expanding demand for this organic acid. Suitable substrates and production factors for high and cost-efficient CA production are discussed. Downstream processes and production systems are also reviewed.

E. Carsanba
Faculty of Agriculture, Food Engineering Department, Cukurova University, Adana, Turkey

Altinozu Agricultural Sciences Vocational School, Mustafa Kemal University, Hatay, Turkey

S. Papanikolaou
Department of Food Science & Human Nutrition, Agricultural University of Athens, Athens, Greece
e-mail: spapanik@aua.gr

P. Fickers
Gembloux Agro-Bio Tech Microbial Processes and Interactions (MiPI), University of Liège, Gembloux, Belgium
e-mail: pfickers@uliege.be

B. Agirman · H. Erten (✉)
Faculty of Agriculture, Food Engineering Department, Cukurova University, Adana, Turkey
e-mail: bagirman@cu.edu.tr; herten@cu.edu.tr

A. Sibirny (ed.), *Non-conventional Yeasts: from Basic Research to Application*,
https://doi.org/10.1007/978-3-030-21110-3_4

Keywords Yeast · *Yarrowia lipolytica* · Citric acid · Citric acid biosynthesis · Factors

4.1 Introduction

4.1.1 Citric Acid

Citric acid ($C_6H_8O_7.H_2O$, 2-hydroxy-1.2.3-propanetricarboxylic acid), an intermediate of the tricarboxylic acid (TCA) cycle, is the most commonly used organic acid in the industry and is naturally found in many plants, animal tissues, and physiological fluids (Cavallo et al. 2017; Rzechonek et al. 2019; Timoumi et al. 2018). Citric acid (CA) is found in lemons and limes (approximately 4–8%) and in other varieties of fruits and vegetables such as oranges, tangerines, raspberries, and strawberries, however, in lower amount (1–3%). Human body tissues such as blood, bones, semen, urine, and human milk have a CA content varying from 10–25 ppm to 2000–4000 ppm (Apelblat 2014). CA can be also produced by microorganisms through being the intermediate compound of the TCA cycle during aerobic metabolism (Cavallo et al. 2017; Timoumi et al. 2018).

CA is a colourless compound that is highly soluble in water (1.48 g/mL at 20 °C) with a density of 1.665 g/cm^3 (anhydrous at 18 °C), a molecular weight of 192.12 g/mol, a melting temperature range of 135–152 °C, and a decomposition temperature of 248 °C. CA has three carboxyl groups, with three dissociation constants: $pKa1 = 3.13$, $pKa2 = 4.76$, and $pKa3 = 6.39$. In aqueous solutions, its salts can be formed by binding citrate ions with many metals. The anhydrous crystal form of CA can be obtained when the crystallization is applied at a temperature over 36.6 °C. In addition, it has biodegradable, eco-friendly, economical, and generally recognized as safe (GRAS) properties (Apelblat 2014; Cavallo et al. 2017; Dhillon et al. 2011).

4.1.2 Importance of Citric Acid in Industry

Recently, CA has attracted increasing interest due to its important role in food, pharmaceuticals, and other industries. The minimum CA selling price (MCSP) has been reported as $0.603 per kilogram (Hou and Bao 2018). Its global production rate reached more than two million tons annually in 2015 with an expected increment of 3.7% per year until 2020 (Cavallo et al. 2017; Ciriminna et al. 2017). Approximately 70% of this production has applications in the food industry, 12% in the pharmaceutical industry, and 18% in other sectors (cosmetic, chemical, metallurgy, textile, paper, waste recycling, mineral fertilizer, agriculture industries, etc.). CA is generally used in anhydrous, monohydrate, and salt forms, such as sodium citrate, calcium citrate, and potassium citrate (Dhillon et al. 2011; Kamzolova and Morgunov 2017; Kieliszek et al. 2017). As a food additive, it is employed as an acidifier, antioxidant, emulsifier, flavouring agent, stabilizer, and preservative in many foods such as beverages, wine, cider drinks, jellies, deserts, jams, cheese, ice cream, confectionary, canned foods, fish, fats, and animal or vegetable oils. Due to its GRAS status, biodegradability, biocompatibility, and non-toxicity as well as its environmentally friendly properties, CA is also used in manufacturing pharmaceutical products where it is used as an acidulant in astringent products, an excipient in tablets, an antimicrobial agent in CA-coated manganese ferrite nanoparticles, and a sparkling agent in effervescent products (in combination with bicarbonate) as well as an anticoagulant in blood transfusion. In addition, it has many applications in the cosmetic, chemical, and detergent industries. Its high buffer capacity promotes its usage in soaps, detergents, and cleaning products. In analytical chemistry, it is used as an eluting agent in ion exchange chromatography. Furthermore, it is used in textiles, printing, paper, leather and construction industries, and in the production of reagents and inks as well as in some metal plating and cleaning technologies. Moreover, it has applications in agriculture as a micronutrient for plants and as an antimicrobial for phytopathogens (Apelblat 2014; Cavallo et al. 2017; Ciriminna et al. 2017; Dhillon et al. 2011; Kamzolova and Morgunov 2017; Morgunov et al. 2017).

4.2 Citric Acid Production in History

CA was first discovered by the alchemist Jabir Ibn Hayyan (also known as Geber) in the eighth century. In 1784, Swedish chemist Carl Wilhelm Scheele isolated and crystallized CA from lemon juice. Later, in 1834, German chemist Justus von Liebig defined it as hydroxytricarboxylic acid (Apelblat 2014; Cavallo et al. 2017; Mattey 1992).

Besides isolation from lemon and lime, it was also synthesized from glycerol with a series of reactions by Grimaux and Adam in 1880. Then, processes for the chemical synthesis of CA were discovered by Haller and Held in 1890, Dunschmann and Pechmann in 1891, and Lawrence in 1897, respectively (Apelblat 2014).

First, commercial production of CA began in the middle of the 1800s in England with isolation from Italian lemons and limes. After the discovery of Carl Wehmer in 1893 that CA could be produced from sugar by using the mould *Penicillium*, it was produced at an industrial scale by the microbial cultivation process. However, this attempt failed due to high microbial contamination during cultivation. In 1913, CA production by *Sterigmatoocystis nigra* (*Aspergillus niger*) was patented by Zahorsky (US Patent No. 1065358). Later, most CA was produced at a high rate by using strains of *Aspergillus niger* at low pH in a medium containing sucrose and small amounts of inorganic salts according to the findings of the food chemist James Currie and the microbiologist Charles Thom in the USA. CA was produced based on this microbiological process in North America, Europe, and Russia until World War II, and it is still actually in use in countries such as China, India, Brazil, Japan, etc. Molasses, sugarcane, and sugar beet were initially preferred as substrates for *A. niger* cultivations due to their high sugar content and ready-to-use properties. Later, inexpensive substrates such as straw, hydrolysed starch, cellulose, and waste products were also used as low-cost substrates. Surface, submerged, continuous, immobilized, yeast-based, and koji processes were applied to CA production (Apelblat 2014; Cavallo et al. 2017).

Although the use of *A. niger* for CA production has a wide range of industrial applications, it has several disadvantages such as the requirement for many processing stages and the limitation of carbon sources and environment harmfulness via the accumulation of a large amount of solid and liquid waste. As a consequence, alternative microorganisms such as yeast and bacteria were used for CA production. After 1965, yeast-based production of CA was developed, and n-alkanes were fermented especially by *Candida* species (*C. oleophila*, *C. intermedia*, *C. fibriae*, *C. parapsilosis*, *C. catenulate*, *C. guilliermondii*, *C. tropicalis*, *C. parapsilosis*, *C. zeylanoides*) (Apelblat 2014; Cavallo et al. 2017; Kieliszek et al. 2017; Max et al. 2010; Papagianni 2007). Later, other yeasts such as *Debaromyces*, *Brettanomyces*, *Endomycopsis*, *Endomyces*, *Hansenula*, *Kloeckera*, *Pichia*, *Rhodotorula*, *Torulopsis*, *Trichosporon*, *Torula*, *Zygosaccharomyces*, *Saccharomyces* species, and *Yarrowia lipolytica* were used as potential CA producers. Bacteria such as *Aerobacter*, *Alcaligenes*, *Arthrobacter*, *Achromobacter*, *Brevibacterium*, *Bacillus*, *Corynebacterium*, *Micrococcus*, *Klebsiella*, *Pseudomonas*, and *Nocardia* spp. have been also studied for CA production (Cavallo et al. 2017; Dhillon et al. 2011; Max et al. 2010; Papagianni 2007).

The use of yeasts rather than moulds for CA production has several advantages such as a wide range of substrate usage and tolerance to high substrate concentrations, salt concentrations, metal ions, and extreme pH, as well as temperature conditions, which enable them to grow on a wide range of unrefined substrates and waste products. Yeasts also have higher production rates compared to moulds. They are easier to cultivate and more convenient for a continuous culture process than moulds. In addition, yeasts can also be genetically modified in order to enhance CA production. However, the production of a high quantity of isocitric acid (ICA) which is an undesired by-product is the main disadvantage of using yeasts over moulds for CA synthesis (Cavallo et al. 2017; Dhillon et al. 2011; Kieliszek et al. 2017; Rywinska et al. 2013; Show et al. 2015; Timoumi et al. 2018).

Recently, new culture techniques, microorganisms, and substrates for the production of CA have been examined. Nonconventional yeast *Y. lipolytica* has been considered to have high potential to produce CA from low-cost substrates. There were numerous research studies concerning the use of *Y. lipolytica* to produce CA from industrial waste or by-products such as rapeseed oil (Kamzolova et al. 2007; Kamzolova et al. 2011), olive mill wastewater (Dourou et al. 2016; Sarris et al. 2017), hydrolysate of pretreated straw cellulose (Liu et al. 2015b), waste cooking oil (Liu et al. 2018; Liu et al. 2015a), pineapple waste (Imandi et al. 2008), extract of Jerusalem artichoke tubers (Wang et al. 2013), crustacean waste (Magdouli et al. 2017), whey (Arslan et al. 2016; Taskin et al. 2015), crude glycerol from biodiesel production (Ferreira et al. 2016b; Mitrea et al. 2017; Rywinska et al. 2011), carrot juice, and celery by-products (Urak et al. 2015; Yalcin 2012).

4.2.1 Yarrowia lipolytica

Nonconventional ascomycetous yeast, *Y. lipolytica* (previously known as *Saccharomycopsis*, *Candida*, or *Endomycopsis lipolytica)* is assigned to the Dipodascacea family and the class *Hemiascomycetes* (Barth and Gaillardin 1997; Fickers et al. 2005; Kurtzman and Fell 1998). It is phylogenetically distant from *Saccharomyces cerevisiae*. Currently, several *Candida* species such as *C. galli*, *C. osloensis*, *C. hollandica*, *C. yakushimensisl*, *C. deformans*, and *C. alimentaria* have been linked to the *Yarrowia* clade. The genus name *Yarrowia* was first recommended by in an acknowledgement of David Yarrow from Delft Microbiology Laboratory where it was identified. The species name *lipolytica* refers to the lipid-hydrolysing capability of this yeast. Strains of *Y. lipolytica* are found naturally in hydrophobic substrate-containing environments and can be isolated from dairy (yoghurt, kefir, and cheese), shrimp salads, poultry, meat (sausages), soy sauce, soil, sewage, oil polluted environments, or hypersaline and marine environments (Zinjarde et al. 2014). *Y. lipolytica* has been considered to be a non-pathogenic microorganism and has been given generally regarded as safe (GRAS) status by the American Food and Drug Administration (FDA) (Fickers et al. 2005; Madzak 2015).

Strains of this yeast are obligate aerobes and generally grow in the temperature range of 24–33 °C. They have tolerance to salt, metal ions, and difficult environmental factors such as low temperatures and pH. The species *Y. lipolytica* is a dimorphic yeast that forms pseudohyphae, septate hyphae, and budding cells which depend on the strain used and environmental conditions such as temperature and pH. Moreover, nitrogen and carbon sources or the presence of specific compounds in culture medium can induce mycelium formation (Barth and Gaillardin 1997).

Y. lipolytica has the ability to degrade proteins and lipids due to the extracellular proteolytic and lipolytic activities. It accumulates extracellular protease (1–2 g/L) in protein-rich media as well as lipase in lipid-rich media. Other extracellular enzymes produced by this organism are phosphatases, RNases, and esterases which could support its growth under diverse conditions. Many biotechnological products can be produced by this yeast: single-cell oil (SCO), single-cell proteins (SCP), organic acids (mainly CA and to a lesser extent acetate, α-ketoglutarate, and pyruvate), lipases, polyols (to a lesser extent mannitol and mostly erythritol), surfactants, γ-decalactone, emulsifiers, and other related end products that make it an excellent model in the biotechnological field (Zinjarde et al. 2014).

Although fructose, glucose, and mannose can be used directly as substrate by *Y. lipolytica*, some other di- and tri-saccharides are degraded outside of the cell since they cannot pass easily through the membrane of the yeast. For instance, sucrose cannot be hydrolysed by *Y. lipolytica* due to the absence of the enzyme invertase in its metabolism (Coelho et al. 2010). Ethanol can be used as substrate in concentrations up to 3% but cannot be synthesized by this yeast. Higher concentrations of ethanol also have a toxic effect on yeast. In addition, organic acids such as acetic, citric, lactic, malic, propionic, and succinic acid can be used as the carbon source, whereas butyric, sorbic, and propionic acid can inhibit its growth. Moreover, sodium acetate at concentrations higher than 1% inhibits cell growth (Rodrigues and Pais 2000).

The amount of oxygen available in the growth medium is essential for this microorganism due to its aerobic nature. Oxygen is usually provided in the growth medium by using a compressed air supply in bioreactors or by agitation of shake-flask culture. Furthermore, the use of sufficient concentrations of iron and perfluorocarbons has significant effects on the improvement of oxygen transfer rate (Amaral et al. 2006).

The genome of *Y. lipolytica* has been sequenced, and the application of metabolic engineering approaches to this yeast has increased. For instance, genetically engineered strains of *Y. lipolytica* have been used to increase CA production (Forster et al. 2007; Fu et al. 2016; Holz et al. 2009, 2011; Tan et al. 2016). From these strains, a maximum CA concentration of 111.1 g/L and a yield of 0.93 g/g could be produced (Fu et al. 2016).

4.3 The Biochemistry of Citric Acid Biosynthesis

The formation of CA in the cell is a very complex process. CA is defined as an organic acid obtained at the end of many metabolic and morphological changes (Anastassiadis et al. 2002). The first studies on CA biochemistry focused on the mould *A. niger*, and the results showed that CA is an intermediate product in the Krebs cycle (TCA). The formation of CA depends on the regulation of the enzyme synthesis involved in each step of the TCA cycle and the cofactors involved in the activation of these enzymes. Metal ions provide cofactor formation and regulate enzyme activation (Angumeenal and Venkappayya 2013).

CA can also be generated by yeast as a result of an abnormality in the TCA cycle (Fig. 4.1). Its production is explained by Gaden Type 2 culture, in which CA is generated indirectly from energy metabolism and it is not a growth-associated product. For instance, at the beginning of culture microorganism growth is along with high substrate consumption and little or no CA production, while at a later stage, cell growth decreases, substrate utilization is still high, and CA is generated (Gaden 2000).

The key parameter for CA production by yeast is a deficiency in the culture medium of nitrogen content, since CA accumulation begins after consumption of all available nitrogen.

Therefore, the sine qua non prerequisite in order for the biosynthesis of CA to be carried out is that there must be sufficiently high initial molar ratios imposed into the culture medium (Papanikolaou and Aggelis 2009). In the above-mentioned culture environments, several metabolic and physiological changes are detected (Ratledge 1988). Therefore, in nitrogen-limited (carbon-excess) media, the nitrogen source is rapidly depleted, but the carbon source remains in surplus. In the presence of excess carbon found in the form of glucose, glycerol, or likewise metabolised

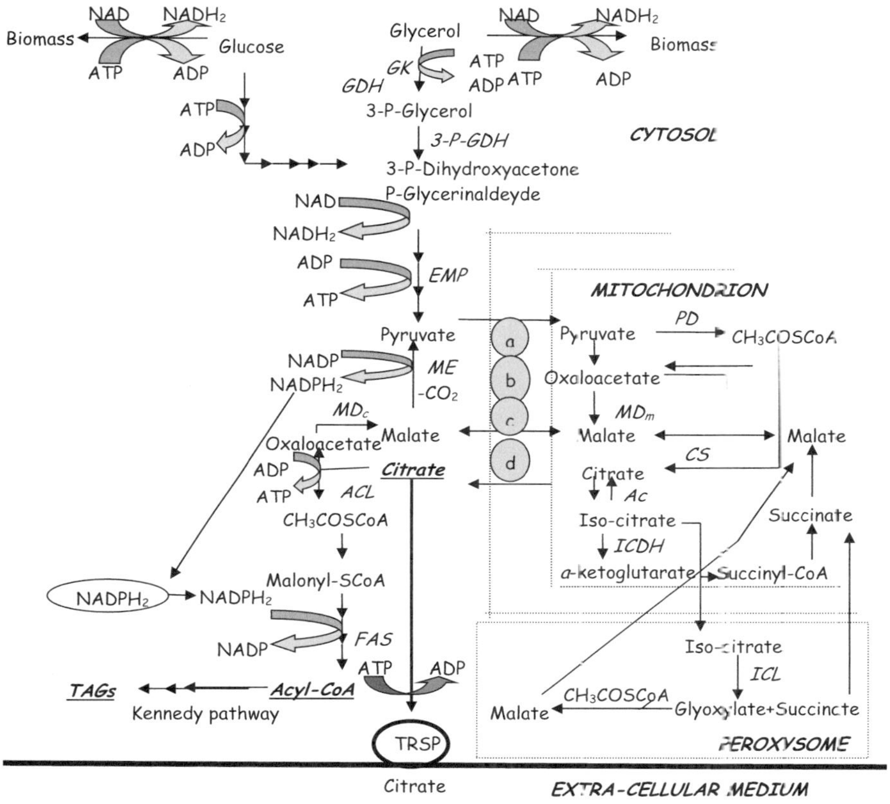

Fig. 4.1 Pathways involved in the intermediate metabolism of glycerol and glucose in order for microbial lipid and citric acid to be synthesized in *Yarrowia lipolytica* yeast. Mitochondrial transport systems, **a**, **b**, **c**, interlinked pyruvate-malate translocase systems; **d**, citrate-malate translocase. Enzymes: *ACL* ATP-citrate lyase, *FAS* fatty acid synthase enzymatic complex, *ICDH* iso-citrate dehydrogenase, *MDc* malate dehydrogenase (cytosolic), *MDm* malate dehydrogenase (mitochondrial), *ME* NADPH+-malic enzyme, *PD* pyruvate dehydrogenase, *CS* citrate synthase, *ICL* isocitrate lyase, *GK* glycerol kinase, *GDH* 3-P-glycerol dehydrogenase. Figure adapted from Ratledge (1987), Ratledge and Wynn (2002), Papanikolaou and Aggelis (2009), Papanikolaou and Aggelis (2011), and Carsanba et al. (2018)

compounds (i.e. polysaccharides, acetic acid, ethanol, etc.) and in conditions of nitrogen limitation, the catalytic growth rate decreases quickly, whereas the rate of carbon assimilation reduces progressively. This results in the favoured channelling of the carbon flux towards either the secretion of CA (or other low-molecular-weight compounds) into the growth medium, or storage lipid neo-synthesis, leading to so-called de novo lipid accumulation (Papanikolaou and Aggelis 2011; Ratledge and Wynn 2002). In these situations, *Y. lipolytica* produces high quantities of TCA cycle intermediates such as CA and ICA, which are not later used by the TCA cycle. Indeed, because of nitrogen depletion, a fast reduction of the concentration of intracellular AMP (adenosine monophosphate) takes place, since *Y. lipolytica* cleaves the existing AMP so as to obtain NH_4^+ ions, crucial for its maintenance. The level of AMP is regulated by the activity of enzyme AMP deaminase as follows:

$$AMP \rightarrow inosine5' - monophosphate + NH_3$$

The extreme reduction of intracellular AMP concentration stops the TCA cycle task; NAD + - (and in several circumstances also NADP+-) isocitrate dehydrogenase loses its activity, since it is allosterically activated by the concentrations of AMP and ADP, and accordingly, isocitric and therefore the CA that is found in equilibrium with isocitrate is accumulated inside the mitochondria (Papanikolaou et al. 2004; Ratledge 1987; Ratledge and Wynn 2002). When the intra-mitochondrial CA quantity reaches a critical level, citrate passes through the cytoplasm in exchange with malate. Finally, for the case of microorganisms that are capable of storing substantial quantities of lipid under nitrogen-limited conditions (the so-called oleaginous microorganisms) (Carsanba et al. 2018; Papanikolaou and Aggelis 2011), CA is cleaved by the ATP citrate lyase (ACL), the enzyme-key characteristic of the lipid production (Athenaki et al. 2018; Ratledge 1987; Ratledge and Wynn 2002). Citric acid cleavage yields oxaloacetate and acetyl-CoA. So, CA must be available in the cytoplasm of the cell where fatty acid (FA) synthesis takes place. Consequently, due to this irregular growth, CA (and, in fact, acetyl-CoA) becomes the substrate amenable for the intracellular accumulation of FAs. In fact, the subsequent acetyl-CoA is carboxylated by acetyl-CoA carboxylase (ACC1) to form malonyl-CoA, the substrate for the biosynthesis of acyl-CoA esters, which occurs by virtue of the action of the enzymatic complex of FA synthase. In the next stage, acyl-CoA esters are incorporated, mainly by virtue of the Kennedy pathway, into the triacylglycerols (TAGs) that are the most common form of lipophilic compounds found in the oleaginous microorganisms (Athenaki et al. 2018; Carsanba et al. 2018; Papanikolaou and Aggelis 2011; Ratledge and Wynn 2002).

In the case of non-lipid producing microorganisms (therefore with absent or with poor regulation of the enzyme ACL), the microbial response upon the imposed nitrogen limitation is the (significant in several cases) secretion of low-molecular-weight compounds (mostly CA and, to lesser extent, mannitol and/or erythritol) into the medium (Athenaki et al. 2018; Chatzifragkou et al. 2011; Papanikolaou and Aggelis 2011). As far as the production of CA is concerned, as previously indicated, mostly yeasts of the species *Y. lipolytica* and fungi of the species *A. niger* are implicated, whereas the poor regulation of ACL can biochemically explain the (noticeable in various cases) biosynthesis and accumulation into the culture medium of CA, since CA itself is the substrate of ACL action (Carsanba et al. 2018; Papanikolaou and Aggelis 2009; Ratledge and Wynn 2002). On the other hand, the synthesis of polyols as a microbial response to nitrogen limitation imposed into the medium has not yet been completely elucidated on a biochemical and molecular level (Rywinska et al. 2013). Moreover, despite poor or even negative regulation of ACL, non-oleaginous microorganisms under the same nitrogen-limited growth conditions do not produce lipid due to the three to four times lesser intra-mitochondrial CA concentration than that of the oleaginous strains (Athenaki et al. 2018). Finally, in the oleaginous strains, the rate of citrate efflux in the existence of L-malate is about 2.5 times bigger in comparison to the non-oleaginous strains (Ratledge and Wynn 2002).

This has led to the conclusion that for high CA production, the activity of CS is required to be substantially high, whereas the activities of other TCA enzymes should be kept low (Kamzolova et al. 2011).

Evidently, ACL activity should equally be substantially low. Furthermore, the aconitate hydratase (AH) enzyme which catalyses the production of ICA in the TCA cycle immediately after CA synthesis changes the CA/ICA ratio negatively (Kamzolova et al. 2015a). Therefore, for high CA production, ACL and AH activities are also required to be low. Based on the previously indicated analysis, many researchers have studied the overexpression or interruption of the genes encoding the relevant enzymes (Forster et al. 2007; Fu et al. 2016; Holz et al. 2009, 2011; Tan et al. 2016).

It should finally be indicated that the production of CA and/or microbial oil during the growth of *Y. lipolytica* on glycerol, glucose, or other sugars looks like strain-dependent, since in some circumstances, in flask or bioreactor processes, mainly microbial oil is accumulated (lipid in dry cell weight [DCW] >35%) (Fontanille et al. 2012; Papanikolaou and Aggelis 2002; Tsigie et al. 2012; Tsigie et al. 2011), while in other applications, CA (and possibly polyols) is produced with quite low levels of lipid (e.g. <22% in DCW) being accumulated (Imandi et al. 2007; Levinson et al. 2007; Makri et al. 2010; Papanikolaou et al. 2002, 2008a, 2009, 2013, 2017; Rymowicz et al. 2009; Rymowicz et al. 2006; Sarris et al. 2011, 2017). Moreover, in several cases in which low lipid accumulation somehow occurs in several *Y. lipolytica* strains cultivated on glucose or similarly catabolized compounds under conditions favouring lipid storage, growth in shake-flask, or fed-batch bioreactor experiments, this has caused consecutive production of extracellular CA and intracellular lipid. In the beginning of nitrogen limitation, some lipid accumulation was initiated (i.e. maximum lipid in DCW values ranging between 20–25% w/w), and afterwards lipid quantity decreased gradually, even though major amount of substrate remained unconsumed into the medium. The intracellular lipid turnover (degradation) period coincided with the production of CA in significant quantities into the culture medium (Makri et al. 2010; Papanikolaou et al. 2009, 2017; Sarris et al. 2011).

The pathways involved in the intermediate cellular catabolism and the synthesis of intracellular lipids or extracellular CA are illustrated in Fig. 4.1.

As mentioned in the previous sections, *Y. lipolytica* is a microorganism that is capable of performing noticeable breakdown of fatty materials, and in several instances, mostly under nitrogen-limited conditions, these substances can be transformed into CA (Kamzolova et al. 2005). When fat substances are used as microbial substrates, the pathways that could be incorporated in the fat turnover process, regardless of the use of eukaryotes or prokaryotes, are as follows:

(a) Lipase-catalysed hydrolysis of the TAGs used as substrates.

(b) Incorporation of the free extracellular aliphatic chains inside the cells or mycelia and fungal pellets. The rate of incorporation of the several exogenous FAs can vary and is associated with the FA composition of the fatty mixture used. In the case of the yeast *Y. lipolytica*, a discrimination (namely, a negative selectivity) against the FA C18:0 is observed, while the unsaturated FAs are incorporated with higher rates (Papanikolaou and Aggelis 2010; Papanikolaou et al. 2001; Tzirita et al. 2018).
(c) Potential reactions of bio-modification of the FAs formerly involved inside the microbial cells (i.e. desaturation and/or elongation).
(d) Reactions incorporating numerous FA chains inside the stored microbial TAGs.
(e) Partial or total assimilation of the acyl-CoA units, via the β-oxidation pathway.
(f) Catabolism of the accumulated acetyl-CoA through the Krebs cycle (formation of energy and ATP) (Athenaki et al. 2018; Fickers et al. 2005; Papanikolaou and Aggelis 2011; Tzirita et al. 2018). In the case of trials performed under nitrogen-limited conditions, the acetyl-CoA units generated through the β-oxidation process cannot be further catabolized through the TCA cycle (see previously indicated biochemical events) and will either be subjected to anabolic reactions through the glyoxylic acid by-pass (reactions of glyconeogenesis), or the formed CA will be secreted into the culture medium (Athenaki et al. 2018; Kamzolova et al. 2005, 2007; Papanikolaou and Aggelis 2011).

4.4 Factors Affecting Citric Acid Production by *Y. lipolytica*

4.4.1 Type of Strain Used

One of the main factors for CA production is type of strain used (Yalcin et al. 2010a). There are several wild types of *Y. lipolytica* strains in different collections such as W, NRRL, ATCC, NCIM, LGAM, VKM, UFLA, NBRC, NCYC, and ACA-YC. Various inbred strains were obtained by different groups from the French (W29), German (H222) and American (CBS 6124–2) strains (Barth and Gaillardin 1997; Nicaud 2012). As mentioned previously, wild-type *Y. lipolytica* strains can produce CA in nitrogen deficiency conditions; however a high level of CA is generally obtained by using mutant strains (Cavallo et al. 2017). Several researchers have achieved a high CA/ICA production value by using acetate or aconitase mutants or by overexpressing the IL encoding gene (Forster et al. 2007; Holz et al. 2009, 2011; Kamzolova et al. 2015b; Yalcin et al. 2010b). Moreover, the CA production capacities of mutant strains were higher than wild-type strains (Anastassiadis et al. 2008; Rywinska et al. 2010b). In general, wild-type strain CA levels reached up to 10 g/L, while mutant strains are capable of producing more than 100 g/L of CA (Cavallo et al. 2017).

4.4.2 Medium Composition

4.4.2.1 Carbon Source

Concentration and the type of carbon source used for CA production are important and predominantly affect yield and CA productivity rates. It is known that wild, mutant, and recombinant strains of *Y. lipolytica* can use a wide variety of carbon sources such as edible oils, n-alkanes, animal fats, alcohols (ethanol and methanol), molasses, starch hydrolysates, glucose, fructose, different agricultural waste residues (pineapple waste, cooking oil, straw cellulose, kiwi fruit peel, apple pomace, corn steep liquor, grape pomace, carob pod, whey, kumara, olive mill wastewater, okra, Jerusalem artichoke tuber extract, inulin), and glycerol, either pure or raw (Abghari and Chen 2017; Arslan et al. 2016; Crolla and Kennedy 2001, 2004b; Darvishi et al. 2009; Imandi et al. 2007; Imandi et al. 2008; Kamzolova et al. 2011; Levinson et al. 2007; Liu et al. 2010; Liu et al. 2015a, b; Papanikolaou et al. 2002, 2006, 2008b; Rzechonek et al. 2019; Taskin et al. 2015; Venter et al. 2004; Wang et al. 2013). Various carbon sources used for CA production by *Y. lipolytica* strains are listed in Table 4.1.

Although the most favourable substrates for CA production by *Aspergillus niger* are sucrose and molasses due to high CA yield and productivity, glucose and glycerol are mainly used for CA production by *Y. lipolytica*. However, invert sugar and molasses can also be used as carbon sources for CA production by *Y. lipolytica* (Wojtatowicz et al. 1991). While it is indicated that the assimilation of fructose by some yeasts is low, Yalcin et al. (2009) showed that the highest CA concentration of 65.1 g/L and the maximum specific CA production rate of 0.0179 g/g h^{-1} were obtained by using an initial fructose content of 200 g/L and 100 g/L, respectively. According to these results, it can be said that an excess concentration of substrate has to be used in order to achieve high CA production. Generally, 150–250 g/L initial substrate usages that promote CA production, and substrate concentrations lower than 50 g/L decrease the accumulation of CA (Antonucci et al. 2001; Kubicek and Karaffa 2001). In addition, it was stated that higher glucose concentration stimulates the glucose transport system (Kubicek and Karaffa 2001). Moreover, sucrose was used as another alternative carbon source, and a recombinant *Y. lipolytica* strain could produce 80 g/L CA from sucrose with a yield and productivity of 0.57 g/g and 1.1 g/L h^{-1}, respectively (Moeller et al. 2013).

Besides glucose, fructose, and sucrose, ethanol was also reported as a carbon source for CA production in submerged continuous cultures (Arzumanov et al. 2000; Finogenova et al. 2002). A CA production of 105 g/L with a yield 88.3% was obtained using an ethanol fed-batch process with a 50% feed every 3 days (Arzumanov et al. 2000). However, the authors noted that ethanol concentration should not exceed 1.2 g/L and culture conditions should be kept at pH 4.5 and 28 °C.

Edible oils such as rapeseed and sunflower oils have been used as substrate for CA production by *Y. lipolytica* strains. Kamzolova et al. (2011) obtained a CA

Table 4.1 Citric acid production from different carbon sources by *Y. lipolytica* strains

Yeast strain	Citric acid (g/L)	Citric acid yield (g/g)	Initial glucose concentration (g/L)	Carbon source	Nitrogen source	Culture mode	References
Y. lipolytica SWJ-1b	31.7	0,4	80	Cooking waste oil	Ammonium sulphate	Batch	Liu et al. (2015a)
Y. lipolytica N1	14.4–19.2	n	0.01–1.0	Ethanol	Ammonium sulphate	Continues	Finogenova et al. (2002)
Y. lipolytica 57	65.1	0.38	200	Fructose	Ammonium chloride, yeast extract	Batch	Yalcin et al. (2009)
Y. lipolytica 57	49.23	0.33	150	Fructose in whey based medium	Ammonium chloride, yeast extract	Batch	Yalcin et al. (2009)
Candida lipolytica Y-1095	13.6–78.5	0.50–0.79	50–150	Glucose	Ammonium chloride, yeast extract	Batch	Rane and Sims (1993)
Candida oleophila ATCC 20177	57.8	n	250	Glucose	Ammonium chloride	Batch	Anastassiadis and Rehm (2005)
Candida oleophila ATCC 20177	50–80	n	120–209	Glucose	Ammonium chloride	Batch and fed-batch	Anastassiadis et al. (2002)
Candida oleophila ATCC 20177	167	n	336	Glucose	Ammonium chloride	Continues	Anastassiadis and Rehm (2006)
Y. lipolytica A-101	34.3	0.84	92	Glucose	Ammonium chloride, yeast extract	Repeated batch	Rymowicz et al. (1993)
Y. lipolytica W29	49	0.85	60	Glucose	Ammonium sulphate, yeast extract	Batch	Papanikolaou et al. (2009)
Y. lipolytica A-101-1.14	>80	0.93	400 (mL/L)	Glucose hydrol	Ammonium chloride, yeast extract	Batch	Wojtatowicz et al. (1991)

Y. lipolytica ACA-DC 50109	28.9	0.82	65	Glucose in olive mill waste water	Ammonium sulphate, yeast extract	Batch	
Y. lipolytica NBRC 1658	38.88	0.38	100	Glucose in whey based medium	Ammonium chloride, yeast extract	Batch	(Yalcin et al. 2009)
Y. lipolytica 57	32.09	0.48	78.30–85.16	Grape must	Ammonium chloride, yeast extract	Batch	Yalcin et al. (2009)
Y. lipolytica SWJ-1b	42.4	0.43	n	Hydrolysate of pretreated straw cellulose	Ammonium sulphate	Fed-batch	Liu et al. (2015b)
Y. lipolytica Transformat 87	77.9	n	100	Inulin	Ammonium sulphate, yeast extract	Batch	Liu et al. (2010)
Y. lipolytica 30	68.3	0,91	84.3	Jerusalem artichoke tubers extract	Ammonium sulphate	Batch	Wang et al. (2013)
Candida lipolytica Y-1095	9.8	n	100–150	N-Paraffin	Urea	Batch	Crolla and Kennedy (2001)
Candida lipolytica Y-1095	42	0.8–1.0	150	N-Paraffin	Urea	Fed-batch	Crolla and Kennedy (2004a, b)
Y. lipolytica ACA-YC 5033	18.1	0.51	35.7	Olive mill waste water	Ammonium sulphate, yeast extract	Batch	Sarris et al. (2011)
Y. lipolytica NCIM 3589	202.35 (g/kg)	n	n	Pineapple waste	Yeast extract	Solid state	Imandi et al. (2008)
Y. lipolytica Wratislavia AWG7	157.5	0.6	300	Pure and raw glycerol	Ammonium chloride, yeast extract	Fed-batch	Rywinska et al. (2010a)

(continued)

Table 4.1 (continued)

Yeast strain	Citric acid (g/L)	Citric acid yield (g/g)	Initial glucose concentration (g/L)	Carbon source	Nitrogen source	Culture mode	References
Y. lipolytica 57	32.8	0.21	160	Pure glycerol	Ammonium chloride, yeast extract	Batch	
Y. lipolytica NRRL YB-423	21.6	0.54	40	Pure glycerol	Ammonium sulphate, yeast extract	Batch	Levinson et al. (2007)
Y. lipolytica 187/1	135	1.55	5	Rapeseed oil	Ammonium sulphate	Batch	Kamzolova et al. (2005)
Y. lipolytica NG40/UV7	175	1.5	20	Rapeseed oil	Ammonium sulphate, yeast extract	Batch	Kamzolova et al. (2011)
Y. lipolytica 1.31	124.5	0.62	200	Raw glycerol	n	Batch	Rymowicz et al. (2006)
Y. lipolytica LGAM S(7)1	33–35	0.42–0.44	80–120	Raw glycerol	Ammonium sulphate, yeast extract	Batch	Papanikolaou et al. (2002)
Y. lipolytica NCIM 3589	77.39	n	54.4	Raw glycerol	Yeast extract	Batch	Imandi et al. (2007)
Y. lipolytica H222-S4(p67ICL1) T5	80	0.57	150	Sucrose	Ammonium sulphate	Repeated fed-batch	Moeller et al. (2013)
Y. lipolytica UOFS Y-1701	18.7	n	30	Sunflower oil	Ammonium chloride, yeast extract	Batch	Venter et al. (2004)
Y. lipolytica H222	53–97	0.39–0.69	150	Glucose	Ammonium chloride	Fed-batch	
Y. lipolytica VKM Y-2373	80–85	0.70–0.75	Pulsed addition 30	Glucose	Ammonium sulphate	Fed-batch	Kamzolova and Morgunov (2017)

production of 175 g/L using oil as a carbon source. Moreover, in another study using sunflower oil (30 g/L) and acetate (10 g/L) as carbon sources, a CA production of 18.7 g /L was obtained (Venter et al. 2004).

Raw glycerol, a by-product of biodiesel industry, has been used as carbon source for CA production. It has been estimated that 1 kg of raw glycerol is generated when 10 kg of biodiesel is produced (Rymowicz et al. 2010). Due to its low cost, high conversion yield of CA, and it being the best carbon source, it has been used for CA production in many research studies (Imandi et al. 2007; Levinson et al. 2007; Morgunov et al. 2013; Papanikolaou et al. 2002; Rymowicz et al. 2006, 2010; Rywinska et al. 2009, 2010a, 2012, 2013; Rywinska and Rymowicz 2010; Rzechonek et al. 2019). Rymowicz et al. (2006) obtained a highest CA titre of 124.5 g/L from an initial raw glycerol concentration of 200 g/L from an acetate mutant of *Y. lipolytica* in batch bioreactor culture. Later, fed-batch culture was applied, and a maximum CA production of 157.5 g/L from 300 g/L initial glycerol concentration was achieved (Rywinska et al. 2010a). Moreover, some studies have reported that glycerol was a better carbon source for *Y. lipolytica* for CA production than glucose when both carbon substrates were present in the medium. This was explained by the excess energy created by glycerol (C3) reducing the activity of the C6 pathway during glycolysis, because glycerol could not stop C6 transporters since their carriers are different (Papanikolaou et al. 2002).

Recently, researchers have sought inexpensive carbon sources such as agricultural residues and waste for CA production. For instance, a Jerusalem artichoke tuber extract was used for CA production by Wang et al. (2013) and 68.3 g/L CA with a 0.91 g/g yield was obtained. In another study using olive mill wastewater enriched with glucose (containing 65 g/L initial sugar) medium, 28.9 g/L CA was produced (Papanikolaou et al. 2008b). In addition, Imandi et al. (2008) employed a solid-state culture for CA production made from pineapple waste and obtained CA of 202.35 g/kg ds (g CA produced/kg of dried pineapple waste as substrate). Lately, the hydrolysate of pretreated straw cellulose and waste cooking oil was also used by Liu et al. (2015b) and Liu et al. (2015a) as carbon sources for CA production, resulting in 42.4 g/L of CA produced from the hydrolysate of pretreated straw cellulose and 31.7 g/L of CA produced from waste cooking oil.

Although agricultural residues and waste are inexpensive carbon sources for CA production, it should be noted that these substrates could contain trace metals which can inhibit the growth of microorganisms and the production of CA. Therefore, deionization or some chemical pretreatment methods should be applied before the use of these substrates (Roukas and Kotzekidou 1997; Yalcin et al. 2010a).

Recently, molecular methods and recombinant technology have often been used for *Y. lipolytica* strains in order to improve yield and CA productivity. With the application of recombinant strains, most of the carbon sources such as sucrose, inulin, etc. can be used for CA production (Liu et al. 2010; Rakicka et al. 2016).

4.4.2.2 Nitrogen Source

The type of nitrogen source and its concentration in culture media are critical factors affecting CA production. A nitrogen source is mainly required for growth of the microorganism and self-maintenance. A high concentration of nitrogen in the medium can have a negative effect on CA productivity (Soccol et al. 2006). The key factor in CA accumulation is the exhaustion of nitrogen content in the culture medium and the limitation of cell growth due to excess carbon and nitrogen deficiency (Fickers et al. 2005). After nitrogen exhaustion during the stationary phase, yeast metabolic activity still continues with carbon source assimilation and CA production. Generally, ammonium salts (ammonium chloride, ammonium sulphate, ammonium nitrate), potassium nitrate, sodium nitrate, peptone, yeast extract, urea, malt extract, and corn steep liquor were used as a nitrogen source for CA production. Amongst these nitrogen sources, the most convenient were reported as yeast extract and ammonium chloride for CA production for the yeasts *Y. lipolytica*, *C. paratropicalis*, and *C. guilliermondii* (Rane and Sims 1996). In a study of CA production by *C. lipolytica*, 1.5 g/L of ammonium chloride as a nitrogen source was found to be the optimum concentration for the highest CA production (Hamissa et al. 1981). In another study on CA production from glycerol, a yeast extract of 0.2682 g/L was estimated as the optimum concentration of a nitrogen source (Imandi et al. 2007). Moreover, Yalcin et al. (2010a) stated that CA production was increased by the addition of 2 g/L of ammonium chloride. It was also reported that the required nitrogen concentration for CA production ranged between 0.1 and 0.4 g/L (Goncalves et al. 2014).

It should be noted that this requirement of limiting the nitrogen content in the medium can be a problem for CA production using high nitrogen content waste. For instance, in a study performed with whey as the carbon source, the addition of an extra nitrogen source to the medium led to decreased CA production (Yalcin et al. 2010a). Although many researchers have attempted to establish the optimum nitrogen concentration for CA production, some authors have recommended that the C/N ratio is more determinant and important for the CA/ICA ratio produced (Levinson et al. 2007; Ochoa-Estopier and Guillouet 2014; Papanikolaou et al. 2002).

4.4.2.3 Trace Elements

Phosphate as another crucial source for CA production should also be limited in quantity in the culture medium. Higher concentrations of phosphate can promote cell growth and lower CA production. Potassium dihydrogen phosphate with a concentration of 0.1% was reported as the most suitable phosphate source for CA production (Grewal and Kalra 1995). Higher concentrations of phosphate in the medium cause an increase in the production of certain sugar acids and the stimulation of growth (Grewal and Kalra 1995; Max et al. 2010; Vandenberghe et al. 1999).

It was reported that trace elements of zinc, iron, manganese, copper, and magnesium have an effect on CA production (Soccol et al. 2006). These elements are required in trace quantity for inducing cell growth to obtain high CA yields (Angumeenal and Venkappayya 2013). For instance, in a study with a manganese concentration of more than 2 mg/L, a decrease of 20% in CA production was observed. Regarding iron, it was reported that the limitation of iron inactivates the aconitase enzyme which catalyses CA degradation within the TCA cycle (Kubicek and Karaffa 2001). Anastassiadis and Rehm (2005) reported that biomass formation was increased with addition of iron to a glucose medium in *C. oleophila* fermentation, whereas Finogenova et al. (2002) indicated that iron and zinc limitation causes cell growth without CA production, while the addition of zinc increased CA production. In the same study, it was also stated that the optimum iron concentration in the medium for high CA production from ethanol should be in the range of 0.2–2.5 mg/g and not more than 7.0 mg/g which inhibits CA production. In another study, Yalcin et al. (2010a) also found that CA production was inhibited and cell growth increased with the addition of iron, copper, and zinc sulphate to the culture medium and the optimum zinc sulphate concentration was found to be 0.008 g/L for the highest CA production by *Y. lipolytica* 57.

4.4.3 Effect of Cultivation Conditions

CA production in submerged cultures is mainly affected by pH, temperature, aeration, and agitation parameters of bioreactors.

4.4.3.1 pH

The pH of the medium is one of the most important parameters for CA production by yeasts, and it should be well-characterized before CA production since each yeast strain shows different CA yield at the defined pH. Generally, if yeast is used for CA production, the initial pH of the medium should be higher than 5 since pH values below 5 show a negative effect on CA production. It was indicated that the production of some poly-alcohols such as erythritol, arabitol, and mannitol increased at pH levels below 5, thus lowering CA production (Mattey 1992). In addition, CA production and its transportation from cells were inhibited at low pH values. For instance, in a study on CA production by *C. oleophile* in continuous culture, the active CA transport system was found to be affected by the pH of medium (Anastassiadis and Rehm 2005). It was also indicated that growth, biomass composition, and CA synthesis depend on pH and the highest CA production was found at a pH of 5 (Anastassiadis and Rehm 2005). In another study performed with *Y. lipolytica* 57 and *Y. lipolytica* NBRC 1658 grown on a glucose medium, a pH range of 5.2–7 increased CA production (Yalcin et al. 2010a). On the other hand, Kamzolova et al. (2008) examined the CA/ICA ratio at different pH levels. They reported that the production of these two acids at pH 4.5 was almost equal and that ICA production increased at pH 6. Moreover, Tomaszewska et al. (2014) reported that high concentrations of CA were produced by using strain Wratislavia 1.31 (70.6 g/L) and Wratislavia AWG7 (85.7 g/L) at pH 5.0 and pH 5.5, respectively, while CA production was inhibited by lowering pH to around 3, and efficient erythritol synthesis took place. Also, in another study, a threefold increase in CA production by *Y. lipolytica* CBS 2073 and W29 was observed when the pH of the medium was dropped from 7 to 5 (Ferreira et al. 2016b). Recently, the optimum pH for CA production by *Y. lipolytica* DSM 3286 was reported as pH 5.5 (Egermeier et al. 2017). It is therefore important to examine the effect of pH on CA production before starting cultivation. Even at the same pH values, different strains can produce different titres of CA and CA/ICA ratios.

4.4.3.2 Temperature

Another important parameter to consider for CA production is temperature. Growth of cells, biomass, and product formation depend on the temperature of the culture medium. Various strains can behave in different ways at the same temperature, and the effect of temperature on yeast should be examined before scaling up the cultivation. The effect of temperature on CA production and biomass formation by *C. lipolytica* was studied, and a temperature range of 26–30 °C was found to be optimal (Crolla and Kennedy 2001). In another study performed using *C. lipolytica* Y 1095, the optimum temperature for CA production was determined to be 27 °C (Rane and Sims 1993). Moreover, the optimal temperature for CA and biomass production by *C. oleophila and Y. lipolytica* 57 was reported as 35 °C and 30 °C, respectively (Anastassiadis and Rehm 2006; Yalcin et al. 2010a). Also, Morgunov et al. (2013) reported that while the maximum CA production rate by *Y. lipolytica* NG40/UV7 was obtained at 28 °C, twofold reductions were observed when temperature decreased to 26 °C or increased to 32 °C.

4.4.3.3 Dissolved Oxygen

Since CA production is an aerobic process, it is obviously affected by aeration (Kubicek and Karaffa 2001). Biochemically, this type of culture requires oxygen to generate CA by *A. niger* as well as the other yeast strains (Kamzolova et al. 2003). It was reported that higher CA production was obtained by high oxygen transfer to yeast cells (Rywinska et al. 2012). Availability of oxygen was a crucial parameter in the growth of *Y. lipolytica*, substrate uptake, and CA synthesis (Workman et al. 2013). High dissolved oxygen (DO) concentration generally results in yeast form in *Y. lipolytica* rather than in the mycelia or pseudomycelia forms (Bellou et al. 2014). However, low DO levels can cause the non-activation of citrate synthase enzyme which directly affects CA production (Rywinska et al. 2012). Different types of aeration devices have been constructed to obtain high oxygen transfer rates (Soccol et al. 2006). The DO ratio in the culture medium generally depends on agitation, aeration, and at the same time the viscosity of the medium. The agitation process increases the area of the dispersed air bubbles in the culture medium that provide a high oxygen transfer rate. However, high agitation can lead to high shear stress on cell walls and the interface between cells and the insoluble substrate. Therefore, agitation speed should be examined before culture is scaled up to achieve the optimum oxygen transfer rate (Crolla and Kennedy 2004a, b). Some authors reported that increasing the agitation speed from 400 to 800 rpm and to 1000 rpm had a positive effect on CA production in small-scale culture (Crolla and Kennedy 2004b; Rywinska et al. 2012). It was also stated that increasing the aeration rate from 0.18 vvm (volume or air/ volume of medium/minute) to 0.6 vvm increased CA concentration (Rywinska et al. 2012). In addition, DO concentration affected the citrate/isocitrate concentration ratio, and it was reported that the ratio of citrate/isocitrate concentration produced by *C. tropicalis* increased with high DO concentration in the medium. In another study performed with *Y. lipolytica* 704, decreasing the DO concentration from 60–95% to 28–30% inhibited CA production (Kamzolova et al. 2003). DO values ranging from 50 to 80% of saturation have been reported as optimal for CA production (Morgunov et al. 2013). This saturation range can be achieved by agitation of 800–900 rpm and aeration of 0.24–036 vvm in a small-scale bioreactor (Rywinska et al. 2012). Moreover, when the initial oxygen volumetric mass transfer coefficient (KLa), which depends on air flow and agitation speed, was raised from 7 up to 55 h^{-1}, sevenfold and eightfold increases in CA production by *Y. lipolytica* W29 were reported (Ferreira et al. 2016a). Kamzolova et al. (2003) reported that iron concentration in the medium had an effect on the oxygen requirements for CA production and suggested that iron-enriched culture could be considered for production of industrial CA to reduce the high oxygen level requirement and its cost.

4.5 Production Systems for CA Production

CA is generally produced by using a batch system. However, fed-batch, repeated-batch, and continuous systems can be also used to increase the productivity and yield of CA. The advantage of using these systems rather than batch systems is the ability to increase substrate concentration.

Rymowicz et al. (2008) studied CA and erythritol production from glycerol in a fed-batch system and obtained a highest CA concentration of 110 g/L after 168 h of fed-batch culture with an initial glycerol concentration of 150 g/L and a total glycerol concentration of 250 g/L. Moreover, two fed-batch systems were also employed by Rywinska et al. (2010a) who added a total of 200 g/L glycerol in pulses in the first system and 300 g/L in the second with a constant feeding rate. Although the highest productivity in this study was achieved by using 200 g/L glycerol in pulses, a high CA concentration of 155–157 g/L was obtained by using a 300 g/L carbon source. In addition, Morgunov et al. (2013) reported a maximum CA concentration of 112 g/L and yield of 0.90 g/g under optimal conditions of a fed-batch culture by applying a pulsed addition of raw glycerol from 20 to 80 g/L.

In order to increase CA productivity, repeated-batch and continuous processes were also studied (Kamzolova et al. 2015b; Moeller et al. 2010, 2013; Ochoa-Estopier and Guillouet 2014; Rywinska et al. 2011; Rywinska and Rymowicz 2010). From these studies, a continuous process was reported as a convenient system for high productivity rather than batch culture (Rywinska et al. 2010a). On the other hand, immobilized cells in repeated-batch systems or continuous airlift bioreactors were also employed by several researchers (Arslan et al. 2016; Kautola et al. 1991; Rymowicz et al. 1993).

4.6 Downstream Processes

CA is generally recovered from culture medium by the classical acid precipitation method, in which it is precipitated with $Ca(OH)_2$ or calcium salt ($CaCO_3$) at high temperatures of 85–90 °C, resulting in calcium citrate. Then, precipitate is cleaned with water to eliminate impurities (sugar and other medium components), and it is subsequently treated with sulphuric acid to transform the calcium citrate to calcium sulphate and CA. Calcium sulphate can then be separated from the solution by filtration since it is not soluble at room temperature. Finally, CA can be decolourized with activated charcoal and later purified by using ion exchange chromatography or other purification techniques (solvent extraction or electrodialysis). Citric acid concentration can be adjusted by employing evaporation, crystallization, or drying methods (Anastassiadis et al. 2008; Cavallo et al. 2017; Dhillon et al. 2011). Wang et al. (2013) could recover 67.2% of CA in the supernatant of the culture and purified 96% of CA in crystal form by using the precipitation method. Rywinska et al. (2010b) reported that the precipitation of CA was difficult when ICA was present in the culture supernatant and it was mainly dependent on the proportion of calcium used for precipitation. For instance, the addition of 3.3 parts of $Ca(OH)_2$ per culture medium promoted precipitation of CA, while the addition of 5.3 parts of $Ca(OH)_2$ resulted in precipitation of ICA (Cavallo et al. 2017).

Aside from the classical precipitation method, solvent extraction method has also been applied to recover and purify CA from unwanted impurities and residues. Organic solvents such as butyl alcohol, acetone, tributyl phosphate, as well as certain amines were used for this purpose. Amongst these, amine extraction has been concluded to be the most convenient. However, these processes have not been found practical in relation to the inhibition of crystallization due to incomplete removal of impurities and unsatisfactory yield (Anastassiadis et al. 2008).

As another method for CA purification, the electrodialysis (ED) field has been used. Electrodeionization, laboratory ED, two-phase electro-electrodialysis, and bipolar membranes have been proposed by several researchers (Luo et al. 2004, 2017; Sun et al. 2017). On the other hand, other methods involving a combination of techniques including the use of ion exchange or weak basic resins or moving beds have also been applied (Cavallo et al. 2017). These techniques can be easily integrated with continuous culture. Moreover, an in situ product recovery technique has been also studied to improve the yield and productivity of CA during the culture process. In this technique, templates are supplemented to the culture medium as a specific surface over where the solute is firstly crystallized (Dhillon et al. 2011).

4.7 Conclusion

Due to numerous functional properties such as high solubility in water, chelating and buffering capacities, pleasant acid taste, biodegradability, biocompatibility, environmentally friendly properties, as well as GRAS, CA has been commonly used in the food, pharmaceutical, and chemical industries. Although industrial CA production is generally carried out by using the mould *A. niger*, recent studies have showed that using the yeast *Y. lipolytica* for CA production over moulds is very promising by reason of numerous advantages such as a wide range of substrate usage (unrefined substrates and waste products), good tolerance to high substrate concentrations and metal ions, as well as extreme conditions such as low temperatures and pH and easier cultivation and convenience for genetic modifications. An increasing number of recent patents on CA production by *Y. lipolytica* also demonstrates this interest. Therefore, alternative CA production at an industrial level by *Y. lipolytica* will play an important role in the near future.

The current review highlights recent studies on CA production by *Y. lipolytica*. The factors affecting CA production such as type of strain, medium composition (carbon, nitrogen, trace elements, and their concentrations), and culture conditions as well as recovery and purification methods have been discussed.

This review can contribute to further studies on CA production by *Y. lipolytica*, particularly when agro-industrial by-products or residues are selected as cheap carbon sources for reducing the cost of CA production. For this purpose, selection of strain, medium composition, and environmental conditions as well as downstream processes can be configured for high productivity according to given information.

References

Abghari A, Chen S (2017) Engineering *Yarrowia lipolytica* for enhanced production of lipid and citric acid. Fermentation 3(3):34

Amaral PFF, Lehocky M, Barros-Timmons AMV, Rocha-Leão MHM, Coelho MAZ, Coutinho JAP (2006) Cell surface characterization of *Yarrowia lipolytica* IMUFRJ 50682. Yeast 23(12):867–877

Anastassiadis S, Rehm HJ (2005) Continuous citric acid secretion by a high specific pH dependent active transport system in yeast *Candida oleophila* ATCC 20177. Electron J Biotechnol 8(2):146–161

Anastassiadis S, Rehm HJ (2006) Citric acid production from glucose by yeast *Candida oleophila* ATCC 20177 under batch, continuous and repeated batch cultivation. Electron J Biotechnol 9(1):26–39
Anastassiadis S, Aivasidis A, Wandrey C (2002) Citric acid production by Candida strains under intracellular nitrogen limitation. Appl Microbiol Biotechnol 60(1–2):81–87
Anastassiadis S, Morgunov IG, Kamzolova SV, Finogenova TV (2008) Citric acid production patent review. Recent Pat Biotechnol 2(2):107–123
Angumeenal AR, Venkappayya D (2013) An overview of citric acid production. Lwt-Food Sci Technol 50(2):367–370
Antonucci S, Bravi M, Bubbico R, Di Michele A, Verdone N (2001) Selectivity in citric acid production by *Yarrowia lipolytica*. Enzym Microb Technol 28(2–3):189–195
Apelblat A (2014) Citric acid. Springer, Cham
Arslan NP, Aydogan MN, Taskin M (2016) Citric acid production from partly deproteinized whey under non-sterile culture conditions using immobilized cells of lactose positive and cold-adapted *Yarrowia lipolytica* B9. J Biotechnol 231:32–39
Arzumanov TE, Shishkanova NV, Finogenova TV (2000) Biosynthesis of citric acid by *Yarrowia lipolytica* repeat-batch culture on ethanol. Appl Microbiol Biotechnol 53(5):525–529
Athenaki M, Gardeli C, Diamantopoulou P, Tchakouteu SS, Sarris D, Philippoussis A, Papanikolaou S (2018) Lipids from yeasts and fungi: physiology, production and analytical considerations. J Appl Microbiol 124(2):336–367
Barth G, Gaillardin C (1997) Physiology and genetics of the dimorphic fungus *Yarrowia lipolytica*. FEMS Microbiol Rev 19(4):219–237
Bellou S, Makri A, Triantaphyllidou IE, Papanikolaou S, Aggelis G (2014) Morphological and metabolic shifts of *Yarrowia lipolytica* induced by alteration of the dissolved oxygen concentration in the growth environment. Microbiology-Sgm 160:807–817
Carsanba E, Papanikolaou S, Erten H (2018) Production of oils and fats by oleaginous microorganisms with an emphasis given to the potential of the nonconventional yeast *Yarrowia lipolytica*. Crit Rev Biotechnol 38(8):1230–1243
Cavallo E, Charreau H, Cerrutti P, Foresti ML (2017) *Yarrowia lipolytica*: a model yeast for citric acid production. FEMS Yeast Res 17(8)
Chatzifragkou A, Makri A, Belka A, Bellou S, Mavrou M, Mastoridou M, Mystrioti P, Onjaro G, Aggelis G, Papanikolaou S (2011) Biotechnological conversions of biodiesel derived waste glycerol by yeast and fungal species. Energy 36(2):1097–1108
Ciriminna R, Meneguzzo F, Delisi R, Pagliaro M (2017) Citric acid: emerging applications of key biotechnology industrial product. Chem Cent J 11:22
Coelho MAZ, Amaral P, Belo I (2010) *Yarrowia lipolytica*: an industrial workhorse. Appl Microbiol Microb Biotechnol 2:930–944
Crolla A, Kennedy KJ (2001) Optimization of citric acid production from *Candida lipolytica* Y-1095 using n-paraffin. J Biotechnol 89(1):27–40
Crolla A, Kennedy K (2004a) In-line mixing for production of citric acid by *Candida lipolytica* grown on n-paraffins. J Chem Technol Biotechnol 79(7):720–728
Crolla A, Kennedy KJ (2004b) Fed-batch production of citric acid by *Candida lipolytica* grown on n-paraffins. J Biotechnol 110(1):73–84
Darvishi F, Nahvi I, Zarkesh-Esfahani H, Momenbeik F (2009) Effect of plant oils upon lipase and citric acid production in *Yarrowia lipolytica* yeast. J Biomed Biotechnol 7
Dhillon GS, Brar SK, Verma M, Tyagi RD (2011) Recent advances in citric acid bio-production and recovery. Food Bioprocess Technol 4(4):505–529
Dourou M, Kancelista A, Juszczyk P, Sarris D, Bellou S, Triantaphyllidou IE, Rywinska A, Papanikolaou S, Aggelis G (2016) Bioconversion of olive mill wastewater into high-added value products. J Clean Prod 139:957–969
Egermeier M, Russmayer H, Sauer M, Marx H (2017) Metabolic flexibility of *Yarrowia lipolytica* growing on glycerol. Front Microbiol 8:49–49

Ferreira P, Lopes M, Mota M, Belo I (2016a) Oxygen mass transfer impact on citric acid production by *Yarrowia lipolytica* from crude glycerol. Biochem Eng J 110:35–42
Ferreira P, Lopes M, Mota M, Belo I (2016b) Oxygen transfer rate and pH as major operating parameters of citric acid production from glycerol by *Yarrowia lipolytica* W29 and CBS 2073. Chem Pap 70(7):869–876
Fickers P, Benetti PH, Wache Y, Marty A, Mauersberger S, Smit MS, Nicaud JM (2005) Hydrophobic substrate utilisation by the yeast *Yarrowia lipolytica*, and its potential applications. FEMS Yeast Res 5(6–7):527–543
Finogenova TV, Kamzolova SV, Dedyukhina EG, Shishkanova NV, Il'chenko AP, Morgunov IG, Chernyavskaya OG, Sokolov AP (2002) Biosynthesis of citric and isocitric acids from ethanol by mutant *Yarrowia lipolytica* N 1 under continuous cultivation. Appl Microbiol Biotechnol 59(4–5):493–500
Fontanille P, Kumar V, Christophe G, Nouaille R, Larroche C (2012) Bioconversion of volatile fatty acids into lipids by the oleaginous yeast *Yarrowia lipolytica*. Bioresour Technol 114:443–449
Forster A, Jacobs K, Juretzek T, Mauersberger S, Barth G (2007) Overexpression of the ICL1 gene changes the product ratio of citric acid production by *Yarrowia lipolytica*. Appl Microbiol Biotechnol 77(4):861–869
Fu G-Y, Lu Y, Chi Z, Liu G-L, Zhao S-F, Jiang H, Chi Z-M (2016) Cloning and characterization of a pyruvate carboxylase gene from *Penicillium rubens* and overexpression of the genein the yeast *Yarrowia lipolytica* for *Enhanced Citric Acid Production*. Mar Biotechnol 18(1):1–14
Gaden EL (2000) Fermentation process kinetics (reprinted from J of biochemical and microbiological technology and engineering, vol 1, pg 413, 1959). Biotechnol Bioeng 67(6):629–635
Goncalves FAG, Colen G, Takahashi JA (2014) *Yarrowia lipolytica* and its multiple applications in the biotechnological industry. Sci World J 14
Grewal HS, Kalra KL (1995) Fungal production of citric-acid. Biotechnol Adv 13(2):209–234
Hamissa FA, Abouzeid AZA, Redwan AA (1981) Fermentative production of citric-acid by yeasts. Agricult Wastes 3(1):21–33
Holz M, Forster A, Mauersberger S, Barth G (2009) Aconitase overexpression changes the product ratio of citric acid production by *Yarrowia lipolytica*. Appl Microbiol Biotechnol 81(6):1087–1096
Holz M, Otto C, Kretzschmar A, Yovkova V, Aurich A, Pötter M, Marx A, Barth G (2011) Overexpression of alpha-ketoglutarate dehydrogenase in *Yarrowia lipolytica* and its effect on production of organic acids. Appl Microbiol Biotechnol 89(5):1519–1526
Hou WL, Bao J (2018) Simultaneous saccharification and aerobic fermentation of high titer cellulosic citric acid by filamentous fungus *Aspergillus niger*. Bioresour Technol 253:72–78
Imandi SB, Bandaru VR, Somalanka SR, Garapati HR (2007) Optimization of medium constituents for the production of citric acid from byproduct glycerol using Doehlert experimental design. Enzym Microb Technol 40(5):1367–1372
Imandi SB, Bandaru VVR, Somalanka SR, Bandaru SR, Garapati HR (2008) Application of statistical experimental designs for the optimization of medium constituents for the production of citric acid from pineapple waste. Bioresour Technol 99(10):4445–4450
Kamzolova SV, Morgunov IG (2017) Metabolic peculiarities of the citric acid overproduction from glucose in yeasts *Yarrowia lipolytica*. Bioresour Technol 243:433–440
Kamzolova SV, Shishkanova NV, Morgunov IG, Finogenova TV (2003) Oxygen requirements for growth and citric acid production of *Yarrowia lipolytica*. FEMS Yeast Res 3(2):217–222
Kamzolova SV, Morgunov IG, Aurich A, Perevoznikova OA, Shishkanova NV, Stottmeister U, Finogenova TV (2005) Lipase secretion and citric acid production in *Yarrowia lipolytica* yeast grown on animal and vegetable fat. Food Technol Biotechnol 43(2):113–122
Kamzolova SV, Finogenova TV, Lunina YN, Perevoznikova OA, Minachova LN, Morgunov IG (2007) Characteristics of the growth on rapeseed oil and synthesis of citric and isocitric acids by *Yarrowia lipolytica* yeasts. Microbiology 76(1):20–24
Kamzolova SV, Finogenova TV, Morgunov IG (2008) Microbiological production of citric and isocitric acids from sunflower oil. Food Technol Biotechnol 46(1):51–59

Kamzolova SV, Lunina JN, Morgunov IG (2011) Biochemistry of citric acid production from rapeseed oil by *Yarrowia lipolytica* yeast. J Am Oil Chemists Soc 88(12):1965–1976

Kamzolova SV, Lunina YN, Allayarov RK, Puntus IF, Laptev IA, Samoilenko VA, Morgunov IG (2015a) Biosynthesis of isocitric acid by the yeast *Yarrowia lipolytica* and its regulation. Appl Biochem Microbiol 51(2):249–254

Kamzolova SV, Vinokurova NG, Lunina JN, Zelenkova NF, Morgunov IG (2015b) Production of technical-grade sodium citrate from glycerol-containing biodiesel waste by *Yarrowia lipolytica*. Bioresour Technol 193:250–255

Kautola H, Rymowicz W, Linko YY, Linko P (1991) Production of citric acid with immobilized *Yarrowia lipolytica*. Appl Microbiol Biotechnol 35(4):447–449

Kieliszek M, Kot AM, Bzducha-Wrobel A, Blazejak S, Gientka I, Kurcz A (2017) Biotechnological use of Candida yeasts in the food industry: a review. Fungal Biol Rev 31(4):185–198

Kubicek, C., Karaffa, L. 2001. Organic acids. in: *Basic Biotechnology*, Cambridge University Press Cambridge, pp. 305–315

Kurtzman C, Fell JW (1998) The yeasts – a taxonomic study. Elsevier Science

Levinson WE, Kurtzman CP, Kuo TM (2007) Characterization of *Yarrowia lipolytica* and related species for citric acid production from glycerol. Enzym Microb Technol 41(3):292–295

Liu XY, Chi Z, Liu GL, Wang F, Madzak C, Chi ZM (2010) Inulin hydrolysis and citric acid production from inulin using the surface-engineered *Yarrowia lipolytica* displaying inulinase. Metab Eng 12(5):469–476

Liu XY, Lv JS, Xu JX, Zhang T, Deng YF, He JL (2015a) Citric acid production in *Yarrowia lipolytica* SWJ-1b yeast when grown on waste cooking oil. Appl Biochem Biotechnol 175(5):2347–2356

Liu XY, Lv JS, Zhang T, Deng YF (2015b) Citric acid production from hydrolysate of pretreated straw cellulose by *Yarrowia lipolytica* SWJ-1b using batch and fed-batch cultivation. Prep Biochem Biotechnol 45(8):825–835

Liu XY, Lv JS, Xu JX, Xia J, He AY, Zhang T, Li XQ, Xu JM (2018) Effects of osmotic pressure and pH on citric acid and erythritol production from waste cooking oil by *Yarrowia lipolytica*. Eng Life Sci 18(6):344–352

Luo GS, Shan XY, Qi X, Lu YC (2004) Two-phase electro-electrodialysis for recovery and concentration of citric acid. Sep Purif Technol 38(3):265–271

Luo HP, Cheng X, Liu GL, Zhou YJ, Lu YB, Zhang RD, Li X, Teng WK (2017) Citric acid production using a biological electrodialysis with bipolar membrane. J Membr Sci 523:122–128

Madzak C (2015) *Yarrowia lipolytica*: recent achievements in heterologous protein expression and pathway engineering. Appl Microbiol Biotechnol 99(11):4559–4577

Magdouli S, Guedri T, Tarek R, Brar SK, Blais JF (2017) Valorization of raw glycerol and crustacean waste into value added products by *Yarrowia lipolytica*. Bioresour Technol 243:57–68

Makri A, Fakas S, Aggelis G (2010) Metabolic activities of biotechnological interest in *Yarrowia lipolytica* grown on glycerol in repeated batch cultures. Bioresour Technol 101(7):2351–2358

Mattey M (1992) The production of organic acids. Crit Rev Biotechnol 12(1–2):87–132

Max B, Salgado JM, Rodriguez N, Cortes S, Converti A, Dominguez JM (2010) Biotechnological production of citric acid. Braz J Microbiol 41(4):862–875

Mitrea L, Trif M, Catoi AF, Vodnar DC (2017) Utilization of biodiesel derived glycerol for 1,3-PD and citric acid production. Microb Cell Factories 16:190

Moeller L, Grunberg M, Zehnsdorf A, Strehlitz B, Bley T (2010) Biosensor online control of citric acid production from glucose by *Yarrowia lipolytica* using semicontinuous fermentation. Eng Life Sci 10(4):311–320

Moeller L, Zehnsdorf A, Aurich A, Barth G, Bley T, Strehlitz B (2013) Citric acid production from sucrose by recombinant *Yarrowia lipolytica* using semicontinuous fermentation. Eng Life Sci 13(2):163–171

Morgunov IG, Kamzolova SV, Lunina JN (2013) The citric acid production from raw glycerol by *Yarrowia lipolytica* yeast and its regulation. Appl Microbiol Biotechnol 97(16):7387–7397

Morgunov IG, Kamzolova SV, Dedyukhina EG, Chistyakova TI, Lunina JN, Mironov AA, Stepanova NN, Shemshura ON, Vainshtein MB (2017) Application of organic acids for plant protection against phytopathogens. Appl Microbiol Biotechnol 101(3):921–932

Nicaud JM (2012) Yarrowia lipolytica. Yeast 29(10):409–418

Ochoa-Estopier A, Guillouet SE (2014) D-stat culture for studying the metabolic shifts from oxidative metabolism to lipid accumulation and citric acid production in *Yarrowia lipolytica*. J Biotechnol 170:35–41

Papagianni M (2007) Advances in citric acid fermentation by *Aspergillus niger*: biochemical aspects, membrane transport and modeling. Biotechnol Adv 25(3):244–263

Papanikolaou S, Aggelis G (2002) Lipid production by *Yarrowia lipolytica* growing on industrial glycerol in a single-stage continuous culture. Bioresour Technol 82(1):43–49

Papanikolaou S, Aggelis G (2009) Biotechnological valorization of biodiesel derived glycerol waste through production of single cell oil and citric acid by *Yarrowia lipolytica*. Lipid Technol 21(4):83–87

Papanikolaou S, Aggelis G (2010) *Yarrowia lipolytica*: a model microorganism used for the production of tailor-made lipids. Eur J Lipid Sci Technol 112(6):639–654

Papanikolaou S, Aggelis G (2011) Lipids of oleaginous yeasts. Part I: biochemistry of single cell oil production. Eur J Lipid Sci Technol 113(8):1031–1051

Papanikolaou S, Chevalot I, Komaitis M, Aggelis G, Marc I (2001) Kinetic profile of the cellular lipid composition in an oleaginous *Yarrowia lipolytica* capable of producing a cocoa-butter substitute from industrial fats. Antonie Van Leeuwenhoek 80(3–4):215–224

Papanikolaou S, Muniglia L, Chevalot I, Aggelis G, Marc I (2002) *Yarrowia lipolytica* as a potential producer of citric acid from raw glycerol. J Appl Microbiol 92(4):737–744

Papanikolaou S, Sarantou S, Komaitis M, Aggelis G (2004) Repression of reserve lipid turnover in *Cunninghamella echinulata* and *Mortierella isabellina* cultivated in multiple-limited media. J Appl Microbiol 97(4):867–875

Papanikolaou S, Galiotou-Panayotou M, Chevalot I, Komaitis M, Marc I, Aggelis G (2006) Influence of glucose and saturated free-fatty acid mixtures on citric acid and lipid production by *Yarrowia lipolytica*. Curr Microbiol 52(2):134–142

Papanikolaou S, Fakas S, Fick M, Chevalot I, Galiotou-Panayotou M, Komaitis M, Marc I, Aggelis G (2008a) Biotechnological valorisation of raw glycerol discharged after bio-diesel (fatty acid methyl esters) manufacturing process: production of 1,3-propanediol, citric acid and single cell oil. Biomass Bioenergy 32(1):60–71

Papanikolaou S, Galiotou-Panayotou M, Fakas S, Komaitis M, Aggelis G (2008b) Citric acid production by *Yarrowia lipolytica* cultivated on olive-mill waste water-based media. Bioresour Technol 99(7):2419–2428

Papanikolaou S, Chatzifragkou A, Fakas S, Galiotou-Panayotou M, Komaitis M, Nicaud J-M, Aggelis G (2009) Biosynthesis of lipids and organic acids by *Yarrowia lipolytica* strains cultivated on glucose. Eur J Lipid Sci Technol 111(12):1221–1232

Papanikolaou S, Beopoulos A, Koletti A, Thevenieau F, Koutinas AA, Nicaud JM, Aggelis G (2013) Importance of the methyl-citrate cycle on glycerol metabolism in the yeast *Yarrowia lipolytica*. J Biotechnol 168(4):303–314

Papanikolaou S, Kampisopoulou E, Blanchard F, Rondags E, Gardeli C, Koutinas AA, Chevalot I, Aggelis G (2017) Production of secondary metabolites through glycerol fermentation under carbon-excess conditions by the yeasts *Yarrowia lipolytica* and *Rhodosporidium toruloides*. Eur J Lipid Sci Technol 119(9):1600507

Rakicka M, Lazar Z, Rywinska A, Rymowicz W (2016) Efficient utilization of inulin and glycerol as fermentation substrates in erythritol and citric acid production using *Yarrowia lipolytica* expressing inulinase. Chem Pap 70(11):1452–1459

Rane KD, Sims KA (1993) Production of citric acid by *Candida lipolytica* Y1095 - effect of glucose-concentration on yield and productivity. Enzym Microb Technol 15(8):646–651

Rane KD, Sims KA (1996) Citric acid production by *Yarrowia lipolytica*: effect of nitrogen and biomass concentration on yield and productivity. Biotechnol Lett 18(10):1139–1144

Ratledge C (1987) Lipid biotechnology: a wonderland for the microbial physiologist. J Am Oil Chem Soc 64(12):1647–1656

Ratledge, C. 1988. Biochemistry, stoichiometry, substrates and economics in: *Single Cell Oil*, (Ed.) R.S. Moreton, Longman Scientific & Technical, Longman House Burnt Mill, Harlow and Essex, , pp. 33–70

Ratledge C, Wynn JP (2002) The biochemistry and molecular biology of lipid accumulation in oleaginous microorganisms. Adv Appl Microbiol 51:1–51

Rodrigues G, Pais C (2000) The influence of acetic and other weak carboxylic acids on growth and cellular death of the yeast *Yarrowia lipolytica*. Food Technol Biotechnol 38(1):27–32

Roukas T, Kotzekidou P (1997) Pretreatment of date syrup to increase citric acid production. Enzym Microb Technol 21(4):273–276

Rymowicz W, Kautola H, Wojtatowicz M, Linko Y-Y, Linko P (1993) Studies on citric acid production with immobilized *Yarrowia lipolytica* in repeated batch and continuous air lift bioreactors. Appl Microbiol Biotechnol 39(1):1–4

Rymowicz W, Rywinska A, Zarowska B, Juszczyk P (2006) Citric acid production from raw glycerol by acetate mutants of *Yarrowia lipolytica*. Chemical Papers-Chemicke Zvesti 60(5):391–394

Rymowicz W, Rywinska A, Gladkowski W (2008) Simultaneous production of citric acid and erythritol from crude glycerol by *Yarrowia lipolytica* Wratislavia K1. Chem Pap 62(3):239–246

Rymowicz W, Rywinska A, Marcinkiewicz M (2009) High yield production of erythritol from raw glycerol in fed-batch cultures of *Yarrowia lipolytica*. Biotechnol Lett 31(3):377–380

Rymowicz W, Fatykhova AR, Kamzolova SV, Rywinska A, Morgunov IG (2010) Citric acid production from glycerol-containing waste of biodiesel industry by *Yarrowia lipolytica* in batch, repeated batch, and cell recycle regimes. Appl Microbiol Biotechnol 87(3):971–979

Rywinska A, Rymowicz W (2010) High yield production of citric acid by *Yarrowia lipolytica* on glycerol in repeated batch bioreactors. J Ind Microbiol Biotechnol 37(5):431–435

Rywinska A, Rymowicz W, Zarowska B, Wojtatowicz M (2009) Biosynthesis of citric acid from glycerol by acetate mutants of *Yarrowia lipolytica* in fed batch fermentation. Food Technol Biotechnol 47(1):1–6

Rywinska, A., Rymowicz, W., Marcinkiewicz, M. 2010a. Valorization of raw glycerol for citric acid production by *Yarrowia lipolytica* yeast. Electron J Biotechnol, 13(4):9–10

Rywinska A, Rymowicz W, Zarowska B, Skrzypinski A (2010b) Comparison of citric acid production from glycerol and glucose by different strains of *Yarrowia lipolytica*. World J Microbiol Biotechnol 26(7):1217–1224

Rywinska A, Juszczyk P, Wojtatowicz M, Rymowicz W (2011) Chemostat study of citric acid production from glycerol by *Yarrowia lipolytica*. J Biotechnol 152(1–2):54–57

Rywinska A, Musial I, Rymowicz W, Zarowska B, Boruczkowski T (2012) Effect of agitation and aeration on the citric acid production by *Yarrowia lipolytica* grown on glycerol. Prep Biochem Biotechnol 42(3):279–291

Rywinska A, Juszczyk P, Wojtatowicz M, Robak M, Lazar Z, Tomaszewska L, Rymowicz W (2013) Glycerol as a promising substrate for *Yarrowia lipolytica* biotechnological applications. Biomass Bioenergy 48:148–166

Rzechonek DA, Dobrowolski A, Rymowicz W, Mironczuk AM (2019) Aseptic production of citric and isocitric acid from crude glycerol by genetically modified *Yarrowia lipolytica*. Bioresour Technol 271:340–344

Sarris D, Galiotou-Panayotou M, Koutinas AA, Komaitis M, Papanikolaou S (2011) Citric acid, biomass and cellular lipid production by *Yarrowia lipolytica* strains cultivated on olive mill waste water based media. J Chem Technol Biotechnol 86(11):1439–1448

Sarris D, Stoforos NG, Mallouchos A, Kookos IK, Koutinas AA, Aggelis G, Papanikolaou S (2017) Production of added value metabolites by *Yarrowia lipolytica* growing in olive mill wastewater-based media under aseptic and non-aseptic conditions. Eng Life Sci 17(6):695–709

Show PL, Oladele KO, Siew QY, Zakry FAA, Lan JCW, Ling TC (2015) Overview of citric acid production from *Aspergillus niger*. Front Life Sci 8(3):271–283

Soccol CR, Vandenberghe LPS, Rodrigues C, Pandey A (2006) New perspectives for citric acid production and application. Food Technol Biotechnol 44(2):141–149

Sun XH, Lu HX, Wang JY (2017) Recovery of citric acid from fermented liquid by bipolar membrane electrodialysis. J Clean Prod 143:250–256

Tan MJ, Chen X, Wang YK, Liu GL, Chi ZM (2016) Enhanced citric acid production by a yeast *Yarrowia lipolytica* over expressing a pyruvate carboxylase gene. Bioprocess Biosyst Eng 39(8):1289–1296

Taskin M, Saghafian A, Aydogan MN, Arslan NP (2015) Microbial lipid production by cold adapted oleaginous yeast *Yarrowia lipolytica* B9 in non-sterile whey medium. Biofuels Bioprodu Biorefining-Biofpr 9(5):595–605

Timoumi A, Guillouet SE, Molina-Jouve C, Fillaudeau L, Gorret N (2018) Impacts of environmental conditions on product formation and morphology of *Yarrowia lipolytica*. Appl Microbiol Biotechnol 102(9):3831–3848

Tomaszewska L, Rakicka M, Rymowicz W, Rywinska A (2014) A comparative study on glycerol metabolism to erythritol and citric acid in *Yarrowia lipolytica* yeast cells. FEMS Yeast Res 14(6):966–976

Tsigie YA, Wang CY, Truong CT, Ju YH (2011) Lipid production from *Yarrowia lipolytica* Po1g grown in sugarcane bagasse hydrolysate. Bioresour Technol 102(19):9216–9222

Tsigie YA, Wang C-Y, Kasim NS, Diem Q-D, Huynh L-H, Ho Q-P, Truong C-T, Ju Y-H (2012) Oil production from *Yarrowia lipolytica* Po1g using rice bran hydrolysate. J Biomed Biotechnol 2012:10

Tzirita M, Papanikolaou S, Chatzifragkou A, Quilty B (2018) Waste fat biodegradation and biomodification by *Yarrowia lipolytica* and a bacterial consortium composed of *Bacillus spp.* and *Pseudomonas putida*. Eng Life Sci 18(12):932–942

Urak S, Yeniay O, Karasu-Yalcin S (2015) Optimization of citric acid production from a carrot juice based medium by *Yarrowia lipolytica* using response surface methodology. Ann Microbiol 65(2):639–649

Vandenberghe LPS, Soccol CR, Pandey A, Lebeault JM (1999) Microbial production of citric acid. Braz Arch Biol Technol 42(3):263–276

Venter T, Kock JLF, Botes PJ, Smit MS, Hugo A, Joseph M (2004) Acetate enhances citric acid production by *Yarrowia lipolytica* when grown on sunflower oil. Syst Appl Microbiol 27(2):135–138

Wang LF, Wang ZP, Liu XY, Chi ZM (2013) Citric acid production from extract of Jerusalem artichoke tubers by the genetically engineered yeast *Yarrowia lipolytica* strain 30 and purification of citric acid. Bioprocess Biosyst Eng 36(11):1759–1766

Wojtatowicz M, Rymowicz W, Kautola H (1991) Comparison of different strains of the yeast *Yarrowia lipolytica* for citric acid production from glucose hydrol. Appl Biochem Biotechnol 31(2):165–174

Workman M, Holt P, Thykaer J (2013) Comparing cellular performance of *Yarrowia lipolytica* during growth on glucose and glycerol in submerged cultivations. AMB Express 3:58

Yalcin SK (2012) Enhancing citric acid production of *Yarrowia lipolytica* by mutagenesis and using natural media containing carrot juice and celery byproducts. Food Sci Biotechnol 21(3):867–874

Yalcin SK, Bozdemir MT, Ozbas ZY (2009) A comparative study on citric acid production kinetics of two *Yarrowia lipolytica* strains in two different media. Indian J Biotechnol 8(4):408–417

Yalcin SK, Bozdemir M, Ozbas Z (2010a) Citric acid production by yeasts: fermentation conditions, process optimization and strain improvement. Curr Res Technol Educ Top Appl Microbiol Microb Bitechnol 27:1374–1382

Yalcin SK, Bozdemir MT, Ozbas ZY (2010b) Effects of different fermentation conditions on growth and citric acid production kinetics of two *Yarrowia lipolytica* strains. Chem Biochem Eng Q 24(3):347–360

Zinjarde S, Apte M, Mohite P, Kumar AR (2014) *Yarrowia lipolytica* and pollutants: interactions and applications. Biotechnol Adv 32(5):920–933

Chapter 5
Effective Technologies for Isolating Yeast Oxido-Reductases of Analytical Importance

Galina Z. Gayda, Olha M. Demkiv, Halyna M. Klepach, Mykhailo V. Gonchar, and Marina Nisnevitch

Abstract Microbial enzymes have gained interest for their widespread use in industries and medicine due to their stability, catalytic activity, and low-cost production, compared to plant and animal analogues. Microbial enzymes are capable of degrading toxic chemical compounds of industrial and domestic wastes by degradation or via conversion to nontoxic products. Enzyme technology broadly involves production, isolation, purification, and use of enzymes in various industries (e.g., food, medicine, agriculture, chemicals, pharmacology). The development of simple technologies for obtaining highly purified novel enzymes is an actual task for biotechnology and enzymology. This chapter presents a review of the main achievements in the elaboration of modern techniques for obtaining recombinant and novel enzymes. The results of a series of the authors' investigations into the development of novel enzymatic approaches, including biosensors, for determination of practically important analytes are summarized. The described methods are related to isolation of highly purified yeast oxido-reductases: alcohol oxidase, flavocytochrome b_2, glycerol dehydrogenase, and formaldehyde dehydrogenase. The enzymes were isolated from selected or recombinant yeast cells using the simple and effective technologies developed by the authors.

Keywords Yeasts · Oxido-reductases · Isolation and purification · Analytical application

---

G. Z. Gayda (✉) · O. M. Demkiv · M. V. Gonchar
Institute of Cell Biology, NAS of Ukraine, Lviv, Ukraine

H. M. Klepach
Drohobych Ivan Franko State Pedagogical University, Drohobych, Ukraine

M. Nisnevitch
Department of Chemical Engineering, Biotechnology and Materials, Ariel University, Ariel, Israel
e-mail: marinan@ariel.ac.il

A. Sibirny (ed.), *Non-conventional Yeasts: from Basic Research to Application*,
https://doi.org/10.1007/978-3-030-21110-3_5

5.1 Introduction

Enzymes are used in many different spheres of human activity, and this use is increasing rapidly due to reduced processing time, low energy input, cost effectiveness, nontoxic, and eco-friendly characteristics (Singh et al. 2016). Enzymatic reactions are the basis of many production processes, and application of microbial enzymes has been widely used since the early twentieth century. Enzymes are necessary in genetic engineering and biotechnology, and in particular for developing the ethanol fuel technology, in various industries (e.g., food, agriculture, chemicals, pharmaceuticals), and in medicine. Enzymes also play an important role in analytical applications (Reyes-De-Corcuera and Powers 2017; Chapman et al. 2018). Analytical systems which contain microbial enzymes possess high selectivity and sensitivity and are widely used in analytical laboratories of the food and microbiological industries, as well as for clinical diagnostics.

The driving force for the growth of the enzymatic test-systems and biosensors market is the need for security control of the environment and food, as well as health status due to the aging population and its related disorders, growth of chronic and infectious diseases, and environmental disasters. Furthermore, people care about their health and want to live comfortably and have fresh air, clean water, and high-quality food. These factors resulted in the development of various services for environmental monitoring of toxic compounds, as well as the emergence of new areas of medicine, including personalized medicine that requires high-performance noninvasive portable test-systems for application in the clinic and at home.

Production of recombinant proteins is a rapidly growing area of research and development. This area emerged in the early 1980s with the development of genetic engineering tools, which represented a compelling alternative to protein extraction from natural sources. Over the years, a high level of heterologous protein was made possible in a variety of cell factories (hosts) ranging from the bacteria *Escherichia coli* to mammalian cells (Vieira Gomes et al. 2018; Owczarek et al. 2019). It is worth mentioning that the production of recombinant biopharmaceutical proteins is a multibillion-dollar industry. The global market for industrial enzymes was valued at USD 4.61 billion in 2016 and is projected to grow at a compound annual growth rate of 5.8% from 2017 and reach USD 6.30 billion by 2022 (Singh et al. 2016; Pharmaion 2019; Feed enzymes market 2019; Industrial Enzymes Market 2019). This branch of science and industry requires rationally chosen cell factories (hosts) and cost-efficient protein isolation protocols.

Of the established hosts, yeasts combine the advantages of unicellular organisms such as fast biomass growth and relatively easy genetic manipulations, with eukaryotic features such as correct post-translational modifications of recombinant proteins and efficient secretory pathways (Rueda et al. 2016; Sibirny 2017; Baghban et al. 2018; Huang et al. 2018a, b; Ekas et al. 2019). This is the reason why yeasts have, in recent decades, become attractive hosts for the production of heterologous proteins, enzymes, organic compounds, biopharmaceuticals, etc. (Daly and Hearn 2005; Idiris et al. 2010; Mattanovich et al. 2014; Singh et al. 2016; Stasyk 2017; Vogl et al. 2018; Yang and Zhang 2018a, b; Ekas et al. 2019; Owczarek et al. 2019).

Analysis of the literature data and our own research experience indicates that the following general approaches are employed for obtaining highly purified yeast enzymes:

1. Screening or gene engineering of the effective yeast strain as a producer of the target enzyme(s)
2. Optimization of cell cultivation conditions for achieving the highest specific activity of a target enzyme in the cells
3. Optimization of disruption conditions for producing intracellular enzymes
4. Development of effective methods for enzyme isolation and purification
5. Selection of methods for enzyme concentration and stabilization during storage

In the current chapter, we focused on the main achievements in the elaboration of modern techniques for isolation and purification of recombinant enzymes. Our investigations into effective technologies for obtaining yeast oxido-reductases of analytical importance are related to two aspects. One is the development of a cost-effective scheme for obtaining several enzymes from the same yeast source. The other is to summarize our previous results on isolation and purifications methods as well as the analytical application of some oxido-reductases for use in bioanalyses.

5.2 Non-conventional Yeasts as Hosts for Production of Heterological Proteins/Enzymes

Yeasts combine the ease of genetic manipulation and fermentation of cells with the capability of secreting and modifying foreign proteins according to a general scheme. Their rapid growth, microbiological safety, and high-density fermentation in simplified medium have a high impact, particularly in the large-scale industrial production of recombinant proteins (Singh et al. 2016; Sibirny 2017; Ekas et al. 2019).

Historically, *Saccharomyces cerevisiae* was the dominant yeast host for heterologous protein production (Muller et al. 1998; Kim et al. 2016; Chen et al. 2018). Lately, other yeasts, including non-conventional ones have emerged as advantageous cell factories. Non-conventional yeasts are considered as convenient expression platforms and promising industrial producers of recombinant proteins of academic and industrial interest (Reiser et al. 1990; Sudbery 1996; Stasyk 2017; Rebello et al. 2018). The yeasts *Kluyveromyces lactis*, *K. marxianus*, *Scheffersomyces stipitis*, *Yarrowia lipolytica*, and *Schizosaccharomyces pombe*, as well as the methylotrophic yeasts *Ogataea* (*Hansenula*) *polymorpha* and *Komagataella phaffii* (*Pichia pastoris*), have been developed as eukaryotic hosts because of their desirable phenotypes, including thermotolerance, assimilation of diverse carbon sources, and secretion of high quantities of proteins (Gellissen et al. 2005; Wagner and Alper 2016; Weninger et al. 2016; Juturu and Wu 2018; Vandermies and Fickers 2019).

Several value-added products – vaccines (Smith et al. 2012; Liu et al. 2014; Bredell et al. 2016; Xiao et al. 2018), mammalian proteins of pharmaceutical interest (Thömmes et al. 2001; Mack et al. 2009; Bawa et al. 2014; Arias et al. 2017; Walker and Pretorius 2018), therapeutic proteins (Griesemer et al. 2014; Kim et al. 2015; Love et al. 2018; Pobre et al. 2018; Zepeda et al. 2018a, b; Baghban et al. 2018; Owczarek et al. 2019), enzymes (Curvers et al. 2001; Hemmerich et al. 2014; Engleder et al. 2018; Vogl et al. 2018; Liu and Zhu 2018; Juturu and Wu 2018), industrial proteins (Singh et al. 2016; Shang et al. 2017; Barrero et al. 2018; Baghban et al. 2018), food additives, bio-renewable value-added chemicals, and biofuels (Jullessen et al. 2015; Kim et al. 2016; Białkowska 2016; Dmytruk et al. 2017; Semkiv et al. 2017; Porro and Branduardy 2017; Avalos et al. 2017; Passoth 2017; Rahman et al. 2017; Xu 2018; Ekas et al. 2019) were generated by the above-mentioned yeasts.

Yeast expression systems are economical and do not contain pyrogenic or viral inclusions. Unlike prokaryotic systems, the eukaryotic subcellular organization of yeasts enables them to carry out many of the post-translational folding, processing, and modification events required to produce "authentic" and bioactive mammalian proteins (Buckholz and Gleeson 1991, Domínguez et al. 1998; Talebkhan et al. 2016; Stasyk 2017). However, secretory expression of heterologous proteins in yeasts is often subject to several bottlenecks that limit yield.

Yeast engineering by genetic modification has been the most useful and effective method for overcoming the drawbacks in yeast secretion pathways (Muller et al. 1998; Daly and Hearn 2005; Pobre et al. 2018; Vogl et al. 2018; Thomas et al. 2018; Bao et al. 2018; Chang et al. 2018; Nora et al. 2019). Metabolic engineering and synthetic biology methods are promising for production of native and transgenic enzymes and proteins on an industrial scale (Krivoruchko et al. 2011; Fletcher et al. 2016; Fernandes et al. 2016; Yang and Zhang 2018a; Theron et al. 2018; Ekas et al. 2019). Methanol-free mutant strains were constructed as alternatives to the traditional system, as a result of synthetic biology tools for reprogramming the cellular behavior of methylotrophic yeasts. The creation of a methanol-free induction system for eliminating the potential risks of methanol and for achieving enhanced recombinant protein production efficiency has been reviewed (Shen et al. 2016).

The methylotrophic yeasts *K. phaffii* (*Pichia pastoris*) and *O. polymorpha* are the most effective sources of recombinant proteins (Sudbery 1996; Berg et al. 2013; Rajamanickam et al. 2017; Peña et al. 2018; Sibirny 2017). The methylotrophic yeast *O. polymorpha* has remarkable thermotolerance. This yeast is therefore successfully used as a cell factory for the production of thermostable enzymes and proteins (Gellissen et al. 2005; Krasovska et al. 2007). Many of the expression platforms, including circular plasmids with the *P. pastoris*-specific autonomously replicating sequence (*PARS1*), were developed in order to facilitate genetic manipulation for plasmid replication and distribution (Song et al. 2003; Sturmberger et al. 2016; Nakamura et al. 2018; Portela et al. 2018).

Yeasts with beneficial native phenotypes and genetically modified cells were therefore selected and used for the heterologous production of recombinant enzymes, proteins, and other products. A detailed and comprehensive description of methylotrophic yeasts as producers of recombinant proteins was published by Stasyk (2017).

5.3 Scheme for Obtaining Several Yeast Enzymes from the Same Source

We carried out a research in order to develop a cost-effective approach for simultaneous isolation and purification of several enzymes with practical importance from the same yeast source. A mutant strain of the thermotolerant methylotrophic yeast *O. polymorpha* C-105 (*gcr1 catX*), overproducer of yeast alcohol oxidase (AO), was selected as the producer of enzymes (Gonchar et al. 1990). This strain has impairment in glucose catabolite repression of AO synthesis, is catalase-defective, and is able to overproduce AO in glucose medium. Contrary to *P. pinus*, the yeast *O. polymorpha* C-105 can generate a single AO isoform, due to the presence of the AO-coding gene only in its genome (Cregg et al. 1985, 1989; Gunkel et al. 2004). However, cultivation of *O. polymorpha* C-105 cells in glucose medium prevents the formation of multiple forms of AO which can be generated in methanol-containing medium as a result of chemical modification of the AO protein by formaldehyde (FA).

Yeast cells were cultivated in mineral medium with 1% glucose and 0.2% yeast extract up to the middle of the exponential growth phase (Gonchar et al. 2002). The cells were washed, resuspended in 30 mM phosphate buffer, pH 7.5 (PB); supplemented with protease inhibitors; and disrupted with glass beads in a planetary disintegrator. The pellet of disrupted cells was removed by centrifugation, and the supernatant (cell-free extract) was used for a two-step ammonium sulfate fractionation (at 40 and 60% saturation).

Pellet 60% (see scheme in Fig. 5.1) was dissolved in 30 mM PB and placed on a column with ion-exchange sorbent DEAE-Toyopearl 650 M (Shleev et al. 2006; Sigawi et al. 2011). The sorbent was washed step by step with 30–200 mM PB; the target enzyme was eluted with 0.25 M PB and collected by fractions. Each fraction was tested for AO activity and protein concentration. AO activity was determined at 30 °C by the rate of hydrogen peroxide formation in reaction with methanol as monitored by the peroxidative oxidation of *o*-dianisidine in the presence of horseradish peroxidase (Gonchar et al. 2001). Chromatographic fractions with the highest specific activity of AO were combined and analyzed electrophoretically in PAG under denaturation conditions.

The resultant AO preparations with a specific activity up to 20 U per mg protein were fourfold purer than in pellet 60% and sevenfold purer than in the cell-free extract. The yield of purified AO from 1 L of yeast culture was about 800 U, which is 30% of the initial activity in the cell-free extract. The purified enzyme was stable, but still not homogeneous in SDS-PAG. AO preparations were stored as a suspension in 70% saturated ammonium sulfate (with an activity of 200 U mL^{-1} suspension) at −10 °C, without any remarkable decrease in activity over a period of 2 years. Before use, the enzyme suspension was centrifuged, and the enzyme precipitate was dissolved in 0.05 M phosphate buffer, pH 7.6.

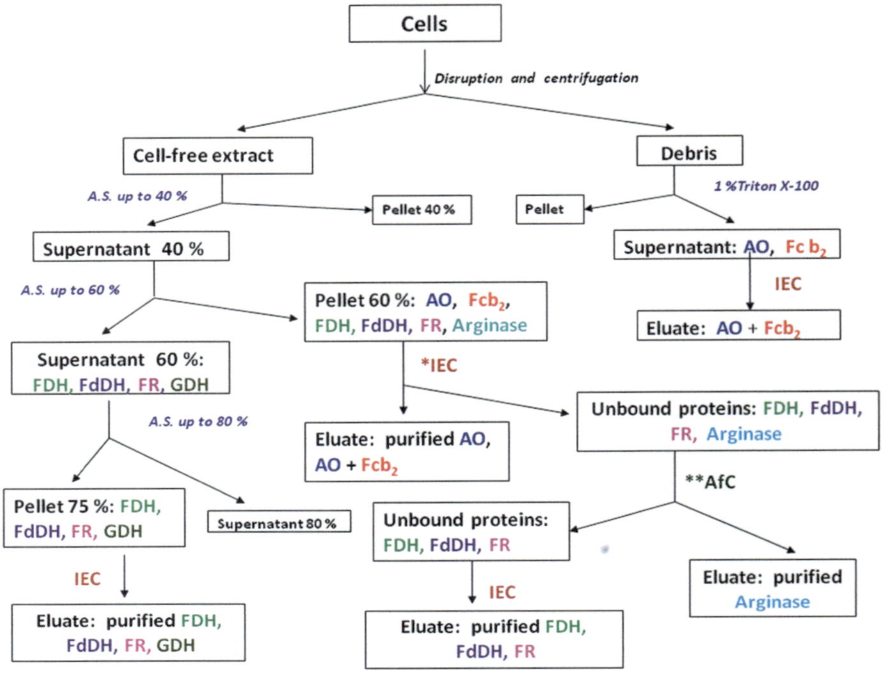

Fig. 5.1 Scheme of obtaining several enzymes from the *O. polymorpha* C-105 cells. Abbreviations: $*$ IEC, ion-exchange chromatography on DEAE-Toyopearl M-650; $**$AfC, affinity chromatography on Arg-containing sorbent

We tested additionally the activity of flavocytochrome b_2 (Gaida et al. 2003) during AO purification in order to obtain the highly purified AO preparations (up to 40 U per mg protein) that were homogeneous during electrophoresis in SDS-PAG (see Fig. 5.1). The proposed approach for optimizing the technology for obtaining the target enzyme with the highest purity is characterized by the following points:

1. Simplification of express qualitative or semiquantitative enzyme activity assays for target enzyme identification in a mixture of different proteins
2. Visualization of enzymatic activity in native PAG for testing the presence of target and waste enzymes in each stage of the purification procedure (from the initial cell-free extract to the final chromatographic fractions)

The examples of express qualitative or semiquantitative assays for AO and Fcb_2 are presented in Fig. 5.2.

A cost-effective technology for obtaining several enzymes from the same yeast producer was thus proposed. This technology demonstrated (Fig. 5.1) the possibility of isolating and chromatographically purifying several yeast enzymes from *O. polymorpha* C-105 cells: alcohol oxidase (AO), flavocytochrome b_2 (Fc b_2), glycerol dehydrogenase (GDH), methyl aminooxidase (AMO), arginase, formaldehyde dehydrogenase (FdDH), formate dehydrogenase (FDH), and formaldehyde reductase (FR). The best yields were observed for AO and Fc b_2. Activities of other enzymes (Fig. 5.1) were tested as described earlier: for GDH by Synenka et al. (2015), for AMO by Krasovska et al. (2006), for arginase by Stasyuk et al. (2013), and for FdDH, FDH, alcohol dehydrogenase (ADH), methyl formiate synthase (MFS) and FR by Demkiv et al. (2011).

The proposed scheme for enzymes, presented in Fig. 5.1, was also applied to other yeast producers (see Table 5.1).

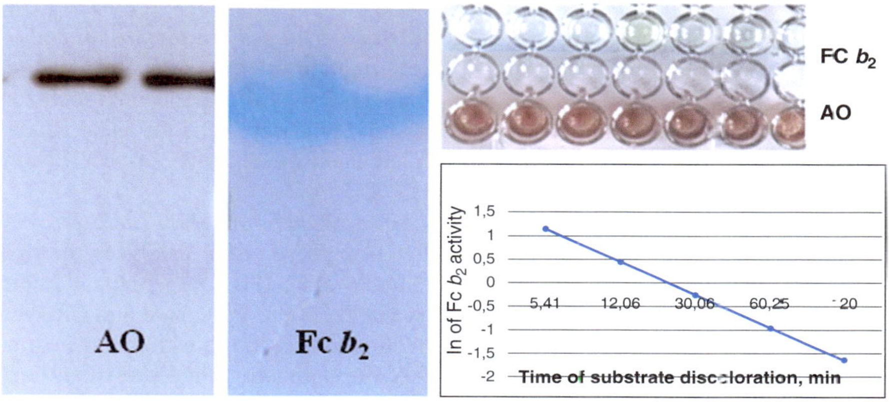

Fig. 5.2 Visualization of enzymatic activities in native PAG and screening tests for AO and Fc b_2 assays

Table 5.1 Enzymes of analytical importance and their producers

Enzyme	Yeast strain – the source of enzyme	Reference
Yeast alcohol oxidase (AO)	Mutant *O. polymorpha* C-105 (*gcr1 catX*)	Gonchar et al. (1990), Shleev et al. (2006)
Yeast alcohol oxidase (mAO)	Mutant *O. polymorpha* CA4 and CA2	Dmytruk et al. (2007)
Yeast formaldehyde dehydrogenase (FdDH)	Recombinant *O. polymorpha* NCYC 495 (*leu*1-1)	Demkiv et al. (2005, 2007, 2011)
Yeast glycerol dehydrogenase (GDH)	Recombinant *Saccharomyces cerevisiae W303* (*HpGDH*)	Nguyen and Nevoigt (2009), Synenka et al. (2015)
Yeast flavocytochrome (Fc b_2)	*O. polymorpha 356*	Gaida et al. (2003), Smutok et al. (201[illegible])
Yeast amine oxidase (AMO)	Recombinant *O. polymorpha* CBS4732	Krasovska et al. (2006)
Human liver arginase I (arginase)	Recombinant *O. polymorpha* NCYC 495-*pGAP1-HsARG1-(leu2car1 Sc:LEU2)*	Stasyuk et al. (2013)

5.4 Oxido-reductases of Analytical Importance

5.4.1 Alcohol Oxidase

Alcohol oxidases (alcohol: O_2 oxidoreductase; EC 1.1.3.x) are flavoenzymes that catalyze the oxidation of alcohols to the corresponding carbonyl compounds with a concomitant release of hydrogen peroxide. Based on substrate specificity, alcohol oxidases may be categorized broadly into four different groups, namely, (a) short-chain alcohol oxidase, (b) long-chain alcohol oxidase, (c) aromatic alcohol oxidase, and (d) secondary alcohol oxidase (Ozimek et al. 2005; Leferink et al. 2008; Goswami et al. 2013; Romero and Gadda 2014; Pickl et al. 2015; van Berkel 2018; Sützl et al. 2018). The sources reported for these enzymes are mostly limited to bacteria, yeasts (Sahm and Wagner 1973; Sagiroglu and Altay 2006; Shleev et al. 2006; Koch et al. 2016; Vonck et al. 2016; Mangkorn et al. 2019), fungi (Janssen and Ruelius 1968; Bringer et al. 1979; Kondo et al. 2008, Isobe et al. 2009; Hernández-Ortega et al. 2012), plants (Panadare and Rathod 2018), insects (Sperry and Sen 2001), and mollusks (Grewal et al. 2000).

Alcohol oxidase (EC 1.1.3.13) also known as methanol oxidase (AO) is a key enzyme of methanol methabolism in methylotrophic yeasts; it catalyzes the first step of methanol oxidation to formic acid (Mincey et al. 1980; Eggeling and Sahm 1980; Sibirny et al. 1988; Lusta et al. 2000; Gadda 2008; Wongnate and Chaiyen 2013; Dijkman et al. 2013; Liu et al. 2018). In addition to the physiological substrate methanol, AO can typically oxidize also short aliphatic primary alcohols consisting of up to four carbons.

AO is a flavoprotein with flavin adenine dinucleotide (FAD) as a prosthetic group, non-covalently, but very tightly bound with apoenzyme. Native protein is an octamer of approximately 600 kDa composed of eight identical FAD-containing subunits (Bringer et al. 1979; Mincey et al. 1980; Boteva et al. 1999; Gunkel et al. 2004; Ozimek et al. 2005; Isobe et al. 2009; van der Klei et al. 1991). Only octameric enzyme has catalytic activity. The mechanism of oligomerization into catalytical active octamers, as well as the role of AO octameric structure in catalysis, is not yet elucidated.

Although AO was discovered 50 years ago, its tertiary structure (for *Pichia pastoris* or *Komagataella phaffii*) was elucidated only in 2016 using crystallography and cryoelectron microscopy (Koch et al. 2016; Vonck et al. 2016).

AO shows a high structural homology to other members of the GMC ("glucose-methanol-choline") family of oxidoreductases, which share a conserved FAD binding domain but have different substrate specificities (Gvozdev et al. 2012; Dijkman et al. 2013; Romero and Gadda 2014; Pickl et al. 2015; Sützl et al. 2018; Liu et al. 2018). The preference of AO for small alcohols is explained by the presence of conserved bulky aromatic residues near the active site. Compared to the other GMC enzymes, AO contains a large number of amino acid inserts, the longest being 75 residues. These segments are found at the periphery of the monomer and make extensive inter-subunit contacts which are responsible for the very stable octamer. A short surface helix forms contacts between two octamers, explaining the tendency of AO to form crystals in the peroxisomes (Vonck et al. 2016).

The crystal structure analysis of the methanol oxidase from *P. pastoris* was described (Koch et al. 2016). The crystallographic phase problem was solved by means of molecular replacement in combination with initial structure rebuilding using Rosetta model completion and relaxation against an averaged electron density map. The subunit arrangement of the homo-octameric AO differs from that of octameric vanillyl alcohol oxidase and other dimeric or tetrameric AOs, due to the insertion of two large protruding loop regions and an additional C-terminal extension in AO. In comparison to other AOs, the active site cavity of AO is significantly reduced in size, which could explain the observed preference for methanol as substrate. All AO subunits of the structure reported here harbor a modified FAD, which contains an arabityl chain instead of a ribityl chain attached to the isoalloxazine ring.

The recently described AO from the white-rot basidiomycete *Phanerochaete chrysosporium* (PcAOX) was reported to feature very mild activity on glycerol. PcAOX was expressed in *Escherichia coli* in high yields and displayed high thermostability. Steady-state kinetics revealed that PcAOX is highly active toward methanol, ethanol, and propanol-1 (k_{cat} = 18; 19 and 11 s^{-1}, respectively), but showed a very limited activity toward glycerol (k_{obs} = 0.2 s^{-1}at 2 M substrate). The crystal structure of the homo-octameric PcAOX was determined at a resolution of 2.6 Å. The catalytic center is a remarkable solvent-inaccessible cavity located at the re-side of the flavin cofactor. Its small size explains the observed preference for methanol and ethanol as best substrates (Nguyen et al. 2018). The catalytic center is a remarkable solvent-inaccessible cavity located at the re-side of the flavin cofactor. Its small size explains the observed preference for methanol and ethanol as best substrates (Nguyen et al. 2018).

In our research, we have obtained a stable highly purified AO (up to a specific activity of 40 U mg^{-1}) from the overproducing strain of the yeast *O. polymorpha* C-105 *(gcr1 catX)* with impaired glucose-induced catabolite repression of AO synthesis and completely devoid of catalase (see Sect. 5.3). Such purity permitted to obtain the enzyme in crystalline form. The crystals of highly purified AO were obtained by different methods, including crystallization in space weightless conditions. X-ray study followed by the calculations for AO complexes with competitive inhibitors structures was performed. Comparative analysis of X-ray database for AO structure with the known protein structures of some oxidases was done, and the model of AO subunit tertiary structure was proposed (Fig. 5.3). Some recommendations for site-specific mutagenesis in AO gene for obtaining enzyme with significantly decreased affinity to ethanol compared to the wild-type AO have been done (Gayda et al. unpublished data).

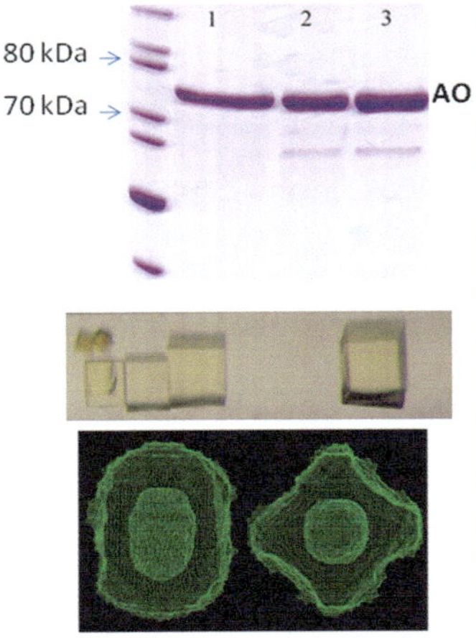

Stages of procedure	AO activity		Yield, %	Factor of purification
	Total, U	Specific, U mg^{-1} protein		
Cell-free extract	1500	3.5	100	1
Two-step fractionation with AS	1200	5.0	80	1.4
Ion-exchange chromatography, pH 7.0, unbound proteins	300	15–20	20	4-6
Ion-exchange chromatography, pH 8.8, eluates	150	25–30	10	7-8
	75	35 - 40	5	10-11

Fig. 5.3 Isolation and purification of AO from *O. polymorpha* C-105 cells

5.4.2 *NAD*$^{+}$*- and Glutathione-Dependent Formaldehyde Dehydrogenase*

Formaldehyde (FA) is a natural metabolite found in tissues, cells, and body fluids. It is present in fruits, vegetables, meat, and fish. FA is also a large-scale product, used extensively in industry. FA is very toxic and is an extremely active chemical compound which causes modifications of bioorganic molecules in living organisms. That is why this dangerous compound is monitored in environmental, industrial, and medical laboratories. Enzymatic methods for valid selective determination of FA are based on using FA-selective enzymes, including formaldehyde dehydrogenase (FdDH). The aim of our research was to obtain highly purified FdDH and develop analytical approaches for FA assay. For this aim, yeast engineering for construction of FdDH-overproducing strains was carried out.

The *O. polymorpha FLD*1 gene with its own promoter was inserted into the integrative plasmid pYT1 (Demkiv et al. 2005) containing the *LEU2* gene of *Saccharomyces cerevisiae* (as a selective marker) in order to construct strains of *O. polymorpha* that overproduce thermostable NAD+- and glutathione-dependent FdDH. The constructed vector was used for multicopy integration of the target gene into the *O. polymorpha* genome by transformation of *leu 1-1* (Demkiv et al. 2005, 2011; Sibirny et al. 2011b) and *leu 2-2* recipient cells (both *leu* alleles were complemented by *S. cerevisiae* gene *LEU2*).

Selection of FdDH-overproducing strains was carried out simultaneously by leucine prototrophy and by resistance to elevated FA concentrations in the medium. Of more than 150 integrative Leu+ transformants with higher resistance to FA (up to 10–12 mM on solid plates), 14 stable clones which were resistant to 15–20 mM FA on plates were selected and studied in greater detail. The growth characteristics of selected clones in the liquid medium are shown in Fig. 5.4. All transformants

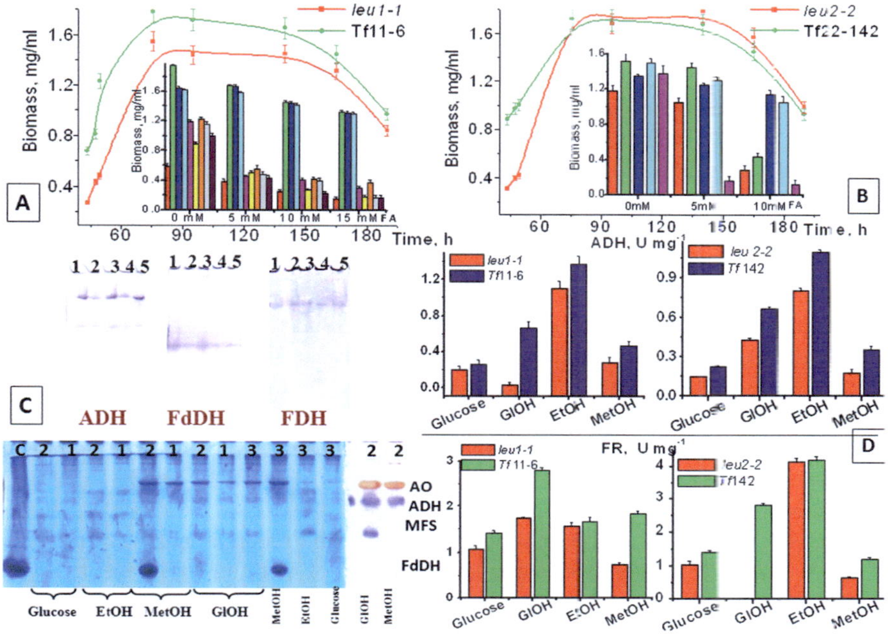

Fig. 5.4 Effect of cultivation conditions on growth and specific activity of FA-utilizing enzymes in cell free extracts. A, B, inductive effect of FA on growth of parental and recombinant cells in a medium with 1% methanol; C, visualization of enzymes activities in native PAG. At a top: *Tf* 11-6 cells, cultivated in medium with MetOH, $(NH_4)_2SO_4$, FA (1); MetOH, $(NH_4)_2SO_4$ (2); MetOH, CH_3NH_3HCl (3); MetOH, NH_4Cl (4); GlOH, NH_4Cl, FA (5). At a bottom: activities of AO, ADH, MFS, and FdDH in *leu 1-1* (1), *Tf* 11-6 (2), and *Tf* 22-142 (3) cells in comparison with purified FdDH (c); D, activities of ADH and FR in parental and recombinant cells.

grew better and were more resistant to elevated FA content in liquid medium with 1% methanol, compared to the recipient strains. A detailed description of this research has been done by Demkiv et al. (2005, 2011). Finally, FdDH specific activities were tested in cell-free extracts (CE) of the best selected FA-resistant Leu-prototrophic transformants (Fig. 5.4).

Tf 11-6 and Tf 22-142 were the most effective recombinant strains with the highest FdDH activity (up to 4.0 U mg^{-1}), which was four- to fivefold higher compared to the parental strains, *leu* 1-1 and *leu* 2-2, respectively. These transformants were characterized and chosen as sources for FdDH production.

Southern dot-blot analysis showed that genomes of the stable recombinant yeast clones contain 6–8 copies of the target *FLD1* gene (Demkiv et al. 2011). The recombinant yeast strain Tf 11-6 contains more than eight copies of the integrated plasmid, as opposed to one copy in the parental strain, probably due to use of the double-gene-containing plasmid *pHp(FLD1)*$_2$ and its tandem aggregation into the genome of the recipient strain.

The influence of growth medium composition on FdDH concentration was studied for the best two strains, Tf 11-6 and Tf 22-142, in order to optimize cultivation conditions for obtaining the highest enzyme yield. FdDH activity in cell-free extract was shown to be dependent on a carbon source. Cultivation in medium with 1% methanol resulted in significant levels of the enzyme synthesis for both tested strains (Fig. 5.4). This is in accordance with the literature data (Hartner and Glieder 2006; van der Klei et al. 1991; Eggeling and Sahm 1980).

We demonstrated that the addition of FA to the methanol medium stimulated synthesis of FdDH. The target enzyme activity was 6.2 U mg^{-1} under experimentally determined optimal conditions (with methanol as a carbon source, methylamine as a nitrogen source, and 5 mM FA as an additional inducer of FdDH synthesis). This is 1.6-fold higher than under normal growth conditions. The addition of 10 mM FA to the optimal culture medium resulted in a FdDH activity of 8.3 U mg^{-1}, which is twofold higher than in the medium without FA. The strong correlation between FA concentration in the medium and FdDH activity in cultivated cells of recombinant yeast strain Tf-11-6 demonstrates the important role of FA as an inducer of FdDH synthesis (Fig. 5.4).

The enzyme was isolated from a cell-free extract of the recombinant overproducing Tf 11-6 strain. Cells were cultivated in 1% methanol medium supplemented with 5 mM FA for 20 h (Demkiv et al. 2007). A simple scheme for FdDH isolation and purification from the recombinant strain by two-step column chromatography on an anion-exchange sorbent was proposed, resulting in a FdDH preparation with specific activity of 27 U mg^{-1} protein.

In the first step, cell-free extract (CE) was applied to the sorbent, equilibrated by PB (pH 8.0). The fraction of unabsorbed proteins, which contained FdDH, was diluted with water (1:3). Tris base solution was added to adjust the pH to 8.8, and the final solution was applied to the same column (the second step), previously washed with 1 M NaCl and equilibrated with 40 mM Tris buffer, pH 8.8 (TB). The enzyme was eluted with 0.1 M NaCl in the initial TB buffer, and the specific activity of FdDH was assayed in each fraction. The fractions of eluate with enzyme activity higher than 10 U mg^{-1} and devoid of AO activity were combined, and dithiothreitol (DTT) up to 2 mM and ammonium sulfate (up to 80% saturation, pH 8.0, at 0 °C) were added. After incubation at 0 °C for 1 h, the enzyme was collected by centrifugation, and the pellet was resuspended in a minimal volume of ammonium sulfate solution (80% saturation) in 40 mM TB with 2 mM DTT.

The specific activity of resulted FdDH was 27 U mg^{-1}. For comparison, the specific activities of commercially available FdDH preparations from *P. putida* and from the yeast *C. boidinii* are 3–5 U mg^{-1} and 17–20 U mg^{-1}, respectively. The purity of the isolated enzyme preparation was controlled by PAG electrophoresis under denaturation conditions according to Laemmly (Demkiv et al. 2007; Sibirny et al. 2011b).

It was reported that the predicted *FLD1* gene product (Fld1p) is a protein of 380 amino acids (Baerends et al. 2002). Since the molecular mass of native FdDH from various methanol-utilizing yeasts was estimated to be between 80 and 85 kDa, the isolated thermostable NAD^+- and GSH-dependent FdDH can be assumed to be dimeric. The molecular mass of the FdDH subunit, estimated by SDS-electrophoresis, was shown to be approximately 40 kDa, which is similar to the 41 kDa found for *C. boidinii* (Yurimoto 2009).

Optimal pH and pH stability of the enzyme were evaluated by incubation in the appropriate buffer at room temperature for 60 min. The optimal pH was found to be in the range of 7.5–8.5, and the highest stability of FdDH was observed at pH 7.0–8.5.

Values of the Michaelis-Menten constant (K_M) for FA and NAD^+ calculated for this enzyme are close to the K_M for the wild-type enzyme. The effect of several inhibitors on the enzymatic properties was studied. The bivalent cations Zn^{2+}, Cu^{2+}, and Mn^{2+} were shown to inhibit FdDH activity, as did the ionic detergent SDS. According to the literature, enzymes from two other yeasts *P. pastoris* and *C. boidini* (Allais et al. 1983; Kato 1990; Patel et al. 1983), were also inhibited in a similar manner.

A limited number of publications on the isolation and characterization of formaldehyde reductase (FdR) and the absence of a corresponding commercial preparation of the enzyme led us to screen potential microbial FdR producers among wild-type and recombinant strains of the yeast *O. polymorpha*. The gene-engineered integrative transformants with the highest FA resistance originating from *leu* 2-2 were shown to overproduce NADH-dependent FdR upon cultivation on 1% ethanol or glycerol. The best integrative clone, T22-126, was chosen as a source for FdR isolation, and optimal cultivation conditions for the highest yield of FR were established (Demkiv et al. 2011). The simple scheme for isolation of FR (Gayda et al. 2008b) from *O. polymorpha* T22-126 yeast cells and chromatographic purification of the enzyme on anion-exchange sorbent was proposed, resulting in *electrophoretically homogeneous* enzyme preparations.

The enzymatic methods and analytical kits for FA assay were developed based on FdDH (Demkiv et al. 2007; Gayda et al. 2008a, 2015; Sibirny et al. 2011a, b). In methylotrophic yeasts, FdDH catalyses the oxidation of FA to formic acid under simultaneous reduction of NAD^+ to NADH. The proposed enzymatic method is based on the photometric detection of a colored product, formazan, which is formed from nitrotetrazolium blue in a reaction coupled with FdDH-catalyzed oxidation of FA in the presence of an artificial mediator, PMS (Demkiv et al. 2007).

Purified preparations of FdDH were also used for construction of FA-selective electrochemical biosensors. Several FA-selective FdDH-based biosensors with different types of signal detection were developed and described in detail (Nikitina et al. 2007; Ali et al. 2007; Demkiv et al. 2008; Gayda et al. 2008a; Sibirny et al. 2011a, b). All constructed biosensors were characterized by high storage and good operational stability, high sensitivity, broad dynamic range, and low applied potential compared to known biosensors. A comparative analysis of different FA-sensitive biosensors was presented in reviews (Sibirny et al. 2011a, b; Gayda et al. 2015).

5.4.3 NAD⁺-Dependent Glycerol Dehydrogenase

Analysis of glycerol (GlOH) is important in clinical diagnostics for assessing the level of triglycerides in obesity and metabolic disorders, in particular lipid metabolism, that cause the development of type II diabetes and cardiovascular disease and in the wine industry for controlling wine quality during the production process. Simple methods for monitoring GlOH content (as a by-product of this technology) are currently in high demand, due to the growth in biodiesel production (Gerpen 2005; Talebian-Kiakalaieh et al. 2018). Additionally, GlOH assay is necessary for new technologies of GlOH conversion to valuable chemical products, including dihydroxyacetone (DHA) (Li et al. 2010; Cho et al. 2015; Kumar and Park 2018; Oh et al. 2018).

A key component of enzymatic kits for glycerol assay are glycerol-selective enzymes, including glycerol dehydrogenase (GDH). GDH (EC 1.1.1.6) is synthesized by mammalian tissues and microorganisms, including bacteria and yeasts (Yamada et al. 1982; Gartner and Kopperschlager 1984; Ruzheinikov et al. 2001; Yamada-Onodera et al. 2002). Three types of enzymes have been described: NAD^+-dependent GDH that converts GlOH to DHA and vice versa, $NADP^+$-dependent GDH that catalyzes the conversion between GlOH and DHA (glycerol 2-dehydrogenase, EC 1.1.1.156), and $NADP^+$-dependent GDH that catalyzes the conversion of GlOH on glyceraldehyde (EC 1.1.1.72). GDH is a promising enzyme for the development of analytical methods for assaying GlOH and other alcohols as well as for GlOH conversion. Commercial preparations of GDH from *Cellulomonas speciaes*, *Enterobacter aerogenes*, and *Bacillus megaterium* are present on the enzymes market (Sigma products).

The aim of our work was to obtain a thermotolerant NAD^+-dependent GDH of *O. polymorpha* (previously *H. polymorpha*) and to investigate its properties in order to develop a reliable and sensitive glycerol assay. We used the recombinant *Saccharomyces cerevisiae* strain that harbors the gene *HpGDH* (Mallinder et al. 1992). This strain was created for the biotransformation of glycerol to DHA (Nguyen and Nevoigt 2009). It contains an extrachromosomal multicopy plasmid p424GDH with an integrated *O. polymorpha GDH* gene under a strong constitutive glyceraldehyde-3-phosphate dehydrogenase (*GPH*) promoter. During cultivation of the cells in a medium with glucose, GDH of *O. polymorpha* was constitutively expressed, and this enzyme oxidized glycerol to DHA. GDH overexpression resulted in DHA extrusion into the extracellular liquid up to 100 mg L^{-1}, which is 60-fold higher than in the wild-type strain.

Optimal conditions for cell cultivation were studied (Fig. 5.5).

The influence of the growth medium composition on biomass and enzyme yield for the chosen producer was studied in order to obtain the highest GDH yield (Fig. 5.5A-E). GDH activity in cell-free extract was dependent on a carbon source. Cultivation in a medium containing 1% GlOH and 0.1% glucose resulted in considerable levels of enzyme activity (Fig. 5.5C). It is noteworthy that two GDH

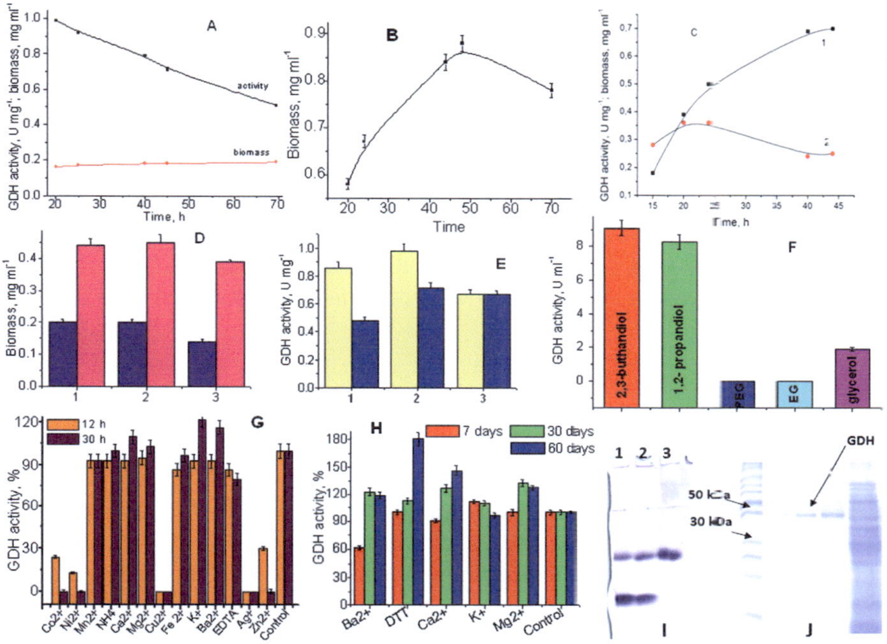

Fig. 5.5 Isolation and characterization of GDH. Effect of cultivation conditions on cells growth (A–D) and specific activity of GDH in cell-free extracts (A, C, E) of producing cells. Properties of the purified GDH: substrate specificity (F), effect of metallic ions (G), storage stability (H), visualization in native PAG after incubation at 40 °C, 50 °C, and 60 °C during 10 min (I), control of purity and estimation of molecular mass of enzyme's subunit in SDS-PAG (J).

isoforms were visualized in native PAG when cells were cultivated in a medium with GlOH and glucose. However, only one thermostable form of GDH was found in cells cultivated in GlOH without glucose (Fig. 5.5I).

A simple scheme for *O. polymorpha* GDH isolation from a cell-free extract of the recombinant strain *S. cerevisiae* was developed. It includes desintegration of the cells by vigorous vortexing followed by two-stage ion-exchange chromatography of cell-free extract on DEAE-Toyopearl M-650 (at pH 8.0 and pH 8.8). This approach was proposed earlier for purification of recombinant FdDH (Demkiv et al. 2007). As a result, a highly purified (tenfold) enzyme preparation with specific activity of 34 U mg^{-1} of protein and 10 % yield was obtained (Fig. 5.5J). For comparison, specific activities of commercially available GDH preparations are 50 U mg^{-1} for enzyme from *Cellulomonas* sp., 15 U mg^{-1} of solid for *B. megaterium*, and 20–80 U mg^{-1} of protein – for *E. aerogenes*. The purity of the isolated enzyme preparation was controlled by PAG electrophoresis under denaturation conditions according to Laemmly. Molecular weight of enzyme's subunit was shown to be 40 kDa (Fig. 5.5J).

The enzymatic method for GlOH assay based on GDH was developed (Gayda et al. 2013a, b). In yeasts, GDH catalyzes the oxidation of GlOH to DHA under simultaneous reduction of NAD^+ to NADH. The proposed enzymatic method includes photometric detection of a colored product, formazan, which is formed from nitrotetrazolium blue in a reaction coupled with GDH-catalyzed oxidation of GlOH in the presence of an artificial mediator, PMS. The same approach was used for the development of an enzymatic method for FA assay (Demkiv et al. 2007). The optimal conditions for an effective reaction were determined. Calibration graphs for GlOH estimation, using a GDH-based method, are presented in Fig. 5.5. This method was used for analysis of GlOH in real samples of commercial wines (see Sect. 5.5.2).

5.4.4 *Flavocytochrome* b_2

L-Lactate is an important metabolite in glucose metabolism. Monitoring lactate levels is a useful indicator of the balance between tissue oxygen demand and utilization and is useful when studying cellular and animal physiology. Lactate detection plays a significant role in healthcare, and food industries and is specially necessitated in conditions like hemorrhage, respiratory failure, hepatic disease, sepsis, and tissue hypoxia.

For L-lactate analysis a lot of physicochemical and chemical methods have been proposed: spectrophotometry, fluorimetry, pH potentiometric measurements, and amperometric biosensors based on O_2 and H_2O_2 electrodes. The available methods include enzymatic approaches which generally use NAD^+-dependent lactate dehydrogenase (LDH) from animal muscle or heart and bacterial lactate oxidase. These classic approaches as well as modern methods were described in detail in the last years (Rassaei et al. 2014; Sharma et al. 2017; Rosati et al. 2018; Rathee et al. 2016; Dagar and Pundir 2018; Bollella et al. 2019). Most of these methods need a lot of time and previous labour-consuming procedures such as filtration, chromatography, deproteinization, etc. On the other hand, most of them require an expensive equipment or are non-selective.

We proposed using the yeast L-lactate-cytochrome *c*-oxidoreductase (EC 1.1.2.3; flavocytochrome b_2, Fc b_2) of the thermotolerant methylotrophic yeast *O. polymorpha* as a promising biocatalyst for enzymatic-chemical analytical methods and amperometric biosensors. Fc b_2 is a tetramer of identical subunits, where each subunit contains FMN- and haem-binding domains. The enzyme exhibits absolute specificity for L-lactate, but application of Fc b_2 from baker's yeast in bioanalytics is hampered by its instability and difficulties in purification of the enzyme. We used the following stages for obtaining Fc b_2:

1. Screening potential yeast producers in order to choose the best source of thermostable Fc b_2

2. Optimizing cultivation conditions in order to achieve the maximal yield of the enzyme
3. Testing different cell disruption methods in order to obtain a stable enzyme with the highest yield
4. Developing a simple scheme for Fc b_2 isolation, chromatographic purification, and stabilization

The highly purified target enzyme was used for developing enzymatic-chemical and biosensor methods for L-lactate assay. Screening of 16 yeast species was carried out in order to choose the most effective producer of the stable form of FC b_2. For this aim, a method of visualization of the activity of Fc b_2 in electrophoretograms was used. This method was based on the interaction between ferrocyanide (generated during the enzymatic reaction) and Fe^{3+}, resulting in the formation of intensely colored precipitates of Berlin blue (Gaida et al. 2003). The main advantages of this method were its high sensitivity (less than 0.005 U Fc b_2 was detected within a suitable time period) and the stability of the dye formed. The method developed can be used for determining Fcb_2 activity in cell-free extracts (e.g., in the selection of Fc b_2 producers) and monitoring chromatographic purification of proteins, as well as in other cases associated with Fc b_2 assessment. *O. polymorpha*, *Kluyveromyces lactis*, and *Rhodotorula pilimanae*, which exhibited the highest specific activities in cell-free extracts, were chosen as the best producers of Fc b_2 (Fig. 5.6). A study of the enzyme's thermostability in the cell-free-extracts revealed that only Fc b_2 from *O. polymorpha* remains active after heating for 10 min at 60 °C or 3 min at 70 °C (Smutok et al. 2006c).

To obtain the target enzyme, *O. polymorpha* cells were cultivated in flasks to the beginning of the stationary growth phase at 30 °C on a shaker with strong aeration (240 rpm) in a mineral medium containing 1% glucose, 0.2% L-lactate sodium, and 0.05% yeast extract. Freshly grown cells were collected by centrifugation, washed, lyophilized, and kept at –20 °C until used (Smutok et al. 2006a).

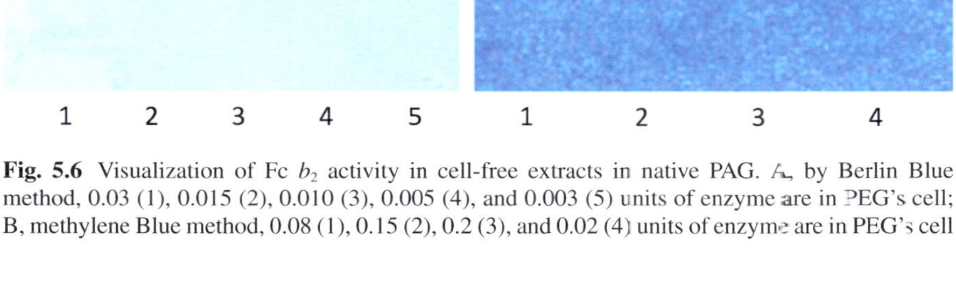

Fig. 5.6 Visualization of Fc b_2 activity in cell-free extracts in native PAG. A, by Berlin Blue method, 0.03 (1), 0.015 (2), 0.010 (3), 0.005 (4), and 0.003 (5) units of enzyme are in PEG's cell; B, methylene Blue method, 0.08 (1), 0.15 (2), 0.2 (3), and 0.02 (4) units of enzyme are in PEG's cell

Cell-free extract (CE) was obtained by incubating lyophilized cells in a lysing mixture with 10% *n*-butanol (Gaida et a. 2003). CE was separated from the cell debris by centrifugation, cell fragments were washed twice with a lysing mixture, and the supernatants were combined (C1). It is worth mentioning that cell fragments in the C1 contained a significant amount of Fc b_2. The pellets were therefore also extracted with 1% Triton X-100 in 50 mM PB, pH 7.5, and the supernatants were combined (C2).

It was demonstrated that the specific activity and stability of Fc b_2 was higher in the C1 and C2 extracts than in the CE. Furthermore, use of Triton X-100 allows an extraction of up to 95% of the total Fc b_2 activity from the cell debris (C1+C2). The enzyme was purified from the (C1+C2) extracts with a total activity of 60 U by column chromatography on the anion-exchange sorbent DEAE-Toyopearl 650 M (TSK-Gel, Japan). The enzyme was eluted by 15% (of saturation at 0 °C) ammonium sulfate in 50 mM PB, pH 7.5, containing L-lactate. Monitoring of enzyme activity and purity was carried out by estimation of the Fc b_2 specific activity in each fraction and by PAG-electrophoresis under native (Gaida et al. 2003) and denaturation conditions according Lemmly.

The most active fractions were combined and treated with ammonium sulfate up to 70% saturation. The highest specific activity of FC b_2 in some fractions was 20 U·mg^{-1} protein; the yield was 10%. Precipitation with ammonium sulfate allows to purify the target enzyme additionally (1.5-fold) up to 30 U mg^{-1} (Smutok et al. 2006a).

Fc b_2 preparation isolated from the methylotrophic yeast *O. polymorpha* 356 has been chosen as a biorecognition element in biosensor's construction (Smutok et al. 2005, 2006a, 2011, 2017; Goriushkina et al. 2009) as well as for the development of enzymatic-chemical methods for L-lactate assay (Gonchar et al. 2009; Smutok et al. 2013).

5.5 Bioanalytical Application of the Isolated Oxido-reductases

Enzymes AO, FdDH, GDH, and FC b_2, isolated from the cells *O. polymorpha* C-105 and other yeast producers, including recombinant cells (see Table 5.1), were used as biocatalysts for the development of analytical approaches to determine correspondent analytes (Fig. 5.7).

5.5.1 *AO, FdDH, and FC* $\mathbf{b}_2$

The functional characteristics of the constructed amperometric biosensors for analysis of practically important analytes, based on the purified yeast oxido-reductases, are summaried in Table 5.2. The main advantage of the developed

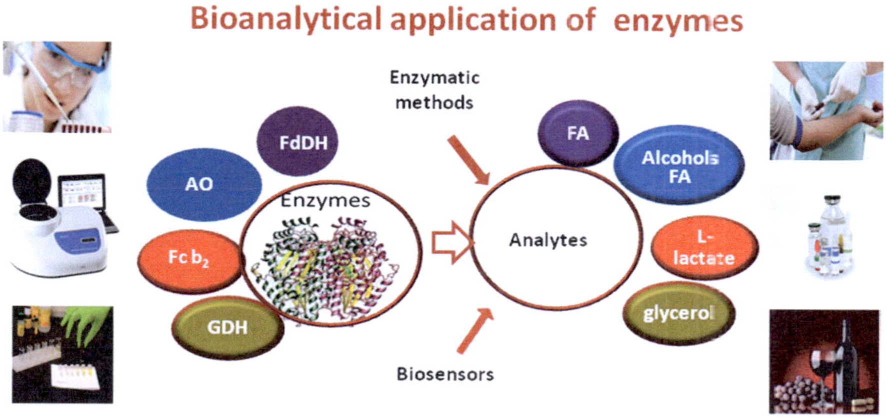

Fig. 5.7 Bioanalytical application of the isolated enzymes

Table 5.2 Analytical characteristics of the developed amperometric biosensors

Analyte	Biomembrane	Linearity range, mM	Response time, s (95%)	Stability, days (50%)	K_M^{app}, mM	Reference
Ethanol	AO/PO	Up to 1.8	45	16	1.94 ± 0.37	Smutok et al. (2006b, 2011)
Ethanol	Mutated AO/[a]PO	Up to 4.0	[b]ND	14	5.4 ± 0.8	Dmytruk et al. (2007)
Ethanol	AO-[c]nAu-enriched/[d]p-cells/PO	Up to 2.0	20	30	1.93 ± 0.08	Karkovska et al. (2017)
Ethanol	AO/PO-like nanozyme	0.01–0.25	15	12	1.99 ± 0.08	Stasyuk et al. (2019)
Methanol	AO/PO-[e]gnPd	Up to 0.8	55	3	0.64 ± 0.02	Gayda et al. (2019)
FA	FdDH/diaphorase	0.05–0.5	ND	1	ND	Nikitina et al. (2007)
FA	NAD/FdDH/glutathione	Up to 20.0	165	3	119 ± 5.54	Demkiv et al. (2008)
FA	AO/PO	Up to 4.0	ND	16	3.1	Sibirny et al. (2011b)
L-Lactate	Fc b_2	Up to 0.5	6	16	0.52 ± 0.02	Smutok et al. (2005)
L-Lactate	nAu-FC b_2 on Au-electrode	0.3–2.0	20	34	0.79 ± 0.03	Smutok et al. (2017)
L-Lactate	Fc b_2 -nAu-enriched/p-cells/PO	0.3–2.7	5	ND	ND	Karkovska et al. (2015, 2017)

Remarks: [a]*PO* horseradish peroxidase, [b]*ND* not determined, [c]*nAu* gold nanoparticles, [d]*p-cells* permeabilized yeast cells, [e]*gnPd* Pd nanoparticles obtained via "green" synthesis

enzyme-based biosensors is a simple procedure of sample preparation: neither pre-treatment of the samples nor their derivatization is required.

Highly stable and sensitive amperometric biosensors on primary alcohols and FA were developed using AO isolated from thermotolerant methylotrophic yeast *O. polymorpha* as biorecognition elements (Smutok et al. 2006b, Shkotova et al. 2006; Smutok et al. 2011; Sigawi et al. 2011, 2014). To construct bi-enzyme sensor, immobilization of AO was performed by means of electrodeposition paints; horse-radish peroxidase (PO) and Os-complex modified polymer were used to decrease the working potential (Smutok et al. 2006b). Mono-enzyme AO-based biosensor was also developed by electrochemical deposition of the Resydrol polymer, conjugated with AO (Shkotova et al. 2006). To facilitate electron transfer between the enzyme and the electrode surface, electroactive polymers were used in biosensor's construction. Both biosensors demonstarted a good reproducibility and operational and storage stability, so they were used for ethanol assay in real alcoholic beverages. For optimization of the electrochemical communication between the immobilized enzymes and the electrode surface, a variety of sensor architectures were tested. Bioanalytical properties of the most effective AO-/HRP-based biosensor were investigated (Table 5.2). The best biosensor with architecture *HRP/Os-Ap59//AOX/CP9* was applied for the determination of ethanol in wine samples (Smutok et al. 2006b).

For construction of highly selective biosensors on FA, FdDH being highly selective to FA was used (see Table 5.2). The developed biosensors on FA show high sensitivity and selectivity to FA and good operational and storage stability. The reagentless biosensor on FA with fixation of all sensor components in a bioactive layer on the transducer surface was proposed (Demkiv et al. 2008). This biosensor was designed to prevent any leakage of the low-molecular and free-diffusing cofactors of FdDH, thus enabling FA determination without addition of the cofactors to the analyte solution. A validity of this biosensor for FA analysis in real samples was approved by testing formalin-containing commercial goods.

A number of amperometric L-lactate-selective biosensors were developed using Fcb_2 and the enzyme-producing yeast cells (Table 5.2). Different immobilization methods and low-molecular free-diffusing redox mediators were tested for optimizing the electrochemical communication between the immobilized enzyme and the electrode surface. The possibility of direct electron transfer from the reduced form of Fc b_2 to carbon electrodes was evaluated. The bioanalytical properties of Fc b_2-based biosensors, such as signal rise time, dynamic range, dependence of the sensor output on the pH value, temperature, and storage stability, were investigated, and the proposed biosensor demonstrated a very fast response and a high selectivity for L-lactate determination (Smutok et al. 2005, 2006a; Goriushkina et al. 2009). The proposed biosensor was successfully tested for L-lactate analysis on the samples of commercial wines.

Combining nanobiotechnology with electrochemical enzyme-based biosensors has become a crucially novel strategy for the development of simple and reliable monitoring systems for food quality and safety. Nanomaterials also endow electrochemical biosensors with device miniaturization and high sensitivity and specificity. They, therefore, have a great potential for on-site food safety assessment (Nikolelis and Nikoleli 2016, Gonchar et al. 2017; Lv et al. 2018).

The improved biosensors were created using a combination of genetic technology and nanotechnology approaches, namely, by overexpression of the corresponding gene in the recombinant yeast cells and by the transfer of enzyme-bound gold nanoparticles (Fc*b*$_2$-nAu and AO-nAu) into the cells (Smutok et al. 2005; Karkovska et al. 2017). The resulted biosensors were shown to possess the high sensitivity and the fast response. A novel mono-enzyme AO-based nanobiosensor on ethanol was constructed with the usage of peroxidase-like PtRu-nanoparticles. This biosensor, being rather stable and very sensitive, was successfully tested on the several real samples of wines (Stasyuk et al. 2019). Recently, we demonstrated the possibility of developing reagentless AO-based amperometric biosensors using nanoparticles of noble metals, synthesized via "green synthesis" in the presence of extracellular metabolites of the yeast *O. polymorpha*. It was shown that AO-based electrode, modified with green-synthesized Pd nanoparticles, although having a lower sensitivity to methanol, reveals a broader linear range of detection and a higher storage stability, compared with unmodified control electrode. Such bioelectrode characteristics are desirable for enzymes, possessing a very high sensitivity for their substrates, because in such cases the tested samples must be very diluted, which is problematic for online analysis of real samples (Gayda et al. 2019).

To achieve the excellent characteristics of enzyme-based sensor, the usage of gold electrode, modified by Fc*b*$_2$-nAu claster, was proposed recently (Smutok et al. 2017). This biosensor was shown to demonstrate a ninefold higher sensitivity to L-lactate and a wider linear range in comparison with the characteristics for free enzyme, immobilized on the same electrode.

Enzymatic-chemical methods for ethanol (Gonchar et al. 2001; Pavlishko et al. 2005), FA (Demkiv et al. 2007; Gayda et al. 2008a; Sibirny et al. 2011a, b; Sigawi et al. 2011), and L-lactate (Gonchar et al. 2009; Smutok et al. 2013) determination were described also earlier. All these methods were successfully tested on the real samples of food products, beverages, as well as biological liquids (Smutok et al. 2011; Pavlishko et al. 2005; Sibirny et al. 2011a, b).

5.5.2 GDH-Based Methods for Glycerol Assay in Wines

The general principles of amperometric detection of glycerol (GlOH) using enzyme-based biosensors were previously reviewed in detail (Goriushkina et al. 2010; Smutok et al. 2011; Synenka et al. 2015; Mahadevan and Fernando 2016). The described methods are based on NADH and mediators-aid registration, on the use of oxygen and hydrogen peroxide electrodes, conductive organic salts, and wiring electrodes. GlOH assay using recombinant yeast GDH was not investigated as thoroughly. We therefore focus on demonstrating the applicability of the proposed enzymatic-chemical method to GlOH assay in real samples: wines.

GlOH is an important by-product of glycolysis and is quantitatively one of the major components of wine (Nieuwoudt et al. 2002). GlOH positively influences the taste of table wines, giving them viscosity, sweetness, and softness. The production of GlOH is closely linked to the availability of the fermentable sugars presented in musts.

We developed an enzymatic-chemical method for GlOH assay based on spectrophotometric detection of solubilized formazan (see Sect. 5.4.2) which is generated in the reaction of nitrotetrazolium blue with NADH, a product of GDH-catalyzed oxidation of GlOH. The validity of the proposed method was tested on commercial wines and compared with a referent method. A standard addition test was used in order to evaluate the negative influence of wine components on enzyme-catalyzed reactions. The results of graphical estimation of the GlOH content in the samples of some commercial wines are presented in Fig. 5.8. It was shown that the estimated values are in good correlation with the data of reference methods, as well as with literature data.

The GlOH content, usually formed by *Saccharomyces cerevisiae* in wine, varies between 1.36 and 11 g/L (15–120 mM) (*http://www.lallemandwine.com/wp-content/uploads/2014/12/Wine-Expert-120321-WE-Glycerol-and-WInemaking.pdf*). Higher GlOH levels are generally considered as improving wine quality. The mean GlOH concentrations in dry red (10.49 g L^{-1}), dry white (6.82 g L^{-1}), and noble late harvest wines (15.55 g L^{-1}) were found to be associated with considerable variation within each respective style (Nieuwoudt et al. 2002). GlOH content is greater in wines from must that was processed with sulfite (Goold et al. 2017; Belda et al. 2017; Rankine and Bridson 1974; Remize et al. 2003) and also in wines made from grapes

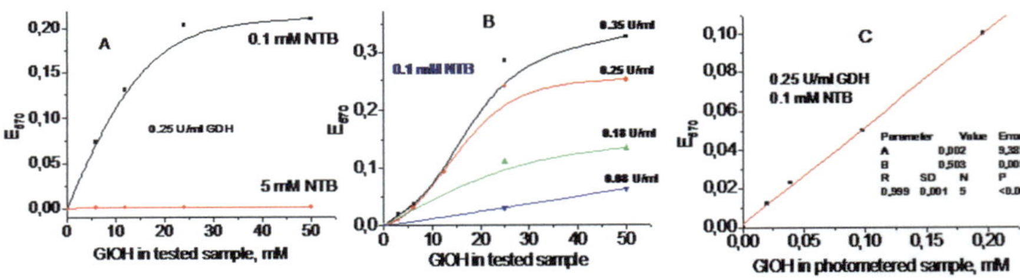

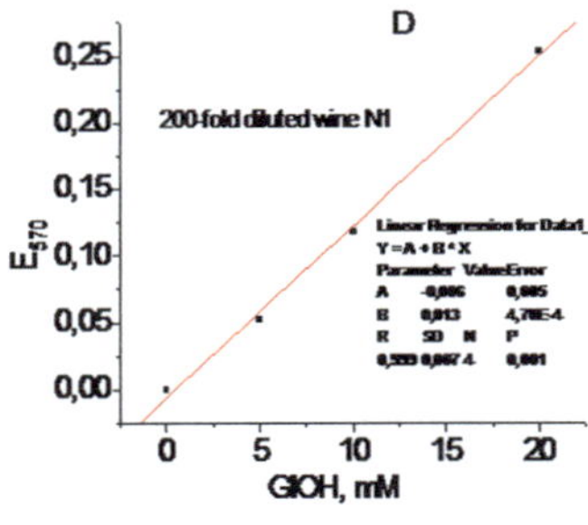

Sample of wine	Glycerol concentration, mM	
	Reference method	GDH-based method
1	8.6±0.73	8.8±0.81
2	7.7±0.29	7.9±0.73
3	5.2±0.04	5.5±0.37
4	5.8±0.26	6.4±0.27

Fig. 5.8 Glycerol dehydrogenase as biocatalyst enzymatic-chemical method for glycerol assay

affected by *Botrytis cinerea*, the “noble mold” – up to 30 g L^{-1} or 330 mM (Nieuwoudt et al. 2002). Thus, comparison of the measured and expected contents of GlOH enables confirmation or challenging the originality of tested wines.

5.6 Conclusions

Enzymes possess high selectivity and sensitivity and are thus widely used in analytical test-systems for control of the environment and food, as well as for clinical diagnostics. Obtaining a wide range of target enzymes on an industrial scale from different sources, including recombinant microorganisms, is an urgent problem of biotechnology and enzymology. This chapter presents the main achievements in the elaboration of modern techniques for recombinant enzymes isolation from selected or recombinant yeasts. The results of a series of the authors’ investigations of these problems are summarized.

Some steps are necessary for isolating a highly purified, stable, and active yeast enzyme: selection or construction of the effective yeast producer, optimization of its cultivation conditions for achievement of the highest specific activity of enzyme in a cell-free extract, and development of an effective technology for target enzyme purification. The scheme for simultaneous isolation of several thermostable yeast enzymes from cells of the thermotolerant methylotrophic yeast *O. polymorpha* followed by their chromatographic purification using ion-exchange sorbent was proposed. The possibility of obtaining alcohol oxidase (AO), flavocytochrome b_2 (Fc b_2), glycerol dehydrogenase (GDH), methylamine oxidase (AMO), arginase, formaldehyde dehydrogenase (FdDH), formate dehydrogenase (FDH), and formaldehyde reductase (FR) from the cell-free extract of the same yeast source was demonstrated.

The highly purified yeast oxido-reductases (AO, FdDH, GDH and Fc b_2) were isolated from *O. polymorpha* cells overproducing alcohol oxidase, as well as from other special yeast producers, including recombinant cells. The target enzymes were characterized and used as biocatalysts for the development of analytical methods for assaying primary alcohols, formaldehyde, glycerol, and L-lactate, respectively.

References

Ali MB, Gonchar M, Gayda G, Paryzhak S, Maaref MA, Jaffrezic-Renault N, Korpan Y (2007) Formaldehyde-sensitive sensor based on recombinant formaldehyde dehydrogenase using capacitance versus voltage measurements. Biosens Bioelectron 22(12):2790–2795

Allais JJ, Louktibi A, Baratti J (1983) Oxidation of methanol by the yeast *Pichia pastoris*. Purification and properties of the formaldehyde dehydrogenase. Agric Biol Chem 47(7):1509–1516

Arias CAD, Marques DAV, Malpiedi LP, Maranhão AQ, Parra DAS, Converti A, Junior AP (2017) Cultivation of *Pichia pastoris* carrying the scFv anti LDL (-) antibody fragment. Effect of pre-culture carbon source. Braz J Microbiol 48(3):419–426

Avalos J, Nordzieke S, Parra O, Pardo-Medina J, Carmen Limon M (2017) Carotenoid Production by Filamentous Fungi and Yeasts. In: Sibirny A. (eds) Biotechnology of Yeasts and Filamentous Fungi. Springer, Cham pp 225–279

Baerends RJ, Sulter GJ, Jeffries TW (2002) Molecular characterization of the *Hansenula polymorpha FLD1* gene encoding formaldehyde dehydrogenase. Yeast 19:37–42

Baghban R, Farajnia S, Ghasemi Y, Mortazavi M, Zarghami N, Samadi N (2018) New developments in Pichia pastoris expression system, review and update. Curr Pharm Biotechnol 19(6):451–467

Bao J, Huang M, Petranovic D, Nielsen J (2018) Balanced trafficking between the ER and the Golgi apparatus increases protein secretion in yeast. AMB Express 8:37. https://doi.org/10.1186/s13568-018-0571-x

Barrero JJ, Casler JC, Valero F, Ferrer P, Glick BS (2018) An improved secretion signal enhances the secretion of model proteins from *Pichia pastoris*. Microb Cell Factories 17:161. https://doi.org/10.1186/s12934-018-1009-5

Bawa Z, Routledge SJ, Jamshad M, Clare M, Sarkar D, Dickerson I, Ganzlin M, Poyner DR, Bill RM (2014) Functional recombinant protein is present in the pre-induction phases of *Pichia pastoris* cultures when grown in bioreactors, but not shake-flasks. Microb Cell Factories 13(1):127. https://doi.org/10.1186/s12934-014-0127-y

Belda I, Ruiz J, Beisert B, Navascués E, Marquina D, Calderón F, Rauhut D, Benito S, Santos A (2017) Influence of Torulaspora delbrueckii in varietal thiol (3-SH and 4-MSP) release inwine sequential fermentations. Int J Food Microbiol 257:183–191

Berg L, Strand TA, Valla S, Brautaset T (2013) Combinatorial mutagenesis and selection to understand and improve yeast promoters. Biomed Res Int 2013:Article ID 926985, 9 pages. https://doi.org/10.1155/2013/926985

Białkowska AM (2016) Strategies for efficient and economical 2,3-butanediol production: new trends in this field. World J Microbiol Biotechnol 32(12):200. https://doi.org/10.1007/s11274-016-2161-x

Bollella P, Sharma S, Cass AEG, Antiochia R (2019) Microneedle-based biosensor for minimally-invasive lactate detection. Biosens Bioelectron 123:152–159

Boteva R, Visser AJ, Filippi B, Vriend G, Veenhuis M, van der Klei IJ (1999) Conformational transitions accompanying oligomerization of yeast alcohol oxidase, a peroxisomal flavoenzyme. Biochemistry 38(16):5034–5044

Bredell H, Smith JJ, Prins WA, Görgens JF, van Zyl WH (2016) Expression of rotavirus VP6 protein: a comparison amongst Escherichia coli, Pichia pastoris and Hansenula polymorpha. FEMS Yeast Research 16 (2) fow001, https://doi.org/10.1093/femsyr/fow001

Bringer S, Sprey B, Sahm H (1979) Purification and properties of alcohol oxidase from *Poria contigua*. Eur J Biochem 101(2):563–570

Buckholz RG, Gleeson MA (1991) Yeast systems for the commercial production of heterologous proteins. Biotechnology (NY) 9(11):1067–1072

Chang CH, Hsiung HA, Hong KL, Huang CT (2018) Enhancing the efficiency of the *Pichia pastoris AOX1* promoter via the synthetic positive feedback circuit of transcription factor Mxr1. BMC Biotechnol 18(1):81. https://doi.org/10.1186/s12896-018-0492-4

Chapman J, Ismail AE, Dinu CZ (2018) Industrial applications of enzymes: recent advances, techniques, and outlooks. Catalysts 8(238):1–26

Chen B, Lee HL, Heng YC, Chua N, Teo WS, Choi WJ, Leong SSJ, Foo JL, Chang MW (2018) Synthetic biology toolkits and applications in *Saccharomyces cerevisiae*. Biotechnol Adv 36(7):1870–1881

Cho S, Kim T, Woo HM, Kim Y, Lee J, Um Y (2015) High production of 2,3-butanediol from biodiesel-derived crude glycerol by metabolically engineered *Klebsiella oxytoca* M1. Biotechnol Biofuels 8:146. https://doi.org/10.1186/s13068-015-0336-6

Cregg JM, Barringer KJ, Hessler AY, Madden KR (1985) *Pichia pastoris* as a host system for transformations. Mol Cell Biol 5(12):3376–3385

Cregg JM, Madden KR, Barringer KJ, Thill GP, Stillman CA (1989) Functional characterization of the two alcohol oxidase genes from the yeast *Pichia pastoris*. Mol Cell Biol 9(3):1316–1323

Curvers S, Brixius P, Klauser T, Thömmes J, Weuster-Botz D, Takors R, Wandrey C (2001) Human chymotrypsinogen B production with *Pichia pastoris* by integrated development of fermentation and downstream processing. Part 1. Fermentation. Biotechnol Prog 17(3):495–502

Dagar K, Pundir CS (2018) Dataset on fabrication of an improved L-lactate biosensor based on lactate oxidase/cMWCNT/CuNPs/PANI modified PG electrode. Data Brief 17:1163–1167

Daly R, Hearn MT (2005) Expression of heterologous proteins in *Pichia pastoris*: a useful experimental tool in protein engineering and production: review. J Mol Recognit 18(2):119–138

Demkiv OM, SYa P, Krasovs'ka OS, Stasyk OV, Gayda GZ, Sibirny AA, Gonchar MV (2005) Construction of methylotrophic yeast *Hansenula polymorpha* strains over-producing formaldehyde dehydrogenase. Biopolymers and Cell 21(6):525–530. (in Ukrainian)

Demkiv OM, Paryzhak SY, Gayda GZ, Sibirny VA, Gonchar MV (2007) Formaldehyde dehydrogenase from the recombinant yeast *Hansenula polymorpha*: isolation and bioanalytic application. FEMS Yeast Res 7:1153–1159

Demkiv O, Smutok O, Paryzhak S, Gayda G, Sultanov Y, Guschin D, Shkil H, Schuhmann W, Gonchar M (2008) Reagentless amperometric formaldehyde-selective biosensors based on the recombinant yeast formaldehyde dehydrogenase. Talanta 76(4):837–846

Demkiv OM, Paryzhak SY, Ishchuk OP, Gayda GZ, Gonchar MV (2011) Activities of the enzymes of formaldehyde catabolism in recombinant strains of *Hansenula polymorpha*. Mirobiology 80(3):307–313

Dijkman WP, de Gonzalo G, Mattevi A, Fraaije MW (2013) Flavoprotein oxidases: classification and applications: review. Appl Microbiol Biotechnol 97(12):5177–5188

Dmytruk KV, Smutok OV, Ryabova OB, Gayda GZ, Sibirny VA, Schuhmann W, Gonchar MV, Sibirny AA (2007) Isolation and characterization of mutated alcohol oxidases from the yeast *Hansenula polymorpha* with decreased affinity toward substrates and their use as selective elements of an amperometric biosensor. BMC Biotechnol 7(7):1–7

Dmytruk KV, Kurylenko OO, Ruchala J, Abbas CA, Sibirny AA (2017) Genetic Improvement of Conventional and Nonconventional Yeasts for the Production of First- and Second-Generation Ethanol. In: Sibirny A. (eds) Biotechnology of Yeasts and Filamentous Fungi. Springer, Cham pp 1–38

Domínguez A, Fermiñán E, Sánchez M, González FJ, Pérez-Campo FM, García S, Herrero AB, San Vicente A, Cabello J, Prado M, Iglesias FJ, Choupina A, Burguillo FJ, Fernández-Lago L, López MC (1998) Non-conventional yeasts as hosts for heterologous protein production. Int Microbiol 1(2):131–142

Eggeling L, Sahm H (1980) Direct enzymatic assay for Alcohol Oxidase, Alcohol Dehydrogenase, and Formaldehyde Dehydrogenase in colonies of *Hansenula polymorpha*. Appl Environ Microbiol 39(1):268–269

Ekas H, Deaner M, Alper HS (2019) Recent advancements in fungal-derived fuel and chemical production and commercialization. Curr Opin Chem Biol 57:1–9

Engleder M, Horvat M, Emmerstorfer-Augustin A, Wriessnegger T, Gabriel S, Strohmeier G, Weber H, Müller M, Kaluzna I, Mink D, Schürmann M, Pichler H (2018) Recombinant expression, purification and biochemical characterization of kievitone hydratase from *Nectria haematococca*. PLoS One 13(2):e0192653. https://doi.org/10.1371/journal.pone.0192653

Fernandes FJ, López-Estepa M, Querol-García J, Vega MC (2016) Production of protein complexes in non-methylotrophic and methylotrophic Yeasts: nonmethylotrophic and methylotrophic Yeasts. Adv Exp Med Biol 896:137–153

Fletcher E, Krivoruchko A, Nielsen J (2016) Industrial systems biology and its impact on synthetic biology of yeast cell factories. Biotechnol Bioeng 113(6):1164–1170

Gadda G (2008) Hydride transfer made easy in the reaction of alcohol oxidation catalyzed by flavin-dependent oxidases: review. Biochemist 47(52):13745–13753

Gaida GZ, Stel'mashchuk SY, Smutok OV, Gonchar MV (2003) A new method of visualization of the enzymatic activity of flavocytochrome b_2 in electrophoretograms. Appl Biochem Microbiol 39(2):221–223

Gartner G, Kopperschlager G (1984) Purification and Properties of Glycerol Dehydrogenase from Candida valida. Microbiology 130: 3225–3233

Gayda G, Demkiv O, Gonchar M, Paryzhak S, Schuhmann W (2008a) Recombinant formaldehyde dehydrogenase and gene-engineered methylotrophic yeasts as bioanalitycal instruments for assay of toxic formaldehyde. In: Evangelista V et al (eds) Algal toxins: nature, occurrence, effect and detection, NATO science for peace and security series A: Chemistry and biology. Springer, Dordrecht, pp 311–333

Gayda G, Paryzhak S, Demkiv O, Ksheminska H, Gonchar M (2008b) Formaldehyde reductase from formaldehyde-resistant gene-engineered methylotrophic yeast *Hansenula polymorpha:* isolation and characterization: 12th International Congress on Yeast, Kyiv, Ukraine. August 11–15, p 304

Gayda G, Pavlishko H, Stasyuk N, Ivash M, Bilek M, Broda D, Gonchar M (2013a) Application of recombinant glycerol dehydrogenase for glycerol assay in wines. 5 th Polish-Ukrainian Weigl conference on Microbiology. Chernivtsi, Ukraine May 23–25, p 72

Gayda G, Stasyuk N, Demkiv O, Klepach H, Broda D, Sibirny A, Gonchar M (2013b) New enzymatic methods for wine assay. International conference Biocatalysis-2013: fundamentals and applications. Moscow, Russia, July 2–5, pp 71–72

Gayda G, Demkiv O, Klepach H, Gonchar M, Levy-Halaf R, Wolf D, Nisnevitch M (2015) Formaldehyde: detection and biodegradation (Chapter 6). In: Patton A (ed) Formaldehyde: synthesis, applications and potential health effects. Nova Science Publishers, Inc., New York, pp 117–142. ISBN: 978-1-63482-412-5

Gayda GZ, Demkiv OM, Stasyuk NYe, Serkiz RYa, Lootsik MD, Errachid A, Gonchar MV, Nisnevitch M (2019) Metallic nanoparticles obtained via "green" synthesis as a platform for biosensor construction. Appl Sci 9:720. https://doi.org/10.3390/app9040720

Gellissen G, Kunze G, Gaillardin C, Cregg JM, Berardi E, Veenhuis M, van der Klei I (2005) New yeast expression platforms based on methylotrophic *Hansenula polymorpha* and *Pichia pastoris* and on dimorphic *Arxula adeninivorans* and *Yarrowia lipolytica* – a comparison. EMS Yeast Res 5(11):1079–1096

Gerpen JV (2005) Biodiesel processing and production. Fuel Proces Technol 86 (10): 1097–1107

Gonchar MV, Ksheminska GP, Hladarevska NM, Sibirny AA (1990) Catalase-minus mutants of methylotrophic yeast Hansenula polymorpha impaired in regulation of alcohol oxidase synthesis. In: Lachowicz TM (Ed) Genetics of Respiratory Enzymes in Yeasts. Wroclaw University Press, Wroclaw, Poland, pp. 222–228

Gonchar MV, Maidan MM, Pavlishko HM, Sibirny AA (2001) A New Oxidase-Peroxidase Kit for Ethanol Assays in Alcoholic Beverages. Food technol. Biotechnol 39(1):37–42

Gonchar M, Maidan M, Korpan Y, Sibirny V, Kotylak Z, Sibirny A (2002) Metabolically engineered methylotrophic yeast cells and enzymes as sensor biorecognition elements. FEMS Yeast Res 2:307–314

Gonchar M, Smutok O, Os'mak H (2009) Flavocytochrome b_2-based enzymatic composition, method and kit for L-lactate. Patent Application PCT/US2008/069637 Publ.WO/2009/009656: http://www.wipo.int/pctdb/en/wo.jsp?WO=2009009656

Gonchar M, Smutok O, Karkovska M, Stasyuk N, Gayda G (2017) Yeast-based biosensors for clinical diagnostics and food control. In: "Biotechnology of Yeasts and Filamentous Fungi" (Ed. A.A. Sibirny). Springer, P. 392–400

Goold HD, Kroukamp H, Williams TC, Paulsen IT, Varela C, Pretorius IS (2017) Yeast's balancing act between ethanol and glycerol production in low-alcohol wines. Microb Biotechnol 10(2):264–278

Goriushkina TB, Orlova AP, Smutok OV, Gonchar MV, Soldatkin AP, Dzyadevych SV (2009) Application of L-lactate-cytochrome c-oxidoreductase for development of amperometric biosensor for L -lactate determination. Biopolymers and Cell 25(3):194–202

Goriushkina TB, Shkotova LV, Gayda GZ, Klepach HM, Gonchar MV, Soldatkin AP, Dziyadevych SV (2010) Amperometric biosensor based on glycerol oxidase for glycerol determination. Sensor Actuat B Chem 144:361–367

Goswami P, Chinnadayyala SS, Chakraborty M, Kumar AK, Kakoti A (2013) An overview on alcohol oxidases and their potential applications: review. Appl Microbiol Biotechnol 97(10):4259–4275
Grewal N, Parveen Z, Large A, Perry C, Connock M (2000) Gastropod mollusc aliphatic alcohol oxidase: subcellular localisation and properties. Comp Biochem Physiol B: Biochem Mol Biol 125(4):543–554
Griesemer M, Young C, Robinson AS, Petzold L (2014) BiP clustering facilitates protein folding in the endoplasmic reticulum. PLoS Comput Biol 10(7):e1003675. https://doi.org/10.1371/journal.pcbi.1003675
Gunkel K, van Dijk R, Veenhuis M, van der Klei IJ (2004) Routing of *Hansenula polymorpha* alcohol oxidase: an alternative peroxisomal protein-sorting machinery. Mol Biol Cell 15(3):1347–1355
Gvozdev AR, Tukhvatullin IA, Gvozdev RI (2012) Quinone-dependent alcohol dehydrogenases and FAD-dependent alcohol oxidases: review. Biochem Mosc 77(8):843–856
Hartner FS, Glieder A (2006) Regulation of methanol utilisation pathway genes in yeasts. Microb Cell Factories 5(39):1–21
Hemmerich J, Adelantado N, Barrigón JM, Ponte X, Hörmann A, Ferrer P, Kensy F, Valero F (2014) Comprehensive clone screening and evaluation of fed-batch strategies in a microbioreactor and lab scale stirred tank bioreactor system: application on *Pichia pastoris* producing *Rhizopus oryzae* lipase. Microb Cell Factories 13(1):36. https://doi.org/10.1186/1475-2859-13-36
Hernández-Ortega A, Ferreira P, Martínez AT (2012) Fungal aryl-alcohol oxidase: a peroxide-producing flavoenzyme involved in lignin degradation: review. Appl Microbiol Biotechnol 93(4):1395–1410
Feed enzymes market. http://www.marketsandmarkets.com/market-reports/feed-enzyme-market-1157.html. Accessed on 19 Feb 2019
Industrial Enzymes Market. http://www.marketsandmarkets.com/Market-Reports/industrial-enzymes-market-237327836.html. Accessed on 19 Feb 2019
Pharmaion. India Industrial Enzymes Market Forecast and Opportunities, 2020 http://www.pharmaion.com/report/india-industrial-enzymes-market-forecast-and-opportunities-2020/10.html. Accessed on 19 Feb 2019
Sigma products. https://www.sigmaaldrich.com/catalog/product/sigma
Huang M, Wang G, Qin J, Petranovic D, Nielsen J (2018a) Engineering the protein secretory pathway of *Saccharomyces cerevisiae* enables improved protein production. Proc Natl Acad Sci U S A 115(47):E11025–E11032
Huang M, Joensson HN, Nielsen J (2018b) High-throughput microfluidics for the screening of yeast libraries. Methods Mol Biol 1671:307–317
Idiris A, Tohda H, Kumagai H, Takegawa K (2010) Engineering of protein secretion in yeast: strategies and impact on protein production. Appl Microbiol Biotechnol 86(2):403–417
Isobe K, Takahashi T, Ogawa J, Kataoka M, Shimizu S (2009) Production and characterization of alcohol oxidase from *Penicillium purpurescens* AIU 063. J Biosci Bioeng 107(2):108–112
Janssen FW, Ruelius HW (1968) Alcohol oxidase, a flavoprotein from several *Basidiomycetes species*. Crystallization by fractional precipitation with polyethylene glycol. Biochim Biophys Acta 151(2):330–342
Jullessen D, David F, Pfleger B, Nielsen J (2015) Impact of synthetic biology and metabolic engineering on industrial production of fine chemicals. Biotechnol Adv 33(7):1395–1402
Juturu V, Wu JC (2018) Heterologous protein expression in *Pichia pastoris*: latest research progress and applications. Chembiochem 19(1):7–21
Karkovska M, Smutok O, Stasyuk N, Gonchar M (2015) L-lactate-selective microbial sensor based on flavocytochrome b_2-enriched yeast cells using recombinant and nanotechnology approaches. Talanta 144:1195–1200
Karkovska MI, Stasyuk NYe, Gayda GZ, Smutok OV, Gonchar MV (2017) Nanomaterials in construction of biosensors of biomedical purposes. In: Stoika R (Ed) Multifunctional nanomaterials for biology and medicine: molecular design, synthesis, and application. pp. 165–177 (in Ukrainian)

Kato N (1990) Formaldehyde dehydrogenase from methylotrophic yeasts. Methods Enzymol 188:455–459

Kim H, Yoo SJ, Kang HA (2015) Yeast synthetic biology for the production of recombinant therapeutic proteins. FEMS Yeast Res 15(1):1–16

Kim JW, Kim J, Seo SO, Kim KH, Jin YS, Seo JH (2016) Enhanced production of 2,3-butanediol by engineered *Saccharomyces cerevisiae* through fine-tuning of pyruvate decarboxylase and NADH oxidase activities. Biotechno Biofuels 9:265. https://doi.org/10.1186/s13068-016-0677-9

Koch C, Neumann P, Valerius O, Feussner I, Ficner R (2016) Crystal structure of alcohol oxidase from *Pichia pastoris*. PLoS One 11(2):e0149846. https://doi.org/10.1371/journal.pone.0149846. eCollection

Kondo T, Morikawa Y, Hayashi N (2008) Purification and characterization of alcohol oxidase from *Paecilomyces variotii* isolated as a formaldehyde-resistant fungus. Appl Microbiol Biotechnol 77(5):995–1002

Krasovska OS, Babiak LIa Nazarko TY, Stasyk OG, Danysh TV, Gayda GZ, Stasyk OV, Gonchar MV, Sybirny AA (2006) Construction of yeast *Hansenula polymorpha* overproducing amine oxidase as bioselective element of sensors for biogenic amines. In: El'skaya AV, Pokhodenko VD (eds) Investigations on sensor systems and technologies, vol 1. IMBG of NAS of Ukraine, Kyiv, pp 141–148

Krasovska OS, Stasyk OG, Nahorny VO, Stasyk OV, Granovski N, Kordium VA, Vozianov OF, Sibirny AA (2007) Glucose-induced production of recombinant proteins in *Hansenula polymorpha* mutants deficient in catabolite repression. Biotechnol Bioeng 97(4):858–870

Krivoruchko A, Siewers V, Nielsen J (2011) Opportunities for yeast metabolic engineering: lessons from synthetic biology. Biotechnol J 6(3):262–276

Kumar V, Park S (2018) Potential and limitations of Klebsiella pneumoniae as a microbial cell factory utilizing glycerol as the carbon source. Biotechnol Adv 36(1):150–167

Leferink NG, Heuts DP, Fraaije MW, van Berkel WJ (2008) The growing VAO flavoprotein family: review. Arch Biochem Biophys 474(2):292–301

Li M, Li M, Wu J, Liu X, Lin J, Wei D, Chen H (2010). Enhanced production of dihydroxyacetone from glycerol by overexpression of glycerol dehydrogenase in an alcohol dehydrogenase-deficient mutant of Gluconobacter oxydans. Bioresour Technol 101 (21): 8294-8299

Liu W-C, Zhu P (2018) Demonstration-scale high-cell-density fermentation of *Pichia pastoris*. Recombinant Glycoprotein Production – 2018 Methods and Protocols Part of the Methods in Molecular Biology book series. MIMB 1674:109–116

Liu C, Yang X, Yao Y, Huang W, Sun W, Ma Y (2014) Diverse expression levels of two codon-optimized genes that encode human papilloma virus type 16 major protein L1 in *Hansenula polymorpha*. Biotechnol Lett 36(5):937–945

Liu J, Wu S, Li Z (2018) Recent advances in enzymatic oxidation of alcohols: review. Curr Opin Chem Biol 43:77–86

Löbs AK, Schwartz C, Wheeldon I (2017) Genome and metabolic engineering in non-conventional yeasts: current advances and applications. Synth Syst Biotechnol 2(3):198–207

Love KR, Dalvie NC, Love JC (2018) The yeast stands alone: the future of protein biologic production. Curr Opin Biotechnol 53:50–58

Lusta KA, Leonovitch OA, Tolstorukov II, Rabinovich YM (2000) Constitutive biosynthesis and localization of alcohol oxidase in the ethanol-insensitive catabolite repression mutant ecr1 of the yeast *Pichia methanolica*. Biochem Mosc 65(5):604–608

Lv M, Liu Y, Geng J, Kou X, Xin Z, Yang D (2018) Engineering nanomaterials-based biosensors for food safety detection. Biosens Bioelectron 106:122–128

Mack M, Wannemacher M, Hobl B, Pietschmann P, Hock B (2009) Comparison of two expression platforms in respect to protein yield and quality: *Pichia pastoris* versus *Pichia angusta*. Protein Expr Purif 66(2):165–171

Mahadevan A, Fernando S (2016) An improved glycerol biosensor with an Au-FeS-NAD-glycerol-dehydrogenase anode. Biosens Bioelectron 92:417–424

Mallinder P, Pritchard A, Moir A (1992) Cloning and characterization of a gene from Bacillus stearothermophilus var. non-diastaticus encoding a glycerol dehydrogenase. Gene 110 (1): 9–16
Mangkorn N, Kanokratana P, Roongsawang N, Laobuthee A, Laosiripojana N, Champreda V (2019) Synthesis and characterization of *Ogataea thermomethanolica* alcohol oxidase immobilized on barium ferrite magnetic microparticles. J Biosci Bioeng 127(3):265–272
Mattanovich D, Sauer M, Gasser B (2014) Yeast biotechnology: teaching the old dog new tricks. Microb Cell Factories 13(1):34. https://doi.org/10.1186/1475-2859-13-34
Mincey T, Tayrien G, Mildvan AS, Abeles RH (1980) Presence of a flavin semiquinone in methanol oxidase. Proc Natl Acad Sci U S A 77(12):7099–7101
Müller S, Sandal T, Kamp-Hansen P, Dalbøge H (1998) Comparison of expression systems in the yeasts *Saccharomyces cerevisiae, Hansenula polymorpha, Klyveromyces lactis, Schizosaccharomyces pombe* and *Yarrowia lipolytica.* Cloning of two novel promoters from *Yarrowia lipolytica*. Yeast 14(14):1267–1283
Nakamura Y, Nishi T, Noguchi R, Ito Y, Watanabe T, Nishiyama T, Aikawa S, Hasunuma T, Ishii J, Okubo Y, Kondo A (2018) A stable, autonomously replicating plasmid vector containing *Pichia pastoris Centromeric* DNA. Appl Environ Microbiol 84(15):e02882–e02817. https://doi.org/10.1128/AEM.02882-17
Nguyen H, Nevoigt E (2009) Engineering of Saccharomyces cerevisiae for the production of dihydroxyacetone (DHA) from sugars: a proof of concept. Metab Eng 11: 335–346
Nguyen QT, Romero E, Dijkman WP, de Vasconcellos SP, Binda C, Mattevi A, Fraaije MW (2018) Structure-based engineering of *Phanerochaete chrysosporium* alcohol oxidase for enhanced oxidative power toward glycerol. Biochemistry 57(43):6209–6218
Nieuwoudt HH, Prior BA, Pretorius S, Bauer FF (2002) Glycerol in South African table wines: an assessment of its relationship to wine quality. Afr J Enol Vitic 23(1):22–30
Nikitina O, Shleev S, Gayda G, Demkiv O, Gonchar M, Gorton L, Csöregi E, Nistor M (2007) Bi-enzyme biosensor based on NAD^+- and glutathione-dependent recombinant formaldehyde dehydrogenase and diaphorase for formaldehyde assay. Sensors Actuators B 125:1–9
Nikolelis DP, Nikoleli GP (2016) Biosensors for security and bioterrorism applications. Springer Int Publi, Switzerland ISBN 978-3-319-28926-7
Nora LC, Westmann CA, Martins-Santana L, Alves LF, Monteiro LMO, Guazzaroni ME, Silva-Rocha R (2019) The art of vector engineering: towards the construction of next-generation genetic tools. Microb Biotechnol 12(1):125–147
Oh BR, Lee SM, Heo SY, Seo JW, Kim CH (2018) Efficient production of 1,3-propanediol from crude glycerol by repeated fed-batch fermentation strategy of a lactate and 2,3-butanediol deficient mutant of *Klebsiella pneumoniae*. Microb Cell Factories 17(1):92. https://doi.org/10.1186/s12934-018-0921-z
Owczarek B, Gerszberg A, Hnatuszko-Konka K (2019) A Brief Reminder of Systems of Production and Chromatography-Based Recovery of Recombinant Protein Biopharmaceuticals. Biomed Res Int 2019:4216060. doi: 10.1155/2019/4216060. eCollection 2019
Ozimek P, Veenhuis M, van der Klei IJ (2005) Alcohol oxidase: a complex peroxisomal, oligomeric flavoprotein. FEMS Yeast Res 5(11):975–983
Panadare D, Rathod KV (2018) Extraction and purification of polyphenol oxidase: a review. Biocatal Agric Biotechnol 14:431–437
Passoth V (2017) Lipids of Yeasts and Filamentous Fungi and Their Importance for Biotechnology. In: Sibirny A. (eds) Biotechnology of Yeasts and Filamentous Fungi. Springer, Cham pp 149–204
Patel RN, Hou CN, Derelanko P (1983) Microbial oxidation of methanol: purification and properties of formaldehyde dehydrogenase from a *Pichia sp.* NRRL-Y-11328. Arch Biochem Biophys 221(1):135–142
Pavlishko HM, Ryabinina OV, Zhilyakova TA, Sakharov IYu, Gerzhikova VG, Gonchar MV (2005) Oxidase-peroxidase method of ethanol assay in fermented musts and wine products. Appl Biochem Microbiol 41 (6): 604–609
Peña DA, Gasser B, Zanghellini J, Steiger MG, Mattanovich D (2018) Metabolic engineering of *Pichia pastoris*. Metab Eng 50:2–15

Pickl M, Fuchs M, Glueck SM, Faber K (2015) The substrate tolerance of alcohol oxidases: review. Appl Microbiol Biotechnol 99(16):6617–6642
Pobre KFR, Poet GJ, Hendershot LM (2018) The endoplasmic reticulum (ER) chaperone BiP is a master regulator of ER functions: getting by with a little help from ERdj friends. J Biol Chem 294(6):2098–2108
Porro D, Branduardi P (2017) Production of Organic Acids by Yeasts and Filamentous Fungi. In: Sibirny A. (eds) Biotechnology of Yeasts and Filamentous Fungi. Springer, Cham pp 205–223
Portela RMC, Vogl T, Ebner K, Oliveira R, Glieder A (2018) *Pichia pastoris* alcohol oxidase 1 (*AOX1*) core promoter engineering by high resolution systematic mutagenesis. Biotechnol J 13(3):e1700340. https://doi.org/10.1002/biot.201700340
Rahman MS, Xu CC, Ma K, Nanda M, Qin W (2017) High production of 2,3-butanediol by a mutant strain of the newly isolated *Klebsiella pneumoniae* SRP2 with increased tolerance towards glycerol. Int J Biol Sci 13(3):308–318
Rajamanickam V, Metzger K, Schmid C, Spadiut O (2017) A novel bi-directional promoter system allows tunable recombinant protein production in *Pichia pastoris*. Microb Cell Factories 16:152. https://doi.org/10.1186/s12934-017-0768-8
Rankine BC, Bridson DA (1974) Glycerol in australian wines and factors influencing its formation. http://www.ajevonline.org/content/ajev/22/1/6.full.pdf
Rassaei L, Olthuis W, Tsujimura S, Sudhölter EJ, van den Berg A (2014) Lactate biosensors: current status and outlook. Anal Bioanal Chem 406(1):123–137
Rathee K, Dhull V, Dhull R, Singh S (2016) Biosensors based on electrochemical lactate detection: a comprehensive review. Biochem Biophys Rep 11(5):35–54
Rebello S, Abraham A, Madhavan A, Sindhu R, Binod P, Karthika Bahuleyan A, Aneesh EM, Pandey A (2018) Non-conventional yeast cell factories for sustainable bioprocesses. FEMS Microbiol Lett 365(21). https://doi.org/10.1093/femsle/fny222
Reiser J, Glumoff V, Kälin M, Ochsner U (1990) Transfer and expression of heterologous genes in yeasts other than *Saccharomyces cerevisiae*. Adv Biochem Eng Biotechnol 43:75–102
Remize F, Cambon B, Barnavon L, Dequin S (2003) Glycerol formation during wine fermentation is mainly linked to Gpd1p and is only partially controlled by the HOG pathway. Yeast 20: 1243–1253
Reyes De Corcuera JI, Powers JR (2017) Application of Enzymes in Food Analysis. In: Food Analysis, Nielsen, S.S. Editor, Springer, pp.469–486
Romero E, Gadda G (2014) Alcohol oxidation by flavoenzymes: review. Biomol Concepts 5(4):299–318
Rosati G, Gherardi G, Grigoletto D, Marcolin G, Cancellara P, Mammucari C, Scaramuzza M, Toni AD, Reggiani C, Rizzuto R, Paccagnella A (2018) Lactate Dehydrogenase and Glutamate Pyruvate Transaminase biosensing strategies for lactate detection on screen-printed sensors. Catalysis efficiency and interference analysis in complex matrices: from cell cultures to sport medicine. Sensing and Bio-Sensing Research 21: 54–64
Rueda F, Gasser B, Sánchez-Chardi A, Roldán M, Villegas S, Puxbaum V, Ferrer-Miralles N, Unzueta U, Vázquez E, Garcia-Fruitós E, Mattanovich D, Villaverde A (2016) Functional inclusion bodies produced in the yeast Pichia pastoris Microbial Cell Factories 15:166 https://doi.org/10.1186/s12934-016-0565-9
Ruzheinikov SN, Burke J, Sedelnikova S, Baker PJ, Taylor R, Bullough PA, Muir NM, Gore MG, Rice DW (2001) Glycerol Dehydrogenase. Structure 9 (9): 789–802
Sagiroglu A, Altay V (2006) Bioconversion of methanol to formaldehyde. II. By purified methanol oxidase from modified yeast, *Hansenula polymorpha*. Prep Biochem Biotechnol 36(4):321–332
Sahm H, Wagner F (1973) Microbial assimilation of methanol. The ethanol- and methanol-oxidizing enzymes of the yeast *Candida boidinii*. Eur J Biochem 36(1):250–256
Semkiv M, Dmytruk K, Abbas C (2017) Biotechnology of Glycerol Production and Conversion in Yeasts. In: Sibirny A. (eds) Biotechnology of Yeasts and Filamentous Fungi. Springer, Cham 8 pp117–148

Shang T, Si D, Zhang D, Liu X, Zhao L, Hu C, Fu Y, Zhang R (2017) Enhancement of thermoalkaliphilic xylanase production by *Pichia pastoris* through novel fed-batch strategy in high cell-density fermentation. BMC Biotechnol 17(1):55. https://doi.org/10.1186/s12896-017-0361-6

Sharma S, Saeed A, Johnson C, Gadegaard N, Cass AE (2017) Rapid, low cost prototyping of transdermal devices for personal healthcare monitoring. Sens Bio-Sens Res 13:104–108

Shen W, Xue Y, Liu Y, Kong C, Wang X, Huang M, Cai M, Zhou X, Zhang Y, Zhou M (2016) A novel methanol-free *Pichia pastoris* system for recombinant protein expression. Microb Cell Factories 15(1):178. https://doi.org/10.1186/s12934-016-0578-4

Shkotova LV, Soldatkin AP, Gonchar MV, Schuhmann W, Dzyadevych SV (2006) Amperometric biosensor for ethanol detection based on alcohol oxidase immobilised within electrochemically deposited Resydrol film. Mater Sci Eng C 26:411–414

Shleev SV, Shumakovich GP, Nikitina OV, Morozova OV, Pavlishko HM, Gayda GZ, Gonchar MV (2006) Purification and characterization of alcohol oxidase from a genetically constructed over-producing strain of the methylotrophic yeast *Hansenula polymorpha*. Biochem Mosc 71(3):245–250

Sibirny AA (Ed) (2017) Book: Biotechnology of yeasts and filamentous fungi. Springer, ISBN 978-3-319-58829-2

Sibirny AA, Titorenko VI, Gonchar MV, Ubiyvovk VM, Ksheminskaya GP, Vitvitskaya OP (1988) Genetic control of methanol utilization in yeasts. J Basic Microbiol 28(5):293–319

Sibirny V, Demkiv O, Klepach H, Honchar T, Gonchar M (2011a) Alcohol oxidase- and formaldehyde dehydrogenase-based enzymatic methods for formaldehyde assay in fish food products. Food Chem 127:774–779

Sibirny V, Demkiv O, Sigawi S, Paryzhak S, Klepach H, Korpan Y, Smutok O, Nisnevich M, Gayda G, Nitzan Y, Puchalski C, Gonchar M (2011b) Formaldehyde oxidizing enzymes and genetically modified yeast *Hansenula polymorpha* cells in monitoring and removal of formaldehyde. In: Einschlag FSG (Ed) Waste Water – Evaluation and Management. ISBN 978-953-307-233-3. InTech (Croatia) (6):115-154

Sigawi S, Smutok O, Demkiv O, Zakalska O, Gayda G, Nitzan Y, Nisnevitch M, Gonchar M (2011) Immobilized formaldehyde-metabolizing enzymes from *Hansenula polymorpha* for removal and control of airborne formaldehyde. J Biotechnol 153:138–144

Sigawi S, Smutok O, Demkiv O, Gayda G, Vus B, Nitzan Y, Gonchar M, Nisnevitch M (2014) Detection of waterborne and airborne formaldehyde: from amperometric chemosensing to a visual biosensor based on alcohol oxidase. Materials 7: 1055–1068

Singh R, Kumar M, Mitta A, Mehta PK (2016) Microbial enzymes: industrial progress in 21st century: review. 3 Biotech 6:174. https://doi.org/10.1007/s13205-016-0485-8

Smith JJ, Burke A, Bredell H, van Zyl WH, Görgens JF (2012) Comparing cytosolic expression to peroxisomal targeting of the chimeric L1/L2 (ChiΔH-L2) gene from human papillomavirus type 16 in the methylotrophic yeasts *Pichia pastoris* and *Hansenula polymorpha*. Yeast 29(9):385–393

Smutok O, Gayda G, Gonchar M, Schuhmann W (2005) A novel L-lactate-selective biosensor based on the use of flavocytochrome b_2 from methylotrophic yeast *Hansenula polymorpha*. Biosens Bioelectron 20:1285–1290

Smutok O, Gayda G, Shuhmann W, Gonchar M (2006a) Development of L-lactate-selective biosensors based on thermostable yeast L-lactate: cytochrome c-oxidoreductase. In: El'skaya AV, Pokhodenko VD (eds) Investigations on sensor systems and technologies. IMBG of NAS of Ukraine, Kyiv, pp 39–45

Smutok O, Ngounou B, Pavlishko H, Gayda G, Gonchar M, Schuhmann W (2006b) A reagentless bienzyme amperometric biosensor based on alcohol oxidase/peroxidase and an Os-complex modified electrodeposition paint. Sensors Actuators B Chem 113(2):590–598

Smutok OV, Os'mak GS, Gaida GZ, Gonchar MV (2006c) Screening of yeasts producing stable L-lactate cytochrome *c* oxidoreductase and study of the regulation of enzyme synthesis. Microbiology (Moscow) 75(1):20–24

Smutok O, Gayda G, Dmytruk K, Klepach H, Nisnevitch M, Sibirny A, Puchalski C, Broda D, Schuhmann W, Gonchar M, Sibirny V (2011) Amperometric biosensors for Lactate, Alcohols, and Glycerol assays in clinical diagnostics. Chapter 20. In: Serra PA (Ed) Biosensors – emerging materials and applications. ISBN 978-953-307-328-6. InTech (Croatia) pp 401–446

Smutok O, Karkovska M, Smutok H, Gonchar M (2013) Flavocytochrome b_2-based enzymatic method of L-lactate assay in food products. The Scientific World Journal 2013: Article ID 461284, 6 pages. https://doi.org/10.1155/2013/461284

Smutok O, Karkovska M, Serkiz Ya, Vus B, Čenas N, Gonchar M (2017) A Novel Mediatorless Biosensor Based On Flavocytochrome b2 Immobilized Onto Gold Nanoclusters For Non-invasive L-lactate Analysis Of Human Liquids. Sensor & Actuators B 250: 469–475

Song H, Li Y, Fang W, Geng Y, Wang X, Wang M, Qiu B (2003) Development of a set of expression vectors in *Hansenula polymorpha*. Biotechnol Lett 25(23):1999–2006

Sperry AE, Sen SE (2001) Farnesol oxidation in insects: evidence that the biosynthesis of insect juvenile hormone is mediated by a specific alcohol oxidase. Insect Biochem Mol Biol 31(2):171–178

Stasyk O (2017) Methylotrophic yeasts as producers of recombinant proteins. In: Sibirny AA (Ed) Biotechnology of yeasts and filamentous fungi. ISBN 978-3-319-58829-2. Springer, pp 325–350. https://doi.org/10.1007/978-3-319-58829-2

Stasyuk N, Gaida G, Gonchar M (2013) L-arginine assay with the use of arginase I. Applied Biochemistry and Microbiology (Moscow) 49(5):529–534

Stasyuk N, Gayda G, Zakalskiy A, Zakalska O, Serkiz R, Gonchar M (2019) Amperometric biosensors based on oxidases and PtRu nanoparticles as artificial peroxidase. Food Chemistry 285: 213–220

Sturmberger L, Chappell T, Geier M, Krainer F, Day KJ, Vide U, Trstenjak S, Schiefer A, Richardson T, Soriaga L, Darnhofer B, Birner-Gruenberger R, Glick BS, Tolstorukov I, Cregg J, Madden K, Glieder A (2016) Refined *Pichia pastoris* reference genome sequence. J Biotechnol 235:121–131

Sudbery PE (1996) The expression of recombinant proteins in yeasts. Curr Opin Biotechnol 7(5):517–524

Sützl L, Laurent CVFP, Abrera AT, Schütz G, Ludwig R, Haltrich D (2018) Multiplicity of enzymatic functions in the CAZy AA3 family. Appl Microbiol Biotechnol 102(6):2477–2492

Synenka M, Gayda G, Klepach H, Ivash M, Gonchar M (2015) Isolation and characterization of recombinant yeast glyceroldehydrogenase. ScienceRise 8(13):7–11. (In Ukrainian). https://doi.org/10.15587/2313-8416.2015.48369

Talebian-Kiakalaieh A, Amin NAS, Najaafi N, Tarighi S (2018) A review on the catalytic acetalization of bio-renewable glycerol to fuel additives. Front Chem 6(573). https://doi.org/10.3389/fchem.2018.00573

Talebkhan Y, Samadi T, Samie A, Barkhordari F, Azizi M, Khalaj V, Mirabzadeh E (2016) Expression of granulocyte colony stimulating factor (GCSF) in *Hansenula polymorpha*. Iran J Microbiol 8(1):21–28

Theron CW, Berrios J, Delvigne F, Fickers P (2018) Integrating metabolic modeling and population heterogeneity analysis into optimizing recombinant protein production by *Komagataella (Pichia) pastoris*. Appl Microbiol Biotechnol 102(1):63–80

Thomas AS, Krikken AM, de Boer R, Williams C (2018) *Hansenula polymorpha* Aat2p is targeted to peroxisomes via a novel Pex20p-dependent pathway. FEBS Lett 592(14):2466–2475

Thömmes J, Halfar M, Gieren H, Curvers S, Takors R, Brunschier R, Kula MR (2001) Human chymotrypsinogen B production from *Pichia pastoris* by integrated development of fermentation and downstream processing. Part 2. Protein recovery. Biotechnol Prog 17(3):503–512

van Berkel WJH (2018) Special issue: Flavoenzymes (Editorial). Molecules 23(8):1957. https://doi.org/10.3390/molecules23081957

van der Klei IJ, Harder W, Veenhuis M (1991) Biosynthesis and assembly of alcohol oxidase, a peroxisomal matrix protein in methylotrophic yeasts: a review. Yeast 7(3):195–209

Vandermies M, Fickers P (2019) Bioreactor-Scale Strategies for the Production of Recombinant Protein in the Yeast Yarrowia lipolytica. Microorganisms 7(2) h pii: E40. doi: 10.3390/microorganisms7020040

Vieira Gomes AM, Souza Carmo T, Silva Carvalho L, Mendonça Bahia F, Farachin NS (2018) Comparison of yeasts as hosts for recombinant protein production. Microorganisms 6(2):E38. https://doi.org/10.3390/microorganisms6020038

Vogl T, Gebbie L, Palfreyman RW, Speight R (2018) Effect of plasmid design and type of integration event on recombinant protein expression in *Pichia pastoris*. Appl Environ Microbiol 84(6):e02712–e02717. https://doi.org/10.1128/AEM.02712-17

Vonck J, Parcej DN, Mills DJ (2016) Structure of alcohol oxidase from *Pichia pastoris* by cryo-electron microscopy. PLoS One 11(7):e0159476. https://doi.org/10.1371/journal.pone.0159476

Wagner JM, Alper HS (2016) Synthetic biology and molecular genetics in non-conventional yeasts: current tools and future advances. Fungal Genet Biol 89:126–136

Walker RSK, Pretorius IS (2018) Applications of yeast synthetic biology geared towards the production of biopharmaceuticals. Genes (Basel) 9(7):e340. https://doi.org/10.3390/genes9070340

Weninger A, Hatzl AM, Schmid C, Vogl T, Glieder A (2016) Combinatorial optimization of CRISPR/Cas9 expression enables precision genome engineering in the methylotrophic yeast *Pichia pastoris*. J Biotechnol 235:139–149

Wongnate T, Chaiyen P (2013) The substrate oxidation mechanism of pyranose 2-oxidase and other related enzymes in the glucose-methanol-choline superfamily: review. FEBS J 280(13):3009–3027

Xiao Y, Zhao P, Du J, Li X, Lu W, Hao X, Dong B, Yu Y, Wang L (2018) High-level expression and immunogenicity of porcine circovirus type 2b capsid protein without nuclear localization signal expressed in *Hansenula polymorpha*. Biologicals 51:18–24

Xu P (2018) Production of chemicals using dynamic control of metabolic fluxes. Curr Opin Biotechnol 53:12–19

Yamada H, Nagao A, Nishise H, Tani Y (1982) Glycerol Dehydrogenase from Cellulomonas sp. NT 3060: Purification and Characterisation. Agnc Bioi Chem 46 (9): 1333–1339

Yamada-Onodera K, Yamamoto H, Emoto E, Kawahara N, Tani Y (2002) Characterisation of glycerol dehydrogenase from a methylotrophic yeast, Hansenula polymorpha DI-1, and its gene cloning. Acta Biotechnol 22: 337–353

Yang Z, Zhang Z (2018a) Engineering strategies for enhanced production of protein and bio-products in *Pichia pastoris*: review. Biotechnol Adv 36(1):182–195

Yang Z, Zhang Z (2018b) Recent advances on production of 2, 3-butanediol using engineered microbes. Biotechnol Adv. https://doi.org/10.1016/j.biotechadv.2018.03.019, in press

Yurimoto H (2009) Molecular basis of methanol-inducible gene expression and its application in the methylotrophic yeast *Candida boidinii*. Biosci Biotechnol Biochem 73(4):793–800

Zepeda AB, Figueroa CA, Pessoa A, Farías JG (2018a) Free fatty acids reduce metabolic stress and favor a stable production of heterologous proteins in *Pichia pastoris*. Braz J Microbiol 49(4):856–864

Zepeda AB, Pessoa A Jr, Farías JG (2018b) Carbon metabolism influenced for promoters and temperature used in the heterologous protein production using *Pichia pastoris* yeast. Braz J Microbiol 49(Suppl 1):119–127

Chapter 6
Glutathione Metabolism in Yeasts and Construction of the Advanced Producers of This Tripeptide

Olena O. Kurylenko, Kostyantyn V. Dmytruk, and Andriy Sibirny

Abstract Glutathione is the most abundant non-protein thiol compound of the most living organisms able to protect cells from nutritional, environmental, and oxidative stresses. Due to the antioxidative properties, glutathione is widely used as an active ingredient of drugs, food, and cosmetic products. Microbial synthesis using yeasts is currently the most common method for the commercial production of glutathione. Construction of glutathione overproducers in yeasts by metabolic engineering approaches and optimization of the technology for its production has potential to satisfy the increasing industrial demand in this tripeptide. This review summarizes the current knowledge of physiological functions and practical applications of glutathione as well as illustrates strategies for its efficient production. The potential of the methylotrophic yeast *Ogataea polymorpha* as a glutathione producer is also discussed.

Keywords Glutathione · Yeasts · Glutathione overproducers · Metabolic engineering · *Ogataea (Hansenula) polymorpha*

6.1 Introduction

Glutathione (γ-L-glutamyl-L-cysteinyl-glycine) is an abundant tripeptide that is synthesized in two consecutive ATP-dependent reactions from three precursor amino acids (L-glutamic acid, L -cysteine, L-glycine). Glutathione is known as an important cellular redox buffer for maintaining the reducing thiol-disulfide balance due to its antioxidative property caused by the thiol group of the cysteine moiety. More than 90% of microbial, plant, and mammalian cell glutathione is present in the reduced form (designated as GSH), while the oxidized form (designated as GSSG)

O. O. Kurylenko (✉) · K. V. Dmytruk
Institute of Cell Biology, NAS of Ukriaine, Lviv, Ukraine

A. Sibirny
Institute of Cell Biology, NAS of Ukriaine, Lviv, Ukraine

Department of Microbiology and Biotechnology, University of Rzeszow, Rzeszow, Poland

A. Sibirny (ed.), *Non-conventional Yeasts: from Basic Research to Application*,
https://doi.org/10.1007/978-3-030-21110-3_6

only accounts for 10%. Both forms are located in the cytoplasm as well as in the mitochondrial membrane. The ratio between reduced and oxidized forms as concentration of GSH depends significantly also on activities of enzymes involved in GSH synthesis and degradation (Spector et al. 2001). Glutathione deficiency in humans can be associated with several medical disorders caused by oxidative stress, poisoning, or compromised immune system. These disorders lead to such diseases as cancer, cataracts, liver cirrhosis, neurodegenerative diseases, gastrointestinal inflammations, and hemolytic anemia. Therefore, glutathione is widely used in medical, cosmetic, and food industries as an active ingredient of drugs, food, and cosmetic products to alleviate dangerous oxidative processes and scavenge toxic compounds. It is also used to strengthen skin whitening and repair due to its antiaging effect and as support medication under cancer therapies (Bachhawat et al. 2009; Townsend et al. 2003).

Although glutathione can be produced by chemical and enzymatic synthesis, microbial production using yeasts, such as *Saccharomyces cerevisiae* and *Candida utilis*, is currently the most common method for the commercial production as a result of high biomass density cultivation with maximum possible glutathione content at the point of biomass harvesting. Different strategies have been proposed to increase intracellular glutathione accumulation, including optimization of cultivation processes and improvement of producer strains. Strains obtained by both random mutagenesis and by genetic engineering techniques are used. Design and construction of novel glutathione producers are of great importance for improvement of the production of this biotechnologically and medically important tripeptide. The development of efficient producer strains requires understanding of the molecular mechanisms controlling intracellular glutathione level in the yeast cells. This chapter describes the role of glutathione in cellular metabolism and discusses the progress in construction of the efficient glutathione producers in yeasts.

Yeast *Ogataea* (*Hansenula*) *polymorpha* is considered as a rich source of glutathione, due to the role of this tripeptide in detoxifications of key intermediates of methanol metabolism accumulated during methylotrophic growth (formaldehyde and hydrogen peroxide). Similar to *S. cerevisiae*, *O. polymorpha* is characterized by simple cultivation mode in inexpensive growth media, well-established genetic tools; besides, there is substantial experience on its industrial cultivation (Kim et al. 2013; Riley et al. 2016). This review also compiles the current knowledge about glutathione metabolism in yeast *O. polymorpha* as well as successful approaches to improve glutathione production in this yeast species.

6.2 Glutathione Biosynthesis, Catabolism, Regulation, Recycling, and Transport

Glutathione was discovered as a substance extracted from yeast by ethanol in 1888 and named philothion. The molecular structure of GSH was elucidated in 1921 (Penninckx and Elskens 1993). Since then three different pathways for synthesis of

GSH was described. Most common pathway of GSH synthesis includes two ATP-dependent enzymatic steps: first, formation of γ-glutamylcysteine from glutamic acid and cysteine and second, GSH formation from γ-glutamylcysteine and glycine. First reaction is catalyzed by γ-glutamylcysteine synthetase EC 6.3.2.2 (GCS, also known as glutamate-cysteine ligase, Gsh1 in eukaryotes, and GshA in prokaryotes, encoded by *GSH1* and *gshA*, respectively). GCSs in yeast and bacteria have a single polypeptide (Dalton et al. 2004). In contrast, these enzymes in fruit flies, rodents, and humans are composed of a catalytic (GCLC) and modifier (GCLM) subunits, which are encoded by different genes (Dalton et al. 2004; Gipp et al. 1992; Gipp et al. 1995; Huang et al. 1993a, b; Yan and Meister 1990). GCLC displays catalytic activity and is subjected to feedback inhibition by GSH (Seelig et al. 1984). GCLM is enzymatically inactive but responsible for decrease Km of enzyme for glutamic acid and increase Ki for GSH (Huang et al. 1993a, b). Hence, the holoenzyme to lesser extent is inhibited by GSH than GCLC. The activity of GCS is feedback-inhibited by GSH, but not by GSSG. Such type of regulation prevents hyperaccumulation of GSH, which is of physiological significance (Richman and Meister 1975). Second reaction of GSH synthesis is catalyzed by glutathione synthetase EC 6.3.2.3 (Gsh2 in eukaryotes and GshB in prokaryotes, encoded by *GSH2* and *gshB*, respectively). Structural and functional characteristics of bacterial glutathione synthetase reveal that the enzyme functions as a tetramer (Yamaguchi et al. 1993). The mammalian, yeast, and plant glutathione synthetases are active as dimers with identical subunits (Gogos and Shapiro 2002; Jez et al. 2004). Glutathione synthetase is not subject to feedback inhibition by GSH (Oppenheimer et al. 1979).

Recently, a single bifunctional enzyme γ-glutamylcysteine synthetases/glutathione synthetase (GshF, encoded by *gshF*) able to perform synthesis of GSH was found in bacteria (Janowiak and Griffith 2005; Vergauwen et al. 2006). GshF orthologs consisting of a γ-glutamylcysteine ligase (GshA) domain fused to an ATP-grasp domain were identified in 20 gram-positive and gram-negative bacteria. Remarkably, 95% of these bacteria are mammalian pathogens. These bacteria are *Listeria innocua*, *Clostridium perfringens*, *Pasteurella multocida*, *Streptococcus gordonii*, *S. mutans*, *S. sanguinis*, *S. sobrinus*, *S. suis*, *S. thermophilus*, *S. uberis*, *Enterococcus faecalis*, *E. faecium*, and *Lactobacillus plantarum* (gram-positive bacteria) and *Actinobacillus pleuropneumoniae*, *Actinobacillus actinomycetemcomitans*, *Haemophilus somnus*, *Mannheimia succiniciproducens*, and *Desulfotalea psychrophila* (Gopal et al., 2005). GshF was described in details in *Listeria monocytogenes*, *Streptococcus agalactiae*, and *Pasteurella multocida*. No significant feedback inhibition of GshF by GSH was detected. It was also shown that GshF is not subjected to inhibition by cystamine, despite the fact that cystamine is a strong inactivator of mammalian but not *E. coli* GCS (Kelly et al. 2002; Lebo and Kredich 1978).

The natural compensatory pathway for GSH synthesis was revealed in yeast and bacteria lacking GCSs (Spector et al. 2001; Veeravalli et al. 2011). γ-Glutamyl kinase is the first enzyme in proline biosynthetic pathway (encoded by *PRO1* in eukaryotes or *proB* in prokaryotes). The enzyme catalyzes the formation of γ-glutamyl phosphate. This compound, if accumulated, could nonenzymatically react with cysteine to form γ-glutamylcysteine, and then glutathione synthetase catalyzes GSH formation from γ-glutamylcysteine and glycine.

GSH degradation plays the important role in GSH homeostasis and recycling of nutrients. GSH degrades by γ-glutamyltranspeptidase, which transfers the γ-glutamyl moiety to acceptor substrates such as water (hydrolysis) or amino acids and peptides (transpeptidation reaction) (Meister and Anderson 1983). During hydrolysis reaction γ-glutamyltranspeptidase catalyzes the cleavage of GSH to glutamate and cysteinyl glycine (Cys-Gly) (Fig. 6.1).

γ-Glutamyltranspeptidase enzyme is conserved among prokaryotes and eukaryotes (Verma et al. 2015). Eukaryotic γ-glutamyltranspeptidases are localized on the plasma membrane (Wickham et al. 2011), while bacterial enzymes are soluble, unlike the membrane-anchored eukaryotic enzymes, but localized in the periplasm (Suzuki et al. 1986). γ-Glutamyltranspeptidases are able to cleave GSH, GSSG, and various γ-glutamyl substrates (e.g., GSH-S conjugates; Wickham et al. 2011).

In the yeast *S. cerevisiae*, GSH is degraded to constituent amino acids glutamate, cysteine, and glycine by consecutive action of γ-glutamyltranspeptidase (Cis2/Ecm38) and cysteinylglycine dipeptidase (Dug1; Fig. 6.1). Cis2 is a dimeric enzyme consisting of two different subunits (Penninckx and Jaspers 1985). The expression of *CIS2* gene is regulated by the nitrogen source. It was shown that ammonium ions repress *CIS2* gene expression, while nitrogen starvation leads to derepression of the gene (Penninckx et al. 1980). The expression of *CIS2* is dependent from transcription activator GATA factors. Under condition of nitrogen limitation, GATA factors Gat1 and Gln3 are required for *CIS2* expression (Springael and Penninckx 2003).

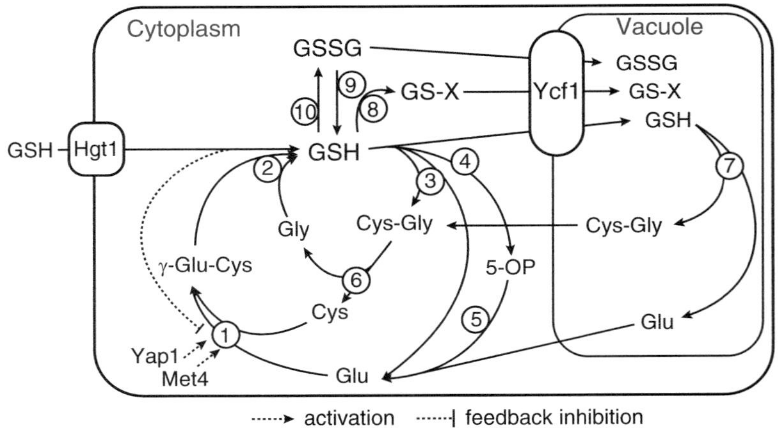

Fig. 6.1 The glutathione recycling in yeast. (*1*) γ-Glutamylcysteine synthetase, (*2*) glutathione synthetase, (*3*) Dug enzyme, (*4*) ChaC2 enzyme, (*5*) 5-oxoprolinase, (*6*) cysteinylglycine dipeptidase Dug1, (*7*) γ-glutamyltranspeptidase, (*8*) glutathione S-transferase, (*9*) dehydroascorbate reductase, glutathione peroxidases, and (*10*) glutathione reductases, Hgt1, high-affinity glutathione transporter; Ycf1, glutathione exporter from cytoplasm to vacuole; Yap1, redox-sensitive transcriptional activator; Met4, leucine-zipper transcriptional activator. (Modified from Bachhawat et al. 2017)

CIS2 is also slightly activated by sulfur limitation (Penninckx and Elskens 1993). Mutants with truncated γ-glutamyltranspeptidase (*ggt1Δ* in *O. polymorpha* and *cis2Δ* in *S. cerevisiae*) revealed importance of this enzyme for detoxification of electrophilic xenobiotics in yeasts (Ubiyvovk et al. 2006). GSH degradation by the action of γ-glutamyltranspeptidase is localized in the vacuole, at least under sulfur and nitrogen starvation (Elskens et al. 1991; Mehdi and Penninckx 1997). Alternative cytosolic GSH degradation pathway, which is independent of γ-glutamyltranspeptidase, was identified in *S. cerevisiae* (Kumar et al. 2003). This so-called DUG pathway contains three metalloproteins Dug1, Dug2, and Dug3 (Ganguli et al. 2007). Dug1 catalyzes hydrolysis of Cys-Gly to cysteine and glycine. Dug1 is a highly specific Cys-Gly dipeptidase. Dug1 is a homodimer enzyme. Deletion of *DUG1* gene resulted in intracellular accumulation of Cys-Gly with concomitant increase of GSH toxicity. Human Cys-Gly dipeptidase gene *CNDP2* can complement yeast *dug1Δ* mutant (Kaur et al. 2009). Dug2 and Dug3 formed heterodimeric Dug enzyme, catalyzing GSH degradation into glutamate and Cys-Gly. Dug enzyme degraded GSH under stress conditions. The catalytic residues responsible for hydrolase activity are located on Dug3, but interaction of Dug3 and Dug2 is necessary for enzymatic activity (Kaur et al. 2012). It was shown that *DUG2* and *DUG3* genes, but not *DUG1*, are derepressed under condition of sulfur limitation. The Dug enzyme was found in fungi and yeast (Bachhawat and Kaur 2017).

Several new enzymes able to GSH degradation were identified. These are ChaC1 enzyme found among higher eukaryotes, the ChaC2 enzyme found from bacteria to man, and the RipAY enzyme represented in some bacteria (Bachhawat and Kaur 2017). ChaC1 hydrolyzes GSH leading to the formation of Cys-Gly and 5-oxoproline. ChaC1 catalyzes degradation of GSH but not GSSG in cytosol (Tsunoda et al. 2014). ChaC1 is upregulated under endoplasmic reticulum stress and amino acid starvation. ChaC2 is also a cytosolic enzyme and acts specifically on GSH to form Cys-Gly and 5-oxoproline. ChaC2 can degrade GSH up to 20-fold less efficiently than ChaC1 (Kaur et al. 2017). ChaC2 is a more ancestral and probably more primitive enzyme as compared to ChaC1. ChaC2 is a constitutively expressed protein functioning primarily in a housekeeping capacity.

RipAY is a GSH-degrading enzyme discovered in a plant pathogenic bacterium *Ralstonia solanacearum*. This enzyme is injected into the host during pathogenesis (Genin and Denny 2012). RipAY possessed very high catalytic efficiency resulting severe alteration in GSH homeostasis in host. RipAY degrades GSH similarly as ChaC enzymes and forms Cys-Gly and 5-oxoproline (Fujiwara et al. 2016).

The ChaC family of GSH-degrading enzymes acts on GSH to yield 5-oxoproline and Cys-Gly. The 5-oxoproline in turn is converted to glutamate by ATP-dependent oxoprolinase (Van Der Werf et al. 1971). Glutamate then can be used for GSH synthesis. The Cys-Gly dipeptide is degraded by peptidase to generate cysteine and glycine, which can be incorporated into GSH (Meister 1988) (Fig. 6.1).

GSH can be used as an endogenous source of sulfur and nitrogen under condition of deprivation of these elements in cells. It was shown that intracellular GSH content depends on the availability of nitrogen and sulfur sources (Elskens et al. 1991; Mehdi and Penninckx 1997). The nitrogen starvation induced a relocation of more than 90% of GSH to the central vacuole in *S. cerevisiae* cell. At the same time, γ-glutamyltranspeptidase was derepressed promoting GSH degradation to provide the cells with glutamate, which further can be converted to other nitrogen-containing compounds required for cellular biosynthetic processes. The nitrogen starvation stimulates *de novo* synthesis of GSH. The nitrogen-response elements were identified in the promoter regions of *GSH1* and *GSH2* genes, leading to their derepression under condition of nitrogen deficiency. Activation of both GSH synthesis and its degradation increases the cellular turnover of the tripeptide (Mehdi and Penninckx 1997). The sulfur starvation in *S. cerevisiae* also increases the turnover rate of GSH, directing the liberated cysteine to protein synthesis and synthesis of other sulfur-containing intracellular compounds. The content of the sulfur amino acids was significantly decreased in GSH-deficient mutants of *S. cerevisiae* as compared to that of the wild-type strain cultivating on the same medium. GSH plays the role of a sulfur storage compound. Intracellular GSH concentration decreased until it reached lower limit of about 10% of its normal value under condition of sulfur starvation (Elskens et al. 1991).

Glutathione redox balance is defined as the ratio of reduced to oxidized glutathione. This redox balance is an important indicator of cell oxidative stress status. Decreased GSH/GSSG ratio in cells leads to strong damages caused by oxidative stress. Increased GSH/GSSG ratio and high intracellular GSH content are toxic due to nonselective glutathionylation of proteins with following interruption of metabolic processes (Bachhawat et al. 2009). For instance, it was shown that overexpression of high-affinity glutathione transporter Hgt1/Opt1 in *S. cerevisiae* resulted in severalfold higher intracellular GSH level and subsequent increase sensitivity to exogenous glutathione as compared to parental strain. GSH toxicity is a consequence of the accumulation of GSH in such a strain. GSH toxicity was also detected in *glr1Δ* strain with truncated glutathione reductase having altered redox balance (Srikanth et al. 2005).

Two factors, such as the intracellular concentration of the precursor amino acids (glutamate, cysteine, and glycine) and ATP, are crucial factors for glutathione synthesis in cell. However, more important factors for glutathione synthesis are sufficient expression level of glutathione biosynthetic genes *GSH1* and *GSH2* as well as activities of corresponding enzymes (Bachhawat et al. 2009). Glutathione biosynthetic reactions are regulated at three different regulatory mechanisms: transcriptional regulation, posttranslational regulation, and substrate level regulation.

Transcriptional regulation of the *GSH1* gene is occurred by the action of two transcriptional activators Yap1 and Met4, which are involved in regulation of sulfur assimilatory pathways. GSH depletion resulted in accumulation of oxidized thioredoxins and subsequent Yap1/Met4-dependent transcriptional activation of *GSH1* gene (Wheeler et al. 2003). Yap1 is responsible for oxidative stress tolerance in yeast. The transcriptional activity of Yap1 is regulated by its cellular localization. Under oxidative stress conditions, an intramolecular disulfide bond is formed in Yap1, and this conformational change directs it in the nucleus (Gulshan et al. 2005). When intracellular concentration of GSH is raised or oxidative stress is passed, Yap1 becomes inactive due to reduction of intramolecular disulfide bond by thioredoxin. Then conformational changed Yap1 is exported from the nucleus to cytoplasm by the nuclear export protein Crm1 (Yan et al. 1998).

It was shown that Yap1 similar to *GSH1* regulates *GSH2*, since *GSH2* mRNA levels and *GSH2-lacZ* reporter gene expression increased in *S. cerevisiae* strains carrying the multicopy *YAP1* gene. Promoters of both *GSH1* and *GSH2* genes of *S. cerevisiae* contain Yap1 response element sequence (TTAC/GTAA) (Sugiyama et al. 2000).

Transcriptional regulation of GSH biosynthesis is upregulated by transcriptional factor Met4. Met4 is a transcriptional activator involved in regulation of the assimilation of extracellular sulfate into the sulfur-containing amino acids (methionine and cysteine) (Thomas and Surdin-Kerjan 1997). Considering this, Met4 has an indirect effect on GSH biosynthesis by regulating the supply of cysteine.

The promoter of *GSH1* gene contains binding sites for Cbf1 protein, located extremely close to the Yap1 binding site (Dormer et al. 2000). DNA-binding protein Cbf1 is a negative regulator of *GSH1* expression, since *GSH1* expression is strongly increased in a *cbf1* mutant. Under conditions of glutathione and cysteine depletion, Met4 interacts with Cbf1 to overcome *GSH1* repression, and Yap1 is able to activate *GSH1* expression. Consequently, in a *met4* mutant, Cbf1 repression is constitutive, and no induction of *GSH1* expression after glutathione depletion has occurred. On the other hand, higher glutathione and cysteine concentrations promote Met4 ubiquitination and inactivation, while Cbf1 represses Yap1 activation of *GSH1* gene by occupation of its promoter (Wheeler et al. 2003).

Posttranslational regulation of GSH biosynthesis occurred by nonallosteric feedback inhibition of Gsh1 activity by increased intracellular GSH concentrations (Soltaninassab et al. 2000). The kinetic studies of mammalian Gsh1 revealed that Ki value of this enzyme is 2 mM for GSH feedback inhibition, which is consistent with the physiological concentration of GSH in those cells (Richman and Meister 1975). Such posttranslational regulation directed to maintain accurate control of GSH biosynthesis in yeast, which is crucial for retaining a proper redox balance in the cell. High intracellular concentrations of GSH stop further accumulation of the tripeptide by the feedback inhibition of *GSH1* gene expression as well as enzymatic activity of Gsh1. At the same time, these regulatory mechanisms ensure acceleration of GSH biosynthesis when its intracellular concentration dropped below the physiolcgical level.

It was shown that cysteine is the main limiting precursor for glutathione accumulation (Wen et al. 2006). Supplementation of yeast cultural media with cysteine significantly increases intracellular GSH concentration (Wang et al. 2007; Liang et al. 2008). Consequently, cysteine biosynthesis pathway as well as sulfur assimilation pathway is crucial for glutathione biosynthesis and could limit glutathione production.

Cysteine can be imported from the media or synthesized from serine and homocysteine. In *S. cerevisiae*, cystathionine β-synthase (Cys4) catalyzes the formation of cystathionine from serine and homocysteine. Cystathionine then converted to cysteine by the action of cystathionine γ-lyase (Cys3) (Ono et al. 1988). Intracellular cysteine regulates sulfur assimilation pathway leading to cysteine biosynthesis

(Thomas and Surdin-Kerjan 1997). It was shown that Met4 upregulates genes responsible for assimilation of inorganic sulfur and for the synthesis of sulfur-containing amino acids. Genes *CYS3* and *CYS4* are among them. Met4 is downregulated through ubiquitination when intracellular cysteine level increased providing indirect regulation of glutathione biosynthesis by supplementation of limiting precursor cysteine (Wheeler et al. 2003).

Glutathione is a powerful antioxidant. Hydrogen peroxide and lipid peroxide generated during aerobic metabolism can be decomposed by glutathione peroxidase to the water and the corresponding less toxic hydroxy fatty acids and oxygen using GSH as a cofactor. Glutathione peroxidase operates in the cytosol, mitochondria, and peroxisomes. During reactions catalyzing by glutathione peroxidase, GSH is oxidized to GSSG. GSSG is reduced back to GSH by glutathione reductase, using NADPH as a cofactor, produced mainly in the oxidative part of pentose phosphate pathway by glucose-6-phosphate dehydrogenase and 6-phosphogluconate dehydrogenase (Rosemeyer 1987). It was shown that expression of *ZWF1* gene encoding glucose-6-phosphate dehydrogenase is inversely correlated to intracellular GSH levels suggesting that GSH replenishment may depend on *ZWF1* expression (Salvemini et al. 1999). Both glutathione peroxidase and glutathione reductase form so-called redox cycle (Lu 2009). Glutathione peroxidase or other enzyme glutathione S-transferase reduce organic peroxides (ROOH) to the corresponding alcohols (ROH) using GSH as a cofactor. Based on the dependency for selenium, glutathione peroxidases can be divided into two types, selenium-dependent and selenium-independent.

Glutathione peroxidase Pmp20 of the methylotrophic yeast *Candida boidinii* is localized in the inner side of peroxisomal membrane. Pmp20 homologs of mammals and lower eukaryotes contain putative peroxisome targeting signal type 1. Catalytic activity and dimerization of Pmp20 of *C. boidinii* depended on the only cysteine residue (Cys53). The *pmp20Δ* strain was unable to grow on methanol as a carbon and energy source. During incubation in methanol medium, *pmp20Δ* did not accumulate hydrogen peroxide, suggesting that the main function of Pmp20 is the decomposition of reactive oxygen species generated at peroxisomal membrane surface, *like* lipid hydroperoxides, rather than to decompose hydrogen peroxide. A physiological level of GSH was detected in peroxisomal fraction of *C. boidinii, suggesting* physiological role for Pmp20 as an antioxidant enzyme within peroxisomes rich in reactive oxygen species (Horiguchi et al. 2001).

Glutathione reductase is an essential enzyme that recycles GSSG back to the GSH. Glutathione reductase is highly conserved and possesses a high degree of similarity between its three-dimensional structures from *E. coli, S. cerevisiae*, and human (Karplus and Schulz 1987; Mittl and Schulz 1994; Yu and Zhou 2007). In prokaryotes, glutathione reductase is localized in the periplasmic space, connected with the inner membrane, and it can be secreted to the extracellular environment (Couto et al. 2015; *Oliveira* et al. 2015). In eukaryotes, this enzyme is found in the cytoplasm, nucleus, mitochondria (Couto et al. 2013; Ge et al. 2014; *Outten and Culotta* 2004), endoplasmic reticulum (Chakravarthi et al. 2006), and lysosomes (Chiang and Maric 2011). In plants, glutathione reductase is found in chloroplasts

(*Ge* et al. 2014). A single gene expresses more than one form of glutathione reductase in yeast and human. Different forms of the enzyme find their localization in the cytoplasm or in different organelles. However, two genes encoding glutathione reductase was identified in plants. Insects and kinetoplastids (a group of protozoa, including *Plasmodium* and *Trypanosoma*) do not express glutathione reductase or even glutathione biosynthetic enzymes. Instead, they express the thioredoxin system or the trypanothione system (Couto et al. 2016).

Glutathione S-transferases (GST) are enzymes responsible for the detoxification of reactive chemical species, including xenobiotics, through conjugation to GSH. GST are found in many prokaryotes and eukaryotes (Townsend et al. 2003). GST are suggested as a marker in cancer development and considered as a target in antitumor therapy (Dong et al. 2018; Noguti et al. 2012; Sawers et al. 2014).

There are seven proteins Grx1, Grx2, Gtt1, Gtt2, Gto1, Gto2, and Gto3 revealing glutathione S-transferase activity in *S. cerevisiae*. Grx1 and Grx2 are glutaredoxins and are active as glutathione peroxidases. At the same time, Grx1 and Grx2 possessed glutathione S-transferase activity. Gtt1 and Gtt2 are responsible for the majority of intracellular glutathione S-transferase activity. Deletion analysis displays that Grx1 and Grx2 have an overlapping function with Gtt1 and Gtt2. Multiple mutants with deletions of *GRX1*, *GRX2*, *GTT1*, and *GTT2* genes show increased sensitivity to stress conditions, including exposure to xenobiotics, heat, and oxidants (Collinson and Grant 2003). Gto1, Gto2, and Gto3 display similarities with human Omega class glutathione S-transferases. These enzymes are active as thiol transferases (glutaredoxins) (Garcerá et al. 2006).

GSH is involved in protein redox signaling. Under oxidative stress conditions, cysteine residues in proteins can be oxidized, leading to sulfenic acid formation. Sulfenic acid residues are able to react with GSH producing glutathionylated proteins. Such glutathionylated residues in frame of proteins can be reduced back to cysteine by glutaredoxins. Glutaredoxins belong to a GSH-dependent oxidoreductase enzyme, which catalyze reversible thiol-disulfide exchange reactions between protein thiols and the abundant thiol pool GSSG/GSH. In fact, glutaredoxins protect protein thiols from irreversible oxidation. They regulate protein activities under different conditions, playing significant role in cell signaling and redox homeostasis (Xiao et al. 2019). Moreover, such a mechanism prevents loss of GSH under oxidative conditions, leading to accumulation of GSSG. Increase of GSSG intracellular content may shift the redox equilibrium, which is determined by ratio between GSH and GSSG (Forman et al. 2009). Stabilization of the redox equilibrium can be achieved via decrease of intracellular GSSG content. GSSG can react with a protein sulfhydryl group (Protein-SH), resulting to the formation of a mixed disulfide (Protein-SSG). GSSG can be reduced to GSH or actively transported out of the cell.

Glutathione synthesis occurs in cytosol. Majority of this tripeptide retained in cytosol but also found in other organelles, like mitochondria, chloroplast, nucleus, endoplasmic reticulum, and vacuoles. There are numerous works on glutathione transporters providing glutathione transport across the plasma membrane in bacteria, yeasts, animals, and plants as well as into the different organelles in eukaryotes (Bachhawat et al. 2013).

Multisubunit protein encoding by *YliABCD* operon in *E. coli* belongs to the prokaryotic periplasmic binding-protein-dependent ABC transporter family. YliABCD transporter is involved in GSH import and together with γ-glutamyltranspeptidase forms way of extracellular GSH utilization as a sole sulfur source (Suzuki et al. 2005). Heterodimeric ABC-type transporter CydDC mediates GSH efflux in *E. coli* (Pittman et al. 2005).

Hgt1 of *S. cerevisiae* is specific high affinity glutathione transporters representing the oligopeptide transporter family (Bourbouloux et al. 2000). Deletion of the *HGT1* gene stops uptake of radioactive GSH. Orthologues of Hgt1 of *S. cerevisiae* in other yeast species were identified as glutathione transporters. Orthologues of Hgt1 in *Schizosaccharomyces pombe* and *Candida albicans* are induced under conditions of sulfur limitation in similar manner to that of Hgt1 in *S. cerevisiae* (Bachhawat et al. 2013). Expression of *HGT1* gene is under control of transcriptional activator Met4, responsible for regulation of sulfur amino acid pathway (Wiles et al. 2006). GSH export in yeast occurred through multidrug resistance-associated proteins Ycf1 and Bpt1 (Li et al. 1996; Sharma et al. 2002). GSH can be transported from the cytoplasm to the vacuole by Ycf1 (Rebbeor et al. 1998). There are also Gex1 and Gex2, a vacuolar and plasma membrane glutathione pumps, that belong to major facilitator superfamily (Dhaoui et al. 2011). GSH export in yeast also facilitated by Gxa1p from the ABC protein family (Kiriyama et al. 2012).

Transport of glutathione in mammalian cells across the plasma membrane is necessary to maintain intracellular and interorgan homeostasis of glutathione. It was shown the existence of sodium-dependent and -independent glutathione transport systems in different tissues (Iantomasi et al. 1997; Kannan et al. 1999; Lash and Jones 1983). Glutathione uptake in mammalian cells occurred indirectly by degradation of this tripeptide by γ-glutamyltranspeptidase outside at the surface, reimport of amino acids (cysteine, glycine, and glutamate) and dipeptides (Cys-Gly) into the cytoplasm, and finally intracellular resynthesis of GSH (Griffith and Meister 1979). Mpr1 and Mpr2 are the multidrug resistance-associated proteins. They participate in GSH export in mammals. Cells overexpressing *MRP1* and *MRP2* lead to increased efflux of GSH and lower intracellular level of GSH (Zaman et al. 1995; Rebbeor et al. 2002). Neighboring neurons in mammalian contain gap junctions. These channels are responsible for controlling transport of small molecules, including GSH (Rana and Dringen 2007).

High affinity and low affinity glutathione transport systems were detected in plants (Schneider et al. 1992). High affinity glutathione transporters are still not identified, while several low affinity glutathione transporters have been reported. Opt1 of *Arabidopsis thaliana* belongs to the oligopeptide transporter family. Opt1 provided low affinity GSH import, confirming by radioactive uptake assay in yeast. Opt1 of *A. thaliana* is also able to transport GSH conjugates (Koh et al. 2002; Cagnac et al. 2004). Several chloroplast-localized Clt proteins are responsible for low affinity GSH efflux from chloroplast to the cytosol (Maughan et al. 2010). It is interesting to note that plant chloroplast is able to own GSH synthesis, since they possess corresponding enzymes (Zechmann 2014). Mrp1 and Mrp2 from the multidrug resistance protein family are involved in efflux of GSH, GSSG, and GSH conjugates in to the plant vacuoles (Liu et al. 2001; Martinoia et al. 1993).

6.3 Practical Applications of Glutathione

Due to its physiological properties, there is an increasing demand for the production of glutathione on an industrial scale. Both GSH and GSSG are widely used in medical, cosmetic, and food industries as an active ingredient of drugs, food, and cosmetic products to protect cells against oxidative damage and scavenge toxic compounds. The estimated global annual production of pure crystalline glutathione and glutathione-enriched yeast extract (15% GSH) exceeds 200 and 800 tons, respectively, sold at a price of 300 and 150 USD/kg. It is expected that glutathione market will exceed 9 billion USD by the year 2019 (Marz 2014; Orumets et al. 2012). Glutathione market is segmented on the basis of geographical location as the USA, China, Europe, Japan, India, and Southeast Asia, and the USA currently remains the dominant region in this market. The top manufacturers in the glutathione market include Kyowa Hakko Bio, Viva, Solgar, Tatiomax Glutathione, Amy Myers MD, Lypo-Spheric, Carlson, Parchem Fine & Specialty Chemicals, Jarrow Formulas, NOW, Max Potency, CCL Advanced, Brandon Sciences, Omnia, Ivory Caps, Puritans Pride, and Swanson (https://www.millioninsights.com/industry-reports/glutathione-market).

An imbalance of glutathione in humans is linked to a wide range of pathologies, for instance, diabetes mellitus, neurodegenerative diseases, HIV, cancer, cataracts, liver cirrhosis, aging, pulmonary diseases, gastrointestinal, and pancreatic inflammation, which are mainly based on disturbed protection against oxidative stress on the cellular level. Hence, glutathione can serve as a potential drug or drug precursor for specific treatments as well as biomarker for detection of disorders (Perricone et al. 2009; Wu et al. 2004). The major cause for several neurodegenerative diseases is the aging-related free radical-dependent oxidative damages of the brain cells. Therefore, age-dependent changes in the levels of glutathione in the brain are assumed to play a role in the development of those diseases. The severe depletion of glutathione level was revealed in the brain of patients with Parkinson's disease (Schulz et al. 2000).

Glutathione deficiency generally leads to impaired liver and kidney functions and also reduced resistance to different stress or infectious agents. In the liver, glutathione is involved in detoxification of widespread natural and synthetic acidic toxins, including cigarette smoke and exhaust gas from motor vehicles, as well as heavy metal ions (copper, cadmium, zinc, silver), forming a water-soluble complex which is ultimately excreted to urine as waste (Townsend et al. 2003).

Chronic ethanol feeding leads to the selective reduction of mitochondrial glutathione due to the partial inactivation of a specific mitochondrial membrane transport protein involved in translocation of glutathione from the cytoplasm to the mitochondria. As a consequence hepatocytes become more sensitive to oxidative stress leading to the alcoholic liver cirrhosis. Glutathione also acts as immune booster being required for replication and activation of the T-lymphocytes and polymorphonuclear leukocytes (Maher 2005; Orumets 2012).

Reactive oxygen species (ROS; H_2O_2, O_2^-, OH^-, etc.) are produced by aerobic cells under physiological conditions. The increased concentration of free radicals in cells can result in mutations in DNA, damage of proteins of plasma/organelle membranes in the cell, which promote the development of cancer. Multiple repair mechanisms remove DNA lesions and ultimately degrade misfolded proteins. Antioxidant mechanisms include enhanced expression of molecules that can modulate ROS accumulation by both enzymatic and nonenzymatic means. Enzymatic antioxidant responses include increased activities of superoxide dismutase or catalase (Bansal and Simon 2018; Wang et al. 2018). In addition, other small redox protein-encoding genes that generate glutathione, thioredoxins, heme oxygenases, and peroxiredoxins are also stimulated. Thus, the limitation of glutathione supply, decreased glutathione synthetic rates, increased glutathione degradation, or lack of GST enzyme activity can reduce physiological protection against carcinogenesis due to inability to detoxify carcinogens and is associated with an increased risk toward a variety of cancers (e.g., lung, colon, and bladder cancer). ROS production is significantly increased in cancer cells because of mitochondrial dysfunction, altered metabolism, and frequent genetic mutations, resulting in an accumulation of large amounts of oxidized protein, DNA, and lipids. Therefore, as an adaptive response, cancer cells harbor elevated levels of ROS-scavenging molecules, especially level of glutathione is often high in tumor cells before treatment, and there is a correlation between elevated levels of intracellular glutathione/sustained glutathione-mediated redox activity and resistance to prooxidant anticancer therapy (Hatem et al., 2017). Additionally, ROS are able to regulate the potential of cancer cells to metastasize to distant locations. Data supporting this indicated that elevated glutathione promotes metastasis in both melanoma and liver cancer (Bansal and Simon 2018; Carretero et al. 1999; Huang et al. 2001). Therefore, antioxidants, including glutathione, may exert dual effects through ROS scavenging and/or regulation of the redox signaling pathway: reducing cancer risk in certain genetic backgrounds but promoting cancer initiation, progression, and metastasis under many other circumstances (Gaucher et al. 2018).

The depletion of glutathione is one of the valuable strategies to increase the sensitivity of cancer cells to radiations and toxic drugs, especially for aggressive and/or metastatic cancers. Possible strategies for the depletion of cytosolic and mitochondrial glutathione level include inhibiting glutathione synthesis; increasing the efflux of cytosolic glutathione; creating a shortage of cysteine, glutamate, and glycine; and inhibiting glutathione transport into mitochondria. As the rate-limiting enzyme in glutathione synthesis, GCS has been an anticancer drug target for more than 30 years. The classical drug used to inhibit GCS activity and therefore to reduce the intracellular glutathione level is BSO (L-buthionine sulfoximine). BSO can trigger apoptosis as a single agent and in combination with arsenic trioxide (As_2O_3) in solid tumors and acute promyelocytic leukemia (APL) cells (Hatem et al. 2017; Maeda et al. 2004). In patients with various types of cancer (ovarian, lung, breast, and colon cancer; melanoma), continuous infusion of BSO resulted in consistent and profound glutathione depletion in tumors (<10% of pretreatment value) without significant

BSO-related toxicity (Hatem et al. 2017). Several other drugs known to reduce cellular glutathione levels are currently used in clinical trials to improve efficacy of targeted therapy; for example, disulfiram induces melanoma cell apoptosis by shifting the ratio of GSH/GSSG toward its oxidized state and is now being used in phase I/II clinical research in metastatic melanoma (Bansal and Simon 2018; Conticello et al. 2012).

High doses of vitamin C can also cause the depletion of glutathione and NADPH, and ultimately ATP, inducing an energetic crisis. As vitamin C is oxidized to dehydroascorbate (DHA) in cell culture media and, subsequently, imported into cells by the glucose transporter GLUT1, glutathione is used as a reducing agent for the conversion of DHA into vitamin C (Yun et al. 2015).

Another approach is based on the detoxifying role of glutathione toward xenobiotics, such as phenethyl isothiocyanate (PEITC), known as a natural compound with chemopreventive and anticancer activity found in consumable cruciferous vegetables. The selective toxicity of PEITC may be explained by generally high glutathione levels in cancer cells. Available data from preclinical and clinical studies suggest PEITC to be among the promising anticancer agents available from natural sources (Gupta et al. 2014; Hatem et al. 2017).

GSH synthesis can also be modulated through the regulation of intracellular cysteine levels. Sulfasalazine is known as anti-inflammatory and immune-modulatory drug that can inhibit the cystine/glutamate antiporter responsible for the export of glutamate in exchange for cystine. Several studies have shown that sulfasalazine used alone, or in combination with anticancer drugs, inhibits the growth of a large spectrum of human cancer cells in cell culture and/or mouse models (Lo et al. 2010; Narang et al. 2007). The targeting of glutamate, required for *de novo* GSH synthesis and for cells to acquire cystine through the Xc-antiporter, represents another strategy to alter intracellular glutathione levels. Also glutamate analogs (e.g., acivicin, L-azaserine, 6-diazo-5-oxol-norleucine, and boronate derivatives) are used as the potent inhibitors of γ-glutamyltranspeptidase, which provide cells with an additional source of cysteine and cystine (Hatem et al. 2017).

GST is primarily responsible for binding to xenobiotics and GSH, forming conjugates that are then secreted extracellularly. There are primarily three types of GSTs: cytosolic, mitochondrial, and nuclear. Specifically, these GSTs bind both the substrate and GSH at different sites of the enzyme, eventually activating the thiol group of GSH to enable nucleophilic attack on the substrate (Bansal and Simon 2018). GSTs are involved in cellular biotransformation of electrophilic compounds and are usually overexpressed in cancer cells leading to increased detoxification of anticancer agents and the development of drug resistance. The GSH-GST-drug conjugate after formation is effluxed out of the cell via multiple resistance-associated protein (MRP1) transporters. MRP1 accumulation is often increased in drug-resistant cancer cells along with GST. GST inhibitors are, therefore, potential cancer therapeutic targets. Different agents targeting GST have been applied, which decrease GST catalytic activity (TKL199, ezatiostat HCl), or are GST-activated prodrugs, converting an inactive prodrug into a cytotoxic species (canfosfamide HCl) (Allocati et al. 2018; Corso and Acco 2018; Hatem et al., 2017).

The most effective anticancer strategy is rational drug combination directed against multiple targets, as it decreases the risk of cancer drug resistance and allows the use of lower therapeutic doses. This approach might be particularly useful in cancer cells that have become adapted to stress and are resistant to anticancer agents. A better understanding of the mechanisms of redox regulation in cancer cells and the development of effective, specific, and clinically safe inhibitors should offer new perspectives and successful strategies for cancer treatment (Bansal and Simon 2018; Hatem et al., 2017).

However, a great majority of diseases (diabetes, cardiovascular diseases, HIV/AIDS, sepsis, cystic fibrosis, stroke, and brain disorders such as Alzheimer's and Parkinson's diseases or schizophrenia) are associated with a decrease in glutathione, combined with various oxidative stress states. Similarly, it has been demonstrated that the glutathione antioxidant defenses of the body decrease linearly with age. As a result, therapeutic strategies to restore the glutathione pool are needed, ideally through oral administration for such chronic conditions. This is especially challenging because of its high degradation rate in the gastrointestinal tract through bacterial and epithelial catalysis by γ-glutamyltranspeptidase. Therefore, the approaches to increase glutathione availability are intensively developed based on its chemistry or drug delivery technology (Gaucher et al. 2018; Orumets 2012).

Glutathione is widely used as a natural antioxidant supplement for oral consumption or cosmetic application. However, the bioavailability of pure glutathione obtained by oral or intravenous administration is still controversial. Also, little is known about the average daily intake of glutathione from food, the concentration of glutathione in various food sources, or the importance of dietary glutathione in prevention of diseases. The estimated daily intake from food has been suggested to be 150 mg of glutathione per day. An oral acute toxicity study of glutathione in mice found that the lethal dose 50 (LD50) was more than 5 g/kg, indicating that glutathione is nontoxic (Hagen et al. 1990).

Any short-term effects were not observed after oral application of 0.15 mmol/kg of glutathione as a single dose. In general, the activity of the hepatic γ-glutamyltranspeptidase, which cleaves tripeptide is believed to prevent an increase in circulating glutathione (Buonocore et al. 2016; Witschi et al. 1992). However, the superiority of sublingual usage of glutathione compared to oral as well as successful oral usage of γ-glutamylcysteine as an alternative to glutathione was recently reported (Schmacht et al. 2017a). Intravenous administration has demonstrated numerous beneficial effects in treatment of cancer, heavy metal overload and Parkinson's disease, and reduction in acute pain from a rheumatoid arthritis and decreased the toxicity of several chemotherapy strategies (Hassan et al. 2001; Sechi et al. 1996). The serum half-life of glutathione after intravenous administration was shown to be less than 2 min. A major problem with reliable determination of bioavailability of glutathione is related to its low concentrations in the plasma due to rapid relocation to different organs (kidneys, liver). The transit of orally administered glutathione to tissues is thought to occur via uptake from the plasma into the cells, export from enterocytes into the blood, and absorption from the intestinal lumen. Such gastrointestinal transport of glutathione appears to be via nonenergy-requiring, sodium-independent, carrier-mediated diffusion. Various epithelial cells, such as alveolar cells, enterocytes, endothelial cells, and retinal pigmented epithelial cells, are capable of uptaking exogenous glutathione. However, glutathione transport capacity in most cells is rather low, and thus the absorption of glutathione is poor. Thus, increasing plasma concentrations by oral administration has been shown to increase the availability of glutathione for transport into these tissues (Buonocore et al. 2016; Witschi et al. 1992). Glutathione is resistant to most gastric proteases due to its distinctive structure (γ-glutamyl peptide bond in glutamylcysteine moiety) but may be partially inactivated by gastric peptidases. Two complementary approaches might increase glutathione bioavailability: one is based on chemistry and the other is based on drug delivery technology.

The first strategy is the usage of glutathione analogs and precursors for avoiding the problem of its poor absorption and transport. However, cysteine cannot be administered directly because of its toxicity and instability; therefore, precursors of cysteine are used. For instance, N-acetylcysteine is an acetylated analog of cysteine able to cross the cell membrane and be rapidly deacetylated inside the cell, allowing the released cysteine to increase intracellular glutathione concentration by *de novo* synthesis. GSH esters, mainly mono- and dimethylesters, due to high hydrophobicity and less sensitivity toward degradation by γ-glutamyltranspeptidase, can cross the cell membrane more easily, and glutathione can be liberated in the cells by the activity of esterases, but no conclusive data on their ability to restore the glutathione pool in humans are currently reported in the literature (Gaucher et al. 2018). Furthermore, other cysteine or glutathione precursors have been evaluated with varying results: L-methionine; S-adenosylmethionine (SAMe) (Lieber 2002); L-2-oxothiazolidine-4-carboxylate (OTC, procysteine), which is enzymatically converted to cysteine within liver cells; 2-(RS)-n-propylthiazolidine-4(R)-carboxylic-acid (PTCA); D-ribose-L-cysteine; L-cysteine-glutathione mixed disulfide (Oz et al. 2007); and γ-glutamylcysteine (Zarka and Bridge 2017).

The second strategy for improvement of the bioavailability of glutathione (or its derivatives) is to use a drug delivery system. For oral administration glutathione can be encapsulated into liposomes (Rosenblat et al. 2007), water-in-oil microemulsions (Wen et al. 2013), pellets of montmorillonite and glutathione (Baek et al. 2012), polymeric nanoparticles and microparticles prepared with natural or synthetic polymers (Naji-Tabasi et al. 2017), hydrogels (Mandracchia et al. 2011), and mucoadhesive films for sublingual delivery (Chen et al. 2015), as well as tablets in combination with L-cystine, vitamin C, and selenium (Buonocore et al. 2016). The administration of the tablet containing 250 mg of glutathione to 15 healthy volunteers (both sexes, 20–40 years old) resulted in the significantly increased glutathione level after 30 and 60 minutes in blood. The additional benefit of such combination aside from improving bioavailability is hiding the undesirable organoleptic (odor and flavor) properties of the thiol drugs (Buonocore et al. 2016).

The glutathione can also be chemically linked to the carrier surface (instead of being passively encapsulated into the carrier). However, only few works have reported the synthesis of glutathione conjugates with a preservation of its antioxidant property by avoiding the formation of disulfide. Two main strategies were led, grafting glutathione either to polymers used as raw material for nanoparticle preparation or to preformed nanoparticles. Glutathione can be functionally conjugated directly to chitosan and polyethylene glycol (Trapani et al. 2010) or to gold nanoparticles (AuNPs) via a linker lipoic acid, which was previously shown to passivate the surface of nanoparticle and limit the access of the thiol function to the gold core (Luo et al. 2016). However, very few of published works have moved from research into clinic applications, especially for oral glutathione supplementation (apart from precursors/prodrugs of glutathione such as N-acetylcysteine and one GSH-containing orobuccal tablet). Nevertheless, the collaboration between the chemical and galenic approaches seems nevertheless to offer promising opportunities in the future (Fraternale et al. 2017; Gaucher et al. 2018).

Glutathione is now investigated as a molecular tool to specifically deliver drugs to the brain or to obtain controlled drug release in the intracellular compartment. One of the major pharmaceutical challenges is drug delivery to the central nervous system as the passage of macromolecules, and 98% of small molecules is prevented by the blood-brain barrier under physiological conditions. The conjugation of glutathione on several pharmaceutical forms safely enhanced the delivery of various encapsulated drugs and nucleic acids to the brain. The GSH-PEG liposomes (G-technology), a liposomal system with a polyethylene glycol (PEG) coating modified with GSH, were successfully applied to deliver carboxyfluorescein (an autoquenched fluorescent tracer) by brain endothelial cells after intraperitoneal or intravenous administration to rats (Rip et al. 2014). This technology was used to deliver amyloid-targeting antibody fragments to the brain in a mouse model of Alzheimer's disease after intravenous administration. In rats, brain-specific uptake of the model drug ribavirin encapsulated into GSH-PEG liposomes was positively correlated with increasing amounts of GSH coating and involved a receptor-mediated mechanism. These GSH-PEG liposomes have also demonstrated brain targeting as well as therapeutic efficacy in murine models of brain cancer (2B3–101; drug: doxorubicin) and neuroinflammation (2B3–201; drug: methylprednisolone). The 2B3–201 product has recently completed a phase I trial in healthy volunteers, while the 2B3–101 product has completed a phase I/IIa trial in patients with various forms of brain cancer (ClinicalTrials.gov Identifier: NCT01386580) and is currently being tested in a phase II trial in patients with breast cancer and leptomeningeal metastases (ClinicalTrials.gov Identifier: NCT01818713) (Brandsma et al. 2014; Gaillard et al. 2014; Gaucher et al. 2018).

GSH coating is also under investigation to obtain the brain delivery of drugs encapsulated into nanoparticles. Recently, this strategy has also been proved to be successful in vitro for gene delivery to brain endothelial cells, thus opening another wide field of application (Gaucher et al. 2018).

Another big field of glutathione application is cosmetic industry (Weschawalit et al. 2017; Wu et al. 2004). It is widely used for composition of different cosmetic products, such as emulsifiers, oily substances, and moisturizers, primarily to enhance the whitening effect on the skin and also to remove or prevent pimple marks. Due to its antioxidant properties, glutathione has gained attention in cosmetic industry as a possible component for fighting against skin aging and wrinkles and for ensuring antisun skin products (Bachhawat et al. 2009).

Aging is related to progression of oxidative damage in the cells and is accompanied by decrease in the GSH/GSSG ratio. The GSH/GSSG ratio was revealed to remain constant until the age of 45 years and declined linearly thereafter, suggesting that glutathione metabolism fails to keep up with oxidizing events and stress in the beginning of late middle age (Jones et al. 2002).

In vitro experiments have demonstrated that glutathione is related to melanogenesis. Its antimelanogenic properties result from a variety of mechanisms including stimulation of pheomelanin synthesis, its antioxidant effects, and interference with intracellular trafficking of melanogenic enzymes. The supplementation of reduced form of glutathione (500 mg/d) had a skin-lightening efficacy in humans. Moreover, the usage of both forms of glutathione (250 mg/d) was well tolerated by humans and resulted in increased skin elasticity, both sun-exposed and sun-protected skin (Weschawalit et al. 2017). The different GSH-containing cosmetic products are available on the market, e.g., Biomimetic Fluid Tears+ (supplied by Dermalab, New Zealand), Liposystem Complex® L-Glutatione Ridotto 13% (supplied by I.R.A. Istituto Ricerche Applicate, Italy), KALILIGHT (supplied by Kalichem, Italy), NanoWhite (supplied by Mibelle Biochemistry, Switzerland), L-Glutathione Reduced (supplied by SMA Collaboratives, USA).

Glutathione has found application as an ingredient in a variety of food products, including baked goods, beverages, breakfast cereals, condiments cheeses, dairy product analogs, fats and oils, sauces, and meat. Depending on the type of baked product, the dough requires a certain combination of strength, extensibility, and tolerance. Reducing agents are a type of dough conditioners used to break the disulfide cross-links between the cysteine sulfhydryl groups of different cereal proteins, in particular gluten proteins, formed upon mixing flour with water (Lagrain et al. 2007; Verheyen et al. 2015). These disulfide bonds increase dough strength and decrease its extensibility that can be detrimental to the quality of a number of products, for example, pizzas, tortillas, crackers, and hard biscuits. Both cysteine and glutathione are often used to induce the reduction of the disulfide cross-links in the dough with high-speed processes, to reduce mix time, lower energy input, and improve machinability as well as the bread loaf volume. While the number of reduced disulfide cross-links is directly proportional to that of cysteine molecules added, considerably less glutathione is required to have the same effect due to the wheat glutathione reductase present in flour, which converts GSSG to GSH with free SH groups that can participate in further disulfide reduction reactions (Orumets 2012). Therefore, GSH-enriched yeasts are used to alter disulfide bonds in the protein network of wheat dough with the effect of dough weakening and modified baking properties. Within the European Union, food additives have to be declared according to Annex II of Regulation (EC) No. 1333/2008 with so-called E-numbers. The advantage of inactivated glutathione-enriched yeasts compared to cysteine (E920) is no need of declaration with an E-number (Schmacht et al. 2017a).

The application of glutathione has gained increased interest in the production of fermented beverages, and in particular, winemaking. The quality of wine depends on its numerous constituents, the presence/absence and the concentration of a wide range of compounds. Grape phenols are the most susceptible compounds to oxidation in wine, which after oxidation can result in a decreased aroma loss and browning. After pressing the grape must, the hydroxycinnamates, especially caftaric acid, are the first phenolic compounds to be oxidized by the grape polyphenol oxidases. This oxidation leads to the formation of respective o-quinones which can further polymerize and interact with various aroma compounds (thiols) resulting in browning of wines and aroma losses. Moreover, nonenzymatic oxidation of diphenols caffeic acid and catechin can also occur during wine aging. Glutathione plays an important role in preventing the oxidative processes in white wines and can also improve their maturation potential. It interrupts the oxidation mechanism by trapping the caftaric acid in the form of 2-S-glutathionylcaftaric acid, also known as grape reaction product, which is chemically stable, colorless, and odorless compounds (Kritzinger et al. 2012; Li et al. 2008; Orumets 2012). The different areas of glutathione application are also illustrated in Table 6.1.

6.4 Microbial Production of Glutathione

The three methods are known for glutathione production: chemical synthesis, in vitro enzymatic synthesis, and in vivo microbial synthesis.

The process of chemical synthesis was commercialized in the 1950s, however, was not favored for industrial production because of its complexity. Moreover, the end product was an optically inactive (racemic) mixture of the D- and L-isomers. As only the L-isomer is physiologically active, an optical resolution was required to separate the two isomers which increased the process cost even higher (Li et al. 2004).

Enzymatic methods for glutathione production use either free or immobilized enzymes in a bioreactor system with added precursor amino acids, ATP, and Mg^{2+} as a cofactor for the two biosynthetic reactions. Enzymes from various microorganisms (*S. cerevisiae*, *E. coli*, *Proteus mirabilis*, *P. vulgaris*) are used. Depending on the microorganisms where the enzymes are derived from the process, conditions may vary. For example, in the case of yeast-derived enzymes (Gsh1 and Gsh2), an optimal temperature 30 °C or 35 °C and pH 7.3–7.5 have been used. With *E. coli* enzymes (gshA and gshB), a process at 37 °C and pH 7 has been preferred. The requirement for ATP makes the enzymatic process difficult to scale up, because it is economically impractical to add it on an industrial scale. For economic feasibility highly efficient ATP regeneration systems are required, in which ATP-requiring reactions are coupled with ATP producing reactions. The drawbacks of this method are high cost of the precursors and the need for ATP regeneration, but a relatively high glutathione concentrations (up to 9 g/L) without process-related by-products can be achieved (Li et al. 2004; Zhang et al. 2017).

Currently, the most common method for the commercial production of glutathione is in vivo synthesis using different microorganisms followed by the extraction

Table 6.1 Different applications of glutathione and related products (Schmacht et al. 2017a)

Industry	Application	Mode of action	Reference(s)
Beverage	Beer	Stabilization of taste, antioxidant	Chen et al. (2015)
Beverage	GSH-enriched yeast in wine	Stabilization of taste, antioxidant	Ortiz-Julien (2012); Mezzetti et al. (2014) and Mao et al. (2016)
Beverage	Extract of reduced GSH in yellow rice wine	Antioxidant	Xia et al. (2019)
Food	Dough modifier	Substitution of L-cysteine, dough relaxation	en.angleyeast.com, organobalance.de, Perrone et al. (2005)
Food	Tasting agent	Taste modulator	Ueda et al. (2014)
Food	GSH-enriched nutrient porridge	Skin protection	Zhang et al. (2016)
Food	Rice noodles	Nourishing faces, skin protection	Lyu (2015)
Food	Yogurt	Antioxidant, aging-delaying effects	Ye et al. (2016)
Food	Chewing gum	Dispelling freckles, beautifying skin	Wang et al. (2016)
Food, packaging	Food packaging film with enhanced GSH stability	Protection of oxidation	Lee et al. (2015)
Feed	Feed supplement	Animal health enhancing	Perrone et al. (2005)
Feed	GSH-enriched yeast as feed supplement	Growth promotion	Xu et al. (2015)
Health	Functional capsules, dietary supplement	Antioxidant, detoxification	Crum (2011), europharmausa.com
Health	GSH-containing capsules	Alleviating hangover, liver protection	Chu (2013), Fu (2015)
Cosmetics	GSH-enriched yeast extract supplement	Inhibition of melanin production	Nakagawa et al. (2016)
Cosmetics	Whitening cosmetic, skin cream	Skin whitener, antiaging drug	Wang (2014) and Watanabe et al. (2014)
Cosmetics	Tooth gel, mouth rinse	Reduction of ROS	Hersh (1999)
Medicine	Eyedropper, eye health supplement	Prevention of treatment of eye-related conditions or diseases	Thiermann (2010) and Saucedo and Ambati (2016)
Biotechnology	Cell culture media	Antioxidant	Sigmaaldrich.com

and purification of the product. Several microorganisms accumulate glutathione and can thus be applied for glutathione production, among which the yeasts, *S. cerevisiae* or *Candida utilis*, are the most widely used. Yeasts are fast growing organisms able to accumulate high intracellular concentrations of glutathione, can be grown to high cell densities on inexpensive substrates, and are easy to handle on a large scale.

However, the synthesis and intracellular accumulation of glutathione are controlled by means of various complex molecular mechanisms (Li et al. 2004; Schmacht et al. 2017a).

The concentration of glutathione in wild-type *S. cerevisiae* strains is reported to occur within the range of 0.1–1.0% of the dry weight (dwt) of cells, depending on growth conditions and the characteristics of individual strains (Bachhawat et al. 2009). Different strategies have been proposed to increase glutathione accumulation in yeast. The application of the producer strains with increased glutathione biosynthetic capacity, obtained either by random mutagenesis or by means of genetic engineering techniques, combined with supplying the precursor amino acids, particularly cysteine, in the cultivation media, has been suggested the main strategies for increase of glutathione production (Schmacht et al. 2017a, b; Wang et al. 2007).

The medium composition is a key factor to achieve high cell densities and high intracellular glutathione accumulation (Li et al. 2004). The optimal production conditions can be found using design of experiments (DoE) approaches or application of alternative substrates (Cha et al. 2004; Zhang et al. 2007). For instance, Yoshida et al. (2011) used an engineered *S. cerevisiae* strain, which expressed amylases for direct utilization of starch. Recently, Schmacht et al. (2017b) developed a chemically defined medium based on yeast's elemental composition, which was suited for high cell density cultivation and high intracellular glutathione concentrations at the same time. The fermentation medium was improved in a way that yeast extract as a complex ingredient could be avoided from the medium. Ammonium sulfate, which serves both as a nitrogen and sulfur source, was beneficial for both biomass and glutathione production.

In most cases, complex media are used for glutathione production as their preparation is simple and all nutritional demands are covered with few ingredients. However, chemically defined media exhibit several advantages, such as high process reproducibility, robustness, a less challenging and faster scale-up from laboratory to industrial scale. The sulfur content in the medium is believed to be a key factor for glutathione production, and the low sulfur availability in the medium can result in consumption of glutathione by the cells as a sulfur source via the γ-glutamyl-cycle (Lorenz et al. 2016). Regarding bioeconomy aspects there is still potential to optimize production media to achieve high reproducibility as well as low-cost complex compounds.

The most popular strategy to increase glutathione production was the single-shot addition of cysteine, glutamate, and glycine as precursors for its synthesis. Additionally, combinations of other inducing amino acids, such as methionine, serine, or cysteine ethyl ester, were also applied (Lorenz et al. 2016). Cysteine was discovered as a key amino acid for glutathione production (Alfafara et al. 1992a; Wang et al. 2012). A suitable L-cysteine addition strategy should be developed to increase glutathione production without causing growth inhibition. Alfafara et al. (1992b) found that single-shot addition of L-cysteine was better than continuous addition, where the concentration of L-cysteine was constant. Therefore, a mass balance model-based feeding strategy in which L-cysteine was added in a single-shot manner to a culture entering the glutathione production phase resulted in approximately twofold increased production rate (Alfafara et al. 1992b).

The second important prerequisite for the maximal intracellular glutathione content is a suitable cultivation strategy. The most common strategy for glutathione production in an industrial scale is the fed-batch mode for high cell densities, although, alternatives such as chemostat or repeated fed-batch approaches are in the focus of research as well. In the process of glutathione production, there are two dominant strategies of the control, such as feedforward control based on the C-source feeding profile and feedback control based on the ethanol concentration. Both these strategies are directed to prevent the formation of by-products, especially ethanol during usage of glucose under aerobic conditions (Schmacht et al. 2017a).

The choice of suitable production strain is a crucial parameter for an efficient glutathione accumulation. Usually, strains already exhibiting high intracellular glutathione concentrations are preferred for further optimization. The optimization of the strain may be achieved by means of evolutionary or metabolic engineering. Overexpression of *GSH1* and *GSH2* genes has been the first attempts to improve glutathione biosynthesis in the cells. However, in most reported studies, the overexpression of either *GSH1* or *GSH2* (or both) in yeast has not led to more than twofold increase in intracellular glutathione level (Bachhawat et al. 2009; Grant et al. 1997). This can be due to feedback inhibition on glutathione biosynthetic reactions, excretion of excess of tripeptide from the cells, or its degradation of by γ-glutamyltranspeptidase (Ecm38p in yeast). The engineering of sulfur metabolism in *S. cerevisiae* has been shown as a more valuable method to increase glutathione production than the simple addition of cysteine to a yeast culture. Such metabolic engineering of a sulfate assimilation reactions allowed to avoid cell growth inhibition related to externally added cysteine toxicity. The intracellular glutathione content increased up to 1.2- and 1.4-fold higher than that of the parental strain by overexpression of adenylylsulfate kinase (*MET14*) and PAPS reductase (*MET16*), the genes involved in sulfate assimilation pathway (Hara et al. 2012).

For the isolation of glutathione overproducers, several selection strategies have been widely described in the literature. The physical or chemical mutagenesis methods used included UV, X-radiation, γ-radiation, and N-methyl-N′-nitro-N-nitrosoguanidine treatment, as well as resistance to highly toxic compounds such as ethionine, 1,2,4-triazole and sodium cyanide (Li et al. 2004). The mechanism for the selection, in most cases, was to disrupt or release the feedback inhibition of Gsh1 by glutathione. For example, Lai et al. (2008) were able to select a mutant resistant to N-methyl-N′-nitro-N-nitrosoguanidine, sodium azide, 1.2.4-triazole, and methylglyoxal, which accumulated glutathione in fourfold higher concentrations (32–34 mg/g dry cell weight) than the respective wild-type strain. The mutant was stable and retained the ability to produce glutathione with high yield even after 30 generations and after 3-year storage. The major problem with random mutagenesis is that the mutations are uncontrolled and may cause growth inhibition or various undesired side effects. The elucidation of biological mechanisms causing the higher glutathione accumulation in mutants obtained by random mutagenesis could assist in the generation glutathione overproducers by means of genetic engineering. Evolutionary engineering of *S. cerevisiae* strain using acrolein as a selection agent resulted in intracellular glutathione concentration of 5.9% under lab-scale bioreactor conditions (Patzschke et al. 2015). The high cell densities and high intracellular glutathione contents were achieved by genetic engineering of nonconventional yeast *Komagataella phaffii* (*Pichia pastoris*) and *Ogataea* (*Hansenula*) *polymorpha* (Fei et al. 2009; Ubiyvovk et al. 2011a).

However, the usage of such genetically modified organisms (GMOs) may be applied for bulk glutathione production but is limited for food applications and is a subject to obligatory declaration (2001/18/EG) in the European Union as consumers usually do not accept GMOs in food products. In general, with increased adaption of the strain properties to high glutathione productivity in a certain fermentation process, the stability of strains under large-scale conditions becomes more challenging (Schmacht et al. 2017a).

The extraction of glutathione from yeast biomass is another important process step of its production. Extraction with hot water (80–90 °C) is commonly used in industrial production of glutathione (Li et al. 2004). Application of different solvents has also been studied. The extraction of glutathione using 25% (v/v) ethanol at room temperature for 60 minutes was also described (Xiong et al. 2009). Such extraction process has many advantages, such as lesser consumption of energy and nondestruction of the cells resulting in a less protein extraction which reduce the complexity and cost of the purification process. However, the solvent extraction does not have significant advantages with regard to extraction yield. Alternatively, induced biological excretion of glutathione from cells can be used. Under normal growth conditions, intracellular glutathione from the yeast cytosol has been shown to be secreted at low levels and then is taken up again by the glutathione transporter (Opt1). Knockout of *OPT1* gene in *S. cerevisiae* leads to threefold higher levels of glutathione in the extracellular medium. A recent genome wide study has tried to identify genes which, upon disruption, would lead to increased glutathione secretion into the intracellular medium (Perrone et al. 2005). The ATP-dependent permease Adp1 was identified as a novel glutathione export ABC protein (Gxa1) in *S. cerevisiae* based on the homology of the protein sequence with known human glutathione export ABC protein (ABCG2). Overexpression of this *GXA1* gene improved the extracellular glutathione production by up to 2.3-fold as compared to the platform host strain lacking the glutathione degradation protein γ-glutamyl transpeptidase (encoded by the *CIS2* gene) and glutathione uptake protein Opt1. Moreover, combinatorial overexpression of the *GXA1* gene and the genes involved in glutathione synthesis in the platform host strain increased the extracellular glutathione production by up to 17.1-fold (Kiriyama et al. 2012).

Another possible way to increase glutathione secretion from the cells is by modulating the fermentation conditions. A low pH (pH 1.5) has proved to lead to increased secretion of glutathione (Nie et al. 2005). Low concentrations of surfactants added to fermentation have also been used to achieve higher levels of tripeptide in the extracellular medium without affecting significantly the growth and viability of the cells. Glutathione secreted into the medium has been found to be predominantly in the reduced form at a GSH:GSSG ratio of 25–30:1 (Bachhawat et al. 2009).

Glutathione produced by fermentation using yeasts is safer for food products than the same compound produced by other microorganisms, such as recombinant bacteria. However, bacteria can produce a significantly higher amount of glutathione in a shorter time of cultivation.

In *E. coli*, overexpression of *gshA* and *gshB* resulted in the increased activities of the corresponding enzymes in 10- and 14-fold, respectively; however, the intracellular glutathione concentration increased only 1.3-fold as compared to the wild-type strain. Nevertheless, the strain was an excellent glutathione biosynthesis system, producing up to 5 g/L of extracellular glutathione in the presence of the three precursor amino acids (Gushima et al. 1983). Stimulatory effect on glutathione production was observed in recombinant *E. coli* when only 9 mM L-cysteine as precursor amino acid was added to the culture at 12 h (Li et al. 2014).

An extremely high intracellular concentration of glutathione (140 mM) upon addition of 5 mM L-cysteine was achieved in *Lactococcus lactis* expressing the genes *gshA* and *gshB* from *E. coli* (Li et al. 2005). The highest intracellular glutathione content achieved in a bacterium could be explained by the fact that *L. lactis* is naturally not capable of synthesizing glutathione and consequently lacks γ-glutamyltranspeptidase activity as well as the feedback inhibition on biosynthetic reactions.

Incomparably high concentrations of 5870–15,210 mg/L of glutathione were achieved using genetically engineered strains of *E. coli* that exhibited different bifunctional glutathione synthetases, GshF, which avoid product inhibition of the GCS (Wang et al. 2016; Zhang et al. 2016). However, problems can occur as bacteria like *E. coli* may include endotoxins; therefore the selection and development of the production strain have to be carefully considered with regard to the later application of the product and the final production process (Schmacht et al. 2017a).

Therefore, several attempts to express *gshF* genes were made for different yeast species. The GshF proteins, from *L. monocytogenes*, *S. agalactiae*, and *L. plantarum*, were each cloned and expressed in *P. pastoris*; however, only GshF from *L. monocytogenes* displayed significant protein expression and catalytic activity in this methylotrophic yeast species and resulted in reduced accumulation of the intermediate metabolite γ-glutamylcysteine (Ge et al. 2012).

The bifunctional glutathione synthetase gene from *S. thermophilus* integrated into ribosomal DNA of *S. cerevisiae* at high copy number resulted in threefold higher glutathione contents accumulated in the recombinant strain as compared to the reference strain (Qiu et al. 2015).

The putative gshF from *A. pleuropneumoniae* resynthesized according to codon usage of *S. cerevisiae* was functionally expressed in *S. cerevisiae* under control of its GAP promoter. Moreover, two fusion proteins Gsh2-Gsh1 and Pro1-GshB were constructed to increase the two-step coupling efficiency of glutathione synthesis. The coexpression of three proteins in *S. cerevisiae* resulted in the highest glutathione accumulation as compared to the expression of each of these proteins separately.

For comparison, *gshF* gene from the nonpathogenic lactic acid bacterium *Streptococcus thermophilus* was overexpressed in tobacco plant. The leaves of plant transformants accumulated 30 times elevated amounts of glutathione relative to that of the nontransformed parental plants (Liedschulte et al. 2010).

A compilation of different strategies for the enhancement of biomass and glutathione concentration in different microorganisms is given in Table 6.2.

6.5 *Ogataea (Hansenula) polymorpha* as a Promising Glutathione Producer

During the last decades, the methylotrophic yeast *O. polymorpha* has gained increasing interest both for basic research and biotechnological applications, such as characterizing the mechanisms of thermotolerance, peroxisome homeostasis, production of numerous heterologous proteins, and high-temperature alcoholic fermentation (Dmytruk et al. 2016, 2017; Ubiyvovk et al. 2011a). The industrial significance of *O. polymorpha* relies on several technologically interesting features such as the ability to grow at high cell densities in bioreactors, capacity to use methanol as sole carbon source, and the availability of strong both regulated and constitutive promoters allowing to reach high product yields. Similar to *S. cerevisiae*, *O. polymorpha* is characterized by simple cultivation mode in inexpensive growth media, well-established genetic tools and experience on industrial cultivation. Additionally, availability of the complete genome sequence and established proteome and transcriptome databases renders *O. polymorpha* as a suitable organism for metabolic engineering to modify and improve particular biosynthetic pathways (Kim et al. 2013; Riley et al. 2016). This yeast species is highly resistant to different stress conditions, induced by heavy metals, xenobiotics, and different pollutants. The methylotrophic yeast *O. polymorpha* is considered to be a rich source of glutathione, due to the role of this tripeptide in detoxifications of key intermediates of methanol metabolism, formaldehyde, hydrogen peroxide, and alkyl hydroperoxides, accumulated during methylotrophic growth (Hartner and Glieder 2006).

In order to obtain a properly functioning system that protects cells against oxidative stress, sufficient glutathione is necessary as a cosubstrate in the reaction mediated by glutathione peroxidases. Glutathione is also important for the formation of *S*-HMG (*S*-hydroxymethyl glutathione), formaldehyde conjugate that serves as an intermediate metabolite in the catabolism of methanol (Fig. 6.2). Thus, the amount of glutathione available for catabolic and antioxidative pathways is important for yeast methylotrophy (Yurimoto et al. 2011). Glutathione is required for metabolism of methanol by detoxification of formaldehyde, hydrogen peroxide, and alkyl hydroperoxides using formaldehyde dehydrogenase, formaldehyde reductase, and glutathione peroxidase reactions in yeasts. Methanol-induced cells of the wild-type strains, and especially of formaldehyde dehydrogenase- and formaldehyde

Table 6.2 The known approaches for enhancement of biomass and glutathione concentration in the cells of different yeast species. Modified from Schmacht et al. (2017a)

Studied substances/parameters	Strain	Process mode	Biomass (max.) (g/L)	Glutathione (max.) (mg/L)	(%)	References
Feedforward/feedback control system	*S. cerevisiae* KY 6186	Fed-batch	63.8	2360	3.7	Sakato and Tanaka (1992)
Precursors: Cys, Glu, Gly, Ser	*S. cerevisiae* T65	Fed-batch	133.0	1875	1.4	Wen et al. (2004)
Precursors: Cys, Glu, Gly	*S. cerevisiae* T65	Fed-batch	133.3	2190	1.6	Wen et al. (2006)
RQ, precursors: Cys, Glu, Gly	*S. cerevisiae* G-14	Fed-batch	132.0	2020	1.5	Wang et al. (2007)
Feedback control of glucose feeding rate, precursor: Cys	*S. cerevisiae* GE-2	Fed-batch	110.0	2280	2.1	Shang et al. (2008)
Precursors: Cys, Glu, Gly	*S. cerevisiae* 26–2	Fed-batch	105.0	2250	2.1	Wang et al. (2012)
RQ-feeding, precursors: Cys, Glu, Gly	*S. cerevisiae* T65	Fed-batch	126.0	2100	1.7	Xiong et al. (2009)
Cell permeabilization via detergents, precursors: Cys, Gly	*S. cerevisiae* GB Italy (commercial)	Batch	100.0	3200	3.2	Rollini et al. (2010)
Precursors: Cys, Glu, Gly, ATP addition	*C. utilis* WSH 02–08	Fed-batch	100.6	2043	2.0	Liang et al. (2008)
Precursors: Cys, Glu, ATP addition, SDS addition	*C. utilis* WSH 02–08	Fed-batch	101.0	2485	2.5	Liang et al. (2010)
Precursors: Cys, Glu, ATP addition, coupled with H_2O_2 stresses	*C. utilis* WSH 02–08	Fed-batch	112.3	2448	2.2	Wang et al. (2010)
GMO strain, precursors: Cys, Glu, Gly	*Pichia pastoris* D18	Fed-batch	98.2	4150	4.2	Fei et al. (2009)
Precursor: Cysteine ethyl ester	*Saccharomyces bayanus* Sa-00645	Batch	21.9	853	3.9	Lorenz et al. (2016)

(continued)

Table 6.2 (continued)

Studied substances/parameters	Strain	Process mode	Biomass (max.) (g/L)	Glutathione (max.) (mg/L)	Glutathione (max.) (%)	References
GMO strain	*Hansenula polymorpha* Moxp-GSH2	Fed-batch	72.0	2270	3.2	Ubiyvovk et al. (2011a, b)
GMO strain with bifunctional glutathione synthetase gshF from *Streptococcus thermophilus*, precursors: Cys, Glu, Gly	*Escherichia coli* BL21(pUC18-gshF)	Fed-batch	31.9	15,210	47.7	Wang et al. (2015)

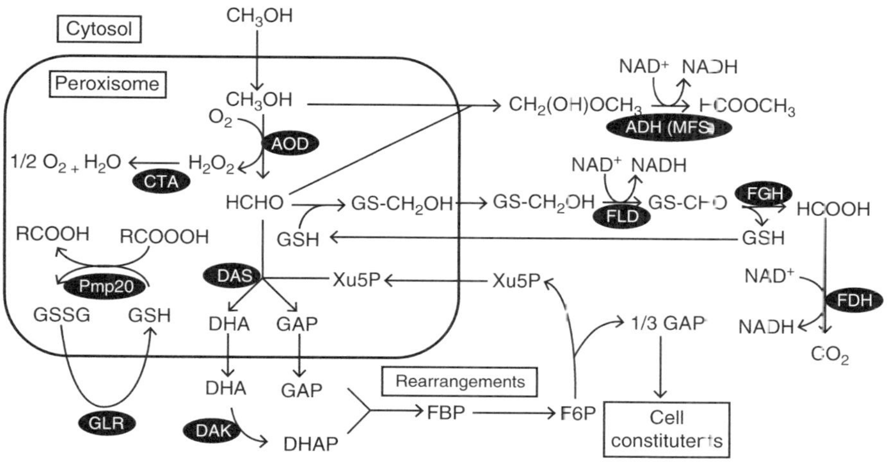

Fig. 6.2 Methanol metabolism in methylotrophic yeasts. Enzymes: *ADH (MFS)* alcohol dehydrogenase (methylformate-synthesizing enzyme), *AOD* alcohol oxidase, *CTA* catalase, *DAK* dihydroxyacetone kinase, *DAS* dihydroxyacetone synthase, *FDH* formate dehydrogenase, *FGH* S-formylglutathione hydrolase, *FLD* formaldehyde dehydrogenase, *GLR* glutathione reductase, *Pmp20* peroxisome membrane protein which has glutathione peroxidase activity. Abbreviations: *DHA* dihydroxyacetone, *DHAP* dihydroxyacetone phosphate, *F6P* fructose 6-phosphate, *FBP* fructose 1,6-bisphosphate, *GAP* glyceraldehyde 3-phosphate, *GS-CH2OH* S-hydroxymethyl glutathione, *GS-CHO* S-formylglutathione, *GSH* reduced form of glutathione, *GSSG* oxidized form of glutathione, *RCOOOH* alkyl hydroperoxide, *Xu5P* xylulose 5-phosphate. (Yurimoto et al. 2011)

reductase-deficient mutants, accumulated elevated levels of both formaldehyde and glutathione (Ubiyvovk et al. 2006).

GSH-deficient mutants of *O. polymorpha* failed to grow on methanol due to toxic accumulation of formaldehyde (Pocsi et al. 2004). Formaldehyde is present in wastewaters, and it is often accompanied by methanol and by chemical formaldehyde derivatives which are difficult to biodegrade. Although new alternative industrial technologies tend to reduce the risk of formaldehyde contamination, the use of this chemical is still widespread in developing countries, and it causes severe environmental problems. Formaldehyde itself is a highly reactive compound, toxic to living organisms (Kaszycki et al. 2001). Most of formaldehyde does not exist in vivo in a free state but is bound to endogenous nucleophiles, such as glutathione or tetrahydrofolate. It is one of the main indoor air pollutants, present in tobacco smoke, furniture, industrial adhesives, and varnishes (Achkor et al. 2003).

The mechanisms involved in the regulation of glutathione biosynthesis in methylotrophic yeast remain obscure though they have a fundamental function in methylotrophic growth. Search of genes required for the biosynthesis of glutathione in *O. polymorpha* was done by Ubiyvovk et al. (2002) by functional complementation of GSH-deficient (*gsh*) mutants. A defect of the first gene of glutathione biosynthetic pathway, *GSH1*, which encodes GCS in the yeast *S. cerevisiae*, led to inability of *gsh* mutants to grow in synthetic media without exogenous glutathione and caused significantly elevated resistance to N-methyl-N-nitro-N-nitrosoquanidine (MNNG) and increased sensitivity to cadmium ions (Ubiyvovk et al. 2002). Mutants of *O. polymorpha* unable to grow on the media with methanol or multicarbon substrates without exogenous glutathione were divided into two complementation groups, designated as *gsh1* and *gsh2*. The *O. polymorpha GSH2* gene cloned from *O. polymorpha* gene library by complementation of the *gsh2* mutation was shown to be a homolog of *S. cerevisiae GSH1* gene coding for GCS (Ubiyvovk et al. 2002).

The *O. polymorpha GSH1* gene, which restored the glutathione level, MNNG sensitivity, cadmium resistance, and growth on minimal GSH-deficient medium of *O. polymorpha gsh1* mutant, displayed homology to the *S. cerevisiae MET1* gene, encoding S-adenosyl-L-methionine uroporphyrinogen III transmethylase responsible for the biosynthesis of sulfite reductase cofactor, sirohaem (Ubiyvovk et al. 2011b). It is known that the cell requirement for sulfur can be fulfilled by the uptake of sulfur-containing amino acids, glutathione, or by assimilation of sulfate (through consequent reduction to sulfite and sulfide) into organic compounds such as cysteine and/or homocysteine. Similar to the *S. cerevisiae met1* mutant, the *O. polymorpha gsh1/met1* mutant did not grow in minimal medium with sulfate and sulfite as a sole sulfur source. However, in contrast to the *S. cerevisiae met1* mutant that satisfied nutritional needs in sulfur with S-amino acids and their derivatives, the null *O. polymorpha gsh1/met1* mutant displayed negligible ability to assimilate methionine and S-adenosylmethionine and reduced growth activity on S-adenosylhomocysteine and homocysteine as a sole sulfur source. In sulfur-free medium supplemented with cysteine or glutathione, *O. polymorpha gsh1/met1* mutant showed total growth restoration. Differences in growth phenotype of *met1* mutants from *S. cerevisiae* and *O. polymorpha* could be explained by distinctions in sulfate assimilation pathways (on the step of H_2S incorporation in cysteine or/and homocysteine) (Ubiyvovk et al. 2011b).

Indeed, recently the detailed sulfur metabolic pathway of *O. polymorpha* was reconstructed and revealed the absence of de novo synthesis of homocysteine from inorganic sulfur in this yeast (Sohn et al. 2014). Thus, the direct biosynthesis of cysteine from sulfide is the only pathway of synthesizing sulfur amino acids from inorganic sulfur in *O. polymorpha*, despite the presence of both directions of transsulfuration pathway. In this pathway, sulfide is condensed with O-acetylserine to generate cysteine in a process catalyzed by cysteine synthase (OAS pathway). In contrast, *S. cerevisiae* sulfide is condensed with O-acetylhomoserine to generate homocysteine, which can be converted to cystathionine and then to cysteine (Fig. 6.3).

Interestingly, analysis of *O. polymorpha* genome indicates that *O. polymorpha* does not have an ORF coding for O-acetylhomoserine sulfhydrylase or homocysteine synthase (encoded by *MET17* or *MET25* in *S. cerevisiae*), which catalyzes the synthesis of homocysteine from homoserine by incorporation of sulfide. Considering that homocysteine is the central molecule for the biosynthesis of sulfur amino acids in all organisms studied, the absence of de novo synthesis of homocysteine from inorganic sulfur might be an exceptional feature of *O. polymorpha*. Moreover, only cysteine, but no other sulfur amino acid, was able to repress the expression of a

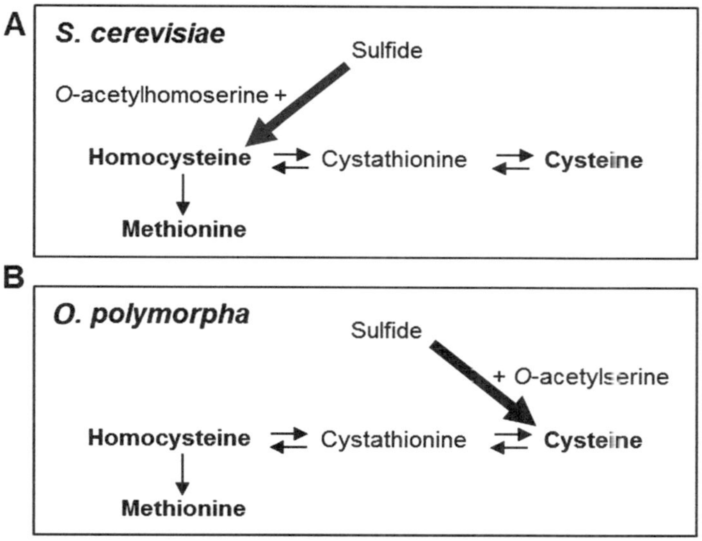

Fig. 6.3 Different pathways of sulfur incorporation into carbon chains and transsulfuration in yeast *S. cerevisiae* (**a**) and *O. polymorpha* (**b**). Inorganic sulfide can be combined with O-acetylhomoserine or O-acetylserine to produce homocysteine and cysteine, respectively. (Modified from Sohn et al. 2014)

subset of sulfur genes, suggesting its central and exclusive role in the control of *O. polymorpha* sulfur metabolism (Sohn et al. 2014) (Fig. 6.4).

It was also demonstrated that GCS activity was not regulated by the sulfur source (sulfate ions, cysteine, or glutathione) in *O. polymorpha* NCYC495 *leu1–1* wild-type strain. Besides, it was shown that GCS activity did not correlate with the increased cellular GSH + GSSG levels in *O. polymorpha* wild-type strains of different genetic lines, indicating that the cellular level of glutathione is balanced by complex processes of transport, degradation, as well as of biosynthesis of glutathione precursor, cysteine (Ubiyvovk et al. 2011b).

Glutathione is important for electrophilic xenobiotic vacuolar accumulation and extrusion from baker's yeast *S. cerevisiae* and methylotrophic yeast *O. polymorpha* (Ubiyvovk et al. 2006). *GGT1* gene of the methylotrophic yeast *O. polymorpha* appears to be a structural and functional homolog of *S. cerevisiae* CIS2/ECM38 gene encoding γ-glutamyltranspeptidase, which catalyzes the transfer of the γ-glutamyl moiety of glutathione and γ-glutamyl compounds to amino acids and also the hydrolytic release of L-glutamate from γ-glutamyl compounds, glutathione, and its S-substituted derivatives. Yeast cells of null *ggt1* mutant as well as the wild-type strain were able to utilize exogenous glutathione as a sole sulfur source and grow in media with different nitrogen sources (ammonium or glutamate). Therefore, yeast *O. polymorpha*, similar to *S. cerevisiae* (Kumar et al. 2003), probably possesses an alternative glutathione degradation pathway to supply it with sulfur and nitrogen from glutathione. It was shown that γ-glutamyltranspeptidases of both *O.*

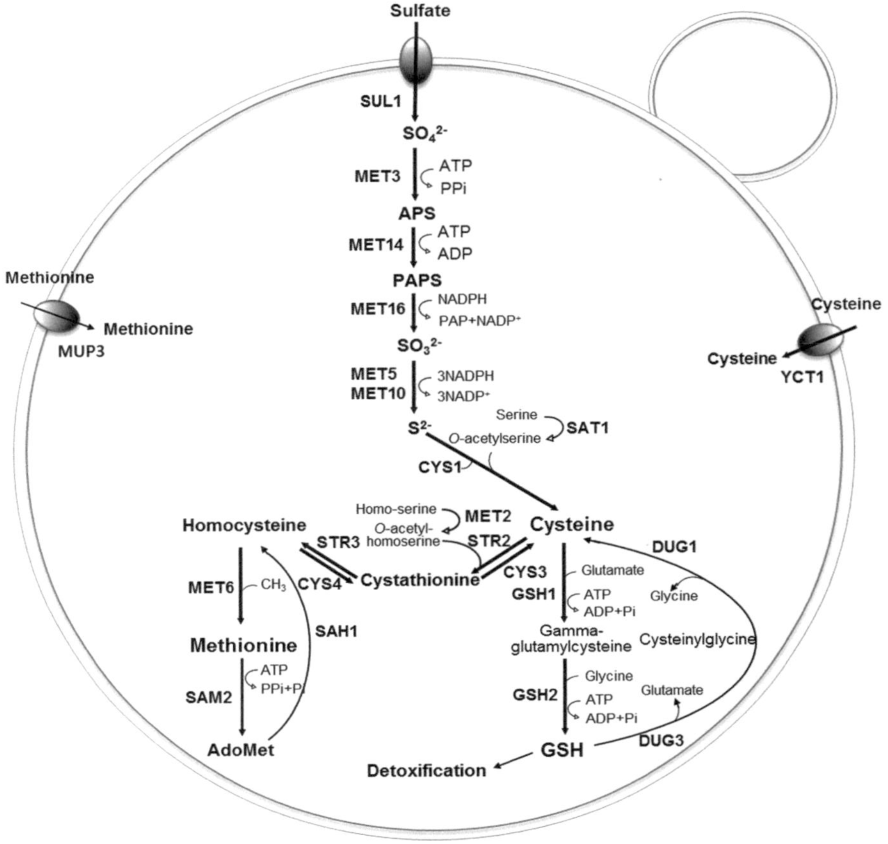

Fig. 6.4 Schematic representation of a reconstructed sulfur pathway of *O. polymorpha*. Genes involved in sulfur metabolism and regulation: *SUL1* sulfate permease, *MET3* ATP sulfurylase, *MET14* adenylylsulfate kinase, *MET16* 3′-phosphoadenylsulfate reductase, *MET5* sulfite reductase beta subunit, *MET10* subunit alpha of assimilatory sulfite reductase, *MET2* homoserine-O-acetyltransferase, *SAT1* serine-O-acetyltransferase, *STR2* cystathionine gamma-synthase, *STR3* cystathionine beta-lyase, *CYS1* cysteine synthase, *CYS3* cystathionine gamma-lyase, *CYS4a, CYS4b, CYS4c* cystathionine beta-synthase, *GSH1* gamma glutamylcysteine synthetase, *GSH2* glutathione synthetase, *GTT1* glutathione *S*-transferase, *TTR1* glutaredoxin, *HYR1* glutathione peroxidase, *MET6* cobalamin-independent methionine synthase, *SAM2 S*-adenosylmethionine sythetase, *SAH1 S*-adenosyl-L-homocysteine hydrolase, *MET4* bZIP transcription factor Met4. (Modified from Sohn et al. 2014)

polymorpha and *S. cerevisiae* are involved in detoxification of electrophilic xenobiotics, as the corresponding mutants appeared to be defective in the disappearance of the fluorescent vacuolar complex of GSH with xenobiotic bimane and the further diffuse distribution of this complex in the cytosol. Metabolism of electrophilic xenobiotics in the yeasts *O. polymorpha* and *S. cerevisiae* probably occurs through a γ-glutamyltranspeptidase-dependent mercapturic acid pathway of GSH-xenobiotic detoxification, similar to that known for mammalian cells, with cysteine-xenobiotics and/or N-acetylcysteine-xenobiotics as the end products (Ubiyvovk et al. 2006).

Glutathione was identified as a sole Cd_2^+ intracellular chelator which suggests that sequestration of this heavy metal in *O. polymorpha* occurs similar to that found in *S. cerevisiae* but differently to *Schizosaccharomys pombe* and *Candida glabrata* which both synthesize phytochelatins. It was found that *gsh2* mutants, impaired in the first step of glutathione biosynthesis, are characterized by increase in net Cd_2^+ uptake by the cells, whereas *gsh1/met1* and *ggt1* mutants impaired in sulfate assimilation and glutathione catabolism, respectively, lost the ability to accumulate Cd_2^+ intracellularly. Apparently *O. polymorpha*, similar to *S. cerevisiae*, forms a Cd-GSH complex in the cytoplasm, which in turn regulates Cd_2^+ uptake. The storage of Cd_2^+ ions in *S. cerevisiae* is controlled by the Ycf1 protein, which is responsible for sequestration of bis(glutathionato)cadmium (GS-Cd-SG), as well as GSH conjugates with electrophilic compounds (GS-X) complexes from cytosol into the vacuole. Cd_2^+ tolerance in *S. cerevisiae* is also rendered by Yap1 transcription factor, which positively regulates the expression of *YCF1* gene as well as *GSH1* and *GLR1* genes coding for GCS and glutathione reductase, respectively (Blazhenko et al. 2006).

During the last decades the metabolic engineering approaches and optimization of the cultivation regimes were applied for improvement of glutathione production in the yeast *O. polymorpha*. The glutathione producing capacity were studied in *O. polymorpha* DL-1 wild-type strain and several recombinant strains depending on parameters of cultivation (dissolved oxygen tension, pH, stirrer speed), carbon substrate (glucose, methanol), and type of overexpressed genes involved in glutathione biosynthetic pathway during batch and fed-batch fermentations. The high-cell-density cultivation of *O. polymorpha* DL-1 wild-type strain led to accumulation of a large amount of glutathione under dual control of substrate feeding (exponential feeding rate mode combined with feedback control of substrate feed by DO set point using on-off regime). Under optimized conditions of glucose fed-batch cultivation, the engineered strains accumulated more than 2250 mg of intracellular glutathione per liter of culture medium, which slightly exceeds the known before maximal glutathione production in *S. cerevisiae* of 2020 mg/L (however, last number was obtained with addition of amino acids to cultural medium, which increases the costs of the aimed product) (Wang et al. 2007). Moreover, one of the constructed *O. polymorpha* recombinant strains accumulated five times higher titer of extracellular glutathione as compared to the best level achieved before for yeasts in methanol medium. In *O. polymorpha*, the glutathione pool was raised by overexpression of *MET4* gene, the central regulatory gene of sulfur metabolism Therefore, the positive effect of the separate multicopy overexpression of *GSH2* and *MET4* genes under control of their native promoters on glutathione production titer was demonstrated (Ubiyvovk et al. 2011a). Moreover, overexpression of *MET4* gene in the background of overexpressed *GSH2* gene resulted in fivefold increased glutathione production during shake flask cultivation as compared to the wild-type strain, reaching 2167 mg/L. During bioreactor cultivation, glutathione accumulation by obtained recombinant strain was fivefold increased relative to that by the parental strain with overexpressed only *GSH2* gene, on the first 25 hours of batch cultivation in mineral medium. Obtained results suggest involvement of Met4 transcriptional activator in regulation of glutathione synthesis in the methylotrophic yeast *O. polymorpha*. The transcription of the genes encoding proteins involved in the synthesis of sulfur-containing amino acids, known as the *MET* regulon, is rapidly induced if methionine and cysteine are limiting for growth. This induction is mediated by the transcription factor Met4, whose function requires different combinations of the cofactors Cbf1, Met28, Met31, and Met32 (Lee et al. 2010; Sadhu et al. 2014). Both Cbf1 and Met31/Met32 binding sites were found in the promoter of *S. cerevisiae GSH1* gene, encoding GCS (Lee et al. 2010). Met4 is also known to regulate the response of *S. cerevisiae GSH1* to cadmium stress, which additionally requires Yap1 (Dormer et al. 2000). In *O. polymorpha* homolog of *S. cerevisiae* Met4 a potential CDEI motifs were revealed similar to Cbf1 binding sites involved in expression of *S. cerevisiae GSH1* gene under cadmium stress (Yurkiv et al. 2018).

As shown in Grabek-Lejko et al. (2011), glutathione level is related to alcoholic fermentation. During the study, several *O. polymorpha* strains with increased glutathione level were tested: the ones expressing either *GSH2* (involved in glutathione biosynthesis) or *MET4* (gene involved in the regulation of sulfur metabolism) from a multicopy vector as well as mutants defective in genes involved in glutathione synthesis and degradation (*gsh2*, *ggt1*). The ethanol production of *O. polymorpha* strains overexpressing genes *MET4* or *GSH2* was two- and threefold higher, respectively, as compared to the wild-type strain which produced about 1.5% of ethanol during fermentation of glucose. Higher ethanol accumulation was accompanied by enhanced glucose consumption by the corresponding strains. Strains defective in glutathione synthesis (*gsh2*) or degradation (*ggt1*) did not differ from the parental wild-type strain regarding ethanol accumulation. The *ggt1* mutation did not lead to any change in the cellular glutathione pool, and the defect in the *gsh2* mutants was counterbalanced by glutathione supplemented to the growth medium (at 0.1 mM concentration) (Grabek-Lejko et al. 2011).

Design and construction of novel glutathione producers are of great importance for improvement of the production of this biotechnologically and medically important tripeptide. One of the approaches for further increase of the intracellular glutathione level in *O. polymorpha* is engineering of GCS protein for alleviation from feedback inhibition normally exerted by glutathione. GCS feedback inhibition engineering may be carried out on the isolated *O. polymorpha* mutants defective both in glutathione uptake and secretory pathways which possess elevated glutathione pool. It is known that *S. cerevisiae* mutants defective in glutathione transport and in secretory pathways overaccumulate glutathione in the medium (Perrone et al. 2005). Another approach for increased intracellular glutathione pool is overexpression of genes involved in the cysteine biosynthesis, which is a key precursor of glutathione, or identification of the genes involved in the regulation of glutathione biosynthesis in yeast *O. polymorpha*.

6.6 Concluding Remarks

Glutathione has been defined as the intracellular redox buffer, and since its discovery understanding of other cellular functions has been the subject of intensive research. Glutathione is an essential molecule for the most living cells. The mouse knockout of GCS, encoding the rate-limiting enzyme in glutathione synthesis, is indeed embryonic lethal, and both yeast and mammalian cells lacking it are unable to grow without exogenous glutathione supplementation. However, this requirement is not probably linked to its redox properties as only trace amounts of glutathione are needed for viability in both yeast and mammals. Furthermore, the lack of glutathione may be compensated by the increased activity of the thioredoxin antioxidant pathway to maintain appropriate cellular redox state. Glutathione essential nature in yeasts has been traced to its requirement for the cytosolic assembly of iron-sulfur clusters (Fe-S), which are the prosthetic groups of many essential proteins. Glutathione requirement for Fe-S assembly are not excluded for mammals too, but it is also important for assisting the phospholipid hydroperoxidase GPx4, the activity of which is essential for preventing ferroptosis, a newly described non-apoptotic cell death pathway mediated by lipid peroxides, and requiring traces of iron (Toledano and Huang 2017). However, the reason why cellular glutathione is present in such high millimolar amounts inside the cells is not definitely known. Recent studies revealed significance of glutathione in the different compartments of the eukaryotic cell, such as mitochondrial matrix and intermembrane space, as well as endoplasmatic reticulum and peroxisomes (Calabrese et al. 2017; Delaunay-Moisan et al. 2017; Horiguchi et al. 2001).

The multiple functions of glutathione in cells result in its significance to prevent various pathophysiological states, opening a means for the use of glutathione as a therapeutic agent. Current studies are focused on the improvement of the bioavailability and targeting ability of glutathione in vivo using nanotechnology approaches.

Therefore, the field of glutathione production is of high importance in research and industry due to the broad applicability of glutathione itself and glutathione-enriched yeasts for food and health products. Historically, mainly yeasts were used as microbial producers; however, within the last years, metabolically engineered bacteria were constructed that show high glutathione production rate. Nowadays there is still potential for improvement of microbial glutathione production by using the most appropriate glutathione producer strain, low-priced raw materials, and developed efficient process strategy.

References

Achkor H, Diaz M, Fernandez RM, Biosca JA, Pare's X, Martinez MC (2003) Enhanced formaldehyde detoxification by overexpression of glutathione-dependent formaldehyde dehydrogenase from *Arabidopsis*. Plant Physiol 132:2248–2255

Alfafara CG, Kanda A, Shioi T, Shimizu H, Shioya S, Suga K (1992a) Effect of amino acids on glutathione production by *Saccharomyces cerevisiae*. Appl Microbiol Biotechnol 36:538–540
Alfafara CG, Miura K, Shimizu H, Shioya S, Suga K (1992b) Cysteine addition strategy for maximum glutathione production in fed-batch culture of *Saccharomyces cerevisiae*. Appl Microbiol Biotechnol 37:141–146
Allocati N, Masulli M, Di Ilio C, Federici L (2018) Glutathione transferases: substrates, inhibitors and pro-drugs in cancer and neurodegenerative diseases. Oncogenesis 7, 8
Bachhawat AK, Kaur A (2017) Glutathione degradation. Antioxid Redox Signal 27:1200–1216
Bachhawat A, Gangul D, Kaur J, Kasturia N, Thakur N, Kaur H, Kumar A, Yadav A (2009) Glutathione production in yeast. In: Satyanarayana T, Kunze G (eds) Yeast biotechnology: diversity and applications. Springer, Dordrecht, pp 259–280
Bachhawat AK, Thakur A, Kaur J, Zulkifli M (2013) Glutathione transporters. Biochim Biophys Acta 1830:3154–3164
Baek M, Choy JH, Choi SJ (2012) Montmorillonite intercalated with glutathione for antioxidant delivery: synthesis, characterization, and bioavailability evaluation. Int J Pharm 425:29–34
Bansal A, Simon MC (2018) Glutathione metabolism in cancer progression and treatment resistance. J Cel Biol 217:2291–2298
Blazhenko OV, Zimmermann M, Kang HA, Bartosz G, Penninckx MJ, Ubiyvovk VM, Sibirny AA (2006) Accumulation of cadmium ions in the methylotrophic yeast *Hansenula polymorpha*. BioMetals 19:593–599
Bourbouloux A, Shahi P, Chakladar A, Delrot S, Bachhawat AK (2000) Hgt1p, a high affinity glutathione transporter from the yeast *Saccharomyces cerevisiae*. J Biol Chem 275:13259–13265
Brandsma D, Kerklaan BM, Diéras V, Altintas S, Anders CK, Ballester MA, Gelderblom H, Soetekouw PMMB, Gladdines W, Lonnqvist F et al (2014) Phase 1/2a study of glutathione pegylated liposomal doxorubicin (2b3-101) in patients with brain metastases (BM) from solid tumors or recurrent high grade gliomas (HGG). Ann Oncol 16:v159–v167
Buonocore D, Grosini M, Giardina S, Michelotti A, Carrabetta M, Seneci A, Marzatico F (2016) Bioavailability study of an innovative orobuccal formulation of glutathione. Oxid Med Cell Longev 2:1–7
Cagnac O, Bourbouloux A, Chakrabarty D, Zhang MY, Delrot S (2004) AtOPT6 transports glutathione derivatives and is induced by primisulfuron. Plant Physiol. 135:1378–1387
Calabrese G, Morgan B, Riemer J (2017) Mitochondrial glutathione: regulation and functions. Antioxid Redox Signal 27(15):1162–1177
Carretero J, Obrador E, Anasagasti MJ, Martin JJ, Vidal-Vanaclocha F, Estrela JM (1999) Growth-associated changes in glutathione content correlate with liver metastatic activity of B16 melanoma cells. Clin Exp Metastasis 17:567–574
Cha J, Park J, Jeon B, Lee Y, Cho Y (2004) Optimal fermentation conditions for enhanced glutathione production by *Saccharomyces cerevisiae* FF-8. J Microbiol 42:51–55
Chakravarthi S, Jessop CE, Bulleid NJ (2006) The role of glutathione in disulphide bond formation and endoplasmic-reticulum-generated oxidative stress. Embo Rep 7:271–275
Chen G, Bunt C, Wen J (2015) Mucoadhesive polymers-based film as a carrier system for sublingual delivery of glutathione. J Pharm Pharmacol 67:26–34
Chiang HS, Maric M (2011) Lysosomal thiol reductase negatively regulates autophagy by altering glutathione synthesis and oxidation. Free Radic Biol Med. 51:688–699
Chu D (2013) Combination containing complex nucleoside, glutathione and yeast extract and application of combination in aspects of alleviating hangover and protecting liver. China Patent Application CN102886042 (A)
Collinson EJ, Grant CM (2003) Role of yeast glutaredoxins as glutathione S-transferases. J Biol Chem 278:22492–22497
Conticello C, Martinetti D, Adamo L, Buccheri S, Giuffrida R, Parrinello N, Lombardo L, Anastasi G, Amato G, Cavalli M et al (2012) Disulfiram, an old drug with new potential therapeutic uses for human hematological malignancies. Int J Cancer 131:2197–2203

Corso CR, Acco A (2018) Glutathione system in animal model of solid tumors: from regulation to therapeutic target. Crit Rev Oncol Hematol 128:43–57
Couto N, Malys N, Gaskell SJ, Barber J (2013) Partition and turnover of glutathione reductase from Saccharomyces cerevisiae: a proteomic approach. J Proteome Res 12 2885–2894
Couto N, Schooling SR, Dutcher JR, Barber J (2015) Proteome profiles of outer membrane vesicles and extracellular matrix of Pseudomonas aeruginosa biofilms. J Proteome Res 14:4207–4222
Couto N, Wood J, Barber J (2016) The role of glutathione reductase and related enzymes on cellular redox homoeostasis network. Free Radic Biol Med 95 27–42
Crum A (2011) Nutritional or therapeutic compositions and methods to increase bodily glutathione levels. United States Reissued Patent USRE42645 (E)
Dalton TP, Chen Y, Schneider SN, Nebert DW, Shertzer HG (2004) Genetically altered mice to evaluate glutathione homeostasis in health and disease. Free Rad Biol Med 37:1511–1526
Delaunay-Moisan A, Ponsero A, Toledano MB (2017) Reexamining the function of glutathione in oxidative protein folding and secretion. Antioxid Redox Signal 27:1178–1199
Dhaoui M, Auchere F, Blaiseau PL, Lesuisse E, Landoulsi A, Camadro JM, Haguenauer-Tsapis R, Belgareh-Touze N (2011) Gex1 is a yeast glutathione exchanger that interferes with pH and redox homeostasis. Mol Biol Cell 22:2054–2067
Dmytruk K, Kurylenko O, Ruchala J, Ishchuk O, Sibirny A (2016) Development of the thermotolerant methylotrophic yeast *Hansenula polymorpha* as efficient ethanol producer. In: Satyanarayana T, Kunze G (eds) Yeast diversity in human welfare. Springer, Singapore, pp 257–282
Dmytruk K, Kurylenko O, Ruchala J, Abbas C, Sibirny A (2017) Genetic improvement of conventional and nonconventional yeasts for the production of first- and second-generation ethanol. In: Sibirny A (ed) Biotechnology of yeasts and filamentous fungi. Springer, Cham, pp 1–38
Dong SC, Sha HH, Xu XY, Hu TM, Lou R, Li H, Wu JZ, Dan C, Feng J (2018) Glutathione S-transferase π: a potential role in antitumor therapy. Drug Des Devel Ther 12:3535–3547
Dormer UH, Westwater J, McLaren NF, Kent NA, Mellor J, Jamieson DJ (2000) Cadmium-inducible expression of the yeast *GSH1* gene requires a functional sulfur-amino acid regulatory network. J Biol Chem 275:32611–32616
Elskens MT, Jaspers CJ, Penninckx MJ (1991) Glutathione as an endogenous sulphur source in the yeast *Saccharomyces cerevisiae*. J Gen Microbiol 137:637–644
Fei L, Wang Y, Chen S (2009) Improved glutathione production by gene expression in *Pichia pastoris*. Bioprocess Biosyst Eng 32:729–735
Forman HJ, Zhang H, Rinna A (2009) Glutathione: overview of its protective roles, measurement, and biosynthesis. Mol Asp Med 30:1–12
Fraternale A, Brundu S, Magnani M (2017) Glutathione and glutathione derivatives in immunotherapy. Biological Chemistry 398:261–275
Fu D (2015) Yeast glutathione nutrition preparation. China Patent CN103750344 (B)
Fujiwara S, Kawazoe T, Ohnishi K, Kitagawa T, Popa C, Valls M, Genin S, Nakamura K, Kuramitsu Y, Tanaka N (2016) RipAY, a plant pathogen effector protein, exhibits robust γ-glutamyl cyclotransferase activity when stimulated by eukaryotic thioredoxins. J Biol Chem 291:6813–6830
Gaillard PJ, Kerklaan BM, Aftimos P, Altintas S, Jager A, Gladdines W, Lonnqvist F, Soetekouw P, Verheul H, Awada A et al (2014) Abstract CT216: Phase I dose escalating study of 2B3-101, glutathione PEGylated liposomal doxorubicin, in patients with solid tumors and brain metastases or recurrent malignant glioma. Cancer Res 74:CT216
Ganguli D, Kumar C, Bachhawat AK (2007) The alternative pathway of glutathione degradation is mediated by a novel protein complex involving three new genes in *Saccharomyces cerevisiae*. Genetics 175:1137–1151
Garcerá A, Barreto L, Piedrafita L, Tamarit J, Herrero E (2006) *Saccharomyces cerevisiae* cells have three Omega class glutathione S-transferases acting as 1-Cys thiol transferases. Biochem J 398:187–196
Gaucher C, Boudier A, Bonetti J, Clarot I, Leroy P, Parent M (2018) Glutathione: antioxidant properties dedicated to nanotechnologies. Antioxidants 7:62

Ge S, Zhu T, Li Y (2012) Expression of bacterial GshF in *Pichia pastoris* for glutathione production. Appl Environ Microbiol 78:5435–5439
Ge C, Spånning E, Glaser E, Wieslander Å (2014) Import determinants of organelle-specific and dual targeting peptides of mitochondria and chloroplasts in *Arabidopsis thaliana*. Mol Plant 7:121–136
Genin S, Denny TP (2012) Pathogenomics of the *Ralstonia solanacearum* species complex. Annu Rev Phytopathol 50:67–89
Gipp JJ, Chang C, Mulcahy RT (1992) Cloning and nucleotide sequence of a full-length cDNA for human liver γ -glutamylcysteine synthetase. Biochem Biophys Res Comm 185:29–35
Gipp JJ, Bailey HH, Mulcahy RT (1995) Cloning and sequence of the cDNA for the light subunit of human liver γ -glutamylcysteine synthetase and relative mRNA levels for heavy and light subunits in human normal tissues. Biochem Biophys Res Comm 206:584–589
Gogos A, Shapiro L (2002) Large conformational changes in the catalytic cycle of glutathione synthase. Structure 10:1669–1676
Gopal S, Borovok I, Ofer A, Yanku M, Cohen G, Goebel W, Aharonowitz Y (2005) A multidomain fusion protein in *Listeria monocytogenes* catalyzes the two primary activities for glutathione biosynthesis. J Bacteriol 187:3839–3847
Grabek-Lejko D, Kurylenko O, Sibirny V, Ubiyvovk V, Penninckx M, Sibirny A (2011) Alcoholic fermentation by wild-type *Hansenula polymorpha* and *Saccharomyces cerevisiae* versus recombinant strains with an elevated level of intracellular glutathione. J Ind Microbiol Biotechnol 38:1853–1859
Grant CM, MacIver FH, Dawes IW (1997) Glutathione synthetase is dispensable for growth under both normal and oxidative stress conditions in the yeast Saccharomyces cerevisiae due to an accumulation of the dipeptide γ-glutamylcysteine. Mol Biol Cell 8:1699–1707
Griffith OW, Meister A (1979) Translocation of intracellular glutathione to membrane-bound gamma-glutamyl transpeptidase as a discrete step in the gamma-glutamyl cycle: glutathionuria after inhibition of transpeptidase. PNAS USA. 76:268–272
Gulshan K, Rovinsky SA, Coleman ST, Moye-Rowley WS (2005) Oxidant-specific folding of Yap1p regulates both transcriptional activation and nuclear localization. J Biol Chem 280:40524–40533
Gupta P, Wright SE, Kim SH, Srivastava SK (2014) Phenethyl isothiocyanate: a comprehensive review of anticancer mechanisms. Biochim Biophys Acta 1846:405–424
Gushima H, Miya T, Murata K, Kimura A (1983) Construction of glutathione producing strains of Escherichiacoli B by recombinant DNA techniques. Appl Biochem Biotechnol 5:43–52
Hagen TM, Wierzbicka GT, Sillau AH, Bowman BB, Jones DP (1990) Bioavailability of dietary glutathione: effect on plasma concentration. Am J Physiol 259:G524–G529
Hara KY, Kiriyama K, Inagaki A, Nakayama H, Kondo A (2012) Improvement of glutathione production by metabolic engineering the sulfate assimilation pathway of *Saccharomyces cerevisiae*. Appl Microbiol Biotechnol 94:1313–1319
Hartner FS, Glieder A (2006) Regulation of methanol utilisation pathway genes in yeasts. Microb Cell Fact 5:39–60
Hassan MQ, Hadi RA, Al-Rawi ZS, Padron VA, Stohs SJ (2001) The glutathione defense system in the pathogenesis of rheumatoid arthritis. J Appl Toxicol 21:69–73
Hatem E, El Banna N, Huang ME (2017) Multifaceted roles of glutathione and glutathione-based systems in carcinogenesis and anticancer drug resistance. Antioxid Redox Signal 27:1217–1234
Hersh T (1999) Intra-oral antioxidant preparations. United States Patent Application US5906811 (A)
Horiguchi H, Yurimoto H, Kato N, Sakai Y (2001) Antioxidant system within yeast peroxisome. Biochemical and physiological characterization of CbPmp20 in the methylotrophic yeast *Candida boidinii*. J Biol Chem 276:14279–14288
Huang C, Anderson ME, Meister A (1993a) Amino acid sequence and function of the light subunit of rat kidney γ-glutamylcysteine synthetase. J Biol Chem 268:20578–20583

Huang C, Chang L, Anderson ME, Meister A (1993b) Catalytic and regulatory properties of the heavy subunit of rat kidney γ -glutamylcysteine synthetase. J Biol Chem 268:19675–19680
Huang ZZ, Chen C, Zeng Z, Yang H, Oh J, Chen L, Lu SC (2001) Mechanism and significance of increased glutathione level in human hepatocellular carcinoma and liver regeneration. FASEB J 15:19–21
Iantomasi T, Favilli F, Marraccini P, Magaldi T, Bruni P, Vincenzini MT (1997) Glutathione transport system in human small intestine epithelial cells. Biochim Biophys Acta 1330:274–283
Janowiak BE, Griffith OW (2005) Glutathione synthesis in *Streptococcus agalactiae*. One protein accounts for γ-glutamylcysteine synthetase and glutathione synthetase activities. J Biol Chem 280:11829–11839
Jez JM, Cahoon RE, Chen S (2004) *Arabidopsis thaliana* glutamate-cysteine ligase: functional properties, kinetic mechanism, and regulation of activity. J Biol Chem 279:33463–33470
Jones DP, Mody VC Jr, Carlson JL, Lynn MJ, Sternberg P Jr (2002) Redox analysis of human plasma allows separation of pro-oxidant events of aging from decline in antioxidant defenses. Free Radic Biol Med 33:1290–1300
Kannan R, Mittur A, Bao Y, Tsuruo T, Kaplowitz N (1999) GSH transport in immortalized mouse brain endothelial cells: evidence for apical localization of a sodium-dependent GSH transporter. J Neurochem 73:390–399
Karplus PA, Schulz GE (1987) Refined structure of glutathione reductase at 1.54 Å resolution. J Mol Biol 195:701–729
Kaszycki P, Tyszka M, Malec P, Kołoczek H (2001) Formaldehyde and methanol biodegradation with the methylotrophic yeast *Hansenula polymorpha*. An application to real wastewater treatment. Biodegradation 12:169–177
Kaur H, Kumar C, Junot C, Toledano MB, Bachhawat AK (2009) Dug1p Is a Cys-Gly peptidase of the gamma-glutamyl cycle of *Saccharomyces cerevisiae* and represents a novel family of Cys-Gly peptidases. J Biol Chem 284:14493–14502
Kaur H, Ganguli D, Bachhawat AK (2012) Glutathione degradation by the alternative pathway (DUG pathway) in *Saccharomyces cerevisiae* is initiated by (Dug2p-Dug3p)2 complex, a novel glutamine amidotransferase (GATase) enzyme acting on glutathione. J Biol Chem 287:8920–8931
Kaur A, Gautam R, Srivastava R, Chandel R, Kumar A, Karthikeyan S, Bachhawat AK (2017) The structure of ChaC2: an enzyme for slow turnover of cytosolic glutathione. J Biol Chem 292:638–651
Kelly BS, Antholine WE, Griffith OW (2002) *Escherichia coli* gamma-glutamylcysteine synthetase. Two active site metal ions affect substrate and inhibitor binding. J Biol Chem 277:50–58
Kim OC, Suwannarangsee S, Oh DB, Kim S, Seo JW, Kim CH, Kang HA, Kim JY, Kwon O (2013) Transcriptome analysis of xylose metabolism in the thermotolerant methylotrophic yeast *Hansenula polymorpha*. Bioprocess Biosyst Eng 36:1509–1518
Kiriyama K, Hara KY, Kondo A (2012) Extracellular glutathione fermentation using engineered Saccharomyces cerevisiae expressing a novel glutathione exporter. Appl Microbiol Biotechnol 96:1021–1027
Koh S, Wiles AM, Sharp JS, Naider FR, Becker JM, Stacey G (2002) An oligopeptide transporter gene family in *Arabidopsis*. Plant Physiol 128:21–29
Kritzinger EC, Bauer FF, du Toit WJ (2012) Role of glutathione in winemaking: a review. J Agric Food Chem 61:269–277
Kumar C, Sharma R, Bachhawat AK (2003) Utilization of glutathione as an exogenous sulfur source is independent of gamma-glutamyl transpeptidase in the yeast *Saccharomyces cerevisiae*: evidence for an alternative glutathione degradation pathway. FEMS Microbiol Lett 219:187–194
Lagrain B, Thewissen BG, Brijs K, Delcour JA (2007) Impact of redox agents on the extractability of gluten proteins during bread making. J Agric Food Chem 55:5320–5325

Lai J, Lee S, Hsieh C, Hwang C, Liao C (2008) *Saccharomyces cerevisiae* strains for hyper-producing glutathione and γ-glutamylcysteine and processes of use. United States Patent 7371557
Lash H, Jones DP (1983) Transport of glutathione by renal basal-lateral membrane vesicles. Biochem Biophys Res Commun 112:55–60
Lebo R, Kredich N (1978) Inactivation of human γ-glutamylcysteine synthetase by cystamine. Demonstration and quantification of enzyme ligand complexes. J Biol Chem 253:2615–2523
Lee TA, Jorgensen P, Bognar AL, Peyraud C, Thomas D, Tyers M (2010) Dissection of combinatorial control by the Met4 transcriptional complex. Mol Biol Cell 21:456–469
Lee HG, Koo SH, Lim DY, Kim ES, Yoon HS, Lee JS, Kim GH (2015) Method for the production of food packaging film with enhanced glutathione stability. Republic of Korea Patent KR101492471 (B1)
Liang G, Mo Y, Du G (2010) Optimization of sodium dedecyl sulfate (SDS) addition coupled with adenosine triphosphate (ATP) regeneration for glutathione overproduction in high density cultivation of *Candida utilis*. Enzym Microb Technol 46(6):526–533
Li ZS, Szczypka M, Lu YP, Thiele DJ, Rea PA (1996) The yeast cadmium factor protein (YCF1) is a vacuolar glutathione S-conjugate pump. J Biol Chem 271:6509–6517
Li Y, Wei G, Chen J (2004) Glutathione: a review on biotechnological production. Appl Microbiol Biotechnol 66:233–242
Li Y, Hugenholtz J, Sybesma W, Abee T, Molenaar D (2005) Using *Lactococcus lactis* for glutathione overproduction. Appl Microbiol Biotechnol 67:83–90
Li H, Guo A, Wang H (2008) Mechanisms of oxidative browning of wine. Food Chem 108:1–13
Li Z, Ye Q, Li W, Zhang S (2014) Method for producing glutathione by fermentation of recombinant *Escherichia coli*. China Patent CN102586369
Liang G, Du G, Chen J (2008) Enhanced glutathione production by using low-pH stress coupled with cysteine addition in the treatment of high cell density culture of *Candida utilis*. Lett appl microbial 46:507–512
Lieber CS (2002) S-Adenosyl-L-methionine and alcoholic liver disease in animal models: implications for early intervention in human beings. Alcohol 27:173–177
Liedschulte V, Wachter A, Zhigang A, Rausch T (2010) Exploiting plants for glutathione (GSH) production: uncoupling GSH synthesis from cellular controls results in unprecedented GSH accumulation. Plant Biotechnol J 8:807–820
Liu G, Sanchez-Fernandez R, Li ZS, Rea PA (2001) Enhanced multispecificity of arabidopsis vacuolar multidrug resistance-associated protein-type ATP-binding cassette transporter, AtMRP2. J Biol Chem 276:8648–8656
Lo M, Ling V, Low C, Wang YZ, Gout PW (2010) Potential use of the anti-inflammatory drug, sulfasalazine, for targeted therapy of pancreatic cancer. Curr Oncol 17:9–16
Lorenz E, Schmacht M, Senz M (2016) Evaluation of cysteine ethyl ester as efficient inducer for glutathione overproduction in *Saccharomyces* spp. Enzyme Microb Technol 93–94:122–131
Lu SC (2009) Regulation of glutathione synthesis. Mol Asp Med 30:42–59
Luo M, Boudier A, Clarot I, Maincent P, Schneider R, Leroy P (2016) Gold nanoparticles grafted by reduced glutathione with thiol function preservation. Colloid Interface Sci Commun 14:8–12
Lyu C (2016) Rice noodles capable of nourishing faces and protecting skins and preparation method of rice noodles. China Patent Application CN105410629 (A)
Maeda H, Hori S, Ohizumi H, Segawa T, Kakehi Y, Ogawa O, Kakizuka A (2004) Effective treatment of advanced solid tumors by the combination of arsenic trioxide and L-buthionine-sulfoximine. Cell Death Differ 11:737–746
Maher P (2005) The effects of stress and aging on glutathione metabolism. Ageing Res Rev 4:288–314
Mandracchia D, Denora N, Franco M, Pitarresi G, Giammona G, Trapani G (2011) New biodegradable hydrogels based on inulin and α,β-polyaspartylhydrazide designed for colonic drug delivery: In Vitro release of glutathione and oxytocin. J Biomater Sci Polym Ed 22:313–328

Martinoia E, Grill E, Tommasini R, Kreuz K, Amrhein N (1993) ATP-dependent glutathione S-conjugate 'export' pump in the vacuolar membrane of plants. Nature 364:247–249
Marz U (2014) Yeasts, yeast extracts, autolysates and related products: the global market, https://www.bccresearch.com/market-research/chemicals/yeast-yeast-extracts-autolysates-products-chm053b.html
Maughan SC, Pasternak M, Cairns N, Kiddle G, Brach T, Jarvis R, Haas F, Nieuwland J, Lim B, Muller C, Salcedo-Sora E, Kruse C, Orsel M, Hell R, Miller AJ, Bray P, Foyer CH, Murray JA, Meyer AJ, Cobbett CS (2010) Plant homologs of the *Plasmodium falciparum* chloroquine-resistance transporter, PfCRT, are required for glutathione homeostasis and stress responses. PNAS USA 107:2331–2336
Mehdi K, Penninckx MJ (1997) An important role for glutathione and γ-glutamyl-transpeptidase in the supply of growth requirements during nitrogen starvation of the yeast *Saccharomyces cerevisiae*. Microbiology 143:1885–1889
Meister A (1988) Glutathione. In: Arias IM, Jakoby WB, Popper H, Schachter D, Shafritz DA (eds) The liver: biology and pathobiology, 2nd edn. Raven Press, New York, pp 401–417
Meister A, Anderson ME (1983) Glutathione. Annu Rev Biochem 52:711–760
Mezzetti F, De Vero L, Giudici P (2014) Evolved wine strains with enhanced glutathione production obtained by an evolution-based strategy. FEMS Yeast Res 14(6):977–987
Mittl PR, Schulz GE (1994) Structure of glutathione reductase from *Escherichia coli* at 1.86 Å resolution: comparison with the enzyme from human erythrocytes. Protein Sci 3:799–809
Naji-Tabasi S, Razavi SMA, Mehditabar H (2017) Fabrication of basil seed gum nanoparticles as a novel oral delivery system of glutathione. Carbohydr Polym 157:1703–1713
Nakagawa T, Kaji N, Tokuriki M (2016) Use of yeast extract including glutathione as melanin production inhibitor. Taiwan Patent Application TW201620396 (A)
Narang VS, Pauletti GM, Gout PW, Buckley DJ, Buckley AR (2007) Sulfasalazine-induced reduction of glutathione levels in breast cancer cells: enhancement of growth-inhibitory activity of Doxorubicin. Chemotherapy 53:210–217
Nie W, Wei G, Du G, Li Y, Chen J (2005) Enhanced intracellular glutathione synthesis and excretion capability of *Candida utilis* by using a low pH-stress strategy. Lett Appl Microbiol 40:378–384
Noguti J, Barbisan LF, Cesar A, Dias Seabra C, Choueri RB, Ribeiro DA (2012) Review: in vivo models for measuring placental glutathione-S-transferase (GST-P7-7) levels: a suitable biomarker for understanding cancer pathogenesis. In Vivo 26:647–650
Oliveira P, Martins NM, Santos M, Couto NA, Wright PC, Tamagnini P (2015) The Anabaena sp. PCC 7120 exoproteome: taking a peek outside the box. Life 5:130–163
Ono B, Shirahige Y, Nanjoh A, Andou N, Ohue H, Ishino-Arao Y (1988) Cysteine biosynthesis in *Saccharomyces cerevisiae*: mutation that confers cystathionine β-synthase deficiency. J Bacteriol 170:5883–5889
Oppenheimer L, Wellner VP, Griffith OW, Meister A (1979) Glutathione synthetase. Purification from rat kidney and mapping of the substrate binding sites. J Biol Chem 254:5184–5190
Ortiz-Julien A (2012) Method for preventing defective ageing of white wines. United States Patent US8268372 (B2)
Orumets K (2012) Molecular mechanisms controlling intracellular glutathione levels in baker's yeast and a random mutagenized glutathione over-accumulating isolate. PhD Thesis, Tallinn university of technology, Tallinn, Estonia
Orumets K, Kevvai K, Nisamedtinov I, Tamm T, Paalme T (2012) *YAP1* over-expression in *Saccharomyces cerevisiae* enhances glutathione accumulation at its biosynthesis and substrate availability levels. Biotechnol J 7:566–568
Outten CE, Culotta VC (2004) Alternative start sites in the *Saccharomyces cerevisiae GLR1* gene are responsible for mitochondrial and cytosolic isoforms of glutathione reductase. J Biol Chem 279:7785–7791
Oz HS, Chen TS, Nagasawa H (2007) Comparative efficacies of 2 cysteine prodrugs and a glutathione delivery agent in a colitis model. Transl Res J Lab Clin Med 150:122–129

Patzschke A, Steiger MG, Holz C, Lang C, Mattanovich D, Sauer M (2015) Enhanced glutathione production by evolutionary engineering of *Saccharomyces cerevisiae* strains. Biotechnol J 10:1719–1726

Penninckx MJ, Elskens MT (1993) Metabolism and functions of glutathione in micro-organisms. Adv Microb Physiol 34:239–301

Penninckx MJ, Jaspers CJ (1985) Molecular and kinetic properties of purified γ-glutamyl transpeptidase from yeast (*Saccharomyces cerevisiae*). Phytochemistry 24:1913–1918

Penninckx MJ, Jaspers C, Wiame JM (1980) Glutathione metabolism in relation to the amino-acid permeation systems of the yeast *Saccharomyces cerevisiae*. Occurrence of γ-glutamyltranspeptidase: its regulation and the effects of permeation mutations on the enzyme cellular level. Eur J Biochem 104:119–123

Perricone C, De Carolis C, Perricone R (2009) Glutathione: a key player in autoimmunity. Autoimmun Rev 8:697–701

Perrone GG, Grant CM, Dawes IW (2005) Genetic and environmental factors influencing glutathione homeostasis in *Saccharomyces cerevisiae*. Mol Biol Cell 16:218–230

Pittman MS, Robinson HC, Poole RK (2005) A bacterial glutathione transporter (*Escherichia coli* CydDC) exports reductant to the periplasm. J Biol Chem 280:32254–32261

Pocsi N, Prade RA, Penninckx MJ (2004) Glutathione, altruistic metabolite in fungi. Adv In Microb Physiol 49:2–76

Qiu Z, Deng Z, Tan H, Zhou S, Cao L (2015) Engineering the robustness of *Saccharomyces cerevisiae* by introducing bifunctional glutathione synthase gene. J Ind Microbiol Biotechnol 42:537–542

Rana S, Dringen R (2007) Gap junction hemichannel-mediated release of glutathione from cultured rat astrocytes. Neurosci Lett 415:45–48

Rebbeor JF, Connolly GC, Dumont ME, Ballatori N (1998) ATP-dependent transport of reduced glutathione on YCF1, the yeast orthologue of mammalian multidrug resistance-associated proteins. J Biol Chem 273:33449–33454

Rebbeor JF, Connolly GC, Ballatori N (2002) Inhibition of Mrp2- and Ycf1p-mediated transport by reducing agents: evidence for GSH transport on rat Mrp2. Biochim Biophys Acta 1559:171–178

Richman PG, Meister A (1975) Regulation of gamma-glutamycysteine synthetase b nonallosteric feedback inhibition by glutathione. J Biol Chem 250:1422–1426

Riley R, Haridas S, Wolfe KH, Lopes MR, Hittinger CT, Goker M, Salamov AA, Wisecaver JH, Long TM, Calvey CH et al (2016) Comparative genomics of biotechnologically important yeasts. Proc Natl Acad Sci U S A 113:9882–9887

Rip J, Chen L, Hartman R, van den Heuvel A, Reijerkerk A, van Kregten J, van der Boom B, Appeldoorn C, de Boer M, Maussang D et al (2014) Glutathione PEGylated liposomes: pharmacokinetics and delivery of cargo across the blood-brain barrier in rats. J Drug Target 22:460–467

Rollini M, Musatti A, Manzoni M (2010) Production of glutathione in extracellular form by *Saccharomyces cerevisiae*. Process Biochem 45(4):441–445

Rosemeyer MA (1987) The biochemistry of glucose-6-phosphate dehydrogenase, 6-phosphogluconate dehydrogenase and glutathione reductase. Cell Biochem Funct 5:79–95

Rosenblat M, Volkova N, Coleman R, Aviram M (2007) Anti-oxidant and anti-atherogenic properties of liposomal glutathione: studies in vitro, and in the atherosclerotic apolipoprotein E-deficient mice. Atherosclerosis 195:61–68

Sadhu MJ, Moresco JJ, Zimmer AD, Yates JR, Rine J (2014) Multiple inputs control sulfur-containing amino acid synthesis in *Saccharomyces cerevisiae*. Mol Biol Cell 25:1653–1665

Sakato K, Tanaka H (1992) Advanced control of glutathione fermentation process. Biotechnol Bioeng 40(8):904–912

Salvemini F, Franzé A, Iervolino A, Filosa S, Salzano S, Ursini MV (1999) Enhanced glutathione levels and oxidoresistance mediated by increased glucose-6-phosphate dehydrogenase expression. J Biol Chem 274:2750–2757

Saucedo AC, Ambati BK (2016) Eye health supplement. United States Patent Application US20160030502 (A1)
Sawers L, Ferguson MJ, Ihrig BR, Young HC, Chakravarty C, Wolf CR, Smith G (2014) Glutathione S-transferase P1 (GSTP1) directly influences platinum drug chemosensitivity in ovarian tumour cell lines. Br J Cancer 111:1150–1158
Schmacht M, Lorenz E, Senz M (2017a) Microbial production of glutathione. World J Microbiol 33:106
Schmacht M, Lorenz E, Stahl U, Senz M (2017b) Medium optimization based on yeast's elemental composition for glutathione production in *Saccharomyces cerevisiae*. J Biosci Bioeng 123:555–561
Schneider A, Martini N, Rennenberg H (1992) Reduced glutathione (GSH) transport into cultured tobacco cells. Plant Physiol Biochem 30:29–38
Schulz JB, Lindenau J, Seyfried J, Dichgans J (2000) Glutathione, oxidative stress and neurodegeneration. Eur J Biochem 267:4904–4911
Sechi G, Deledda MG, Bua G, Satta WM, Deiana GA, Pes GM, Rosati G (1996) Reduced intravenous glutathione in the treatment of early Parkinson's disease. Prog Neuro-Psychopharmacol Biol Psychiatry 20:1159–1170
Seelig GF, Simondsen RP, Meister A (1984) Reversible dissociation of γ -glutamylcysteine synthetase into two subunits. J Biol Chem 259:9345–9347
Shang F, Wang Z, Tan T (2008) High-cell-density cultivation for co-production of ergosterol and reduced glutathione by *Saccharomyces cerevisiae*. Appl Microbiol Biotechnol 77(6):1233–1240
Sharma KG, Mason DL, Liu G, Rea PA, Bachhawat AK, Michaelis S (2002) Localization, regulation, and substrate transport properties of Bpt1p, a *Saccharomyces cerevisiae* MRP-type ABC transporter. Eukaryot Cell 1:391–400
Sohn MJ, Yoo SJ, Oh DB, Kwon O, Lee SY, Sibirny AA, Kang HA (2014) Novel cysteine-centered sulfur metabolic pathway in the thermotolerant methylotrophic yeast *Hansenula polymorpha*. PLoS One 9:e100725
Soltaninassab SR, Sekhar KR, Meredith MJ, Freeman ML (2000) Multifaceted regulation of γ-glutamylcysteine synthetase. J Cel Physiol 182:163–170
Spector D, Labarre J, Toledano MB (2001) A genetic investigation of the essential role of glutathione mutations in the proline biosynthesis pathway are the only suppressors of glutathione auxotrophy in yeast. J Biol Chem 276:7011–7016
Springael J, Penninckx MJ (2003) Nitrogen-source regulation of yeast γ-glutamyl transpeptidase synthesis involves the regulatory network including the GATA zinc-finger factors Gln3, Nil1/Gat1 and Gzf3. Biochem J 371:589–595
Srikanth CV, Vats P, Bourbouloux A, Delrot S, Bachhawat AK (2005) Multiple cis-regulatory elements and the yeast sulphur regulatory network are required for the regulation of the yeast glutathione transporter, Hgt1p. Curr Genet 47:345–358
Sugiyama K, Izawa S, Inoue Y (2000) The Yap1p-dependent induction of glutathione synthesis in heat shock response of Saccharomyces cerevisiae. J Biol Chem 275:15535–15540
Suzuki H, Kumagai H, Tochikura T (1986) Gamma-glutamyltranspeptidase from *Escherichia coli* K-12: formation and localization. J Bacteriol 168:1332–1335
Suzuki H, Koyanagi T, Izuka S, Onishi A, Kumagai H (2005) The yliA, -B, -C, and –D genes of *Escherichia coli* K-12 encode a novel glutathione importer with an ATP-binding cassette. J Bacteriol 187:5861–5867
Thiermann M (2010) Non-surgical method for treating cataracts in mammals including man. United States Patent US7776364 (B2)
Thomas D, Surdin-Kerjan Y (1997) Metabolism of sulfur amino acids in *Saccharomyces cerevisiae*. Microbiol Mol Biol Rev 61:503–532
Toledano MB, Huang ME (2017) The unfinished puzzle of glutathione physiological functions, an old molecule that still retains many enigmas. Antioxid Redox Signal 27:1127–1129
Townsend DM, Tew KD, Tapiero H (2003) The importance of glutathione in human disease. Biomed Pharmacother 57:145–155

Trapani A, Lopedota A, Franco M, Cioffi N, Ieva E, Garcia-Fuentes M, Alonso MJ (2010) A comparative study of chitosan and chitosan/cyclodextrin nanoparticles as potential carriers for the oral delivery of small peptides. Eur J Pharm Biopharm 75:26–32

Tsunoda S, Avezov E, Zyryanova A, Konno T, Mendes-Silva L, Melo EP, Harding HP, Ron D (2014) Intact protein folding in the glutathione-depleted endoplasmic reticulum implicates alternative protein thiol reductants. Elife 3:e03421

Ubiyvovk VM, Nazarko TY, Stasyk OG, Sohn MJ, Kang HA, Sibirny AA (2002) *GSH2*, a gene encoding gamma-glutamylcysteine synthetase in the methylotrophic yeast *Hansenula polymorpha*. FEMS Yeast Res 2:327–332

Ubiyvovk VM, Blazhenko OV, Gigot D, Penninckx M, Sibirny AA (2006) Role of gamma-glutamyltranspeptidase in detoxification of xenobiotics in the yeasts *Hansenula polymorpha* and *Saccharomyces cerevisiae*. Cell Biol Int 30:665–671

Ubiyvovk VM, Ananin VM, Malyshev AY, Kang HA, Sibirny AA (2011a) Optimization of glutathione production in batch and fed-batch cultures by the wild-type and recombinant strains of the methylotrophic yeast *Hansenula polymorpha* DL-1. BMC Biotechnol 11:8

Ubiyvovk VM, Blazhenko OV, Zimmermann M, Sohn MJ, Kang HA (2011b) Cloning and functional analysis of the *GSH1/MET1* gene complementing cysteine and glutathione auxotrophy of the methylotrophic yeast *Hansenula polymorpha*. Ukr Biokhim Zh 83:67–81

Ueda Y, Yonemitsu M, Tsubuku T, Sakaguchi M, Miyajima R (2014) Flavor characteristics of glutathione in raw and cooked foodstuffs. Biosci Biotechnol Biochem 61(12):1977–1980

Van Der Werf P, Orlowski M, Meister A (1971) Enzymatic conversion of 5-oxo-L-proline (L-pyrrolidone carboxylate) to L-glutamate coupled with cleavage of adenosine triphosphate to adenosine diphosphate, a reaction in the γ -glutamyl cycle. Proc Natl Acad Sci 68:2982–2985

Veeravalli K, Boyd D, Iverson BL, Beckwith J, Georgiou G (2011) Laboratory evolution of glutathione biosynthesis reveals natural compensatory pathways. Nat Chem Biol 7:101–105

Vergauwen B, De Vos D, Van Beeumen JJ (2006) Characterization of the bifunctional γ-glutamate–cysteine ligase/glutathione synthetase (GshF) of *Pasteurella multocida*. J Biol Chem 281:4380–4394

Verheyen C, Albrecht A, Herrmann J, Strobl M, Jekle M, Becker T (2015) The contribution of glutathione to the destabilizing effect of yeast on wheat dough. Food Chem 173:243–249

Verma VV, Gupta R, Goel M (2015) Phylogenetic and evolutionary analysis of functional divergence among Gamma glutamyl transpeptidase (GGT) subfamilies. Biol Direct 10:49

Wang Z (2014) Whitening cosmetic and preparation method thereof. China Patent Application CN104027293 (A)

Wang Z, Tan T, Song J (2007) Effect of amino acids addition and feedback control strategies on the high-cell-density cultivation of *Saccharomyces cerevisiae* for glutathione production. Bioresour Technol 42:108–111

Wang B, Liang G, Zhou Q, Xie J, Mo Y (2010) Combined amino acids modulation with H2O2 stress for glutathione overproduction in *Candida utilis*. Afr J Biotechnol 9:5399–5406

Wang M, Sun J, Xue F, Shang F, Wang Z, Tan T (2012) The effect of intracellular amino acids on GSH production by high-cell density cultivation of Saccharomyces cerevisiae. Appl Biochem Biotechnol 168:198–205

Wang C, Zhang J, Wu H, Li Z, Ye Q (2015) Heterologous gshF gene expression in various vector systems in *Escherichia coli* for enhanced glutathione production. J Biotechnol 214:63–68

Wang D, Wang C, Wu H, Li Z, Ye Q (2016) Glutathione production by recombinant *Escherichia coli* expressing bifunctional glutathione synthetase. J Ind Microbiol Biotechnol 43:45–53

Wang YR, Branicky N, Noë A, Hekimi S (2018) Superoxide dismutases: dual roles in controlling ROS damage and regulating ROS signaling. J Cell Biol 217:1915–1928

Watanabe F, Hashizume E, Chan GP, Kamimura A (2014) Skin-whitening and skin-condition-improving effects of topical oxidized glutathione: a double-blind and placebo-controlled clinical trial in healthy women. Clin Cosmet Investig Dermatol:267

Wen S, Zhang T, Tan T (2004) Utilization of amino acids to enhance glutathione production in *Saccharomyces cerevisiae*. Enzym Microb Technol 35(6–7):501–507

Wen S, Zhang T, Tan T (2006) Maximizing production of glutathione by amino acid modulation and high-cell-density fed-batch culture of *Saccharomyces cerevisiae*. Process Biochem 41:2424–2428

Wen J, Du Y, Li D, Alany R (2013) Development of water-in-oil microemulsions with the potential of prolonged release for oral delivery of L-glutathione. Pharm Dev Technol 18:1424–1429

Weschawalit S, Thongthip S, Phutrakool P, Asawanonda P (2017) Glutathione and its antiaging and antimelanogenic effects. Clin Cosmet Investig Dermatol 10:147–153

Wheeler GL, Trotter EW, Dawes IW, Grant CM (2003) Coupling of the transcriptional regulation of glutathione biosynthesis to the availability of glutathione and methionine via the Met4 and Yap1 transcription factors. J Biol Chem 278:49920–49928

Wickham S, West MB, Cook PF, Hanigan MH (2011) Gamma-glutamyl compounds: substrate specificity of gamma-glutamyl transpeptidase enzymes. Anal Biochem 414:208–214

Wiles AM, Cai H, Naider F, Becker JM (2006) Nutrient regulation of oligopeptide transport in *Saccharomyces cerevisiae*. Microbiology 152:3133–3145

Witschi A, Reddy S, Stofer B, Lauterburg BH (1992) The systemic availability of oral glutathione. Eur J Clin Pharmacol 43:667–669

Wu G, Fang Y, Yang S, Lupton JR, Turner ND (2004) Glutathione metabolism and its implications for health. Int J Nutr 134:489–492

Xiao Z, La Fontaine S, Bush AI, Wedd AG (2019) Molecular mechanisms of glutaredoxin enzymes: versatile hubs for thiol-disulfide exchange between protein thiols and glutathione. J Mol Biol 431:158–177

Xiong Z, Guo M, Guo Y, Chu J, Zhuang Y, Zhang S (2009) Efficient extraction of intracellular reduced glutathione from fermentation broth of *Saccharomyces cerevisiae* by ethanol. Bioresour Technol 100:1011–1014

Xu G, Zhang L, Chen Y, Cheng W, Guo Y, Wei Y, Liang J, Tan W (2015) *Saccharomyces cerevisiae* microbial preparation with high yield of glutathione and preparation method thereof. China Patent Application CN104286415 (A)

Yamaguchi H, Kato H, Hata Y, Nishioka T, Kimura A, Oda J, Katsube Y (1993) Three-dimensional structure of the glutathione synthetase from *Escherichia coli* B at 2.0 A resolution. J Mol Biol 229:1083–1100

Yan N, Meister A (1990) Amino acid sequence of rat kidney γ-glutamylcysteine synthetase. J Biol Chem 265:1588–1593

Yan C, Lee LH, Davis LI (1998) Crm1p mediates regulated nuclear export of a yeast AP-1-like transcription factor. EMBO J 17:7416–7429

Ye C, Liu Y, Wang Y, Li H, Wang K (2016) Glutathione beautifying yogurt and production method thereof. China Patent Application CN105685226 (A)

Yoshida H, Arai S, Hara KY, Yamada R, Ogino C, Fukuda H, Kondo A (2011) Efficient and direct glutathione production from raw starch using engineered *Saccharomyces cerevisiae*. Appl Microbiol Biotechnol 89:1417–1422

Yu J, Zhou CZ (2007) Crystal structure of glutathione reductase Glr1 from the yeast *Saccharomyces cerevisiae*. Proteins: Struct Funct Bioinf 68:972–979

Yun J, Mullarky E, Lu C, Bosch KN, Kavalier A, Rivera K, Roper J, Chio II, Giannopoulou EG, Rago C, Muley A, Asara JM, Paik J, Elemento O, Chen Z, Pappin DJ, Dow LE, Papadopoulos N, Gross SS, Cantley LC (2015) Vitamin C selectively kills KRAS and BRAF mutant colorectal cancer cells by targeting GAPDH. Science 350:1391–1396

Yurimoto H, Oku M, Sakai Y (2011) Yeast methylotrophy: metabolism, gene regulation and peroxisome homeostasis. Int J of Microbiol 2011:1–8

Yurkiv M, Kurylenko O, Vasylyshyn R, Dmytruk K, Fickers P, Sibirny A (2018) Gene of the transcriptional activator *MET4* is involved in regulation of glutathione biosynthesis in the methylotrophic yeast *Ogataea (Hansenula) polymorpha*. FEMS Yeast Res 18(2)

Zaman GJ, Lankelma J, van Tellingen O, Beijnen J, Dekker H, Paulusma C, Oude Elferink RP, Baas F, Borst P (1995) Role of glutathione in the export of compounds from cells by the multidrug-resistance-associated protein. PNAS USA 92:7690–7694

Zarka MH, Bridge WJ (2017) Oral administration of γ-glutamylcysteine increases intracellular glutathione levels above homeostasis in a randomised human trial pilot study. Redox Biol 11:631–636

Zechmann B (2014) Compartment-specific importance of glutathione during abiotic and biotic stress. Front Plant Sci 5:566

Zhang T, Wen S, Tan T (2007) Optimization of the medium for glutathione production in *Saccharomyces cerevisiae*. Process Biochem 42:454–458

Zhang J, Quan C, Wang C, Wu H, Li Z, Ye Q (2016) Systematic manipulation of glutathione metabolism in *Escherichia coli* for improved glutathione production. Microb Cell Factories 15:38

Zhang X, Wu H, Huang B, Li Z, Ye Q (2017) One-pot synthesis of glutathione by a two-enzyme cascade using a thermophilic ATP regeneration system. J Biotechnol 241:163–169

Chapter 7
Nitrogen Assimilation Pathways in Budding Yeasts

Tomas Linder

Abstract The element nitrogen is an essential macronutrient for all living organisms. Like other microorganisms, budding yeasts (phylum Ascomycota, subphylum Saccharomycotina) have evolved a versatile enzymatic toolbox for the extraction of nitrogen from a wide array of nitrogen-containing compounds. This chapter will review our current knowledge of pathways and enzymes involved in the assimilation of individual categories of nitrogen compounds including ammonia, nitrate, amino acids, amides, amines, purines, pyrimidines as well as aromatic and heterocyclic nitrogen compounds. The genes encoding the corresponding enzymes are listed whenever possible. Since the ability to assimilate specific categories of nitrogen compounds continue to be used for classification of budding yeasts, the taxonomic context of the occurrence of individual pathways and enzymes is emphasized throughout. Current as well as possible future biotechnology applications of budding yeast nitrogen assimilation pathways and enzymes are also discussed.

Keywords Metabolism · Nitrogen · Orphan enzyme · Orphan pathway · Yeast

7.1 Introduction

The element nitrogen is integral to proteins and nucleic acids as well as some lipids, carbohydrates, and other biomolecules. Therefore, the acquisition of nitrogen from the external environment is essential for all living organisms. Microorganisms are particularly adept at extracting nitrogen from a wide spectrum of nitrogen-containing organic substrates. Organic nitrogen sources are typically assimilated by one of two pathways. The first involves a deamination reaction, whereby the nitrogen atom is separated from the carbon backbone, predominantly in the form of ammonia (NH_3). The second pathway is a transamination reaction, where the nitrogen is transferred

T. Linder (✉)
Department of Molecular Sciences, Swedish University of Agricultural Sciences, Uppsala, Sweden
e-mail: tomas.linder@slu.se

A. Sibirny (ed.), *Non-conventional Yeasts: from Basic Research to Application*,
https://doi.org/10.1007/978-3-030-21110-3_7

to acceptor substrate such as an α-ketoacid. The transfer of an amino group (-NH_2) to the α-position of α-ketoglutarate to produce L-glutamate is perhaps the most common transamination reaction that occurs during nitrogen assimilation in nature.

The ability of any one microorganism to assimilate particular nitrogenous compounds depends on possessing the requisite metabolic pathway to enable, for example, deamination or transamination. With regard to nitrogen assimilation by fungi, the common baker's yeast *Saccharomyces cerevisiae* is by far the most intensely studied organism (Cooper 1982). However, compared to many other species of budding yeasts (phylum Ascomycota, subphylum Saccharomycotina), the spectrum of nitrogen sources assimilated by *S. cerevisiae* is relatively narrow (Large 1986). The genus *Saccharomyces* belongs to a subgroup of the family Saccharomycetaceae known to have undergone a whole-genome duplication (WGD) event (Kellis et al. 2004). This subgroup of the Saccharomycetaceae is therefore commonly referred to as the "WGD clade" and, in addition to the genus *Saccharomyces*, also includes the genera *Kazachstania*, *Nakaseomyces*, *Naumovozyma*, *Tetrapisispora*, and *Vanderwaltozyma*. The species of the WGD clade appear to share the relatively limited nitrogen assimilation profile that is observed in *S. cerevisiae*.

The study of nitrogen assimilation pathways that do not occur in *S. cerevisiae* has long lagged behind due to a lack of genetic tools and genomic data. In the past few years, however, a substantial number of genomes from non-*S. cerevisiae* species (often referred to as "nonconventional yeasts") have been sequenced, which will facilitate the connection of phenotype (in this case the ability to use a particular nitrogen source) to individual enzymes and the genes that encode them.

This chapter will give an overview of known nitrogen assimilation pathways among the budding yeasts and give brief descriptions of the enzymes involved as well as list the genes that encode these enzymes, if their identity is known. For the sake of space, this chapter will not provide an exhaustive review of all previous research on each and every pathway but will instead cite key studies. The author recommends previous reviews on yeast nitrogen assimilation that provide a more comprehensive background of previous research (Cooper 1982; Large 1986; Messenguy et al. 2006). Nitrogen assimilation pathways and their associated enzymes will be discussed in a taxonomic context whenever possible since the author has previously reported a clear lineage-specific distribution of most nitrogen assimilation pathways among budding yeasts (Linder 2019a). Although the present chapter will focus on budding yeasts, selected references will be made to other fungal taxonomic groups whenever it is deemed appropriate. The present chapter will not discuss cell surface transporters for dedicated import of nitrogen sources into the cell interior from the external environment. Likewise, regulation of nitrogen assimilation pathways and individual enzymes by transcriptional, posttranslational, or allosteric mechanisms will not be discussed in this chapter. Studies of regulatory mechanisms of yeast nitrogen metabolism have predominantly been carried out in *S. cerevisiae*, and the author would refer the reader to a number of excellent reviews on the topic (Broach 2012; Cooper 2002; Ljungdahl 2009; Magasanik and Kaiser 2002; Messenguy et al. 2006; Wong et al. 2008).

7.2 Assimilation of Ammonia

Ammonia (NH_3) is one of the simplest nitrogen substrates used as a nitrogen source by budding yeasts. Ammonia is also a product of the catabolism of several classes of organic nitrogen compounds including urea (Sect. 7.4), L-asparagine (Sect. 7.5.4), D-amino acids (Sect. 7.5.8), aliphatic amides (Sect. 7.6), aliphatic amines (Sect. 7.7), purines (Sect. 7.8), cytosine (Sect. 7.9.1), and dihydropyrimidines (Sect. 7.9.2).

Ammonia can be used directly in a small number of amination reactions for the synthesis of L-glutamate, L-glutamine, guanosine 5'-phosphate (GMP) and the diphthamide posttranslational modification of eukaryotic translation elongation factor 2 (encoded by the *EFT1* gene in yeast). However, the vast majority of biosynthetic pathways for nitrogen-containing biomolecules in yeast first require ammonia to be converted into L-glutamate and L-glutamine, which are then used as amino group donors in downstream anabolic transamination reactions (Magasanik 2003). The enzyme $NADP^+$-dependent glutamate dehydrogenase (EC 1.4.1.4; encoded by the *GDH1* gene) catalyzes the amination of α-ketoglutarate to form L-glutamate (Fig. 7.1a) (Moye et al. 1985; Nagasu and Hall 1985), while the enzyme glutamine synthetase (EC 6.3.1.2; encoded by the *GLN1* gene) (Minehart and Magasanik 1992; Mitchell 1985) catalyzes the amination of L-glutamate to form L-glutamine (Fig. 7.1b) (Hachimori et al. 1974). A second copy of the *GDH1* gene (termed *GDH3*) has been described in *S. cerevisiae* (Avendaño et al. 1997) and appears to play a distinct physiological role in nitrogen metabolism (DeLuna et al. 2001, 2005). This duplication of the *GDH1* gene is possibly unique to the genus

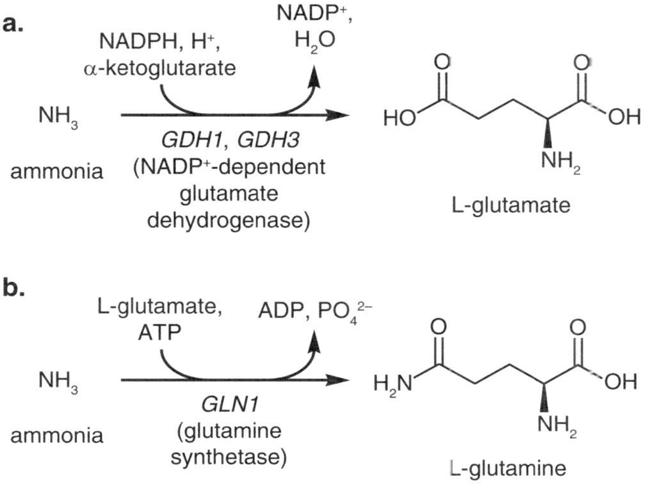

Fig. 7.1 Pathways for assimilation of ammonia in budding yeasts. (**a**) Assimilation of ammonia via $NADP^+$-dependent glutamate dehydrogenase. (**b**) Assimilation of ammonia via glutamine synthetase

Saccharomyces among the budding yeasts and has been functionally linked to the particular fermentative lifestyle of this genus (Magasanik 2003).

Carbon and nitrogen availability appears to determine the proportions of L-glutamate and L-glutamine synthesized from ammonia by NADP$^+$-dependent glutamate dehydrogenase and glutamine synthetase, respectively (Holmes et al. 1991; Lacerda et al. 1992; Magasanik 2003). Under conditions of glucose excess, *S. cerevisiae* will assimilate ammonia predominantly through reductive amination of α-ketoglutarate to L-glutamate catalyzed by the NADP$^+$-dependent glutamate dehydrogenase Gdh1 despite the higher energetic cost (Magasanik 2003). Conversely, ammonia assimilation in *S. cerevisiae* during glucose limitation proceeds predominantly through L-glutamine synthesis by Gln1 followed by L-glutamate synthesis by glutamate synthase (Lacerda et al. 1992) (see Sect. 7.5.1). Ammonia assimilation also proceeds through L-glutamine synthesis by Gln1 when extracellular concentrations of ammonia are low since Gln1 has higher affinity for ammonia than Gdh1.

7.3 Assimilation of Nitrate and Nitrite

The ability to assimilate inorganic nitrogen sources other than ammonia is thought to be rare among budding yeasts (Choudary and Rao 1984; Norkrans 1969). Nitrate (NO_3^-) assimilation by budding yeasts appears to be restricted to the families Phaffomycetaceae, Pichiaceae, and Trichomonascaceae as well as unassigned genera within the CUG^{Leu2} and CUG^{Ala} clades (Krassowski et al. 2018; Linder 2019a; Shen et al. 2018). Nitrate is assimilated by a two-step reduction via nitrite (NO_2^-) to produce ammonia (Fig. 7.2).

The reduction of nitrate into nitrite is carried out by the enzyme nitrate reductase (EC 1.7.1.2/EC 1.7.1.3), which is encoded by the *YNR1* gene (Avila et al. 1995). In addition to the presence of flavin adenine dinucleotide (FAD) and heme prosthetic groups, nitrate reductase is one of the few enzymes currently known in budding yeasts that require the molybdenum cofactor (MoCo) for its activity (Hipkin et al. 1993; Kay et al. 1990; Truong et al. 1991). The restricted taxonomic distribution of the MoCo biosynthetic pathway among budding yeasts therefore explains the limited number of yeasts possessing the nitrate assimilation pathway (Linder 2019a; Shen et al. 2018). The redox cofactor requirements for yeast nitrate reductase appear to differ between species. The nitrate reductase in *Blastobotrys adeninivorans* (family

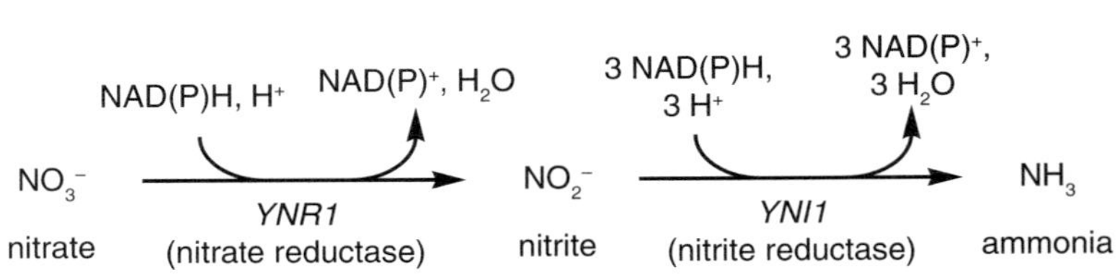

Fig. 7.2 Pathway for assimilation of nitrate and nitrite in budding yeasts

Trichomonascaceae) is specific for NADPH (Böer et al. 2009), while the nitrate reductases of *Candida boidinii* (family Pichiaceae), *Cyberlindnera jadinii* (family Phaffomycetaceae), and *Ogataea polymorpha* (family Pichiaceae) can use both NADH and NADPH (Choudary et al. 1986; Gromes et al. 1991; Pignocchi et al. 1998).

Nitrite is further reduced to ammonia by the FAD-containing enzyme nitrite reductase (EC 1.7.1.4), which is encoded by the *YNI1* gene (Brito et al. 1996; García-Lugo et al. 2000). The redox cofactor specificity for budding yeast nitrite reductases has not been studied in a comprehensive manner. Thus far, it is known that the *B. adeninivorans* nitrite reductase is specific for NADPH (Böer et al. 2009), while the *C. jadinii* nitrite reductase has been shown to use NADPH but has not been tested for NADH (Sengupta et al. 1996).

There are a small number of yeast genomes that only contain the nitrite reductase (Shen et al. 2018; Vigliotta et al. 2007) and therefore can assimilate nitrite but not nitrate. This polyphyletic group of yeasts includes species that lack the MoCo biosynthetic pathway (Linder 2019a), which suggests that the *YNI1* gene was acquired through horizontal gene transfer in these cases. In budding yeast species that possess the complete nitrate assimilation pathway, the *YNR1* and *YNI1* genes are located in a gene cluster together with nitrate transporters and regulatory transcription factors (Böer et al. 2009; Brito et al. 1996; García-Lugo et al. 2000; Linder 2019a).

7.4 Assimilation of Urea

Unlike other fungi, budding yeasts lack urease (EC 3.5.1.5) and instead deaminate urea using a bifunctional urea amidolyase (Roon and Levenberg 1972; Sumrada and Cooper 1982a), which is encoded by the *DUR1,2* gene (Cooper et al. 1980). The budding yeast urea amidolyase is an ATP-dependent biotin-containing enzyme composed of a C-terminal urea carboxylase domain (EC 6.3.4.6) preceded by an allophanate hydrolase domain (EC 3.5.1.54) at the N-terminus. Consequently, urea is catabolized in two steps through an allophanate intermediate (Whitney and Cooper 1972) (Fig. 7.3).

The budding yeast bifunctional urea amidolyase is thought to have originated from a gene fusion event between a duplicated copy of the fungal methylcrotonyl CoA carboxylase (*mccA*) gene and an allophanate hydrolase of unknown origin (Navarathna et al. 2010). Unlike the simple hydrolytic reaction catalyzed by urease,

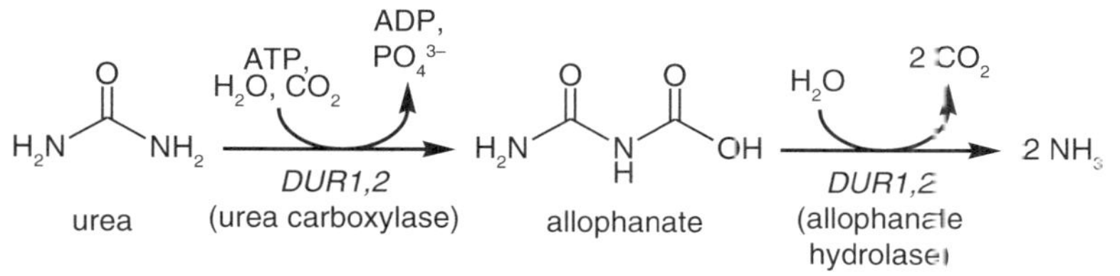

Fig. 7.3 Pathway for assimilation of urea in budding yeasts

urea deamination by urea amidolyase requires an input in the form of ATP. It has been suggested that this was a trade-off to enable the evolutionary ancestor of budding yeasts to no longer rely on the external availability of nickel ions (Ni^{2+}), which are required for urease activity, and therefore be able to eject the genes required for Ni^{2+} sensing and import (Navarathna et al. 2010). This might give budding yeasts a competitive advantage in nickel-poor environments. However, the reliance on urea amidolyase has instead resulted in an increased demand for biotin during assimilation of urea or nitrogen sources that proceed through a urea intermediate (Di Carlo et al. 1953).

7.5 Assimilation of Amino Acids

7.5.1 Assimilation of L-Glutamate and L-Glutamine

L-glutamate and L-glutamine are the main amino group donors in yeast anabolic pathways. Even so, a handful of anabolic amination reactions in yeast require ammonia (Sect. 7.2). Ammonia can be produced from L-glutamate through deamination (Fig. 7.4a) catalyzed by the enzyme NAD^+-dependent glutamate dehydrogenase (EC 1.4.1.2) (Hemmings 1980; Roon and Even 1973), which is encoded by the *GDH2* gene (Middelhoven et al. 1978; Miller and Magasanik 1990). Ammonia production from L-glutamine first requires α-ketoglutarate-dependent conversion of L-glutamine into two molecules of L-glutamate (Fig. 7.4b) by the enzyme glutamate

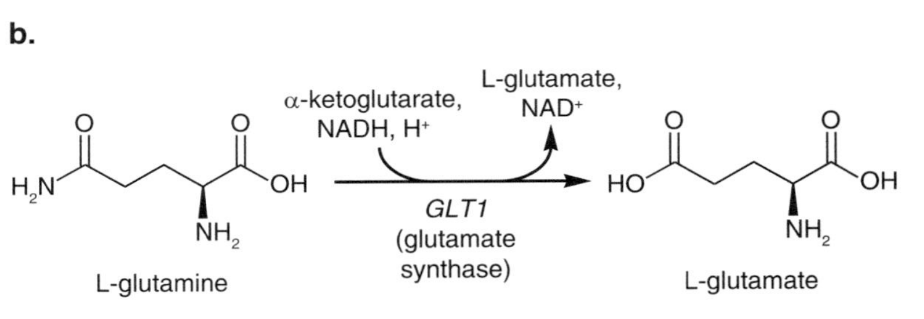

Fig. 7.4 Pathways for assimilation of L-glutamate and L-glutamine in budding yeasts. (**a**) Assimilation of L-glutamate via NADH-dependent glutamate dehydrogenase. (**b**) Assimilation of L-glutamine via glutamate synthase

synthase (EC 1.4.1.14) (Roon et al. 1974), which is encoded by the *GLT1* gene (Filetici et al. 1996; Folch et al. 1989). L-glutamate can subsequently be deaminated by the Gdh2 NAD⁺-dependent glutamate dehydrogenase as described above.

In addition to the requirement for ammonia, interconversion of L-glutamate and L-glutamine is also required under conditions when either of the two amino acids is provided as the sole nitrogen source. When L-glutamate is the sole nitrogen source, the conversion of L-glutamate to L-glutamine is carried out by glutamine synthetase (Sect. 7.2) with the ammonia co-substrate provided by the Gdh2 NAD⁺-dependent glutamate dehydrogenase. Conversely, when L-glutamine is the sole nitrogen source, the conversion of L-glutamine to L-glutamate is carried out by glutamate synthase as described above. Glutamate synthase can also play a role in ammonia assimilation in concert with glutamine synthase as described in Sect. 7.2 (Holmes et al. 1991; Lacerda et al. 1992; Magasanik 2003).

7.5.2 Assimilation of Linear, Branched, and Aromatic N_1 L-α-Amino Acids

L-Amino acids that only contain one molar equivalent of nitrogen in the form of the α-amino group are predominantly catabolized through transamination by pyridoxal 5'-phosphate (PLP)-dependent transaminases (EC 2.6.1). Typically, the transamination reaction transfers the α-amino group from the L-amino acid to an α-ketoacid acceptor substrate (predominantly α-ketoglutarate), which produces the corresponding α-ketoacid and L-amino acid (predominantly L-glutamate). In many cases, transaminases can play a role in both L-amino acid assimilation and biosynthesis. In the latter case, L-glutamate serves as a substrate rather than a product.

L-Alanine is assimilated through transamination by the enzyme alanine transaminase (EC 2.6.1.2), which is encoded by the *ALT1* gene (García-Campusano et al. 2009). Most species belonging to the WGD clade of the family Saccharomycetaceae possess at least one additional copy of alanine transaminase, which is encoded by the *ALT2* gene. The *S. cerevisiae ALT2* gene has been shown to be regulated by L-alanine availability although the resulting protein appears to lack enzymatic activity (Peñalosa-Ruiz et al. 2012). L-Aspartate is assimilated through transamination by the enzyme aspartate transaminase (EC 2.6.1.1), which exists as both mitochondrial and cytosolic isoenzymes in budding yeasts (encoded by the *AAT1* and *AAT2* genes, respectively) (Chéret et al. 1993; Morin et al. 1992; Verleur et al. 1997). Branched chain amino acids such as L-leucine, L-isoleucine, and L-valine are assimilated through transamination by the enzyme branched chain amino acid transaminase (EC 2.6.1.42). This enzyme activity exists as mitochondrial and cytosolic isoenzymes (encoded by the genes *BAT1* and *BAT2*, respectively) (Eden et al. 1996; Kispal et al. 1996) in species belonging to the WGD clade of the family Saccharomycetaceae as well as in *Yarrowia lipolytica* (family Dipodascaceae) (Bondar et al. 2005) while existing as a single enzyme in some other species of budding yeasts (Colón et al. 2011; Montalvo-Arredondo et al. 2015). The branched chain amino acid transaminase can also catalyze the transamination of other L-amino acids such as L-methionine (Bondar et al. 2005).

Aromatic amino acids such as L-phenylalanine, L-tyrosine, and L-tryptophan are assimilated through transamination by either of the enzymes aromatic amino acid transaminases I and II (EC 2.6.1.57 and EC 2.6.1.58, respectively) (Kradolfer et al. 1982), which are encoded by the genes *ARO8* and *ARO9*, respectively (Iraqui et al. 1998; Urrestarazu et al. 1998). Aromatic amino acid transaminase I in *S. cerevisiae* can also transaminate nonaromatic L-amino acids in vitro such as L-leucine, L-methionine, and L-α-aminoadipate (Urrestarazu et al. 1998). Both aromatic amino acid transaminases are able to use a range of α-ketoacid amino group acceptors in *S. cerevisiae* (Kradolfer et al. 1982). In *S. cerevisiae*, the α-ketoacids produced from the catabolism of branched and aromatic amino acids cannot be further assimilated and are instead converted to their corresponding long-chain alcohols – so-called fusel oils (Hazelwood et al. 2008).

L-Serine and L-threonine are not assimilated through transamination but are instead both deaminated by the enzyme serine/threonine dehydratase (EC 4.2.1.13/4.2.1.19; encoded by the *CHA1* gene) (Petersen et al. 1988) to produce ammonia and either pyruvate or α-ketobutyrate, respectively.

7.5.3 Assimilation of L-Arginine, L-Ornithine, and L-Proline

The amino acids L-arginine, L-ornithine, and L-proline share a common catabolic pathway in yeast (Brandriss and Magasanik 1980; Middelhoven 1964) (Fig. 7.5a). The first step involves the hydrolysis of the guanidino group of L-arginine to produce L-ornithine and urea, which is catalyzed by the enzyme arginase (EC 3.5.3.1; encoded by the *CAR1* gene) (Jauniaux et al. 1982; Sumrada and Cooper 1982b, 1984). The second step involves the transfer of the γ-amino group from L-ornithine to α-ketoglutarate, which produces L-glutamate and L-glutamate γ-semialdehyde. This reaction is catalyzed by the PLP-dependent enzyme L-ornithine transaminase (EC 2.6.1.13), which is encoded by the *CAR2* gene (Degols et al. 1987). L-Glutamate γ-semialdehyde spontaneously dehydrates to form L-Δ^1-pyrroline-5-carboxylate, which is then reduced to L-proline by the enzyme NADPH-dependent L-Δ^1-pyrroline-5-carboxylate reductase (EC 1.5.1.2), which is encoded by the *PRO3* gene (Brandriss and Falvey 1992). L-Proline is subsequently reoxidized to L-Δ^1-pyrroline-5-carboxylate by the enzyme FAD-dependent proline oxidase (EC 1.5.99.8), which is encoded by the *PUT1* gene (Brandriss and Magasanik 1979; Wang and Brandriss 1986). L-Δ^1-Pyrroline-5-carboxylate is then further oxidized to L-glutamate by the enzyme NAD^+-dependent L-Δ^1-pyrroline-5-carboxylate dehydrogenase (EC 1.5.1.12), which is encoded by the *PUT2* gene (Brandriss and Magasanik 1979; Krzywicki and Brandriss 1984).

Car1, Car2, and Pro3 are all cytosolic enzymes (Brandriss and Magasanik 1981; Jauniaux et al. 1978), while Put1 and Put2 are localized to the mitochondria

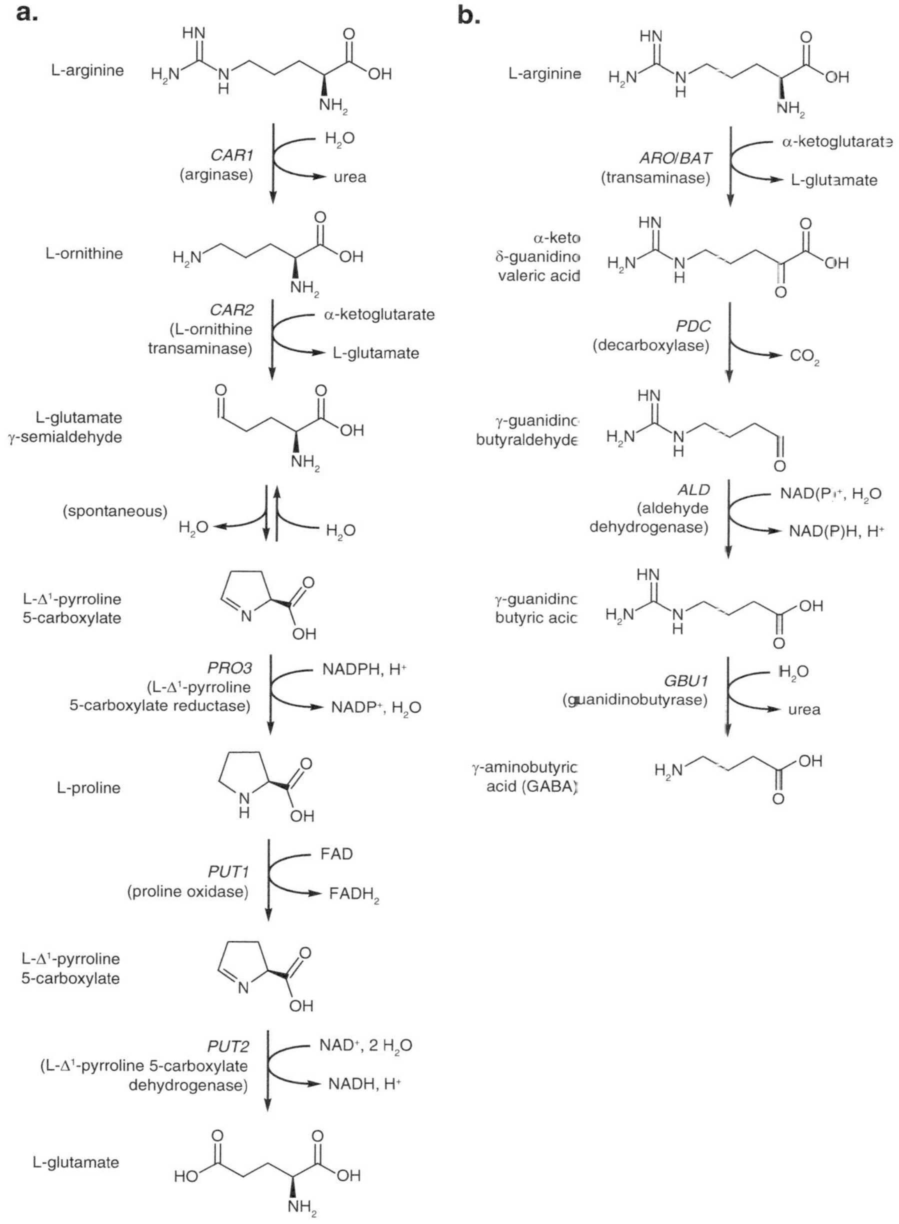

Fig. 7.5 Pathways for assimilation of L-arginine, L-ornithine, and L-proline in budding yeasts. (**a**) Conventional super-pathway for assimilation of L-arginine, L-ornithine, and L-proline in budding yeasts. (**b**) Alternative pathway for assimilation of L-arginine via a γ-guanidinobutyric acid intermediate

(Brandriss and Krzywicki 1986; Wang and Brandriss 1987). As both Pro3 and Put2 share L-Δ^1-pyrroline-5-carboxylate as a substrate, the separation of the two enzymes into two distinct cellular compartments avoids futile cycling between L-proline and L-Δ^1-pyrroline-5-carboxylate (Brandriss and Magasanik 1981).

An alternative pathway for L-arginine catabolism has recently been described in *Kluyveromyces lactis* (family Saccharomycetaceae) (Romagnoli et al. 2014). In *K. lactis*, L-arginine catabolism can also proceed by transamination of the α-amino group to produce α-keto-δ-guanidinovaleric acid, which is then subsequently decarboxylated into γ-guanidinobutyraldehyde and then reduced to γ-guanidinobutyric acid (Fig. 7.5b). The enzyme guanidinobutyrase (EC 3.5.3.7; encoded the *GBU1* gene in *K. lactis*) then hydrolyzes γ-guanidinobutyric acid to produce the ω-amino acid γ-aminobutyric acid (GABA) and urea (see Sects. 7.5.9 and 7.4, respectively, for subsequent catabolic steps). The *GBU1* gene is fairly common among budding yeasts but is missing in the genomes of the Saccharomycetaceae WGD clade. Heterologous expression of the *K. lactis GBU1* gene in *S. cerevisiae* can partially restore the ability to utilize L-arginine as a nitrogen source in a *Δcar1* genetic background (Romagnoli et al. 2014), which demonstrates that the upstream enzymes for the production of γ-guanidinobutyric acid from L-arginine are present in *S. cerevisiae*.

7.5.4 Assimilation of L-Asparagine

L-Asparagine catabolism in yeast involves the hydrolysis of the terminal amide group by the enzyme asparaginase (EC 3.5.1.1) to produce ammonia and L-aspartate (Fig. 7.6). The *S. cerevisiae* genome is known to encode two asparaginase isoenzymes (Kim et al. 1988; Sinclair et al. 1994) – asparaginase I (encoded by the *ASP1* gene) and asparaginase II (encoded by four copies of the *ASP3* gene). Asparaginase I is a constitutively expressed cytosolic enzyme (Jones and Mortimer 1973), while asparaginase II is associated with the extracellular face of the cell wall and is induced under conditions of nitrogen limitation (Dunlop and Roon 1975). The isoenzymes also differ somewhat in substrate specificity with asparaginase I being specific for L-asparagine, while asparaginase II can also hydrolyze the terminal amide group of D-asparagine (Dunlop et al. 1976). The *ASP1* gene appears to be

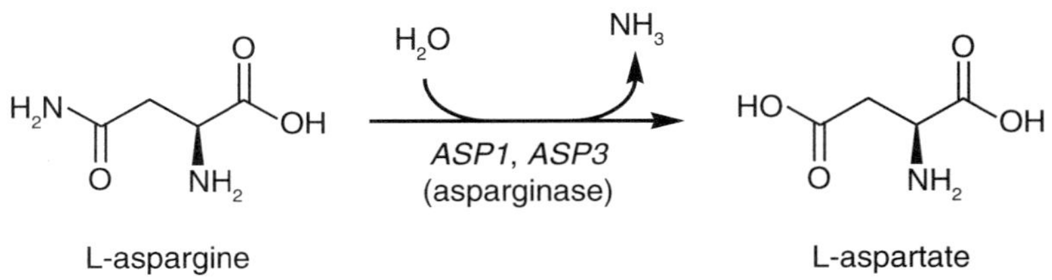

Fig. 7.6 Pathway for assimilation of L-asparagine in budding yeasts

ubiquitous among budding yeasts, while the *ASP3* gene occurs sporadically and is thought to have been recently acquired by *S. cerevisiae* through horizontal gene transfer from a species of the genus *Wickerhamomyces* (League et al. 2012).

7.5.5 Assimilation of L-Histidine

The ability to utilize L-histidine as a nitrogen source does not appear to be ubiquitous among budding yeasts (Brunke et al. 2014). Unlike archaea, bacteria, animals, and most fungi, budding yeasts lack the enzyme histidinase (EC 4.3.1.3), which deaminates histidine to produce ammonia and urocanic acid (Hall 1952). Instead, L-histidine assimilation in the yeasts *Candida albicans* (family Debaryomycetaceae) and *Nakaseomyces glabrata* (family Saccharomycetaceae) has been shown to employ the aromatic L-amino acid transaminase Aro8, which is thought to produce L-glutamate and imidazole pyruvate from L-histidine and α-ketoglutarate (Brunke et al. 2014; Rząd et al. 2018). Although *S. cerevisiae* fails to grow with L-histidine as sole nitrogen source, it does appear that L-histidine can still be transaminated by this species (Brunke et al. 2014). It has been suggested that the resulting α-ketoacid, imidazole pyruvate, can probably not be metabolized further by *S. cerevisiae* and therefore inhibits growth. The metabolic fate of imidazole pyruvate in *C. albicans* and *N. glabrata* is currently unknown.

7.5.6 Assimilation of L-Lysine

L-Lysine can be utilized as a nitrogen source by the majority of budding yeast but not by *S. cerevisiae* and other species belonging to the WGD clade of the family Saccharomycetaceae (Brady 1965). The catabolism of L-lysine as a nitrogen source remains poorly understood, and thus far only one gene has successfully been identified as encoding one of the previously described enzyme activities involved in L-lysine catabolism (Beckerich et al. 1994). Based on biochemical studies, a total of three separate pathways have so far been reported for assimilation of L-lysine as a nitrogen source, and these pathways may display some degree of redundancy in individual species (Hammer et al. 1991a).

The first described pathway for L-lysine catabolism in budding yeasts proceeds by initial acetylation of the ε-amino group of L-lysine followed by transamination of the α-amino group (Fig. 7.7a). This pathway has been described in *Candida tropicalis* (family Debaryomycetaceae) (Large and Robertson 1991), *Candida maltosa* (family Debaryomycetaceae) (Schmidt et al. 1988b; Schmidt and Bode 1992), *Cyberlindnera saturnus* (family Phaffomycetaceae), and *Y. lipolytica* (Gaillardin et al. 1976; Rothstein 1965). The resulting α-keto-ε-acetamidocaproic acid is then decarboxylated to form δ-acetamidovaleraldehyde, which is then oxidized to δ-acetamidovaleric acid (Large and Robertson 1991; Rothstein 1965).

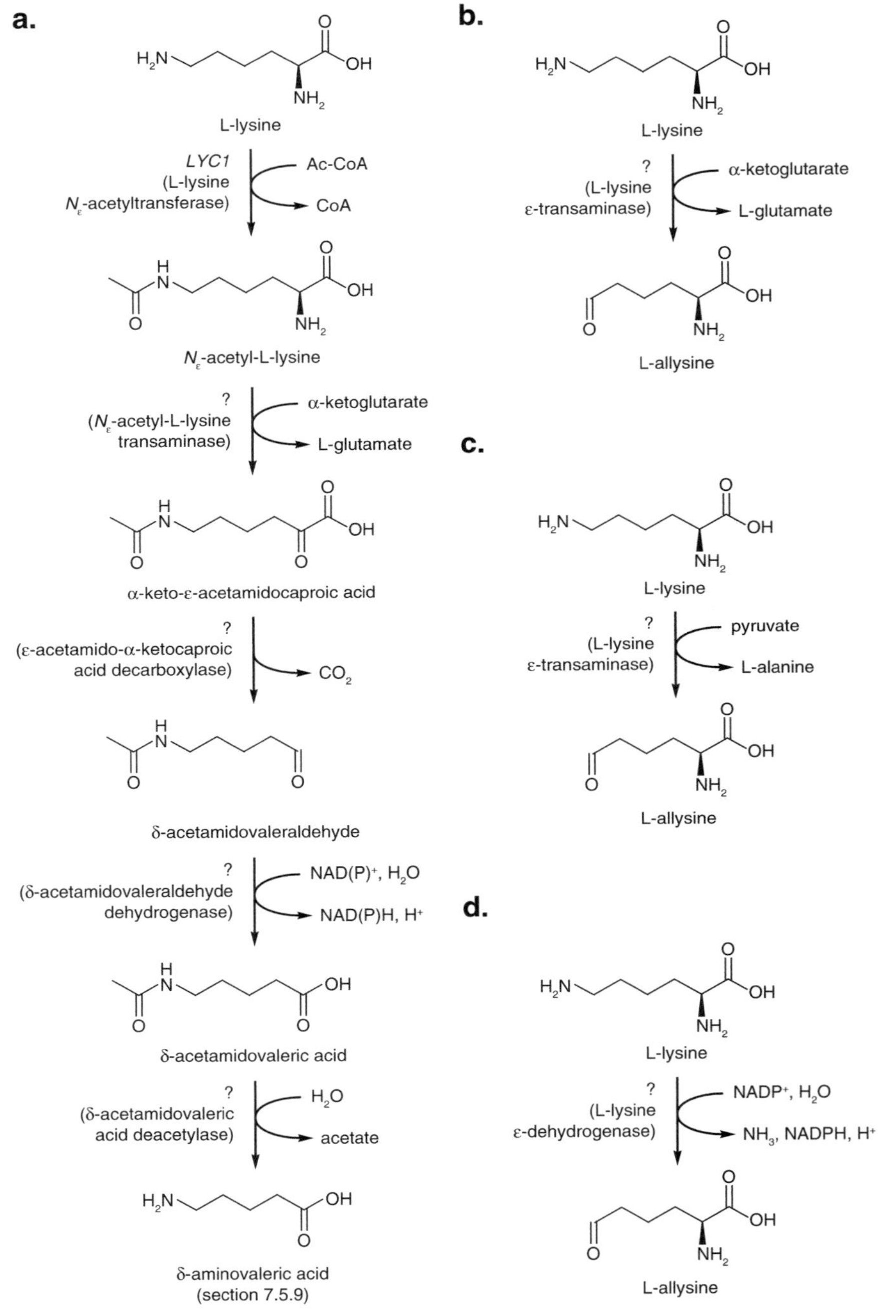

Fig. 7.7 Pathways for assimilation of L-lysine in budding yeasts. (**a**) Assimilation of L-lysine via initial acetylation of the ε-amino group by L-lysine *N*ε-acetyltransferase (encoded by the *LYC1* gene). (**b**) Assimilation of L-lysine via initial transamination of the ε-amino group by L-lysine:α-ketoglutarate ε-transaminase. (**c**) Assimilation of L-lysine via initial transamination of the ε-amino group by L-lysine:pyruvate ε-transaminase. (**d**) Assimilation of L-lysine via initial deamination of the ε-amino group by L-lysine ε-dehydrogenase

δ-Acetamidovaleric acid is then thought to be deacetylated to produce the ω-amino acid δ-aminovaleric acid, which can subsequently be transaminated to produce L-glutamate from α-ketoglutarate (Der Garabedian 1986) (see Sect. 7.5.9 regarding ω-amino acid transamination).

The second described pathway for L-lysine catabolism in budding yeasts involves transamination of the ε-amino group of L-lysine, which has been reported in *C. jadinii* and *Meyerozyma guilliermondii* (family Debaryomycetaceae) with either α-ketoglutarate or puruvate as the amino group acceptors to produce L-glutamate or L-alanine, respectively (Fig. 7.7b, c) (Hammer and Bode 1992; Schmidt et al. 1988a). The third pathway, which was first described in *C. albicans*, an $NADP^+$-dependent dehydrogenase, catalyzes the oxidative deamination of the ε-amino group of L-lysine (Hammer et al. 1991b) (Fig. 7.7d). Both these latter pathways produce L-allysine (L-α-aminoadipic acid δ-semialdehyde), which is then thought to be oxidized to L-α-aminoadipic acid, followed by transamination of the α-amino group to an α-ketoacid acceptor such as α-ketoglutarate to ultimately form L-glutamate (Hammer et al. 1991b). There have so far been no complementary genetic studies to identify the genes encoding the enzymes responsible for these two latter pathways, and it is uncertain to what degree all three pathways may coexist in a single species of budding yeast.

7.5.7 Assimilation of Glycine and Its N*-Methylated Variants*

Glycine is the simplest proteinogenic amino acid. *N*-Methylated species of glycine, such as glycine betaine (*N,N,N*-trimethylglycine), are used by many organisms as compatible solutes to increase cellular stress tolerance. Glycine and its *N*-methylated variants sarcosine (*N*-methylglycine), *N,N*-dimethylglycine, and glycine betaine form a metabolic axis connected to choline (Fig. 7.8) in many organisms, but it is unclear to what extent these reactions occur among budding yeasts. Glycine betaine synthesis has so far not been reported among budding yeast although it has recently been reported that two species of the genus *Wickerhamiella* (family Trichomonascaceae) may have acquired putative choline oxidases through horizontal gene transfer (Shen et al. 2018), which could be involved in glycine betaine synthesis from choline. The utilization of *N*-methylated variants of glycine as nitrogen sources by budding yeasts is rare, while the utilization of glycine itself as a nitrogen source is more common but not universal among budding yeasts (Linder 2014; Middelhoven et al. 1991).

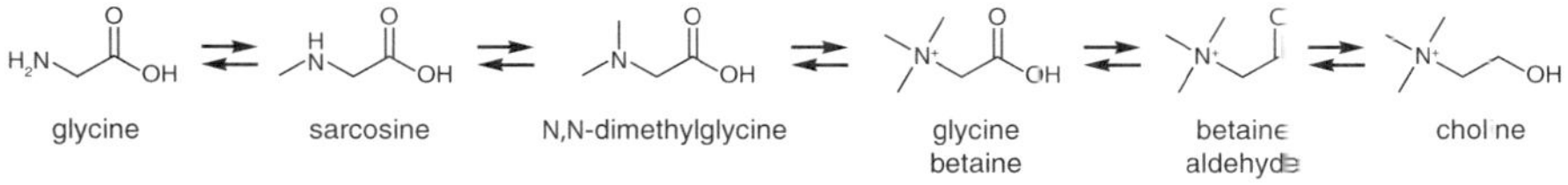

Fig. 7.8 The glycine-choline metabolic axis

S. cerevisiae has the ability to assimilate glycine as a nitrogen source through the tetrahydrofolate (THF)-dependent glycine cleavage system (EC 2.1.2.10) (McNeil et al. 1997; Sinclair and Dawes 1995), possibly coupled to the activity of the enzyme serine hydroxymethyltransferase (EC 2.1.2.1). The glycine cleavage system consists of four proteins (encoded by the genes *GCV1*, *GCV2*, *GCV3*, and *LPD1*, respectively) (McNeil et al. 1997; Nagarajan and Storms 1997; Ross et al. 1988; Sinclair et al. 1996), which together decarboxylate glycine to produce ammonia and carbon dioxide with concurrent reduction of NAD^+ to NADH as well as conversion of THF to 5,10-methylene-THF (5,10-CH_2-THF; Fig. 7.9a). The enzyme serine hydroxymethyltransferase (the mitochondrial and cytosolic isoenzymes encoded by the genes *SHM1* and *SHM2*, respectively) (McNeil et al. 1994) can convert glycine into L-serine while simultaneously regenerating THF from 5,10-CH_2-THF (Fig. 7.9b). L-Serine can subsequently be deaminated by the serine/threonine dehydratase Cha1 (see Sect. 7.5.2). Although both the glycine cleavage system and serine hydroxymethyltransferase appear to be ubiquitous among budding yeasts, the ability to assimilate glycine clearly is not. This strongly suggests the presence of additional pathways for glycine assimilation pathways among the budding yeasts. Glycine assimilation in the yeast *C. boidinii* has been shown to induce catalase (Haywood and Large 1981), which would suggest that an oxidase-like enzyme is involved. Metabolomic analysis of *S. cerevisiae* cultivated with isotope-labeled glycine as the sole nitrogen source has revealed that glycine can be also be assimilated through an as yet identified pathway, which produces glyoxylate as a by-product (Villas-Bôas et al. 2005). The existence of at least one additional glycine catabolic pathway in *S. cerevisiae* that does not involve the glycine cleavage system is also supported by genetic data (Sinclair and Dawes 1995).

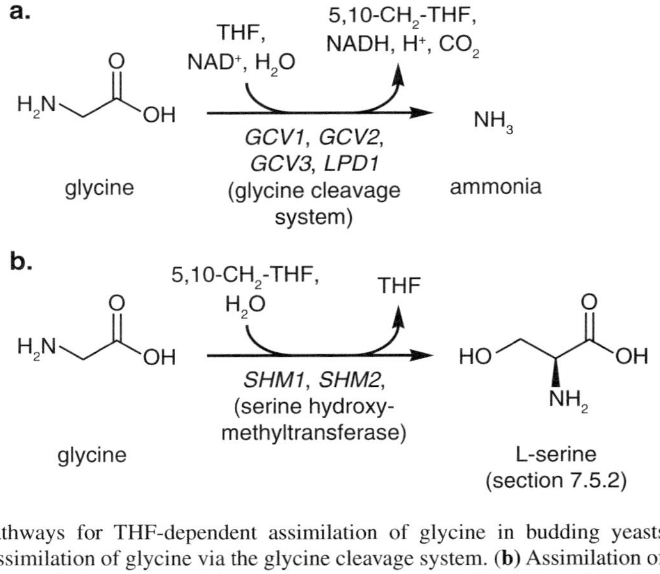

Fig. 7.9 Pathways for THF-dependent assimilation of glycine in budding yeasts. (**a**) THF-dependent assimilation of glycine via the glycine cleavage system. (**b**) Assimilation of glycine via serine hydroxymethyltransferase with concurrent regeneration of THF from 5,10-CH_2-THF

7.5.8 Assimilation of D-Amino Acids

The assimilation of D-amino acids is common among budding yeasts but is absent in *S. cerevisiae* and related species (LaRue and Spencer 1967a). The main catabolic process responsible for D-amino acid assimilation in budding yeasts is catalyzed by the enzyme D-amino acid oxidase (EC 1.4.3.3), which oxidatively deaminates D-amino acids to produce the corresponding α-keto acid, ammonia, and hydrogen peroxide (Fig. 7.10a). D-Amino acid oxidases contain a non-covalently bound FAD moiety and localize to the peroxisome (Berg and Rodden 1976; Yurimoto et al. 2000, 2001; Zwart et al. 1983a). Budding yeast D-amino acid oxidases appear to use a noncanonical C-terminal type 1 peroxisome targeting signal (PTS1) for peroxisomal localization (González et al. 1997; Klompmaker et al. 2010; Yurimoto et al. 2000). There are at least two types of D-amino acid oxidases in budding yeasts, which are encoded by the paralogous genes *DAO1* and *DAO2*, respectively (González et al. 1997; Klompmaker et al. 2010; Yurimoto et al. 2000). The Dao1 isoenzyme appears to be specific for neutral D-amino acids (Berg and Rodden 1976; Yurimoto et al. 2001), while the Dao2 isoenzyme specifically deaminates acidic D-amino acids (Klompmaker et al. 2010) and is therefore commonly referred to as D-aspartate oxidase. Pyruvate carboxylase (encoded by the *PYC1* gene) is required for full Dao1 activity in *Komagataella pastoris* (Klompmaker et al. 2010) and possibly other species as well. The *DAO1* gene was previously shown to be dispensable for utilization of D-alanine as a nitrogen source in *C. boidinii* (Yurimoto et al. 2000), which suggests some degree of genetic redundancy in D-alanine assimilation in this species.

Deamination of D-serine (and to a lesser extent, D-threonine) can also be catalyzed by PLP/zinc-dependent D-serine dehydratase (EC 4.3.1.18; Fig. 7.10b), which is encoded by the *DSD1* gene (Ito et al. 2008). The Dsd1 enzyme was first biochemically characterized in *S. cerevisiae*, which is notable since this yeast does not

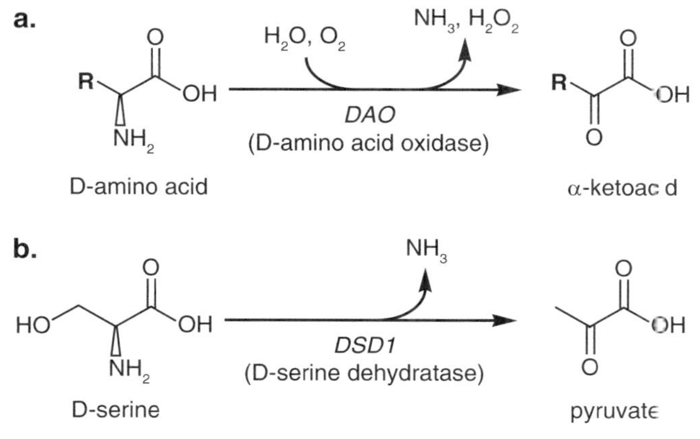

Fig. 7.10 Pathways for assimilation of D-amino acids in budding yeasts. (**a**) Assimilation of D-amino acids via D-amino acid oxidase. (**b**) Assimilation of D-serine via D-serine dehydratase

utilize D-serine as a nitrogen source. This could indicate that D-serine dehydratase may not be involved in D-amino acid assimilation but instead plays a role in D-amino acid detoxification. A putative proline racemase (EC 5.1.1.4), which catalyzes the interconversion of D- and L-proline, has been identified in the genome of *Candida parapsilosis* (family Debaryomycetaceae) and appears to be the result of a recent horizontal gene transfer event from bacteria (Fitzpatrick et al. 2008).

7.5.9 Assimilation of ω-Amino Acids

ω-Amino acids are defined by the structural formula $H_2N\text{-}(CH_2)_n\text{-}COOH$ and are metabolic intermediates in the catabolism of dihydropyrimidines, diamines (including L-lysine), and polyamines (Gojković et al. 2001; Large and Robertson 1988; Rothstein 1965). GABA is also produced from decarboxylation of L-glutamate, which is catalyzed by the enzyme glutamate decarboxylase (EC 4.1.1.15; encoded by the *GAD1* gene) in response to external stress (Coleman et al. 2001). Although the α-amino acid glycine is technically also an ω-amino acid, the catabolism of glycine is discussed separately in Sect. 7.5.7. ω-Amino acids are assimilated by transamination to produce the corresponding dicarboxylic acid semialdehyde (Fig. 7.11). The dicarboxylic acid semialdehyde is then typically converted to either a dicarboxylic acid or an ω-hydroxycarboxylic acid (Andersson Rasmussen et al. 2014; Bach et al. 2009; Cao et al. 2014). The initial transamination reaction is catalyzed by a class III PLP-dependent transaminase (EC 2.6.1.19).

Thus far, only three genes encoding ω-amino acid transaminases have been described in budding yeasts. The *UGA1* gene encodes GABA transaminase, which was first described in *S. cerevisiae* (André and Jauniaux 1990; Ramos et al. 1985). A gene encoding β-alanine transaminase (*PYD4*) has been described in *Lachancea kluyveri* (family Saccharomycetaceae) (Andersen et al. 2007). Both enzymes use α-ketoglutarate as an amino group acceptor to produce L-glutamate (Garabedian et al. 1986; Schnackerz et al. 2008). The *S. cerevisiae* genome contains the *UGA1*

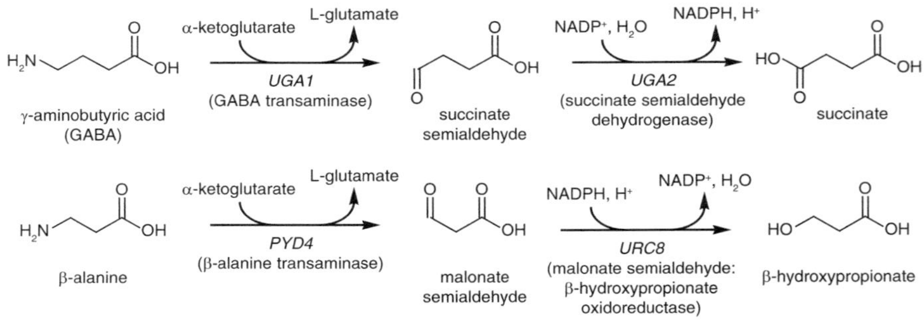

Fig. 7.11 Pathways for assimilation of ω-amino acids in budding yeasts

gene but not *PYD4* and consequently is unable to utilize β-alanine as a nitrogen source (Di Carlo et al. 1952), while the *L. kluyveri* genome contains both the *UGA1* and *PYD4* genes and can therefore utilize both ω-amino acids as nitrogen sources (Andersen et al. 2007). A third gene encoding a putative β-alanine transaminase was recently described in the yeast *Scheffersomyces stipitis* (family Debaryomycetaceae) (Linder 2019b). Phylogenetic analysis of budding yeast class III transaminases demonstrated that the *S. stipitis* protein does not form a monophyletic clade with the *L. kluyveri* Pyd4 protein, which could suggest that these two genes may have originated by independent duplications of an ancestral *UGA1* gene (Linder 2019b).

A previous biochemical study of purified GABA transaminase from *M. guilliermondii* demonstrated some degree of activity toward β-alanine, δ-aminovaleric acid, and ε-aminocaproic acid as well (Garabedian et al. 1986). An additional ω-amino acid transaminase has been described in *M. guilliermondii*, which in pure form preferentially transaminates δ-aminovaleric acid with some activity toward GABA (Der Garabedian 1986). However, the identities of the corresponding genes of both enzymes in the *M. guilliermondii* genome have not been established to date.

The $NADP^+$-dependent enzyme succinate semialdehyde dehydrogenase (EC 1.2.1.79; encoded by the *UGA2* gene) is responsible for the oxidation of succinate semialdehyde to succinate (Cao et al. 2014; Ramos et al. 1985). Succinate can subsequently be assimilated as a carbon source through the citric acid cycle. Production of γ-hydroxybutyrate from succinate semialdehyde has also been reported in GABA-utilizing *S. cerevisiae* cultures (Bach et al. 2009). It involves an as yet unidentified succinate semialdehyde dehydrogenase. The *L. kluyveri URC8* gene encodes an NADPH-dependent malonate semialdehyde dehydrogenase (EC 1.2.1.15), which reduces malonate semialdehyde to β-hydroxypropionate (Andersson Rasmussen et al. 2014).

7.6 Assimilation of Aliphatic Nitriles and Amides

Organic compounds with nitrile functional groups (R-C≡N) are found in nature, predominantly in the form of cyanogenic plant antifeedant compounds. Synthetic nitrile compounds are also used extensively in industrial chemical synthesis. The simplest nitrile – cyanide (HCN) – is abundant in mining waste and as such poses a considerable environmental hazard. The ability to catabolize nitriles would be expected in phytopathogenic lineages among the budding yeasts. It is also conceivable that budding yeasts that colonize the digestive tracts of herbivorous insects would be under direct selection to maintain the ability to metabolize and detoxify plant antifeedant nitrile compounds. The utilization of aliphatic as well as aromatic nitriles as nitrogen sources has been reported in a number of species of budding yeasts including *Cyberlindnera fabianii* (family Phaffomycetaceae) (Brewis et al. 1995), *Debaryomyces hansenii* (family Debaryomycetaceae) (Linardi et al. 1996), *Lachancea thermotolerans* (family Saccharomycetaceae) (Prasad et al. 2005), *M. guilliermondii* (Dias et al. 2000) as well as a number of species belonging to the polyphyletic genus *Candida* (Rezende et al. 1999).

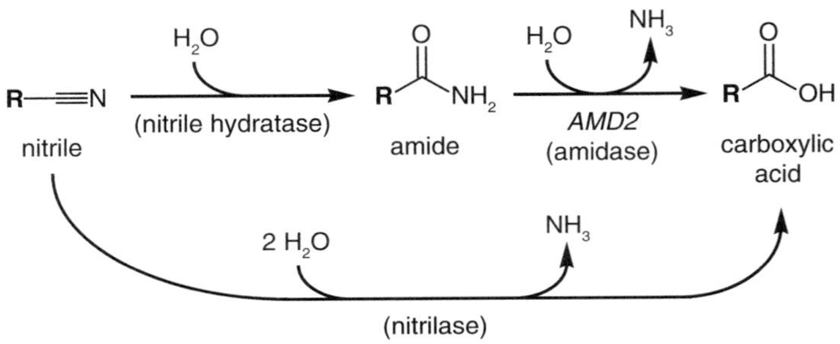

Fig. 7.12 Proposed pathways for assimilation of aliphatic amides and nitriles in budding yeasts

Previous work on microbial catabolism of aliphatic nitriles have identified two catabolic routes toward nitrile deamination (Fig. 7.12). One route involves the enzyme nitrilase (EC 3.5.5.1), which deaminates aliphatic nitriles through a single hydrolytic step to produce ammonia and the corresponding carboxylic acid. A second route toward nitrile deamination employs a two-step process, whereby the enzyme nitrile hydratase (EC 4.2.1.84) initially hydrolyses the nitrile substrate to produce the corresponding aliphatic amide followed by a second hydrolytic step catalyzed by an amidase (EC 3.5.1.4) to produce ammonia and the corresponding carboxylic acid. Cyanide hydratases (EC 4.2.1.66) are a subclass of nitrile hydratases that convert cyanide to formamide. These enzymes are often found in phytopathogenic fungi for the detoxification of cyanide released by cyanogenic plants (Nolan et al. 2003; Wang and van Etten 1992). Although no nitriliases or nitrile hydratases have been characterized in budding yeasts thus far, previously characterized cyanide hydratases from filamentous ascomycete fungi do have unambiguous homologs in the genomes of a number of budding yeast species (not shown). Interestingly, homologs of the downstream enzyme formamidase (EC 3.5.1.49), which hydrolyzes formamide into formic acid, have also been identified in the genomes of budding yeasts (Blandin et al. 2000). However, no biochemical or genetic characterization of a putative yeast formamidase has been carried out to date. The ability to utilize acetamide and larger aliphatic amides as nitrogen sources appears widespread among budding yeasts (Mira-Gutiérrez et al. 1995). Curiously, although *S. cerevisiae* cannot utilize aliphatic amides as nitrogen sources (Shepherd and Piper 2010), the *S. cerevisiae* genome does contain a putative amidase gene named *AMD2* (Chang and Abelson 1990).

7.7 Assimilation of Aliphatic Amines

For the purpose of this chapter, the word "amine" will be used to specifically refer to a category of organic compounds containing amino groups that is distinct from other categories of amino group containing compounds such as amino acids or aliphatic amides. Amines are in turn classified according to whether they contain one amino group (monoamines), two amino groups (diamines), or more than two amino groups (polyamines). The majority of studies on amine assimilation in budding yeasts have focused on aliphatic amines with linear or branched alkyl side chains. Amines with cyclic side chains will therefore not be considered here, while amines with aromatic side chains (e.g., anilines) will be discussed in Sect. 7.10. An individual amino group within an amine can be either mono-, di-, tri-, or tetraalkylated, and the corresponding compounds are therefore classified as either a primary (1°), secondary (2°), tertiary (3°), or quaternary (4°) amines, respectively.

7.7.1 Assimilation of Aliphatic Monoamines

Aliphatic monoamines are ubiquitous throughout nature and adopt a wide range of roles in cellular function. Ethanolamine and choline are both common functional groups of phospholipids, while zwitterionic monoamine compatible solutes such as choline *O*-phosphate, choline *O*-sulfate, trimethylamine *N*-oxide, and taurine function as protection against environmental stress. Amine-nitrogen is extracted through deamination of primary amines, which means that any multi-alkylated amino group must first be dealkylated to its primary amine form before deamination can occur.

Deamination of primary monoamines ($R\text{-}CH_2\text{-}NH_2$) in budding yeasts is performed by copper-containing amine oxidases (EC 1.4.3.6) to release ammonia, hydrogen peroxide, and the corresponding alkylaldehyde. There are two main isoforms of amine oxidases in budding yeasts – methylamine oxidase and benzylamine oxidase (Green et al. 1982; Haywood and Large 1981) – which are encoded by the paralogous genes *AMO1* and *AMO2*, respectively. The Amo1 methylamine oxidase, which contains an N-terminal type 2 peroxisomal targeting signal (PTS2) composed of the nonapeptide motif RLXXXXX$^{H}/_{Q}$L and localizes to the peroxisome (Faber et al. 1994), displays higher affinity toward short-chain aliphatic amines (Haywood and Large 1981). The Amo2 benzylamine oxidase, which is assumed to be cytosolic, has higher affinity toward amines with longer and/or bulkier side chains (Haywood and Large 1981). Amine oxidases are completely absent in the genomes of the WGD clade of the family Saccharomycetaceae (Linder 2019a), which is in agreement with the inability of this group of yeasts to utilize monoamines as a nitrogen sources (Linder 2014; van der Walt 1962). The methylotrophic yeast *K. pastoris* has been reported to possess a third type of amine oxidase known as lysyl oxidase (EC 1.4.3.13) (Kucha and Dooley 2001), which deaminates a broad range of substrates including lysine amino groups within polypeptides (Tur and Lerch 1988).

Secondary (di-alkylated) and tertiary (tri-alkylated) monoamines can also be assimilated by some budding yeasts (Linder 2014; van Dijken and Bos 1981), which involves stepwise dealkylation into primary monoamines. Biochemical studies of the dealkylation of tertiary and secondary amines indicate that two distinct enzyme activities are involved, which have characteristics similar to that of cytochrome P450 monooxygenases (Fattakhova et al. 1991; Green and Large 1983, 1984). The genetic identities of these two enzyme activities remain to be established although a recent genetic study in the yeast *S. stipitis* suggested that the enzymes involved are not cytochrome P450 monooxygenases (Linder 2018).

Choline is one of the few quaternary (tetra-alkylated) amines that is known to be assimilated by budding yeasts (Linder 2014; van Dijken and Bos 1981). The initial dealkylation reaction, which produces trimethylamine and glycolaldehyde (Mori et al. 1988) (Fig. 7.13), is thought to be catalyzed by a putative choline monooxygenase that is encoded by the *CMO1* gene (Linder 2014). Trimethylamine is then demethylated via dimethylamine and methylamine before ammonia is released by amine oxidase (Zwart et al. 1983b). The yeast Cmo1 protein has yet to be biochemically characterized, but the homologous plant protein, which is responsible for converting choline into betaine aldehyde, requires both oxygen and ferredoxin (Brouquisse et al. 1989). It is notable that the yeast *Lipomyces starkeyi* (family Lipomycetaceae) can utilize choline as a nitrogen source despite lacking the *CMO1* gene (Linder 2014). One possibility is that this species, which sits at the base of the budding yeast subphylum (Shen et al. 2016), employs an alternative pathway for choline assimilation more reminiscent of filamentous ascomycete fungi (Lambou et al. 2013).

Assimilation of trimethylamine *N*-oxide as a nitrogen source has been reported in the yeasts *C. boidinii* (Green and Large 1983) and *Sporopachydermia cereana* (Whitfield and Large 1986; Whitfield and Large 1987). Trimethylamine *N*-oxide

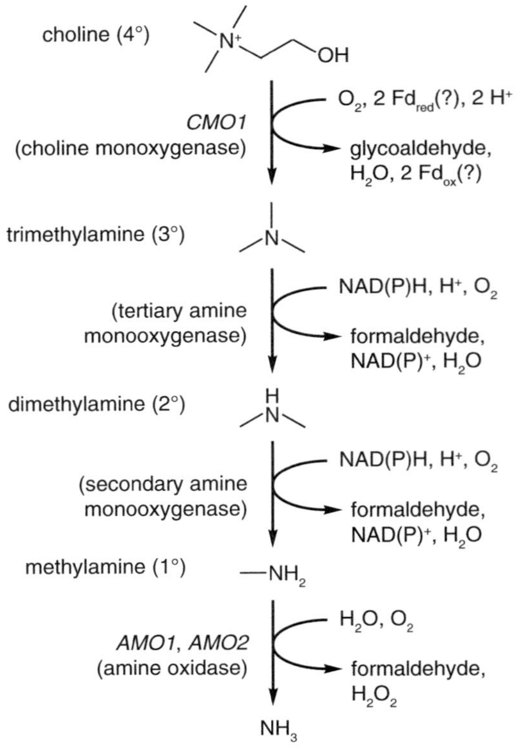

Fig. 7.13 Proposed pathway for assimilation of choline in budding yeasts

catabolism is thought to proceed through initial reduction to trimethylamine, which is catalyzed by an as yet unidentified NAD(P)H-dependent reductase (Whitfield and Large 1987).

7.7.2 Assimilation of Aliphatic Diamines and Polyamines

Aliphatic diamines (e.g., putrescine and cadaverine) and polyamines (spermine and spermidine; Fig. 7.14a) play essential roles in cellular metabolism (Chattopadhyay et al. 2003; White et al. 2003). A number of budding yeasts have the ability to assimilate diamines and polyamines as nitrogen sources although the ability to assimilate polyamines as nitrogen sources appears to be rarer (Brady 1965).

Compared to aliphatic monoamines, the assimilation of aliphatic diamines and polyamines poses some additional challenges. Deamination of amines ($R-CH_2-NH_2$) gives rise to aldehydes (R-CHO), which can form imines with other amines ($R-CH=NH-CH_2-R'$) through nonenzymatic condensation (Green et al. 1983; Gunasekaran and Gunasekaran 1999; Haywood and Large 1985). Consequently, deamination of organic substrates containing more than one amino group, such as diamines and polyamines, will result in reaction products where amino and carbonyl groups are present within the same molecule. This in turn can lead to the formation of either cyclic or polymeric products that may be toxic to the cell. Budding yeasts appear to have solved this problem through initial acetylation of one amino group prior to the first deamination step (Haywood and Large 1985). Thus far, the only amine *N*-acetyltransferases who have been assigned to genes belong to *S. cerevisiae*, which lacks the ability to assimilate diamines and polyamines. However, polyamines are also catabolized during synthesis of, for example, pantothenic acid, which raises the possibility that the individual amine *N*-acetyltransferases may be involved in both diamine and polyamine assimilations as well as in pantothenic acid biosynthesis. The *S. cerevisiae PAA1* gene encodes an amine *N*-acetyltransferase specific for the acetylation of putrescine, spermidine, and spermine (Liu et al. 2005) with lower activity toward monoamines (Ganguly et al. 2001). Two paralogous acetyltransferases encoded by the *HPA2* and *HPA3* genes in *S. cerevisiae* have also been shown to acetylate putrescine, spermidine, and spermine (Sampath et al. 2013).

Following acetylation of one amino group, catabolism of aliphatic diamines and polyamines proceeds by different routes. The free amino group of *N*-acetylated aliphatic diamines is deaminated by a monoamine oxidase to produce the corresponding acetamidoaldehyde, which is then oxidized to an acetamidoalkanoic acid by an acetamidoaldehyde dehydrogenase (Fig. 7.14b) (Gillyon et al. 1987). The acetamidoalkanoic acid is subsequently deacetylated by an acetamidoalkanoic acid deacetylase (Haywood and Large 1986) to produce the corresponding ω-amino acid, which is then further catabolized to extract the second nitrogen atom (Sect. 7.5.9).

The catabolism of the polyamines spermine and spermidine can proceed through a variety of routes (Fig. 7.14c). Before deamination can occur, both polyamines

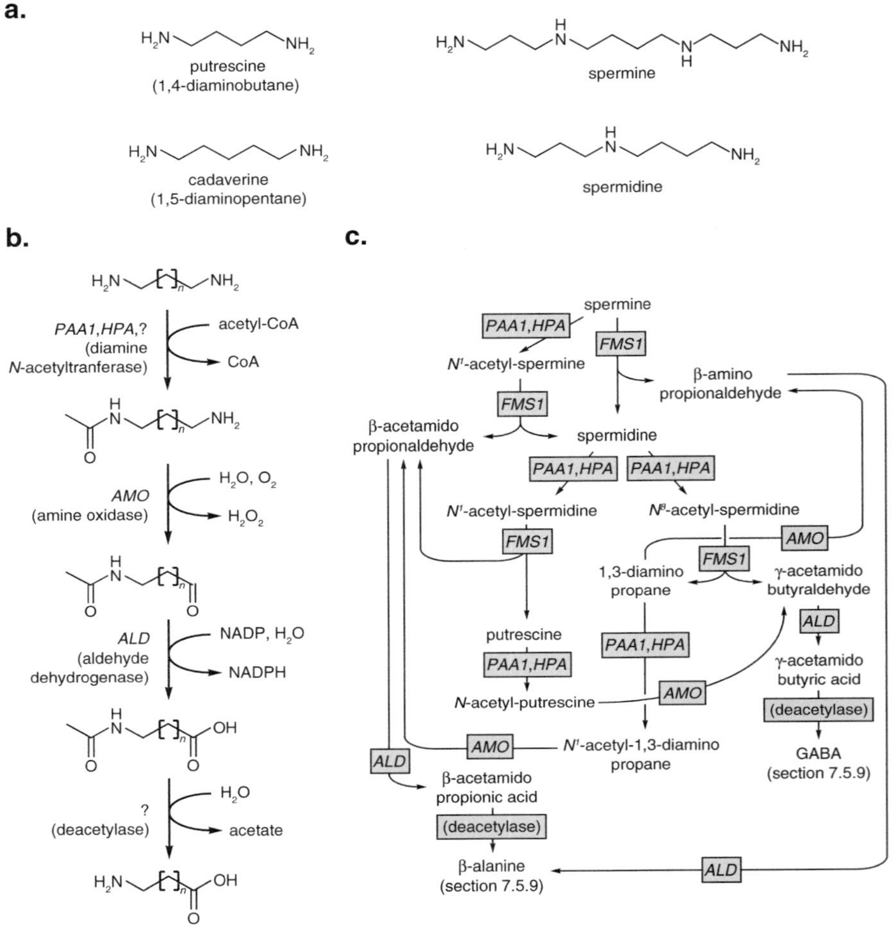

Fig. 7.14 Pathways for assimilation of aliphatic di- and polyamines in budding yeasts. (**a**) Structural formulae for common aliphatic di- and polyamines. (**b**) Generic pathway for assimilation of diamines in budding yeasts. (**c**) Possible routes for polyamine assimilation in budding yeasts. Proposed enzymes are identified by gray rectangles bearing the corresponding gene name (*FMS1* polyamine oxidase, *PAA1*/*HPA* di-/polyamine *N*-acetyltransferase, *AMO* amine oxidase, *ALD* aldehyde dehydrogenase, unassigned deacetylase)

must first be cleaved to produce mono- and diamines. The cleavage of polyamines is catalyzed by the FAD-containing enzyme polyamine oxidase (EC 1.5.3.17), which is encoded by the *FMS1* gene (Chattopadhyay et al. 2003; Landry and Sternglanz 2003). Polyamine oxidase can cleave both acetylated and non-acetylated spermine as well as acetylated spermidine but not non-acetylated spermidine (Landry and Sternglanz 2003). Cleavage of spermine, whether acetylated or not, produces spermidine and either β-aminopropionaldehyde or β-acetamidopropionaldehyde (Adachi et al. 2010; Landry and Sternglanz 2003). β-aminopropionaldehyde can be oxidized to β-alanine by aldehyde dehydrogenase (White et al. 2003), followed by transamination of β-alanine (Sect. 7.5.9). Spermidine is subsequently acetylated at either terminal amino group, producing *N*[1]- or *N*[8]-acetylspermidine, respectively, although the *S. cerevisiae* Fms1 polyamine oxidase preferentially cleaves *N*[1]-acetylspermidine (Landry and Sternglanz 2003). The cleavage products of *N*[1]-acetylspermidine are β-acetamidopropionaldehyde and putrescine (1,4-diaminobutane), while the cleavage products of *N*[8]-acetylspermidine are 1,3-diaminopropane and γ-acetamidobutyraldehyde (Landry and Sternglanz 2003). The resulting acetamidoaldehydes and diamines are then further catabolized as described previously.

7.8 Assimilation of Purines, Uric Acid, and Allantoin

Budding yeasts assimilate purines, uric acid, and allantoin through a joint, modular super-pathway (Fig. 7.15a, c). The first pathway module converts the purines adenine and guanine into xanthine, which in turn is converted to uric acid. The assimilation of adenine and guanine commences with hydrolytic deamination to produce hypoxanthine and xanthine, respectively. Adenine deaminase (EC 3.5.4.2) is encoded by the *AAH1* gene (Deeley 1992), while guanine deaminase (EC 3.5.4.3) is encoded by the *GUD1* gene (Saint-Marc and Daignan-Fornier 2004). Both purine deaminases also participate in purine salvage pathways and are therefore not unique to yeasts that assimilate purines as nitrogen sources. Hypoxanthine is converted into xanthine through the xanthine oxidase activity (EC 1.17.3.2) of the bifunctional enzyme xanthine oxidoreductase (Corte and Stirpe 1972), which is encoded by the *XAN1* gene (Linder 2019a). Xanthine oxidoreductase can also further convert xanthine into uric acid through its xanthine dehydrogenase activity (EC 1.17.1.4). At present, only the xanthine oxidoreductase of *B. adeninivorans* has been biochemically characterized (Jankowska et al. 2013). Xanthine oxidoreductase is one of few known MoCo-dependent enzymes among budding yeasts, another being nitrate reductase (Sect. 7.3). However, unlike nitrate reductase, xanthine oxidoreductase requires the sulfurized form of MoCo, and the *XAN1* gene has been shown to co-occur with a putative MoCo sulfurase (encoded by the *MOC6* gene) in budding yeast genomes (Linder 2019a; Shen et al. 2018).

Due to the restricted taxonomic distribution of the MoCo biosynthetic pathway among budding yeasts, most species lack xanthine oxidoreductase and only possess the downstream modules of the super-pathway (Linder 2019a). However, many species of budding yeast also possess a putative α-ketoglutarate-dependent xanthine dioxygenase (EC 1.14.11.48), which is encoded by the *XAN2* gene (Linder 2019a). Xanthine dioxygenase appears to be an enzyme unique to fungi and has been shown to convert xanthine into uric acid in the filamentous ascomycete *Aspergillus nidulans* and the fission yeast *Schizosaccharomyces pombe* (Cultrone et al. 2005;

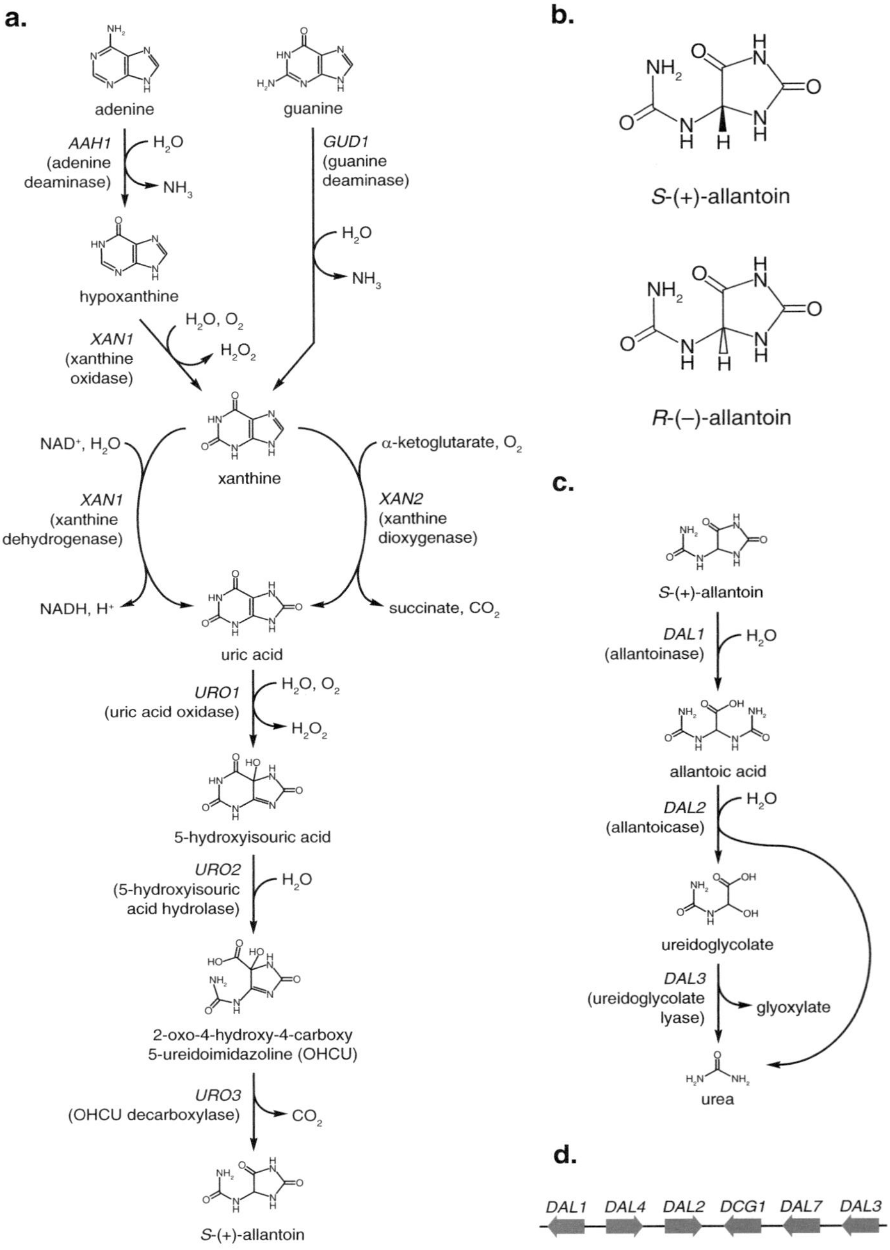

Fig. 7.15 Pathways for assimilation of purines, uric acid, and allantoin in budding yeasts. (**a**) The purine and uric acid catabolic modules. (**b**) The structural formulae of the *S*-(+)- and *R*-(−)-enantiomers of allantoin. (**c**) The allantoin catabolic module. (**d**) The *DAL* gene cluster in *S. cerevisiae*. Each gene is represented by an arrow with the direction of the arrow indicating whether the gene is located on the sense (rightward) or antisense (leftward) strand. Genes and intergenic regions are not drawn to scale

Montero-Morán et al. 2007). The budding yeast xanthine dioxygenase remains to be characterized genetically and biochemically, but the co-occurrence of the *XAN2* gene with downstream genes in the genomes of budding yeasts suggests that it plays the same function as previously reported in filamentous ascomycetes and fission yeasts (Linder 2019a).

The second module of the purine catabolic super-pathway involves the conversion of uric acid to *S*-(+)-allantoin in three enzymatic steps. The first step is carried out by the enzyme uric acid oxidase (EC 1.7.3.3; encoded by the *URO1* gene) (Koyama et al. 1996), which converts uric acid to 5-hydroxyisouric acid (Ramazzina et al. 2006). 5-Hydroxyisouric acid is hydrolyzed to 2-oxo-4-hydroxy-4-carboxy-5-ureidoimidazoline (OHCU) by the enzyme 5-hydroxyisouric acid hydrolase (EC 3.5.2.17) (Ramazzina et al. 2006), which is encoded by the *URO2* gene (Lee et al. 2013). OHCU is subsequently converted to *S*-(+)-allantoin by OHCU decarboxylase (EC 4.1.1.97) (Ramazzina et al. 2006), which is encoded by the *URO3* gene (Lee et al. 2013). The *URO2* and *URO3* genes have at present been characterized only in the basidiomycete yeast *Cryptococcus neoformans* (Lee et al. 2013) but are believed to participate in uric acid catabolism in budding yeasts as well (Linder 2019a).

It is notable that OHCU decarboxylase specifically produces the *S*-(+)-allantoin enantiomer (Ramazzina et al. 2006). Allantoin can also be formed spontaneously from 5-hydroxyisourate but will produce a racemic mixture of the *S*-(+)- and *R*-(−)-allantoin enantiomers (Fig. 7.15b) (Okumura et al. 1976; Ramazzina et al. 2006). Budding yeasts possess a putative allantoin racemase (EC 5.1.99.3) encoded by the *DCG1* gene (Yoo and Cooper 1991), which would interconvert the *S*-(+)- and *R*-(−)-allantoin enantiomers. An allantoin racemase has previously been purified in *C. jadinii* (Okumura and Yamamoto 1978), but the identity of the protein has not been established although it is assumed to be the product of the *DCG1* gene.

The third and final module of the purine catabolic super-pathway consists of three enzymatic steps that convert one molecule of *S*-(+)-allantoin into two molecules of urea (Fig. 7.15c). While the initial catabolism of uric acid takes place in the peroxisome, subsequent catabolism of *S*-(+)-allantoin is localized to the cytosol (Large et al. 1990). *S*-(+)-Allantoin is hydrolyzed to allantoic acid by the enzyme allantoinase (EC 3.5.2.5), which is encoded by the *DAL1* gene (Buckholz and Cooper 1991). Allantoic acid is further hydrolyzed to ureidoglycolate and urea by the enzyme allantoicase (EC 3.5.3.4), which is encoded by the *DAL2* gene (Yoo and Cooper 1991). Ureidoglycolate is then ultimately cleaved to glyoxylate and urea by the enzyme ureidoglycolate lyase (EC 4.3.2.3), which is encoded by the *DAL3* gene (Yoo et al. 1985). The two urea equivalents produced from the catabolism of allantoin is then further metabolized as described previously (Sect. 7.4).

The three *DAL* genes are scattered in the majority of budding yeast genomes (Linder 2019a; Wong and Wolfe 2005). However, in the WGD clade of the Saccharomycetaceae, the *DAL* genes are organized in a cluster (Fig. 7.15d) (Cooper et al. 1979; Lawther et al. 1974) of recent evolutionary origin (Wong and Wolfe 2005), which also includes associated genes such as the previously mentioned putative allantoin racemase encoded by the *DCG1* gene as well as an allantoin permease and a malate synthase (encoded by the *DAL4* and *DAL7* genes, respectively).

7.9 Assimilation of Pyrimidines and Dihydropyrimidines

Unlike many other eukaryotes, budding yeasts do not assimilate pyrimidines through a dihydropyrimidine intermediate (Andersen et al. 2008) although the budding yeast pyrimidine assimilation pathway has yet to be fully resolved. However, a number of budding yeasts can assimilate dihydropyrimidines as nitrogen sources through a separate pathway that is homologous to that of other eukaryotes (Gojković et al. 2000, 2001). This indicates that dihydropyrimidines are an ecologically relevant nitrogen source for budding yeasts. There is some degree of taxonomic overlap between the two pathways (LaRue and Spencer 1968), but this does not appear to be the result of common metabolic intermediates or enzymatic activities (Linder 2019a).

7.9.1 Assimilation of Pyrimidines

In budding yeasts, the assimilation of cytosine proceeds through hydrolytic deamination to produce uracil and ammonia and is carried out by the enzyme cytosine deaminase (EC 3.5.4.1), which is encoded by the *FCY1* gene (Erbs et al. 1997). The utilization of thymine as a nitrogen source is notably rarer than the utilization of either cytosine or uracil (LaRue and Spencer 1968), and it remains unclear whether thymine assimilation proceeds through the uracil assimilation pathway.

Studies of the uracil catabolic pathway have so far been restricted to the yeast *L. kluyveri* (Andersen et al. 2008), where seven loci have thus far been shown to participate in uracil catabolism. These loci have been assigned gene names *URC1* through *URC6* as well as *URC8* (Andersen et al. 2008; Andersson Rasmussen et al. 2014; Björnberg et al. 2010). The products of genes *URC1* and *URC4* genes (which encode a putative cyclohydrolase and a putative ribosyl-urea degrading enzyme, respectively) are unique to the uracil catabolic pathway (Linder 2019a). The *URC3* and *URC5* genes have been shown to correspond to the urea amidolyase-encoding gene *DUR1,2* (Andersen et al. 2008), which suggests that uracil catabolism proceeds through a urea intermediate. The *L. kluyveri URC6* gene encodes a putative uracil phosphoribosyltransferase, which is similar to the *S. cerevisiae FUR1* gene product, while the *URC2* gene encodes a putative transcription factor similar to the gene product of *S. cerevisiae* gene *YDR520C* (Andersen et al. 2008). The *L. kluyveri URC8* gene encodes an NADPH-dependent short-chain dehydrogenase and is thought to be responsible the conversion the catabolic by-product malonate semialdehyde into β-hydroxypropionate (Andersson Rasmussen et al. 2014). The *URC8* gene is also thought play a role in β-alanine assimilation (Sect. 7.5.9).

7.9.2 Assimilation of Dihydropyrimidines

Budding yeasts have been shown to assimilate the dihydropyrimidines 5,6-dihydrouracil, 5,6-dihydrothymine, and hydantoin (Gojković et al. 2000; LaRue and Spencer 1968). Dihydropyrimidines are initially hydrolyzed to produce the corresponding ureido acid (Fig. 7.16). This reaction is catalyzed by the enzyme dihydropyrimidinase (EC 3.5.2.2), which is encoded by the *PYD2* gene (Gojković et al. 2000). The *L. kluyveri* Pyd2 dihydropyrimidinase has been shown to hydrolyze hydantoin with less efficiency than 5,6-dihydrouracil or 5,6-dihydrothymine (Lohkamp et al. 2006).

The ureido acids produced through hydrolysis of dihydropyrimidines are hydrolyzed further to produce the corresponding ω-amino acid, ammonia, and carbon dioxide (Fig. 7.16). This reaction is catalyzed by the enzyme β-ureidopropionase/β--alanine synthase (EC 3.5.1.6), which is encoded by the *PYD3* gene (Gojković et al. 2001). Despite the name, this enzyme is also responsible for hydrolyzing the ureido group of β-ureidoisobutyric acid and hydantoic acid. The resulting ω-amino acids and ammonia are further assimilated as described in Sects. 7.2, 7.5.7, and 7.5.9.

7.10 Assimilation of Aromatic and Heterocyclic Nitrogen Compounds

Aromatic and heterocyclic nitrogen compounds such as anilines, pyridines, triazines, and imidazoles are commonly found among anthropogenic xenobiotics such as pesticides and pharmaceuticals. There are a number of reports on the utilization of aromatic and heterocyclic nitrogen compounds as nitrogen sources by budding yeasts. The majority of these reports feature species of the soil-associated genus

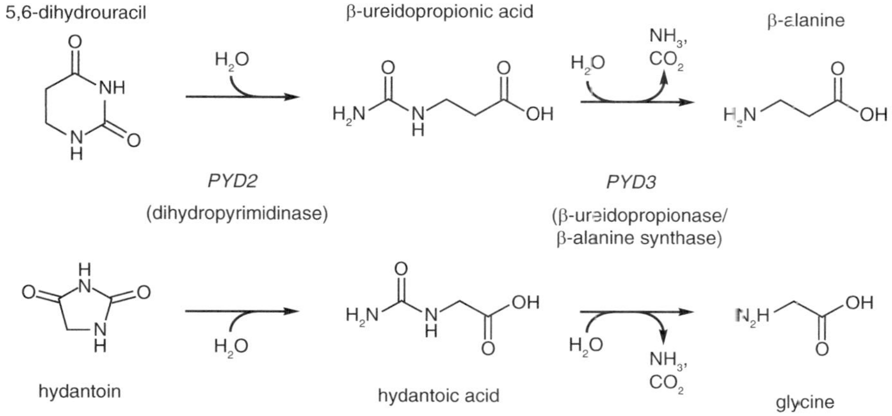

Fig. 7.16 Pathway for assimilation of dihydropyrimidines in budding yeasts

Lipomyces (family Lipomycetaceae), which forms a basal branch of the subphylum Saccharomycotina (Shen et al. 2016). *Lipomyces* species have so far been shown to utilize a variety of aromatic and heterocyclic nitrogen-containing herbicides as nitrogen sources including paraquat (Carr et al. 1985; Hata et al. 1986), picloram (Sadowsky et al. 2009), and a number of triazine herbicides (Nishimura et al. 2002). The utilization of imidazole compounds as nitrogen sources has been reported in the Lipomycetaceae species *L. starkeyi* and *Lipomyces lipoferus* as well as in *Trigonopsis variabilis* (family Trigonopsidaceae) and *Starmerella gropengiesseri* (family Trichomonascaceae) (LaRue and Spencer 1967b). A strain of *C. tropicalis* isolated from granular sludge has been reported to utilize aniline as a source of both carbon and nitrogen (Wang et al. 2011). The utilization of the heterocyclic nitrogen compound urotropin appears to be fairly common among budding yeasts (Middelhoven and van Doesburg 2007). At the time of writing, little is known about the enzyme activities involved in the catabolism of aromatic and heterocyclic nitrogen compounds or the identity of the genes that encode these enzymes.

7.11 Biotechnological Applications of Yeast Nitrogen Assimilation Pathways

Pathways for the assimilation of nitrogen in yeast have a number of potential applications in biotechnology. One such potential application is the use of yeasts for the degradation of nitrogen-containing xenobiotics such as pesticides and pharmaceuticals. As described previously (Sect. 7.10), the basal genus *Lipomyces* is particularly adept at catabolizing aromatic and heterocyclic nitrogen compounds, which could be applied in bioremediation of contaminated environments.

There have been a number of instances where nitrogen assimilation genes have used as selection markers for the generation of transgenic yeast strains. The yeast cytosine deaminase gene *FCY1* has been developed into a dual selection marker since the gene is essential for utilization of cytosine as a nitrogen source, while the deletion mutant is resistant to 5-fluorocytosine (Hartzog et al. 2005). If the *FCY1* gene can be removed after genomic integration, for example, through excision by a recombinase followed by counterselection on medium containing 5-fluorocytosine, the gene can be reused multiple times. Similarly, the *amdS* amidase gene from *A. nidulans* has been developed into a dual selection marker, which enables *S. cerevisiae* to use acetamide as a nitrogen source but can also be selected against using fluoroacetamide (Solis-Escalante et al. 2013). However, the inability to use acetamide as nitrogen source is mostly restricted the WGD clade of the family Saccharomycetaceae. Therefore, the *amdS* gene is of limited use to other yeast species that already possess endogenous amidases. Considering the currently limited knowledge of nitrogen assimilation pathways among budding yeasts, there likely remain many other nitrogen assimilation genes that could be appropriated as selection markers for a wider range of yeast species.

Another approach to employ nitrogen assimilation pathways in a biotechnological context is to engineer production strains to enable assimilation of xenobiotic or ecologically rare nitrogen sources. This approach, termed "robust operation by utilization of substrate technology" (ROBUST), confers a competitive advantage to the engineered strains over potential contaminant organisms under conditions where the selected nitrogen compounds are provided as either the main or only nitrogen source (Shaw et al. 2016).

7.12 Concluding Remarks

It is clear that much still remains to be learned about nitrogen assimilation pathways in budding yeasts – in particular, those pathways that are unique to nonconventional yeasts. The two main "known unknowns" of budding yeast nitrogen assimilation pathways can be divided into "orphan" enzymes and "orphan" pathways. The term orphan enzyme describes an enzyme that has only been described in terms of its biochemical properties. However, its sequence – and therefore its genetic identity – remains unknown (Hanson et al. 2009). Relevant examples of budding yeast orphan enzymes include a number of activities that are involved in the assimilation L-lysine (Sect. 7.5.6) and the cytochrome P450 monooxygenase-like activities responsible for the dealkylation of secondary and tertiary aliphatic amines (Sect. 7.7.1). Conversely, the only known characteristic of an orphan pathway is the identity of the substrate that it assimilates with little or no information about metabolic intermediates or the enzymes involved in assimilation. Budding yeast orphan pathways include the alternative pathway for glycine catabolism (Sect. 7.5.7) as well as catabolism of aromatic and heterocyclic nitrogen sources (Sect. 7.10). The recent wealth of budding yeast genome sequences will enable the identification of some genes within orphan pathways by simply correlating their presence in a specific genome with the ability to assimilate a particular nitrogen source (Linder 2014). "Omics" approaches such as transcriptomics, proteomics, and metabolomics will also be instrumental in the identification and characterization of orphan enzymes and pathways.

This chapter has not discussed the ecological context of individual nitrogen assimilation pathways. However, it would seem likely that the ability to catabolize particular nitrogen sources is key for the colonization of different ecological niches (Shiraishi et al. 2015). The fact that many species of budding yeasts are associated with insects and other invertebrates (Ganter 2006) – in many cases as intestinal endosymbionts (Suh et al. 2003, 2005; Woolfolk and Inglis 2004; Zhang et al. 2003) – it would be informative to investigate whether individual nitrogen assimilation pathways can influence not only the ability of the yeast to colonize the host but also confer tangible benefits to it. For instance, budding yeast endosymbionts may assist in the detoxification of potentially hazardous nitrogen-containing metabolites in the diet of the host organism (Martin et al. 2018).

The sensing of individual external nitrogen compounds and how the corresponding assimilation pathways are regulated have not been discussed in this chapter. Our current knowledge of how nitrogen assimilation pathways are regulated in budding yeasts predominantly comes from studies in *S. cerevisiae* (Broach 2012; Cooper 2002; Ljungdahl 2009; Magasanik and Kaiser 2002; Messenguy et al. 2006; Wong et al. 2008). It therefore follows that the regulation of nitrogen assimilation pathways that are absent in *S. cerevisiae* is far less well understood. Analyses of promoter sequences from nitrogen assimilation genes specific to nonconventional yeasts have identified a number of "orphan" regulatory sequence motifs, i.e., short conserved DNA elements that are likely to serve as binding sites for as yet unidentified transcriptional regulatory proteins (Linder 2019a). Consequently, like all other aforementioned aspects of budding yeast nitrogen assimilation, the regulatory aspect remains a fertile ground for new discoveries.

Acknowledgments The author wishes to thank Professor Terrance G. Cooper (University of Tennessee) and Professor Peter J. Large (University of Hull) for commenting on the text.

References

Adachi MS, Torres JM, Fitzpatrick PF (2010) Mechanistic studies of the yeast polyamine oxidase Fms1: kinetic mechanism, substrate specificity, and pH dependence. Biochemistry 49:10440–10448

Andersen G, Andersen B, Dobritzsch D, Schnackerz KD, Piškur J (2007) A gene duplication led to specialized gamma-aminobutyrate and beta-alanine aminotransferase in yeast. FEBS J 274:1804–1817

Andersen G, Björnberg O, Polakova S, Pynyaha Y, Rasmussen A, Møller K, Hofer A, Moritz T, Sandrini MP, Merico AM, Compagno C, Akerlund HE, Gojković Z, Piškur J (2008) A second pathway to degrade pyrimidine nucleic acid precursors in eukaryotes. J Mol Biol 380:656–666

Andersson Rasmussen A, Kandasamy D, Beck H, Crosby SD, Björnberg O, Schnackerz KD, Piškur J (2014) Global expression analysis of the yeast *Lachancea* (*Saccharomyces*) *kluyveri* reveals new *URC* genes involved in pyrimidine catabolism. Eukaryot Cell 13:31–42

André B, Jauniaux JC (1990) Nucleotide sequence of the yeast *UGA1* gene encoding GABA transaminase. Nucleic Acids Res 18:3049

Avendaño A, Deluna A, Olivera H, Valenzuela L, Gonzalez A (1997) *GDH3* encodes a glutamate dehydrogenase isozyme, a previously unrecognized route for glutamate biosynthesis in *Saccharomyces cerevisiae*. J Bacteriol 179:5594–5597

Avila J, Pérez MD, Brito N, González C, Siverio JM (1995) Cloning and disruption of the *YNR1* gene encoding the nitrate reductase apoenzyme of the yeast *Hansenula polymorpha*. FEBS Lett 366:137–142

Bach B, Meudec E, Lepoutre JP, Rossignol T, Blondin B, Dequin S, Camarasa C (2009) New insights into γ-aminobutyric acid catabolism: evidence for γ-hydroxybutyric acid and polyhydroxybutyrate synthesis in *Saccharomyces cerevisiae*. Appl Environ Microbiol 75:4231–4239

Beckerich JM, Lambert M, Gaillardin C (1994) *LYC1* is the structural gene for lysine N-6-acetyl transferase in yeast. Curr Genet 25:24–29

Berg PC, Rodden FA (1976) Purification of D-amino acid oxidase from *Trigonopsis variabilis*. Anal Biochem 71:214–222

Björnberg O, Vodnala M, Domkin V, Hofer A, Rasmussen A, Andersen G, Piškur J (2010) Ribosylurea accumulates in yeast *urc4* mutants. Nucleosides Nucleotides Nucleic Acids 29:433–437

Blandin G, Ozier-Kalogeropoulos O, Wincker P, Artiguenave F, Dujon B (2000) Genomic exploration of the hemiascomycetous yeasts: 16. *Candida tropicalis*. FEBS Lett 487:91–94

Böer E, Schröter A, Bode R, Piontek M, Kunze G (2009) Characterization and expression analysis of a gene cluster for nitrate assimilation from the yeast *Arxula adeninivorans*. Yeast 26:83–93

Bondar DC, Beckerich JM, Bonnarme P (2005) Involvement of a branched-chain aminotransferase in production of volatile sulfur compounds in *Yarrowia lipolytica*. Appl Environ Microbiol 71:4585–4591

Brady BL (1965) Utilization of amino compounds by yeasts of the genus *Saccharomyces*. Antonie Van Leeuwenhoek 31:95–102

Brandriss MC, Falvey DA (1992) Proline biosynthesis in *Saccharomyces cerevisiae*: analysis of the *PRO3* gene, which encodes Δ^1-pyrroline-5-carboxylate reductase. J Bacteriol 174:3782–3788

Brandriss MC, Krzywicki KA (1986) Amino-terminal fragments of Δ^1-pyrroline-5-carboxylate dehydrogenase direct β-galactosidase to the mitochondrial matrix in *Saccharomyces cerevisiae*. Mol Cell Biol 6:3502–3512

Brandriss MC, Magasanik B (1979) Genetics and physiology of proline utilization in *Saccharomyces cerevisiae*: enzyme induction by proline. J Bacteriol 140:498–503

Brandriss MC, Magasanik B (1980) Proline: an essential intermediate in arginine degradation in *Saccharomyces cerevisiae*. J Bacteriol 143:1403–1410

Brandriss MC, Magasanik B (1981) Subcellular compartmentation in control of converging pathways for proline and arginine metabolism in *Saccharomyces cerevisiae*. J Bacteriol 145:1359–1364

Brewis EA, van Der Walt JP, Priora BA (1995) The utilization of aromatic, cyclic and heterocyclic nitriles by yeasts. Syst Appl Microbiol 18:338–342

Brito N, Avila J, Perez MD, Gonzalez C, Siverio JM (1996) The genes *YNI1* and *YNR1*, encoding nitrite reductase and nitrate reductase respectively in the yeast *Hansenula polymorpha*, are clustered and co-ordinately regulated. Biochem J 317:89–95

Broach JR (2012) Nutritional control of growth and development in yeast. Genetics 192:73–105

Brouquisse R, Weigel P, Rhodes D, Yocum CF, Hanson AD (1989) Evidence for a ferredoxin-dependent choline monooxygenase from spinach chloroplast stroma. Plant Physiol 90:322–329

Brunke S, Seider K, Richter ME, Bremer-Streck S, Ramachandra S, Kiehntopf M, Brock M, Hube B (2014) Histidine degradation via an aminotransferase increases the nutritional flexibility of *Candida glabrata*. Eukaryot Cell 13:758–765

Buckholz RG, Cooper TG (1991) The allantoinase (*DAL1*) gene of *Saccharomyces cerevisiae*. Yeast 7:913–923

Cao J, Singh NK, Locy RD (2014) Characterization of the recombinant succinic semi-aldehyde dehydrogenase from *Saccharomyces cerevisiae*. Yeast 31:411–420

Carr RJ, Bilton RF, Atkinson T (1985) Mechanism of biodegradation of paraquat by *Lipomyces starkeyi*. Appl Environ Microbiol 49:1290–1294

Chang TH, Abelson J (1990) Identification of a putative amidase gene in yeast *Saccharomyces cerevisiae*. Nucleic Acids Res 18:7180

Chattopadhyay MK, Tabor CW, Tabor H (2003) Spermidine but not spermine is essential for hypusine biosynthesis and growth in *Saccharomyces cerevisiae*: spermine is converted to spermidine *in vivo* by the *FMS1*-amine oxidase. Proc Natl Acad Sci U S A 100:13869–13874

Chéret G, Pallier C, Valens M, Diagnan-Fornier B, Fukuhara H, Bolotin-Fukuhara M, Sor F (1993) The DNA sequence analysis of the *HAP4-LAP4* region on chromosome XI of *Saccharomyces cerevisiae* suggests the presence of a second aspartate aminotransferase gene in yeast. Yeast 9:1259–1265

Choudary PV, Rao GR (1984) Molecular analysis of inorganic nitrogen assimilation in yeasts. Arch Microbiol 138:183–186

Choudary PV, Deobagkar DN, Rao GR (1986) Partial purification and properties of the assimilatory nitrate reductase of the food yeast *Candida utilis*. Microbios 47:135–147
Coleman ST, Fang TK, Rovinsky SA, Turano FJ, Moye-Rowley WS (2001) Expression of a glutamate decarboxylase homologue is required for normal oxidative stress tolerance in *Saccharomyces cerevisiae*. J Biol Chem 276:244–250
Colón M, Hernández F, López K, Quezada H, González J, López G, Aranda C, González A (2011) *Saccharomyces cerevisiae* Bat1 and Bat2 aminotransferases have functionally diverged from the ancestral-like *Kluyveromyces lactis* orthologous enzyme. PLoS One 6:e16099
Cooper TG (1982) Nitrogen metabolism in Saccharomyces cerevisiae. In: Strathern JN, Jones EW, Broach JR (eds) The molecular biology of the yeast Saccharomyces: metabolism and gene expression. Cold Spring Harbor Laboratory Press, Cold Spring Harbor, pp 39–99
Cooper TG (2002) Transmitting the signal of excess nitrogen in *Saccharomyces cerevisiae* from the Tor proteins to the GATA factors: connecting the dots. FEMS Microbiol Rev 26:223–238
Cooper TG, Gorski M, Turoscy V (1979) A cluster of three genes responsible for allantoin degradation in *Saccharomyces cerevisiae*. Genetics 92:383–396
Cooper TG, Lam C, Turoscy V (1980) Structural analysis of the *dur* loci in *S. cerevisiae*: two domains of a single multifunctional gene. Genetics 94:555–580
Corte ED, Stirpe F (1972) The regulation of rat liver xanthine oxidase. Involvement of thiol groups in the conversion of the enzyme activity from dehydrogenase (type D) into oxidase (type O) and purification of the enzyme. Biochem J 126:739–745
Cultrone A, Scazzocchio C, Rochet M, Montero-Morán G, Drevet C, Fernández-Martín R (2005) Convergent evolution of hydroxylation mechanisms in the fungal kingdom: molybdenum cofactor-independent hydroxylation of xanthine via α-ketoglutarate-dependent dioxygenases. Mol Microbiol 57:276–290
Deeley MC (1992) Adenine deaminase and adenine utilization in *Saccharomyces cerevisiae*. J Bacteriol 174:3102–3110
Degols G, Jauniaux JC, Wiame JM (1987) Molecular characterization of transposable-element-associated mutations that lead to constitutive L-ornithine aminotransferase expression in *Saccharomyces cerevisiae*. Eur J Biochem 165:289–296
DeLuna A, Avendano A, Riego L, Gonzalez A (2001) NADP-glutamate dehydrogenase isoenzymes of *Saccharomyces cerevisiae*. Purification, kinetic properties, and physiological roles. J Biol Chem 276:43775–43783
DeLuna A, Quezada H, Gómez-Puyou A, González A (2005) Asparaginyl deamidation in two glutamate dehydrogenase isoenzymes from *Saccharomyces cerevisiae*. Biochem Biophys Res Commun 328:1083–1090
Der Garabedian PA (1986) Candida δ-aminovalerate:α-ketoglutarate aminotransferase: purification and enzymologic properties. Biochemistry 25:5507–5512
Der Garabedian PA, Lotti AM, Vermeersch JJ (1986) 4-Aminobutyrate:2-oxoglutarate aminotransferase from *Candida*. Purification and properties. Eur J Biochem 156:589–596
Di Carlo FJ, Schultz AS, Kent AM (1952) On the mechanism of pyrimidine metabolism by yeasts. J Biol Chem 199:333–343
Di Carlo FJ, Schultz AS, Kent AM (1953) The mechanism of allantoin catabolism by yeast. Arch Biochem Biophys 44:468–474
Dias JC, Rezende RP, Rosa CA, Lachance MA, Linardi VR (2000) Enzymatic degradation of nitriles by a *Candida guilliermondii* UFMG-Y65. Can J Microbiol 46:525–531
Dunlop PC, Roon RJ (1975) L-asparaginase of *Saccharomyces cerevisiae*: an extracellular enzyme. J Bacteriol 122:1017–1024
Dunlop PC, Roon RJ, Even HL (1976) Utilization of D-asparagine by *Saccharomyces cerevisiae*. J Bacteriol 125:999–1004
Eden A, Simchen G, Benvenisty N (1996) Two yeast homologs of ECA39, a target for c-Myc regulation, code for cytosolic and mitochondrial branched-chain amino acid aminotransferases. J Biol Chem 271:20242–20245

Erbs P, Exinger F, Jund R (1997) Characterization of the *Saccharomyces cerevisiae FCY1* gene encoding cytosine deaminase and its homologue *FCA1* of *Candida albicans*. Curr Genet 31:1–6

Faber KN, Keizer-Gunnink I, Pluim D, Harder W, Ab G, Veenhuis M (1994) The N-terminus of amine oxidase of *Hansenula polymorpha* contains a peroxisomal targeting signal. FEBS Lett 357:115–120

Fattakhova AN, Ofitserov EN, Garusov AV (1991) Cytochrome P-450-dependent catabolism of triethanolamine in *Rhodotorula mucilaginosa*. Biodegradation 2:107–113

Filetici P, Martegani MP, Valenzuela L, González A, Ballario P (1996) Sequence of the *GLT1* gene from *Saccharomyces cerevisiae* reveals the domain structure of yeast glutamate synthase. Yeast 12:1359–1366

Fitzpatrick DA, Logue ME, Butler G (2008) Evidence of recent interkingdom horizontal gene transfer between bacteria and *Candida parapsilosis*. BMC Evol Biol 8:181

Folch JL, Antaramián A, Rodríguez L, Bravo A, Brunner A, González A (1989) Isolation and characterization of a *Saccharomyces cerevisiae* mutant with impaired glutamate synthase activity. J Bacteriol 171:6776–6781

Gaillardin C, Fournier P, Sylvestre G, Heslot H (1976) Mutants of *Saccharomycopsis lipolytica* defective in lysine catabolism. J Bacteriol 125:48–57

Ganguly S, Mummaneni P, Steinbach PJ, Klein DC, Coon SL (2001) Characterization of the *Saccharomyces cerevisiae* homolog of the melatonin rhythm enzyme arylalkylamine *N*-acetyltransferase (EC 2.3.1.87). J Biol Chem 276:47239–47247

Ganter PF (2006) Yeast and invertebrate associations. In: Péter G, Rosa C (eds) Biodiversity and Ecophysiology of Yeasts. The Yeast handbook. Springer, Berlin/Heidelberg, pp 303–370

García-Campusano F, Anaya VH, Robledo-Arratia L, Quezada H, Hernández H, Riego L, González A (2009) *ALT1*-encoded alanine aminotransferase plays a central role in the metabolism of alanine in *Saccharomyces cerevisiae*. Can J Microbiol 55:368–374

García-Lugo P, González C, Perdomo G, Brito N, Avila J, de La Rosa JM, Siverio JM (2000) Cloning, sequencing, and expression of *H.a.YNR1* and *H.c.YNI1*, encoding nitrate and nitrite reductases in the yeast *Hansenula anomala*. Yeast 16:1099–1105

Gillyon C, Haywood GW, Large PJ, Nellen B, Robertson A (1987) Putrescine breakdown in the yeast *Candida boidini*: subcellular location of some of the enzymes involved and properties of two acetamidoaldehyde dehydrogenases. J Gen Microbiol 133:2477–2485

Gojković Z, Jahnke K, Schnackerz KD, Piškur J (2000) *PYD2* encodes 5,6-dihydropyrimidine amidohydrolase, which participates in a novel fungal catabolic pathway. J Mol Biol 295:1073–1087

Gojković Z, Sandrini MP, Piškur J (2001) Eukaryotic β-alanine synthases are functionally related but have a high degree of structural diversity. Genetics 158:999–1011

González FJ, Montes J, Martin F, López MC, Fermiñán E, Catalán J, Galán MA, Domínguez A (1997) Molecular cloning of *TvDAO1*, a gene encoding a D-amino acid oxidase from *Trigonopsis variabilis* and its expression in *Saccharomyces cerevisiae* and *Kluyveromyces lactis*. Yeast 13:1399–1408

Green J, Large PJ (1983) Oxidation of dimethylamine and trimethylamine in methazotrophic yeasts by microsomal mono-oxygenases sensitive to carbon monoxide. Biochem Biophys Res Commun 113:900–907

Green J, Large PJ (1984) Subcellular localization and properties of partially purified dimethylamine and trimethylamine mono-oxygenase activities in *Candida utilis*. J Gen Microbiol 130:2577–2588

Green J, Haywood GW, Large PJ (1982) More than one amine oxidase is involved in the metabolism of primary amines supplied as nitrogen source. J Gen Microbiol 128:991–995

Green J, Haywood GW, Large PJ (1983) Serological differences between the multiple amine oxidases of yeasts and comparison of the specificities of the purified enzymes from *Candida utilis* and *Pichia pastoris*. Biochem J 211:481–493

Gromes R, Schwartz H, Heinrich M, Johannssen W (1991) Nitrate reductase from yeast: cultivation, partial purification and characterization. Appl Microbiol Biotechnol 35:491–495

Gunasekaran M, Gunasekaran U (1999) Partial purification and properties of putrescine oxidase from *Candida guilliermondii*. Appl Biochem Biotechnol 76:229–336

Hachimori A, Ito M, Samejima T (1974) Some properties of glutamine synthetase from baker's yeast. J Biochem 76:1075–1081

Hall DA (1952) Histidine α-deaminase and the production of urocanic acid in the mammal. Biochem J 51:499–504

Hammer T, Bode R (1992) Purification and characterization of an inducible L-lysine: 2-oxoglutarate 6-aminotransferase from *Candida utilis*. J Basic Microbiol 32:21–27

Hammer T, Bode R, Schmidt H, Birnbaum D (1991a) Distribution of three lysine-catabolizing enzymes in various yeast species. J Basic Microbiol 31:43–49

Hammer T, Bode R, Birnbaum D (1991b) Occurrence of a novel yeast enzyme, L-lysine ε-dehydrogenase, which catalyses the first step of lysine catabolism in *Candida albicans*. Microbiology 137:711–715

Hanson AD, Pribat A, Waller JC, de Crécy-Lagard V (2009) 'Unknown' proteins and 'orphan' enzymes: the missing half of the engineering parts list – and how to find it. Biochem J 425:1–11

Hartzog PE, Nicholson BP, McCusker JH (2005) Cytosine deaminase MX cassettes as positive/negative selectable markers in *Saccharomyces cerevisiae*. Yeast 22:789–798

Hata S, Shirata K, Takagishi H (1986) Degradation of paraquat and diquat by the yeast *Lipomyces starkeyi*. J Gen Appl Microbiol 32:193–202

Haywood GW, Large PJ (1981) Microbial oxidation of amines. Distribution, purification and properties of two primary-amine oxidases from the yeast *Candida boidinii* grown on amines as sole nitrogen source. Biochem J 199:187–201

Haywood GW, Large PJ (1985) The occurrence, subcellular localization and partial purification of diamine acetyltransferase in the yeast *Candida boidinii* grown on spermidine or putrescine as sole nitrogen source. Eur J Biochem 148:277–283

Haywood GW, Large PJ (1986) 4-acetamidobutyrate deacetylase in the yeast *Candida boidinii* grown on putrescine or spermidine as sole nitrogen source and its probable role in polyamine catabolism. J Gen Microbiol 132:7–14

Hazelwood LA, Daran JM, van Maris AJ, Pronk JT, Dickinson JR (2008) The Ehrlich pathway for fusel alcohol production: a century of research on *Saccharomyces cerevisiae* metabolism. Appl Environ Microbiol 74:2259–2266

Hemmings BA (1980) Purification and properties of the phospho and dephospho forms of yeast NAD-dependent glutamate dehydrogenase. J Biol Chem 255:7925–7932

Hipkin CR, Kau DA, Cannons AC (1993) Further characterization of the assimilatory nitrate reductase from the yeast *Candida nitratophila*. J Gen Microbiol 139:473–478

Holmes AR, McNaughton GS, More RD, Shepherd MG (1991) Ammonium assimilation by *Candida albicans* and other yeasts: a ^{13}N isotope study. Can J Microbiol 37:226–232

Iraqui I, Vissers S, Cartiaux M, Urrestarazu A (1998) Characterisation of *Saccharomyces cerevisiae ARO8* and *ARO9* genes encoding aromatic aminotransferases I and II reveals a new aminotransferase subfamily. Mol Gen Genet 257:238–248

Ito T, Hemmi H, Kataoka K, Mukai Y, Yoshimura T (2008) A novel zinc-dependent D-serine dehydratase from *Saccharomyces cerevisiae*. Biochem J 409:399–406

Jankowska DA, Trautwein-Schult A, Cordes A, Hoferichter P, Klein C, Bode R, Baronian K, Kunze G (2013) *Arxula adeninivorans* xanthine oxidoreductase and its application in the production of food with low purine content. J Appl Microbiol 115:796–807

Jauniaux JC, Urrestarazu LA, Wiame JM (1978) Arginine metabolism in *Saccharomyces cerevisiae*: subcellular localization of the enzymes. J Bacteriol 133:1096–1107

Jauniaux JC, Dubois E, Vissers S, Crabeel M, Wiame JM (1982) Molecular cloning, DNA structure, and RNA analysis of the arginase gene in *Saccharomyces cerevisiae*. A study of cis-dominant regulatory mutations. EMBO J 1:1125–1131

Jones GE, Mortimer RK (1973) Biochemical properties of yeast L-asparaginase. Biochem Genet 9:131–146

Kay CJ, Barber MJ, Solomonson LP, Kau D, Cannons AC, Hipkin CR (1990) Spectroscopic, thermodynamic and kinetic properties of *Candida nitratophila* nitrate reductase. Biochem J 272:545–548

Kellis M, Birren BW, Lander ES (2004) Proof and evolutionary analysis of ancient genome duplication in the yeast *Saccharomyces cerevisiae*. Nature 428:617–624

Kim KW, Kamerud JQ, Livingston DM, Roon RJ (1988) Asparaginase II of *Saccharomyces cerevisiae*. Characterization of the *ASP3* gene. J Biol Chem 263:11948–11953

Kispal G, Steiner H, Court DA, Rolinski B, Lill R (1996) Mitochondrial and cytosolic branched-chain amino acid transaminases from yeast, homologs of the myc oncogene-regulated Eca39 protein. J Biol Chem 271:24458–24464

Klompmaker SH, Kilic A, Baerends RJ, Veenhuis M, van der Klei IJ (2010) Activation of a peroxisomal *Pichia pastoris* D-amino acid oxidase, which uses D-alanine as a preferred substrate, depends on pyruvate carboxylase. FEMS Yeast Res 10:708–716

Koyama Y, Ichikawa T, Nakano E (1996) Cloning, sequence analysis, and expression in *Escherichia coli* of the gene encoding the *Candida utilis* urate oxidase (uricase). J Biochem 120:969–973

Kradolfer P, Niederberger P, Hütter R (1982) Tryptophan degradation in *Saccharomyces cerevisiae*: characterization of two aromatic aminotransferases. Arch Microbiol 133:242–248

Krassowski T, Coughlan AY, Shen XX, Zhou X, Kominek J, Opulente DA, Riley R, Grigoriev IV, Maheshwari N, Shields DC, Kurtzman CP, Hittinger CT, Rokas A, Wolfe KH (2018) Evolutionary instability of CUG-Leu in the genetic code of budding yeasts. Nat Commun 9:1887

Krzywicki KA, Brandriss MC (1984) Primary structure of the nuclear *PUT2* gene involved in the mitochondrial pathway for proline utilization in *Saccharomyces cerevisiae*. Mol Cell Biol 4:2837–2842

Kucha JA, Dooley DM (2001) Cloning, sequence analysis, and characterization of the 'lysyl oxidase' from *Pichia pastoris*. J Inorg Biochem 83:193–204

Lacerda VA, Marsden A, Ledingham WM (1992) Ammonia utilization in *S. cerevisiae* under chemostatic growth. Appl Biochem Biotechnol 32:15–21

Lambou K, Pennati A, Valsecchi I, Tada R, Sherman S, Sato H, Beau R, Gadda G, Latgé JP (2013) Pathway of glycine betaine biosynthesis in *Aspergillus fumigatus*. Eukaryot Cell 12:853–863

Landry J, Sternglanz R (2003) Yeast Fms1 is a FAD-utilizing polyamine oxidase. Biochem Biophys Res Commun 303:771–776

Large PJ (1986) Degradation of organic nitrogen compounds by yeasts. Yeast 2:1–34

Large PJ, Robertson A (1988) The subcellular location of 4-aminobutyrate aminotransferase in *Candida boidinii* and its probable role in the breakdown of putrescine and spermidine. Yeast 4:149–153

Large PJ, Robertson A (1991) The route of lysine breakdown in *Candida tropicalis*. FEMS Microbiol Lett 66:209–213

Large PJ, Waterham HR, Veenhuis M (1990) Subcellular location of the enzymes of purine breakdown in the yeast *Candida famata* grown on uric acid. FEMS Microbiol Lett 72:303–307

LaRue TA, Spencer JF (1967a) The utilization of D-amino acid by yeast. Can J Microbiol 13:777–788

LaRue TA, Spencer JFT (1967b) The utilization of imidazoles by yeasts. Can J Microbiol 13:789–794

LaRue TA, Spencer JF (1968) The utilization of purines and pyrimidines by yeasts. Can J Microbiol 14:79–86

Lawther RP, Riemer E, Chojnacki B, Cooper TG (1974) Clustering of the genes for allantoin degradation in *Saccharomyces cerevisiae*. J Bacteriol 119:461–468

League GP, Slot JC, Rokas A (2012) The *ASP3* locus in *Saccharomyces cerevisiae* originated by horizontal gene transfer from *Wickerhamomyces*. FEMS Yeast Res 12:859–863

Lee IR, Yang L, Sebetso G, Allen R, Doan TH, Blundell R, Lui EY, Morrow CA, Fraser JA (2013) Characterization of the complete uric acid degradation pathway in the fungal pathogen *Cryptococcus neoformans*. PLoS One 8:e64292

Linardi VR, Dias JCT, Rosa CA (1996) Utilization of acetonitrile and other aliphatic nitriles by a *Candida famata* strain. FEMS Microbiol Lett 144:67–71

Linder T (2014) *CMO1* encodes a putative choline monooxygenase and is required for the utilization of choline as the sole nitrogen source in the yeast *Scheffersomyces stipitis* (syn. *Pichia stipitis*). Microbiology 160:929–940

Linder T (2018) Genetic redundancy in the catabolism of methylated amines in the yeast *Scheffersomyces stipitis*. Antonie Van Leeuwenhoek 111:401–411

Linder T (2019a) A genomic survey of nitrogen assimilation pathways in budding yeasts (subphylum Saccharomycotina). Yeast. https://doi.org/10.1002/yea.3364

Linder T (2019b) Phenotypical characterisation of a putative ω-amino acid transaminase in the yeast *Scheffersomyces stipitis*. Arch Microbiol 201:185–192

Liu B, Sutton A, Sternglanz R (2005) A yeast polyamine acetyltransferase. J Biol Chem 280:16659–16664

Ljungdahl PO (2009) Amino-acid-induced signalling via the SPS-sensing pathway in yeast. Biochem Soc Trans 37:242–247

Lohkamp B, Andersen B, Piškur J, Dobritzsch D (2006) The crystal structures of dihydropyrimidinases reaffirm the close relationship between cyclic amidohydrolases and explain their substrate specificity. J Biol Chem 281:13762–13776

Magasanik B (2003) Ammonia assimilation by *Saccharomyces cerevisiae*. Eukaryot Cell 2:827–829

Magasanik B, Kaiser CA (2002) Nitrogen regulation in *Saccharomyces cerevisiae*. Gene 290:1–18

Martin E, Varotto Boccazzi I, De Marco L, Bongiorno G, Montagna M, Sacchi L, Mensah P, Ricci I, Gradoni L, Bandi C, Epis S (2018) The mycobiota of the sand fly *Phlebotomus perniciosus*: involvement of yeast symbionts in uric acid metabolism. Environ Microbiol 20:1064–1077

McNeil JB, McIntosh EM, Taylor BV, Zhang FR, Tang S, Bognar AL (1994) Cloning and molecular characterization of three genes, including two genes encoding serine hydroxymethyltransferases, whose inactivation is required to render yeast auxotrophic for glycine. J Biol Chem 269:9155–9165

McNeil JB, Zhang F, Taylor BV, Sinclair DA, Pearlman RE, Bognar AL (1997) Cloning, and molecular characterization of the *GCV1* gene encoding the glycine cleavage T-protein from *Saccharomyces cerevisiae*. Gene 186:13–20

Messenguy F, André B, Dubois E (2006) Diversity of nitrogen metabolism among yeast species: regulatory and evolutionary aspects. In: Péter G, Rosa C (eds) Biodiversity and Ecophysiology of Yeasts. The Yeast handbook. Springer, Berlin/Heidelberg, pp 123–153

Middelhoven WJ (1964) The pathway of arginine breakdown in *Saccharomyces cerevisiae*. Biochim Biophys Acta 93:650–652

Middelhoven WJ, van Doesburg W (2007) Utilization of hexamethylenetetramine (urotropine) by bacteria and yeasts. Antonie Van Leeuwenhoek 91:191–196

Middelhoven WJ, van Eijk J, van Renesse R, Blijham JM (1978) A mutant of Saccharomyces cerevisiae lacking catabolic NAD-specific glutamate dehydrogenase. Growth characteristics of the mutant and regulation of enzyme synthesis in the wild-type strain. Antonie Van Leeuwenhoek 44:311–320

Middelhoven WJ, de Jong IM, de Winter M (1991) *Arxula adeninivorans*, a yeast assimilating many nitrogenous and aromatic compounds. Antonie Van Leeuwenhoek 59:129–137

Miller SM, Magasanik B (1990) Role of NAD-linked glutamate dehydrogenase in nitrogen metabolism in *Saccharomyces cerevisiae*. J Bacteriol 172:4927–4935

Minehart PL, Magasanik B (1992) Sequence of the *GLN1* gene of *Saccharomyces cerevisiae*: role of the upstream region in regulation of glutamine synthetase expression. J Bacteriol 174:1828–1836

Mira-Gutiérrez J, Garcia-Martos P, Mira-Gordillo AJ (1995) Identification of yeasts by hydrolysis of amides. Mycoses 38:101–106
Mitchell AP (1985) The *GLN1* locus of *Saccharomyces cerevisiae* encodes glutamine synthetase. Genetics 111:243–258
Montalvo-Arredondo J, Jiménez-Benítez Á, Colón-González M, González-Flores J, Flores-Villegas M, González A, Riego-Ruiz L (2015) Functional roles of a predicted branched chain aminotransferase encoded by the *LkBAT1* gene of the yeast *Lachancea kluyveri*. Fungal Genet Biol 85:71–82
Montero-Morán GM, Li M, Rendòn-Huerta E, Jourdan F, Lowe DJ, Stumpff-Kane AW, Feig M, Scazzocchio C, Hausinger RP (2007) Purification and characterization of the Fe^{II}- and α-ketoglutarate-dependent xanthine hydroxylase from *Aspergillus nidulans*. Biochemistry 46:5293–5304
Mori N, Shirakawa K, Uzura K, Kitamoto Y, Ichikawa Y (1988) Formation of ethylene-glycol and trimethylamine from choline by *Candida tropicalis*. FEMS Microbiol Lett 51:41–44
Morin PJ, Subramanian GS, Gilmore TD (1992) *AAT1*, a gene encoding a mitochondrial aspartate aminotransferase in *Saccharomyces cerevisiae*. Biochim Biophys Acta 1171:211–214
Moye WS, Amuro N, Rao JK, Zalkin H (1985) Nucleotide sequence of yeast *GDH1* encoding nicotinamide adenine dinucleotide phosphate-dependent glutamate dehydrogenase. J Biol Chem 260:8502–8508
Nagarajan L, Storms RK (1997) Molecular characterization of *GCV3*, the *Saccharomyces cerevisiae* gene coding for the glycine cleavage system hydrogen carrier protein. J Biol Chem 272:4444–4450
Nagasu T, Hall BD (1985) Nucleotide sequence of the *GDH* gene coding for the NADP-specific glutamate dehydrogenase of *Saccharomyces cerevisiae*. Gene 37:247–253
Navarathna DH, Harris SD, Roberts DD, Nickerson KW (2010) Evolutionary aspects of urea utilization by fungi. FEMS Yeast Res 10:209–213
Nishimura K, Yamamoto M, Nakagomi T, Takiguchi Y, Naganuma T, Uzuka Y (2002) Biodegradation of triazine herbicides on polyvinylalcohol gel plates by the soil yeast *Lipomyces starkeyi*. Appl Microbiol Biotechnol 58:848–852
Nolan LM, Harnedy PA, Turner P, Hearne AB, O'Reilly C (2003) The cyanide hydratase enzyme of *Fusarium lateritium* also has nitrilase activity. FEMS Microbiol Lett 221:161–165
Norkrans B (1969) Hydroxylamine as the sole nitrogen source for growth of some *Candida* sp. Acta Chem Scand 23:1457–1459
Okumura I, Yamamoto T (1978) Enzymic racemization of allantoin. J Biochem 84:891–895
Okumura I, Kondo K, Miyake Y, Itaya K, Yamamoto T (1976) Stereospecificity of conversion of uric acid into allantoic acid by enzymes of *Candida utilis*. J Biochem 79:1013–1019
Peñalosa-Ruiz G, Aranda C, Ongay-Larios L, Colon M, Quezada H, Gonzalez A (2012) Paralogous *ALT1* and *ALT2* retention and diversification have generated catalytically active and inactive aminotransferases in *Saccharomyces cerevisiae*. PLoS One 7:e45702
Petersen JG, Kielland-Brandt MC, Nilsson-Tillgren T, Bornaes C, Holmberg S (1988) Molecular genetics of serine and threonine catabolism in *Saccharomyces cerevisiae*. Genetics 119:527–534
Pignocchi C, Berardi E, Cox BS (1998) Nitrate reduction and the isolation of Nit- mutants in *Hansenula polymorpha*. Microbiology 144:2323–2330
Prasad S, Sharma DR, Bhalla TC (2005) Nitrile- and amide-hydrolysing activity in *Kluyveromyces thermotolerans* MGBY 37. World J Microbiol Biotechnol 21:1447–1450
Ramazzina I, Folli C, Secchi A, Berni R, Percudani R (2006) Completing the uric acid degradation pathway through phylogenetic comparison of whole genomes. Nat Chem Biol 2:144–148
Ramos F, el Guezzar M, Grenson M, Wiame JM (1985) Mutations affecting the enzymes involved in the utilization of 4-aminobutyric acid as nitrogen source by the yeast *Saccharomyces cerevisiae*. Eur J Biochem 149:401–404
Rezende RP, Dias JC, Rosa CA, Carazza F, Linardi VR (1999) Utilization of nitriles by yeasts isolated from a Brazilian gold mine. J Gen Appl Microbiol 45:185–192

Romagnoli G, Verhoeven MD, Mans R, Fleury Rey Y, Bel-Rhlid R, van den Broek M, Seifar RM, Ten Pierick A, Thompson M, Müller V, Wahl SA, Pronk JT, Daran JM (2014) An alternative, arginase-independent pathway for arginine metabolism in *Kluyveromyces lactis* involves guanidinobutyrase as a key enzyme. Mol Microbiol 93:369–389

Roon RJ, Even HL (1973) Regulation of the nicotinamide adenine dinucleotide- and nicotinamide adenine dinucleotide phosphate-dependent glutamate dehydrogenases of *Saccharomyces cerevisiae*. J Bacteriol 116:367–372

Roon RJ, Levenberg B (1972) Urea amidolyase. I. Properties of the enzyme from *Candida utilis*. J Biol Chem 247:4107–4113

Roon RJ, Even HL, Larimore F (1974) Glutamate synthase: properties of the reduced nicotinamide adenine dinucleotide-dependent enzyme from *Saccharomyces cerevisiae*. J Bacteriol 118:89–95

Ross J, Reid GA, Dawes IW (1988) The nucleotide sequence of the *LPD1* gene encoding lipoamide dehydrogenase in *Saccharomyces cerevisiae*: comparison between eukaryotic and prokaryotic sequences for related enzymes and identification of potential upstream control sites. J Gen Microbiol 134:1131–1139

Rothstein M (1965) Intermediates of lysine dissimilation in the yeast, *Hansenula saturnus*. Arch Biochem Biophys 111:467–476

Rząd K, Milewski S, Gabriel I (2018) Versatility of putative aromatic aminotransferases from *Candida albicans*. Fungal Genet Biol 110:26–37

Sadowsky MJ, Koskinen WC, Bischoff M, Barber BL, Becker JM, Turco RF (2009) Rapid and complete degradation of the herbicide picloram by *Lipomyces kononenkoae*. J Agric Food Chem 57:4878–4882

Saint-Marc C, Daignan-Fornier B (2004) *GUD1* (*YDL238c*) encodes *Saccharomyces cerevisiae* guanine deaminase, an enzyme expressed during post-diauxic growth. Yeast 21:1359–1363

Sampath V, Liu B, Tafrov S, Srinivasan M, Rieger R, Chen EI, Sternglanz R (2013) Biochemical characterization of Hpa2 and Hpa3, two small closely related acetyltransferases from *Saccharomyces cerevisiae*. J Biol Chem 288:21506–21513

Schmidt H, Bode R (1992) Characterization of a novel enzyme, N^6-acetyl-L-lysine: 2-oxoglutarate aminotransferase, which catalyses the second step of lysine catabolism in *Candida maltosa*. Antonie Van Leeuwenhoek 62:285–290

Schmidt H, Bode R, Birnbaum D (1988a) A novel enzyme, L-lysine pyruvate aminotransferase, catalyses the first step of lysine catabolism in *Pichia guilliermondii*. FEMS Microbiol Lett 49:203–206

Schmidt H, Bode R, Birnbaum D (1988b) Lysine degradation in *Candida maltosa*: occurrence of a novel enzyme, acetyl-CoA: L-lysine N-acetyltransferase. Arch Microbiol 150:215–218

Schnackerz KD, Andersen G, Dobritzsch D, Piškur J (2008) Degradation of pyrimidines in *Saccharomyces kluyveri*: transamination of beta-alanine. Nucleosides Nucleotides Nucleic Acids 27:794–799

Sengupta S, Shaila MS, Rao GR (1996) Purification and characterization of assimilatory nitrite reductase from *Candida utilis*. Biochem J 317:147–155

Shaw AJ, Lam FH, Hamilton M, Consiglio A, MacEwen K, Brevnova EE, Greenhagen E, LaTouf WG, South CR, van Dijken H, Stephanopoulos G (2016) Metabolic engineering of microbial competitive advantage for industrial fermentation processes. Science 353:583–586

Shen XX, Zhou X, Kominek J, Kurtzman CP, Hittinger CT, Rokas A (2016) Reconstructing the backbone of the Saccharomycotina yeast phylogeny using genome-scale data. G3 6:3927–3939

Shen XX, Opulente DA, Kominek J, Zhou X, Steenwyk JL, Buh KV, Haase MAB, Wisecaver JH, Wang M, Doering DT, Boudouris JT, Schneider RM, Langdon QK, Ohkuma M, Endoh R, Takashima M, Manabe RI, Čadež N, Libkind D, Rosa CA, DeVirgilio J, Hulfachor AB, Groenewald M, Kurtzman CP, Hittinger CT, Rokas A (2018) Tempo and mode of genome evolution in the budding yeast subphylum. Cell 175:1533–1545

Shepherd A, Piper PW (2010) The Fps1p aquaglyceroporin facilitates the use of small aliphatic amides as a nitrogen source by amidase-expressing yeasts. FEMS Yeast Res 10:527–534

Shiraishi K, Oku M, Kawaguchi K, Uchida D, Yurimoto H, Sakai Y (2015) Yeast nitrogen utilization in the phyllosphere during plant lifespan under regulation of autophagy. Sci Rep 5:9719
Sinclair DA, Dawes IW (1995) Genetics of the synthesis of serine from glycine and the utilization of glycine as sole nitrogen source by *Saccharomyces cerevisiae*. Genetics 140:1213–1222
Sinclair K, Warner JP, Bonthron DT (1994) The *ASP1* gene of *Saccharomyces cerevisiae*, encoding the intracellular isozyme of L-asparaginase. Gene 144:37–43
Sinclair DA, Hong SP, Dawes IW (1996) Specific induction by glycine of the gene for the P-subunit of glycine decarboxylase from *Saccharomyces cerevisiae*. Mol Microbiol 19:611–623
Solis-Escalante D, Kuijpers NG, Bongaerts N, Bolat I, Bosman L, Pronk JT, Daran JM, Daran-Lapujade P (2013) *amdSYM*, a new dominant recyclable marker cassette for *Saccharomyces cerevisiae*. FEMS Yeast Res 13:126–139
Suh SO, Marshall CJ, McHugh JV, Blackwell M (2003) Wood ingestion by passalid beetles in the presence of xylose-fermenting gut yeasts. Mol Ecol 12:3137–3145
Suh SO, McHugh JV, Pollock DD, Blackwell M (2005) The beetle gut: a hyperdiverse source of novel yeasts. Mycol Res 109:261–265
Sumrada RA, Cooper TG (1982a) Urea carboxylase and allophanate hydrolase are components of a multifunctional protein in yeast. J Biol Chem 257:9119–9127
Sumrada RA, Cooper TG (1982b) Isolation of the *CAR1* gene from *Saccharomyces cerevisiae* and analysis of its expression. Mol Cell Biol 2:1514–1523
Sumrada RA, Cooper TG (1984) Nucleotide sequence of the *Saccharomyces cerevisiae* arginase gene (*CAR1*) and its transcription under various physiological conditions. J Bacteriol 160:1078–1087
Truong HN, Meyer C, Daniel-Vedele F (1991) Characteristics of *Nicotiana tabacum* nitrate reductase protein produced in *Saccharomyces cerevisiae*. Biochem J 278:393–397
Tur SS, Lerch K (1988) Unprecedented lysyloxidase activity of *Pichia pastoris* benzylamine oxidase. FEBS Lett 238:74–76
Urrestarazu A, Vissers S, Iraqui I, Grenson M (1998) Phenylalanine- and tyrosine-auxotrophic mutants of *Saccharomyces cerevisiae* impaired in transamination. Mol Gen Genet 257:230–237
van der Walt JP (1962) Utilization of ethylamine by yeasts. Antonie Van Leeuwenhoek 28:91–96
van Dijken JP, Bos P (1981) Utilization of amines by yeasts. Arch Microbiol 128:320–324
Verleur N, Elgersma Y, Van Roermund CW, Tabak HF, Wanders RJ (1997) Cytosolic aspartate aminotransferase encoded by the *AAT2* gene is targeted to the peroxisomes in oleate-grown *Saccharomyces cerevisiae*. Eur J Biochem 247:972–980
Vigliotta G, Di Giacomo M, Carata E, Massardo DR, Tredici SM, Silvestro D, Paolino M, Pontieri P, Del Giudice L, Parente D, Alifano P (2007) Nitrite metabolism in *Debaryomyces hansenii* TOB-Y7, a yeast strain involved in tobacco fermentation. Appl Microbiol Biotechnol 75:633–645
Villas-Bôas SG, Åkesson M, Nielsen J (2005) Biosynthesis of glyoxylate from glycine in *Saccharomyces cerevisiae*. FEMS Yeast Res 5:703–709
Wang SS, Brandriss MC (1986) Proline utilization in *Saccharomyces cerevisiae*: analysis of the cloned *PUT1* gene. Mol Cell Biol 6:2638–2645
Wang SS, Brandriss MC (1987) Proline utilization in *Saccharomyces cerevisiae*: sequence, regulation, and mitochondrial localization of the *PUT1* gene product. Mol Cell Biol 7:4431–4440
Wang P, van Etten HD (1992) Cloning and properties of a cyanide hydratase gene from the phytopathogenic fungus *Gloeocercospora sorghi*. Biochem Biophys Res Commun 187:1048–1054
Wang D, Zheng G, Wang S, Zhang D, Zhou L (2011) Biodegradation of aniline by *Candida tropicalis* AN1 isolated from aerobic granular sludge. J Environ Sci 23:2063–2068
White WH, Skatrud PL, Xue Z, Toyn JH (2003) Specialization of function among aldehyde dehydrogenases: the *ALD2* and *ALD3* genes are required for β-alanine biosynthesis in *Saccharomyces cerevisiae*. Genetics 163:69–77
Whitfield D, Large PJ (1986) Enzymes metabolizing dimethylamine, trimethylamine and trimethylamine *N*-oxide in the yeast *Sporopachydermia cereana* grown on amines as sole nitrogen source. FEMS Microbiol Lett 35:99–105

Whitfield D, Large PJ (1987) Assimilatory reduction of trimethylamine *N*-oxide in the yeast *Sporopachydermia cereana*. Appl Microbiol Biotechnol 26:277–282

Whitney PA, Cooper TG (1972) Urea carboxylase and allophanate hydrolase. Two components of adenosine triphosphate:urea amido-lyase in *Saccharomyces cerevisiae*. J Biol Chem 247:1349–1353

Wong S, Wolfe KH (2005) Birth of a metabolic gene cluster in yeast by adaptive gene relocation. Nat Genet 37:777–782

Wong KH, Hynes MJ, Davis MA (2008) Recent advances in nitrogen regulation: a comparison between *Saccharomyces cerevisiae* and filamentous fungi. Eukaryot Cell 7:917–925

Woolfolk SW, Inglis GD (2004) Microorganisms associated with field-collected *Chrysoperla rufilabris* (Neuroptera: Chrysopidae) adults with emphasis on yeast symbionts. Biol Control 29:155–168

Yoo HS, Cooper TG (1991) Sequences of two adjacent genes, one (*DAL2*) encoding allantoicase and another (*DCG1*) sensitive to nitrogen-catabolite repression in *Saccharomyces cerevisiae*. Gene 104:55–62

Yoo HS, Genbauffe FS, Cooper TG (1985) Identification of the ureidoglycolate hydrolase gene in the *DAL* gene cluster of *Saccharomyces cerevisiae*. Mol Cell Biol 5:2279–2288

Yurimoto H, Hasegawa T, Sakai Y, Kato N (2000) Physiological role of the D-amino acid oxidase gene, *DAO1*, in carbon and nitrogen metabolism in the methylotrophic yeast *Candida boidinii*. Yeast 16:1217–1227

Yurimoto H, Hasegawa T, Sakai Y, Kato N (2001) Characterization and high-level production of D-amino acid oxidase in *Candida boidinii*. Biosci Biotechnol Biochem 65:627–633

Zhang N, Suh SO, Blackwell M (2003) Microorganisms in the gut of beetles: evidence from molecular cloning. J Invertebr Pathol 84:226–233

Zwart KB, Overmans FH, Harder W (1983a) The role of peroxisomes in the metabolism of D-alanine in the yeast *Candida utilis*. FEMS Microbiol Lett 19:225–231

Zwart KB, Veenhuis M, Harder W (1983b) Significance of yeast peroxisomes in the metabolism of choline and ethanolamine. Antonie Van Leeuwenhoek 49:369–385

Chapter 8
Microtubules in Non-conventional Yeasts

Hiromi Maekawa and Douglas R. Drummond

Abstract Microtubules polymerise from tubulin proteins and play a significant role in the growth and proliferation of eukaryotic cells. In yeasts, most studies on microtubules and tubulins have utilised the budding yeast *Saccharomyces cerevisiae* and fission yeast *Schizosaccharomyces pombe* model systems. However, more recently interest in the microtubules of other non-conventional yeast and fungal species has increased, both for investigation of biological processes such as fungal evolution and for applications such as developing antifungal drugs. We review the microtubule cytoskeleton and its role in yeast and fungal cellular processes in vivo and the tubulin proteins found in yeast cells and their study in vitro, together with the recent advances in cryoEM leading to detailed molecular structures of yeast microtubules. We examine what is known about the microtubule cytoskeleton in non-conventional yeasts and highlight the significant differences, as well as many conserved aspects, in their microtubule biology compared to the two model yeasts. Finally, we discuss the potential role of microtubules as drug targets for treatment of yeast and fungal infections.

Keywords microtubule · tubulin · cytoskeleton · yeast · SPB · MTOC

8.1 Introduction

The microtubule (MT) cytoskeleton is a key component of eukaryotic cells with important roles in intracellular trafficking, chromosome segregation during mitosis and meiosis, nuclear positioning and cell polarisation. The MTs are dynamic structures assembled from α- and β-tubulin protein subunits whose growth is nucleated by γ-tubulin-containing complexes in microtubule-organising centres (MTOCs).

The yeasts *Saccharomyces cerevisiae* and *Schizosaccharomyces pombe* have been used extensively as model organisms to study microtubules and their role in

H. Maekawa (✉) · D. R. Drummond (✉)
Faculty of Agriculture, Kyushu University, Fukuoka, Japan
e-mail: hmaekawa@agr.kyushu-u.ac.jp; d.drummond@agr.kyushu-u.ac.jp

A. Sibirny (ed.), *Non-conventional Yeasts: from Basic Research to Application*,
https://doi.org/10.1007/978-3-030-21110-3_8

eukaryotic cell biology. These studies have also been extended to the molecular level in order to better understand the biochemistry of tubulin proteins, their assembly into dynamic microtubules and the role of microtubule-associated proteins in regulating these processes. These in vitro studies have exploited the well-developed technology in yeasts for gene manipulation and expression and, for example, enable mutagenesis studies which are difficult with the traditional mammalian sources of tublin protein used for biochemical analysis. However, despite these studies in the model yeasts, our knowledge of the MTs in other non-conventional yeasts (NCYs) and non-model fungal systems has tended to lag behind. More recently this has been changing through a combination of advances in technology such as genome sequencing that has increased our knowledge of the tubulins and MT-interacting proteins present in NCYs and a recognition of their usefulness for investigating biological processes such as fungal evolution and cell division and proliferation. Increasing awareness of the role many NCYs play in causing infections has also led to interest in applications such as developing antifungal drugs.

We focus on the MT cytoskeleton, its regulation and function. We first describe the microtubule cytoskeleton, its formation, organisation and role in yeast and fungal cellular processes. Then we examine the tubulin proteins present in yeast cells and the study of yeast tubulin and microtubules in vitro, together with the recent advances in cryoEM leading to detailed molecular structures of yeast microtubules. Finally, we discuss tubulin as a drug target for treatment of yeast and fungal infections. In each case we review what is known from the well-studied *S. cerevisiae* budding and *S. pombe* fission yeast model systems and compare this to the NCYs and other non-model fungi, highlighting the similarities and differences, as well as where further studies are required to fill the gaps in our existing knowledge of the NCYs.

8.2 Spindle Pole Body

In the model yeasts *S. cerevisiae* and *S. pombe*, the mitotic spindle is organised from structures called spindle pole bodies (SPBs), which are microtubule-organising centres (MTOCs) equivalent to centrosomes in mammals (Fig. 8.1). The structures of SPBs in *S. cerevisiae* and *S. pombe* have been investigated intensively. Based on electron microscopy studies, both *S. cerevisiae* and *S. pombe* SPBs are cylindrical multilayered structures with a lateral extension that spans across the nuclear membrane called the half-bridge (Fig. 8.2) (Cavanaugh and Jaspersen 2017). In *S. cerevisiae*, SPBs are embedded in the nuclear envelope throughout the cell cycle. Standard EM shows outer, central and inner plaques. Finer analyses by electron tomography or cryoEM reveal two further intermediate layers IL1 and IL2 (O'Toole et al. 1999). The central plaque is embedded in the nuclear envelope and tethers the outer and inner plaques. The outer and inner plaques function as sites for microtubule (MT) nucleation and anchoring, whilst the half-bridge is important as a cytoplasmic microtubule (cMT)-anchoring site during G1 and also as the site for assembly of a new SPB during SPB duplication (Rüthnick and Schiebel 2016; Cavanaugh and Jaspersen 2017).

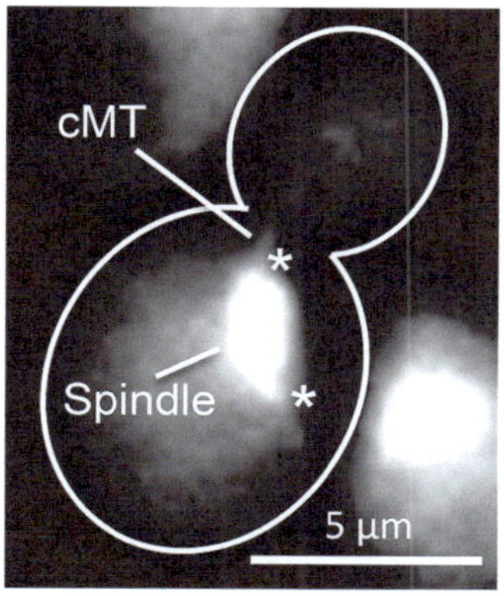

Fig. 8.1 Microtubule organisation in *S. cerevisiae*. Metaphase spindle and cytoplasmic MTs were visualised using α-tubulin fused to fluorescent protein (CFP-Tub1). The outline of the cell is illustrated by white lines. Asterisks mark the SPBs. Cytoplasmic (astral) MT emanating from the SPB close to the bud neck enters the bud, whilst cMT from the other SPB extends towards the mother cell cortex. Scale bar, 5 µm

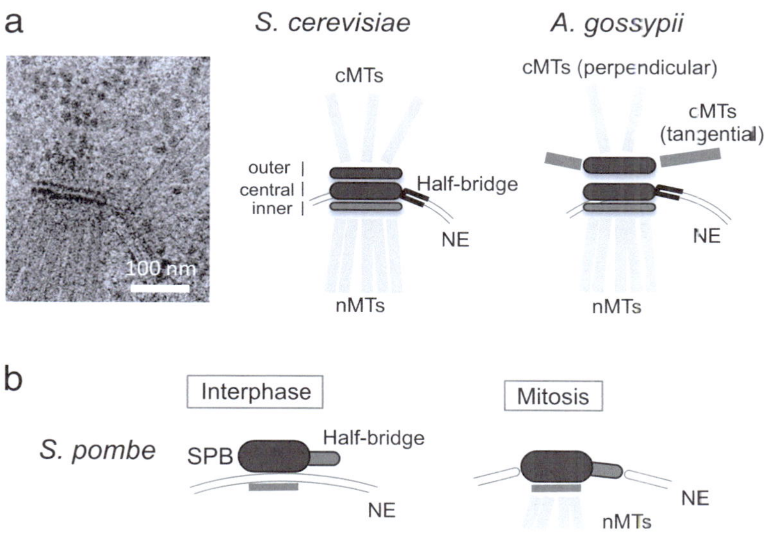

Fig. 8.2 Schematics of SPBs in *S. cerevisiae*, *A. gossypii* and *S. pombe*. (**a**, left) Electron micrograph of an SPB in the cell 20 min after release from an α-factor arrest showing the layered ultrastructure of the SPB, nuclear and cytoplasmic MTs. Scale bar, 100 nm. Kindly provided by Dr. Annett Neuner, University of Heidelberg, Germany. (**a**, middle and right) Schematics of SPB layer structures. The schematic of *A. gossypii* SPB is based on (Gibeaux et al. 2012). (**b**) *S. pombe* SPB is inserted into the nuclear envelope only during mitosis

A similar layered SPB structure is found in *Ashbya gossypii* multinucleated hyphae (Lang et al. 2010a, b). AgSPBs are embedded in the nuclear envelope throughout the cell cycle. In Saccharomycetes, SPBs are the only MTOC and organise nuclear and cMTs at all stages of the cell cycle. Although SPBs in *S. cerevisiae* and *A. gossypii* are structurally similar, a difference in cMT attachment to the SPB is observed (Fig. 8.2a). In *S. cerevisiae*, cMTs attach perpendicularly to the cytoplasmic layer (called the outer plaque) (Cavanaugh and Jaspersen 2017). In contrast, in *A. gossypii* there are two geometries, with both perpendicular and tangential cMT attachments to the outer plaque (Lang et al. 2010b). Perpendicularly attached cMTs are similar to *S. cerevisiae* cMTs and contact with the cell cortex. Tangentially attached MTs are usually very long and are thought to be important for bypassing other nuclei within the hypha. The tangential attachment is unique to *A. gossypii* and has not been observed in other Saccharomycetes species. The *A. gossypii* SPB has a longer distance between the outer plaque and the layer underneath compared to *S. cerevisiae*, which is proposed to account for the tangential attachment of cMTs (Lang et al. 2010a). Although *A. gossypii* grow exclusively as hyphae, it is phylogenetically much closer to *S. cerevisiae* than to other filamentous fungi such as *A. nidulans* and *N. crassa* (Kellis et al. 2004; Dietrich et al. 2004). *A. gossypii* diverged from the common ancestor with *S. cerevisiae* before the whole genome duplication (Wolfe and Shields 1997). So far, at least 18 proteins have been identified as constitutive components of SPBs in *S. cerevisiae*, all of which are encoded in the *A. gossypii* genome, albeit with low amino acid sequence identities (Lang et al. 2010a). The minus ends of MTs are anchored to the SPB by γ-tubulin and γ-tubulin receptor proteins, with Spc110 anchoring nuclear MTs to the inner plaque and Spc72 anchoring cMTs to the outer plaque (Rout and Kilmartin 1990; Kilmartin and Goh 1996; Spang et al. 1996; Knop and Schiebel 1997, 1998; Pereira et al. 1998, 1999; Nguyen et al. 1998). Nuclear MTs only anchor to Spc110, which is an integral component of the inner plaque. In contrast, although cMTs are tethered to Spc72, which appears to be an integral component of the outer plaque during most of the cell cycle, during G1 Spc72 can also bind to the half-bridge or bridge structure to organise cMTs (Rout and Kilmartin 1990; Pereira et al. 1999). Spc72 directly binds to the outer plaque protein Nud1 which in turn binds to a coiled-coil protein Cnm67, whilst the half-bridge association is mediated by binding to Kar1 (Brachat et al. 1998; Gruneberg et al. 2000; Schaerer et al. 2001; Muller et al. 2005; Cavanaugh and Jaspersen 2017). In *S. cerevisiae*, Spc72 is essential for survival in strain S288C but non-essential in W303 backgrounds. In W303 cells with *spc72Δ*, which are viable, nuclear segregation is mediated by short cMTs and a microtubule motor protein (Knop and Schiebel 1998; Chen et al. 1998; Souès and Adams 1998; Hoepfner et al. 2002). In contrast, although cMTs are mostly lost in an *A. gossypii spc72Δ* mutant, it is viable, despite having no short cMTs and only a few long tangential cMTs. *A. gossypii spc72Δ* mutants have a severe defect in nuclear oscillation, but fast nuclear bypassing is still observed in a small subpopulation of nuclei (Lang et al. 2010a). As expected, based upon cMT regulation in *S. cerevisiae*, the half-bridge association of cMTs was also observed in *nud1Δ* and *cnm67Δ* in *A. gossypii*, and these cMTs support nuclear migration and bypassing (Lang et al. 2010a; Brachat et al.

1998; Adams and Kilmartin 1999; Gruneberg et al. 2000). The SPB layered structure and its permanent insertion into the nuclear envelop appear to be conserved in Saccharomycetes since SPBs in the methylotrophic yeast *Ogataea polymorpha*, which is more distantly related to *S. cerevisiae* than *A. gossypii* or *Candida albicans*, also share these features (Cavanaugh and Jaspersen 2017; Maekawa et al. 2017). Regulation, however, may be more diverse. In *O. polymorpha*, the outer plaque was clearly observed, by electron microscopy, in SPBs during mitosis but not in G1 cells, suggesting that the outer plaque disintegrates as cells exit from mitosis and enter into the next cell cycle (Fig. 8.3) (Maekawa et al. 2017). Unlike in *S. cerevisiae* and *A. gossypii*, OpSpc72 is not a constitutive outer plaque component. Recruitment to the SPB occurs only shortly before anaphase onset, in a polo-like kinase-dependent manner. Some of the *O. polymorpha* SPB proteins may have more divergent amino acid sequences. So, although proteins that localise to the outside of the SPB such as γ-tubulin complex, Spc110, Spc72 and Nud1 are conserved, the proteins forming the SPB internal structures could not be identified in homology searches of the *O. polymorpha* genome, suggesting very low sequence homology. Also, OpSPB may not have the second Spc72 binding site at the half-bridge, which may be explained by the absence of a Kar1 orthologue (Maekawa et al. 2017). Thus it is not clear whether all SPB components are conserved amongst all Saccharomycetes species.

The *S. pombe* SPB has a multilayered structure similar to *S. cerevisiae* (Fig. 8.2b), but forms less distinct ellipsoidal structures in EM micrographs. As in *S. cerevisiae*, a cytoplasmic bridge extends from the SPB and connects the duplicated SPBs; however, the SpSPB undergoes more pronounced cell cycle-dependent structural alterations. The interphase SPB is not inserted into the nuclear envelope but rather locates in the cytoplasm close to the nuclear envelope and is connected to the electron-dense materials on the nuclear side of the nuclear envelope (Ding et al. 1997). MTs are nucleated in the cytoplasm and associate with SPBs laterally when no nuclear MTs are formed. At the beginning of mitosis, a fenestrated region of nuclear envelope forms and SPBs become inserted into the nuclear envelope and nucleate spindle MTs (Tanaka and Kanbe 1986). SPB components are mostly conserved with a few exceptions. The central plaque components in *S. cerevisiae*, Spc42 and Spc29, are absent in the *S. pombe* genome. Instead, the *S. pombe*-specific Ppc89 protein is proposed to fulfil their function (Rosenberg et al. 2006; Cavanaugh and Jaspersen 2017). Also absent are the half-bridge component Kar1 and the Mps2 and Nbp1 proteins connecting the central plaque and the half-bridge (Schramm et al. 2000). Another fission yeast *Schizosaccharomyces japonicus* shares similar overall SPB structure and behaviour during the cell cycle with *S. pombe*, although the SjSPB differs in that it remains on the fenestra of the nuclear envelope during mitosis (Horio et al. 2002).

There have been few studies on the SPB in other species. In the black yeast *Exophiala dermatitidis*, which belongs to the Pezizomycotina subphyla of Ascomycetes, SPBs are disk-shaped elements locating on the nuclear outer membrane with electron-dense materials on the inner face of the nuclear envelope, as in *S. pombe* (Yamaguchi et al. 2002). Nuclear microtubules are attached before

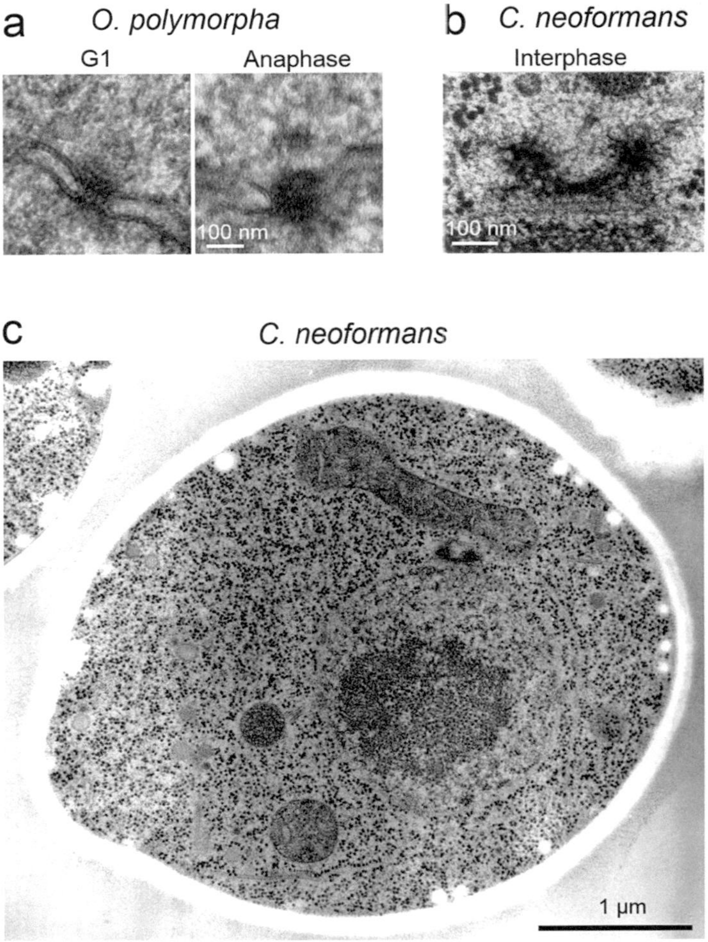

Fig. 8.3 Electron micrographs of NCY SPBs. (**a**) SPBs in G1 and anaphase cells of *O. polymorpha*. Reproduced from (Maekawa et al. 2017) under a creative commons attribution licence 4.0 (CC BY). (**b** and **c**) SPB in G1 cell (**b**) and G1 cell (**c**) of *C. neoformans*. Kindly provided by Dr. Masashi Yamaguchi, Chiba University, Japan. Reproduced from (Yamaguchi et al. 2009) with permission

embedding of the SPB into the nuclear membrane at mitosis. SPB duplication occurs in G1 before bud emergence; therefore, two disk-shaped SPBs, connected by a bridge-like structure, are observed during most of interphase. Before mitotic prophase, the SPBs become more than two times larger than in interphase. In *Cryptococcus neoformans*, a basidiomycetous human pathogen, dynamic changes in SPB structure were observed by electron microscopy (Yamaguchi et al. 2009). Interphase SPBs are either dumbbell or bar shaped and then grow more globular in shape during mitosis. Based on the limited number of species where SPB structures have been examined, embedding into the nuclear envelope may be a common feature in *Saccharomycetes*, whilst in the Pezizomycotina and Taphrinomycotina subphyla of Ascomycota and in Basidiomycota, SPBs locate outside the nuclear envelope during interphase and are then inserted into the nuclear envelope for mitosis (Yamaguchi et al. 2002, 2009; Horio et al. 2002; Cavanaugh and Jaspersen 2017; Maekawa et al. 2017; Lang et al. 2010a, b).

8.3 MT Nucleation by the γ-Tubulin Complex and Its Regulators

MT nucleation primarily depends on γ-tubulin, a member of the tubulin superfamily and an essential protein found in all eukaryotic cells (Rosselló et al. 2018). The γ-tubulin associates with the conserved γ-tubulin ring complex protein 2 (GCP2) and GCP3 to form a complex called the γ-tubulin small complex (γ-TuSC) (Fig. 8.4).

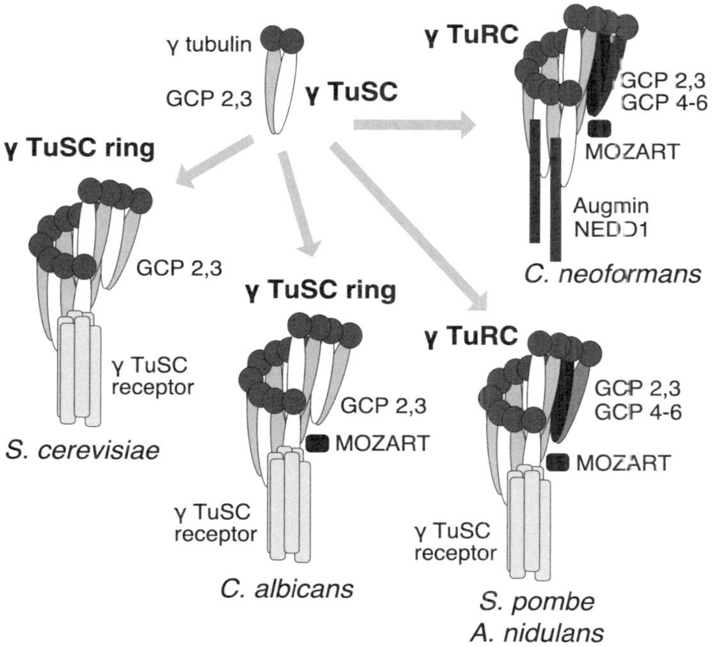

Fig. 8.4 Schematics of the main components of the γ-tubulin complexes found in different yeasts and fungi. The γ-TuSC and γ-TuRC rings form a template for microtubule nucleation. Components such as MOZART help stabilise ring formation. The γ-TuSC receptors promote ring formation and anchor the complex. Other molecules such as Augmin also anchor and localise the complex in the cell. See the main text for a detailed description. Complex components are not drawn to scale, the interactions are speculative and for clarity only a few representative molecules are shown

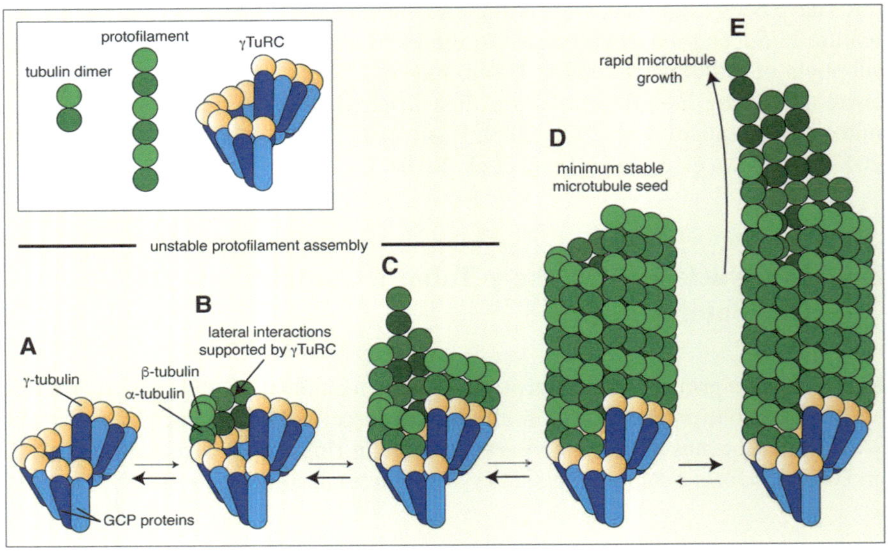

Fig. 8.5 Nucleation of microtubules by the γ-tubulin complex. A model of templated microtubule nucleation. (**a**) γ-Tubulin is held in a helical structure in the γTuRC. (**b**, **c**) αβ-Tubulin heterodimers assemble endwise onto the γ-tubulin template, which helps stabilise the incomplete MT lattice. Extension of the MT (**d**) forms a complete, stable lattice structure for normal rapid MT elongation (**e**). (Reproduced from (Tovey and Conduit 2018) under a creative commons attribution licence 4.0 (CC BY))

The γ-TuSC, additional GCP proteins and other proteins form a larger γ-tubulin ring complex (γ-TuRC) that forms a template for microtubule nucleation in most organisms (Fig. 8.5) (Sanchez and Feldman 2016; Roostalu and Surrey 2017; Tovey and Conduit 2018). In contrast, *S. cerevisiae* uses the most basic MT nucleation machinery (Fig. 8.6), and the γ-TuSC consisting of γ-tubulin, Spc97/GCP2 and Spc98/GCP3 is able to form a ring-like complex by self-oligomerisation in vitro. In physiological conditions, the presence of γ-TuSC receptor promotes the formation and stability of the γ-TuSC filament structure and nucleation activity. The SPB inner plaque component Spc110 is the exclusive γ-TuSC receptor in the nucleus, whilst the outer plaque component Spc72 is the only receptor in the cytoplasm. Because of the role of γ-tubulin receptors in MT nucleation activity, their cellular localisation probably explains the restriction of MT nucleation to only the SPB in *S. cerevisiae*. Furthermore, MT nucleation is temporally regulated through Spc110 phosphorylation in the nucleus during the cell cycle. Although they have significantly divergent amino acid sequences, Spc110-like proteins containing CM1 (centrosomin motif 1) domain and PACT (pericentrin-AKAP450 centrosomal targeting) domain and Spc72-like proteins containing CM1 domain and MASC domain are present in Ascomycota including *S. pombe*, *C. albicans*, *O. polymorpha* and *A. nidulans,* as well as in Basidiomycota such as *C. neoformans* (Maekawa et al. 2017; Lin et al. 2014; Flory et al. 2002). The molecular activity of these Spc110-like proteins to promote oligomerisation of γ-TuSC complexes has not been investigated.

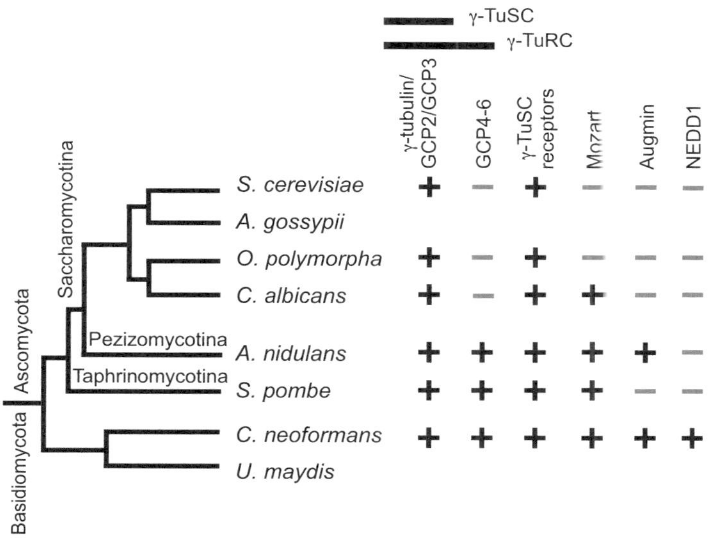

Fig. 8.6 Conservation of γ-tubulin complex proteins and recruiting factors in yeasts and fungi. Presence and absence are based on Lin et al. (2014, 2016b) but are unknown in *A. gossypii* and *U. maydis*. The phylogenetic tree is not to scale

In many organisms, γ-TuSC complexes together with GCP4–6 proteins and another conserved protein MOZART assemble into γ-tubulin ring complexes (γ-TuRC). Genomes of fungal species outside the *Saccharomycetes*, such as *S. pombe*, *A. nidulans* and *C. neoformans*, all encode GCP4–6 proteins and all encode MOZART (Anders et al. 2006; Fujita et al. 2002; Venkatram et al. 2004; Xiong and Oakley 2009; Lin et al. 2014). The γ-TuRC assembly is crucial for centrosomal MT nucleation and spindle assembly in humans and for recruitment to specific cellular locations in *Drosophila* (Izumi et al. 2008; Bouissou et al. 2009; Bahtz et al. 2012). However, GCP4–6 proteins in *Drosophila* and fungi are not essential for viability, and their deletion does not result in the loss of MT nucleation activity. In *S. pombe*, unlike its γ-TuSC mutants, deletions removing Gfh1/GCP4, Mod21/GCP5 and Alp16/GCP6 reduce both the number of cMTs and interphase MTOC activity, but do not cause major mitotic defects (Anders et al. 2006; Fujita et al. 2002; Venkatram et al. 2004). MOZART is conserved in all organisms that carry GCP4–6 orthologues. However, although the Saccharomycetes *S. cerevisiae* and *C. albicans* both lack GCP4–6 orthologues and MOZART is absent in *S. cerevisiae*, MOZART is present in *C. albicans*. In *C. albicans* MOZART/Mzt1 can directly interact with the conserved CM1 domain of Spc72 and Spc110, as well as the N-terminus of the γ-TuSC component CaSpc98, in an in vitro reconstitution system (Lin et al. 2016b). CaMzt1 increases the affinity of the γ-TuSC complex for the γ-tubulin complex receptors and cooperates with the γ-tubulin complex receptors to oligomerise γ-TuSC complex into the ring structure that has MT nucleation activity (Lin et al. 2016b). The receptor stimulation of γ-TuSC complex oligomerisation activity is higher in *S. cerevisiae* than *C. albicans*, which may compensate for the lack of MOZART protein in *S. cerevisiae*. In *S. cerevisiae* the γ-TuSC with Spc110 protein can form ring structures with 13 γ-tubulins, which are thought to form the template for microtubule formation (Kollman et al. 2010, 2015). Also, at least in vitro, the receptor Spc72 and TOG protein Stu2 also increase the efficiency of *S. cerevisiae* γ-TuSC microtubule nucleation (Gunzelmann et al. 2018). It is proposed that MOZART was lost during evolution in some budding yeast species, and compensation mechanisms for the loss of MOZART, probably through structural alteration of receptors, may have then evolved in these yeasts (Lin et al. 2016a). In other organisms, MOZART is involved in the recruitment of γ-TuRC to various MTOCs (Cota et al. 2017; Tovey and Conduit 2018).

In *S. pombe*, GCP4–6 orthologues (Gfh1/GCP4, Mod21/GCP5, Alp16/GCP6) were identified as nonessential components of the γ-TuSC complex (Fujita et al. 2002; Venkatram et al. 2004; Anders et al. 2006). All single mutants are viable and exhibit no major mitotic phenotypes. However, a reduced number of cMTs, detachment of cMTs from the SPB and lower nucleation activity in interphase MTOCs were observed. In contrast, MOZART/Mzt1 is essential for cell viability. As in *C. albicans*, SpMzt1 directly interacts with the N-terminus of Alp6/GCP3. SpMzt1 is not required for γ-tubulin complex formation but is important for the recruitment of the γ-tubulin complex to the SPB (Masuda et al. 2013; Dhani et al. 2013). Amongst noncore GCP components, Alp16/GCP6 localises to the SPB independently of Gfh1 and Mod21 and recruits the γ-tubulin complex to the SPB during mitosis. This effect is synergistic to Mzt1 function and may promote the assembly of the ring structure of the γ-tubulin complex at the SPB (Masuda and Toda 2016). At non-centrosomal MTOCs, Spc72-like protein Mto1 which contains CM1 and MASC domains, together with Mto2, is responsible for the nucleation activity (Lynch et al. 2014).

The recruitment of γ-TuRC to MTOCs is an important step to form an appropriate microtubule network and is regulated differently depending on organism, cell type and stage of the cell cycle. Assembly, recruitment or activation of γ-TuRC may be regulated through phosphorylation of the γ-tubulin complex receptors (Lin et al. 2014; Friedman et al. 1996, 2001; Fong et al. 2018; Tovey and Conduit 2018). In *S. pombe* the γ-TuRC may also be regulated by the activity of the motor protein Pkl1, which can remove the γ-tubulins (Olmsted et al. 2013). Additional proteins are also known to be involved in the γ-TuRC recruitment to specific MTOCs. In humans, NEDD1 and the Augmin complex mediate the attachment of γ-TuRC to the lattice of spindle MTs and is required for MT-dependent MT nucleation during mitosis (Lin et al. 2014; Tovey and Conduit 2018; Lüders et al. 2006; Haren et al. 2006; Lecland and Lüders 2014). Augmin subunits are absent in ascomycetous yeast including *S. cerevisiae*, *C. albicans* and *S. pombe*, but some subunits are present in the genome of the filamentous fungus *A. nidulans* (Edzuka et al. 2014; Lin et al. 2014; Farache et al. 2018). Although the homologue of Augmin subunit 6 localises at the SPB and on the mitotic spindle, no role in the recruitment of the γ-tubulin complex is reported (Edzuka et al. 2014). Basidiomycetous species, including *C. neoformans*, also have some of the Augmin subunits and NEDD1-like proteins in their genome.

MT nucleation activity, independent of the γ-TuRC, is reported for the TOG domain protein XMAP215/Stu2 and TPX2 (Roostalu et al. 2015; Flor-Parra et al. 2018; Lüders 2018). An XMAP215/Stu2 homologue is conserved in yeast and fungi as *S. cerevisiae* Stu2 and *A. nidulans* AlpA, and both localise to the SPB (Enke et al. 2007; Lin et al. 2014). Although interaction between TOG domain proteins and γ-TuRC is found in both animal cells and yeast, the details differ. XMAP215 binds to γ-tubulin through its C-terminus, whilst *S. cerevisiae* Stu2 indirectly interacts with the γ-tubulin complex through its receptor Spc72 (Usui et al. 2003; Thawani et al. 2018). *S. pombe* Alp14 also co-immunoprecipitates with the γ-TuRC (Flor-Parra et al. 2018); however, its exact binding site is not known. It is proposed that TOG domain proteins, associated with the γ-TuRC through their C-terminus, promote the addition of tubulin dimers (Lüders 2018). TPX2 is also important for microtubule nucleation in animal systems, and although TPX2 is absent in Ascomycota, it is conserved in a subset of basidiomycetous species; however, it is not known if the TPX2 homologue is involved in microtubule nucleation (Lin et al. 2014). Accumulating evidence suggests that XMAP215 and TPX2 not only nucleate MTs in cooperation with γ-TuRC, but can also do so in the absence of γ-TuRC (Roostalu et al. 2015; Woodruff et al. 2017). In yeast, *S. cerevisiae* Stu2 nucleates MTs from unattached kinetochores in the nucleus independently of γ-TuSC (Gandhi et al. 2011). Interaction of Stu2 with Spc72 is direct and does not require the γ-TuSC, and Stu2 with only Spc72 present can form asters in vitro (Gunzelmann et al. 2018). However, under normal circumstances, the γ-TuSC-dependent nucleation is dominant, and the γ-TuSC-independent MT nucleation becomes visible only when γ-tubulin is depleted. The presence of TPX2 as well as NEDD1 and Augmin may result from the semi-open mode of mitosis in these fungi (Lin et al. 2014; Farache et al. 2018).

8.4 Regulation of Nuclear Microtubules for Chromosome Segregation

The mitotic spindle consists of two sets of nuclear MTs: kinetochore MTs and interpolar microtubules. In many organisms, bipolar mitotic spindle formation requires the functions and interactions of multiple kinesin motors, particularly from the kinesin-5 and the kinesin-14 families (kinesin-5 = human KIF11, *Xenopus* Eg5, *S. cerevisiae* Cin8 and Kip1, *S. pombe* Cut7, *A. nidulans* BimC; kinesin-14 = human KIFC1, *Drosophila melanogaster* Ncd, *S. pombe* Pkl1 and Klp2, *S. cerevisiae* Kar3) (Schoch et al. 2003). Kinesin-5 is a plus-end-directed motor, which forms a homotetramer with two pairs of motor domains (Hentrich and Surrey 2010). Each motor domain interacts with a microtubule in an antiparallel orientation, which cross-links antiparallel MTs and slides them to generate outward forces that drive spindle pole separation. Interestingly, bidirectional motility in vitro has been demonstrated in fungal kinesin-5, although its contribution to cellular functions is unclear (Roostalu et al. 2011; Edamatsu 2014; Gerson-Gurwitz et al. 2011; Britto et al. 2016; Shapira et al. 2017). Kinesin-14 is a minus-end-directed motor, which can cross-link and bundle parallel MTs, provides inward forces and also has the ability to focus spindle poles (She and Yang 2017). The 'tug of war' between kinesin-5 and kinesin-14 is a common mechanism to maintain a bipolar spindle structure. In *S. pombe*, mutations in the kinesin-5 Cut7 lead to mitotic arrest with a monopolar spindle (Hagan and Yanagida 1990). Similar phenotypes are observed in other organisms including human cells (Kapoor et al. 2000). Kinesin-14 Pkl1 localises to the SPB and mediates spindle anchoring by recruiting the proteins Msd1 and the mitotic-specific SPB component Wdr8 (Syrovatkina and Tran 2015). The other *S. pombe* kinesin-14 Klp2 localises to the spindle, and an antagonistic relationship between kinesin-5 and kinesin-14 motors is supported by the observations that the lethality of *cut7Δ* is rescued by deletion of *pkl1*, and the unfocused minus-end phenotype of *pkl1Δ* is suppressed by *cut7Δ* (Syrovatkina and Tran 2015). Pkl1 has a motor-independent activity to directly inhibit microtubule nucleation at the γ-TuRC, which is also counteracted by Cut7 (Olmsted et al. 2013, 2014). In *S. cerevisiae*, the major kinesin-5 Cin8 and the minor kinesin-5 Kip1 provide a pushing force which slides antiparallel MTs apart and contributes to spindle formation, stability and elongation (Saunders and Hoyt 1992). In addition, Cin8 constitutes a length-dependent MT depolymerase activity for metaphase chromosome congression by kinetochore MTs (Gardner et al. 2008). The proposed model is that, since the kinesin-5 Cin8 walks to the MT plus end and promotes catastrophe, long MTs, which have more binding sites for Cin8, accumulate more Cin8 at their plus ends causing catastrophe, whilst shorter MTs are less likely to accumulate sufficient Cin8 at the MT plus end to prevent MT growth. This model is supported by experiments in *C. albicans*, which has one of the smallest known mitotic spindles (McCoy et al. 2015). In *S. cerevisiae*, the antagonistic motor to Cin8 and Kip1 activity is the kinesin-14 Kar3, which was originally isolated as a karyogamy mutant (Meluh and Rose 1990). Unlike other kinesin-14 motors that form homodimers, Kar3 forms heterodimeric complexes with a nonmotor subunit of either Cik1 or Vik1 (Barrett et al. 2000; Duan et al. 2012). The same sets of heterodimers, Kar3-Cik1 and Kar3-Vik1, are reported in the closely related *C. glabrata* (Joshi et al. 2013). The binding partner determines the localisation and function of Kar3. Thus, Kar3-Cik1 localises to the spindle MTs or the plus ends of MTs during mitosis and remains in the nucleus during interphase (Hepperla et al. 2014; Sproul et al. 2005), whilst Kar3-Vik1 is recruited to the spindle poles during mitosis (Manning et al. 1999). The Kar3 motor's role in karyogamy is fulfilled only by Kar3-Cik1 (Manning et al. 1999). Kinesin-14 monomeric motors are non-processive and the processive movement of Kar3 depends on its dimerisation with Cik1 or Vik1 (Mieck et al. 2015). The kinesin-14 KlpA in *A. nidulans* exhibits either processive motility towards the plus end of MTs or minus-end-directed motility, dependent on the experimental conditions (Popchock et al. 2017). The N-terminal nonmotor tail domain is proposed to function in switching directionality. Whether this feature is shared by other kinesin-14 motors is unclear.

In *S. cerevisiae*, consistent with the SPB being embedded into the nuclear envelope throughout the cell cycle, nuclear MTs are retained throughout the cell cycle. In unbudded cells, equivalent numbers of nuclear MTs and chromosomes were observed by electron microscopy, and each kinetochore is captured by one nuclear microtubule (Winey et al. 1995). In mitotic cells, there are also additional nuclear interpolar MTs (O'Toole et al. 1999). In *C. albicans* which has regional centrosomes of 3–5-kb length, a kinetochore also attaches one microtubule (Joglekar et al. 2008). In contrast, the human kinetochore covers 10 Mb of centromeric DNA and attaches to 15–20 MTs (McDonald et al. 1992). Single MT attachment is also not common in most fungal kinetochores. In *A. gossypii*, which probably has point centromeres of ~ 180 bp, compared to ~ 125 bp in *S. cerevisiae*, EM analysis shows 2–3 MTs are associated with each kinetochore (Gibeaux et al. 2012), whilst *S. pombe*, with regional centromeres spanning 40–100 kb, has 2–4 MTs attached to each kinetochore (Ding et al. 1993). The link between centromere DNA and the plus end of nuclear MTs is mediated by the kinetochore-microtubule attachment containing the highly conserved KMN network, comprising KNL1, Mis12 and Ndc80 complexes (Freitag 2016; Hara and Fukagawa 2018). In addition, the Dam1-DASH complex consisting of 10 proteins associates with kinetochores and is essential for viability. It is proposed to form rings around the MTs and stabilise the kinetochore-microtubule attachment in *S. cerevisiae* (Westermann et al. 2005; Thomas et al. 2017). It functions in spindle integrity and sister chromatid bi-orientation. Interestingly, whilst the Dam1 complex is also essential in *C. albicans*, it is dispensable in *S. pombe*, although chromosome mis-segregation is increased in the mutants (Sanchez-Perez et al. 2005). There appears to be a correlation between the Dam1 complex being essential and having only a single microtubule attachment to each kinetochore throughout the cell cycle (Thakur and Sanyal 2011). Importantly, Dam1 complex is absent in higher eukaryotes, which makes the Dam1 complex an attractive target for antifungal drug development.

8.5 Organisation of Cytoplasmic Microtubules and Nuclear Migration

Yeasts are generally highly polarised and thus cell division has asymmetric features. Therefore, like many eukaryotic cells, positioning and regulating the orientation of the mitotic spindle are crucial for successful distribution of chromosomes. Some fungi switch between the yeast form of proliferating cycle and filamentous growth. These organisms deal with both the same issues as budding yeast in their yeast form as well specific problems in their hyphal form, such as regulation of nuclear spacing in multinucleated hyphae. Interphase nuclear positioning and mitotic spindle orientation largely rely on the cytoplasmic microtubules (cMTs) and their interactions with cortical factors (Fig. 8.7). In *S. cerevisiae*, the SPB is the only MTOC and organises cMTs from the cytoplasmic side of the SPB throughout the cell cycle. No apparent difference has been reported between interphase cMTs and mitotic astral MTs, and cMTs stably associate with the SPB throughout the cell cycle and with similar dynamics in both G1 and metaphase cells (Maddox et al. 2000; Kosco et al.

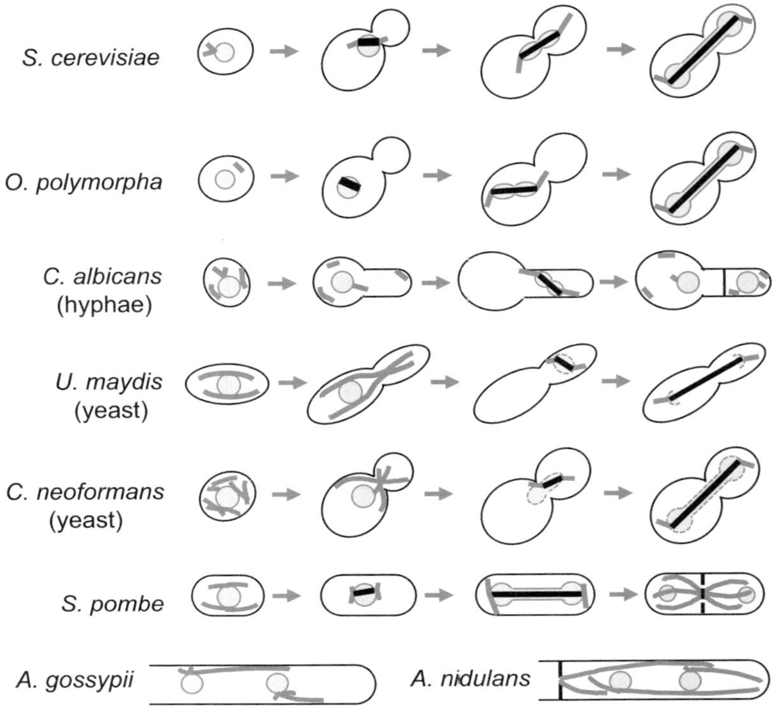

Fig. 8.7 Microtubule organisation in yeasts and fungi. Nuclear and cytoplasmic MTs during the cell cycle in *S. cerevisiae*, *O. polymorpha*, *C. albicans*, *C. neoformans*, *U. maydis* and *S. pombe* and interphase MTs in *A. gossypii* and *A. nidulans*. Thick black lines, nuclear MTs (nMTs); grey lines, cMTs. Not to scale

2001). In small budded cells, cMTs already orient the SPB towards the bud through extensive contact with the bud neck. As cells enter anaphase, the SPB that is closest to the bud neck migrates into the bud. Once the SPB enters the bud, cMTs emanating from that SPB maintain the contact with the bud cell cortex, whilst the other SPB still remaining in the mother cell organises cMTs that contact the mother cell cortex. Since SPB position is tightly regulated from an early stage of the cell cycle, the destiny of each SPB is usually determined soon after the SPBs are duplicated and the spindle is formed. The spindle is positioned close to the bud neck in the mother cell body and oriented along the mother-bud axis before the onset of anaphase (Fig. 8.8). This nuclear positioning in the cell cycle is common amongst most of the budding yeast in Ascomycota, with only a few exceptions (Martin et al. 2004; Maekawa et al. 2017). In *C. albicans* and *O. polymorpha*, the nucleus is not strongly biased towards the bud neck prior to anaphase. This is because the metaphase spindle is not positioned close to the bud neck and oriented along the mother-bud axis. Therefore, the anaphase spindle elongates initially in a misoriented direction. The misoriented anaphase spindle is in most of cases quickly corrected and inserted into

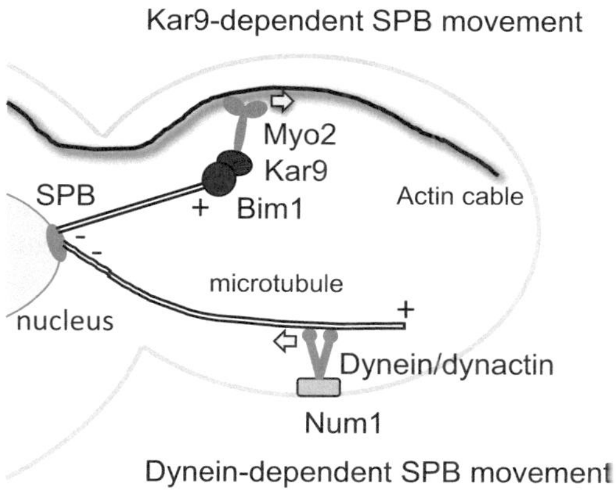

Fig. 8.8 Kar9-dependent and dynein-dependent spindle orientation pathways. (Upper) Association of Kar9 with the plus end of cMTs depends on the interaction with Bim1. Kar9 also interacts with myosin V (Myo2) that walks along the actin cable towards the bud tip, which moves the cMT and attached SPB towards the bud. (Lower) Dynein/dynactin is anchored at the cortical Num1 site and walks on the cMT towards the minus end, which pulls the SPB towards the bud. Arrows indicate the direction of motor movement along the actin cable and cMT

the bud successfully, without any apparent delay of anaphase progression (Martin et al. 2004; Maekawa et al. 2017). In *O. polymorpha*, this phenotype is explained by the lack of stable cMTs from G1 to metaphase. Emergence of cMTs is not frequent, and when emerged they are not maintained long enough to move SPBs. Unstable attachment of cMTs is probably due to the significantly reduced amount of the γ-tubulin complex receptor Spc72 at SPBs during early stages of the cell cycle, since when OpSpc72 is overexpressed Spc72 is targeted to SPBs and organises cMTs, even in interphase. In such cells, metaphase spindles are positioned close to the bud neck and oriented along the mother-bud axis as in *S. cerevisiae*. Whether *C. albicans* cells employ a similar regulation to that in *O. polymorpha* is unknown. In *C. albicans*, free cMTs, which are probably released from the SPB, are frequently observed during interphase. These short cMTs are proposed to function as storage for tubulin protein (Finley and Berman 2005). Free cMTs are also reported in *O. polymorpha* (Maekawa et al. 2017).

In the cylindrical cells of the fission yeast *S. pombe*, interphase MTs are arranged as 3–5 bundles. The ends of each bundle have the MT dynamic plus ends facing the cell ends, and the MTs from each bundle end overlap in a short region containing the antiparallel MT minus ends near the cell centre (Hagan 1998; Drummond and Cross 2000; Sagolla et al. 2003; Carazo-Salas et al. 2005; Hoog et al. 2007). The overlap regions are usually either associated with the SPB or directly with the nuclear envelope, and the equivalent microtubule dynamics at both ends of the bundle, pushing on the cell ends, keep the attached nucleus centred within the cell (Tran et al. 2001). The most prominent difference in the *S. pombe* cMT organisation is the presence of non-centrosomal MTOCs (Sawin and Tran 2006). During mitosis, SPBs are the sole MTOC and organise the spindle MTs and cytoplasmic astral

MTs. In contrast, during interphase, multiple non-centrosomal MTOCs appear in addition to the SPB. Non-centrosomal MTOCs include multiple sites on the nuclear envelope, existing MTs and the cytoplasm. These interphase non-centrosomal MTOCs become inactive at the entry into mitosis. Another type of non-centrosomal MTOC is formed transiently at the cell equator at the end of cell division, which is called an eMTOC. The γ-tubulin complex receptor Mto1, together with Mto2, localises to all types of MTOCs in the cytoplasm, independently of the interaction with the γ-tubulin complex (Janson et al. 2005; Sawin et al. 2004; Samejima et al. 2008; Venkatram et al. 2004, 2005), and their recruitment is a crucial step to form MTOCs and recruit the γ-tubulin complex. Localisation to the SPB and eMTOC is mediated through the conserved MASC domain at the Mto1 C-terminus. Overlapping, but distinct, amino acid sequences within the MASC domain are required for each localisation. The binding site of Mto1 is the unconventional myosin Myp2 at the eMTOC and the ScNud1 homologue Cdc11 at the SPB during mitosis (Samejima et al. 2010). Although SpCdc11 functions as a scaffold for the SPB-associated signalling pathway septation initiation network (SIN), its role in Mto1 recruitment is independent of its function in the SIN (Simanis 2015). Localisation to MTOCs on the nuclear envelope is regulated differently through the N-terminal region of Mto1 and the exportin Crm1 (Bao et al. 2018). The NES-like sequence within the Mto1 N-terminus interacts with exportin Crm1 in the cytoplasm by mimicking a nuclear export cargo. Crm1-Mto1/2 complex is docked to a nuclear pore complex (NPC) in an FG-repeat containing Nup146- and Ran GTP-dependent manner. Whether this non-export function of the nuclear export machinery is conserved in other organisms is currently unknown (Bao et al. 2018). Distribution of Mto1 amongst different non-centrosomal MTOCs is regulated by the J domain co-chaperone Rsp1 (Shen et al. 2019). MT organisation in another fission yeast *S. japonicus* is similar to that in *S. pombe* (Horio et al. 2002). As in *S. pombe* SPBs are likely the only MTOCs during mitosis and nucleate spindle MTs and astral MTs. Although interphase MTOCs may not be present in the cytoplasm, reorganisation of cMTs from the cell equator in late mitotic phase suggests the presence of an eMTOC.

Cytoplasmic MT organisation is very different in the basidiomycetous yeast *C. neoformans* and dimorphic plant pathogen *Ustilago maydis*, which can grow as haploid yeast in culture (Banuett 1995). Nuclear movement during the mitotic cell cycle in basidiomycetous yeast shows striking differences from that in ascomycetous yeast. In Basidiomycota, the nucleus migrates into the bud before nuclear division, and anaphase initiates in the bud. During anaphase, one of the nuclei moves back into the mother cell (Kopecka et al. 2001). Although the spindle is formed from SPBs, the SPB is not the only MTOC and the cMT organisation is more complex. In *U. maydis*, whilst astral MTs are organised from SPBs during mitosis, cMTs are nucleated at many cytoplasmic sites and extend along the long axis in anti-polar orientation in G1. In G2, a limited number of γ-tubulin-containing MTOCs are formed close to the bud neck and cMTs are reorganised into parallel bundles. Additional motile MTOCs are also observed in different cytoplasmic locations. The MT motor dynein plays important roles in positioning of MTOCs at the bud neck as well as overall organisation of polarised cMT bundles. Interestingly, both metaphase SPBs carry astral MTs contacting with the bud cell cortex, which is in stark contrast to *S. cerevisiae* where cMTs from only one of the SPBs establish the interaction with the bud cortex (Straube et al. 2003; Steinberg and Fuchs 2004; Banuett et al. 2008; Fink and Steinberg 2006). A slightly different picture was reported in the basidiomycetous *C. neoformans* and *C. laurentii* (Kopecka et al. 2001; David et al. 2007; Mochizuki 1998). At the beginning of the cell cycle, before bud emergence, a random network of cMTs locates close to the cell cortex. After the bud emerges, bundles of cMTs are inserted into the bud in a parallel orientation as the nucleus migrates into the bud. By the time of spindle formation, cMTs disappear and astral MTs are nucleated from both SPBs. The cMT network is restored as the spindle disassembles at the end of mitosis. Whether and how metaphase spindle orientation is regulated in the bud is unclear. The cMTs in basidiomycetous yeast have some similarities with those in *S. pombe*. Firstly, the MTs are arranged in bundles with the plus ends facing the bud and mother cell tips, with the minus ends originating at a non-SPB MTOC just inside the bud, and secondly, the cMTs go through a dynamic reorganisation at mitotic entry (Kopecka et al. 2001). This may fit with the common requirement for cMTs in cell morphogenesis (Banuett et al. 2008).

In filamentous fungi, cMTs play prominent roles in regulating nuclear position and successful chromosome segregation and also as tracks for rapid hyphal growth and the long-distance transport of vesicles, endosomes, mRNAs and other organelles such as peroxisomes towards the hyphal tip (Horio and Oakley 2005; Riquelme et al. 2018, 2011; Steinberg et al. 2012; Takeshita et al. 2014; Peñalva et al. 2017; Salogiannis and Reck-Peterson 2017; Ishitsuka et al. 2015). Whilst *A. nidulans* γ-tubulin MipA has been intensively studied as the founding member of the γ-tubulin family, understanding of the composition of MTOCs and their regulation remains limited (Oakley et al. 2015; Oakley and Oakley 1989; Xiong and Oakley 2009; Zekert et al. 2010; Horio et al. 1991; Zhang et al. 2017; Riquelme et al. 2018). In *A. nidulans*, SPBs nucleate spindle MTs as well as astral MTs during mitosis. In interphase, SPBs remain active and organise cMTs, which often overlap with cMTs emanating from the SPB of the neighbouring nucleus within the same cell compartment. Additional MTOCs were discovered at septa that contribute to the cMT array during interphase (Konzack et al. 2005). γ-Tubulin is present at both SPBs and septal MTOCs. Similar to the eMTOC in *S. pombe*, the septal MTOCs also contain several SPB outer plaque components including the γ-tubulin receptor ApsB (SpMto1/ScSpc72 homologue), Spa18 (SpMto2 homologue), GcpB and GcpC (homologues of ScSpc97/GCP2 and ScSpc98/GCP3, respectively) and GcpD–F (homologues of GCP4–6), whilst the half-bridge component SfiA and the γ-tubulin receptor at inner plaque PcpA are not detected (Konzack et al. 2005; Zhang et al. 2017). Localisation of ApsB seems to be the key since it binds to γ-tubulin and recruits the γ-TuSC to the septal MTOCs. SPB localisation of ApsA requires the ScNud1 homologue SepK, although the septal-pore-associated protein Spa10 is likely to be the ApsA anchor protein at septal MTOCs (Zhang et al. 2017). The current model of septal MTOC formation is that Spc18 and ApsB bind first to the

Spa10 disc-like structure at the septal pore and then recruit the γ-TuRC. Unlike the eMTOCs of *S. pombe*, which are transient structures at the end of mitosis, the septal MTOC structures are persistent throughout the cell cycle. In *A. nidulans*, a small number of cMTs are maintained during mitosis, which presumably guarantees continuous intracellular transport (Riquelme et al. 2003; Zekert and Fischer 2009). It remains unclear whether septal MTOCs are active during mitosis (Riquelme et al. 2018). In another filamentous fungus *Neurospora crassa*, cortical and cytoplasmic MTs are much more abundant and are present in both polarities, mainly parallel to the long axis of hyphae, forming an extensive network throughout the cytoplasm (Mouriño-Pérez et al. 2006, 2013, 2016; Uchida et al. 2008). Rotational or helical motions of the network were observed using live cell imaging. Non-centrosomal MTOCs are present at the growing tip of hyphae (Mouriño-Pérez et al. 2006). Spa10 was originally identified in a screen for septum-associated disordered proteins in *N. crassa* (Lai et al. 2012). The presence of an MTOC at septa remains unclear, although cMTs were observed entrapped in a septal pore in *N. crassa*. It is presumed that a septal MTOC may not be required since the cell cycle of nuclei in *N. crassa* hyphae is not synchronised, and the cMT network remains intact to support transport (Riquelme et al. 2018).

8.6 Nuclear Position and Migration Are Regulated by Cytoplasmic Microtubules

The position and orientation of mitotic spindles are regulated in such a way as to ensure that each daughter cell inherits only one set of chromosomes and other cellular content as appropriate. This is particularly important in asymmetric cell division, such as that in the budding yeast. In *S. cerevisiae*, chromosome segregation initiates within the mother cell body, followed by migration of one daughter nuclei into the bud. Since the bud neck, the junction between the mother cell and the bud, is the predetermined cytokinesis site, mitotic exit and cytokinesis should take place only after proper segregation of chromosomes across the bud neck. Therefore, coordination of cell cycle progression with nuclear movement is crucial for the success of mitotic cell division. In the fission yeast *S. pombe*, in which the position of the nucleus determines the position of the cell division plane, maintaining the nucleus at the cell centre is essential for producing two daughter cells of equal size, with MTs playing an important role in ensuring this (Tran et al. 2001; Daga et al. 2006). In multinucleated filamentous fungi, proper distribution of interphase nuclei is also important for hyphal and colony growth and the cMTs play critical roles in nuclear movement (Xiang 2018). The dynamic instability of cMTs lets the plus ends of the MTs explore the cytoplasm until they meet binding partners. Attachment of cMTs to cortical factors leads to force generation, which then moves the spindle and the nucleus. Specific sets of microtubule motors and microtubule-associating proteins at the plus end of cMTs play important roles in the force generation and regulation of cMTs. Although nuclear positioning is commonly regulated by cMTs, the detailed mechanisms of nuclear positioning differ depending on the organism and cell type (Pruyne et al. 2004; Xiang 2018). In *S. cerevisiae*, two redundant molecular pathways were identified: one is the Kar9 pathway, which is active at early stages of the cell cycle, and the other the dynein pathway in anaphase (Fig. 8.8). In normal mitosis, the SPB in G1, which is inherited from the previous cell cycle, already has cMTs, which establish contact with the bud cortex soon after the bud emergence. SPB duplication is initiated in late G1 to build a new SPB structure next to the pre-existing old SPB, followed by SPB separation and short spindle formation (Winey and Bloom 2012; Cavanaugh and Jaspersen 2017). The old SPB maintains contact with the bud cortex through cMTs, whilst the acquisition of cMTs on the new SPB is delayed (Shaw et al. 1997; Segal et al. 2000). The short spindle, and therefore the nucleus, moves close to the bud neck within the mother cell body and orientates itself along the mother-bud axis. This configuration ensures efficient insertion of one set of chromosomes into the daughter cell compartment upon spindle elongation in anaphase. The Kar9-dependent mechanism is dominant in spindle positioning and orientation during the early stages of the cell cycle. In anaphase, dynein-dependent sliding of cMTs along the cortex brings the spindle and the nucleus into the bud (Moore and Cooper 2010; Geymonat and Segal 2017).

8.6.1 Nuclear Movement Mediated by Kar9

Kar9 was first reported as one of the proteins required for nuclear fusion during mating (karyogamy) and later shown to be involved in spindle positioning and orientation (Miller and Rose 1998; Kurihara et al. 1994). Kar9 connects the cMTs with actin cables by interacting simultaneously with the MT plus-end-tracking protein (+TIP) Bim1 and the type V myosin Myo2 (Fig. 8.8). As Myo2 motor complexed with Kar9-Bim1 moves along the actin cable towards the bud tip, cMTs linked via the Kar9-Bim1 complex guide the SPB towards the bud (Palmer et al. 1992; Miller et al. 1999; Yin et al. 2000; Hwang et al. 2003). Kar9 is loaded on the cytoplasmic side of the SPB and then transported to the plus end of cMTs together with Bim1 (Maekawa et al. 2003; Liakopoulos et al. 2003). Strikingly, Kar9 preferentially associates with the old SPB. This asymmetric binding is important to ensure that only one of the SPBs acquires a bud-ward destiny. Although the Kar9 pathway moves the spindle and the nucleus along the actin cables, directing them towards the bud tip, insertion of the spindle into the bud is normally prevented until the onset of anaphase. The transient nature of the cMT contacts with the cortex, probably through post-translational modifications of related factors, is an important part of the mechanism (Maekawa and Schiebel 2004; Schweiggert et al. 2016). Another important factor is the SPB movement caused by the capture-shrinkage mechanism involving Bud6. Bud6 is a cortical protein that interacts with the formins and plays a role in actin organisation (Graziano et al. 2011; Moseley et al.

2004). In G1, cMT capture is mediated by Bud6 at the bud cortical domain and the SPB is positioned facing the bud. Later, accumulation at the bud neck helps maintain the metaphase spindle and the nucleus in the mother cell body (Segal et al. 2000, 2002; Adames and Cooper 2000). The Bud6-dependent capture of cMTs is active throughout the cell cycle, which contributes to the positioning of the SPB near the previous cytokinesis site in G1 (Huisman et al. 2004; Segal et al. 2002; Ten Hoopen et al. 2012). The shrinkage of MTs at Bud6 sites is promoted by the depolymerisation activity of Kip3, a member of the kinesin-8 family. When Kip3 detaches from the shrinking plus end of MTs, the MT ends remain connected to the cortex through dynein (Ten Hoopen et al. 2012). Direct interaction of Bud6 with the +TIP Bim1 is necessary to promote cMT capture but dispensable for its role in actin organisation (Segal et al. 2002; Huisman et al. 2004; Delgehyr et al. 2008; Ten Hoopen et al. 2012).

A Kar9 homologue is conserved in saccharomycetous budding yeast and filamentous fungi, but not found in either Taphrinomycotina, including the fission yeast, or in Basidiomycota. Kar9 is likely to be involved in the dynein-independent mechanisms of nuclear movement observed in *C. albicans* and *A. nidulans*, although the precise function of Kar9 in these organisms remains unclear (Xiang et al. 1995; Martin et al. 2004; Finley and Berman 2005). Kar9 in *C. albicans* localises as a cortical dot at the bud tip during the early stages of the cell cycle and then later appears as a series of dots on the mother and bud cortex as well as at the bud neck (Li et al. 2005). It is unclear whether or not Kar9 associates with the plus end of MTs since these localisation studies were performed with GFP-Kar9 expressed from a strong *GAL1* promoter, and only examination of its localisation at endogenous expression levels will clarify if the asymmetric distribution of Kar9 observed in *S. cerevisiae* is also conserved in *C. albicans*. Kar9 may also be responsible for dynein-independent spindle orientation and nuclear movement, albeit the contribution in wild type cells is small (Martin et al. 2004; Finley and Berman 2005). In the filamentous fungus *A. nidulans*, MigA, a Kar9 orthologue, localises to SPBs and the cell cortex, and is involved in spindle positioning during the early stages of mitosis by facilitating cMT contacts with the cell cortex (Manck et al. 2015). However, the molecular mechanism may differ from that in *S. cerevisiae*, since MigA also localises to the mitotic spindle and associates with MTs independently of EbA, the ScEB1 orthologue. Its role in spindle orientation is likely mediated by its interaction with ApsA, the orthologue of the PH domain containing cortical protein ScNum1. This is consistent with the observation that the actin cables are limited to the hyphal tip. The kinesin-8 motor KipB is involved in spindle positioning during mitosis, probably as part of the MigA-dependent mechanism (Rischitor et al. 2004). However, deletion of MigA or KipB does not cause any defects in nuclear distribution in hyphae (Rischitor et al. 2004; Manck et al. 2015). In contrast to other filamentous fungi, a dynein-independent mechanism strongly pulls the nucleus towards the hyphal tips in *A. gossypii*, which is phylogenetically close to *S. cerevisiae* (Alberti-Segui et al. 2001).

8.6.2 Nuclear Movement Mediated by Dynein

A dynein-dependent mechanism is more common in many fungal species. Cytoplasmic dynein is a large protein complex with minus-end-directed MT motor activity. It requires many cofactors such as LIS1 and dynactin for its in vivo functions. In *S. cerevisiae*, dynein accumulation at the plus end of cMTs depends on the CLIP-170 homologue Bik1, the kinesin Kip2, the LIS1 homologue Pac1 and NudEL homologue Ndl1 (Fig. 8.8). Dynein is offloaded at the cell cortex through interaction with the cortical protein Num1 (Heil-Chapdelaine et al. 2000; Farkasovsky and Küntzel 2001; Tang et al. 2009). Whilst Pac1 inhibits dynein motility at the plus end of MTs, Num1 releases the dynein-dynactin complex from the plus end and activates the dynein minus-end-directed movement (Lammers and Markus 2015). Since dynein remains anchored to the cortex, the walk towards the minus end leads to the sliding of the MT and pulling of the SPB attached to the cMT (Moore and Cooper 2010).

A role for dynein in positioning of the spindle and nucleus has also been reported in other yeasts and filamentous fungi including *C. albicans*, *A. gossypii*, *A. nidulans*, *U. maydis* and *Schizophyllum commune*. In *C. albicans*, nuclear movement largely relies on the dynein-dependent mechanism in both the yeast and hyphal forms (Martin et al. 2004; Finley and Berman 2005; Finley et al. 2008). In the yeast form, the nucleus stays away from the bud neck until anaphase. Therefore, spindle elongation in early anaphase occurs in the mother cell without alignment along the mother-bud axis. Then the spindle orientation is efficiently realigned to insert one of the daughter nuclei into the bud, driven mainly by the dynein-dependent mechanism (Martin et al. 2004). In the absence of dynein, nuclear segregation may still succeed when the spindle elongation happens by chance in a small number of cases to occur along the mother-bud axis. However, even though the majority of mitoses occur in the mother cell, a dynein-independent mechanism (probably the Kar9-dependent mechanism) ensures the successful segregation of one nucleus into the bud. As a result, only a small number of dynein-deficient cells carry multiple nuclei (Martin et al. 2004). When a cell starts forming a hypha (germ tube), the hyphal neck (equivalent to the bud neck) is no longer the future cytokinesis site. The septin ring is not formed at the neck but instead at the future site of septation (presumptum) in the germ tube. SPB-anchored cMTs facilitate nuclear migration from the mother cell into the germ tube, approaching but not passing the presumptum before anaphase. This migration is mediated by the dynein-dependent sliding of cMTs. Anaphase spindle elongation pushes one nucleus forward, whilst the other nucleus moves back towards the hyphal neck. After mitotic exit and septation, the nucleus, which was pushed back to the hyphal neck, moves back into the mother cell body. The molecular basis of this nuclear movement is not understood (Finley and Berman 2005). The Bub2 spindle checkpoint prevents cytokinesis until nuclear migration is completed (Finley et al. 2008). Although it is not essential for normal cell division, the checkpoint is crucial when dynein function is compromised. Interestingly, formation of MTs free in the cytoplasm during interphase may depend upon dynein, since dynein-dependent pulling forces may be involved in the detachment of cMTs from SPB (Finley et al. 2008). Although the nuclear movements in the yeast form and in germ tubes appear different, in both, the nucleus is positioned close to the site of the septin ring by dynein activity.

In multinucleate hyphae, the actions of opposing forces are proposed to determine the distribution of nuclei (Plamann et al. 1994). In filamentous fungi, dynein is crucial for nuclear distribution (Xiang and Fischer 2004). However, the molecular detail seems to be different depending on the species. In multinucleated *A. gossypii*, short-range oscillatory movement of nuclei was observed, as well as long-range nuclear bypassing movements, which often change the order of nuclei in hyphae (Alberti-Segui et al. 2001; Gladfelter et al. 2006; Grava et al. 2011). Since dynein is responsible for all nuclear movement in growing hyphae, including nuclear distribution, nuclei become stacked at the hyphal tips in dynein mutants (Grava et al. 2011). Similar to the mechanism in *S. cerevisiae*, the cortical anchor Num1 and dynactin play important roles in nuclear distribution (Grava et al. 2011). However, the impact of mutations in *NUM1* and dynactin genes seems to be milder than that of dynein mutants, suggesting that there may be other cortical factors contributing to the dynein activity. Another possibility is that dynein's function is partly independent of cortical receptors. The current model is supported by a simulation of the nuclear movement in *A. gossypii*, which uses published parameters (Gibeaux et al. 2017). The accumulation and activation of dynein at the plus ends of cMTs are cell cycle regulated and become most prominent during anaphase in *S. cerevisiae*. However, the mitoses in the multinucleated hyphae of *A. gossypii* are asynchronous. LIS1, Bik1 and Kip2 have roles in the plus end accumulation of dynein in *S. cerevisiae*, whilst plus end accumulation in *A. nidulans* and *U. maydis* requires kinesin-1 and dynactin (Xiang et al. 2000; Han et al. 2001; Sheeman et al. 2003; Zhang et al. 2003; Yao et al. 2012; Qiu et al. 2018). In *A. nidulans*, dynein and the Num1 homologue ApsA are important for nuclear distribution and migration in hyphae (Suelmann et al. 1997; Fischer and Timberlake 1995). In germ tubes of the dynein mutant, nuclei were clustered in the spore end, which is the opposite phenotype of *A. gossypii* dynein mutants (Xiang et al. 1994). In *N. crassa* nuclei are clustered, fail to distribute evenly and do not reach the hyphal apex in dynein mutants. However, dynein plays only a minor role in nuclear distribution, and cytoplasmic bulk flow is suggested to be the most important factor (Mouriño-Pérez et al. 2016).

8.7 Nuclear Movement in Sexual Development

Nuclear movement is important for karyogamy (nuclear fusion following mating) and asexual spore development. Amongst the proteins required for karyogamy that have microtubule-related functions, Kar9 and the kinesin-14 family member Kar3 are conserved in Saccharomycetes (Kurihara et al. 1994). Whereas Kar9's role in karyogamy is probably conserved in ascomycetous budding yeast, since it is present in *C. albicans*, there are no apparent Kar9 orthologues in either fission yeast or

Basidiomycota such as *C. neoformans* (Bennett et al. 2005; Lee and Heitman 2012). In *S. cerevisiae*, Kar3 plays a central role in nuclear congression at the SPB where it captures and pulls on MTs attached to the SPB on the other nucleus (Gibeaux et al. 2013). Similarly, Kar3 function in karyogamy is conserved in *C. albicans* and probably in *A. gossypii* (Wasserstrom et al. 2013; Bennett et al. 2005). In *S. pombe*, Klp2, one of its two kinesin-14 family kinesins, and dynein play the major role in nuclear congression (Troxell et al. 2001; Scheffler et al. 2015). Interestingly, in *C. neoformans*, there is no Kar3 orthologue (Lee and Heitman 2012), although nuclear congression is still necessary for sexual development in opposite-sex mating. This may reflect less need for nuclear movement in the karyogamy of *C. neoformans* where two nuclei are already packed into a dikaryotic hyphal compartment (Lee and Heitman 2012).

In *S. pombe*, after nuclear fusion, diploid nuclei undergo rigorous dynein-driven 'horsetail' movements during meiotic prophase, which are important for initial pairing of homologous chromosomes and recombination, as well as other meiotic roles (Chikashige et al. 1994; Yamamoto and Hiraoka 2003; Ding et al. 2004; Chacón et al. 2016). Similar movement is reported in *S. japonicus* (Niki 2014). Cooperation of Num1 (also known as Mcp5) and dynactin complex establishes anchorage of dynein at the cell cortex in *S. pombe* (Yamashita and Yamamoto 2006; Saito et al. 2006; Fujita et al. 2015). Whilst SpNum1 prevents sliding of dynein along the cell cortex, dynactin complex is involved in the regulation of microtubule dynamics. The *NUM1* gene is expressed specifically during meiosis.

8.8 Roles of MTs in Polarised Growth

MTs are involved in cell morphogenesis through the regulation of actin assembly in both cylindrical *S. pombe* and filamentous fungi cells. As such, in *S. pombe* cells and fungal hyphae, microtubule-dependent transport plays an important role in polarised growth at cell end or hyphal tips (Chang and Martin 2009; Riquelme et al. 2018; Tay et al. 2018). This is in contrast to *S. cerevisiae* where actin polarity regulates the orientation of cMTs (Winey and Bloom 2012). *S. pombe* cells have distinct cortical domains at the sides and cell tips throughout the cell cycle and on the division plane at the end of mitosis (Chang and Martin 2009). Polarised cell growth is restricted to the cell tips during interphase, and the cell tips are marked by cell end marker proteins and the enrichment of actin cytoskeleton (Chiou et al. 2017). Microtubule-dependent transport mechanisms deliver cell end markers and deposit them at the cell tips (Mata and Nurse 1997; Behrens and Nurse 2002; Feierbach et al. 2004; Martin et al. 2005; Tatebe et al. 2005). Cytoplasmic MTs elongate until they reach the cell tips where they undergo catastrophe and shrink. The kelch-repeat protein Tea1 and SH3 domain protein Tea4 act as cell end marker proteins, and their localisation to the cell tip is important for maintenance of cell polarity. The association of Tea1/Tea4 with the growing MT plus ends is dependent upon +TIP CLIP-170 homologue Tip1, EB1 homologue Mal3 and kinesin Tea2, which form the

Tea2/Tip1/Mal3 complex that carries the Tea1/Tea4 cargo and walks to the MT plus ends (Browning et al. 2000, 2003; Brunner and Nurse 2000; Busch et al. 2004). Slightly behind the Tea2/Tip1/Mal3 complex, the heterocomplex of two kinesin-8 proteins Klp5 and Klp6 follows (Meadows et al. 2018). The Klp5/Klp6 complex also contains Mcp1, which has similarity to the antiparallel MT binding protein Ase1/PRC1/MAP 65, and the complete complex exhibits MT-destabilising activity. The presence of Tea2/Tip1/Mal3 prevents the accumulation of Klp5/Klp6/Mcp1 MT-destabilising complex at the MT plus end and reduces the MT catastrophe frequency (Meadows et al. 2018). When MTs extend to the cell tips and MT growth is paused, the Klp5/Klp6/Mcp1 complex reaches the plus end and promotes displacement of Tea2/Tip1/Mal3 and MT shrinkage. The Tea1/Tea4 cargo detaches from the motor complex and MTs and is then tethered to the tip cortex where Tea1/Tea4 recruit a formin component For3 to promote actin assembly (Martin et al. 2005). Cell growth is initiated at the 'old end', which existed in the preceding division, in G1. In G2, cells also start growth from the new end (new end take off, NETO). Although the Tea1/Tea4 complex is crucial for polarised growth, its localisation alone is insufficient to promote cell growth, which is evident from the observation that the complex localises to both cell ends even before NETO (Mata and Nurse 1997; Martin et al. 2005). Apparently Tea2-dependent transport and the cargo deposition do not distinguish old and new ends.

A similar mechanism has been reported for the cell end regulation in *A. nidulans*. The plus ends of most cMTs in the tip compartment face towards the cell end and converge to the apex of the hyphal tip (Konzack et al. 2005; Sampson and Heath 2005). Cell end markers Tea1 and Tea4, as well as the kinesin Tea2, are conserved in *A. nidulans* (TeaA, TeaC, KipA, respectively). Analyses of the mutant phenotypes of these genes indicate an *S. pombe*-like system for the regulation of polarised growth. The cell end marker Tea1 is delivered to the cell tip by growing MTs through interaction with the Tea2/Tip1/Mal3 motor complex and then released and anchored at the cell cortex. Then Tea1 recruits other proteins including formin components. Placing of the cell end markers is important for ensuring a straight form of hyphae (Takeshita et al. 2008). In contrast, MTs are not important during germ tube formation and slow hyphal growth (Horio and Oakley 2005). The detailed regulation of MT dynamics by +TIP proteins in *A. nidulans* seems to be different from that in *S. pombe* and other organisms. In *kipAΔ* mutant cells, the normal focussing of MTs at the cell tips is disturbed and a special structure called the Spizenkörper is mispositioned (Konzack et al. 2005). In addition, whilst most of the cMTs fail to grow and do not reach the cell tips in the *clipAΔ* mutant cells, in agreement with the catastrophe-suppressing activity of other CLIP-170 family proteins, the few MTs that do reach the cell tip are more stable than those in wild type cells (Efimov et al. 2006). The precise mechanism normally promoting shrinkage is not understood. Localisation of TeaA at the cell tip influences cMT convergence. TeaA interacts with the XMAP215 homologue AlpA when MTs reach the cell tip and possibly regulates AlpA activity (Takeshita et al. 2013). The MT plus end localisation of +TIP proteins is temperature-dependent, though the functional significance of this remains unclear (Efimov et al. 2006).

Another unique and important role of cMTs in filamentous fungi is the transport of vesicles such as secretory vesicles and early endosomes, which is mediated by actin in the budding yeast. Conventional kinesin-1 transports secretory vesicles and nuclear pore complexes in *U. maydis* (Schuster et al. 2012; Steinberg et al. 2012) and, along with myosin 5 and dynein/dynactin/p25, Rab11 secretory vesicles in *A. nidulans* (Peñalva et al. 2017), whilst dynein and either the kinesin-3 family protein UncA in *A. nidulans* or Kin3 in *U. maydis* have functions in early endosome transport (Peñalva et al. 2017; Seidel et al. 2013; Zekert and Fischer 2009; Wedlich-Söldner et al. 2002; Schuster et al. 2011). Rab11 secretory vesicles require dynein and the p25 subunit of dynactin, but not the adaptor or hook complex for early endosome transport. However, the hook complex, dynein and dynactin are important for the transport of Rab5 vesicles (Riquelme et al. 2018). Such intracellular membrane traffic is not only responsible for delivering membrane components for tip expansion but also transports many other cargos, such as mRNAs and ribosomes that are attached on the surface of endosomes for transport along MTs (Niessing et al. 2018), as well as organelles such as peroxisomes and mitochondria. Extensive studies on MT-dependent transport have been performed in yeast and filamentous fungi (Egan et al. 2012; Haag et al. 2015; Niessing et al. 2018).

8.9 Tubulin Proteins in Yeasts

MTs assemble by endwise polymerisation of tubulin protein heterodimers containing α- and β-tubulins, with assembly in vivo usually nucleated by microtubule-organising centres containing γ-tubulin (Goodson and Jonasson 2018). Members of the tubulin superfamily are found in all eukaryotes, and mammals in particular contain a large number of tubulin isoforms (Ludueña 1998; Khodiyar et al. 2007; Leandro-García et al. 2010). In contrast most fungal and yeast cells have relatively few isoforms (Fig. 8.9), with some, such as *Candida tropicalis,* having only a single α-, β- and γ-tubulin gene (Zhao et al. 2014). Based on sequence comparisons, Zhao et al. (Zhao et al. 2014) have proposed that a common fungal ancestor contained 2 α- and 2 β-tubulin genes with subsequent loss of some isoforms as the fungi diverged during evolution, which was further complicated in some species by the subsequent duplication of some of the remaining genes. Thus both *S. pombe* and *S. cerevisiae* contain two α-tubulin genes thought to have arisen by duplication from a single gene following their divergence as species, together with a single β- and γ-tubulin. Many NCYs have a similar arrangement of tubulin genes, though others vary with *Yarrowia lipolytica* having a single α-, two β- and single γ-tubulins, whilst *Candida albicans* has a single α-, single β- and two γ-tubulins (Zhao et al. 2014). Such complexities of duplication and divergence make it difficult to use tubulin sequences alone to establish fungal phylogenies (Keeling 2003; Zhao et al. 2014), though they may still be of use in the identification of species or determining the phylogeny of closely related species (Kharazi et al. 2018). Tubulin protein sequences are well conserved, and *S. pombe* and *S. cerevisiae* tubulin sequences are as similar to those

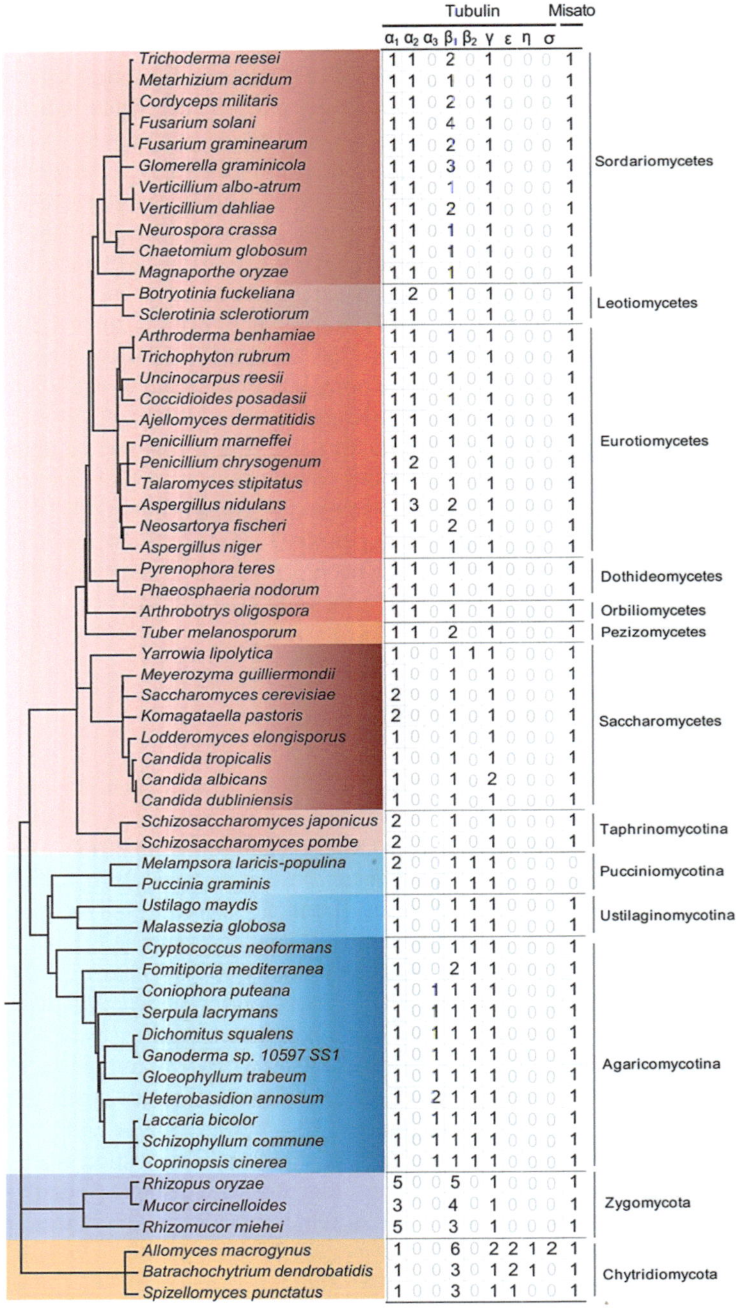

Fig. 8.9 Tubulin genes present in yeast and fungi. A tree based on α-tubulin protein sequences in yeast and fungi, with the number of different tubulin genes present in each. α1, 2 and 3 and β1 and 2 represent different clades. Multiple members within each clade, such as the two α1 genes found in many yeasts, are thought to have arisen through species-specific gene duplications (Zhao et al. 2014). Misato is a tubulin-related gene found in all species. δ-, ε- and η-tubulins are not present in yeasts but are present in *Chytridiomycetes*. (Reproduced from (Zhao et al. 2014). Copyright Springer Nature, reproduced with permission)

of mammals as they are to each other, with about 75% identity between *S. pombe*, *S. cerevisiae* and pig α-tubulins (Schatz et al. 1986a), which may reflect the divergence of the two yeasts from each other over 300 million years ago coupled with a high rate of evolution in *S. pombe* (Sipiczki 2000). This sequence conservation results in core structural and functional features being conserved. Thus mammalian and yeast tubulins can co-assemble to form MTs in vitro (Kilmartin 1981; Schatz et al. 1986a), whilst an α-tubulin gene from *C. albicans* can complement tubulin mutants in *S. cerevisiae* cells (Daly et al. 1997), a human or plant γ-tubulin can function in *S. pombe* cells (Horio and Oakley 1994, 2003), and a chimeric chicken and yeast β-tubulin can incorporate into MTs in cultured mouse cells (Bond et al. 1986). However, despite the basic microtubule assembly function being conserved, significant differences are found in the behaviour of different yeast α-tubulins.

Genetic analysis in both *S. cerevisiae* and *S. pombe* revealed functional differences between the two α-tubulin genes present in each (reviewed in (Huffaker et al. 1987; Yanagida 1987)), with deletion analysis showing that whilst one gene is essential the other is non-essential. *S. cerevisiae* diploids are viable with 50% of normal tubulin α-/β-heterodimer protein levels, though they have an increased sensitivity to microtubule depolymerising drugs (Katz et al. 1990). Modest overexpression of either α-tubulin gene alone or α- plus β-tubulin (but not of β-tubulin alone) is tolerated (Katz et al. 1990); however, high levels of expression in any combination arrest cell division and lead to a loss of viability (Burke et al. 1989; Weinstein and Solomon 1990), which makes overexpression of tubulin genes for tubulin protein production difficult. The single β-tubulin gene is essential in *S. cerevisiae* (Neff et al. 1983), whilst of the α-tubulin genes only *TUB1* is essential, and *TUB3* nulls are viable, though with increased sensitivity to microtubule depolymerising drugs and reduced spore viability (Schatz et al. 1986b). Tub1 α-tubulin protein appears to be incorporated into MTs at a higher level than Tub3 α-tubulin (Schatz et al. 1986a); however, overexpression of *TUB3* can rescue the lethal effect of deletion of *TUB1*, whilst overexpression of *TUB1* can rescue the *TUB3* deletion phenotype (Schatz et al. 1986b). Thus, neither α-tubulin protein isoform has unique functions that are essential for cell viability, and each can be substituted by, at least by an excess of, the other tubulin. Similar results are observed in *S. pombe* (Hiraoka et al. 1984; Toda et al. 1984; Adachi et al. 1986). Adachi et al. (1986) also showed that both α-tubulin proteins are present in *S. pombe* at similar levels in wild type cells, but only expression of the essential *nda2* gene is modulated in cells to compensate for loss of the other non-essential *atb2* α-tubulin, making loss of the *nda2* gene lethal.

Although the genetic analysis in cells suggested no functional difference in the α-tubulin protein isoforms, at least for cell viability, measurements of microtubule dynamics in vitro have found significant differences between the α-tubulin proteins in both *S. cerevisiae* (Bode et al. 2003) and *S. pombe* tubulins (Hussmann et al. 2016). Studies on mammalian and *C. elegans* tubulins also find differences in microtubule dynamics based on the isoform composition of the MTs (Honda et al. 2017; Parker et al. 2018) (Ti et al. 2018) and suggest that at least part of the reason for having multiple isoforms may be to fine-tune the microtubule dynamics in particular cell types (Vemu et al. 2017). Why limited numbers of particular isoforms are maintained in different unicellular yeast species remains unclear. It may, as in mammalian cells, also be a method of fine-tuning the microtubule dynamics, perhaps to enable microtubule dynamics over a wider temperature range than would be possible with a single tubulin isoform. In dimorphic yeast, expression of different tubulin isoforms in the yeast and mycelial forms has been observed in the NCY pathogen *Paracoccidioides brasiliensis* (Silva et al. 2001), where the α2 tubulin is constitutively expressed but the α1 isoform is only expressed in the mycelial form, which may result from different requirements for microtubule dynamics in the two cell types.

8.10 Post-translational Modifications

Tubulins from many species contain a range of post-translational modifications including acetylation, phosphorylation, glutamylation and tyrosination, which can influence microtubule stability and dynamics or the binding of motor proteins such as dyneins to MTs (reviewed by (Song and Brady 2015; Yu et al. 2015; Wloga et al. 2017; Roll-Mecak 2018; Sirajuddin et al. 2014)). However, these modifications appear to be absent from the yeast tubulins studied so far. For example, α-tubulins in many organisms are acetylated on lysine 40 (Sadoul and Khochbin 2016). However, this modification is absent from *S. cerevisiae* and *S. pombe* as their α-tubulins have no lysine residue at position 40 (Alfa and Hyams 1991).

Mammalian tubulins also have a cycle of removal and re-addition of the C-terminal tyrosine residue, which affects the binding of motor and other proteins to MTs and is thought to have a role in directing intracellular traffic along particular MTs (Nieuwenhuis and Brummelkamp 2018). In vitro, using MTs assembled from chimeric yeast and mammalian brain tubulin without a C-terminal tyrosine decreased the velocity of kinesin-1 motor proteins and the rate of microtubule depolymerisation by kinesin-13, though the velocity and processivity of a kinesin-2 motor increased (Sirajuddin et al. 2014). *S. cerevisiae* α-tubulin has a carboxyterminal phenylalanine rather than tyrosine, and the phenylalanine residue is normally present in the yeast cells. Removing the phenylalanine residue affects binding of the clasp protein Bik1 and interferes with microtubule stability and interactions with the cell cortex (Badin-Larçon et al. 2004). *S. pombe* do have a carboxyterminal tyrosine residue in their α-tubulins, but this is also not removed and all of the tubulin in *S. pombe* remains tyrosinated (Alfa and Hyams 1991). Analysis of tubulins from *S. pombe* (Toda et al. 1984; Adachi et al. 1986; Hagan 1998) and *S. cerevisiae* (Schatz et al. 1986a) by 2-D gel electrophoresis shows only the expected number of protein spots corresponding to the tubulin encoding genes, and suggests the absence of extensive modifications causing charge changes such as phosphorylation and glutamylation. This is further supported by mass spectroscopy analysis of purified tubulins from *S. pombe*, which have molecular masses in agreement with those predicated from the gene sequence (Drummond et al. 2011). Although there is no evidence of extensive post-translational modifications in yeasts, low levels of modification have been detected in *S. cerevisiae*. For example, some α-tubulin in *S.*

cerevisiae is modified by addition of the palmate fatty acid to cysteine 374 (Caron et al. 2001). This modification is also found in other membrane-associated tubulins (Wolff 2009), however, its role remains unclear (reviewed in (Song and Brady 2015)). Mutation of Cys374 to Ser, which would prevent palmitoylation, affects astral microtubule length and orientation and mitotic spindle positioning and transition through the bud neck (Caron et al. 2001), but it is unclear if these effects result from the amino acid changes or the loss of palmitoylation. SUMOylated α- and β-tubulins, where the small ubiquitin-like SUMO protein is ligated to proteins through lysine side chains, have also been detected in *S. cerevisiae* (Panse et al. 2004; Wohlschlegel et al. 2004). SUMOylation is thought to have a regulatory role in a range of cellular systems (Han et al. 2018) including the cytoskeleton (Alonso et al. 2015), though the precise function of tubulin SUMOylation remains unclear. Another exception is γ-tubulin, where phosphorylation is important for regulation of microtubule formation (Vogel et al. 2001; Lin et al. 2011). However, at least for the model yeasts, there seems to be less requirement for the complexity of extensive additional post-translational modifications of tubulins compared to multicellular organisms. It remains to be seen if this remains true for the other less well-studied NCYs.

8.11 Purification of Tubulin Protein from Yeast

A soluble extract of mammalian pig brain contains about 26% tubulin protein (Hiller and Weber 1978), which simplifies purification to remove contaminating proteins (Gell et al. 2011). In contrast yeasts such as *S. cerevisiae* and *S. uvarum* (Kilmartin 1981) (Davis et al. 1993) have low concentrations of tubulin (0.05%), and this makes purification of the native tubulin protein more challenging.

Despite continuing reports of tubulin expression in prokaryotic systems and subsequent refolding in vitro (MacDonald et al. 2001, 2003; Oxberry et al. 2001; Jang et al. 2008; Koo et al. 2009; Liu et al. 2018), these methods do not seem generally applicable and they have not been widely adopted. This may be because efficient folding of native tubulin normally depends on chaperones and associated proteins found in eukaryotic cells (Lopez-Fanarraga et al. 2001) including in yeasts (Mori and Toda 2013; Fedyanina et al. 2009; Radcliffe et al. 1999, 2000; Grishchuk and McIntosh 1999; Lacefield and Solomon 2003; Lacefield et al. 2006). So at present the most practical source of tubulin for biochemical experiments is from eukaryotic cells.

Since the initial reports of purifying native *S. cerevisiae* (Clayton et al. 1979; Kilmartin 1981) and *S. pombe* tubulin (Alfa and Hyams 1991), methods have been developed for the large-scale purification of both *S. cerevisiae* (Barnes et al. 1992; Davis et al. 1993) and *S. pombe* (Drummond et al. 2011) tubulins to obtain sufficient tubulin of high enough purity for biochemical and structural analysis. These methods typically require cell cultures of 80L to obtain final yields of about 10 mg of purified tubulin following extensive purification steps. Yield can be improved and purification simplified by increasing the level of expression using an inducible GAL10 promoter to control tubulin expression from multicopy plasmids in *S. cerevisiae* (Bellocq et al. 1992). To simplify purification, His tags have also been attached to the C-terminus of the β-tubulin in *S. cerevisiae* to permit use of affinity chromatography to isolate the tubulin (Gupta Jr et al. 2002, 2003). However, since the His tag was found to interfere with the attachment and function of microtubule motor proteins, an alternative construct with an internal His tag position has been developed (Sirajuddin et al. 2014), where the His tag is inserted into a loop in α-tubulin which tolerates short insertions (Schatz et al. 1987). This loop is also in the interior of the assembled MT and does not interfere with the attachment of motor proteins to the microtubule exterior (Sirajuddin et al. 2014). Since the tubulin genes are essential in yeast, loss of function or dominant deleterious tubulin mutants are difficult to express. One approach is to express untagged tubulin using a combination of plasmid shuffling and a strain containing a β-tubulin which has a highly charged carboxyterminal tag, but can still complement loss of function mutants. This tagged tubulin is then separated from the mutant tubulin of interest by ion exchange chromatography (Uchimura et al. 2006, 2010, 2015). Alternatively, the mutant tubulins can be expressed under the control of an inducible promoter from multicopy plasmids containing the α- and β-tubulins, with the host cell retaining its normal wild type α- and β-tubulins (Johnson et al. 2011; Sirajuddin et al. 2014). In this system the α-tubulin is tagged (Sirajuddin et al. 2014), so pure α-tubulin mutants may be obtained, but β-tubulin mutants will always have a low level of wild type β-tubulin from the genomic tubulin of the host strain. Similar protein expression technology has not been developed in *S. pombe* so *S. cerevisiae* is currently the system of choice.

It is unclear if heterologous tubulins from other species can be expressed using the *S. cerevisiae* system. Attempts to express human tubulin were unsuccessful as it was insoluble, although functional chimeras of *S. cerevisiae* and human tubulin were successfully expressed (Sirajuddin et al. 2014). However, the α-tubulin from *Candida albicans* can complement an *S. cerevisiae* tubulin cold-sensitive α-tubulin mutant (Daly et al. 1997), whilst β-tubulin from *Candida albicans* can complement a β-tubulin gene deletion in *S. cerevisiae* (Smith et al. 1990), so overexpression of functional tubulins from at least some other related NCYs may be possible in *S. cerevisiae*. Alternatively, other host cells may prove more amenable for expression of particular tubulins, as has been found for human tubulin which can be successfully produced in insect cells (Minoura et al. 2013). The NCY *Pichia pastoris* has been used to express tubulin from the amoeba *Reticulomyxa filosa* (Linder et al. 1998), and tubulin from the fungus *Podosphaera xanthii* has been expressed in *S. pombe* (Vela-Corcia et al. 2018), so these also have potential for expression of tubulins from other NCYs.

An alternative approach to obtaining purified NCY tubulin is to use the native tubulin. Although this was potentially difficult in the past, because of the low expression levels and complex purification, the development of a one-step affinity purification method using the tubulin-binding TOG domain of the Stu protein from *S. cerevisiae* (Widlund et al. 2012) has simplified tubulin purification from a variety of native sources including *S. cerevisiae* (Widlund et al. 2012; Munguia et al. 2017) and may prove useful for isolating tubulins from NCYs. However, low levels of tubulin expression in some yeasts and fungi may still prove challenging. Although not possible in all NCYs due to the lethality problems with tubulin overexpression discussed earlier, one approach taken in the filamentous fungus *Aspergillus nidulans* was to introduce additional gene copies to increase the level of tubulin expression and enable purification by conventional methods (Yoon and Oakley 1995). A similar approach was used in *S. pombe* to enhance expression of a single alpha tubulin isoform from the *nda2* gene, by replacing the non-essential *atb2* gene with a second copy of the *nda2* gene (des Georges et al. 2008; Braun et al. 2009). These methods, in combination with TOG purification, may provide an alternative approach when expression in *S. cerevisiae* or another host cell is not possible.

8.12 Yeast Microtubule Polymerisation In Vitro

Purified heterodimers of α- and β-tubulin protein in a buffer containing Mg^{2+} and GTP spontaneously assemble by endwise polymerisation to form MTs. The MTs undergo phases of growth and then a catastrophe event leading to rapid shrinkage, which can then be followed by rescue and repolymerisation of the microtubule in a non-equilibrium process termed dynamic instability (Mitchison and Kirschner 1984). Using suitable microscopy methods such as DIC, dark field or fluorescence to visualise the MTs, rates of growth and shrinkage and the frequency of catastrophe and rescue events can be measured in vitro (Zwetsloot et al. 2018) (Fig. 8.10) and the apparent kinetic rates for the overall assembly process determined (Walker et al. 1988). Although dynamic instability is qualitatively similar in mammalian brain and *S. pombe* tubulin MTs, with both having faster growth at the (presumed) plus end compared to the minus end of the microtubule (Walker et al. 1988; Hussmann et al. 2016), the assembly kinetics are quite distinct, with faster apparent rate constants in mammalian brain tubulin (Walker et al. 1988) compared to *S. pombe* tubulin (Hussmann et al. 2016) or *S. cerevisiae* tubulin MTs (Geyer et al. 2015). The apparent affinity for tubulin heterodimers at the microtubule ends, which determines the minimum concentration for microtubule assembly from seeds, is lower (K_D ~2μM) for *S. pombe* (Hussmann et al. 2016) and *S. cerevisiae* tubulin (0.12 μM) (Geyer et al. 2015) than mammalian brain tubulin (K_D ~ 5μM) (Walker et al. 1988), and assembly of *S. pombe* tubulin or *S. cerevisiae* tubulin (Bode et al. 2003) in vitro is possible at a lower tubulin concentration than with mammalian tubulin.

The estimated total concentration of tubulin in *S. pombe* cells in vivo is 1–5 μM (Hoog et al. 2007; Al-Bassam et al. 2012; Hussmann et al. 2016) but is about tenfold higher in *S. cerevisiae* at ~ 35 μM, though normally only about 3 μM is present as non-polymerised tubulin (Winey and Bloom 2012). Given that both yeast tubulins have similar assembly properties in vitro, the reason for this large concentration difference in vivo is unclear.

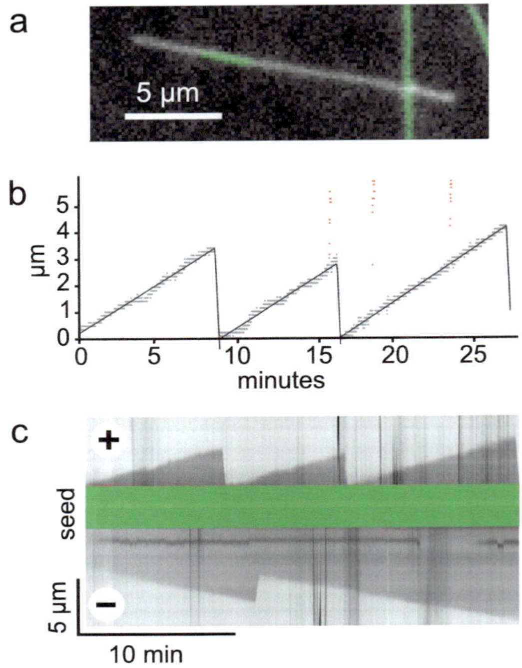

Fig. 8.10 *S. pombe* microtubule dynamics in vitro. (**a**) Merged dark field and fluorescence image of purified unlabelled *S. pombe* tubulin polymerised from stabilised MT seeds containing fluorescently labelled tubulin. This is a single frame from a time-lapse movie recording. (**b**) A plot of microtubule length against time generated from the kymograph plot in (**c**), which is used to determine rates of MT growth and shrinkage and frequency of catastrophic events when growth switches to shrinkage. (**c**) Kymograph plot of the MT length against time, with the position of the stabilised seed shown in the centre. The MT plus (+) and minus (-) ends are indicated. The more dynamic MT plus end grows faster and has more frequent catastrophes than the slower-growing less dynamic MT minus end. Reproduced from (Katsuki et al. 2014) under a Creative Commons Attribution Licence 3.0 (CC BY), with alterations to labelling

MTs assembled from the two isoforms of α-tubulin present in either *S. cerevisiae* (Bode et al. 2003) or *S. pombe* (Hussmann et al. 2016) have different dynamics. In *S. pombe* MTs containing α1 tubulin had slower apparent rate constants for growth and a lower critical concentration, compared to MTs containing both α1 and α2 isoforms (Hussmann et al. 2016). In *S. cerevisiae* both α-tubulin isoforms had similar microtubule growth rates, but the presence of Tub3 α reduced the frequency of catastrophe events and the rate of rapid post-catastrophe depolymerisation, despite Tub3 α only forming ~ 10% of the α-tubulin protein present (Bode et al. 2003). Thus, there are measurable differences in the properties of the tubulin isoforms, although as discussed earlier these do not affect cell viability, at least under laboratory conditions. Recent work suggests that the overall blend of tubulin isoforms in mammalian MTs defines its overall kinetics (Vemu et al. 2017; Parker et al. 2018), and the same appears to be true in yeasts.

Dynamic instability is driven by the hydrolysis of GTP (Hyman et al. 1992; Davis et al. 1994; Sage et al. 1995; Geyer et al. 2015). Upon assembly into MTs GTP in the exchangeable site of β-tubulin undergoes hydrolysis, which is thought to introduce strain into the microtubule lattice and destabilise the MTs. The microtubule is thought to be stabilised by a cap of GTP containing newly added subunits at the ends, with loss of this cap function leading to catastrophe. Studies using *S. pombe* Mal3 protein suggest the cap may be an extended structure with both GTP and GDP. Pi components (Duellberg et al. 2016), though the precise details of how this relates to MT stabilisation and destabilisation are still being determined (Brouhard 2015; Brouhard and Rice 2018). Measurements of GTP and Pi content in *S. cerevisiae* tubulin suggest that the GTP or GDP. Pi cap is much larger than in mammalian brain tubulin, with the microtubule retaining significantly more Pi (Dougherty et al. 1998). The relationship of cap size to the different kinetics found in yeast compared to mammalian MTs is unclear.

These studies show that the yeasts have distinct microtubule dynamics compared to mammalian MTs, despite their extensive similarities in tubulin protein sequences. Tubulin from the filamentous fungus *A. nidulans* also has unusual assembly and stability properties in vitro, which are quite distinct from those of either mammalian or the yeast tubulins, and MT assembly requires high tubulin concentrations and disassembly is resistant to cold and calcium ions, which normally depolymerise MTs (Yoon and Oakley 1995). It will therefore be interesting to determine the extent of variation in tubulin properties amongst the other NCYs and fungi, which at present remain unstudied. The extent to which the requirements of growth in fungi and yeast, such as wide cellular temperature range, or the relative simplicity of unicellular life determine these differences also remains unknown at present.

8.13 Microtubule-Associated Proteins

Microtubule dynamics in vivo are modified from those of MTs assembled from purified tubulin in vitro due to the effect of microtubule-associated proteins (Akhmanova and Steinmetz 2015; Mustyatsa et al. 2017; Goodson and Jonasson 2018), and in vitro studies are a useful complement to genetic and cellular studies to determine the role of individual proteins or groups of proteins (Bieling et al. 2007; Al-Bassam et al. 2012). However, until the isolation of yeast tubulins in sufficient quantity and purity, such studies depended on the use of mammalian brain tubulin and the use of such heterologous tubulins can affect the results obtained (Alonso et al. 2007). Thus Hussmann et al. (Hussmann et al. 2016) observed altered microtubule-promoting activity by the *S. pombe* TOG protein Alp14 on *S. pombe* tubulin compared to earlier studies using brain tubulin (Al-Bassam et al. 2012). Howes et al. found that the *S. cerevisiae* Stu2 +TIP protein would only track the ends of *S. cerevisiae* MTs not human MTs (Howes et al. 2018), whilst Kollman et al. (Kollman et al. 2015) found that nucleating centres were more effective at promoting microtubule polymerisation with tubulin from the same species.

Increasingly, in vitro studies of microtubule-associated proteins are using tubulin from the same species. The effect of these proteins on MTs and their dynamics have been studied using *S. cerevisiae* tubulin together with *S. cerevisiae* +TIP clasp protein Stu1 (Majumdar et al. 2018), Bim1 (Geyer et al. 2015; Howes et al. 2017, 2018) and the kinesin Kip3 depolymerase (Arellano-Santoyo et al. 2017). *S. pombe* tubulin has been used in studies of *S. pombe* Mal3 (des Georges et al. 2008; Katsuki et al. 2009; von Loeffelholz et al. 2017), Tea2 kinesin motor protein (Braun et al. 2009), Alp14 TOG protein (Hussmann et al. 2016) and Cut7 kinesin-5 motor (von Loeffelholz et al. 2019). One recent development is the creation of an *S. cerevisiae* extract to study microtubule dynamics (Bergman et al. 2018), and this may provide a useful intermediate between the simplicity of the in vitro system and full complexity of the cell, particularly for studies of cell cycle-dependent effects on microtubule dynamics. Although, as with tubulin, little is known at present about the biochemical properties of the microtubule-associated proteins in NCYs, it is clear that such studies will require both the MAPs and the tubulin from the same NCY to give meaningful results.

8.14 Structure of Yeast Microtubules

Usually EM images show the growing ends of MTs as flat sheets before they round up to form the microtubule (Atherton et al. 2017, 2018; Guesdon et al. 2016). However, a recent study found that the flared ends of growing MTs (Hoog et al. 2011) contained curved individual protofilaments of αβ-heterodimers joined head to tail rather than flat sheets. This was observed both in vitro and in several cell types in vivo, including *S. cerevisiae* and *S. pombe* cells (McIntosh et al. 2018). These growing microtubule ends are similar to the ends of shrinking MTs (Atherton et al. 2017, 2018; Guesdon et al. 2016; McIntosh et al. 2018).

The tubulin heterodimer lattice of the microtubule contains a variable number of protofilaments when they spontaneously assemble in vitro, but a more regular number in vivo (Atherton et al. 2018). In vivo most cells have MTs with 13 protofilaments (Chaaban and Brouhard 2017), and *S. pombe* cells are also thought to contain 13 protofilament MTs based on the diameter of the MTs (Höög and Antony 2007). In vitro *S. pombe* tubulin MTs have a variable number of protofilaments, but when co-assembled with Mal3 the number is biased to 13 as found in vivo (des Georges et al. 2008; von Loeffelholz et al. 2017), and a similar bias to 13 protofilament microtubules is found when *S. cerevisiae* tubulin is assembled with Bim1 (Howes et al. 2018).

Although high-resolution X-ray crystallography is possible for tubulin complexed with stathmin fragments (Dorleans et al. 2007), imaging of tubulin packed within protofilaments or MTs depends upon cryoEM methods (Downing and Nogales 2010; Nogales 2015; Manka and Moores 2018). Using cryoEM, a structure of *S. pombe* tubulin MTs with Mal3 (*S. pombe* EB protein) bound was determined (des Georges et al. 2008). Development of EM methods with more sensitive imag-

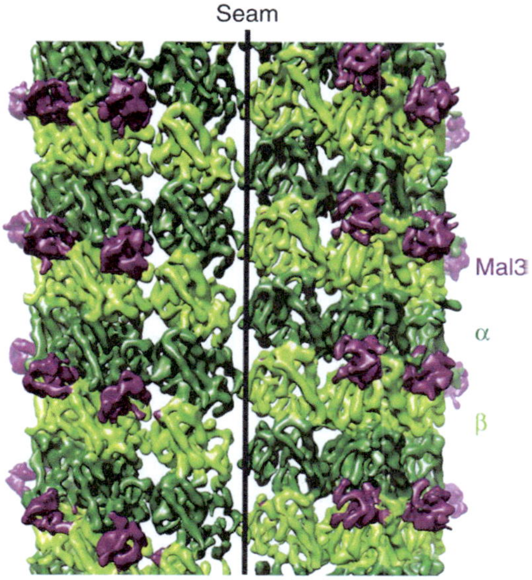

Fig. 8.11 *S. pombe* microtubule structure. 4.6 Å resolution structure determined by cryoEM of the *S. pombe* Mal3 (EB1-like) protein bound to a microtubule assembled from *S. pombe* tubulin. The microtubule has a B lattice structure with mainly α-α and β-β lateral contacts, apart from at the MT seam (indicated by vertical line). Mal3 binds between the protofilaments at the junction of 4 heterodimers, apart from along the seam. Reproduced from (von Loeffelholz et al. 2017), used under a Creative Commons Attribution 4.0 International Licence (CC BY)

ing and particle averaging has led to improvements in the resolution of the images (Manka and Moores 2018), and 4.6 Å resolution images of *S. pombe* tubulin MTs either alone or with bound Mal3 (Fig. 8.11) (von Loeffelholz et al. 2017) or Cut7 kinesin-5 motor (von Loeffelholz et al. 2019) have been created. Structures have also been determined for *S. cerevisiae* tubulin MTs at 3.7–4.0 Å resolution, either alone or with Bim1 (*S. cerevisiae* EB protein) bound (Howes et al. 2017). Unlike *S. pombe* MTs where one Mal3 binds at the interface of the tubulin heterodimers, in *S. cerevisiae* MTs Bim1 can also bind within heterodimers.

CryoEM structures have shown that MTs assembled from mammalian brain tubulin undergo distinct structural changes upon the hydrolysis of GTP to GDP. These changes alter the interface between the heterodimers resulting in an overall compaction of the lattice structure (Alushin et al. 2014; Zhang et al. 2015, 2018). Conversely when non-hydrolysed GTP analogues are used, or MT stabilising drugs such as Taxol are added, the GDP lattice elongates. Thus, there is a good correlation between lattice compaction and microtubule destabilisation in mammalian MTs (Alushin et al. 2014). However, MTs assembled with tubulin from either *S. cerevisiae* (Howes et al. 2017) or *S. pombe* (von Loeffelholz et al. 2017) reveal no large compaction of the lattice on GTP hydrolysis. Furthermore, the structural changes observed were different between the *S. cerevisiae* (Howes et al. 2017) and *S. pombe* MTs (von Loeffelholz et al. 2017), yet both yeast MTs are dynamic.

However, in *S. cerevisiae* MTs binding of Bim1 does induce compaction (Howes et al. 2017), which is not observed with *S. pombe* MTs (von Loeffelholz et al. 2017). These observations emphasise the diversity found in the MTs of yeast tubulins and highlight the current gaps in models of MT dynamic instability mechanisms (Cross 2019). The potential diversity that might be found amongst NCY MTs may help identify the critical features common to all MTs. Certainly the differences in MT mechanism observed between mammalian and yeast tubulins suggest the possibility of developing drugs which might only target a yeast-specific MT mechanism, though much more research will be required to determine if this is a realistic prospect.

8.15 NCYs and Human Disease

Fungal infections are a major health problem in humans, other animals and plants (Fisher et al. 2012; Ghosh et al. 2018), with the NCYs *Candida* and *Cryptococcus* being two of the four fungal genera causing over 90% of the 1.5 million human deaths from fungal infections annually (Brown et al. 2012). The emergence of resistance to existing antifungal drugs makes the identification of suitable drug targets and development of new drugs a major priority for research (Fisher et al. 2018).

NCY infections occur in both humans (Schmidt 2000) and other animals (H_ örrmansdorfer and Bauer 2000; Stewart et al. 2008), often in immunocompromised individuals (Hirschi et al. 2012). Although unusual, even *S. cerevisiae*, despite normally being benign, can be found associated with infections (Cimolai et al. 1987) . There is therefore ongoing interest in developing treatments for these NCY infections (Bassetti et al. 2018).

8.16 Microtubules as Drug Targets

The important role of MTs in the cellular biology of eukaryotes is emphasised by the use of drugs that specifically target MTs, which are active against MTs from a wide variety of species (Dostal and Libusova 2014; Ilan 2018). Such drugs are used in humans to treat cancer and are effective as they can disrupt MT-dependent processes in cells at all stages of the cell cycle not only mitosis (Bates and Eastman 2016). Likewise, in yeasts many cellular processes such as mitosis and meiosis depend on MTs. In *Candida albicans*, which causes candidiasis, the cells undergo a switch from the yeast form to an invasive filamentous form during infection (Li and Nielsen 2017), and in this hyphal form MTs are required for nuclear movement (Finley and Berman 2005).

Benzimidazole fungicides have been used to treat fungal phytopathogens (reviewed by (Davidse 1986)). Benzimidazoles cause a destabilisation and depolymerisation of MTs (Pisano et al. 2000), and *Candida albicans* is sensitive to methyl benzimidazole, which causes aneuploidy (Barton and Gull 1992), and nocodazole, which is effective at inhibiting cell elongation (Yokoyama et al. 1990). However, there are several problems in using such microtubule-targeting drugs in treatment of disease, including drug resistance and toxicity.

8.17 Microtubule Drug Resistance

Resistance to tubulin-targeting drugs has developed in many organisms (Geerts and Gryseels 2000; Giannakakou et al. 2000; Kowal et al. 2016; Jaeger and Carvalho-Costa 2017) including in fungi (reviewed in (Ma and Michailides 2005; Fisher et al. 2018)). Resistance to benzimidazole anti-tubulin fungicides has developed in a wide range of fungal phytopathogens including *Botrytis*, *Gibberella* and *Fusarium* species (Koenraadt et al. 1992; Leroux et al. 2002; Cabanas et al. 2009; Qiu et al. 2012; Liu et al. 2014), which has led to decreased benzimidazole usage (Hollomon 2010). Resistance is usually caused by mutations in the tubulin genes (Foster et al. 1987; Jung and Oakley 1990; Jung et al. 1992; Davidson and Hanson 2006; Chen et al. 2009, 2014; Chen and Zhou 2009; Vela-Corcia et al. 2018). In the model yeasts benzimidazole drugs have been used in studies of the cellular cytoskeleton, and both resistant and supersensitive mutations have been identified in tubulin (Yamamoto 1980; Umesono et al. 1983; Li et al. 1996; Richards et al. 2000).

8.18 Toxicity of Benzimidazole Drugs

Although benzimidazole fungicides and their metabolites such as carbendazim preferentially inhibit assembly of fungal tubulin (Davidse 1986), they can affect other organisms. Benomyl inhibits mammalian brain tubulin polymerisation (Gupta et al. 2004), binds to sheep tubulin and has effects on HeLa cells (Clement et al. 2008), whilst carbendazim inhibits MCF7 breast cancer cells by inhibiting microtubule dynamics, though not by inducing microtubule depolymerisation (Yenjerla et al. 2009). As a result of these effects benzimidazoles have even been suggested as potential treatments for human cancers (Laryea et al. 2010). Benzimidazole (Stringer and Wright 1976; Rombke et al. 2007) and carbendazim (Liu et al. 2012) are also toxic to earthworms, and affect spermatogenesis (Sorour and Larink 2001). Carbendazim also has toxic effects on the testis in birds and mammals (Lim and Miller 1997; Aire 2005) and oogenesis in hamsters (Zuelke and Perreault 1995), and many benzimidazole derivatives also affect targets other than tubulin (Bansal and Silakari 2012). As a result, use of benomyl and carbendazim fungicides is now restricted in many countries (EU pesticides database 2019).

8.19 Development of Benzimidazole Derivatives

Since tubulin-targeting fungicides can be effective in treating infections, there is ongoing interest in developing alternatives which may overcome the limitations of existing compounds. Benzimidazole is being used as the basis to develop new derivatives (reviewed by (Gaba and Mohan 2016)). These are routinely screened against a panel of organisms including *C. albicans*, but it is not always confirmed that they are targeting tubulin alone (Goker et al. 1998; Oren et al. 1999; Ayhan-Kilcigil et al. 1999; Klimesova et al. 2002; Devereux et al. 2004; Ozdemir et al. 2010; Ozkay et al. 2011; Al-Ebaisat 2011; Khabnadideh et al. 2012; Janeczko et al. 2016; El-Gohary and Shaaban 2017; Tuncbilek et al. 2009; Chandrika et al. 2016; El-Gohary and Shaaban 2017; Kaplancikli et al. 2017). Some of these compounds show promising properties, such as lack of effect on human cell lines, that would make them useful as drug treatments for NCY infections (Bauer et al. 2011).

8.20 Development of New Anti-microtubule Drugs from Natural Products

An alternative approach is to develop entirely novel compounds, which are often natural products derived from other organisms, such as the epothilones extracted from myxobacterium. The epothilones bind in the taxane-binding site on β-tubulin, and, unlike paclitaxel-related compounds, stabilise *S. cerevisiae* MTs (Bode et al. 2002) CryoEM molecular structures have been determined of *S. cerevisiae* (Howes et al. 2017) and *S. pombe* MTs (von Loeffelholz et al. 2019) with epothilone bound. Several other compounds have been identified with anti-microtubule activity in NCYs including in *Candida albicans* (Xu et al. 2007). Alteramide A is a macrolactam isolated from a lysobacter that has anti-microtubule activity in *C. albicans* (Ding et al. 2016). Spongistatin 1 is a macrocyclic lactone polyether isolated from a marine sponge, which is active against *Candida*, *Trichosporon* and *Cryptococcus* species, disrupts microtubule formation in *Cryptococcus* and was also shown to be active in vivo against a *Candida* infection in a mouse model system (Pettit et al. 2005).

8.21 Repurposing Existing Drugs

Given the difficulties and costs of developing and gaining approval for novel drugs, there is also interest in the alternative approach of repurposing existing drugs (Corsello et al. 2017). For tubulin-targeting drugs, one issue is cross-species activity as there is natural variation in drug sensitivity in different species (Dostal and Libusova 2014). So, for example, the taxanes, which are widely used in treatment of human cancers, are not effective against *S. cerevisiae* (Gupta Jr et al. 2003) or *Cryptococcus* tubulin (Kopecka and Gabriel 2009; Kopecka 2016). From a range of other drugs normally used for cancer treatment in humans, vincristine was the most effective against *C. albicans* (Routh et al. 2011), methyl benzimidazole-2-ylcarbamate and vincristine were effective against *Cryptococcus* (Kopecka and Gabriel 2009; Kopecka 2016), and dolastatin 10, which binds to human tubulin, was effective against *Candida*, *Cryptococcus* and *Trichosporon* NCYs (Woyke et al. 2001). One obvious difficulty in using cancer drugs is that they were selected for efficacy against human tubulin, which is not a desirable characteristic for a drug aimed at treating an NCY infection in humans. Therefore a more promising approach may be in drugs already used to treat other human pathogens such as anthelminthic benzimidazole drugs, which target MTs (Stojkovic et al. 2009) and are also active against *Cryptococcus* (Cruz et al. 1994; Cruz and Edlind 1997; Nixon et al. 2018). However, these drugs may also have additional non-tubulin targets in *Cryptococcus* (Joffe et al. 2017) and will require further work to determine their mode of action. One of the anthelmintics, flubendazole, has also been shown to be effective in a mouse model of cryptococcal meningitis (Nixon et al. 2018).

8.22 Future Developments

Almost 20 years ago, Kurischko and Swoboda could write that 'it is very astonishing that very little is known about the cytoskeletal proteins of *C. albicans* and *Yarrowia lipolytica*' (Kurischko and Swoboda 2000). Over the past 20 years we have, due in part to developments in genome sequencing, learnt much more about the tubulins and microtubule-associated proteins present in many of the NCYs, though our knowledge of their function is still surprisingly lacking. The same is not true in the model yeasts *S. cerevisiae* and *S. pombe* where there have been many advances in the study of their cytoskeleton both in vivo and in the detailed biochemistry of the purified tubulins and their associated proteins in vitro. The recent development of an *S. cerevisiae* extract system should also enhance the study of cell cycle regulation of the microtubule cytoskeleton in yeasts. High-resolution cryoEM images are now available of both *S. cerevisiae* and *S. pombe* MTs, which together with their biochemical studies make them the best characterised tubulins of any species apart from mammalian brain tubulin. It is therefore all the more striking how little is still currently known about the activity of the tubulin proteins and MTs from the other non-conventional yeasts and fungi. Study of non-mammalian tubulin has long been hampered by various technical problems related in large part to the difficulty of purifying these tubulins. Methods have now been developed that address many of these issues and should now make possible the isolation and study of a much more diverse range of tubulins than was previously practicable. Such studies will be vital in determining the detailed activity of the tubulins from the NCYs and will complement cell biological studies. It is clear that yeast tubulins behave differently to mammalian brain tubulin and that there are also differences in function between the tubulins of *S. cerevisiae* and *S. pombe*. How much more diverse other NCY tubulins may be remains to be determined, but this information will be vital if tubulin-targeting drugs are to have a role in combating NCY disease in humans and other animals.

Acknowledgements We thank Dr. A. Neuner and Dr. M. Yamaguchi for providing electron micrographs.

References

Adachi Y, Toda T, Niwa O, Yanagida M (1986) Differential expressions of essential and nonessential alpha-tubulin genes in Schizosaccharomyces pombe. Mol Cell Biol 6:2168–2178

Adames NR, Cooper JA (2000) Microtubule interactions with the cell cortex causing nuclear movements in Saccharomyces cerevisiae. J Cell Biol 149:863–874

Adams IR, Kilmartin JV (1999) Localization of core spindle pole body (SPB) components during SPB duplication in Saccharomyces cerevisiae. J Cell Biol 145:809–823

Aire TA (2005) Short-term effects of carbendazim on the gross and microscopic features of the testes of Japanese quails (Coturnix coturnix japonica). Anat Embryol 210:43–49

Akhmanova A, Steinmetz MO (2015) Control of microtubule organization and dynamics: two ends in the limelight. Nat Rev Mol Cell Biol 16:711–726

Al-Bassam J, Kim H, Flor-Parra I, Lal N, Velji H, Chang F (2012) Fission yeast Alp14 is a dose-dependent plus end-tracking microtubule polymerase. Mol Biol Cell 23:2878–2890

Alberti-Segui C, Dietrich F, Altmann-Jöhl R, Hoepfner D, Philippsen P (2001) Cytoplasmic dynein is required to oppose the force that moves nuclei towards the hyphal tip in the filamentous ascomycete Ashbya gossypii. J Cell Sci 114:975–986

Al-Ebaisat HS (2011) Synthesis and biological activities of some benzimidazoles derivatives. J Appl Sci Environ Manag 15:451–454

Alfa CE, Hyams JS (1991) Microtubules in the fission yeast Schizosaccharomyces pombe contain only the tyrosinated form of alpha-tubulin. Cell Motil Cytoskeleton 18:86–93

Alonso MC, Drummond DR, Kain S, Hoeng J, Amos L, Cross RA (2007) An ATP gate controls tubulin binding by the tethered head of kinesin-1. Science 316:120–123

Alonso A, Greenlee M, Matts J, Kline J, Davis KJ, Miller RK (2015) Emerging roles of sumoylation in the regulation of actin, microtubules, intermediate filaments, and septins. Cytoskeleton (Hoboken) 72:305–339

Alushin GM, Lander GC, Kellogg EH, Zhang R, Baker D, Nogales E (2014) High-resolution microtubule structures reveal the structural transitions in αβ-tubulin upon GTP hydrolysis. Cell 157:1117–1129

Anders A, Lourenço PCC, Sawin KE (2006) Noncore components of the fission yeast gamma-tubulin complex. Mol Biol Cell 17:5075–5093

Arellano-Santoyo H, Geyer EA, Stokasimov E, Chen GY, Su X, Hancock W, Rice LM, Pellman D (2017) A tubulin binding switch underlies Kip3/Kinesin-8 Depolymerase activity. Dev Cell 42:37–51.e8

Atherton J, Jiang K, Stangier MM, Luo Y, Hua S, Houben K, van Hooff JJE, Joseph AP, Scarabelli G, Grant BJ, Roberts AJ, Topf M, Steinmetz MO, Baldus M, Moores CA, Akhmanova A (2017) A structural model for microtubule minus-end recognition and protection by CAMSAP proteins. Nat Struct Mol Biol 24:931–943

Atherton J, Stouffer M, Francis F, Moores CA (2018) Microtubule architecture in vitro and in cells revealed by cryo-electron tomography. Acta Crystallogr D Struct Biol 74:572–584

Ayhan-Kilcigil G, Tuncbilek M, Altanlar N, Goker H (1999) Synthesis and antimicrobial activity of some new benzimidazole carboxylates and carboxamides. Farmaco 54:562–565

Badin-Larçon AC, Boscheron C, Soleilhac JM, Piel M, Mann C, Denarier E, Fourest-Lieuvin A, Lafanechère L, Bornens M, Job D (2004) Suppression of nuclear oscillations in Saccharomyces cerevisiae expressing Glu tubulin. Proc Natl Acad Sci U S A 101:5577–5582
Bahtz R, Seidler J, Arnold M, Haselmann-Weiss U, Antony C, Lehmann WD, Hoffmann I (2012) GCP6 is a substrate of Plk4 and required for centriole duplication. J Cell Sci 125:486–496
Bansal Y, Silakari O (2012) The therapeutic journey of benzimidazoles: a review. Bioorg Med Chem 20:6208–6236
Banuett F (1995) Genetics of Ustilago maydis, a fungal pathogen that induces tumors in maize. Annu Rev Genet 29:179–208
Banuett F, Quintanilla RH, Reynaga-Peña CG (2008) The machinery for cell polarity, cell morphogenesis, and the cytoskeleton in the Basidiomycete fungus Ustilago maydis-a survey of the genome sequence. Fungal Genet Biol 45(Suppl 1):S3–S14
Bao XX, Spanos C, Kojidani T, Lynch EM, Rappsilber J, Hiraoka Y, Haraguchi T, Sawin KE (2018) Exportin Crm1 is repurposed as a docking protein to generate microtubule organizing centers at the nuclear pore. elife 7:1195
Barnes G, Louie KA, Botstein D (1992) Yeast proteins associated with microtubules in vitro and in vivo. Mol Biol Cell 3:29–47
Barrett JG, Manning BD, Snyder M (2000) The Kar3p kinesin-related protein forms a novel heterodimeric structure with its associated protein Cik1p. Mol Biol Cell 11:2373–2385
Barton RC, Gull K (1992) Isolation, characterization, and genetic analysis of monosomic, aneuploid mutants of Candida albicans. Mol Microbiol 6:171–177
Bassetti M, Righi E, Montravers P, Cornely OA (2018) What has changed in the treatment of invasive candidiasis? A look at the past 10 years and ahead. J Antimicrob Chemother 73:i14–i25
Bates D, Eastman A (2016) Microtubule destabilizing agents: far more than just anti-mitotic anticancer drugs. Br J Clin Pharmacol 83:255–268
Bauer J, Kinast S, Burger-Kentischer A, Finkelmeier D, Kleymann G, Rayyan WA, Schroppel K, Singh A, Jung G, Wiesmuller KH, Rupp S, Eickhoff H (2011) High-throughput-screening-based identification and structure-activity relationship characterization defined (S)-2-(1-aminoisobutyl)-1-(3-chlorobenzyl)benzimidazole as a highly antimycotic agent nontoxic to cell lines. J Med Chem 54:6993–6997
Behrens R, Nurse P (2002) Roles of fission yeast tea1p in the localization of polarity factors and in organizing the microtubular cytoskeleton. J Cell Biol 157:783–793
Bellocq C, Andrey-Tornare I, Paunier Doret AM, Maeder B, Paturle L, Job D, Haiech J, Edelstein SJ (1992) Purification of assembly-competent tubulin from Saccharomyces cerevisiae. Eur J Biochem 210:343–349
Bennett RJ, Miller MG, Chua PR, Maxon ME, Johnson AD (2005) Nuclear fusion occurs during mating in Candida albicans and is dependent on the KAR3 gene. Mol Microbiol 55:1046–1059
Bergman ZJ, Wong J, Drubin DG, Barnes G (2018) Microtubule dynamics regulation reconstituted in budding yeast lysates. J Cell Sci 132:jcs219386
Bieling P, Laan L, Schek H, Munteanu EL, Sandblad L, Dogterom M, Brunner D, Surrey T (2007) Reconstitution of a microtubule plus-end tracking system in vitro. Nature 450:1100–1105
Bode CJ, Gupta ML, Reiff EA, Suprenant KA, Georg GI, Himes RH (2002) Epothilone and paclitaxel: unexpected differences in promoting the assembly and stabilization of yeast microtubules. Biochemistry 41:3870–3874
Bode CJ, Gupta ML, Suprenant KA, Himes RH (2003) The two alpha-tubulin isotypes in budding yeast have opposing effects on microtubule dynamics in vitro. EMBO Rep 4:94–99
Bond JF, Fridovich-Keil JL, Pillus L, Mulligan RC, Solomon F (1986) A chicken-yeast chimeric beta-tubulin protein is incorporated into mouse microtubules in vivo. Cell 44:461–468
Bouissou A, Vérollet C, Sousa A, Sampaio P, Wright M, Sunkel CE, Merdes A, Raynaud-Messina B (2009) {gamma}-Tubulin ring complexes regulate microtubule plus end dynamics. J Cell Biol 187:327–334
Brachat A, Kilmartin JV, Wach A, Philippsen P (1998) Saccharomyces cerevisiae cells with defective spindle pole body outer plaques accomplish nuclear migration via half-bridge-organized microtubules. Mol Biol Cell 9:977–991

Braun M, Drummond DR, Cross RA, McAinsh AD (2009) The kinesin-14 Klp2 organizes microtubules into parallel bundles by an ATP-dependent sorting mechanism. Nat Cell Biol 11:724–730
Britto M, Goulet A, Rizvi S, von Loeffelholz O, Moores CA, Cross RA (2016) Schizosaccharomyces pombe kinesin-5 switches direction using a steric blocking mechanism. Proc Natl Acad Sci U S A 113:E7483–E7489
Brouhard GJ (2015) Dynamic instability 30 years later: complexities in microtubule growth and catastrophe. Mol Biol Cell 26:1207–1210
Brouhard GJ, Rice LM (2018) Microtubule dynamics: an interplay of biochemistry and mechanics. Nat Rev Mol Cell Biol 19:451–463
Brown GD, Denning DW, Gow NAR, Levitz SM, Netea MG, White TC (2012) Hidden killers: human fungal infections. Sci Trans Med 4:165rv13
Browning H, Hayles J, Mata J, Aveline L, Nurse P, McIntosh JR (2000) Tea2p is a kinesin-like protein required to generate polarized growth in fission yeast. J Cell Biol 151:15–28
Browning H, Hackney DD, Nurse P (2003) Targeted movement of cell end factors in fission yeast. Nat Cell Biol 5:812–818
Brunner D, Nurse P (2000) CLIP170-like tip1p spatially organizes microtubular dynamics in fission yeast. Cell 102:695–704
Burke D, Gasdaska P, Hartwell L (1989) Dominant effects of tubulin overexpression in Saccharomyces cerevisiae. Mol Cell Biol 9:1049–1059
Busch KE, Hayles J, Nurse P, Brunner D (2004) Tea2p kinesin is involved in spatial microtubule organization by transporting tip1p on microtubules. Dev Cell 6:831–843
Cabanas R, Castella G, Abarca ML, Bragulat MR, Cabanes FJ (2009) Thiabendazole resistance and mutations in the beta-tubulin gene of Penicillium expansum strains isolated from apples and pears with blue mold decay. FEMS Microbiol Lett 297:189–195
Carazo-Salas RE, Antony C, Nurse P (2005) The kinesin Klp2 mediates polarization of interphase microtubules in fission yeast. Science 309:297–300
Caron JM, Vega LR, Fleming J, Bishop R, Solomon F (2001) Single site alpha-tubulin mutation affects astral microtubules and nuclear positioning during anaphase in Saccharomyces cerevisiae: possible role for palmitoylation of alpha-tubulin. Mol Biol Cell 12:2672–2687
Cavanaugh AM, Jaspersen SL (2017) Big lessons from little yeast: budding and fission yeast centrosome structure, duplication, and function. Annu Rev Genet 51:361–383
Chaaban S, Brouhard GJ (2017) A microtubule bestiary: structural diversity in tubulin polymers. Mol Biol Cell 28:2924–2931
Chacón MR, Delivani P, Tolić IM (2016) Meiotic nuclear oscillations are necessary to avoid excessive chromosome associations. CellReports 17:1632–1645
Chandrika NT, Shrestha SK, Ngo HX, Garneau-Tsodikova S (2016) Synthesis and investigation of novel benzimidazole derivatives as antifungal agents. Bioorg Med Chem 24:3680–3686
Chang F, Martin SG (2009) Shaping fission yeast with microtubules. Cold Spring Harb Perspect Biol 1:a001347–a001347
Chen Y, Zhou MG (2009) Characterization of Fusarium graminearum isolates resistant to both carbendazim and a new fungicide JS399-19. Phytopathology 99:441–446
Chen XP, Yin H, Huffaker TC (1998) The yeast spindle pole body component Spc72p interacts with Stu2p and is required for proper microtubule assembly. J Cell Biol 141:1169–1179
Chen CJ, Yu JJ, Bi CW, Zhang YN, Xu JQ, Wang JX, Zhou MG (2009) Mutations in a beta-tubulin confer resistance of Gibberella zeae to benzimidazole fungicides. Phytopathology 99:1403–1411
Chen Z, Gao T, Liang S, Liu K, Zhou M, Chen C (2014) Molecular mechanism of resistance of Fusarium fujikuroi to benzimidazole fungicides. FEMS Microbiol Lett 357:77–84
Chikashige Y, Ding DQ, Funabiki H, Haraguchi T, Mashiko S, Yanagida M, Hiraoka Y (1994) Telomere-led premeiotic chromosome movement in fission yeast. Science 264:270–273
Chiou J-G, Balasubramanian MK, Lew DJ (2017) Cell polarity in yeast. Annu Rev Cell Dev Biol 33:77–101
Cimolai N, Gill MJ, Church D (1987) Saccharomyces cerevisiae fungemia: case report and review of the literature. Diagn Microbiol Infect Dis 8:113–117

Clayton L, Pogson CI, Gull K (1979) Microtubule proteins in the yeast, Saccharomyces cerevisiae. FEBS Lett 106:67–70

Clement MJ, Rathinasamy K, Adjadj E, Toma F, Curmi PA, Panda D (2008) Benomyl and colchicine synergistically inhibit cell proliferation and mitosis: evidence of distinct binding sites for these agents in tubulin. Biochemistry 47:13016–13025

Corsello SM, Bittker JA, Liu Z, Gould J, McCarren P, Hirschman JE, Johnston SE, Vrcic A, Wong B, Khan M (2017) The drug repurposing hub: a next-generation drug library and information resource. Nat Med 23:405

Cota RR, Teixidó-Travesa N, Ezquerra A, Eibes S, Lacasa C, Roig J, Lüders J (2017) MZT1 regulates microtubule nucleation by linking γTuRC assembly to adapter-mediated targeting and activation. J Cell Sci 130:406–419

Cross RA (2019) Microtubule lattice plasticity. Curr Opin Cell Biol 56:88–93

Cruz MC, Edlind T (1997) Beta-Tubulin genes and the basis for benzimidazole sensitivity of the opportunistic fungus Cryptococcus neoformans. Microbiology 143:2003–2008

Cruz MC, Bartlett MS, Edlind TD (1994) In vitro susceptibility of the opportunistic fungus Cryptococcus neoformans to anthelmintic benzimidazoles. Antimicrob Agents Chemother 38:378–380

Daga RR, Yonetani A, Chang F (2006) Asymmetric microtubule pushing forces in nuclear centering. Curr Biol 16:1544–1550

Daly S, Yacoub A, Dundon W, Mastromei G, Islam K, Lorenzetti R (1997) Isolation and characterization of a gene encoding alpha-tubulin from Candida albicans. Gene 187:151–158

David M, Gabriel M, Kopecká M (2007) Cytoskeletal structures, ultrastructural characteristics and the capsule of the basidiomycetous yeast Cryptococcus laurentii. Antonie Van Leeuwenhoek 92:29–36

Davidse LC (1986) Benzimidazole fungicides: mechanism of action and biological impact. Annu Rev Phytopathol 24:43–65

Davidson RM, Hanson LE (2006) Analysis of b-tubulin gene fragments from Benzimidazole-sensitive and –tolerant Cercospora beticola. J Phytopathol 154:321–328

Davis A, Sage CR, Wilson L, Farrell KW (1993) Purification and biochemical characterization of tubulin from the budding yeast Saccharomyces cerevisiae. Biochemistry 32:8823–8835

Davis A, Sage CR, Dougherty CA, Farrell KW (1994) Microtubule dynamics modulated by guanosine triphosphate hydrolysis activity of beta-tubulin. Science 264:839–842

Delgehyr N, Lopes CSJ, Moir CA, Huisman SM, Segal M (2008) Dissecting the involvement of formins in Bud6p-mediated cortical capture of microtubules in S. cerevisiae. J Cell Sci 121:3803–3814

des Georges A, Katsuki M, Drummond DR, Osei M, Cross RA, Amos LA (2008) Mal3, the Schizosaccharomyces pombe homolog of EB1, changes the microtubule lattice. Nat Struct Mol Biol 15:1102–1108

Devereux M, McCann M, Shea DO, Kelly R, Egan D, Deegan C, Kavanagh K, McKee V, Finn G (2004) Synthesis, antimicrobial activity and chemotherapeutic potential of inorganic derivatives of 2-(4′-thiazolyl)benzimidazole[thiabendazole]: X-ray crystal structures of [Cu(TBZH)2Cl]Cl.H2O.EtOH and TBZH2NO3 (TBZH=thiabendazole). J Inorg Biochem 98:1023–1031

Dhani DK, Goult BT, George GM, Rogerson DT, Bitton DA, Miller CJ, Schwabe JWR, Tanaka K (2013) Mzt1/Tam4, a fission yeast MOZART1 homologue, is an essential component of the γ-tubulin complex and directly interacts with GCP3(Alp6). Mol Biol Cell 24:3337–3349

Dietrich FS, Voegeli S, Brachat S, Lerch A, Gates K, Steiner S, Mohr C, Pöhlmann R, Luedi P, Choi S, Wing RA, Flavier A, Gaffney TD, Philippsen P (2004) The Ashbya gossypii genome as a tool for mapping the ancient Saccharomyces cerevisiae genome. Science 304:304–307

Ding R, McDonald KL, McIntosh JR (1993) Three-dimensional reconstruction and analysis of mitotic spindles from the yeast, Schizosaccharomyces pombe. J Cell Biol 120:141–151

Ding R, West RR, Morphew DM, Oakley BR, McIntosh JR (1997) The spindle pole body of Schizosaccharomyces pombe enters and leaves the nuclear envelope as the cell cycle proceeds. Mol Biol Cell 8:1461–1479

Ding D-Q, Yamamoto A, Haraguchi T, Hiraoka Y (2004) Dynamics of homologous chromosome pairing during meiotic prophase in fission yeast. Dev Cell 6:329–341
Ding Y, Li Y, Li Z, Zhang J, Lu C, Wang H, Shen Y, Du L (2016) Alteramide B is a microtubule antagonist of inhibiting Candida albicans. Biochim Biophys Acta 1860:2097–2106
Dorleans A, Knossow M, Gigant B (2007) Studying drug-tubulin interactions by X-ray crystallography. Methods Mol Med 137:235–243
Dostal V, Libusova L (2014) Microtubule drugs: action, selectivity, and resistance across the kingdoms of life. Protoplasma 251:991–1005
Dougherty CA, Himes RH, Wilson L, Farrell KW (1998) Detection of GTP and Pi in wild-type and mutated yeast microtubules: implications for the role of the GTP/GDP-Pi cap in microtubule dynamics. Biochemistry 37:10861–10865
Downing KH, Nogales E (2010) Cryoelectron microscopy applications in the study of tubulin structure, microtubule architecture, dynamics and assemblies, and interaction of microtubules with motors. Methods Enzymol 483:121–142
Drummond DR, Cross RA (2000) Dynamics of interphase microtubules in Schizosaccharomyces pombe. Curr Biol 10:766–775
Drummond DR, Kain S, Newcombe A, Hoey C, Katsuki M, Cross RA (2011) Purification of tubulin from the fission yeast Schizosaccharomyces pombe. Methods Mol Biol 777:29–55
Duan D, Hnatchuk DJ, Brenner J, Davis D, Allingham JS (2012) Crystal structure of the Kar3-like kinesin motor domain from the filamentous fungus Ashbya gossypii. Proteins 80:1016–1027
Duellberg C, Cade NI, Holmes D, Surrey T (2016) The size of the EB cap determines instantaneous microtubule stability. elife 5:e13470
Edamatsu M (2014) Bidirectional motility of the fission yeast kinesin-5, Cut7. Biochem Biophys Res Commun 446:231–234
Edzuka T, Yamada L, Kanamaru K, Sawada H, Goshima G (2014) Identification of the augmin complex in the filamentous fungus Aspergillus nidulans. PLoS One 9:e101471
Efimov VP, Zhang J, Xiang X (2006) CLIP-170 homologue and NUDE play overlapping roles in NUDF localization in Aspergillus nidulans. Mol Biol Cell 17:2021–2034
Egan MJ, McClintock MA, Reck-Peterson SL (2012) Microtubule-based transport in filamentous fungi. Curr Opin Microbiol 15:637–645
El-Gohary NS, Shaaban MI (2017) Synthesis, antimicrobial, antiquorum-sensing and antitumor activities of new benzimidazole analogs. Eur J Med Chem 137:439–449
Enke C, Zekert N, Veith D, Schaaf C, Konzack S, Fischer R (2007) Aspergillus nidulans Dis1/XMAP215 protein AlpA localizes to spindle pole bodies and microtubule plus ends and contributes to growth directionality. Eukaryot Cell 6:555–562
EU Pesticides Database (2019) http://ec.europa.eu/food/plant/pesticides/eu-pesticides-database
Farache D, Emorine L, Haren L, Merdes A (2018) Assembly and regulation of γ-tubulin complexes. Open Biol 8:170266
Farkasovsky M, Küntzel H (2001) Cortical Num1p interacts with the dynein intermediate chain Pac11p and cytoplasmic microtubules in budding yeast. J Cell Biol 152:251–262
Fedyanina OS, Book AJ, Grishchuk EL (2009) Tubulin heterodimers remain functional for one cell cycle after the inactivation of tubulin-folding cofactor D in fission yeast cells. Yeast 26:235–247
Feierbach B, Verde F, Chang F (2004) Regulation of a formin complex by the microtubule plus end protein tea1p. J Cell Biol 165:697–707
Fink G, Steinberg G (2006) Dynein-dependent motility of microtubules and nucleation sites supports polarization of the tubulin array in the fungus Ustilago maydis. Mol Biol Cell 17:3242–3253
Finley KR, Berman J (2005) Microtubules in Candida albicans hyphae drive nuclear dynamics and connect cell cycle progression to morphogenesis. Eukaryot Cell 4:1697–1711
Finley KR, Bouchonville KJ, Quick A, Berman J (2008) Dynein-dependent nuclear dynamics affect morphogenesis in Candida albicans by means of the Bub2p spindle checkpoint. J Cell Sci 121:466–476

Fischer R, Timberlake WE (1995) Aspergillus nidulans apsA (anucleate primary sterigmata) encodes a coiled-coil protein required for nuclear positioning and completion of asexual development. J Cell Biol 128:485–498
Fisher MC, Henk DA, Briggs CJ, Brownstein JS, Madoff LC, McCraw SL, Gurr SJ (2012) Emerging fungal threats to animal, plant and ecosystem health. Nature 484:186–194
Fisher MC, Hawkins NJ, Sanglard D, Gurr SJ (2018) Worldwide emergence of resistance to antifungal drugs challenges human health and food security. Science 360:739–742
Flor-Parra I, Iglesias-Romero AB, Chang F (2018) The XMAP215 Ortholog Alp14 promotes microtubule nucleation in fission Yeast. Curr Biol: CB 28:1681–1691.e4
Flory MR, Morphew M, Joseph JD, Means AR, Davis TN (2002) Pcp1p, an Spc110p-related calmodulin target at the centrosome of the fission yeast Schizosaccharomyces pombe. Cell Growth Differ: The Mol Biol J Am Assoc Cancer Res 13:47–58
Fong KK, Zelter A, Graczyk B, Hoyt JM, Riffle M, Johnson R, MacCoss MJ, Davis TN (2018) Novel phosphorylation states of the yeast spindle pole body. Biology open 7:bio033647
Foster KE, Burland TG, Gull K (1987) A mutant beta-tubulin confers resistance to the action of benzimidazole-carbamate microtubule inhibitors both in vivo and in vitro. Eur J Biochem 163:449–455
Freitag M (2016) The kinetochore interaction network (KIN) of ascomycetes. Mycologia 108:485–505
Friedman DB, Sundberg HA, Huang EY, Davis TN (1996) The 110-kD spindle pole body component of Saccharomyces cerevisiae is a phosphoprotein that is modified in a cell cycle-dependent manner. J Cell Biol 132:903–914
Friedman DB, Kern JW, Huneycutt BJ, Vinh DB, Crawford DK, Steiner E, Scheiltz D, Yates J, Resing KA, Ahn NG, Winey M, Davis TN (2001) Yeast Mps1p phosphorylates the spindle pole component Spc110p in the N-terminal domain. J Biol Chem 276:17958–17967
Fujita A, Vardy L, Garcia MA, Toda T (2002) A fourth component of the fission yeast gamma-tubulin complex, Alp16, is required for cytoplasmic microtubule integrity and becomes indispensable when gamma-tubulin function is compromised. Mol Biol Cell 13:2360–2373
Fujita I, Yamashita A, Yamamoto M (2015) Dynactin and Num1 cooperate to establish the cortical anchoring of cytoplasmic dynein in S. pombe. J Cell Sci 128:1555–1567
Gaba M, Mohan C (2016) Development of drugs based on imidazole and benzimidazole bioactive heterocycles: recent advances and future directions. Med Chem Res 25:173–210
Gandhi SR, Gierliński M, Mino A, Tanaka K, Kitamura E, Clayton L, Tanaka TU (2011) Kinetochore-dependent microtubule rescue ensures their efficient and sustained interactions in early mitosis. Dev Cell 21:920–933
Gardner MK, Bouck DC, Paliulis LV, Meehl JB, O'Toole ET, Haase J, Soubry A, Joglekar AP, Winey M, Salmon ED, Bloom K, Odde DJ (2008) Chromosome congression by Kinesin-5 motor-mediated disassembly of longer kinetochore microtubules. Cell 135:894–906
Geerts S, Gryseels B (2000) Drug resistance in human helminths: current situation and lessons from livestock. Clin Microbiol Rev 13:207–222
Gell C, Friel CT, Borgonovo B, Drechsel DN, Hyman AA, Howard J (2011) Purification of tubulin from porcine brain. Methods Mol Biol 777:15–28
Gerson-Gurwitz A, Thiede C, Movshovich N, Fridman V, Podolskaya M, Danieli T, Lakämper S, Klopfenstein DR, Schmidt CF, Gheber L (2011) Directionality of individual kinesin-5 Cin8 motors is modulated by loop 8, ionic strength and microtubule geometry. EMBO J 30:4942–4954
Geyer EA, Burns A, Lalonde BA, Ye X, Piedra FA, Huffaker TC, Rice LM (2015) A mutation uncouples the tubulin conformational and GTPase cycles, revealing allosteric control of microtubule dynamics. elife 4:e10113
Geymonat M, Segal M (2017) Intrinsic and extrinsic determinants linking spindle pole fate, spindle polarity, and asymmetric cell division in the budding Yeast S. cerevisiae. Results Probl Cell Differ 61:49–82
Ghosh PN, Fisher MC, Bates KA (2018) Diagnosing emerging fungal threats: a one health perspective. Front Genet 9:article376

Giannakakou P, Gussio R, Nogales E, Downing KH, Zaharevitz D, Bollbuck B, Poy G, Sackett D, Nicolaou KC, Fojo T (2000) A common pharmacophore for epothilone and taxanes: molecular basis for drug resistance conferred by tubulin mutations in human cancer cells. Proc Natl Acad Sci U S A 97:2904–2909

Gibeaux R, Lang C, Politi AZ, Jaspersen SL, Philippsen P, Antony C (2012) Electron tomography of the microtubule cytoskeleton in multinucleated hyphae of Ashbya gossypii. J Cell Sci 125:5830–5839

Gibeaux R, Politi AZ, Nédélec F, Antony C, Knop M (2013) Spindle pole body-anchored Kar3 drives the nucleus along microtubules from another nucleus in preparation for nuclear fusion during yeast karyogamy. Genes Dev 27:335–349

Gibeaux R, Politi AZ, Philippsen P, Nédélec F (2017) Mechanism of nuclear movements in a multinucleated cell. Mol Biol Cell 28:645–660

Gladfelter AS, Hungerbuehler AK, Philippsen P (2006) Asynchronous nuclear division cycles in multinucleated cells. J Cell Biol 172:347–362

Goker H, Tuncbilek M, Ayhan G, Altanlar N (1998) Synthesis of some new benzimidazolecarboxamides and evaluation of their antimicrobial activity. Farmaco 53:415–420

Goodson HV, Jonasson EM (2018) Microtubules and microtubule-associated proteins. Cold Spring Harb Perspect Biol 10:a022608

Grava S, Keller M, Voegeli S, Seger S, Lang C, Philippsen P (2011) Clustering of nuclei in multinucleated hyphae is prevented by dynein-driven bidirectional nuclear movements and microtubule growth control in Ashbya gossypii. Eukaryot Cell 10:902–915

Graziano BR, DuPage AG, Michelot A, Breitsprecher D, Moseley JB, Sagot I, Blanchoin L, Goode BL (2011) Mechanism and cellular function of Bud6 as an actin nucleation-promoting factor. Mol Biol Cell 22:4016–4028

Grishchuk EL, McIntosh JR (1999) Sto1p, a fission yeast protein similar to tubulin folding cofactor E, plays an essential role in mitotic microtubule assembly. J Cell Sci 112:1979–1988

Gruneberg U, Campbell K, Simpson C, Grindlay J, Schiebel E (2000) Nud1p links astral microtubule organization and the control of exit from mitosis. EMBO J 19:6475–6488

Guesdon A, Bazile F, Buey RM, Mohan R, Monier S, Garcia RR, Angevin M, Heichette C, Wieneke R, Tampe R, Duchesne L, Akhmanova A, Steinmetz MO, Chretien D (2016) EB1 interacts with outwardly curved and straight regions of the microtubule lattice. Nat Cell Biol 18:1102–1108

Gunzelmann J, Ruthnick D, Lin TC, Zhang W, Neuner A, Jakle U, Schiebel E (2018) The microtubule polymerase Stu2 promotes oligomerization of the gamma-TuSC for cytoplasmic microtubule nucleation. elife 7:934

Gupta ML Jr, Bode CJ, Thrower DA, Pearson CG, Suprenant KA, Bloom KS, Himes RH (2002) Beta-tubulin C354 mutations that severely decrease microtubule dynamics do not prevent nuclear migration in yeast. Mol Biol Cell 13:2919–2932

Gupta ML Jr, Bode CJ, Georg GI, Himes RH (2003) Understanding tubulin-Taxol interactions: mutations that impart Taxol binding to yeast tubulin. Proc Natl Acad Sci U S A 100:6394–6397

Gupta K, Bishop J, Peck A, Brown J, Wilson L, Panda D (2004) Antimitotic antifungal compound benomyl inhibits brain microtubule polymerization and dynamics and cancer cell proliferation at mitosis, by binding to a novel site in tubulin. Biochemistry 43:6645–6655

H_örrmansdorfer S, Bauer J (2000) Yeast infections in veterinary medicine. In: Ernst JF, Schmidt A (eds) Dimorphism in human pathogenic and apathogenic yeasts. Karger, Basel

Haag C, Steuten B, Feldbrugge M (2015) Membrane-coupled mRNA trafficking in Fungi. Annu Rev Microbiol 69:265–281

Hagan IM (1998) The fission yeast microtubule cytoskeleton. J Cell Sci 111:1603–1612

Hagan I, Yanagida M (1990) Novel potential mitotic motor protein encoded by the fission yeast cut7+ gene. Nature 347:563–566

Han G, Liu B, Zhang J, Zuo W, Morris NR, Xiang X (2001) The Aspergillus cytoplasmic dynein heavy chain and NUDF localize to microtubule ends and affect microtubule dynamics. Curr Biol 11:719–724

Han ZJ, Feng YH, Gu BH, Li YM, Chen H (2018) The post-translational modification, SUMOylation, and cancer (review). Int J Oncol 52:1081–1094

Hara M, Fukagawa T (2018) Kinetochore assembly and disassembly during mitotic entry and exit. Curr Opin Cell Biol 52:73–81

Haren L, Remy M-H, Bazin I, Callebaut I, Wright M, Merdes A (2006) NEDD1-dependent recruitment of the gamma-tubulin ring complex to the centrosome is necessary for centriole duplication and spindle assembly. J Cell Biol 172:505–515

Heil-Chapdelaine RA, Oberle JR, Cooper JA (2000) The cortical protein Num1p is essential for dynein-dependent interactions of microtubules with the cortex. J Cell Biol 151:1337–1344

Hentrich C, Surrey T (2010) Microtubule organization by the antagonistic mitotic motors kinesin-5 and kinesin-14. J Cell Biol 189:465–480

Hepperla AJ, Willey PT, Coombes CE, Schuster BM, Gerami-Nejad M, McClellan M, Mukherjee S, Fox J, Winey M, Odde DJ, O'Toole E, Gardner MK (2014) Minus-end-directed Kinesin-14 motors align antiparallel microtubules to control metaphase spindle length. Dev Cell 31:61–72

Hiller G, Weber K (1978) Radioimmunoassay for tubulin: a quantitative comparison of the tubulin content of different established tissue culture cells and tissues. Cell 14:795–804

Hiraoka Y, Toda T, Yanagida M (1984) The NDA3 gene of fission yeast encodes beta-tubulin: a cold-sensitive nda3 mutation reversibly blocks spindle formation and chromosome movement in mitosis. Cell 39:349–358

Hirschi S, Letscher-Bru V, Pottecher J, Lannes B, Jeung MY, Degot T, Santelmo N, Sabou AM, Herbrecht R, Kessler R (2012) Disseminated Trichosporon mycotoxinivorans, Aspergillus fumigatus, and Scedosporium apiospermum coinfection after lung and liver transplantation in a cystic fibrosis patient. J Clin Microbiol 50:4168–4170

Hoepfner D, Schaerer F, Brachat A, Wach A, Philippsen P (2002) Reorientation of mispositioned spindles in short astral microtubule mutant spc72Delta is dependent on spindle pole body outer plaque and Kar3 motor protein. Mol Biol Cell 13:1366–1380

Hollomon DW (2010) New fungicide modes of action strengthen disease control strategies. Outlooks on Pest Management June:1–5

Honda Y, Tsuchiya K, Sumiyoshi E, Haruta N, Sugimoto A (2017) Tubulin isotype substitution revealed that isotype combination modulates microtubule dynamics in C. elegans embryos. J Cell Sci 130:1652–1661

Höög JL, Antony C (2007) Whole-cell investigation of microtubule cytoskeleton architecture by electron tomography. Methods Cell Biol 79:145–167

Hoog JL, Schwartz C, Noon AT, O'Toole ET, Mastronarde DN, McIntosh JR, Antony C (2007) Organization of interphase microtubules in fission yeast analyzed by electron tomography. Dev Cell 12:349–361

Hoog JL, Huisman SM, Sebo-Lemke Z, Sandblad L, McIntosh JR, Antony C, Brunner D (2011) Electron tomography reveals a flared morphology on growing microtubule ends. J Cell Sci 124:693–698

Horio T, Oakley BR (1994) Human gamma-tubulin functions in fission yeast. J Cell Biol 126:1465–1473

Horio T, Oakley BR (2003) Expression of Arabidopsis gamma-tubulin in fission yeast reveals conserved and novel functions of gamma-tubulin. Plant Physiol 133:1926–1934

Horio T, Oakley BR (2005) The role of microtubules in rapid hyphal tip growth of Aspergillus nidulans. Mol Biol Cell 16:918–926

Horio T, Uzawa S, Jung MK, Oakley BR, Tanaka K, Yanagida M (1991) The fission yeast gamma-tubulin is essential for mitosis and is localized at microtubule organizing centers. J Cell Sci 99:693–700

Horio T, Kimura N, Basaki A, Tanaka Y, Noguchi T, Akashi T, Tanaka K (2002) Molecular and structural characterization of the spindle pole bodies in the fission yeast Schizosaccharomyces japonicus var japonicus. Yeast 19:1335–1350

Howes SC, Geyer EA, LaFrance B, Zhang R, Kellogg EH, Westermann S, Rice LM, Nogales E (2017) Structural differences between yeast and mammalian microtubules revealed by cryo-EM. J Cell Biol 216:2669–2677

Howes SC, Geyer EA, LaFrance B, Zhang R, Kellogg EH, Westermann S, Rice LM, Nogales E (2018) Structural and functional differences between porcine brain and budding yeast microtubules. Cell Cycle 17:278–287
Huffaker TC, Hoyt MA, Botstein D (1987) Genetic analysis of the yeast cytoskeleton. Annu Rev Genet 21:259–284
Huisman SM, Bales OAM, Bertrand M, Smeets MFMA, Reed SI, Segal M (2004) Differential contribution of Bud6p and Kar9p to microtubule capture and spindle orientation in S. cerevisiae. J Cell Biol 167:231–244
Hussmann F, Drummond DR, Peet DR, Martin DS, Cross RA (2016) Alp7/TACC-Alp14/TOG generates long-lived, fast-growing MTs by an unconventional mechanism. Sci Rep 6:20653
Hwang E, Kusch J, Barral Y, Huffaker TC (2003) Spindle orientation in Saccharomyces cerevisiae depends on the transport of microtubule ends along polarized actin cables. J Cell Biol 161:483–488
Hyman AA, Salser S, Drechsel DN, Unwin N, Mitchison TJ (1992) Role of GTP hydrolysis in microtubule dynamics: information from a slowly hydrolyzable analogue, GMPCPP. Mol Biol Cell 3:1155–1167
Ilan Y (2018) Microtubules: from understanding their dynamics to using them as potential therapeutic targets. J Cell Physiol 234:7923–7937
Ishitsuka Y, Savage N, Li Y, Bergs A, Grün N, Kohler D, Donnelly R, Nienhaus GU, Fischer R, Takeshita N (2015) Superresolution microscopy reveals a dynamic picture of cell polarity maintenance during directional growth. Sci Adv 1:e1500947
Izumi N, Fumoto K, Izumi S, Kikuchi A (2008) GSK-3beta regulates proper mitotic spindle formation in cooperation with a component of the gamma-tubulin ring complex, GCP5. J Biol Chem 283:12981–12991
Jaeger LH, Carvalho-Costa FA (2017) Status of benzimidazole resistance in intestinal nematode populations of livestock in Brazil: a systematic review. BMC Vet Res 13:358
Janeczko M, Kazimierczuk Z, Orzeszko A, Niewiadomy A, Krol E, Szyszka R, Maslyk M (2016) In search of the antimicrobial potential of Benzimidazole derivatives. Pol J Microbiol 65:359–364
Jang MH, Kim J, Kalme S, Han JW, Yoo HS, Koo BS, Kim SK, Yoon MY (2008) Cloning, purification, and polymerization of Capsicum annuum recombinant alpha and beta tubulin. Biosci Biotechnol Biochem 72:1048–1055
Janson ME, Setty TG, Paoletti A, Tran PT (2005) Efficient formation of bipolar microtubule bundles requires microtubule-bound gamma-tubulin complexes. J Cell Biol 169:297–308
Joffe LS, Schneider R, Lopes W, Azevedo R, Staats CC, Kmetzsch L, Schrank A, Del Poeta M, Vainstein MH, Rodrigues ML (2017) The anti-helminthic compound Mebendazole has multiple antifungal effects against Cryptococcus neoformans. Front Microbiol 8:535
Joglekar AP, Bouck D, Finley K, Liu X, Wan Y, Berman J, He X, Salmon ED, Bloom KS (2008) Molecular architecture of the kinetochore-microtubule attachment site is conserved between point and regional centromeres. J Cell Biol 181:587–594
Johnson V, Ayaz P, Huddleston P, Rice LM (2011) Design, overexpression, and purification of polymerization-blocked yeast alphabeta-tubulin mutants. Biochemistry 50:8636–8644
Joshi M, Duan D, Drew D, Jia Z, Davis D, Campbell RL, Allingham JS (2013) Kar3Vik1 mechanochemistry is inhibited by mutation or deletion of the C terminus of the Vik1 subunit. J Biol Chem 288:36957–36970
Jung MK, Oakley BR (1990) Identification of an amino acid substitution in the benA, beta-tubulin gene of Aspergillus nidulans that confers thiabendazole resistance and benomyl supersensitivity. Cell Motil Cytoskeleton 17:87–94
Jung MK, Wilder IB, Oakley BR (1992) Amino acid alterations in the benA (beta-tubulin) gene of Aspergillus nidulans that confer benomyl resistance. Cell Motil Cytoskeleton 22:170–174
Kaplancikli ZA, Levent S, Osmaniye D, Saglik BN, Cevik UA, Cavusoglu BK, Ozkay Y, Ilgin S (2017) Synthesis and Anticandidal activity evaluation of new Benzimidazole-Thiazole derivatives. Molecules 22:2051
Kapoor TM, Mayer TU, Coughlin ML, Mitchison TJ (2000) Probing spindle assembly mechanisms with monastrol, a small molecule inhibitor of the mitotic kinesin, Eg5. J Cell Biol 150:975–988

Katsuki M, Drummond DR, Osei M, Cross RA (2009) Mal3 masks catastrophe events in Schizosaccharomyces pombe microtubules by inhibiting shrinkage and promoting rescue. J Biol Chem 284:29246–29250
Katsuki M, Drummond DR, Cross RA (2014) Ectopic A-lattice seams destabilize microtubules. Nat Commun 5:3094
Katz W, Weinstein B, Solomon F (1990) Regulation of tubulin levels and microtubule assembly in Saccharomyces cerevisiae: consequences of altered tubulin gene copy number. Mol Cell Biol 10:5286–5294
Keeling PJ (2003) Congruent evidence from alpha-tubulin and beta-tubulin gene phylogenies for a zygomycete origin of microsporidia. Fungal Genet Biol 38:298–309
Kellis M, Birren BW, Lander ES (2004) Proof and evolutionary analysis of ancient genome duplication in the yeast Saccharomyces cerevisiae. Nature 428:617–624
Khabnadideh S, Rezaei Z, Pakshir K, Zomorodian K, Ghafari N (2012) Synthesis and antifungal activity of benzimidazole, benzotriazole and aminothiazole derivatives. Res Pharm Sci 7:65–72
Kharazi M, Ahmadi B, Makimur K, Farhang A, Kianipour S, Motamedi M, Mirhendi H (2018) Characterization of beta-tubulin DNA sequences within Candida parapsilosis complex. Curr Med Mycol 4:24–29
Khodiyar VK, Maltais LJ, Ruef BJ, Sneddon KM, Smith JR, Shimoyama M, Cabral F, Dumontet C, Dutcher SK, Harvey RJ, Lafanechère L, Murray JM, Nogales E, Piquemal D, Stanchi F, Povey S, Lovering RC (2007) A revised nomenclature for the human and rodent alpha-tubulin gene family. Genomics 90:285–289
Kilmartin JV (1981) Purification of yeast tubulin by self-assembly in vitro. Biochemistry 20:3629–3633
Kilmartin JV, Goh PY (1996) Spc110p: assembly properties and role in the connection of nuclear microtubules to the yeast spindle pole body. EMBO J 15:4592–4602
Klimesova V, Koci J, Pour M, Stachel J, Waisser K, Kaustova J (2002) Synthesis and preliminary evaluation of benzimidazole derivatives as antimicrobial agents. Eur J Med Chem 37:409–418
Knop M, Schiebel E (1997) Spc98p and Spc97p of the yeast gamma-tubulin complex mediate binding to the spindle pole body via their interaction with Spc110p. EMBO J 16:6985–6995
Knop M, Schiebel E (1998) Receptors determine the cellular localization of a gamma-tubulin complex and thereby the site of microtubule formation. EMBO J 17:3952–3967
Koenraadt H, Somerville SC, Jones AL (1992) Characterization of mutations in the beta-tubulin gene of benomyl-resistant field strains of Venturia inaequalis and other plant pathogenic fungi. Phytopathology 82:1348–1354
Kollman JM, Polka JK, Zelter A, Davis TN, Agard DA (2010) Microtubule nucleating gamma-TuSC assembles structures with 13-fold microtubule-like symmetry. Nature 466:879–882
Kollman JM, Greenberg CH, Li S, Moritz M, Zelter A, Fong KK, Fernandez JJ, Sali A, Kilmartin J, Davis TN, Agard DA (2015) Ring closure activates yeast γTuRC for species-specific microtubule nucleation. Nat Struct Mol Biol 22:132–137
Konzack S, Rischitor PE, Enke C, Fischer R (2005) The role of the kinesin motor KipA in microtubule organization and polarized growth of Aspergillus nidulans. Mol Biol Cell 16:497–506
Koo BS, Kalme S, Yeo SH, Lee SJ, Yoon MY (2009) Molecular cloning and biochemical characterization of alpha- and beta-tubulin from potato plants (Solanum tuberosum L.). Plant Physiol Biochem 47:761–768
Kopecka M (2016) Microtubules and actin cytoskeleton of Cryptococcus neoformans as targets for anticancer agents to potentiate a novel approach for new antifungals. Chemotherapy 61:117–121
Kopecka M, Gabriel M (2009) Microtubules and actin cytoskeleton of potentially pathogenic basidiomycetous yeast as targets for antifungals. Chemotherapy 55:278–286
Kopecka M, Gabriel M, Takeo K, Yamaguchi M, Svoboda A, Ohkusu M, Hata K, Yoshida S (2001) Microtubules and actin cytoskeleton in Cryptococcus neoformans compared with ascomycetous budding and fission yeasts. Eur J Cell Biol 80:303–311

Kosco KA, Pearson CG, Maddox PS, Wang PJ, Adams IR, Salmon ED, Bloom K, Huffaker TC (2001) Control of microtubule dynamics by Stu2p is essential for spindle orientation and metaphase chromosome alignment in yeast. Mol Biol Cell 12:2870–2880

Kowal J, Wyrobisz A, Nosal P, Kucharski M, Kaczor U, Skalska M, Sendor P (2016) Benzimidazole resistance in the ovine Haemonchus contortus from southern Poland - coproscopical and molecular findings. Ann Parasitol 62:119–123

Kurihara LJ, Beh CT, Latterich M, Schekman R, Rose MD (1994) Nuclear congression and membrane fusion: two distinct events in the yeast karyogamy pathway. J Cell Biol 126:911–923

Kurischko C, Swoboda RK (2000) Cytoskeletal proteins and morphogenesis in *Candida albicans* and *Yarrowia lipolytica*. In: Ernst JF, Schmidt A (eds) Dimorphism in human pathogenic and apathogenic yeasts. Karger, Basel

Lacefield S, Solomon F (2003) A novel step in beta-tubulin folding is important for heterodimer formation in Saccharomyces cerevisiae. Genetics 165:531–541

Lacefield S, Magendantz M, Solomon F (2006) Consequences of defective tubulin folding on heterodimer levels, mitosis and spindle morphology in Saccharomyces cerevisiae. Genetics 173:635–646

Lai J, Koh CH, Tjota M, Pieuchot L, Raman V, Chandrababu KB, Yang D, Wong L, Jedd G (2012) Intrinsically disordered proteins aggregate at fungal cell-to-cell channels and regulate intercellular connectivity. Proc Natl Acad Sci U S A 109:15781–15786

Lammers LG, Markus SM (2015) The dynein cortical anchor Num1 activates dynein motility by relieving Pac1/LIS1-mediated inhibition. J Cell Biol 211:309–322

Lang C, Grava S, van den Hoorn T, Trimble R, Philippsen P, Jaspersen SL (2010a) Mobility, microtubule nucleation and structure of microtubule-organizing centers in multinucleated hyphae of Ashbya gossypii. Mol Biol Cell 21:18–28

Lang C, Grava S, Finlayson M, Trimble R, Philippsen P, Jaspersen SL (2010b) Structural mutants of the spindle pole body cause distinct alteration of cytoplasmic microtubules and nuclear dynamics in multinucleated hyphae. Mol Biol Cell 21:753–766

Laryea D, Gullbo J, Isaksson A, Larsson R, Nygren P (2010) Characterization of the cytotoxic properties of the benzimidazole fungicides, benomyl and carbendazim, in human tumour cell lines and primary cultures of patient tumour cells. Anti-Cancer Drugs 21:33–42

Leandro-García LJ, Leskelä S, Landa I, Montero-Conde C, López-Jiménez E, Letón R, Cascón A, Robledo M, Rodríguez-Antona C (2010) Tumoral and tissue-specific expression of the major human beta-tubulin isotypes. Cytoskeleton (Hoboken) 67:214–223

Lecland N, Lüders J (2014) The dynamics of microtubule minus ends in the human mitotic spindle. Nat Publ Group 16:770–778

Lee SC, Heitman J (2012) Function of Cryptococcus neoformans KAR7 (SEC66) in karyogamy during unisexual and opposite-sex mating. Eukaryot Cell 11:783–794

Leroux P, Fritz R, Debieu D, Albertini C, Lanen C, Bach J, Gredt M, Chapeland F (2002) Mechanisms of resistance to fungicides in field strains of Botrytis cinerea. Pest Manag Sci 58:876–888

Li Z, Nielsen K (2017) Morphology changes in human fungal pathogens upon interaction with the host. J Fungi (Basel) 3:66

Li J, Katiyar SK, Edlind TD (1996) Site-directed mutagenesis of Saccharomyces cerevisiae beta-tubulin: interaction between residue 167 and benzimidazole compounds. FEBS Lett 385:7–10

Li CR, Wang Y-M, De Zheng X, Liang HY, Tang JCW, Wang Y (2005) The formin family protein CaBni1p has a role in cell polarity control during both yeast and hyphal growth in Candida albicans. J Cell Sci 118:2637–2648

Liakopoulos D, Kusch J, Grava S, Vogel J, Barral Y (2003) Asymmetric loading of Kar9 onto spindle poles and microtubules ensures proper spindle alignment. Cell 112:561–574

Lim J, Miller MG (1997) The role of the benomyl metabolite carbendazim in benomyl-induced testicular toxicity. Toxicol Appl Pharmacol 142:401–410

Lin TC, Gombos L, Neuner A, Sebastian D, Olsen JV, Hrle A, Benda C, Schiebel E (2011) Phosphorylation of the yeast γ-tubulin Tub4 regulates microtubule function. PLoS One 6:e19700

Lin TC, Neuner A, Schlosser YT, Scharf AN, Weber L, Schiebel E (2014) Cell-cycle dependent phosphorylation of yeast pericentrin regulates γ-TuSC-mediated microtubule nucleation. elife 3:e02208
Lin C, Schuster M, Guimaraes SC, Ashwin P, Schrader M, Metz J, Hacker C, Gurr SJ, Steinberg G (2016a) Active diffusion and microtubule-based transport oppose myosin forces to position organelles in cells. Nat Commun 7:11814
Lin TC, Neuner A, Flemming D, Liu P, Chinen T, Jakle U, Arkowitz R, Schiebel E (2016b) MOZART1 and gamma-tubulin complex receptors are both required to turn gamma-TuSC into an active microtubule nucleation template. J Cell Biol 215:823–840
Linder S, Schliwa M, Kube-Granderath E (1998) Expression of Reticulomyxa filosa alpha- and beta-tubulins in Escherichia coli yields soluble and partially correctly folded material. Gene 212:87–94
Liu K, Pan X, Han Y, Tang F, Yu Y (2012) Estimating the toxicity of the weak base carbendazim to the earthworm (Eisenia fetida) using in situ pore water concentrations in different soils. Sci Total Environ 438:26–32
Liu Y, Chen X, Jiang J, Hamada MS, Yin Y, Ma Z (2014) Detection and dynamics of different carbendazim-resistance conferring β-tubulin variants of Gibberella zeae collected from infected wheat heads and rice stubble in China. Pest Manag Sci 70:1228–1236
Liu C, Yao J, Yin J, Xue J, Zhang H (2018) Recombinant α- and β-tubulin from Echinococcus granulosus: expression, purification and polymerization. Parasite 25:62
Lopez-Fanarraga M, Avila J, Guasch A, Coll M, Zabala JC (2001) Review: postchaperonin tubulin folding cofactors and their role in microtubule dynamics. J Struct Biol 135:219–229
Lüders J (2018) XMAP215 joins microtubule nucleation team. Nat Publ Group 20:508–510
Lüders J, Patel UK, Stearns T (2006) GCP-WD is a gamma-tubulin targeting factor required for centrosomal and chromatin-mediated microtubule nucleation. Nat Cell Biol 8:137–147
Ludueña RF (1998) Multiple forms of tubulin: different gene products and covalent modifications. Int Rev Cytol 178:207–275
Lynch EM, Groocock LM, Borek WE, Sawin KE (2014) Activation of the γ-tubulin complex by the Mto1/2 complex. Curr Biol 24:896–903
Ma Z, Michailides TJ (2005) Advances in understanding molecular mechanisms of fungicide resistance and molecular detection of resistant genotypes in phytopathogenic fungi. Crop Prot 24:853–863
MacDonald LM, Armson A, Thompson RC, Reynoldson JA (2001) Expression of Giardia duodenalis beta-tubulin as a soluble protein in Escherichia coli. Protein Expr Purif 22:25–30
MacDonald LM, Armson A, Thompson RC, Reynoldson JA (2003) Characterization of factors favoring the expression of soluble protozoan tubulin proteins in Escherichia coli. Protein Expr Purif 29:117–122
Maddox PS, Bloom KS, Salmon ED (2000) The polarity and dynamics of microtubule assembly in the budding yeast Saccharomyces cerevisiae. Nat Cell Biol 2:36–41
Maekawa H, Schiebel E (2004) Cdk1-Clb4 controls the interaction of astral microtubule plus ends with subdomains of the daughter cell cortex. Genes Dev 18:1709–1724
Maekawa H, Usui T, Knop M, Schiebel E (2003) Yeast Cdk1 translocates to the plus end of cytoplasmic microtubules to regulate bud cortex interactions. EMBO J 22:438–449
Maekawa H, Neuner A, Rüthnick D, Schiebel E, Pereira G, Kaneko Y (2017) Polo-like kinase Cdc5 regulates Spc72 recruitment to spindle pole body in the methylotrophic yeast Ogataea polymorpha. elife 6:e24340
Majumdar S, Kim T, Chen Z, Munyoki S, Tso SC, Brautigam CA, Rice LM (2018) An isolated CLASP TOG domain suppresses microtubule catastrophe and promotes rescue. Mol Biol Cell 29:1359–1375
Manck R, Ishitsuka Y, Herrero S, Takeshita N, Nienhaus GU, Fischer R (2015) Genetic evidence for a microtubule-capture mechanism during polarised growth of Aspergillus nidulans. J Cell Sci 128:3569–3582
Manka SW, Moores CA (2018) Microtubule structure by cryo-EM: snapshots of dynamic instability. Essays Biochem 62:737–751

Manning BD, Barrett JG, Wallace JA, Granok H, Snyder M (1999) Differential regulation of the Kar3p kinesin-related protein by two associated proteins, Cik1p and Vik1p. J Cell Biol 144:1219–1233

Martin R, Walther A, Wendland J (2004) Deletion of the dynein heavy-chain gene DYN1 leads to aberrant nuclear positioning and defective hyphal development in Candida albicans. Eukaryot Cell 3:1574–1588

Martin SG, McDonald WH, Yates JR, Chang F (2005) Tea4p links microtubule plus ends with the formin for 3p in the establishment of cell polarity. Dev Cell 8:479–491

Masuda H, Toda T (2016) Synergistic role of fission yeast Alp16GCP6 and Mzt1MOZART1 in γ-tubulin complex recruitment to mitotic spindle pole bodies and spindle assembly. Mol Biol Cell 27:1753–1763

Masuda H, Mori R, Yukawa M, Toda T (2013) Fission yeast MOZART1/Mzt1 is an essential γ-tubulin complex component required for complex recruitment to the microtubule organizing center, but not its assembly. Mol Biol Cell 24:2894–2906

Mata J, Nurse P (1997) tea1 and the microtubular cytoskeleton are important for generating global spatial order within the fission yeast cell. Cell 89:939–949

McCoy KM, Tubman ES, Claas A, Tank D, Clancy SA, O'Toole ET, Berman J, Odde DJ (2015) Physical limits on kinesin-5-mediated chromosome congression in the smallest mitotic spindles. Mol Biol Cell 26:3999–4014

McDonald KL, O'Toole ET, Mastronarde DN, McIntosh JR (1992) Kinetochore microtubules in PTK cells. J Cell Biol 118:369–383

McIntosh JR, O'Toole E, Morgan G, Austin J, Ulyanov E, Ataullakhanov F, Gudimchuk N (2018) Microtubules grow by the addition of bent guanosine triphosphate tubulin to the tips of curved protofilaments. J Cell Biol 217:2691–2708

Meadows JC, Messin LJ, Kamnev A, Lancaster TC, Balasubramanian MK, Cross RA, Millar JB (2018) Opposing kinesin complexes queue at plus tips to ensure microtubule catastrophe at cell ends. EMBO Rep 19:e46196

Meluh PB, Rose MD (1990) KAR3, a kinesin-related gene required for yeast nuclear fusion. Cell 60:1029–1041

Mieck C, Molodtsov MI, Drzewicka K, van der Vaart B, Litos G, Schmauss G, Vaziri A, Westermann S (2015) Non-catalytic motor domains enable processive movement and functional diversification of the kinesin-14 Kar3. elife 4:1161

Miller RK, Rose MD (1998) Kar9p is a novel cortical protein required for cytoplasmic microtubule orientation in yeast. J Cell Biol 140:377–390

Miller RK, Matheos D, Rose MD (1999) The cortical localization of the microtubule orientation protein, Kar9p, is dependent upon actin and proteins required for polarization. J Cell Biol 144:963–975

Minoura I, Hachikubo Y, Yamakita Y, Takazaki H, Ayukawa R, Uchimura S, Muto E (2013) Overexpression, purification, and functional analysis of recombinant human tubulin dimer. FEBS Lett 587:3450–3455

Mitchison T, Kirschner M (1984) Dynamic instability of microtubule growth. Nature 312:237–242

Mochizuki T (1998) Three-dimensional reconstruction of mitotic cells of *Cryptococcus neoformans* based on serial section electron microscopy. Jpn J Med Mycol 39:123–127

Moore JK, Cooper JA (2010) Coordinating mitosis with cell polarity: molecular motors at the cell cortex. Semin Cell Dev Biol 21:283–289

Mori R, Toda T (2013) The dual role of fission yeast Tbc1/cofactor C orchestrates microtubule homeostasis in tubulin folding and acts as a GAP for GTPase Alp41/Arl2. Mol Biol Cell 24:1713–24, S1

Moseley JB, Sagot I, Manning AL, Xu Y, Eck MJ, Pellman D, Goode BL (2004) A conserved mechanism for Bni1- and mDia1-induced actin assembly and dual regulation of Bni1 by Bud6 and profilin. Mol Biol Cell 15:896–907

Mouriño-Pérez RR, Roberson RW, Bartnicki-Garcia S (2006) Microtubule dynamics and organization during hyphal growth and branching in Neurospora crassa. Fungal Genet Biol 43:389–400

Mouriño-Pérez RR, Linacre-Rojas LP, Román-Gavilanes AI, Lew TK, Callejas-Negrete OA, Roberson RW, Freitag M (2013) MTB-3, a microtubule plus-end tracking protein (+TIP) of Neurospora crassa. PLoS One 8:e70655
Mouriño-Pérez RR, Riquelme M, Callejas-Negrete OA, Galván-Mendoza JI (2016) Microtubules and associated molecular motors in Neurospora crassa. Mycologia 108:515–527
Muller EG, Snydsman BE, Novik I, Hailey DW, Gestaut DR, Niemann CA, O'Toole ET, Giddings TH, Sundin BA, Davis TN (2005) The organization of the core proteins of the yeast spindle pole body. Mol Biol Cell 16:3341–3352
Munguia B, Teixeira R, Veroli V, Melian E, Saldana J, Minteguiaga M, Senorale M, Marin M, Dominguez L (2017) Purification of native M. vogae and H. contortus tubulin by TOG affinity chromatography. Exp Parasitol 182:37–44
Mustyatsa VV, Boyakhchyan AV, Ataullakhanov FI, Gudimchuk NB (2017) EB-family proteins: functions and microtubule interaction mechanisms. Biochemistry (Mosc) 82:791–802
Neff NF, Thomas JH, Grisafi P, Botstein D (1983) Isolation of the beta-tubulin gene from yeast and demonstration of its essential function in vivo. Cell 33:211–219
Nguyen T, Vinh DB, Crawford DK, Davis TN (1998) A genetic analysis of interactions with Spc110p reveals distinct functions of Spc97p and Spc98p, components of the yeast gamma-tubulin complex. Mol Biol Cell 9:2201–2216
Niessing D, Jansen RP, Pohlmann T, Feldbrugge M (2018) mRNA transport in fungal top models. Wiley Interdiscip Rev RNA 9:e1453
Nieuwenhuis J, Brummelkamp TR (2018) The tubulin Detyrosination cycle: function and enzymes. Trends Cell Biol 29:80–92
Niki H (2014) Schizosaccharomyces japonicus: the fission yeast is a fusion of yeast and hyphae. Yeast 31:83–90
Nixon GL, McEntee L, Johnson A, Farrington N, Whalley S, Livermore J, Natal C, Washbourn G, Bibby J, Berry N, Lestner J, Truong M, Owen A, Lalloo D, Charles I, Hope W (2018) Repurposing and reformulation of the Antiparasitic agent Flubendazole for treatment of Cryptococcal meningoencephalitis, a neglected fungal disease. Antimicrob Agents Chemother 62:e01909–e01917
Nogales E (2015) An electron microscopy journey in the study of microtubule structure and dynamics. Protein Sci 24:1912–1919
Oakley CE, Oakley BR (1989) Identification of gamma-tubulin, a new member of the tubulin superfamily encoded by mipA gene of Aspergillus nidulans. Nature 338:662–664
Oakley BR, Paolillo V, Zheng Y (2015) γ-Tubulin complexes in microtubule nucleation and beyond. Mol Biol Cell 26:2957–2962
Olmsted ZT, Riehlman TD, Branca CN, Colliver AG, Cruz LO, Paluh JL (2013) Kinesin-14 Pkl1 targets γ-tubulin for release from the γ-tubulin ring complex (γ-TuRC). Cell Cycle 12:842–848
Olmsted ZT, Colliver AG, Riehlman TD, Paluh JL (2014) Kinesin-14 and kinesin-5 antagonistically regulate microtubule nucleation by γ-TuRC in yeast and human cells. Nat Commun 5:5339
Oren I, Temiz O, Yalcin I, Sener E, Altanlar N (1999) Synthesis and antimicrobial activity of some novel 2,5- and/or 6-substituted benzoxazole and benzimidazole derivatives. Eur J Pharm Sci 7:153–160
O'Toole ET, Winey M, McIntosh JR (1999) High-voltage electron tomography of spindle pole bodies and early mitotic spindles in the yeast *Saccharomyces cerevisiae*. Mol Biol Cell 10:2017–2031
Oxberry ME, Geary TG, Winterrowd CA, Prichard RK (2001) Individual expression of recombinant alpha- and beta-tubulin from Haemonchus contortus: polymerization and drug effects. Protein Expr Purif 21:30–39
Ozdemir I, Temelli N, Gunal S, Demir S (2010) Gold(I) complexes of N-heterocyclic carbene ligands containing benzimidazole: synthesis and antimicrobial activity. Molecules 15:2203–2210
Ozkay Y, Tunali Y, Karaca H, Isikdag I (2011) Antimicrobial activity of a new series of benzimidazole derivatives. Arch Pharm Res 34:1427–1435

Palmer RE, Sullivan DS, Huffaker T, Koshland D (1992) Role of astral microtubules and actin in spindle orientation and migration in the budding yeast, Saccharomyces cerevisiae. J Cell Biol 119:583–593
Panse VG, Hardeland U, Werner T, Kuster B, Hurt E (2004) A proteome-wide approach identifies sumoylated substrate proteins in yeast. J Biol Chem 279:41346–41351
Parker AL, Teo WS, Pandzic E, Vicente JJ, McCarroll JA, Wordeman L, Kavallaris M (2018) Beta-tubulin carboxy-terminal tails exhibit isotype-specific effects on microtubule dynamics in human gene-edited cells. Life Sci Alliance 1:e201800059
Peñalva MA, Zhang J, Xiang X, Pantazopoulou A (2017) Transport of fungal RAB11 secretory vesicles involves myosin-5, dynein/dynactin/p25, and kinesin-1 and is independent of kinesin-3. Mol Biol Cell 28:947–961
Pereira G, Knop M, Schiebel E (1998) Spc98p directs the yeast gamma-tubulin complex into the nucleus and is subject to cell cycle-dependent phosphorylation on the nuclear side of the spindle pole body. Mol Biol Cell 9:775–793
Pereira G, Grueneberg U, Knop M, Schiebel E (1999) Interaction of the yeast gamma-tubulin complex-binding protein Spc72p with Kar1p is essential for microtubule function during karyogamy. EMBO J 18:4180–4195
Pettit RK, Woyke T, Pon S, Cichacz ZA, Pettit GR, Herald CL (2005) In vitro and in vivo antifungal activities of the marine sponge constituent spongistatin. Med Mycol 43:453–463
Pisano C, Battistoni A, Antoccia A, Degrassi F, Tanzarella C (2000) Changes in microtubule organization after exposure to a benzimidazole derivative in Chinese hamster cells. Mutagenesis 15:507–515
Plamann M, Minke PF, Tinsley JH, Bruno KS (1994) Cytoplasmic dynein and actin-related protein Arp1 are required for normal nuclear distribution in filamentous fungi. J Cell Biol 127:139–149
Popchock AR, Tseng K-F, Wang P, Karplus PA, Xiang X, Qiu W (2017) The mitotic kinesin-14 KlpA contains a context-dependent directionality switch. Nat Commun 8:13999
Pruyne D, Legesse-Miller A, Gao L, Dong Y, Bretscher A (2004) Mechanisms of polarized growth and organelle segregation in yeast. Annu Rev Cell Dev Biol 20:559–591
Qiu J, Huang T, Xu J, Bi C, Chen C, Zhou M (2012) Beta-Tubulins in Gibberella zeae: their characterization and contribution to carbendazim resistance. Pest Manag Sci 68:1191–1198
Qiu R, Zhang J, Xiang X (2018) p25 of the dynactin complex plays a dual role in cargo binding and dynactin regulation. J Biol Chem 293:15606–15619
Radcliffe PA, Hirata D, Vardy L, Toda T (1999) Functional dissection and hierarchy of tubulin-folding cofactor homologues in fission yeast [In Process Citation]. Mol Biol Cell 10:2987–3001
Radcliffe PA, Garcia MA, Toda T (2000) The cofactor-dependent pathways for alpha- and beta-tubulins in microtubule biogenesis are functionally different in fission yeast. Genetics 156:93–103
Richards KL, Anders KR, Nogales E, Schwartz K, Downing KH, Botstein D (2000) Structure-function relationships in yeast tubulins. Mol Biol Cell 11:1887–1903
Riquelme M, Fischer R, Bartnicki-García S (2003) Apical growth and mitosis are independent processes in Aspergillus nidulans. Protoplasma 222:211–215
Riquelme M, Yarden O, Bartnicki-Garcia S, Bowman B, Castro-Longoria E, Free SJ, Fleissner A, Freitag M, Lew RR, Mouriño-Pérez R, Plamann M, Rasmussen C, Richthammer C, Roberson RW, Sanchez-Leon E, Seiler S, Watters MK (2011) Architecture and development of the Neurospora crassa hypha – a model cell for polarized growth. Fungal Biol 115:446–474
Riquelme M, Aguirre J, Bartnicki-Garcia S, Braus GH, Feldbrügge M, Fleig U, Hansberg W, Herrera-Estrella A, Kämper J, Kück U, Mouriño-Pérez RR, Takeshita N, Fischer R (2018) Fungal morphogenesis, from the polarized growth of hyphae to complex reproduction and infection structures. Microbiol Mol Biol Rev 82:e00068–e00017
Rischitor PE, Konzack S, Fischer R (2004) The Kip3-like kinesin KipB moves along microtubules and determines spindle position during synchronized mitoses in Aspergillus nidulans hyphae. Eukaryot Cell 3:632–645
Roll-Mecak A (2018) How cells exploit tubulin diversity to build functional cellular microtubule mosaics. Curr Opin Cell Biol 56:102–108

Rombke J, Garcia MV, Scheffczyk A (2007) Effects of the fungicide benomyl on earthworms in laboratory tests under tropical and temperate conditions. Arch Environ Contam Toxicol 53:590–598
Roostalu J, Surrey T (2017) Microtubule nucleation: beyond the template. Nat Rev Mol Cell Biol 18:702–710
Roostalu J, Hentrich C, Bieling P, Telley IA, Schiebel E, Surrey T (2011) Directional switching of the kinesin Cin8 through motor coupling. Science 332:94–99
Roostalu J, Cade NI, Surrey T (2015) Complementary activities of TPX2 and chTOG constitute an efficient importin-regulated microtubule nucleation module. Nat Publ Group 17:1422–1434
Rosenberg JA, Tomlin GC, McDonald WH, Snydsman BE, Muller EG, Yates JR, Gould KL (2006) Ppc89 links multiple proteins, including the septation initiation network, to the core of the fission yeast spindle-pole body. Mol Biol Cell 17:3793–3805
Rosselló CA, Lindström L, Eklund G, Corvaisier M, Kristensson MA (2018) γ-Tubulin–γ-tubulin interactions as the basis for the formation of a meshwork. Int J Mol Sci 19:3245
Rout MP, Kilmartin JV (1990) Components of the yeast spindle and spindle pole body. J Cell Biol 111:1913–1927
Routh MM, Raut JS, Karuppayil SM (2011) Dual properties of anticancer agents: an exploratory study on the in vitro anti-Candida properties of thirty drugs. Chemotherapy 57:372–380
Rüthnick D, Schiebel E (2016) Duplication of the yeast spindle pole body once per cell cycle. Mol Cell Biol 36:1324–1331
Sadoul K, Khochbin S (2016) The growing landscape of tubulin acetylation: lysine 40 and many more. Biochem J 473:1859–1868
Sage CR, Davis AS, Dougherty CA, Sullivan K, Farrell KW (1995) Beta-Tubulin mutation suppresses microtubule dynamics in vitro and slows mitosis in vivo. Cell Motil Cytoskeleton 30:285–300
Sagolla MJ, Uzawa S, Cande WZ (2003) Individual microtubule dynamics contribute to the function of mitotic and cytoplasmic arrays in fission yeast. J Cell Sci 116:4891–4903
Saito TT, Okuzaki D, Nojima H (2006) Mcp5, a meiotic cell cortex protein, is required for nuclear movement mediated by dynein and microtubules in fission yeast. J Cell Biol 173:27–33
Salogiannis J, Reck-Peterson SL (2017) Hitchhiking: a non-canonical mode of microtubule-based transport. Trends Cell Biol 27:141–150
Samejima I, Miller VJ, Groocock LM, Sawin KE (2008) Two distinct regions of Mto1 are required for normal microtubule nucleation and efficient association with the gamma-tubulin complex in vivo. J Cell Sci 121:3971–3980
Samejima I, Miller VJ, Rincon SA, Sawin KE (2010) Fission yeast Mto1 regulates diversity of cytoplasmic microtubule organizing centers. Curr Biol 20:1959–1965
Sampson K, Heath IB (2005) The dynamic behaviour of microtubules and their contributions to hyphal tip growth in Aspergillus nidulans. Microbiology 151:1543–1555
Sanchez AD, Feldman JL (2016) Microtubule-organizing centers: from the centrosome to non-centrosomal sites. Curr Opin Cell Biol 44:93–101
Sanchez-Perez I, Renwick SJ, Crawley K, Karig I, Buck V, Meadows JC, Franco-Sanchez A, Fleig U, Toda T, Millar JBA (2005) The DASH complex and Klp5/Klp6 kinesin coordinate bipolar chromosome attachment in fission yeast. EMBO J 24:2931–2943
Saunders WS, Hoyt MA (1992) Kinesin-related proteins required for structural integrity of the mitotic spindle. Cell 70:451–458
Sawin KE, Tran PT (2006) Cytoplasmic microtubule organization in fission yeast. Yeast 23:1001–1014
Sawin KE, Lourenço PCC, Snaith HA (2004) Microtubule nucleation at non-spindle pole body microtubule-organizing centers requires fission yeast centrosomin-related protein mod20p. Curr Biol 14:763–775
Schaerer F, Morgan G, Winey M, Philippsen P (2001) Cnm67p is a spacer protein of the Saccharomyces cerevisiae spindle pole body outer plaque. Mol Biol Cell 12:2519–2533
Schatz PJ, Solomon F, Botstein D (1986a) Genetically essential and nonessential alpha-tubulin genes specify functionally interchangeable proteins. Mol Cell Biol 6:3722–3733

Schatz PJ, Pillus L, Grisafi P, Solomon F, Botstein D (1986b) Two functional alpha-tubulin genes of the yeast Saccharomyces cerevisiae encode divergent proteins. Mol Cell Biol 6:3711–3721
Schatz PJ, Georges GE, Solomon F, Botstein D (1987) Insertions of up to 17 amino acids into a region of alpha-tubulin do not disrupt function in vivo. Mol Cell Biol 7:3799–3805
Scheffler K, Minnes R, Fraisier V, Paoletti A, Tran PT (2015) Microtubule minus end motors kinesin-14 and dynein drive nuclear congression in parallel pathways. J Cell Biol 209:47–58
Schmidt A (2000) Yeast infections in humans with special emphasis on *Malassezia furfur* and infections caused by "rare" yeast species. In: Ernst JF, Schmidt A (eds) Dimorphism in human pathogenic and apathogenic yeasts. Karger, Basel, pp 36–53
Schoch CL, Aist JR, Yoder OC, Gillian Turgeon B (2003) A complete inventory of fungal kinesins in representative filamentous ascomycetes. Fungal Genet Biol 39:1–15
Schramm C, Elliott S, Shevchenko A, Schiebel E (2000) The Bbp1p-Mps2p complex connects the SPB to the nuclear envelope and is essential for SPB duplication. EMBO J 19:421–433
Schuster M, Kilaru S, Fink G, Collemare J, Roger Y, Steinberg G (2011) Kinesin-3 and dynein cooperate in long-range retrograde endosome motility along a nonuniform microtubule array. Mol Biol Cell 22:3645–3657
Schuster M, Treitschke S, Kilaru S, Molloy J, Harmer NJ, Steinberg G (2012) Myosin-5, kinesin-1 and myosin-17 cooperate in secretion of fungal chitin synthase. EMBO J 31:214–227
Schweiggert J, Stevermann L, Panigada D, Kammerer D, Liakopoulos D (2016) Regulation of a spindle positioning factor at kinetochores by SUMO-targeted ubiquitin ligases. Dev Cell 36:415–427
Segal M, Clarke DJ, Maddox P, Salmon ED, Bloom K, Reed SI (2000) Coordinated spindle assembly and orientation requires Clb5p-dependent kinase in budding yeast. J Cell Biol 148:441–452
Segal M, Bloom K, Reed SI (2002) Kar9p-independent microtubule capture at Bud6p cortical sites primes spindle polarity before bud emergence in Saccharomyces cerevisiae. Mol Biol Cell 13:4141–4155
Seidel C, Moreno-Velásquez SD, Riquelme M, Fischer R (2013) Neurospora crassa NKIN2, a kinesin-3 motor, transports early endosomes and is required for polarized growth. Eukaryot Cell 12:1020–1032
Shapira O, Goldstein A, Al-Bassam J, Gheber L (2017) A potential physiological role for bi-directional motility and motor clustering of mitotic kinesin-5 Cin8 in yeast mitosis. J Cell Sci 130:725–734
Shaw SL, Yeh E, Maddox P, Salmon ED, Bloom K (1997) Astral microtubule dynamics in yeast: a microtubule-based searching mechanism for spindle orientation and nuclear migration into the bud. J Cell Biol 139:985–994
She Z-Y, Yang W-X (2017) Molecular mechanisms of kinesin-14 motors in spindle assembly and chromosome segregation. J Cell Sci 130:2097–2110
Sheeman B, Carvalho P, Sagot I, Geiser J, Kho D, Hoyt MA, Pellman D (2003) Determinants of S. cerevisiae dynein localization and activation: implications for the mechanism of spindle positioning. Curr Biol 13:364–372
Shen J, Li T, Niu X, Liu W, Zheng S, Wang J, Wang F, Cao X, Yao X, Zheng F, Fu C (2019) The J-domain co-chaperone Rsp1 interacts with Mto1 to organize non-centrosomal microtubule assembly. Mol Biol Cell 30:256–267
Silva WP, Soares RBA, Jesuino RSA, Izacc SMS, Felipe MSS, Soares CMA (2001) Expression of alpha tubulin during the dimorphic transition of Paracoccidioides brasiliensis. Med Mycol 39:457–462
Simanis V (2015) Pombe's thirteen - control of fission yeast cell division by the septation initiation network. J Cell Sci 128:1465–1474
Sipiczki M (2000) Where does fission yeast sit on the tree of life. Genome Biol 1:REVIEWS1011
Sirajuddin M, Rice LM, Vale RD (2014) Regulation of microtubule motors by tubulin isotypes and post-translational modifications. Nat Cell Biol 16:335–344
Smith HA, Gorman JW, Koltin Y, Gorman JA (1990) Functional expression of the Candida albicans beta-tubulin gene in Saccharomyces cerevisiae. Gene 90:115–123

Song Y, Brady ST (2015) Post-translational modifications of tubulin: pathways to functional diversity of microtubules. Trends Cell Biol 25:125–136
Sorour J, Larink O (2001) Toxic effects of benomyl on the ultrastructure during spermatogenesis of the earthworm Eisenia fetida. Ecotoxicol Environ Saf 50:180–188
Souès S, Adams IR (1998) SPC72: a spindle pole component required for spindle orientation in the yeast Saccharomyces cerevisiae. J Cell Sci 111:2809–2818
Spang A, Geissler S, Grein K, Schiebel E (1996) Gamma-Tubulin-like Tub4p of Saccharomyces cerevisiae is associated with the spindle pole body substructures that organize microtubules and is required for mitotic spindle formation. J Cell Biol 134:429–441
Sproul LR, Anderson DJ, Mackey AT, Saunders WS, Gilbert SP (2005) Cik1 targets the minus-end kinesin depolymerase kar3 to microtubule plus ends. Curr Biol 15:1420–1427
Steinberg G, Fuchs U (2004) The role of microtubules in cellular organization and endocytosis in the plant pathogen Ustilago maydis. J Microsc 214:114–123
Steinberg G, Schuster M, Theisen U, Kilaru S, Forge A, Martin-Urdiroz M (2012) Motor-driven motility of fungal nuclear pores organizes chromosomes and fosters nucleocytoplasmic transport. J Cell Biol 198:343–355
Stewart AJ, Salazar T, Welles EG (2008) Systemic and pulmonary fungal infections. Compendium on Continuing Education for the Practicing Veterinarian Equine Edition 3:260–272
Stojkovic M, Zwahlen M, Teggi A, Vutova K, Cretu CM, Virdone R, Nicolaidou P, Cobanoglu N, Junghanss T (2009) Treatment response of cystic echinococcosis to benzimidazoles: a systematic review. PLoS Negl Trop Dis 3:e524
Straube A, Brill M, Oakley BR, Horio T, Steinberg G (2003) Microtubule organization requires cell cycle-dependent nucleation at dispersed cytoplasmic sites: polar and perinuclear microtubule organizing centers in the plant pathogen Ustilago maydis. Mol Biol Cell 14:642–657
Stringer A, Wright MA (1976) The toxicity of benomyl and some related 2-substituted benzimidazoles to the earthworm Lumbricus terrestris. Pestic Sci 7:459–464
Suelmann R, Sievers N, Fischer R (1997) Nuclear traffic in fungal hyphae: in vivo study of nuclear migration and positioning in Aspergillus nidulans. Mol Microbiol 25:757–769
Syrovatkina V, Tran PT (2015) Loss of kinesin-14 results in aneuploidy via kinesin-5-dependent microtubule protrusions leading to chromosome cut. Nat Commun 6:7322
Takeshita N, Higashitsuji Y, Konzack S, Fischer R (2008) Apical sterol-rich membranes are essential for localizing cell end markers that determine growth directionality in the filamentous fungus Aspergillus nidulans. Mol Biol Cell 19:339–351
Takeshita N, Mania D, Herrero S, Ishitsuka Y, Nienhaus GU, Podolski M, Howard J, Fischer R (2013) The cell-end marker TeaA and the microtubule polymerase AlpA contribute to microtubule guidance at the hyphal tip cortex of Aspergillus nidulans to provide polarity maintenance. J Cell Sci 126:5400–5411
Takeshita N, Manck R, Grün N, de Vega SH, Fischer R (2014) Interdependence of the actin and the microtubule cytoskeleton during fungal growth. Curr Opin Microbiol 20:34–41
Tanaka K, Kanbe T (1986) Mitosis in the fission yeast Schizosaccharomyces pombe as revealed by freeze-substitution electron microscopy. J Cell Sci 80:253–268
Tang X, Punch JJ, Lee W-L (2009) A CAAX motif can compensate for the PH domain of Num1 for cortical dynein attachment. Cell Cycle 8:3182–3190
Tatebe H, Shimada K, Uzawa S, Morigasaki S, Shiozaki K (2005) Wsh3/Tea4 is a novel cell-end factor essential for bipolar distribution of Tea1 and protects cell polarity under environmental stress in S. pombe. Curr Biol 15:1006–1015
Tay YD, Leda M, Goryachev AB, Sawin KE (2018) Local and global Cdc42 guanine nucleotide exchange factors for fission yeast cell polarity are coordinated by microtubules and the Tea1-Tea4-Pom1 axis. J Cell Sci 131:jcs216580
Ten Hoopen R, Cepeda-García C, Fernández-Arruti R, Juanes MA, Delgehyr N, Segal M (2012) Mechanism for astral microtubule capture by cortical Bud6p priming spindle polarity in S. cerevisiae. Curr Biol 22:1075–1083

Thakur J, Sanyal K (2011) The essentiality of the fungus-specific Dam1 complex is correlated with a one-kinetochore-one-microtubule interaction present throughout the cell cycle, independent of the nature of a centromere. Eukaryot Cell 10:1295–1305

Thawani A, Kadzik RS, Petry S (2018) XMAP215 is a microtubule nucleation factor that functions synergistically with the γ-tubulin ring complex. Nat Publ Group 20:575–585

Thomas GE, Renjith MR, Manna TK (2017) Kinetochore-microtubule interactions in chromosome segregation: lessons from yeast and mammalian cells. Biochem J 474:3559–3577

Ti SC, Alushin GM, Kapoor TM (2018) Human beta-tubulin isotypes can regulate microtubule protofilament number and stability. Dev Cell 47:175

Toda T, Adachi Y, Hiraoka Y, Yanagida M (1984) Identification of the pleiotropic cell division cycle gene NDA2 as one of two different alpha-tubulin genes in Schizosaccharomyces pombe. Cell 37:233–242

Tovey CA, Conduit PT (2018) Microtubule nucleation by γ-tubulin complexes and beyond. Essays Biochem 91:EBC20180028

Tran PT, Marsh L, Doye V, Inoué S, Chang F (2001) A mechanism for nuclear positioning in fission yeast based on microtubule pushing. J Cell Biol 153:397–411

Troxell CL, Sweezy MA, West RR, Reed KD, Carson BD, Pidoux AL, Cande WZ, McIntosh JR (2001) pkl1(+)and klp2(+): two kinesins of the Kar3 subfamily in fission yeast perform different functions in both mitosis and meiosis. Mol Biol Cell 12:3476–3488

Tuncbilek M, Kiper T, Altanlar N (2009) Synthesis and in vitro antimicrobial activity of some novel substituted benzimidazole derivatives having potent activity against MRSA. Eur J Med Chem 44:1024–1033

Uchida M, Mouriño-Pérez RR, Freitag M, Bartnicki-Garcia S, Roberson RW (2008) Microtubule dynamics and the role of molecular motors in Neurospora crassa. Fungal Genet Biol 45:683–692

Uchimura S, Oguchi Y, Katsuki M, Usui T, Osada H, Nikawa J, Ishiwata S, Muto E (2006) Identification of a strong binding site for kinesin on the microtubule using mutant analysis of tubulin. EMBO J 25:5932–5941

Uchimura S, Oguchi Y, Hachikubo Y, Ishiwata S, Muto E (2010) Key residues on microtubule responsible for activation of kinesin ATPase. EMBO J 29:1167–1175

Uchimura S, Fujii T, Takazaki H, Ayukawa R, Nishikawa Y, Minoura I, Hachikubo Y, Kurisu G, Sutoh K, Kon T, Namba K, Muto E (2015) A flipped ion pair at the dynein-microtubule interface is critical for dynein motility and ATPase activation. J Cell Biol 208:211–222

Umesono K, Toda T, Hayashi S, Yanagida M (1983) Cell division cycle genes nda2 and nda3 of the fission yeast Schizosaccharomyces pombe control microtubular organization and sensitivity to anti-mitotic benzimidazole compounds. J Mol Biol 168:271–284

Usui T, Maekawa H, Pereira G, Schiebel E (2003) The XMAP215 homologue Stu2 at yeast spindle pole bodies regulates microtubule dynamics and anchorage. EMBO J 22:4779–4793

Vela-Corcia D, Romero D, de Vicente A, Perez-Garcia A (2018) Analysis of beta-tubulin-carbendazim interaction reveals that binding site for MBC fungicides does not include residues involved in fungicide resistance. Sci Rep 8:7161

Vemu A, Atherton J, Spector JO, Moores CA, Roll-Mecak A (2017) Tubulin isoform composition tunes microtubule dynamics. Mol Biol Cell 28:3564–3572

Venkatram S, Tasto JJ, Feoktistova A, Jennings JL, Link AJ, Gould KL (2004) Identification and characterization of two novel proteins affecting fission yeast gamma-tubulin complex function. Mol Biol Cell 15:2287–2301

Venkatram S, Jennings JL, Link A, Gould KL (2005) Mto2p, a novel fission yeast protein required for cytoplasmic microtubule organization and anchoring of the cytokinetic actin ring. Mol Biol Cell 16:3052–3063

Vogel J, Drapkin B, Oomen J, Beach D, Bloom K, Snyder M (2001) Phosphorylation of gamma-tubulin regulates microtubule organization in budding yeast. Dev Cell 1:621–631

von Loeffelholz O, Venables NA, Drummond DR, Katsuki M, Cross R, Moores CA (2017) Nucleotide- and Mal3-dependent changes in fission yeast microtubules suggest a structural plasticity view of dynamics. Nat Commun 8:2110

von Loeffelholz O, Pena A, Drummond DR, Cross R, Moores CA (2019) 4.5Å Cryo-EM structure of yeast Kinesin-5-microtubule complex reveals a distinct binding footprint and mechanism of drug resistance. J Mol Biol 431:864–872
Walker RA, O'Brien ET, Pryer NK, Soboeiro MF, Voter WA, Erickson HP, Salmon ED (1988) Dynamic instability of individual microtubules analyzed by video light microscopy: rate constants and transition frequencies. J Cell Biol 107:1437–1448
Wasserstrom L, Lengeler KB, Walther A, Wendland J (2013) Molecular determinants of sporulation in Ashbya gossypii. Genetics 195:87–99
Wedlich-Söldner R, Straube A, Friedrich MW, Steinberg G (2002) A balance of KIF1A-like kinesin and dynein organizes early endosomes in the fungus Ustilago maydis. EMBO J 21:2946–2957
Weinstein B, Solomon F (1990) Phenotypic consequences of tubulin overproduction in Saccharomyces cerevisiae: differences between alpha-tubulin and beta-tubulin. Mol Cell Biol 10:5295–5304
Westermann S, Avila-Sakar A, Wang H-W, Niederstrasser H, Wong J, Drubin DG, Nogales E, Barnes G (2005) Formation of a dynamic kinetochore- microtubule interface through assembly of the Dam1 ring complex. Mol Cell 17:277–290
Widlund PO, Podolski M, Reber S, Alper J, Storch M, Hyman AA, Howard J, Drechsel DN (2012) One-step purification of assembly-competent tubulin from diverse eukaryotic sources. Mol Biol Cell 23:4393–4401
Winey M, Bloom K (2012) Mitotic spindle form and function. Genetics 190:1197–1224
Winey M, Mamay CL, O'Toole ET, Mastronarde DN, Giddings TH, McDonald KL, McIntosh JR (1995) Three-dimensional ultrastructural analysis of the *Saccharomyces cerevisiae* mitotic spindle. J Cell Biol 129:1601–1615
Wloga D, Joachimiak E, Fabczak H (2017) Tubulin post-translational modifications and microtubule dynamics. Int J Mol Sci 18:2207
Wohlschlegel JA, Johnson ES, Reed SI, Yates JR (2004) Global analysis of protein sumoylation in Saccharomyces cerevisiae. J Biol Chem 279:45662–45668
Wolfe KH, Shields DC (1997) Molecular evidence for an ancient duplication of the entire yeast genome. Nature 387:708–713
Wolff J (2009) Plasma membrane tubulin. Biochim Biophys Acta 1788:1415–1433
Woodruff JB, Ferreira Gomes B, Widlund PO, Mahamid J, Honigmann A, Hyman AA (2017) The centrosome is a selective condensate that nucleates microtubules by concentrating tubulin. Cell 169:1066–1077
Woyke T, Pettit GR, Winkelmann G, Pettit RK (2001) In vitro activities and postantifungal effects of the potent dolastatin 10 derivative auristatin PHE. Antimicrob Agents Chemother 45:3580–3584
Xiang X (2018) Nuclear movement in fungi. Semin Cell Dev Biol 82:3–16
Xiang X, Fischer R (2004) Nuclear migration and positioning in filamentous fungi. Fungal Genet Biol 41:411–419
Xiang X, Beckwith SM, Morris NR (1994) Cytoplasmic dynein is involved in nuclear migration in Aspergillus nidulans. Proc Natl Acad Sci 91:2100–2104
Xiang X, Roghi C, Morris NR (1995) Characterization and localization of the cytoplasmic dynein heavy chain in Aspergillus nidulans. Proc Natl Acad Sci 92:9890–9894
Xiang X, Han G, Winkelmann DA, Zuo W, Morris NR (2000) Dynamics of cytoplasmic dynein in living cells and the effect of a mutation in the dynactin complex actin-related protein Arp1. Curr Biol 10:603–606
Xiong Y, Oakley BR (2009) In vivo analysis of the functions of gamma-tubulin-complex proteins. J Cell Sci 122:4218–4227
Xu D, Jiang B, Ketela T, Lemieux S, Veillette K, Martel N, Davison J, Sillaots S, Trosok S, Bachewich C, Bussey H, Youngman P, Roemer T (2007) Genome-wide fitness test and mechanism-of-action studies of inhibitory compounds in Candida albicans. PLoS Pathog 3:e92
Yamaguchi M, Biswas SK, Suzuki Y, Furukawa H, Sameshima M, Takeo K (2002) The spindle pole body duplicates in early G1 phase in the pathogenic yeast Exophiala dermatitidis: an ultrastructural study. Exp Cell Res 279:71–79

Yamaguchi M, Biswas SK, Ohkusu M, Takeo K (2009) Dynamics of the spindle pole body of the pathogenic yeast Cryptococcus neoformans examined by freeze-substitution electron microscopy. FEMS Microbiol Lett 296:257–265
Yamamoto M (1980) Genetic analysis of resistant mutants to antimitotic benzimidazole compounds in Schizosaccharomyces pombe. Mol Gen Genet 180:231–234
Yamamoto A, Hiraoka Y (2003) Cytoplasmic dynein in fungi: insights from nuclear migration. J Cell Sci 116:4501–4512
Yamashita A, Yamamoto M (2006) Fission yeast Num1p is a cortical factor anchoring dynein and is essential for the horse-tail nuclear movement during meiotic prophase. Genetics 173:1187–1196
Yanagida M (1987) Yeast tubulin genes. Microbiol Sci 4:115–118
Yao X, Zhang J, Zhou H, Wang E, Xiang X (2012) In vivo roles of the basic domain of dynactin p150 in microtubule plus-end tracking and dynein function. Traffic (Copenhagen, Denmark) 13:375–387
Yenjerla M, Cox C, Wilson L, Jordan MA (2009) Carbendazim inhibits cancer cell proliferation by suppressing microtubule dynamics. J Pharmacol Exp Ther 328:390–398
Yin H, Pruyne D, Huffaker TC, Bretscher A (2000) Myosin V orientates the mitotic spindle in yeast. Nature 406:1013–1015
Yokoyama K, Kaji H, Nishimura K, Miyaji M (1990) The role of microfilaments and microtubules in apical growth and dimorphism of Candida albicans. J Gen Microbiol 136:1067–1075
Yoon Y, Oakley BR (1995) Purification and characterization of assembly-competent tubulin from Aspergillus nidulans. Biochemistry 34:6373–6381
Yu I, Garnham CP, Roll-Mecak A (2015) Writing and Reading the tubulin code. J Biol Chem 290:17163–17172
Zekert N, Fischer R (2009) The Aspergillus nidulans kinesin-3 UncA motor moves vesicles along a subpopulation of microtubules. Mol Biol Cell 20:673–684
Zekert N, Veith D, Fischer R (2010) Interaction of the Aspergillus nidulans microtubule-organizing center (MTOC) component ApsB with gamma-tubulin and evidence for a role of a subclass of peroxisomes in the formation of septal MTOCs. Eukaryot Cell 9:795–805
Zhang J, Li S, Fischer R, Xiang X (2003) Accumulation of cytoplasmic dynein and dynactin at microtubule plus ends in Aspergillus nidulans is kinesin dependent. Mol Biol Cell 14:1479–1488
Zhang R, Alushin GM, Brown A, Nogales E (2015) Mechanistic origin of microtubule dynamic instability and its modulation by EB proteins. Cell 162:849–859
Zhang Y, Gao X, Manck R, Schmid M, Osmani AH, Osmani SA, Takeshita N, Fischer R (2017) Microtubule-organizing centers of Aspergillus nidulans are anchored at septa by a disordered protein. Mol Microbiol 106:285–303
Zhang R, LaFrance B, Nogales E (2018) Separating the effects of nucleotide and EB binding on microtubule structure. Proc Natl Acad Sci U S A 115:E6191–E6200
Zhao Z, Liu H, Luo Y, Zhou S, An L, Wang C, Jin Q, Zhou M, Xu J-R (2014) Molecular evolution and functional divergence of tubulin superfamily in the fungal tree of life. Sci Rep 4:6746
Zuelke KA, Perreault SD (1995) Carbendazim (MBC) disrupts oocyte spindle function and induces aneuploidy in hamsters exposed during fertilization (meiosis II). Mol Reprod Dev 42:200–209
Zwetsloot AJ, Tut G, Straube A (2018) Measuring microtubule dynamics. Essays Biochem 62(6):725–735

Chapter 9
Systematics, Diversity and Ecology of the Genus *Yarrowia* and the Methanol-Assimilating Yeasts

Gábor Péter, Edina Szandra Nagy, and Dénes Dlauchy

Abstract Yeasts have been exploited by humankind for millennia. Currently, numerous yeast species beyond *Saccharomyces cerevisiae* are utilised by the bioindustry, the so-called non-conventional yeasts. Among the non-conventional yeasts, *Yarrowia lipolytica* and some methanol-utilising yeast species occupy an important position. During the past one and half decade, *Yarrowia* has expanded from a monotypic genus including only *Y. lipolytica* to a genus containing 14 species. Similarly, the number of known methanol-utilising yeasts has increased dynamically. Currently, their number exceeds 90, and the majority of them are assigned to the genera *Komagataella*, *Kuraishia*, *Ogataea*, and some of them yet to *Candida*. Methanol-assimilating *Candida* species are related to the genus *Ogataea*.

The most important changes in the systematics of these important yeasts are summarised, and some aspects of their ecology and diversity are discussed as well. Diversity of these yeasts in a few habitats, including a newly recognised one, is briefly introduced. Selective enrichment-based isolation methods for recovering strains of these two exciting yeast groups are discussed as well.

Keywords Biodiversity · Ecology · Methylotrophic yeasts · *Pichia pastoris* · *Komagataella* · *Kuraishia* · *Ogataea* · Phylloplane · Systematics · *Yarrowia lipolytica*

G. Péter (✉) · D. Dlauchy
National Collection of Agricultural and Industrial Microorganisms, Faculty of Food Science, Szent István University, Budapest, Hungary
e-mail: peter.gabor@etk.szie.hu; denes.dlauchy@etk.szie.hu

E. S. Nagy
Department of Brewing and Distilling, Faculty of Food Science, Szent István University, Budapest, Hungary
e-mail: nagy.edina.szandra@etk.szie.hu

A. Sibirny (ed.), *Non-conventional Yeasts: from Basic Research to Application*,
https://doi.org/10.1007/978-3-030-21110-3_9

9.1 Introduction

Yarrowia lipolytica and some methanol-assimilating yeast species, first of all *Ogataea* (*Hansenula*) *polymorpha* and *Komagataella* (*Pichia*) "*pastoris*", have been exploited by biotechnology for a long time. The application of DNA sequencing resulted in unprecedented accuracy and rapidity in yeast identification, creating the basis of more thorough exploration of yeast biodiversity in natural habitats as well as in man-made substrates, e.g. in foods. In several cases, cryptic species have been recognised by molecular techniques. For example, all but one of the currently recognised methylotrophic *Komagataella* species are indistinguishable from each other if standard phenotypic characters are considered. The monotypic genera, like *Yarrowia*, *Komagataella* and *Kuraishia*, have expanded considerably during the latest decades, while *Ogataea*, initially proposed for a few species in 1994 (Yamada et al. 1994) together with the phylogenetically related *Candida* species, has grown to a large clade containing almost 70 species now. Exploring biodiversity by culture-based methods produces also a growing pool of strains of microorganisms, which are available for biotechnological studies. After putting into context by a short historical overview, the most important, recent developments in systematics and exploration of biodiversity of the genus *Yarrowia* and the methanol-assimilating yeasts are summarised, and also some ecological aspects are discussed.

9.2 The Genus *Yarrowia*

Yarrowia has been a monotypic genus for a long time. Its type species, *Y. lipolytica*, is one of the most extensively studied "non-conventional" yeasts. It is widely used in the industry, in molecular biology and in genetic studies. Moreover, it is a model for dimorphism studies. *Y. lipolytica* is capable of producing important metabolites and have an intense secretory activity. This species can be used for such diverse industrial applications that it was called "an industrial workhorse" (Coelho et al. 2010).

Although it may have undesirable contribution to food spoilage, it has several beneficial properties in the food industry. It has desirable effects on the ripening of cheese and meat products, for example it can reduce the ripening time, delay rancidity and preserve the appealing red colour of fermented sausages. Due to the production of aroma compounds, *Y. lipolytica* contributes to superior organoleptic properties. It is a high-quality protein source for feeding livestock, and this species can be used also in the production of biofuels and in bioremediation (Groenewald et al. 2013). *Yarrowia lipolytica* is a biotechnological production host for organic acids (e.g. citric acid, hydrophobic substances such as polyunsaturated fatty acids (PUFAs), aromas and carotenoids), and it is a heterologous production host for pharmaceutical and industrial proteins and enzymes. One of the most important enzymes secreted by this microorganism is lipase which can be exploited for several applications in the detergent, food, pharmaceutical and environmental industries

(Coelho et al. 2010). New developments in the field of some important biotechnological applications of *Y. lipolytica* have recently been summarised (Barth 2013).

In a recent study, *Yarrowia* (*Candida*) *oslonensis* proved to be the best erythritol and mannitol producer among the tested *Yarrowia* strains. Additionally, *Yarrowia* (*Candida*) *hollandica* and *Yarrowia divulgata* are also promising species for sweetener production (Rakicka et al. 2016). Based on the complexity of the *Yarrowia* genus discussed in this chapter, and the promising results with *Yarrowia* species other than *Y. lipolytica*, maybe the expression "a whole stud instead of a single workhorse" is more appropriate.

9.2.1 *History and Systematics of the Genus* Yarrowia

Based on the latest phylogenetic classification, the genus *Yarrowia* is placed in the class Saccharomycetes, in the order Saccharomycetales and in the family Dipodascaceae (Gouliamova et al. 2017). For a long time *Y. lipolytica* has been the only known species in the genus, and its anamorph has been classified in the genus *Candida* as *C. lipolytica*, the type strain of which was isolated in 1921 from expired margarine in the Netherlands. Its examiner, Jacobsen, named it *Torula lipolytica* (nom. nud.) due to its lipolytic activity, although he did not describe the species. In 1928 Harrison assigned the strain to the genus *Mycotorula*, and described it as *M. lipolytica*. Due to its ability to produce pseudohyphae, this species was reassigned to the genus *Candida* in 1942, and a few species described earlier, *Monilia cornealis*, *Proteomyces cornealis*, *Pseudomonilia deformans* and *Candida deformans*, were considered to be conspecific with *C. lipolytica* (Lodder and Kreger-van Rij 1952). In 1949 Kreger-van Rij and Verona described *Candida olea* isolated from olives, which was believed to be different from *C. lipolytica* based on some phenotypic characters, for example, its shorter cells, differently formed pseudohyphae, the absence of true hyphae and because unlike *C. lipolytica*, it formed only a thin membrane in malt extract medium and grew faster in a medium containing ethanol as a sole carbon source (Lodder and Kreger-van Rij 1952). The isolation of four other strains prompted the re-examination of the differences between the two species. Many of these differences were not observed at this time, so it has been concluded that *C. olea* is the same species as *C. lipolytica* (Lodder and Kreger-van Rij 1952).

The teleomorph of *C. lipolytica* was described in 1970 as *Endomycopsis lipolytica* (Wickerham et al. 1970). Shortly after the description of *E. lipolytica* it was pointed out that the genus name *Endomycopsis* is invalid, because it is an obligate synonym of *Saccharomycopsis* (van der Walt and Scott 1971; von Arx 1972). Thus most of the *Endomycopsis* species were reassigned to other genera (van der Walt and Scott 1971), but not *E. lipolytica*. This species was further examined and reclassified in the genus *Saccharomycopsis*, as *S. lipolytica* (Yarrow 1972). However, coenzyme Q-9 system was observed in *S. lipolytica* (Yamada et al. 1976), while other members of the genus possessed coenzyme Q-8 as the predominant ubiquinone, which raised the prospect that this species may be reassigned to another genus. A few years later it was noticed that carbohydrate composition of the cell of *S. lipolytica* differs from that of the majority of the taxa considered because it contained galactose (van der Walt and von Arx 1980). Variable size and shape of the ascospores (spherical, spheroid to ellipsoid, walnut-shaped, saturnoid or angular, navicular, crateriform or galeate with narrow basal brim, apical cap-like appendages, lateral projections may also be present), cylindrical or ovate conidia with scars, positive urease reaction, and high G + C values (49.5–50.2%) were also observed, so this species seemed unique among other *Saccharomycopsis* species. Concerning the above-mentioned differences, the species has been reassigned to the novel monotypic genus *Yarrowia* as *Y. lipolytica*. At that time there was not any other known closely related species, consequently *Y. lipolytica,* the teleomorph of *C. lipolytica* took an isolated position in the Endomycetales (van der Walt and von Arx 1980), currently Saccharomycetales. It was noted that *Y. lipolytica* differs from other yeasts of class Saccharomycetes by several traits, e.g. by a unique nuclear arrangement of rRNA genes (Fournier et al. 1986).

As mentioned above, since 1942 *C. deformans* has been believed to be the same species as *C. lipolytica*. Based on phenotypic characteristics in 1970, it was still considered a variant of *C. lipolytica* as *C. lipolytica* var. *deformans* which was differentiated from *C. lipolytica* var. *lipolytica* by its β-glucoside-assimilating ability (van Uden and Buckley 1970). Based on morphological and physiological characteristics and mating experiments, *C. deformans* was suspected to be the synonym of *Saccharomycopsis* (*Yarrowia*) *lipolytica* (Meyer et al. 1984). In the subsequent edition of *The Yeasts: A Taxonomic Study*, *C. deformans* was still considered to be a synonym of *Y. lipolytica* (Kurtzman 1998).

Since the beginning of the 1990s, considerable efforts have been made to reclassify yeasts based on phylogenetic basis (Bigey et al. 2003; Knutsen et al. 2007; Kurtzman and Robnett 1994, 1995, 1998; Suzuki et al. 1999). These studies provided insight into the phylogenetic placement of *Y. lipolytica* and other species of the *Yarrowia* clade and allowed to recognise the borderlines among species. As a result of comparing SSU rRNA (Suzuki et al. 1999) and partial LSU rRNA gene sequences (Kurtzman and Robnett 1994, 1995, 1998), it was concluded that *Y. lipolytica* is a remote relative of other ascomycete yeasts and in the phylogenetic trees *Y. lipolytica* occupied an isolated position. The phylogenetic placement of *Y. lipolytica* has been variable depending on the taxon sampling and the analysed phylogenetic markers. Multigene analysis based on five loci (Kurtzman and Robnett 2013a) identified *Wickerhamiella domercqiae* as the closest relative of *Y. lipolytica*. In a recent genome-scale reconstruction of phylogeny *Y. lipolytica* and the other *Yarrowia* species included in the analysis formed a clade with *Nadsonia fulvescens* (Shen et al. 2018).

Differences were noticed between the sequences of coding regions of lipase gene of *Y. lipolytica* and *C. deformans*, demonstrating for the first time that they are two distinct species. The extent of nucleotide divergence in the variable D1/D2 region of the large-subunit (26S) ribosomal RNA gene was also compared. This region was found to be sufficiently divergent to confirm that *C. deformans* is a species different from *Y. lipolytica* (Bigey et al. 2003).

In 2004, strains originating from chicken breast and chicken liver were described as *Candida galli*. Phenotypic characteristics of these strains were similar to that of *Y. lipolytica*, however, based on the differences detected along the partial sequences of the small (18S) and large (26S) subunit rRNA genes they obviously represented another species. *C. galli* also possesses some physiological characteristics, which enable differentiation from *Y. lipolytica* (Péter et al. 2004).

In order to re-evaluate the relationship among the members of the *Yarrowia* clade, several strains of *Y. lipolytica* and related species and some unidentified strains resembling *Y. lipolytica* were studied (Knutsen et al. 2007). Profound examinations were carried out on the strains using different techniques, for example, PCR fingerprinting, sequence analysis of internal transcribed spacer (ITS) and D1/D2 region, and mating experiments; furthermore, physiology and morphology of the strains were also studied. Two previously described species, *C. galli* and *C. yakushimensis* (nom. inval.), were also involved in the examinations, and as a result three novel species, *Candida alimentaria*, *Candida hollandica* and *Candida oslonensis*, were described. The possibility of the differentiation based on the divergences in the sequences of ITS and D1/D2 region of LSU rRNA gene of the above-mentioned species had been demonstrated. Significant differences were found among the above-mentioned sequences of *Y. lipolytica* and *C. deformans*, so the earlier conclusion (Bigey et al. 2003) had been confirmed: these yeasts represent two different species. In the latest edition of *The Yeasts: A Taxonomic Study* (Kurtzman et al. 2011), they were already treated as two different species.

The next member added to the genus was *Candida phangngensis* recovered from estuarine water in mangrove forests in Thailand (Limtong et al. 2008). Based on D1/D2 and SSU rRNA gene sequences, it was found to be distinct from the described species of the genus.

Mating and ascosporulation among *C. deformans* strains were observed a few years later, and *Y. deformans* the teleomorph of *C. deformans* was described as well as *Y. yakushimensis*, an anamorphic member of the genus (Groenewald and Smith 2013). *Candida hispaniensis* was also referred to as a member of the *Yarrowia* clade (Kurtzman et al. 2011; Kurtzman 2005).

Yarrowia keelungensis was described based on a strain isolated from sea-surface microlayer. This species successfully degrades petroleum and hydrocarbons (Chang et al. 2013).

Additional three novel species belonging to the genus *Yarrowia* were described in 2013 and 2014 as *Y. divulgata*, *Y. porcina* and *Y. bubula*. Strains originated mainly from minced meat; other isolation sources were river sediment and a bacon processing plant. Ascospores were found in the mixture of some *Y. porcina* strains, so this species became the third teleomorphic member of the genus. The ascospores of *Y. porcina* are embedded in a capsular material, which is a unique characteristic among *Yarrowia* species (Nagy et al. 2013, 2014).

Seven strains from the gut of *Parophonus hirsutulus* (Carabidae) were examined and based on ITS and D1/D2 sequence comparisons it was concluded to be a novel species and was described as *Y. parophonii* (Gouliamova et al. 2017). The DNA sequence divergence between D1/D2 regions of the type strains of *Y. parophonii* and its closest relative in terms of pair-wise sequence similarities, *Y. oslonensis* (16 substitutions) supports the proposal of the new species. However, due to minor variations between the ITS sequences of *Y. parophonii* and *Y. oslonensis* (1 substitution and 1 gap) the status of *Y. parophonii* is worth further studying (see additional discussion below).

Very recently, two strains of a novel yeast species have been isolated from traditional Chinese sauerkraut samples. Phylogenetic analysis based on the concatenated sequences of the ITS regions and D1/D2 domain of the large subunit rRNA gene showed that these strains belong to the *Yarrowia* clade, and they were described as *Y. brassicae* (Liu et al. 2018).

C. bentonensis has earlier been regarded also as a member of the *Yarrowia* clade (Kurtzman et al. 2011; Kurtzman 2005), although its physiological characteristics significantly differ from that of the other members of the clade. Phylogenetic analysis of ITS and D1/D2 sequences of *Y. lipolytica* and related species suggested that *C. bentonensis* is not the member of the *Yarrowia* clade (Nagy et al. 2013).

The Amsterdam declaration on fungal nomenclature recognized the need for an orderly transition to a single-name nomenclatural system for all fungi (Hawksworth et al. 2011). This idea has also been supported by the International Code of Nomenclature for Algae, Fungi and Plants (McNeill et al. 2012). It means that dual nomenclature of pleomorphic fungi has discontinued. One fungus can only have one name (Hawksworth 2011). This had a huge impact on fungal nomenclature and already had an effect on the systematics of the *Yarrowia* clade, as most of the *Candida* species nested in the *Yarrowia* clade has already been transferred to genus *Yarrowia* (Groenewald et al. 2013; Gouliamova et al. 2017) and also several new anamorphic members of the *Yarrowia* clade were already described as *Yarrowia* species.

The phylogenetic relationships among the currently known *Yarrowia* species deduced from the analysis of ITS and LSU D1/D2 sequences are shown in Fig. 9.1.

9.2.2 *Possibilities of Phenotypic Differentiation of the Members of the Genus* Yarrowia

Urease test is a commonly used tool for the discrimination between yeasts of the phyla *Ascomycota* and *Basidiomycota*. The test is negative in case of Ascomycota, except of the Taphrynomycotina subphylum. For a long time *Y. lipolytica* has been regarded as an exception, since it is an ascomycete but still gave a positive urease reaction (Kurtzman 1998). However, in 1987 negative urease reaction was reported from a study examining the type strain of *Y. lipolytica* (Booth and Vishniac 1987). This observation was confirmed and extended to other strains of the species as well, and it was also demonstrated that the colour change observed in Christensen's urea agar is a false-positive reaction. Instead of actual urease activity, it is the result of a non-specific alkalisation of the medium, which can be triggered even without adding urea (Péter and Deák 1991).

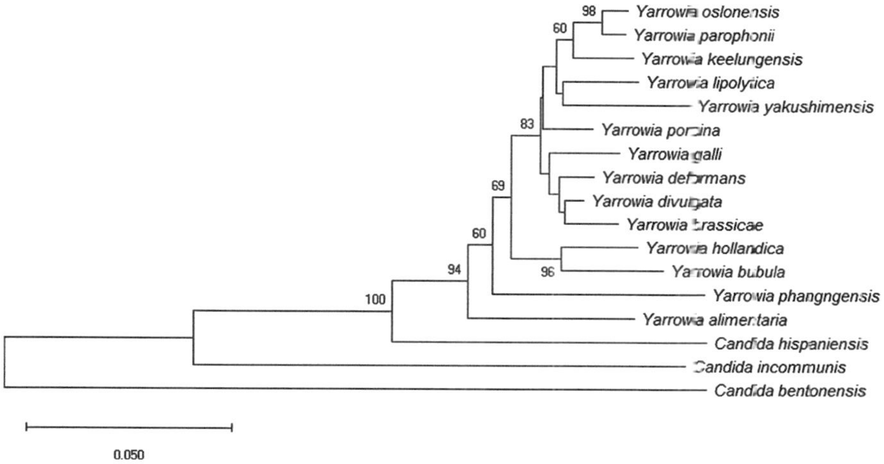

Fig. 9.1 Phylogenetic relationships among the members of the *Yarrowia* clade determined from Neighbour-Joining analysis of concatenated sequences of the ITS regions and LSU rRNA gene D1/D2 domain
The evolutionary history was inferred using the Neighbor-Joining algorithm (Saitou and Nei 1987). Bootstrap percentages (Felsenstein 1985) (1000 replicates) exceeding 50% are given at next to the branches. Bar, 5% nucleotide sequence divergence. The evolutionary distances were computed using the Tamura 3-parameter method (Tamura 1992). All positions containing gaps and missing data were eliminated. There were a total of 562 positions in the final dataset. *Candida bentonensis* was used as the outgroup species. Evolutionary analysis was conducted in MEGA X (Kumar et al. 2018)

Based on the information in the description of *Y.* (*C.*) *galli* (Péter et al. 2004) and in the latest edition of The Yeasts: A Taxonomic Study (Kurtzman et al. 2011), this species is able to grow in vitamin-free medium. This ability since then has been considered as a suitable trait for phenotypic discrimination of this species from the other members of the *Yarrowia* clade (Michely et al. 2013; Galán-Sánchez et al. 2014).

During the course of a comparative physiological analysis of some species of the *Yarrowia* clade among other tests, the assimilation of 31 carbon sources (30 by using API ID32C test, and fructose on a microtitre plate) was investigated (Michely et al. 2013). The type strains of nine species were investigated, and attention has been called to the fact that results from examining only one strain per species do not certainly represent the assimilation spectrum of the given species properly. There is a few contradicting information about the assimilating patterns of the species of the *Yarrowia* clade. The above-mentioned authors refer to *C. hispaniensis* as the only member being able to assimilate trehalose and *Y.* (*Candida*) *galli* to grow in vitamin-free medium. They also refer to *C. hispaniensis* and *Y.* (*Candida*) *oslonensis* as being able to assimilate galactose and sorbose, meanwhile other members of the clade cannot grow or grow only weakly with these carbon compounds. Some of their observations are not fully consistent with those of the Westerdijk Fungal

Biodiversity Institute database or the latest edition of The Yeasts: A Taxonomic Study (Kurtzman et al. 2011).

The phenotypic characteristics of the known *Yarrowia* species are rather similar. Fermentation is absent, nitrate is not assimilated, urease activity is absent, and they can grow in the presence of 0.1% cycloheximide. Their carbon source assimilation patterns are quite similar as well; in addition, several reactions show intraspecific variability. Every known *Yarrowia* species assimilates hexadecane which is an uncommon characteristic among yeasts. Including *Yarrowia* species only about 10% of all known yeasts species can grow with this carbon source (Kurtzman et al. 2011).

Based on the standard phenotypic characteristics applied for the characterisation of strains in yeast taxonomy, not all species of the *Yarrowia* clade can be differentiated from each other. *Yarrowia lipolytica*, *Y. keelungensis*, *Y. deformans*, *Y. porcina* and *Y. divulgata* form a complex group, and they are undistinguishable based on standard phenotypic characteristics. Considering this, information about the physiological characteristics of the members of the *Yarrowia* clade may be controversial and non-representative (Michely et al. 2013). Although some species of the *Yarrowia* clade (10 of 15) according to our current knowledge can be distinguished based on phenotypic characteristics, identification from phenotype is not reliable even in case of the 10 above-noted species, since it does not consider yet unrecognized potential phenotypic variability within species, or unknown novel species of the clade. Based on phenotypic characteristics, *Y. lipolytica* and the above-listed additional four species cannot be differentiated from each other. Therefore, for reliable identification molecular biological methods are needed.

9.2.3 *Differentiation of the Members of the* Yarrowia *Clade Based on Molecular Methods*

As mentioned above, some species of the *Yarrowia* genus cannot be differentiated by conventional methods. They form a species complex, consisting of species indistinguishable based on phenotypic characteristics. Unlike for the majority of other fungi, the D1/D2 domain of LSU rRNA coding gene is still the primary barcoding region used for the identification of yeasts. If there is no clear-cut result based on the D1/D2 sequences, determination of the DNA sequences for other loci, e.g. the ITS, is needed. In the genus *Yarrowia* considerable intraspecific nucleotide variability among the LSU D1/D2 sequences is not unusual (Table 9.1). In these cases the significant variability in the D1/D2 sequences is often coupled with much smaller differences in the ITS region. Bigger sequence variability of the coding D1/D2 region than that of the non-coding ITS region is unexpected. Considering the accepted guidelines (Kurtzman and Robnett 1998; Daniel et al. 2009; Vu et al. 2016), in some cases the comparison of the D1/D2 sequences suggests different species, while that of the ITS regions suggests conspecificity. From two cases where the investigated strains were compared with other molecular methods as well, it seems that in these

Table 9.1 Some intra- and interspecific nucleotide differences detected in the *Yarrowia* genus (Gouliamova et al. 2017; Knutsen et al. 2007; Nagy et al. 2014)

Species (number of strains involved)	LSU D1/D2	ITS
Yarrowia (*Candida*) *alimentaria* (2)	9 (5 sts, 4 gaps)	1 st
Yarrowia porcina (7)	0–7 sts	0–2 sts
Y. oslonensis/parophonii	16 sts	2 (1 st, 1 gap)
Yarrowia parophonii (7)	0	3 sts

cases the sequence variability detected along the ITS regions is decisive (Knutsen et al. 2007; Nagy et al. 2014). A *Y.* (*Candida*) *galli*-like strain was isolated from seawater (Butinar et al. 2011), its LSU D1/D2 sequence differs from the *Y. galli* type strain by 11 nucleotides, and it is referred to as a novel species, although no information has been mentioned about its ITS sequence, which could help to resolve if it is indeed a new species or it is a divergent *Y. galli* strain. The type strain of the recently described *Y. parophonii* (Gouliamova et al. 2017) differs from the type strain of *Y. oslonensis* by 16 substitutions in the D1/D2 region. However, their ITS sequences exhibit merely one substitution and one gap difference. In the meantime, three substitutions can be observed among the ITS sequences of the seven paratype strains of *Y. parophonii*. Based on the significant sequence variation detected along the D1/D2 regions of LSU rDNA, the seven strains were described as a novel species (Gouliamova et al. 2017), although ITS sequence comparisons suggested conspecificity with *Y. oslonensis*.

Microsatellite primed PCR fingerprinting method is commonly used for the discrimination of species or strains within species; it is also applied for the identification of yeast species and in many yeast diversity studies; however, based on a recent study (Ramírez-Castrillón et al. 2014), its reliability can be queried. In another recent study (Nagy 2014a), 219 yeast strains of the *Yarrowia* group were investigated to assess the reliability of microsatellite primed polymerase chain reaction fingerprinting using $(GAC)_5$ primer in the differentiation and identification of the yeast strains of this clade. Using this method, yeast strains of the *Yarrowia* clade from raw meat, raw milk, cheese and cottage cheese formed seven groups. Following MP-PCR fingerprinting-based clustering, strains were identified by sequencing the D1/D2 region of the LSU rDNA of all strains. As each group corresponded to one species, the results supported the reliability of the MP-PCR-based differentiation and identification of the members of the *Yarrowia* genus. Using this method the diversity of *Yarrowia* strains in food has been assessed as well (Nagy 2014a).

9.2.4 Isolation of Yarrowia *Strains*

For the isolation of specific groups of yeasts, selective and differential media can be applied. For the detection of *Y. lipolytica* several methods have been described. It was reported (Lin and Fung 1985) that *Y.* (*Candida*) *lipolytica* can easily be differentiated from other yeast species on YM medium containing either 0.1% Chrystal violet or 0.004% Malachite green, by providing distinctive white, coarsely folded colonies. *Yarrowia lipolytica* was reported to have the unique ability to produce brown pigments from tyrosine which has provided the basis for composing a differential medium for the detection of *Y. lipolytica* (Carreira and Loureiro 1998). At the time of describing these methods *Y. lipolytica* was the only known member of the clade, and since then it became clear that phenotype-based identification is unreliable. Since several species of the *Yarrowia* genus, including *Y. lipolytica*, cannot be differentiated from each other based on phenotypic characteristics, the abovementioned methods are not necessarily suitable for discriminating *Y. lipolytica* from other *Yarrowia* species; however, they may be useful to separate the members of the *Yarrowia* clade from other yeast species. Moreover, bacteria often predominate yeasts; thus the isolation of yeasts along with members of the *Yarrowia* genus can be difficult without suppressing the growth of bacteria.

A method suitable for the isolation of *Yarrowia* strains has been recently reported (Nagy et al. 2013). Before isolation a three-step enrichment procedure was carried out in liquid yeast nitrogen base (YNB) medium prepared with phosphate-citric acid buffer (pH 3.6), supplemented with 0.5% (v/v) hexadecane. The reduced pH served to discourage bacterial growth, and the rationale for using hexadecane was that only about 10% of the known yeast species are able to grow with this compound as a sole carbon source (among the species tested for this character), including members of the *Yarrowia* genus (Kurtzman et al. 2011). Following enrichment serial decimal dilutions and surface plating on Rose-Bengal chloramphenicol (RBC) agar was applied. Following incubation in darkness at 25 °C, representative colonies were isolated, purified and identified by DNA sequencing. According to our experiences obtained during processing of more than 130 samples of minced meet (pork, beef, turkey), raw milk, cheese and cottage cheese, this method is not fully selective, but can efficiently be used to isolate the members of the genus *Yarrowia*.

To enhance the selectivity of the method, the composition of the enrichment medium was later modified. McIlvaine buffer (McIlvaine 1921) containing citric acid was substituted by a buffer containing KH_2PO_4 and phosphoric acid, thus hexadecane was the sole carbon source in the medium. To compare their efficiency, both media were used parallel for processing the same samples and the modified medium performed better than the original one.

9.2.5 Ecology and Natural Occurrence in Foods

Yarrowia lipolytica has pronounced lipolytic and proteolytic activities (Corbo et al. 2001) so it is not surprising that it can often be found in foods with high proportions of fat and/or protein, especially in meat and dairy products (Groenewald et al. 2013). However, its ability to grow under relatively extreme conditions allows *Y. lipolytica* to be found in a wide variety of foods (and other habitats) in a relatively high number. *Yarrowia lipolytica* is one of the most tolerant yeast species to sorbate and benzoate preservatives at pH 5.0. It is able to grow even in the presence of 1000 mg/l sorbate and at a_w 0.89, in the presence of 12.5% NaCl or at pH 2, or in the presence of 50% sucrose (Praphailong and Fleet 1997). *Yarrowia lipolytica* is almost as tolerant against acids as *Zygosaccharomyces bailii*, one of the most resistant food spoilage yeast; it can grow even at pH 10–10.5, and thus it can also be regarded as an alkali-tolerant yeast (Deák 2006); and it can grow and become predominant even at 5 °C in chilled poultry (Ismail et al. 2000).

Yarrowia lipolytica is frequently identified in meat and fish (Deák 2008); it is one of the most frequent species in minced meat (Fleet 1992). It is a frequently isolated species from fresh beef (Liang 1989) and sausages (Viljoen et al 1993). Due to its tolerance against low temperature, high salt-concentrate and low pH, it is one of the most frequent species in fermented sausages (Romano et al. 2006). Based on a study 69% of yeast isolates from fresh beef and 62% of yeast isolates from ham were identified as *Y. lipolytica* (Liang 1989). To the contrary, according to a review, none of the 41 yeast isolates from different kinds of meat were identified as *Y. lipolytica* (Fung and Liang 1990). In frozen chicken meat it was found to be the second most frequently isolated yeast species, with 18% share of the isolates, and it could be isolated from frozen fish as well (Diriye et al. 1993). It can be isolated from irradiated poultry meat and commercial chilled foods (Sinigaglia et al. 1994). Forty-five percent of yeast strains isolated from marinated poultry and 54% of yeast isolates from roasted poultry were identified as *Y. lipolytica*, but this species was also reported to be the most abundant yeast in chicken liver and wings, turkey neck and sausage (Ismail et al. 2000). It can be isolated from Spanish fermented sausages (Encinas et al. 2000), German sausages (Samelis and Sofos 2003), Frankfurters (Drake et al. 1959), Vienna and other sausages (Viljoen et al. 1993), as well as from ham (Liang 1989), biltong (Wolter et al. 2000), salami (Gardini et al. 2001) and fish (Leme et al. 2011). *Yarrowia lipolytica* was found to be the third most frequent yeast species in salami, almost 14% of the isolates belonged to this species (Abunyewa et al. 2000). The second most frequently found species from pastirma, a Turkish dry-cured meat product, was *Y. deformans*; 11% of the yeast isolates were *Y. galli*, 10% of the yeast isolates were *Y. alimentaria* and 5% of them belonged to *Y. lipolytica*. These species were isolated at each sampling point during the production pipeline of this product (Ozturk 2015). Both *Y. lipolytica* and *Y. alimentaria* were isolated from raw meat and bacon (Nielsen et al. 2008).

Several studies report the occurrence of *Y. lipolytica* in a variety of cheeses (mould-ripened, smear-ripened, blue-veined and fresh) produced worldwide. It is often reported to be one of the most frequently isolated yeast species from cheese (Roostita and Fleet 1996; Welthagen and Viljoen 1998; Larpin et al. 2006; Monnet et al. 2010). *Yarrowia lipolytica* can often be isolated from surface-ripened soft cheeses like Camembert and blue-veined cheese, although it was not used as a starter culture (Addis et al. 2001).

At lower frequencies, *Y. lipolytica* has been isolated from raw milk and dairy products, for example, from milk of cow, ewe and water buffalo (Corbo et al. 2001; Chen et al. 2010); it also occurs in fermented milk products such as kefir and yoghurt (Rohm et al. 1992; Gadaga et al. 2000; Lourens-Hattingh and Viljoen 2002;

Fröhlich-Wyder 2003; Viljoen et al. 2003; Bai et al. 2010). *Yarrowia lipolytica* was considered to be the third most frequent yeast species in yoghurt representing 15% of all isolates (Deák 2006). It is also one of the most frequent yeast species in butter (Sinigaglia et al. 1994; Lanciotti et al. 1992; Lopandic et al. 2006); it occurs in cream and margarine (Pitt and Hocking 2009). Due to its key properties offering competitive advantages, such as the tolerance against higher salt concentration and lower temperature, assimilation of lactate and citrate and production of lipolytic and proteolytic enzymes (Sinigaglia et al. 1994; Fleet 1990; Guerzoni et al. 1993, 2001; van den Tempel and Jakobsen 2000; Suzzi et al. 2001), it has the ability to outcompete other yeasts; thus it is able not only to grow in dairy products but even to become predominant (Lanciotti et al. 2005).

Sources of *Y. lipolytica* contamination can be the environment of cheese-making, such as the air, equipment surfaces, hands and aprons and the brine bath. Milk has also been alluded as a potential source of *Y. lipolytica* in cheese (Welthagen and Viljoen 1998). Another study attributed the occurrence of *Y. lipolytica* in cheese to potential contamination of brine used in cheese-making (Bintsis and Robinson 2004).

By reviewing more than 60 publications on the occurrence of *Y. lipolytica* in a variety of different cheeses (mould-ripened, smear-ripened, blue-veined and fresh cheeses), no obvious differences in the prevalence of *Y. lipolytica* were found among cheeses produced from raw or pasteurized milk (Groenewald et al. 2013). Among the different milk sources, cow milk seems to be over-represented, which can be explained by the much larger annual production volumes of cow versus ewe, goat and buffalo milk and cheese. On the other hand, when normalising to the annual production volumes, data seem to suggest a higher prevalence of *Y. lipolytica* in ewe, goat and buffalo cheese as compared to cow cheese (Groenewald et al. 2013).

Some frequent isolation sources of *Yarrowia* species and *C. hispaniensis* are given below (Table 9.2). In view of the fact that several strains of different *Yarrowia* species were isolated from products of warm-blooded animals, it is notable that many of these strains are unable to grow at 35 °C or higher temperature (e.g. *Y. bubula*, *Y. divulgata*, *Y. galli*). The inability of these strains to grow at or above 35 °C suggests that they are secondary contaminants.

9.2.5.1 Diversity in Meat

A recent study revealed rich biodiversity of the *Yarrowia* species in raw meat (Nagy 2014b). According to the published data, updated with some recent results, 135 *Yarrowia* strains were isolated from 66 samples of minced raw meat. These strains were identified at species level by amplifying and sequencing the D1/D2 region of the ribosomal RNA's large subunit coding gene. They were assigned to 8 species. Four of them [*Yarrowia lipolytica, Y. deformans, Y.* (*Candida*) *galli,* and *Y.* (*Candida*) *alimentaria*] were earlier described species, four of them proved to be novel ones, from which three already had been described as *Y. divulgata* (Nagy et al. 2013), *Y. porcina* and *Y. bubula* (Nagy et al. 2014). The predominant *Yarrowia* species recovered from raw meat samples during this study was *Y. deformans* with 33% (48

Table 9.2 Isolation sources of the members of the *Yarrowia* clade

Yarrowia alimentaria	Yoghurt, Norway; cured ham, Norway; bacon before smoking, Denmark; raw meat, Denmark; brine, Denmark; pastirma, Nuve, Ankara, Turkey	Knutsen et al. (2007), Nielsen et al. (2008), and Ozturk (2015)
Yarrowia brassicae	Sauerkraut, Nanyang, Henan Province, China	Liu et al. (2018)
Yarrowia bubula	Minced beef, Hungary; minced pork, Hungary; turkey meat pulp, Hungary	Nagy et al. (2014)
Yarrowia deformans	Raw meat, Hungary; human, Germany; fingernail, Australia; preserved fish, Japan; pastirma, Nuve, Turkey	Kurtzman et al. (2011), Nagy (2014b), and Ozturk (2015)
Yarrowia divulgata	Bacon processing plant, Denmark; chicken liver, Griffin, GA, USA; chicken breast, Hungary; minced beef, Hungary; deep sea water; ocean fish	Nagy et al. (2013)
Yarrowia galli	Chicken liver, Griffin, GA, USA; chicken breast, Griffin, GA, USA; raw meat, Hungary; pastirma, Turkey; seawater, Kongsfjorden coast, western Spitsbergen, Svalbard	Péter et al. (2004), Nagy (2014b), Ozturk (2015), and Butinar et al. (2011)
Yarrowia hollandica	Back of cow, the Netherlands	Knutsen et al. (2007)
Yarrowia keelungensis	Sea-surface microlayer near the Keelung City off the northern coast, Taiwan	Chang et al. (2013)
Yarrowia lipolytica	Milled corn (maize) fibre tailings, Illinois, USA; fingernail, Austria; corneal lesion, Italy; petroleum storage tanks, Iowa, USA; petroleum storage tanks, South Dakota, USA; cold-stored frankfurters, USA; sputum, Cleveland, Ohio, USA; refrigerated meat products, Virginia, USA; refrigerated Canadian bacon, Peoria, Illinois, USA; pastirma, Turkey; rancid margarine, the Netherlands; olives, Italy; lung, Argentina; hydrocarbons, France; soil, the Netherlands; soil, Russia; kerosene aviation fuel, dairy products, USA; wastewater, France; mayonnaise-based salad, Germany; apple juice, Norway; petroleum oil-polluted seawater, near Mumbai, India	Kurtzman et al. (2011), Ozturk (2015), (Pitt and Hocking (2009), Groenewald et al. (2013) and the references therein, Palande et al. (2014)
Yarrowia oslonensis	Yoghurt, Norway	Knutsen et al. (2007)
Yarrowia phangngensis	Estuarine water from mangrove forest, Khao Lumpee-Haad Thaimueang National Park, Phang-Nga Province, Thailand; estuarine water from mangrove forest, Mu Ko Ra-Ko Prathong National Park, Phang-Nga Province, Thailand	Limtong et al. (2008)
Yarrowia parophonii	Gut of *Parophonus hirsutulus* (Carabidae), Bulgaria	Gouliamova et al. (2017)
Yarrowia porcina	Minced beef, Hungary; minced pork, Hungary; sediment of tropical freshwater river, Minas Gerais, Brazil	Nagy et al. (2014)

(continued)

Table 9.2 (continued)

Yarrowia yakushimensis	Gut of Japanese termite (*Hodotermopsis sjoestedti*), Japan	Groenewald and Smith (2013)
Candida hispaniensis	Larva of *Spondylis buprestoides*, conifer forest in Spain	Kurtzman et al. (2011)

strains), and actually only 20% (29 strains) belonged to the species *Y. lipolytica*. Fourteen percent of them were *C. galli* (21 strains), 10% were *C. alimentaria* (14 strains) and 23% of them could be assigned to the 4 new species (Nagy et al. 2013, 2014), 12% to *Y. bubula* (18 strains), 6% to *Y. porcina* (8 strains), 4% to *Y. divulgata* (6 strains) and 1 strain belonged to a yet undescribed *Yarrowia* species.

Based on standard conventional phenotypic tests, 43% of the yeast strains (the strains of *Y. deformans*, *Y. divulgata* and *Y. porcina*) isolated during this study would have been misidentified as *Y. lipolytica*. Based on these results, at least eight species, which are about half of the currently known species of the *Yarrowia* clade, can be isolated from raw meat. Species composition of the genus *Yarrowia* in raw meat is shown in Fig. 9.2.

9.2.5.2 Diversity in Milk and Dairy Products

Contrary to the high diversity of *Yarrowia* species in raw meat, with a few exceptions only *Y. lipolytica* was revealed from raw milk and dairy products (cheese and cottage cheese) during the course of our studies in Hungary. From the 17 yeast strains isolated from raw milk 14 belonged to *Y. lipolytica*, and 3 of them belonged to *Y. bubula*. From cheese only *Y. lipolytica* (12 strains) were isolated, from cottage cheese 61 *Y. lipolytica* strains and only one *C. alimentaria* strain were isolated. Taken together, *Y. lipolytica* dominated over other members of the clade, 96% of the isolates from milk and dairy products belonged to this species (Fig. 9.3).

9.3 Methylotrophic Yeasts

Although methanol-utilising yeasts species amount only to about 5% of the number of species in the whole yeast domain, their biotechnological importance exceeds their proportion. Shortly after the recognition that yeasts are able to utilise methanol as a sole source of carbon and energy, methylotrophic yeasts have been considered as a potential source of single-cell protein. Although considerable effort has been made by several companies and also fermentation technology has greatly improved, the competitiveness with other protein sources was low (Wegner and Harder 1987). Later the interest was turned towards other utilisations of methylotrophic yeasts. Among other applications some species, first of all *Ogataea* (*Hansenula*) *polymorpha* and *Komagataella* spp. (including "*Pichia pastoris*"), have been successfully

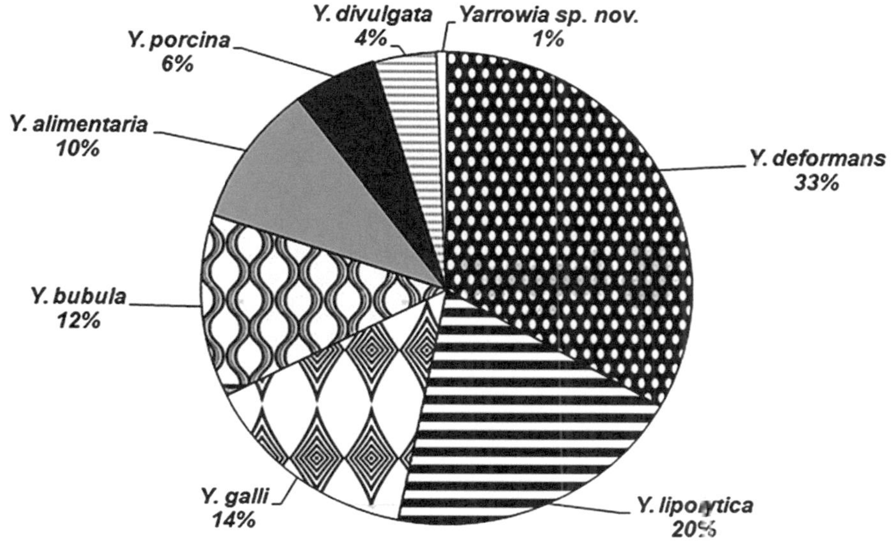

Fig. 9.2 Species composition of *Yarrowia* strains, isolated from raw meat (135 strains from 66 samples)

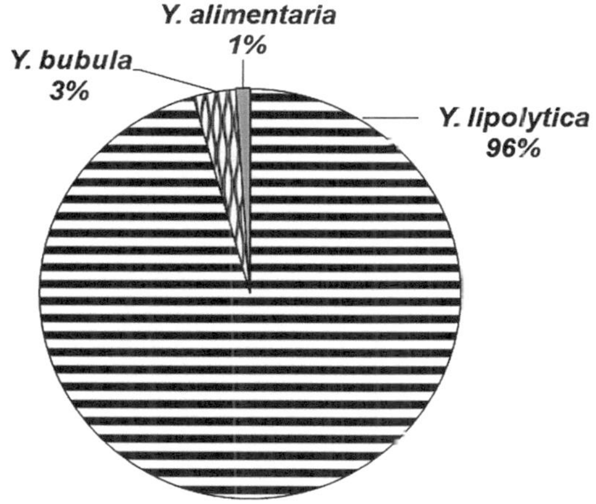

Fig. 9.3 Species composition of *Yarrowia* strains, in raw milk and dairy products (91 strains from 67 samples)

applied for the production of heterologous proteins (Johnson and Echavarri-Erasun 2011). The following sections provide an insight into the recent achievements in the taxonomy, ecology and diversity of this important group of yeasts.

9.3.1 A Half-Century-Old Discovery: The Beginning of the Story

The first report on the utilisation of methanol by yeast was published exactly 50 years ago (Ogata et al. 1969). The first documented methanol-utilising yeast (strain No. 2201) was supposed to be closely related to *Kloeckera* or *Torulopsis* in the first report, but 1 year later it was already referred to as *Kloeckera* sp. (Ogata et al. 1970) and it has been interpreted so until it was reidentified as *Candida boidinii* (Lee and Komagata 1980). The first report on methanol utilisation by yeasts (Ogata et al. 1969) prompted screening of culture collections for this kind of yeasts. However, no methanol-assimilating yeast among the 192 strains was found in an institutional collection (Oki et al. 1972). The screening of the yeast collection of the Centraalbureau voor Schimmelcultures (CBS) also confirmed that methanol assimilation among yeasts is rare (Hazeu et al. 1972). Only 15 methanol-utilising type strains were found among the tested approximately 500 strains representing 422 species including all yeast type strains maintained in CBS at that time. All methanol-assimilating species revealed in CBS belonged to four genera, *Hansenula*, *Pichia*, *Candida* and *Torulopsis*. It was also noted that the majority of methanol utilisers in the CBS collection were recovered from tree bark or from insects living on trees and it was supposed that the number of methanol-assimilating yeasts in this habitat might be enhanced by the presence of lignin, a compound rich in methoxy groups (Hazeu et al. 1972).

Early attempts for isolating methanol-assimilating yeasts were partly fuelled by the demand of strains suitable for the production of single-cell protein (SCP) from methanol. As it will be discussed below, it was noted by several authors that enrichment in methanol-containing media greatly enhances the probability of the isolation of methanol-utilising yeast strains. In the latest five decades the number of known methylotrophic yeast species has increased steadily and the vast majority of the hitherto known methylotrophic yeast species are ascomycetous. Methylotrophic yeasts represent only about 5% of all the known yeast species. In Fig. 9.4 the number of methanol-assimilating yeast species treated in the subsequent editions of the Yeasts: A Taxonomic Study, as well as their currently estimated number are depicted. The increase in the number of methylotrophic yeast species has been accelerated following the establishment of freely accessible database of nucleic acid sequences for ascomycetous yeasts (first of all D1/D2 LSU rRNA gene sequences; (Kurtzman and Robnett 1998)) coupled by the widespread application of sequence-based yeast identification.

9.3.2 Systematics of Methylotrophic Yeasts

The majority of the methanol-assimilating yeasts belong to the subphylum *Saccharomycotina*, and initially they were described first of all as members of the teleomorphic genera *Hansenula* and *Pichia* and the anamorphic ones as *Torulopsis*

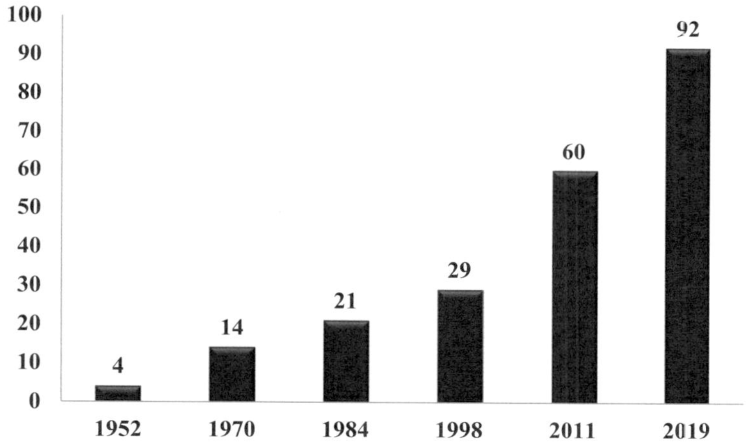

Fig. 9.4 The number of described methanol-assimilating yeast species
1952 – Lodder J & Kreger-van Rij NJW: The Yeasts: A Taxonomic Study (Lodder and Kreger-van Rij 1952)
1970 – Lodder J (ed): The Yeasts: A Taxonomic Study. 2nd ed. (Lodder 1970)
1984 – Kreger-van Rij NJW (ed): The Yeasts: A Taxonomic Study. 3rd ed. (Kreger-van Rij 1984)
1998 – Kurtzman CP & Fell JW (eds): The Yeasts: A Taxonomic Study. 4th ed (Kurtzman and Fell 1998)
2011 – Kurtzman CP, Fell JW, Boekhout T (eds): The Yeasts: A Taxonomic Study. 5th ed. (Kurtzman et al. 2011)
2019 – The estimated number of the currently known methanol-assimilating yeast species

and *Candida*. Following the merger of the two teleomorphic (Kurtzman 1984) and the two anamorphic (Yarrow and Meyer 1978) genera, the majority of ascospore-forming methylotrophic yeast species were transferred to *Pichia*, while the anamorphic ascomycetous methylotrophs became members of the genus *Candida*. However, the polyphyletic nature of these two huge genera has been recognised for a long time (Kurtzman 2011a, b, c; Lachance et al. 2011). The genus *Pichia* has already been reduced to the species phylogenetically related to its type species, *P. membranifaciens* (Kurtzman et al. 2008) and a similar process is in progress in case of the genus *Candida*. *Candida* species phylogenetically not related to *C. vulgaris* (a synonym of *C. tropicalis*) are gradually being reclassified in extant or newly erected genera (Daniel et al. 2014). As a part of the above-noted process, based on partial small and large subunit rRNA sequence analyses new monotypic genera, *Kuraishia* (Yamada et al. 1994) and *Komagataella* (Yamada et al. 1995) were proposed for the methylotrophic yeast species *Pichia capsulata* and *P. pastoris*, respectively while some ascospore-forming methanol-assimilating *Pichia* (former *Hansenula*) species were reassigned to the newly erected *Ogataea* genus (Yamada et al. 1994). These proposals initially have not generally been accepted, because of the low number of the taxa included in the analyses. However, subsequent analyses with wider taxon and phylogenetic marker sampling supported the proposals of the above-noted genera and gradually the methanol-assimilating *Pichia* species has been reassigned to these genera and the recently described phylogenetically related species have already been placed to *Komagataella*, *Kuraishia* and *Ogataea*.

The vast majority of the known *Ogatae* species form hat-shaped ascospores, but *O. salicorniae* (reassigned from *Williopsis* (Kurtzman and Robnett 2010)) forms globose ascospores having an equatorial ledge, while *O. allantospora* is unique in the genus by forming allantoid ascospores. *Ogataea ramenticola* is the only documented heterothallic member of the genus. The fermentation of glucose and other sugars as well as nitrate assimilation are variable. All *Ogataea* species, except *O. salicorniae*, grow with methanol as a sole carbon source (Kurtzman 2011b). The *Ogatae* clade includes also an anamorphic species, *C. ortonii*, which does not grow with methanol (Lachance et al. 2011). Some *Ogataea* species grow at 45–50 °C. If data are available, the predominant ubiquinone in *Ogatae* species is coenzyme Q-7 (Kurtzman 2011b). The quickly expanding *Ogataea* clade in addition to the *Ogataea* species includes phylogenetically related *Candida* species as well. Considering its current species membership, the *Ogataea* clade lacks statistical support in the phylogenetic tree based on the analysis of D1/D2 region of the LSU rRNA gene (Fig. 9.5) and also in the multigene phylogenetic tree published in 2010 (Kurtzman and Robnett 2010). Figure 9.5 includes all the known members of the *Ogataea* clade at the time of writing this chapter. In the D1/D2 tree based on the analysis of the D1/D2 region statistical support is lacking in the majority of the branches. It was also noted earlier (Naumov et al. 2018a) that single marker phylogenetic analyses have low informative value in case of *Ogataea* clade; however, currently no complete dataset for a reliable multigene analysis can be obtained from publicly available databases. Based on D1/D2 phylogenetic analyses (Nagatsuka et al. 2008), the existence of several smaller groups (referred to as clusters) has been revealed with very weak linkage among the species which are currently assigned to genus *Ogataea*. However, even these "clusters" lacked significant statistical support. A multigene tree (Kurtzman and Robnett 2010) was constructed following the analysis of four loci, but the taxon sampling at that time was yet narrower than in case of the current analysis. Contrary to the D1/D2 tree (Fig. 9.5), in the multigene tree several subclades have strong statistical support, but the entire *Ogataea* clade does not possess significant support and the heterogeneity of the genus *Ogataea* has also been noted (Kurtzman and Robnett 2010). Based on previous findings (Kurtzman and Robnett 2010) it was stated that the genus *Ogataea* is heterogeneous and should be split into at least five different genera (Naumov et al. 2018a). The revision of the *Ogataea* clade must be based on multigene (Naumov et al. 2018a) or genome-scale analysis.

The genus *Komagataella* currently includes seven species (Fig. 9.6). They form hat-shaped ascospores in deliquescent asci. All species are homothallic. Glucose in fermented nitrate is not assimilated. The major ubiquinone, if determined, is CoQ-8 (Kurtzman 2011d). The *Komagataella* species, except for *K. kurtzmanii*, are indistinguishable if the standard physiological tests are considered (Naumov et al. 2018b). In the case of *K. pseudopastoris*, it was noted that the growth of this species is inhibited by smaller tannic acid concentration than that of *K. pastoris*, a characteristic correlating to the isolation source of *K. pseudopastoris* (Dlauchy et al.

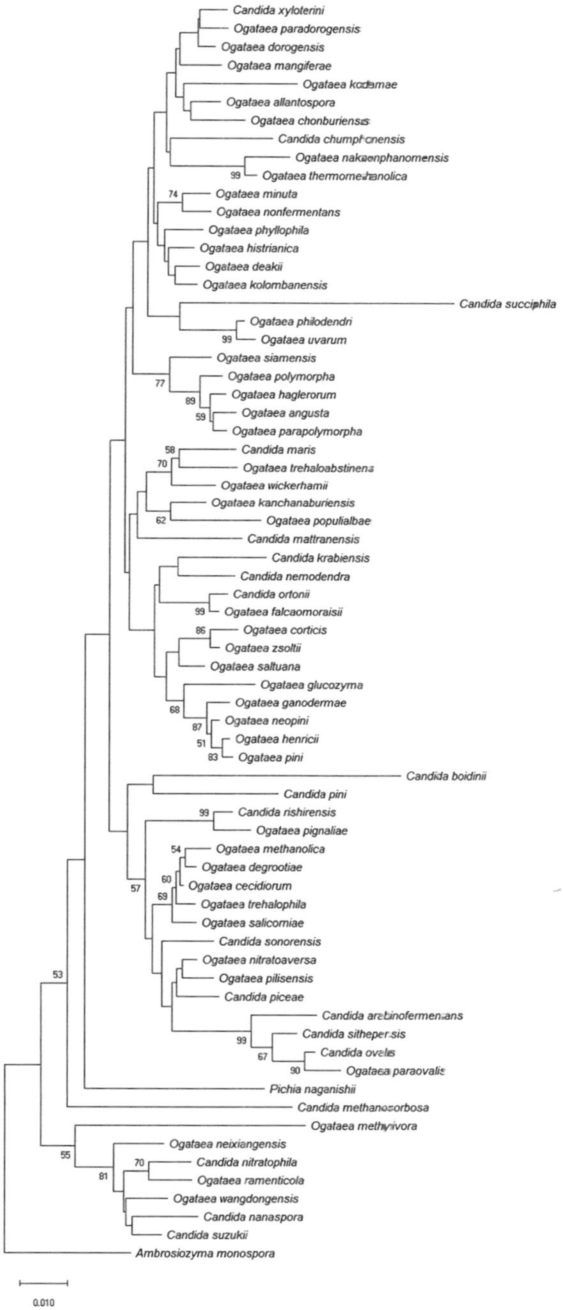

Fig. 9.5 Phylogenetic relationships among the members of the *Ogataea* clade determined from Neighbour-Joining analysis of sequences of LSU rRNA gene D1/D2 domain The evolutionary history was inferred using the Neighbour-Joining algorithm (Saitou and Nei 1987). Bootstrap percentages (Felsenstein 1985) (1000 replicates) exceeding 50% are given at next to the branches. Bar, 1% nucleotide sequence divergence. The evolutionary distances were computed using the Tamura 3-parameter method (Tamura 1992). All positions containing gaps and missing data were eliminated. There were a total of 505 positions in the final dataset. *Ambrosiozyma monospora* was used as the designated outgroup species. Evolutionary analysis was conducted in MEGA X (Kumar et al. 2018)

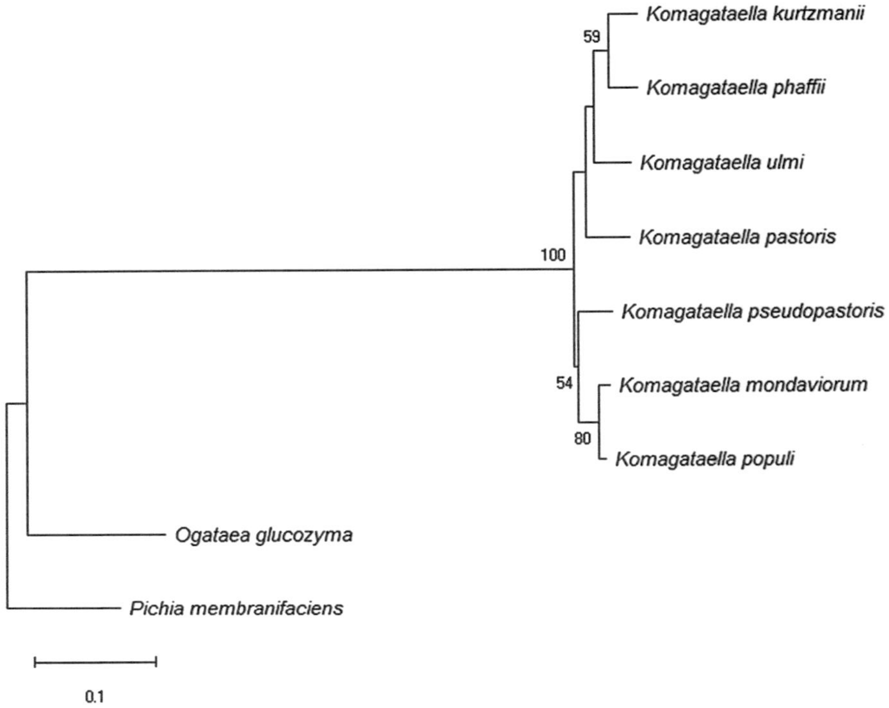

Fig. 9.6 Phylogenetic relationships among the members of the genus *Komagataella* determined from maximum likelihood analysis of concatenated sequences of the ITS regions and LSU rRNA gene D1/D2 domain
The evolutionary history was inferred by using the Maximum Likelihood method and Kimura 2-parameter model (Kimura 1980). The tree with the highest log likelihood is shown. Bootstrap percentages (Felsenstein 1985) (1000 replicates) exceeding 50% are given at next to the branches. Bar, 10% nucleotide sequence divergence. All positions containing gaps and missing data were eliminated. There were a total of 560 positions in the final dataset. *Pichia membranifaciens* was used as the outgroup species. Evolutionary analysis was conducted in MEGA X (Kumar et al. 2018)

2003). Before the spreading of sequence-based yeast identification, accurate species-level identification of the closely related *Komagataella* species has been difficult. It has been shown by gene sequence comparisons that the strain utilised in "*Pichia pastoris* Expression Kit" and supposed to be *K.* (*Pichia*) *pastoris* is actually *K. phaffii* (Kurtzman 2009). Most but not all industrially applied strains labelled as "*Pichia pastoris*" are *K. phaffii* (Gasser and Mattanovich 2018).

The genus *Kuraishia* contains strongly or weakly fermenting species and nitrate is assimilated by the majority of them. The major ubiquinone, if determined, is CoQ-8 or CoQ-9 (Lachance et al. 2011; Péter et al. 2011). *Kuraishia* includes phylogenetically more distantly related species than the genus *Komagataella* (Fig 9.7). In the ITS-D1/D2 tree *K. capsulata*, the type species of the genus, together with *K. molischiana*, *K. borneana* and the recently described *K. mediterranea* form a well-

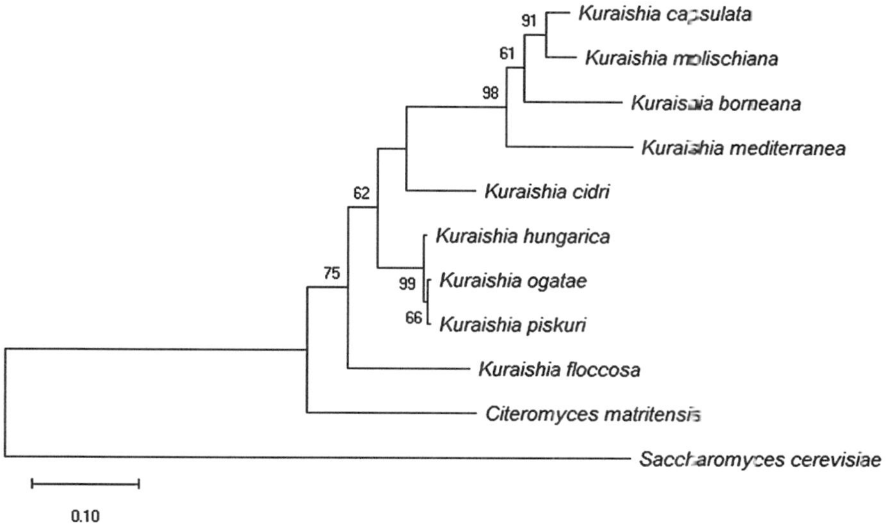

Fig. 9.7 Phylogenetic relationships among the members of the genus *Kuraishia* determined from maximum likelihood analysis of concatenated sequences of the ITS regions and LSU rRNA gene D1/D2 domain
The evolutionary history was inferred by using the maximum likelihood method and Tamura-Nei model (Tamura and Nei 1993). The tree with the highest log likelihood is shown. Bootstrap percentages (Felsenstein 1985) (1000 replicates) exceeding 50% are given at next to the branches. Bar, 10% nucleotide sequence divergence. All positions containing gaps and missing data were eliminated. There were a total of 920 positions in the final dataset. *Saccharomyces cerevisiae* was used as the outgroup species. Evolutionary analysis was conducted in MEGA X (Kumar et al. 2018)

supported core group within the genus. Except of *K. borneana* all species of this core group form ascospores while no ascosporulation has been detected in the other members of the genus. The anamorphic members of the genus, in agreement with the Melbourne Code (McNeill et al. 2012), were transferred to *Kuraishia* as a result of multigene sequence analysis (Kurtzman and Robnett 2014).

From the methanol-assimilating *Candida* species, those which are placed by phylogenetic analyses in the *Ogataea* clade are still waiting for reassignment. Known members of the genus *Komagataella* invariably form ascospores. The major changes in the generic assignments of the methanol-assimilating yeasts in the *Ogatae* clade and in the genera *Komagataella* and *Kuraishia* are summarised in Fig. 9.8.

Although the vast majority of the hitherto known methylotrophic yeast species belong to the above-listed three groups, some additional, phylogenetically unrelated methanol-assimilating species were reported as well. In their case the growth with methanol as a sole carbon source is often weak or slow. In addition to a few ascomycetous yeast species, *Ascoidea asiatica*, *Alloascoidea hylecoeti* (Kurtzman and Robnett 2013b), *C. xylanilytica* (*Spathaspora* clade) (Boonmak et al. 2011), some

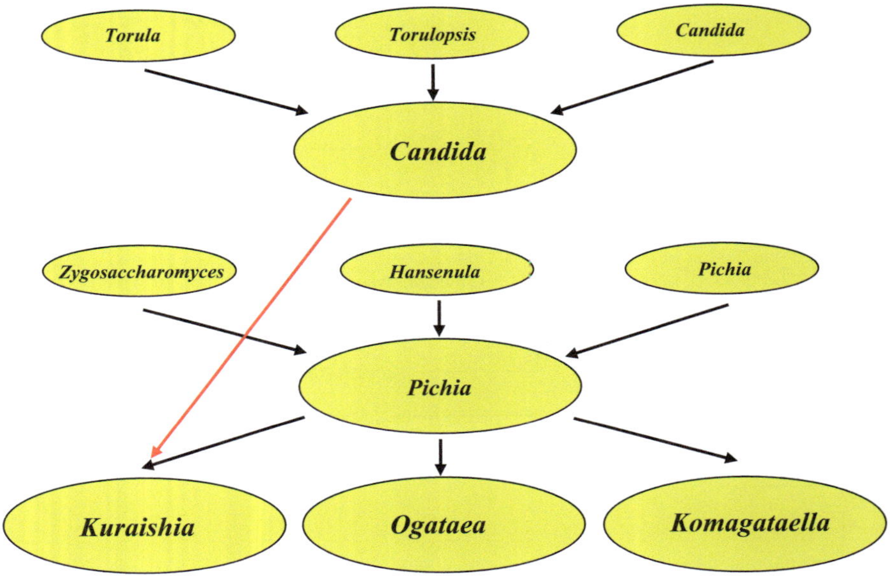

Fig. 9.8 Major changes in the generic assignments of the methanol-assimilating yeasts in subphylum *Saccharomycotina*

basidiomycetous species, *Chionosphaera cuniculicola*, *Pseudozyma flocculosa*, *P. rugulosa* (current name *Moesziomyces bullatus*), *Udeniomyces puniceus*, *U. pyricola* and *Tremella foliacea* (current name *Phaeotremella foliacea*) (Kurtzman et al. 2011) were also reported to be able to utilise methanol.

9.3.3 Isolation of Methylotrophic Yeast Strains: Suitable Methods and Substrates

Reports on the isolation of methanol-assimilating yeast strains appeared soon after the first report of methanol assimilation by yeasts (Ogata et al. 1969) was published. They have in common the application of enrichment in methanol-containing media before the isolation of yeasts. The substrates yielding methanol-utilising yeasts in the first attempts during the 1970s included soil (Asthana et al. 1971; Sahm and Wagner 1972; Levine and Cooney 1973; Volfová and Pilát 1974; van Dijken and Harder 1974), rotten tomato, rotten flower (Oki et al. 1972) and wastewater (Volfová and Pilát 1974). To sum up the experiences of the pioneering efforts, the essential conditions for the isolation of methanol assimilation of yeasts were collected (van Dijken and Harder 1974). According to the list the medium optimally has a pH of 4.5, it contains antibiotics for discouraging the growth of bacteria, and vitamins to fulfil the requirements of some methanol-assimilating yeasts. The suggested concentration of methanol is 0.1–0.5% (v/v), because higher methanol concentration may inhibit the growth of some methanol-assimilating yeast species. According to our experiences, Yeast Nitrogen Base (YNB) supplemented with 0.5% (v/v) methanol is a suitable enrichment medium for methanol-assimilating yeasts. We applied it successfully in a two-step enrichment process for several hundreds of rotten wood, leaf and tree exudate samples (see below). In case of rotten wood and leaves the sample size was usually 10 g, while tree exudates were sampled with sterile cotton swabs. In case of success the second enriched culture was serially diluted and surface plated on Rose-Bengal Chloramphenicol (RBC) agar (Dlauchy et al. 2003). Contrary to the proposal (van Dijken and Harder 1974) and most of the reports published, we have not routinely applied antibiotics for discouraging bacterial growth. Although bacterial cells (usually in minority) were occasionally seen in the second enrichment culture, they never hampered the isolation of yeasts from the (chloramphenicol containing) RBC agar. In some cases, mainly during processing leaf samples, 200 mgL^{-1} chloramphenicol was applied in the enrichment medium to suppress protists feeding on yeasts (Péter et al. 2007). Although methanol-assimilating yeasts can occasionally be isolated from some substrates, e.g. tree exudates or necrotic cactus tissue (Miller et al. 1976), even without enrichment, the application of enrichment in methanol-containing media is a far more effective option, especially in the case of substrates containing low numbers of methylotrophic yeasts.

Soil rich in organic matter was found to be a good source of methylotrophic yeasts (van Dijken and Harder 1974), and numerous methanol-assimilating yeast strains were isolated also from tree bark or from insects living on trees (Hazeu et al. 1972). The isolation sources of methanol-assimilating yeasts obtained from the latest edition of the Yeasts: A Taxonomic Study (Kurtzman et al. 2011), and in case of recently described species, from the papers containing the descriptions are listed in Table 9.3. In many cases the substrate yielding the methylotrophic yeast is a plant organ or tissue or soil which is the "ultimate repository for organic and inorganic materials" (Phaff and Starmer 1987) including plant materials. The occurrence of methylotrophes in soil is probably related to the presence of plant lignin and pectin in soil (Fall and Benson 1996). The association of methylotrophic yeasts with plants has been recognised for a long time. It was supposed already in 1972 that the number of methanol-assimilating yeasts on tree bark might be enhanced by the presence of lignin, a compound rich in methoxy groups (Hazeu et al. 1972). Methanol can be originated from the methoxy groups of lignin (de Koning and Harder 1992). Methanol is formed also by the hydrolysis of the methyl esters of pectin and it was observed that many methylotrophic yeast strains were able to assimilate pectin as well. It was also suggested that methanol-assimilating yeasts play an important role in the decomposition of methanol originated from plant material (Lee and Komagata 1980). Although all hitherto known methanol-assimilating yeast species are facultative methylotrophs, the fact that methanol can be derived from two common constituents of plants may explain the observation that "decaying plant material and soil samples rich in organic matter are particularly good sources for isolation of methylotrophic yeasts" (Harder and Veenhuis 1989).

Table 9.3 Isolation sources of the methylotrophic yeast species assigned to genera *Komagataella*, *Kuraishia* and *Ogataea* and the *Candida* species of the *Ogataea* clade

Species	Isolation source	Reference
Candida arabinofermentans	Insect frass of a dead larch (*Larix* sp., Pinaceae), Alaska; insect frass of a loblolly pine (*Pinus taeda*), Mississippi; insect frass of a longleaf pine (*Pinus australis*), Florida; all from USA	Lachance et al. (2011)
Candida boidinii	Tanning fluid in Spain; sap fluxes of Douglas fir (*Pseudotsuga menziesii*, Pinaceae), California, USA; sap fluxes of California black oak (*Quercus kelloggii*, Fagaceae), California, USA; sap fluxes of northern red oak (*Quercus rubra*), Ontario, Canada; sap fluxes of western cottonwood (*Populus fremontii*, Salicaceae), Arizona, USA; fruit of *Opuntia megacantha* (Cactaceae), Hawai'i Island; necrotic cladodes of *Opuntia megacantha*, Hawai'i Island; fermenting stem of *Clermontia* sp. (Araliaciae) Maui, Hawai'i; sap fluxes of koa (*Acacia koa*, Fabales: Mimosaceae), Hawai'i Island; fruit fly (*Drosophila* sp.), tequila distillery, Mexico; sap flux of guapinol (*Hymenaea courbaril*, Fabales: Caesalpinaceae), Costa Rica; flower of *Heliconia* sp. (Heliconiaceae), Costa Rica; gummy exudate of Guanacaste tree (*Enterolobium cyclocarpum*, Fabales: Mimosaceae), Costa Rica; sap fluxes of breadfruit (*Artocarpus altilis*, Moraceae), Islands of Oahu and Hawai'i; sap flux of koa (*Acacia koa*), Molokai, Hawai'i; sap flux of hemlock (*Tsuga canadensis*, Pinaceae), W.T. Starmer, Tennessee, USA; seawater; soil in South Africa; alpechin, Spain; seawater, Florida, USA; wine, Japan; ginger ale, USA; floor of hospital ward in Finland; soil, the Netherlands; washed soft-drink bottles; tepache, a spicy pineapple drink, Mexico; slimy mud in Japan; soil, Poland	Lachance et al. (2011)
Candida chumphonensis	Surface of plant leaves, Chumphon, Thailand	Koowadjanakul et al. (2011)
Candida krabiensis	Soil, Krabi Province, Thailand	Lachance et al. (2011)
Candida maris	Seawater, Torres Strait, Australia	Lachance et al. (2011)
Candida mattranensis	Surface of plant leaves, Chumphon, Thailand	Koowadjanakul et al. (2011)
Candida methanosorbosa	Sewage, Nagoya, Japan; soil, Japan	Lachance et al. (2011)
Candida nanaspora	Capuchin monkey (*Cebus apella*) in a zoo, France	Lachance et al. (2011)
Candida nemodendra	Tunnels of scolytid beetle (*Xyleborus* sp., Coleoptera: Scolytidae) in Cape beech (*Rapanea melanophloeos*, Myrsinaceae), South Africa	Lachance et al. (2011)

(continued)

Table 9.3 (continued)

Species	Isolation source	Reference
Candida nitratophila	Bark beetle (*Dendroctonus monticolae*, Coleoptera: Scolytidae) in ponderosa pine (*Pinus ponderosa*), California, USA	Lachance et al. (2011)
Candida ovalis	Soil, Japan	Lachance et al. (2011)
Candida piceae	Insect frass in Sitka spruce (*Picea sitchensis*, Pinaceae), Alaska, USA; insect frass, *Picea* sp., Ontario, Canada	Lachance et al. (2011)
Candida pini	Water-logged heart of a pine tree, Sweden	Lachance et al. (2011)
Candida rishirensis	Soil in a pine forest, Rishiri Island, Japan	Nakase et al. (2010)
Candida sithepensis	Soil in Sithep Historical Park, Petchabun Province, Thailand	Lachance et al. (2011)
Candida sonorensis	Tissue of organ-pipe cactus (*Lemaireocereus thurberi*, Cactaceae), Mexico, and many others from cactus tissue or associated *Drosophila* spp. (Diptera: Drosophilidae) or moths (*Lepidoptera*: Pyralidae) from Arizona and Hawai'i, USA, the Bahamas, the Caribbean, Argentina, Brazil and South Africa; tissue of saguaro cactus (*Carnegiea gigantea*), Arizona, and hundreds of others, from cactus tissue or associated insects, in Mexico, the southern USA and Australia	Lachance et al. (2011)
Candida succiphila	Sap of peach tree (*Amygdalus persica*, Rosaceae), Japan; tree exudate, Japan; black knot (*Dibotryon morbosum*, Venturaceae) on chokecherry (*Prunus virginiana*, Rosaceae), Ontario, Canada; soil of oilfield, Japan	Lachance et al. (2011)
Candida suzukii	Bark of an unidentified tree, Taiwan	Lachance et al. (2011)
Candida xyloterini	Body surface and galleries of *Xyloterinus politus*, Wisconsin, USA	Suh and Zhou (2010a)
Komagataella kurtzmanii	Fir flux in the Catalina Mountains, Southern AZ, USA	Naumov et al. (2013)
Komagataella mondaviorum	Cottonwood tree (*Populus deltoides*), Davis, CA, USA; slime flux of (*Quercus* sp.), CA, USA; exudate of black oak (*Quercus kelloggii*), Sierra Nevada, USA; dry frass from willow (*Salix* sp.), WA, USA; flux of hackberry tree (*Celtis* sp.), CA, USA	Naumov et al. (2018b)
Komagataella pastoris	Exudate of a chestnut tree (*Castanea* sp., Fagaceae), France; exudates and rotted wood obtained from Norway; hedge maples (*Acer platanoides*, *A. campestre*, Aceraceae); oaks (*Quercus* spp., Fagaceae), European white birch (*Betula pendula*, Betulaceae), European larch (*Larix decidua*, Pinaceae), willow (*Salix* sp., Salicaceae), European hornbeam (*Carpinus betulus*, Betulaceae); and European beech (*Fagus sylvatica*, Fagaceae), all from Hungary	Kurtzman (2011d)

(continued)

Table 9.3 (continued)

Species	Isolation source	Reference
Komagataella phaffii	Black oak tree (*Quercus kelloggii*), California, USA; sap flux, emory oak (*Quercus emoryi*), Tucson, Arizona, USA; slime flux of an elm tree (*Ulmus* sp.), USA	Kurtzman (2011d)
Komagataella populi	Exudate on a cottonwood tree (*Populus deltoides*), Peoria, Illinois, USA	Kurtzman (2012)
Komagataella pseudopastoris	Rotted wood of willow tree (*Salix alba*), Hungary	Kurtzman (2011d)
Komagataella ulmi	Exudate on an elm tree (*Ulmus americana*), Peoria, Illinois, USA	Kurtzman (2011d)
Kuraishia (*Candida*) *borneonana*	Fruit waste collected in markets in Brunei, Borneo	Sipiczki (2012)
Kuraishia capsulata	Insect frass of a coniferous tree on the shore of Wabatongushi Lake, near Franz, Ontario, Canada; from soil, Finland; rotten wood of black pine (*Pinus nigra*), Hungary; rotten wood of Scotch pine (*Pinus sylvestris*), Hungary; rotten wood of Norway spruce (*Picea abies*), Hungary; lichen, Wyoming, USA; insect frass, black spruce (*Picea mariana*), Ontario, Canada; insect frass, yellow spruce (*Picea rubens*), Ontario, Canada; insect frass, fir (*Tsuga* sp.), Tokyo, Japan; frass, larch (*Larix* sp.), Germany; insect frass, larch, Germany; insect frass, fir (*Tsuga* sp.), Germany; resin, fir, Quebec, Canada	Péter (2011)
Kuraishia (*Candida*) *cidri*	Cider, UK	Lachance et al. (2011)
Kuraishia (*Candida*) *floccosa*	Flux of an oak (*Quercus petrea*, Fagaceae) Pilis Mountain, Hungary; flux of a red oak (*Quercus rubra*), Ontario, Canada	Lachance et al. (2011)
Kuraishia (*Candida*) *hungarica*	Rotten wood of oak (*Quercus* sp., Fagaceaea), Gödöllő, Hungary; rotten wood of oak (*Quercus* sp., Fagaceaea), Pilis Mountains, Hungary	Lachance et al. (2011)
Kuraishia mediterranea	Olive oil sediments, Slovenia; spoiled olive oils, Portugal	Čadež et al. (2017)
Kuraishia molischiana	Used tanning bark; water, at 40 °C in wood-working factory, Sweden; soil, Japan; fruiting body of mushroom (*Suillus* sp.), Hungary; rotten wood of black pine (*Pinus nigra*), Hungary; rotten wood of black locust (*Robinia pseudoacacia*), Hungary; rotten wood of European beech (*Fagus sylvatica*); feces of domesticated rabbit, Hungary; insect frass, red pine (*Pinus resinosa*), Canada; insect frass, loblolly pine (*Pinus taeda*), Georgia; insect frass, juniper (*Juniperus* sp.), Montana, USA; insect frass, ponderosa pine (*Pinus ponderosa*), Washington, USA; insect frass, pine (*Pinus* sp.), Germany; insect frass, white pine (*Pinus strobus*), Wisconsin, USA	Péter (2011)

(continued)

Table 9.3 (continued)

Species	Isolation source	Reference
Kuraishia (*Candida*) *ogatae*	Rotten wood (*Quercus* sp.), Hungary; freshwater (Danube), Hungary	Péter et al. (2009a)
Kuraishia piskuri	Insect frass from tree, Florida, USA	Kurtzman and Robnett (2014)
Ogataea angusta	*Drosophila pseudoobscura*, USA	Suh and Zhou (2010b)
Ogataea allantospora	Leaf of a European hornbean (*Carpinus betulus*, Betulaceae), Pilis Mountains, Hungary; leaf of an oak (*Quercus* sp., Fagaceae), Pilis Mountains, Hungary; leaf of an ash (*Fraxinus* sp., Oleaceae), Pilis Mountains, Hungary	Kurtzman (2011b)
Ogataea chonburiensis	Soil, Chonburi Province, Thailand	Kurtzman (2011b)
Ogataea cecidiorum	Galls induced by sawflies on the leaves of willows in the Losiny Ostrov National Park, Russia	Glushakova et al. (2010)
Ogataea corticis	Tree bark, Japan	Kurtzman (2011b)
Ogataea deakii	Rotten beech wood, Hungary	Čadež et al. (2013)
Ogataea degrootiae	soil, the Netherlands	Groenewald et al. (2018)
Ogataea (*Pichia*) *dorogensis*	Rotted railroad ties, Dorog, Hungary	Kurtzman (2011b)
Ogataea falcaomoraisii	Tree sap exudates of *Sclerolobium* sp. in a natural forest fragment (ipuca), Lago Verde Estate, Tocantins, Brazil; tree sap exudates of *Hymenaea courbarail* from Guanacaste Conservation Area, Santa Rosa Sector, Costa Rica	Kurtzman (2011b)
Ogataea ganodermae	Basidiocarps of *Ganoderma* sp., collected from a tree trunk, Mangshan Mountain, Hunan Province, China	Kurtzman (2011b)
Ogataea glucozyma	Frass in an Engelmann spruce (*Picea engelmannii*, Pinaceae) growing in the Medicine Bow Mountains, Wyoming, USA; southern beech (*Nothofagus* sp., Nothofagaceae), Chile	Kurtzman (2011b)
Ogataea haglerorum	Rotting tissue of *Opuntia phaeacantha*, Arizona, USA	Naumov et al. (2017)
Ogataea henricii	Bird feces collected in the Medicine Bow Mountains, Wyoming, USA; frass from a lodgepole pine (*Pinus contorta*) collected near Markleeville, California, USA; soil, Medicine Bow Mountains, Wyoming, USA	Kurtzman (2011b)
Ogataea histrianica	Olive oil sediments, Slovenia; extra virgin olive oil originating from Italy	Čadež et al. (2013)
Ogataea kanchanaburiensis	External surface of *Mangifera indica* leaves, Kanchanaburi province, Thailand	Limtong et al. (2013)
Ogataea kolombanensis	Olive oil sediments, Slovenia	Čadež et al. (2013)

(continued)

Table 9.3 (continued)

Species	Isolation source	Reference
Ogataea kodamae	Insect infestations of spurge (*Euphorbia ingens*), Groblersdal district; South Africa	Kurtzman (2011b)
Ogataea mangiferae	Mango (*Mangifera indica*) leaves, Brazil	Santos et al. (2015)
Ogataea methanolica (Pichia methanolica)	Soil, Japan; soil, Marion, Illinois, USA	Kurtzman (2011b)
Ogataea methylivora	Rotted tree, Iwate Prefecture, Japan; decayed timber, Japan	Kurtzman (2011b)
Ogataea minuta	Fermenting mushroom (*Mycena pura*); forest soil, Germany	Kurtzman (2011b)
Ogataea naganishii	Exudate from a camellia (*Camellia japonica*, Theaceae), Japan; rotted log, Marion, Illinois, USA; damaged fruit of an Osage orange tree (*Maclura pomifera*, Moraceae) Peoria, Illinois, USA	Kurtzman (2011b)
Ogataea nakhonphanomensis	Tree exudate, Nakhon-Phanom Province, Thailand	Kurtzman (2011b)
Ogataea neixiangensis	Rotting wood, China	Lu et al. (2017)
Ogataea neopini	Frass in a loblolly pine (*Pinus taeda*); frass in a short leaf pine (*Pinus echinata*), Piney Woods, Mississippi, USA	Kurtzman (2011b)
Ogataea nitratoaversa	Leaf of a beech tree (*Fagus sylvatica*, Fagaceae), Pilis Mountains, Hungary; rotted wood, alder tree (*Alnus glutinosa*, Betulaceae), Pilis Mountains, Hungary; rotted wood, beech tree (*Fagus sylvatica*), Bükk Mountains, Hungary	Kurtzman (2011b)
Ogataea nonfermentans	Libby Creek near Laramie, Wyoming, USA	Kurtzman (2011b)
Ogataea paradorogensis	Insect (*Xyleborus* sp.) galleries of a Tabu-no-ki tree (*Machilus thunbergii*, Lauraceae), Japan; insect (*Xyleborus* sp.) galleries in Japanese beech (*Fagus crenata*, Fagaceae), Japan	Kurtzman (2011b)
Ogataea paraovalis	Rotting wood, China	Lu et al. (2017)
Ogataea parapolymorpha	Soil, USA	Suh and Zhou (2010b)
Ogataea philodendri	Frass from tunnels of the Bostrichid beetle (*Xylion adustus*), which was infesting a moribund fig (*Ficus sycomorus*, Moraceae), South Africa; tunnel of a beetle (*Platypus externedentatus*) in the tree Macaranga capensis (Euphorbiaceae), South Africa; tunnel of *P. externedentatus* in a sweet thorn acacia tree (*Acacia karroo*, Fabaceae), South Africa	Kurtzman (2011b)
Ogataea phyllophila	Surface of plant leaves, Chumphon, Thailand	Koowadjanakul et al. (2011)
Ogataea pilisensis	Rotted oak wood (*Quercus* sp., Fagaceae), Pilis Mountains, Hungary	Kurtzman (2011b)

(continued)

Table 9.3 (continued)

Species	Isolation source	Reference
Ogataea pignaliae	Tanning fluid, France; exudate of sessile oak (*Quercus petrea*), Pilis Mountains, Hungary; leaf of agrimony (*Agrimonia eupatoria*), Pilis Mountains, Hungary; leaf of maple (*Acer* sp.), Pilis Mountains, Hungary; unidentified moss, Pilis Mountains, Hungary	Péter et al. (2010)
Ogataea pini	Bark beetle *Dendroctonus brevicomis*, USA; larva of *Dendroctonus monticola* in Western white pine (*Pinus monticola*, Pinaceae), California, USA; curculionidae larva on a monkey-puzzle tree (*Araucaria araucana*, Araucariaceae), Chile; *Hylurgonotus brunneus* on a monkey-puzzle tree, Chile; frass, short-leaf pine (*Pinus echinata*); frass, loblolly pine (*Pinus taeda*); larva of *Dendroctonus monticola* in ponderosa pine (*Pinus ponderosa*); frass, pine (*Pinus* sp.); frass, lodge pole pine (*Pinus contorta*); frass, fir tree (*Abies firma*), Japan; dead birch tree (*Betula* sp., Betulaceae), USA; tunnel of a pine-borer beetle (*Xyleborus torquatus*), South Africa	Kurtzman (2011b)
Ogataea polymorpha	Soil, Brazil; soil, Bagasse, Thailand; soil, Taiwan; soil, Korea; soil, Costa Rica; intestinal tract, swine, Portugal; alpechin, Spain; human knee infection, Worcester, Massachusetts, USA; catheter-related infection, Chicago, Illinois, USA; mix of sugar cane bagasse and soil, Louisiana, USA; spoiled orange juice; human blood, USA	Kurtzman (2011b) and Suh and Zhou (2010b)
Ogataea populialbae	19 strains, all isolated from exudates of white poplar trees (*Populus alba*, Salicaceae) growing in various areas of Hungary	Kurtzman (2011b)
Ogataea ramenticola	Frass, tunnel of an unidentified insect in loblolly pine (*Pinus taeda*, Pinaceae), Piney Woods, Mississippi, USA; frass, tunnel of an unidentified insect in shortleaf pine (*P. echinata*), Piney Woods, Mississippi, USA; frass, tunnel of western pine beetle (*Dendroctonus brevicomis*); Coulter pine (*P. coulteri*), Santa Barbara County, California, USA; insect frass, longleaf pine (*P. palustris*), Wilma, Florida, USA; insect frass, white pine (*P. strobus*), Duluth, Minnesota, USA; insect frass, red pine (*P. resinosa*), Angola, Indiana, USA; 15 strains from insect frass, unidentified pine (*Pinus* sp.), near Gainesville, Florida, USA	Kurtzman (2011b)
Ogataea saltuana	Rotten wood, Hungary; leaves of a sessile oak (*Quercus petraea*), Hungary; gut of an unidentified beetle collected from under the bark of a coniferous tree, Bulgaria	Péter et al. (2011)

(continued)

Table 9.3 (continued)

Species	Isolation source	Reference
Ogataea siamensis	Flower of grape jasmine (*Ervatamia coronaria*, Apocynaceae), Kanchanaburi Province, Thailand; flower of water willow (*Justicia fragilis*, Acanthaceae), Kanchanaburi Province, Thailand; flux of a mango tree (*Manaifera indica*, Anacardiaceae), Kalasin Province, Thailand	Kurtzman (2011b)
Ogataea thermomethanolica	Soil, Pathalung Province, Thailand; soil, Saraburi Province, Thailand	Kurtzman (2011b)
Ogataea trehaloabstinens	Exudate of an oak (*Quercus* sp., Fagaceae); exudate of a European hornbean (*Carpinus betulus*, Betulaceae); exudate of a sessile oak (*Q. petrea*), all from Pilis Mountains, Hungary	Kurtzman (2011b)
Ogataea trehalophila	Slime flux of cottonwood (*Populus trichocarpa*, Salicaceae), California, USA	Kurtzman (2011b)
Ogataea uvarum	Grape (*Vitis vinifera*) bunches, Cutrofiano, Italy	Roscini et al. (2018)
Ogataea wangdongensis	External surface of *Vitis vinifera* leaves, Kanchanaburi province, Thailand	Limtong et al. (2013)
Ogataea wickerhamii	Soil from a swamp in which spruce and broad-leafed trees were growing, Hÿÿttiala Province of Tavastia Australis, Finland	Kurtzman (2011b)
Ogataea zsoltii	Rotted railroad tie, Dorog, Hungary	Kurtzman (2011b)

Some recently described methylotrophic *Ogatae* and *Candida* species were recovered from leaves (Table 9.3) of different plants at different geographic locations at three continents. At first glance this is a surprising sequel, considering that the predominant yeast colonisers of phylloplane revealed by culture-dependent approaches are usually basidiomycetous (Fonseca and Inácio 2006; Limtong and Nasanit 2017; Kemler et al. 2017). It was observed more than 20 years ago that plants emit methanol to the atmosphere from their leaves (MacDonald and Fall 1993; Nemecek-Marshall et al. 1995). The methanol produced in leaves is probably a by-product of pectin metabolism during cell wall synthesis (Fall and Benson 1996). The annual amount of methanol emitted by plants globally was estimated to be 10^8 tons (Galbally and Kirstine 2002). Although the majority of methanol produced in leaves is emitted to the atmosphere through stomata, it is probable that the plant surfaces are enriched in methanol produced inside the leaves. It is also known that leaf methanol supports an abundant population of epiphytic methylotrophic bacteria (Fall and Benson 1996). On leaves of *Arabidopsis thaliana*, methanol was identified as the main available carbon source. The concentration of free methanol on leaf surfaces was estimated to be between 25 and 250 mM and it was also observed that the distribution of methanol is not limited to the area surrounding the stomata but is present on the entire leaf surface (Kawaguchi et al. 2011). Considering these findings, it was reasonable to ask whether leaves could harbour methylotrophic yeasts as well. The application of enrichment in methanol containing media gave a positive answer. Methanol-assimilating yeasts, although in small numbers, are consistently present on the leaves of many plants and can repeatedly be isolated given that suitable method (enrichment) is applied and the sample size is sufficient as well. Our results (see below) indicate that more than half of the leaf samples of trees and herbaceous plants collected in Hungary yielded methanol-assimilating yeasts. Some novel methylotrophic yeast species have also been described recently from leaves: *O. allantospora* (Péter et al. 2007), *O. phyllophila*, *C. chumphonensis*, *C. mattranensis* (Koowadjanakul et al. 2011), *O. kanchanaburiensis*, *O. wangdongensis* (Limtong et al. 2013) and *O. mangiferae* (Santos et al. 2015). Future investigations shall reveal whether methylotrophic yeasts are autochthonous or allochthonous members of the microbial community inhabiting the leaves. Methylotrophic yeasts are occasionally isolated also from sea- and freshwater (Kurtzman 2011b; Lachance 2003).

Among the isolation sources of methylotrophic yeasts are insects, insect frass and insect galleries. As in many other cases insects are supposed to be amid the dispersing vectors of methylotrophic yeasts. Tree exudates, rotten wood, aerial plant surfaces, soil, and suitable substrates for isolation of methanol-assimilating yeasts are inhabited or visited by insects. Insects are also supposed to influence the geographic distribution of some methylotrophic yeast species. It was reported that *K.* (*Pichia*) *pastoris* commonly occurs in tree exudates in western North America and in Europe, but is extremely rare in Japan. The limited distribution of *K. pastoris* was supposed to be correlated with the distribution of specific insects which introduce the yeasts in tree fluxes (Phaff and Starmer 1987; Phaff et al. 1978). It has been shown that large foliar emissions of methanol is induced by caterpillars (of *Euphydryas aurinia*) feeding on devil's-bit (*Succisa pratensis*) leaves. Attacked leaves emitted about five times more methanol than unattacked ones (Peñuelas et al. 2005). The available data suggest that a complex set of interactions among plants, insects and methylotrophic yeasts waits for more detailed studies.

9.3.4 The Occurrence of Methylotrophic Yeasts in Three Plant Substrates

In this section, the diversity of methylotrophic yeasts detected in rotting wood, tree exudates and on leaves collected mainly in Hungary is summarised based on our, partly unpublished, results. The samples were processed with the above-described two-step enrichment in methanol-containing broth. The sample size of rotting wood and leaf in most cases was 10 gram. If several indistinguishable strains were isolated from the same sample, they are reported as single strains. The number of the processed samples, the number and proportion of the samples yielding methanol-assimilating yeasts and the number of methanol-assimilating isolates for each type of substrate are summarised in Table 9.4. Rotting wood samples were obtained from numerous tree species including some unidentified ones. The highest number of rotting wood samples originated from oaks (*Quercus* spp.), beech (*Fagus sylvatica*) and hornbeam (*Carpinus betulus*). In the case of tree exudates, samples were collected from 15 plant genera. Most frequently oaks, poplars (*Populus* spp.), hornbeam and beech exudates were sampled. Leaf samples were collected most often from oaks, hornbeam and beech. More than half of the leaf samples yielded methylotrophic yeast species of the subphylum *Saccharomycotina*.

The methylotrophic yeast species isolated from rotting wood, tree exudates and leaves in their descending order of isolation frequency are listed in Table 9.5. Strikingly, the most frequently isolated methylotrophic yeast species from all substrate types in Hungary was *K. pastoris*. Its proportion exceeded 50% in the case of all substrate categories, but in tree exudates it came out in almost two-thirds of the isolates. *Candida boidinii*, the second species considering its isolation frequency after *K. pastoris*, was regularly isolated from rotting wood and tree exudates but was not detected among the 77 strains recovered from leaves.

Ogataea populialbae was found in tree exudates only, and even within tree exudates, it was recovered exclusively from the exudates of white poplar (*Populus alba*). The high abundance of *O. populialbae* in the exudates of white poplar and its absence in the exudates of other sympatric tree species suggest that the exudates of white poplar provide a selective habitat for the colonising yeasts (Péter et al. 2009b). It was also noted by the same authors that not one strain of *K. pastoris*, which is very common in the exudates of other trees in Hungary, was found in the investigated 31 white poplar exudates but two strains of *K. pseudopastoris* were isolated from this substrate. The selective forces in tree exudates may act either at the level of yeast-vectoring insects or directly on the growth of yeasts inhabiting the tree exudates (Lachance et al. 2001). In case of exudate of white poplar the supposed selective forces as well as the mode of their action are still waiting for exploration.

An additional pattern was observed in the host range of the distribution of *K. pseudopastoris*, which like *C. boidinii* was not found on leaves (Table 9.5). *Komagataella pseudopastoris* was not isolated from rotting wood or exudates of oaks (*Quercus* spp.) known to have high tannic acid content. It was supposed that the reason of the absence of this species on oaks is its sensitivity against tannic acid. The growth of *K. pseudopastoris* was inhibited by lower tannic acid concentration than that of *K. pastoris*, a species common on oaks in Hungary (Dlauchy et al. 2003).

Table 9.4 The efficiency of the enrichment in methanol-containing broth in the recovery of methylotrophic yeasts from different substrates

Substrate	Number of samples	Number and percentage of samples yielding methylotrophic yeasts	Number of isolated methylotrophic yeast strains
Rotting wood	242	134 (55%)	150
Tree exudate	197	137 (70%)	149
Leaf	121	67 (55%)	77

Table 9.5 Species composition of methylotrophic yeasts in rotting wood, tree exudates and leaves

Species	Number of strains isolated from rotting wood (altogether 150)	Number of strains isolated from tree exudates (altogether 149)	Number of strains isolated from leaves (altogether 77)
Komagataella pastoris	81	97	44
Candida boidinii	18	14	–
Ogataea populialbae	–	19	–
Komagataella pseudopastoris (*Pichia pseudopastoris*)[a]	9	4	–
Kuraishia molischiana	3	1	9
Ogataea trehaloabstinens (*Pichia trehaloabstinens*)[a]	4	3	2
Candida succiphila	4	2	2
Ogataea dorogensis (*Pichia dorogensis*)[a]	2	–	5
Ogataea pignaliae	–	2	4
Ogataea allantospora	–	–	5
Ogataea nitratoaversa	4	–	1
Ogataea trehalophila	4	–	1
Ogataea saltuana	3	–	2
Kuraishia capsulata	3	–	–
Ogataea zsoltii (*Pichia zsoltii*)[a]	2	–	1
Kuraishia floccosa (*Candida floccosa*)[a]	–	2	–
Kuraishia cidri	2	–	–
Kuraishia hungarica (*Candida hungarica*)[a]	2	–	–
Kuraishia ogatae (*Candida ogatae*)[a]	2	–	–
Candia nitratophila	1	1	–
Ogataea polymorpha	1	–	–
Ogataea parapolymorpha	1	–	–
Ogataea deakii	1	–	–
Ogataea pini	–	1	–
Ogataea pilisensis (*Pichia pilisensis*)[a]	1	–	–
Kuraishia cidri	–	1	–
Ogataea cf. *trehalophila*	2	1	–
Candida cf. *boidinii*	–	1	–

[a]Described as given in parentheses

Kuraishia molischiana, the anamorphic state of which (*C. molischiana*) has been considered to be conspecific with *K. capsulata* for a long time, was isolated more frequently (13 times) than *K. capsulata* (3 times) from the investigated 560 rotting wood, tree exudate and leaf samples. *Ogataea allantospora*, the first and hitherto only species in the genus *Ogataea* which forms allantoid ascospores, was found until now exclusively on leaves. Despite the extensive sampling several recently described species were recovered from the investigated substrates only a few times; moreover, some of them are represented by single isolates only (Table 9.5). Therefore, speculations on their ecology at this point are premature. Most probably they are allochthonous components of the microbial communities of these substrates and their "real" habitats are to be explored. Listed at the bottom of Table 9.5 are also two undescribed species designated as *C.* cf. *boidinii* and *O.* cf. *trehalophila* which were recovered merely one and three times, respectively.

9.4 Concluding Remarks

The latest few decades have witnessed an accelerating exploring rate of yeast biodiversity, first of all as a result of the application of quick and accurate DNA sequence-based tools including culture-based and culture-independent methods. The number of described yeast species has grown quickly. The number of species has also increased in several biotechnologically important yeast groups, like those discussed above. Analyses of DNA sequences, moving gradually from single locus through multigene to genome-scale approaches, help us to better understand the evolution of yeasts. It can be predicted that metagenomic studies will result in a major advance in the exploration of biodiversity of different habitats, while as a result of ongoing culture-based studies coupled with the application of whole-genome sequencing, increasing amount of well-characterised biological material will be available for future biotechnological studies.

References

Abunyewa AAO, Laing E, Hugo A, Viljoen BC (2000) The population change of yeasts in commercial salami. Food Microbiol 17(4):429–438. https://doi.org/10.1006/fmic.1999.0333

Addis E, Fleet GH, Cox JM, Kolak D, Leung T (2001) The growth, properties and interactions of yeasts and bacteria associated with the maturation of Camembert and blueveined cheeses. Int J Food Microbiol 69(1–2):25–36. https://doi.org/10.1016/S0168-1605(01)00569-4

Asthana H, Humphrey AE, Moritz V (1971) Growth of yeast on methanol as the sole carbon substrate. Biotechnol Bioeng 13(6):923–929. https://doi.org/10.1002/bit.260130615

Bai M, Qing M, Guo Z, Zhang Y, Chen X, Bao Q, Zhang H, Sun T (2010) Occurrence and dominance of yeast species in naturally fermented milk from the Tibetan Plateau of China. Can J Microbiol 56(9):707–714. https://doi.org/10.1139/W10-056

Barth G (ed) (2013) *Yarrowia lipolytica*. Biotechnological applications. Springer, Berlin
Bigey F, Tuery K, Bougard D, Nicaud J, Moulin G (2003) Identification of a triacylglycerol lipase gene family in *Candida deformans*: molecular cloning and functional expression. Yeast 20(3):233–248. https://doi.org/10.1002/yea.958
Bintsis T, Robinson RK (2004) A study of the effects of adjunct cultures on the aroma compounds of Feta-type cheese. Food Chem 88(3):435–441. https://doi.org/10.1016/j.foodchem.2004.01.057
Boonmak C, Limtong S, Jindamorakot S, Am-In S, Yongmanitchai W, Suzuki K, Nakase T, Kawasaki H (2011) *Candida xylanilytica* sp. nov., a xylan-degrading yeast species isolated from Thailand. Int J Syst Evol Microbiol 61(Pt5):1230–1234. https://doi.org/10.1099/ijs.0.021873-0
Booth JL, Vishniac HS (1987) Urease testing and yeast taxonomy. Can J Microbiol 33(5):396–404. https://doi.org/10.1139/m87-069
Butinar L, Strmole T, Gunde-Cimerman N (2011) Relative incidence of ascomycetous yeasts in arctic coastal environment. Microb Ecol 61(4):832–843. https://doi.org/10.1007/s00248-010-9794-3
Čadež N, Dlauchy D, Raspor P, Péter G (2013) *Ogataea kolombanensis* sp. nov., *Ogataea histrianica* sp. nov. and *Ogataea deakii* sp. nov., three novel yeast species from plant sources. Int J Syst Evol Microbiol 63(Pt8):3115–3123. https://doi.org/10.1099/ijs.0.052589-0
Čadež N, Dlauchy D, Tóbiás A, Péter G (2017) *Kuraishia mediterranea* sp. nov., a methanol-assimilating yeast species from olive oil and its sediment. Int J Syst Evol Microbiol 67(11):4846–4850. https://doi.org/10.1099/ijsem.0.002392
Carreira A, Loureiro V (1998) A differential medium to detect *Yarrowia lipolytica* within 24 hours. J Food Mycol 1:3–12
Chang C-F, Chen C-C, Lee C-F, Liu S-M (2013) Identifying and characterizing *Yarrowia keelungensis* sp. nov., an oil-degrading yeast isolated from the sea surface microlayer. A Van Leeuw J Microb 104(6):1117–1123. https://doi.org/10.1007/s10482-013-0033-z
Chen L-S, Ma Y, Maubois J-L, Chen L-J, Liu Q-H, Guo J-P (2010) Identification of yeasts from raw milk and selection for some specific antioxidant properties. Int J Dairy Technol 63(1):47–54. https://doi.org/10.1111/j.1471-0307.2009.00548.x
Coelho MAZ, Amaral PFF, Belo I (2010) *Yarrowia lipolytica*: an industrial workhorse. In: Mendez-Vilas A (ed) Current Research, Technology and Education Topics in Applied Microbiology and Microbial Biotechnology. Formatex Research Center, Badajoz, pp 930–944
Corbo MR, Lanciotti R, Albenzio M, Sinigaglia M (2001) Occurrence and characterization of yeasts isolated from milks and dairy products of Apulia region. Int J Food Microbiol 69(1–2):147–152. https://doi.org/10.1016/S0168-1605(01)00585-2
Daniel HM, Vrancken G, Takrama JF, Camu N, De Vos P, De Vuyst L (2009) Yeast diversity of Ghanaian cocoa bean heap fermentations. FEMS Yeast Res 9(5):774–783. https://doi.org/10.1111/j.1567-1364.2009.00520.x
Daniel HM, Lachance MA, Kurtzman CP (2014) On the reclassification of species assigned to *Candida* and other anamorphic ascomycetous yeast genera based on phylogenetic circumscription. A Van Leeuw J Microb 106(1):67–84. https://doi.org/10.1007/s10482-014-0170-z
de Koning W, Harder W (1992) Methanol-utilizing yeasts. In: Murell JC, Dalton H (eds) Methane and methanol utilizers. Plenum Press, New York, pp 207–244
Deák T (2006) *Candida* and related genera. In: Blackburn C (ed) Food spoilage microorganisms. Woodhead Publishing Limited, Cambridge, pp 334–353
Deák T (2008) Handbook of food spoilage yeasts, 2nd edn. CRC Press, Boca Raton
Diriye EU, Scorzeiti G, Martini A (1993) Methods for the separation of yeast cells from surfaces of processed, frozen foods. Int J Food Microbiol 19(1):27–37. https://doi.org/10.1016/0168-1605(93)90121-V
Dlauchy D, Tornai-Lehoczki J, Fülöp L, Péter G (2003) Pichia (Komagataella) pseudopastoris sp. nov., a new yeast species from Hungary. A Van Leeuw J Microb 83(4):327–332. https://doi.org/10.1023/A:1023318829389
Drake SD, Evans JB, Niven CF (1959) The identity of yeasts in the surface flora of packaged frankfurters. J Food Sci 24(3):243–246. https://doi.org/10.1111/j.1365-2621.1959.tb17268.x

Encinas JP, López-Diáz TM, García-López ML, Otero A, Moreno B (2000) Yeast populations on Spanish fermented sausages. Meat Sci 54(3):203–208. https://doi.org/10.1016/S0309-1740(99)00080-7

Fall R, Benson AA (1996) Leaf methanol – the simplest natural product from plants. Trends Plant Sci 1(9):296–301. https://doi.org/10.1016/S1360-1385(96)88175-0

Felsenstein J (1985) Confidence limits on phylogenies: an approach using the bootstrap. Evolution 39(4):783–791. https://doi.org/10.2307/2408678

Fleet GH (1990) Yeasts in dairy products. J Appl Bacteriol 68:199–211

Fleet GH (1992) Spoilage yeasts. Crit Rev Biotechnol 12(1–2):1–44. https://doi.org/10.3109/07388559209069186

Fonseca Á, Inácio J (2006) Phylloplane yeasts. In: Rosa C, Péter G (eds) Biodiversity and ecophysiology of yeasts. Springer, Berlin, pp 263–301

Fournier P, Gaillardin C, Persuy MA, Klootwijk J, Heerikhuizen H (1986) Heterogeneity in the ribosomal family of the yeast *Yarrowia lipolytica*: genomic organization and segregation studies. Gene 42(3):273–282. https://doi.org/10.1016/0378-1119(86)90231-3

Fröhlich-Wyder MT (2003) Yeasts in dairy products. In: Boekhout T, Robert V (eds) Yeasts in food: beneficial and detrimental aspects. CRC Press, Boca Raton, pp 209–237

Fung DY, Liang C (1990) Critical review of isolation, detection, and identification of yeasts from meat products. Crit Rev Food Sci 29(5):341–379. https://doi.org/10.1080/10408399009527532

Gadaga TH, Mutukumira AN, Narvhus JA (2000) Enumeration and identification of yeasts isolated from Zimbabwean traditional fermented milk. Int Dairy J 10(7):459–466. https://doi.org/10.1016/S0958-6946(00)00070-4

Galán-Sánchez F, Garcia-Agudo L, Garcia-Martos P, Rodrigues-Iglesias M (2014) *Candida galli* as a cause of tinea unguium - molecular characterization of a rare clinical fungal entity. Mycopathologia 178(3–4):303–330. https://doi.org/10.1007/s11046-014-9789-6

Galbally IE, Kirstine W (2002) The production of methanol by flowering plants and the global cycle of methanol. J Atmos Chem 43(3):195–229. https://doi.org/10.1023/A:1020684815474

Gardini F, Suzzi G, Lombardi A, Galgano F, Crudele MA, Andrighetto C, Schirone M, Tofalo R (2001) A survey of yeasts in traditional sausages of southern Italy. FEMS Yeast Res 1(2):161–167. https://doi.org/10.1016/S1567-1356(01)00024-1

Gasser B, Mattanovich D (2018) A yeast for all seasons – is *Pichia pastoris* a suitable chassis organism for future bioproduction? FEMS Microbiol Lett 365(17):1–4. https://doi.org/10.1093/femsle/fny181

Glushakova AM, Maximova IA, Kachalkin AV, Yurkov AM (2010) *Ogataea cecidiorum* sp. nov., a methanol-assimilating yeast isolated from galls on willow leaves. A Van Leeuw J Microb 98(1):93–101. https://doi.org/10.1007/s10482-010-9433-5

Gouliamova DE, Dimitrov RA, Guéorguiev BV, Smith MT, Groenewald M (2017) *Yarrowia parophonii*. Fungal Planet description sheets Personia 39:288–289

Groenewald M, Smith MT (2013) The teleomorph state of *Candida deformans* Langeron & Guerra and description of *Yarrowia yakushimensis* comb. nov. A Van Leeuw J Microb 103(5):1023–1028. https://doi.org/10.1007/s10482-013-9882-8

Groenewald M, Boekhout T, Neuvéglise C, Gaillardin C, van Dijck PWM, Wyss M (2013) *Yarrowia lipolytica*: safety assessment of an oleaginous yeast with a great industrial potential. Crit Rev Microbiol 40(3):187–206. https://doi.org/10.3109/1040841X.2013.770386

Groenewald M, Lombard L, de Vries M, Lopez AG, Smith M, Crous PW (2018) Diversity of yeast species from Dutch garden soil and the description of six novel Ascomycetes. FEMS Yeast Res 18(7). https://doi.org/10.1093/femsyr/foy076

Guerzoni ME, Lancotti R, Marchetti R (1993) Survey of the physiological properties of the most frequent yeasts associated with commercial chilled foods. Int J Food Microbiol 17(4):329–341. https://doi.org/10.1016/0168-1605(93)90203-s

Guerzoni ME, Lanciotti R, Vannini L, Galgano F, Favati F, Gardini F, Suzzi G (2001) Variability of the lipolytic activity in *Yarrowia lipolytica* and its dependence on environmental conditions. It J Food Microbiol 69(1–2):79–89. https://doi.org/10.1016/S0168-1605(01)00575-x

Harder V, Veenhuis M (1989) Metabolism of one-carbon compounds. In: Rose A, Harrison J (eds) Metabolism and physiology of yeasts, 3rd edn. Academic, San Diego, pp 289–316
Hawksworth DL (2011) A new dawn for the naming of fungi: impacts of decisions made in Melbourne in July 2011 on the future publication and regulation of fungal names. MycoKeys 1:7–20. https://doi.org/10.3897/mycokeys.1.2062
Hawksworth DL, Crous PW, Redhead AS, Reynolds DR, Samson RA, Seifert KA, Taylor JW, Wingfield MJ, 80 signers (2011) The Amsterdam declaration on fungal nomenclature. Mycotaxon 2(1):491–500. https://doi.org/10.5598/imafungus.2011.02.01.14
Hazeu W, de Bruyn JC, Bos P (1972) Methanol assimilation by yeasts. Arch Mikrobiol 87(2):185–188. https://doi.org/10.1007/BF00425000
Ismail SAS, Deák T, Abd-Rahman HA, Yassien MA, Beuchat LR (2000) Presence and changes in populations of yeasts on raw and processed poultry products stored at refrigeration temperature. Int J Food Microbiol 62(1–2):113–121. https://doi.org/10.1016/S0168-1605(00)00414-1
Johnson EA, Echavarri-Erasun C (2011) Yeast biotechnology. In: Kurtzman CP, Fell JW, Boekhout T (eds) The yeasts: a taxonomic study, 5th edn. Elsevier, Amsterdam, pp 21–63
Kawaguchi K, Yurimoto H, Oku M, Sakai Y (2011) Yeast methylotrophy and autophagy in a methanol-oscillating environment on growing Arabidopsis thaliana leaves. PLoS One 6:e25257. https://doi.org/10.1371/journal.pone.0025257
Kemler M, Witfeld F, Begerow D, Yurkov A (2017) Phylloplane yeasts in temperate climates. In: Buzzini P, Lachance MA, Yurkov A (eds) Yeasts in natural ecosystems: diversity. Springer, Cham, pp 171–197
Kimura M (1980) A simple method for estimating evolutionary rate of base substitutions through comparative studies of nucleotide sequences. J Mol Evol 16(2):111–120. https://doi.org/10.1007/BF01731581
Knutsen AK, Robert V, Poot A, Epping W, Figge M, Holst-Jensen A, Skaar I, Smith MT (2007) Polyphasic re-examination of *Yarrowia lipolytica* strains and the description of three novel *Candida* species: *Candida oslonensis* sp. nov., *Candida alimentaria* sp. nov. and *Candida hollandica* sp. nov. Int J Syst Evol Micr 57(10):2426–2435. https://doi.org/10.1099/ijs.0.65200-0
Koowadjanakul N, Jindamorakot S, Yongmanitchai W, Limtong S (2011) *Ogataea phyllophila* sp. nov., *Candida chumphonensis* sp. nov. and *Candida mattranensis* sp. nov., three methylotrophic yeast species from phylloplane in Thailand. A Van Leeuw J Microb 100(2):207–217. https://doi.org/10.1007/s10482-011-9579-9
Kreger-van Rij NJW (1984) The yeasts: a taxonomic study, 3rd edn. Elsevier, Amsterdam
Kumar S, Stecher G, Li M, Knyaz C, Tamura K (2018) MEGA X: molecular evolutionary genetics analysis across computing platforms. Mol Biol Evol 35(6):1547–1549. https://doi.org/10.1093/molbev/msy096
Kurtzman CP (1984) Synonymy of the yeast genera *Hansenula* and *Pichia* demonstrated through comparisons of deoxyribonucleic acid relatedness. A Van Leeuw J Microb 50(3):209–217. https://doi.org/10.1007/BF02342132
Kurtzman CP (1998) *Yarrowia* van der Walt & von Arx. In: Kurtzman CP, Fell JW (eds) The yeasts, a taxonomic study, 4th edn. Elsevier, Amsterdam, pp 420–421
Kurtzman CP (2005) New species and a new combination in the *Hyphopichia* and *Yarrowia* yeast clades. Anton Leeuw Int J G 88(2):121–130. https://doi.org/10.1007/s10482-005-2495-0
Kurtzman CP (2009) Biotechnological strains of *Komagataella* (*Pichia*) *pastoris* are *Komagataella phaffii* as determined from multigene sequence analysis. J Ind Microbiol Biot 36(11):1435–1438. https://doi.org/10.1007/s10295-009-0638-4
Kurtzman CP (2011a) Discussion of teleomorphic and anamorphic ascomycetous yeasts and yeast-like taxa. In: Kurtzman CP, Fell JW, Boekhout T (eds) The yeasts: a taxonomic study, 5th edn. Elsevier, Amsterdam, pp 293–307
Kurtzman CP (2011b) *Ogataea* Y. Yamada, K. Maeda & Mikata (1994). In: Kurtzman CP, Fell JW, Boekhout T (eds) The yeasts: a taxonomic study, 5th edn. Elsevier, Amsterdam, p 645–671
Kurtzman CP (2011c) *Pichia* E.C. Hansen (1904) In: Kurtzman CP, Fell JW, Boekhout T (eds) the yeasts: a taxonomic study, 5th edn. Elsevier, Amsterdam, p 685–708

Kurtzman CP (2011d) *Komagataella* Y. Yamada, Matsuda, Maeda & Mikata (1995) In: Kurtzman CP, Fell JW, Boekhout T (eds) the yeasts: a taxonomic study, 5th edn. Elsevier, Amsterdam, p 491–495

Kurtzman CP (2012) *Komagataella populi* sp. nov. and *Komagataella ulmi* sp. nov., two new methanol assimilating yeasts from exudates of deciduous trees. A Van Leeuw J Microb 101(4):859–868. https://doi.org/10.1007/s10482-012-9702-6

Kurtzman CP and Fell JW (eds) (1998) The Yeasts: A Taxonomic Study, 4th edn. Elsevier Science, B.V. Amsterdam

Kurtzman CP, Robnett CJ (1994) Orders and families of ascosporogenous yeasts and yeastlike taxa compared from ribosomal RNA sequence similarities. In: Hawksworth DL (ed) Ascomycete systematics: problems and perspectives in the nineties. Plenum Press, New York, pp 249–258

Kurtzman CP, Robnett CJ (1995) Molecular relationships among hyphal ascomycetous yeasts and yeastlike taxa. Can J Bot 73(S1):824–830. https://doi.org/10.1139/b95-328

Kurtzman CP, Robnett CJ (1998) Identification and phylogeny of ascomycetous yeasts from analysis of nuclear large subunit (26S) ribosomal DNA partial sequences. A Van Leeuw J Microb 73(4):331–371. https://doi.org/10.1023/A:1001761008817

Kurtzman CP, Robnett CJ (2010) Systematics of methanol assimilating yeasts and neighboring taxa from multigene sequence analysis and the proposal of *Peterozyma* gen. nov., a new member of the Saccharomycetales. FEMS Yeast Res 10(3):353–361. https://doi.org/10.1111/j.1567-1364.2010.00625.x

Kurtzman CP, Robnett CJ (2013a) Relationships among Genera of the Saccharomycotyna (Ascomycota) from multigene phylogenetic analysis of type species. FEMS Yeast Res 13(1):23–33. https://doi.org/10.1111/1567-1364.12006

Kurtzman CP, Robnett CJ (2013b) *Alloascoidea hylecoeti* gen. nov., comb. nov., *Alloascoidea africana* comb. nov., *Ascoidea tarda* sp. nov., and *Nadsonia starkeyi-henricii* comb. nov., new members of the Saccharomycotina (Ascomycota). FEMS Yeast Res 13(5):423–432. https://doi.org/10.1111/1567-1364.12044

Kurtzman CP, Robnett CJ (2014) Description of *Kuraishia piskuri* f.a., sp. nov., a new methanol assimilating yeast and transfer of phylogenetically related *Candida* species to the genera *Kuraishia* and *Nakazawaea* as new combinations. FEMS Yeast Res 14(7):1028–1036. https://doi.org/10.1111/1567-1364.12192

Kurtzman CP, Robnett CJ, Basehoar-Powers E (2008) Phylogenetic relationships among species of *Pichia*, *Issatchenkia* and *Williopsis* determined from multigene sequence analysis, and the proposal of *Barnettozyma* gen. nov., *Lindnera* gen. Nov. and *Wickerhamomyces* gen. nov. FEMS Yeast Res 8(6):939–954. https://doi.org/10.1111/j.1567-1364.2008.00419.x

Kurtzman CP, Fell JW, Boekhout T (2011) The yeasts, a taxonomic study, 5th edn. Elsevier, Amsterdam

Lachance MA (2003) The Phaff school of yeast ecology. Int Microbiol 6(3):163–167. https://doi.org/10.1007/s10123-003-0129-9

Lachance MA, Klemens JA, Bowles JM, Janzen DH (2001) The yeast community of sap fluxes of Costa Rican *Maclura* (*Chlorophora*) *tinctoria* and description of two new yeast species, *Candida galis* and *Candida ortonii*. FEMS Yeast Res 1(2):87–92. https://doi.org/10.1111/j.1567-1364.2001.tb00019.x

Lachance MA, Boekhout T, Scorzetti G, Fell JW, Kurtzman CP (2011) *Candida* Berkhout (1923). In: Kurtzman CP, Fell JW, Boekhout T (eds) The yeasts: a taxonomic study, 5th edn. Elsevier, Amsterdam, pp 987–1278

Lanciotti R, Massa S, Guerzoni ME, Di Fabio G (1992) Light butter: natural microbial population and potential growth of *Listeria monocytogenes* and *Yersinia enterocolitica*. Lett Appl Microbiol 15(6):256–258. https://doi.org/10.1111/j.1472-765X.1992.tb00777.x

Lanciotti R, Vannini L, Chaves Lopez C, Gobbetti M, Guerzoni ME (2005) Evaluation of the ability of *Yarrowia lipolytica* to impart strain-dependent characteristics to cheese when used as a ripening adjunct. Int J Dairy Technol 58(2):89–99. https://doi.org/10.1111/j.1471-0307.2005.00197.x

Larpin S, Mondoloni C, Goerges S, Vernoux J-P, Guéguen M, Desmasures N (2006) *Geotrichum candidum* dominates in yeast population dynamics in Livarot, a French redsmear cheese. FEMS Yeast Res 6(8):1243–1253. https://doi.org/10.1111/j.1567-1364.2006.00127.x

Lee J-D, Komagata K (1980) Taxonomic study of methanol-assimilating yeasts. J Gen Appl Microbiol 26(2):133–158. https://doi.org/10.2323/jgam.26.133

Leme FC, de Barros Negreiros MM, Koga FA, de Moraes Gimenes Bosco S, Bagagli E, Haddad V Jr (2011) Evaluation of pathogenic fungi occurrence in traumatogenic structures of freshwater fish. Rev Soc Bras Med Tro 44(2):182–185. doi:https://doi.org/10.1590/S0037-86822011005000007

Levine DW, Cooney CL (1973) Isolation and characterization of a thermotolerant methanol-utilizing yeast. Appl Microbiol 26(6):982–990

Liang C (1989) Rapid presumptive identification of yeasts in meat products, thesis. Kansas State University p 146

Limtong S, Nasanit R (2017) Phylloplane yeasts in tropical climates. In: Buzzini P, Lachance MA, Yurkov A (eds) Yeasts in natural ecosystems: diversity. Springer, Cham, pp 199–223

Limtong S, Youngmanitchai W, Kawasaki H, Seki T (2008) *Candida phangngensis* sp. nov., an anamorphic yeast species in the *Yarrowia* clade, isolated from water in mangrove forests in Phang-Nga Province, Thailand. Int J Syst Evol Micr 58(Pt2):515–519. https://doi.org/10.1099/ijs.0.65506-0

Limtong S, Kaewwichian R, Groenewald M (2013) *Ogataea kanchanaburiensis* sp. nov. and *Ogataea wangdongensis* sp. nov., two novel methylotrophic yeast species from phylloplane in Thailand. A Van Leeuw J Microb 103(3):551–558. https://doi.org/10.1007/s10482-012-9837-5

Lin CCS, Fung DYC (1985) Effect of dyes on the growth of food yeasts. J Food Sci 50(1):241–244. https://doi.org/10.1111/j.1365-2621.1985.tb13319.x

Liu KF, Li XH, Hui FL (2018) *Yarrowia brassicae* f.a., sp. nov., a new yeast species from traditional Chinese sauerkraut. Int J Syst Evol Microbiol 68(6):2024–2027. https://doi.org/10.1099/ijsem.0.002783

Lodder J (ed) (1970) The yeasts: a taxonomic study, 2nd edn. North-Holland Publishing Company, Amsterdam

Lodder J, Kreger-van Rij NJW (1952) The yeasts: a taxonomic study. North-Holland Publishing Company, Amsterdam

Lopandic K, Zelger S, Bánszky LK, Eliskases-Lechner F, Prillinger H (2006) Identification of yeasts associated with milk products using traditional and molecular techniques. Food Microbiol 23(4):341–350. https://doi.org/10.1016/j.fm.2005.05.001

Lourens-Hattingh A, Viljoen BC (2002) Survival of dairy-associated yeasts in yoghurt and yoghurt-related products. Food Microbiol 19(6):597–604. https://doi.org/10.1006/fmic.2002.0515

Lu YF, Wang M, Zheng J, Hui FL (2017) *Ogataea neixiangensis* sp. nov. and *Ogataea paraovalis* f.a., sp. nov., two methanol-assimilating yeast species isolated from rotting wood. Int J Syst Evol Microbiol 67(8):3038–3042. https://doi.org/10.1099/ijsem.0.002075

MacDonald RC, Fall R (1993) Detection of substantial emissions of methanol from plants to the atmosphere. Atmos Environ A-Gen 27(11):1709–1713. https://doi.org/10.1016/0960-1686(93)90233-o

McIlvaine TC (1921) A buffer solution for colorimetric comparison. J Biol Chem 49:183–186

McNeill J, Barrie FR, Buck WR, Demoulin V, Greuter W, Hawksworth, DL, Herendeen PS, Knapp S, Marhold K et al. (2012) International Code of Nomenclature for algae, fungi, and plants (Melbourne Code). Koeltz Scientific Books. http://www.iapt-taxon.org/nomen/main.php. ISBN:9783874294256

Meyer SA, Ahearn DG, Yarrow D (1984) *Candida* Berkhout. In: Kreger-van Rij NJW (ed) The yeasts, a taxonomic study, 3rd edn. Elsevier, Amsterdam, pp 585–849

Michely S, Gaillardin C, Nicaud J-M, Neuvéglise C (2013) Comparative physiology of oleaginous species from the *Yarrowia* clade. PLoS One 8(5).:7. Article number: e63356 https://doi.org/10.1371/journal.pone.0063356

Miller MW, Phaff HJ, Miranda M, Heed WB, Starmer WT (1976) *Torulopsis sonorensis*, a new species of the genus *Torulopsis*. Int J Syst Bacteriol 26:88–91. https://doi.org/10.1099/00207713-26-1-88

Monnet C, Bleicher A, Neuhaus K, Sarthou AS, Leclercq-Perlat M-N, Irlinger F (2010) Assessment of the antilisterial activity of microfloras from the surface of smear-ripened cheeses. Food Microbiol 27(2):302–310. https://doi.org/10.1016/j.fm.2009.11.009

Nagatsuka Y, Saito S, Sugiyama J (2008) *Ogataea neopini* sp. nov. and *O. corticis* sp. nov., with the emendation of the ascomycete yeast genus *Ogataea*, and transfer of *Pichia zsoltii*, *P. dorogensis*, and *P. trehaloabstinens* to it. J Gen Appl Microbiol 54(6):353–365. https://doi.org/10.2323/jgam.54.353

Nagy E (2014a) Differentiation of food spoilage yeast strains of the *Yarrowia* group by microsatellite polymerase chain reaction fingerprinting. J Univ Sci 1(1):6–11. https://doi.org/10.17202/JUSO.2014.1.6

Nagy E (2014b) Isolation and diversity of food spoilage *Yarrowia* yeast strains from meat. Acta Aliment Hung 3(Supplement 1):101–106. https://doi.org/10.1556/AAlim.43.2014.Suppl.15

Nagy E, Niss M, Dlauchy D, Arneborg N, Nielsen DS, Péter G (2013) *Yarrowia divulgata* f.a., sp. nov., a yeast species from animal-related and marine sources. Int J Syst Eol Microbiol 63(Pt12):4818–4823. https://doi.org/10.1099/ijs.0.057208-0

Nagy E, Dlauchy D, Medeiros AO, Péter G, Rosa CA (2014) *Yarrowia porcina* sp. nov. and *Yarrowia bubula* f.a. sp. nov., two yeast species from meat and river sediment. A Van Leeuw J Microb 105(4):697–707. https://doi.org/10.1007/s10482-014-0125-4

Nakase T, Imanishi Y, Ninomiya S, Takashima M (2010) *Candida rishirensis* sp. nov., a novel methylotrophic anamorphic yeast species isolated from soil on Rishiri Island in Japan. J Appl Microbiol 56(2):169–173. https://doi.org/10.2323/jgam.56.169

Naumov GI, Naumova ES, Tyurin OV, Kozlov DG (2013) *Komagataella kurtzmanii* sp. nov., a new sibling species of *Komagataella* (*Pichia*) *pastoris* in accordance with multigene sequence analysis. A Van Leeuw J Microb 104(3):339–347. https://doi.org/10.1007/s10482-013-9956-7

Naumov GI, Naumova ES, Lee CF (2017) *Ogataea haglerorum* sp. nov., a novel member of the species complex, *Ogataea (Hansenula) polymorpha*. Int J Syst Evol Microbiol 67(7):2465–2469. https://doi.org/10.1099/ijsem.0.002012

Naumov GI, Shalamitskiy MY, Naumova ES, Lee CF (2018a) Phylogenetics, biogeography, and ecology of methylotrophic yeasts of the heterogeneous genus *Ogataea*: achievements and prospects. Microbiology 87(4):443–452. https://doi.org/10.1134/S002626171804015x

Naumov GI, Naumova ES, Boundy-Mills KL (2018b) Description of *Komagataella mondaviorum* sp. nov., a new sibling species of *Komagataella* (*Pichia*) *pastoris*. A Van Leeuw J Microb 111(7):1197–1207. https://doi.org/10.1007/s10482-018-1028-6

Nemecek-Marshall M, MacDonald RC, Franzen JJ, Wojciechowsky CL, Fall R (1995) Methanol emission from leaves. Plant Physiol 108(4):1359–1368. https://doi.org/10.1104/pp.108.4.1359

Nielsen DS, Jacobsen T, Jespersen L, Koch AG, Arneborg N (2008) Occurrence and growth of yeasts in processed meat products: implications for potential spoilage. Meat Sci 80(3):919–926. https://doi.org/10.1016/j.meatsci.2008.04.011

Ogata K, Nishikawa H, Ohsugi M (1969) A yeast capable of utilizing methanol. Agr Biol Chem 33(10):1519–1520. https://doi.org/10.1080/00021369.1969.10859497

Ogata K, Nishikawa H, Ohsugi M, Tochikura T (1970) Studies on the production of yeast (II) the cultural conditions of methanol assimilatin yeast *Kloeckera* sp. No. 2201. J Ferment Technol 48:470–477

Oki T, Kouno K, Kitai A, Ozaki A (1972) New yeasts capable of assimilating methanol. J Gen Appl Microbiol 18(4):295–305. https://doi.org/10.2323/jgam.18.295

Ozturk I (2015) Presence, changes and technological properties of yeast species during processing of pastirma, a Turkish dry cured meat product. Food Control 50:76–84. https://doi.org/10.1016/j.foodcont.2014.08.039

Palande AS, Kulkarni SV, León-Ramirez C, Campos-Góngora E, Ruiz-Herrera J, Deshpande MV (2014) Dimorphism and hydrocarbon metabolism in *Yarrowia lipolytica var. indica*. Arch Microbiol 196(8):545–556. https://doi.org/10.1007/s00203-014-0990-2

Peñuelas J, Filella I, Stefanescu C, Llusià J (2005) Caterpillars of *Euphydryas aurinia* (Lepidoptera: Nymphalidae) feeding on *Succisa pratensis* leaves induce large foliar emissions of methanol. New Phytol 167(3):851–857. https://doi.org/10.1111/j.1469-8137.2005.01459.x

Péter G (2011) *Kuraishia* Y. Yamada, Maeda & Mikata (1994). In: Kurtzman CP, Fell JW, Boekhout T (eds) The yeasts: a taxonomic study, 5th edn. Elsevier, Amsterdam, pp 503–506

Péter G, Deák T (1991) On the false positive urease activity of *Yarrowia lipolytica*. A Van Leeuw J Microb 60(1):55–59. https://doi.org/10.1007/BF00580442

Péter G, Dlauchy D, Vasdinyei R, Tornai-Lehoczki J, Deák T (2004) *Candida galli* sp. nov., a new yeast from poultry. Anton Leeuw Int J G 86(2):105–110. https://doi.org/10.1023/B:ANTO.0000036117.03847.bd

Péter G, Tornai-Lehoczki J, Dlauchy D (2007) *Ogataea allantospora* sp. nov., an ascomycetous yeast species from phylloplane. A Van Leeuw J Microb 92(4):443–448. https://doi.org/10.1007/s10482-007-9172-4

Péter G, Tornai-Lehoczki J, Dlauchy D (2009a) *Candida ogatae* sp. nov., an anamorphic member of the Kuraishia clade. FEMS Yeast Res 9(2):328–333. https://doi.org/10.1111/j.1567-1364.2008.00466.x

Péter G, Tornai-Lehoczki J, Dlauchy D (2009b) *Ogataea populialbae* sp. nov., a yeast species from white poplar. FEMS Yeast Res 9(6):936–941. https://doi.org/10.1111/j.1567-1364.2009.00547.x

Péter G, Tornai-Lehoczki J, Dlauchy D (2010) *Ogataea pignaliae* sp. nov., the teleomorph of *Candida pignaliae*. Int J Syst Evol Microbiol 60(Pt10):2496–2500. https://doi.org/10.1099/ijs.0.019174-0

Péter G, Dlauchy D, Tornai-Lehoczki J, Gouliamova D, Kurtzman CP (2011) *Ogataea saltuana* sp. nov., a novel methanol-assimilating yeast species. A Van Leeuw J Microb 100(3):375–383. https://doi.org/10.1007/s10482-011-9592-z

Phaff HJ, Starmer WT (1987) Yeasts associated with plants, insects and soil. In: Rose AH, Harrison JS (eds) The yeasts, vol 1. The biology of yeasts, 2nd edn. Academic, London, pp 123–180

Phaff HJ, Miller MW, Mrak EM (1978) The life of yeasts, 2nd edn. Harvard University Press, Cambridge, MA

Pitt JI, Hocking AD (2009) Fresh and perishable foods. In: Pitt JI, Hocking AD (eds) Fungi and food spoilage, 3rd edn. Springer, London, pp 383–400

Praphailong W, Fleet GH (1997) The effect of pH, sodium chloride, sucrose, sorbate and benzoate on the growth of food spoilage yeasts. Food Microbiol 14(5):459–468. https://doi.org/10.1006/fmic.1997.0106

Rakicka M, Kieroń A, Piotr Hapeta P, Neuveglise-Degouy C, Lazar Z (2016) Sweet and sour potential of yeast from the *Yarrowia* clade. Biomass Bioenergy 92:48–54. https://doi.org/10.1016/j.biombioe.2016.06.004

Ramírez-Castrillón M, Mendes SDC, Inostroza-Ponta M, Valente P (2014) $(GTG)_5$ MSPPCR fingerprinting as a technique for discrimination of wine associated yeasts? PLoS One 9(8):e105870. https://doi.org/10.1371/journal.pone.0105870

Rohm H, Eliskases-Lechner F, Bräuer M (1992) Diversity of yeasts in selected dairy products. J Appl Bacteriol 72(5):370–376. https://doi.org/10.1111/j.1365-2672.1992.tb01848.x

Romano P, Capece A, Jespersen L (2006) Taxonomic and ecological diversity of food and beverage yeasts. In: Querol A, Fleet G (eds) The Yeast Handbook, volume 2, Yeasts in food and beverages. Springer, Berlin/Heidelberg, pp 13–54

Roostita R, Fleet GH (1996) The occurrence and growth of yeasts in Camembert and blueveined cheeses. Int J Food Microbiol 28(3):393–404. https://doi.org/10.1016/0168-1605(95)00018-6

Roscini L, Tristezza M, Corte L, Colabella C, Perrotta C, Rampino P, Robert V, Vu D, Cardinali G, Grieco F (2018) Early ongoing speciation of *Ogataea uvarum* sp. nov. within the grape ecosystem revealed by the internal variability among the rDNA operon repeats. Front Microbiol 3(9):1687. https://doi.org/10.3389/fmicb.2018.01687

Sahm H, Wagner F (1972) Microbial assimilation of methanol. Arch Mikrobiol 84(1):29–42. https://doi.org/10.1111/j.1432-1033.1973.tb02907.x

Saitou N, Nei M (1987) The neighbor-joining method: a new method for reconstructing phylogenetic trees. Mol Biol Evol 24(4):406–425. https://doi.org/10.1093/oxfordjournals.molbev.a040454

Samelis J, Sofos JN (2003) Yeasts in meat and meat products. In: Boekhout T, Robert V (eds) Yeasts in food. CRC Press, Boca Raton, pp 239–266

Santos AR, Faria ES, Lachance MA, Rosa CA (2015) *Ogataea mangiferae* sp. nov., a methylotrophic yeast isolated from mango leaves. Int J Syst Evol Microbiol 65(6):1855–1859. https://doi.org/10.1099/ijs.0.000194

Shen XX, Opulente DA, Kominek J et al (2018) Tempo and mode of genome evolution in the budding yeast subphylum. Cell 175(6):1533–1545. https://doi.org/10.1016/j.cell.2018.10.023

Sinigaglia M, Lanciotti R, Guerzoni ME (1994) Biochemical and physiological characteristics of *Yarrowia lipolytica* strains in relation to isolation source. Can J Microbiol 40(1):54–59. https://doi.org/10.1139/m94-008

Sipiczki M (2012) *Candida borneonana* sp. nov., a methanol-assimilating anamorphic yeast isolated from decaying fruit. Int J Syst Evol Microbiol 62(Pt9):2303–2306. https://doi.org/10.1099/ijs.0.039412-0

Suh SO, Zhou J (2010a) Yeasts associated with the curculionid beetle *Xyloterinus politus*: *Candida xyloterini* sp. nov., *Candida palmyrensis* sp. nov. and three common ambrosia yeasts. Int J Syst Evol Microbiol 60(Pt7):1702–1708. https://doi.org/10.1099/ijs.0.016907-0

Suh SO, Zhou JJ (2010b) Methylotrophic yeasts near *Ogataea* (*Hansenula*) *polymorpha*: a proposal of *Ogataea angusta* comb. nov. and *Candida parapolymorpha* sp. nov. FEMS Yeast Res 10(5):631–638. https://doi.org/10.1111/j.1567-1364.2010.00634.x

Suzuki M, Suh S-O, Sugita T, Nakase T (1999) A phylogenetic study on galactosecontaining *Candida* species based on 18S ribosomal DNA sequences. The J Gen Appl Microbiol 45(5):229–238. https://doi.org/10.2323/jgam.45.229

Suzzi G, Larnote M, Galgano F, Andrighetto C, Lombardi A, Lanciotti R, Guerzoni M (2001) Proteolytic, lipolytic and molecular characterization of *Yarrowia lipolytica* isolated from cheese. Int J Food Microbiol 69(1–2):69–77. https://doi.org/10.1016/S0168-1605(01)00574-8

Tamura K (1992) Estimation of the number of nucleotide substitutions when there are strong transition-transversion and G + C-content biases. Mol Biol Evol 9(4):678–687. https://doi.org/10.1093/oxfordjournals.molbev.a040752

Tamura K, Nei M (1993) Estimation of the number of nucleotide substitutions in the control region of mitochondrial DNA in humans and chimpanzees. Mol Biol Evol 10(3):512–526. https://doi.org/10.1093/oxfordjournals.molbev.a040023

van Uden N, Buckley HR (1970) *Candida* Berkhout. In: Lodder J (ed) The yeasts, a taxonomic study, 2nd edn. North-Holland Publishing Company, Amsterdam, p 893–1087

van den Tempel T, Jakobsen M (2000) The technological characteristics of *Debaryomyces hansenii* and *Yarrowia lipolytica* and their potential as starter cultures for production of Danablu. Int Dairy J 10(4):263–270. https://doi.org/10.1016/S0958-6946(00)00053-4

van der Walt JP, von Arx JA (1980) The yeast genus *Yarrowia* gen. nov. A Van Leeuw J Microb 46(6):517–521. https://doi.org/10.1007/bf00394008

van Dijken JP, Harder W (1974) Optimal conditions for the enrichment and isolation of methanol-assimilating yeasts. J Gen Microbiol 84(2):409–411. https://doi.org/10.1099/00221287-84-2-409

Viljoen BC, Dykes GA, Callis M, von Holy A (1993) Yeasts associated with Vienna sausage packaging. Int J Food Microbiol 18(1):53–62. https://doi.org/10.1016/0168-1605(93)90007-4

Viljoen BC, Lourens-Hattingh A, Ikalafeng B, Péter G (2003) Temperature abuse initiating yeast growth in yoghurt. Food Res Int 36(2):193–197. https://doi.org/10.1016/S0963-9969(02)00138-2

Volfová O, Pilát P (1974) Studies on methanol-oxidizing yeasts. Folia Microbiol 19(5):249–256. https://doi.org/10.1007/bf02878112

von Arx JA (1972) On Endomyces, Endomycopsis and related yeast-like fungi. A Van Leeuw J Microb 38(1):289–309. https://doi.org/10.1007/bf02328100

Vu D, Groenewald M, Szöke S, Cardinali G, Eberhardt U, Stielow B, de Vries M, Verkleij GJM, Crous PW, Boekhout T, Robert V (2016) DNA barcoding analysis of more than 9000 yeast isolates contributes to quantitative thresholds for yeast species and genera delimitation. Stud Mycol 85:91–105. https://doi.org/10.1016/j.simyco.2016.11.007

van der Walt JP, Scott DB (1971) The yeast genus *Saccharomycopsis* Schiönning. Mycopath Mycol Appl 43(3–4):279–288. doi:https://doi.org/10.1007/bf02051747
Wegner GH, Harder W (1987) Methylotrophic yeasts–1986. In: van Verseveld HW, Duine JA (eds) Microbial growth on C1 compounds. Springer, Dordrecht p 131–138
Welthagen JJ, Viljoen BC (1998) Yeast profile in Gouda cheese during processing and ripening. Int J Food Microbiol 41(3):185–194. https://doi.org/10.1016/S0168-1605(98)00042-7
Wickerham LJ, Kurtzman CP, Herman AI (1970) Sexual reproduction in *Candida lipolytica*. Science 167(3921):11–41. https://doi.org/10.1126/science.167.3921.1141
Wolter H, Laing E, Viljoen BC (2000) Isolation and identification of yeasts associated with intermediate moisture meats. Food Technol Biotech 38(1):69–75
Yamada Y, Nojiri M, Matsuyama M, Kondo K (1976) Coenzyme Q system in the classification of the ascosporogenous yeast genera *Debaryomyces*, *Saccharomyces*, *Kluyveromyces*, and *Endomycopsis*. J Gen Appl Microbiol 22(6):325–337. https://doi.org/10.2323/jgam.22.325
Yamada Y, Maeda K, Mikata K (1994) The phylogenetic relationships of the hat-shaped ascospore-forming, nitrate-assimilating *Pichia* species, formerly classified in the genus *Hansenula* Sydow et Sydow, based on the partial sequences of 18S and 26S ribosomal RNAs (Saccharomycetaceae): the proposals of three new genera, *Ogataea*, *Kuraishia*, and *Nakazawaea*. Biosci Biotech Bioch 58(7):1245–1257. https://doi.org/10.1271/bbb.58.1245
Yamada Y, Matsuda M, Maeda K, Mikata K (1995) The phylogenetic relationships of methanol-assimilating yeasts based on the partial sequences of 18S and 26S ribosomal RNAs: the proposal of *Komagataella* gen. nov. (Saccharomycetaceae). Biosci Biotech Biochem 59(3):439–444. https://doi.org/10.1271/bbb.59.439
Yarrow D (1972) Four new combinations in yeasts. A Van Leeuw J Microb 38(1):357–360. https://doi.org/10.1007/bf02328105
Yarrow D, Meyer SA (1978) Proposal for amendment of the diagnosis of the genus *Candida* Berkhout nom. cons. Int J Syst Bacteriol 28(4):611–615. https://doi.org/10.1099/00207713-28-4-611

Chapter 10
Anhydrobiosis in Non-conventional Yeasts

Alexander Rapoport

Abstract Anhydrobiosis is a unique state of living organisms that provides the possibility of surviving conditions of extreme heat and drought with temporary and reversible suspended processes of metabolism. Beyond its ecological importance in nature, it is used for the production of viable dry microbial preparations, including active dry yeast. During the last decade, this state has also been studied in non-conventional yeasts (NCY), mainly because of the necessity of active dry wine yeasts. At the same time, many other modern biotechnological processes are also linked to the use of NCY. Correspondingly, this implies an interest in the long-term maintenance of various species of NCY in a viable dry state. This review describes the results of studies of the dehydration of various NCY species. It demonstrates the very important contribution of these yeasts to the development of our knowledge of this unique phenomenon of nature. The results of the structural and biochemical investigations of NCY have revealed the main cell changes during their dehydration-rehydration, as well as the intracellular protective reactions necessary for the maintenance of yeast cell viability. The synthesis of various intracellular protective compounds was shown in NCY cells under the conditions of their dehydration-rehydration. All of this accumulated information is important for the development of new approaches for the successful production and use of biotechnologically important species of NCY in an active dry state, as well as for the improvement of our knowledge of the mechanisms of anhydrobiosis.

Keywords Anhydrobiosis · Dehydration-rehydration · Drying · Organelle changes · Protective reactions · Biotechnological application

---

A. Rapoport (✉)
Laboratory of Cell Biology, Institute of Microbiology and Biotechnology,
University of Latvia, Riga, Latvia
e-mail: rapoport@mail.eunet.lv

A. Sibirny (ed.), *Non-conventional Yeasts: from Basic Research to Application*,
https://doi.org/10.1007/978-3-030-21110-3_10

10.1 Introduction

Anhydrobiosis is one of the most unique phenomena of nature. It has been developed by various groups of living organisms during their evolution and provides the possibility of surviving and maintaining their viability in extremely unfavourable environmental conditions. Anhydrobiosis is a state in which the metabolism of living organisms is temporarily reversibly suspended as a result of water losses linked with a hot climate and drought, and the corresponding strong desiccation of the cells and tissues. Studies of anhydrobiosis have been conducted for more than 300 years, after its discovery by the great Dutch naturalist Anthony van Leeuwenhoek in 1701, and great progress in our knowledge has been reached in the last 50 years. Nevertheless, we are still far from understanding all the mechanisms of this unique state of life (Crowe 1971, 2015).

Yeasts and other microorganisms are able to transition into anhydrobiosis with a subsequent restoration of active life (Beker and Rapoport 1987; Hernandez-Garcia 2011, 2014; Dupont et al. 2014; Rapoport 2017; Rapoport et al. 2019). Results from the last 50 years of active studies are very interesting and important for basic science, as well as for biotechnology and medicine. This information helps us to understand the role of water in the maintenance of the structure and conformation of various intracellular membranes and macromolecules. It reveals the hidden potential of the main intracellular organelles directed towards the protection of an organisms' viability in extreme environmental conditions, as well as their additional functional value for metabolism. New protective compounds and earlier unknown intracellular protective reactions have been found and may still be revealed in future studies of anhydrobiosis.

Because the yeast cell is a universal model of eukaryotic cells, all the knowledge accumulated from studies of anhydrobiosis can be easily extrapolated to higher organisms and used, for example, in medicine for humans and animals when there is a need to increase the resistance of organisms during various treatments and difficult surgeries. Modern biotechnology broadly uses the phenomenon of anhydrobiosis for practical goals, including obtaining active dry microbial preparations that may be stored in a dry state for long time periods without serious losses to quality or the viability of dehydrated cells. The most well-known application of dry yeast preparations is in baking. Such preparations of active dry yeast or instant yeast that are produced in huge amounts by the yeast industry belong to the species of conventional yeast, *Saccharomyces cerevisiae*. At the same time, in the last decades, there is increasing attention of the possibility of also using non-conventional yeasts (NCY), including their dry preparations, in various fields of modern biotechnology (Rapoport et al. 2016; Rebello et al. 2018). It is important that, in many studies of anhydrobiosis, NCY have been used as the main object of investigations. In some cases, the obtained results were checked and confirmed in later experiments with conventional yeast *S. cerevisiae* or with other species of NCY. This review summarises the main results obtained from research with NCY and the achievements reached in these studies for the development of our knowledge of this unique phenomenon of nature.

10.2 Structural Studies

The initial yeast anhydrobiosis studies at the structural level were linked with the research of NCY. In the first electron microscopy studies of yeast, intracellular changes at their transfer into the state of anhydrobiosis and subsequent rehydration/reactivation were studied in *Endomyces magnusii* cells; the synonyms of this species are *Magnusiomyces magnusii* and *Dipodascus magnusii*. This research showed a decrease in cell sizes during dehydration, which led to the cells transitioning into a state of anhydrobiosis. The strong folding of the plasma membrane, which was revealed in these studies, is undoubtedly a specific organisms' protective reaction directed towards the maintenance of the integrity of this membrane under the conditions of diminishing cell size. These results were obtained using ultra-thin sections of dehydrated yeast cells (Rapoport and Kostrikina 1973). A few years later, these results were confirmed using another electron microscopy approach – freeze-etching (Biryusova and Rapoport 1978). Identical plasma membrane folding results were obtained for various yeast species – *E. magnusii*, *Candida utilis* and *S. cerevisiae*. This research showed that native yeast cells also have small folds of the plasma membrane, which significantly increase during dehydration and lead to the formation of a net of folds. On the basis of these studies, the functional role of the small folds of the plasma membrane in native cells became clear. They are necessary for the protection of the membrane in conditions of hyperosmotic stress and dehydration, providing the possibility for the cells to develop in cases of decreased cell size and to maintain the integrity of the plasma membrane (Rapoport and Kostrikina 1973; Rapoport et al. 1973).

In the same electron microscopy research, it was established that the dehydration of all studied yeast species leads to changes in the shape of the cell's nucleus. Moreover, essential widening of some nucleus' pores appeared in practically all of the cells studied using the method of ultra-thin sections, whereas freeze-etching studies showed definite considerable contraction of the nuclear pores. The analysis of these results concluded that there are serious changes in the lipid component in some areas of nuclear membranes. Thereof, the appropriate chemical fixation of these membrane areas does not take place during the preparation of the samples for electron microscopy studies. In these changed areas of the nuclear membranes, the localisation of mitochondria and/or lipid droplets may be revealed. Mitochondria are necessary in these areas as the source of energy for the subsequent repair of these changes/damages, but lipid droplets may supply reserve lipid material for these repair processes (Rapoport and Kostrikina 1973; Rapoport et al. 1973, 2019; Biryusova and Rapoport 1978).

Another extremely important observation from these studies of anhydrobiosis in NCY is linked to the detection of chromatin condensation in the nucleus in all viable cells. It was concluded that chromatin condensation is one of the most important intracellular protective reactions, which is necessary for the maintenance of cells' viability in a dehydrated state (Rapoport et al. 1973). An additional interesting intracellular protective reaction also linked with the nucleus was found in the studies of anhydrobiosis in *C. utilis* cells. It was revealed that, at the early stages of these cells' dehydration processes, the localisation of condensed chromatin at one of the poles of the nucleus and the synthesis of new nuclear membrane may take place in the cells, which separates the chromatin-containing part of the nucleus from the other part. Then, during these cells' rehydration/reactivation, the chromatin-containing part of the nucleus increases in size, but the part without chromatin is subjected to gradual autolysis and inactivation by active phagosomes. This phenomenon is directed towards the maintenance of the cells' genome under the conditions of strong dehydration and the diminished amount of water necessary for the maintenance of the main nucleus activities (Rapoport et al. 1973).

In all studied yeast species (and later also in *S. cerevisiae*), the mitochondria were revealed as the organelles with minimum changes during the dehydration process. Their main change was linked with the condensation of mitochondrial DNA. It was also shown that, during the subsequent restoration of their activities, they rather quickly start to multiply during the early stages of the reactivation process (Rapoport and Kostrikina 1973).

Interesting observations were made regarding the changes of peroxisomes during the dehydration-rehydration/reactivation of *Candida boidinii* yeast. An electron microscopy study revealed the "disappearance" of the membranes of peroxisomes in some areas inside the packs of peroxisomes. This phenomenon is similar to that revealed in studies of nucleus changes and identifies changes to the lipid components of the membranes of peroxisomes (Rapoport et al., unpublished).

The majority of the intracellular protective reactions found in yeast cells during their transition into anhydrobiosis and subsequent rehydration and reactivation were revealed in studies of various species of NCY. Besides the previously described reactions, linked with the folding of the plasma membrane, condensation of chromatin in the nucleus and mitochondria, separation of the nucleus into chromatin-containing and non-containing parts, an additional unusual protective reaction was found in yeast *E. magnusii*. In some cases, the whole cells may separate into two parts – viable and non-viable. The viable part may be larger than the non-viable part, but may also be smaller. *E. magnusii* yeast contains some nuclei in one and the same cell, and the viable part of the cell, of course, has to contain at least one nucleus. This cell separation takes place during dehydration, when the formation of new plasma membrane and cell walls, which separate the two parts of the cell, also takes place. During the subsequent rehydration/reactivation stages, the viable part of the cell increases in size and multiplies, whereas the non-viable part is subjected to gradual autolysis (Rapoport 1973).

In this chapter of the review, the protective reactions found at the structural level were described, but there are a number of other reactions, such as the synthesis of various protective compounds, which were also revealed in NCY studies; some of these will be presented in the next chapter of this review.

10.3 Physiological Studies

One of the main directions in anhydrobiosis studies is linked with attempts to understand the mechanisms of cell transition in this state. Correspondingly, the reasons for resistance or the increased sensitivity to the dehydration of living organisms should also be investigated. This task may be accomplished by the comparison of yeast cells with different dehydration resistances; NCY are very suitable and convenient for such research.

The first experiments in this direction were performed with various strains of *Saccharomycodes ludwigii* yeast. These yeast strains were isolated from soil, fruits, insects and oak trees (Lachance et al. 1995; Miller and Phaff 1998; Barnett et al. 2000; Vejarano 2018). Up to the present, this yeast still had not been used in biotechnology, in spite of the existence of rather interesting proposals. For example, it was reported that *S. ludwigii* may be used for the production of a "refreshing summer drink" (Romano et al. 1999). Other ideas were linked with its use in the production of vinegar, pre-treatment of wine musts before fermentation for a de-sulphite process and the production of low-alcohol beer (De Francesco et al. 2015; Vejarano 2018). Additionally, some other interesting applications of this yeast were indicated for winemaking: ageing on lees, stabilising red wines, improving the wine aromatic profile and reducing the alcohol content in wine (Romano et al. 2003; Palomero et al. 2009; Kulkarni et al. 2015; Vejarano 2018). An investigation directed towards revealing the possible reasons for yeast resistance or sensitivity to dehydration stress was performed with four strains of *S. ludwigii*, which had related origins (Rapoport et al. 1981). The resistant strain *S. ludwigii* VKM Y-631 is a mutant of strain *S. ludwigii* VKM Y-630, which is not resistant. The *S. ludwigii* VKM Y-639 yeast strain, which is not resistant to dehydration, was obtained from one spore of the four-spore resistant *S. ludwigii* VKM Y-638 strain. The results of this study showed that a considerably higher nucleic acid content in the fractions of the phosphorus compounds was found in the non-resistant strains, compared to that of the resistant cultures. The total amount of nucleic acids and nucleotides in yeast sensitive to drying also noticeably exceeded the corresponding value for resistant organisms. The application of nucleic acids fractionation methods that were available at that time showed that the amount of nucleic acids in hot perchloric acid extract directly correlated with the dehydration resistance of the corresponding strain. These results were the first indication that the nucleic acid properties of yeast organisms may be one of the factors that determines the resistance of the cells to dehydration (Rapoport et al. 1981).

Another comparison was performed with one strain of yeast *Pachysolen tannophilus* CBS 4044 grown in different conditions, which also led to changes in the cells' resistance to dehydration (Krallish et al. 1994, 1997). This yeast species has been known since 1957, when it was described by J. Boidin and J.M. Adzet. Most of these yeast strains were isolated from tree extracts. *P. tannophilus* was the first yeast identified as being capable of producing significant amounts of ethanol from xylose. For many years, it has been used as a model for studies of other yeasts, mediating this conversion (Slininger et al. 1987). In anhydrobiosis studies, this yeast has been used because of its ability to synthesise xylitol during growth in xylose-containing medium. This research was linked with previously obtained results demonstrating that special yeast pre-treatment in solutions with increased osmotic pressure improved the cells' resistance to dehydration and led to the synthesis of polyols, which were able to substitute the main protective compound – trehalose (Rapoport and Beker 1983; Rapoport et al. 1988). The goal of the study with *P. tannophilus* was to verify the possible protective effect of polyols (in this case, xylitol). It was shown that when glucose was fermented by *P. tannophilus*, glycerol, ethanol and acetic acid were formed, whereas when *P. tannophilus* fermented a xylose-containing medium, xylitol was formed, in addition to ethanol and acetic acid, but no glycerol formation was detected in these conditions. The intracellular trehalose content was lower in xylose-grown *P. tannophilus* cells, compared with glucose-grown cells (39 mkg/mg dry weight and 127 mkg/mg dry weight, respectively; Krallish et al. 1994, 1997). It is important to note that the increased synthesis of xylitol in these cells was accompanied with a decrease in trehalose synthesis. Dehydration of this yeast led to a rather significant increase of xylitol in both cases – after growth in glucose- and xylose-containing media; the xylitol content in dry cells reached 88 mkg/g dry weight and 110 mkg/g dry weight, respectively. At the same time, in both cases, the amount of trehalose in these cells increased very slightly, whereas in the case of *S. cerevisiae* yeast, which is not able to synthesise xylitol, drying is followed by a very significant increase in the amount of trehalose – by about two times or sometimes more (Krallish et al. 1994, 1997). It is clear that during dehydration in the case of a simultaneous increase in the additional synthesis of xylitol, there is very moderate synthesis of trehalose. This suggests that xylitol can at least partially perform the same protective functions as trehalose.

To summarise, higher intracellular xylitol content in native and dehydrated cells coincided with higher viability in a dry state. The intracellular trehalose content in this yeast increased very slightly during dehydration and this did not correspond to the enhanced viability of dehydrated cells (Krallish et al. 1997). These experiments also confirmed the idea that polyols may function as efficient intracellular protective compounds. Later, it was concluded that if trehalose is absent, or present in the cells in insufficient quantities, compounds such as sugar alcohols (polyols) can substitute trehalose in at least some protective mechanisms. If the amount of trehalose in cells is still sufficient to protect membranes in a dry state, but not enough to protect proteins, polyols, such as xylitol, sorbitol and inositol, may participate in the protection of macromolecules (Rapoport et al. 2019).

The resistance to dehydration of an osmotolerant yeast *Debaryomyces hansenii* was verified in another study, with the goal of understanding whether there are general mechanisms of cells resistance to the treatments, with similar natures (Khroustalyova et al. 2001). The first isolation of this yeast was from sea water. *D.*

hansenii can also be found in many other habitats with low water activity. For example, it may be found in cheese, meat, wine, beer, fruits, soil and some high sugar-containing products (Norkrans 1966; Tilbury 1980; Garcia-Gonzalez and Ochoa 1999; Barnett et al. 2000). *D. hansenii* is a very prospective and biotechnologically important yeast. It is connected to or may be used in processes such as the synthesis of dairy products, the fermentation of meat products and the synthesis of fine chemicals, including unusual fatty acids or tailor-made lipids and lytic enzymes of commercial interest. It may be used as a therapeutic agent in medicine, a bio-preservative in food fermentation and in other biotechnological processes (Breuer and Harms 2006). The study of the dehydration resistance of *D. hansenii* showed completely unexpected results. After cultivation in basal medium and subsequent dehydration, exponential growth phase cells maintained a very high viability (70–84%), whereas the viability of dehydrated stationary phase cells grown in the same medium was lower (53–67%); results of the high viability of exponential phase cells had not been previously found and the viability of this phase of cells, for example, in the case of *S. cerevisiae*, had never been higher than 30%. Moreover, stationary phase cells were always significantly more resistant than exponential growth phase cells. It should be noted that the synthesis of arabinitol in both the exponential and stationary growth phase cells of *D. hansenii* yeast had been previously found, and in the exponential growth phase cells, the arabinitol content was higher (Adler and Gustafsson 1980). This means that in all studied species – *S. cerevisiae, P. tannophilus* and *D. hansenii* – the synthesis of various polyols coincided with an increase of these yeasts' resistance to dehydration; this confirms the protective functions of these compounds during the cells' transition into the state of anhydrobiosis. Further experiments demonstrated that growth at high salinity improves dehydration tolerance. Cultivation of this yeast in a medium with 10% NaCl led to an additional increase in their viability in a dry state; in some experiments this reached 88–90% for cells from both the exponential and stationary growth phases (Khroustalyova et al. 2001). The main conclusion from this study was that osmotolerant yeast can be very resistant to dehydration. Further research of this yeast's resistance to dehydration was performed by us, together with Prof. John Crowe (University of California, Davis), revealing serious differences in the plasma membrane characteristics of *D. hansenii* from those of baker's yeast, *S. cerevisiae*. The membranes of dehydrated *D. hansenii* had lower phase transition temperatures for membrane lipids, and this temperature was practically the same for both exponential and stationary growth phase cells. In the case of *S. cerevisiae* cells, these temperatures were different in both growth phases, with temperatures being significantly higher for exponential growth phase cells (Rapoport et al. 1995). These variations are most likely linked with the different chemical content of the plasma membranes in these yeasts.

A comparison of various strains of baker's yeast, *S. cerevisiae*, showed that, like osmotolerant strains, thermotolerant yeasts also possess high resistance to dehydration (Rapoport et al. 2014). Recently, it appears that psychrotolerant yeasts are also very resistant to dehydration-rehydration stress. In one study, six strains of two yeast species – *Solicoccozyma terricola* and *Naganishia albida* – were revealed as highly resistant to this treatment. The average viability of dehydrated cells in all six studied strains exceeded 78%, and these results were achieved after fast rehydration in water. The viability of *S. terricola* strains was at least 93% (Khroustalyova et al. 2019). The application of preliminary gradual rehydration of the dry cells of all six investigated strains led to an additional increase in the viability of reactivated cells, which reached almost 100% in the case of *S. terricola* and 85–90% in the case of *N. albida* (Khroustalyova et al. 2019). These experiments also revealed a very high resistance of the plasma membranes of these cells. As a result, the permeability of the plasma membranes of dehydrated cells during fast rehydration was at an unusually low level (6.3–9.8%). An additional decrease in plasma membrane permeability was reached as a result of the preliminary gradual rehydration of dehydrated cells in water vapour. In such experiments, the permeability of plasma membranes decreased to 4.2–8.4% (Khroustalyova et al. 2019). At such low plasma membrane permeabilities, there was no direct correlation of this parameter with cell viability. This new information led to the hypothesis that "other factors may play an important role in determining the resistance of yeast cells under conditions of dehydration-rehydration", besides the resistance of the plasma membrane (Khroustalyova et al. 2019). Thus, psychrotolerant yeasts that are usually isolated in polar and subpolar environments are also extremely resistant to dehydration stress.

To summarise, by reviewing the studies of the high resistance of osmotolerant, thermotolerant and psychrotolerant yeasts to dehydration, the value that NCY studies add to our understanding of anhydrobiosis can be seen; these have shown that extremophilic organisms are also always able to successful (with a very high viability rate) transition into the state of anhydrobiosis.

The experimental use of various species of NCY has also revealed the role of some additional protective compounds, besides the well-known importance of trehalose. The first group of these protective compounds is the different polyols (shown above), which probably strengthen trehalose protection or may substitute it. Another important group is antioxidants. Recent studies have supported previous proofs of the high importance of glutathione in yeast dehydration stress (Espindola et al. 2003; Franca et al. 2005). Experiments with recombinant strains of *Ogataea (Hansenula) polymorpha* yeast, directed towards the improvement of their resistance to dehydration, led to unexpected results. Usually, recombinant yeast strains are very sensitive to dehydration, and their viability rate after drying is rather low. A study of a genetically engineered strain that exhibits an increased production of glutathione showed that, in this case, there was no difference in the resistance of the parental and recombinant strains (Kulikova-Borovikova et al. 2018). With the goal of determining whether the unusual resistance of a recombinant strain is characteristic of other recombinant strains of this species, the genetically engineered strain of *O. polymorpha*, constructed for the production of ethanol, was investigated under the same conditions. The obtained results demonstrated that this strain exhibits low resistance to dehydration-rehydration, as has been shown previously for other recombinant yeast strains. This means that the resistance of a glutathione-overproducing recombinant strain to dehydration-rehydration is unusual for other genetically engineered strains. The authors of this research hypothesised that these results are linked to the overproduction of glutathione in recombinant strain cells, which facilitates the maintenance of their viability during the drying procedure (Kulikova-Borovikova et al. 2018). In additional experiments, the authors revealed that the plasma membrane permeability of dry cells of the recombinant strain during their fast rehydration was lower than that of the wild-type strain, despite their similar viability. These results indicate that the increased synthesis of glutathione in the cells of recombinant strains improves the resistance of the plasma membrane during dehydration (Kulikova-Borovikova et al. 2018). The higher amount of glutathione likely decreases the detrimental effects linked with the possible peroxidation and de-esterification of membrane lipids during the process of cell drying.

An additional interesting study in this direction was recently performed using three other strains of NCY: *Torulaspora delbrueckii*, *Metschnikowia pulcherrima* and *Lachancea thermotolerans*. These species may be rather interesting for the wine industry (Câmara et al. 2019). Existing information shows that these yeasts may be important for obtaining the necessary colour profile, the anthocyanin composition, the complexity of volatile compounds and the aroma profile of wine (Gschaedler 2017; Chen et al. 2018). Thus, *T. delbrueckii* produces higher alcohols and esters, which may change and improve the aromatic profile and quality of wine (Chen and Liu 2016). *M. pulcherrima* is also able to synthesise higher alcohols, esters and terpenols and improve the fruity and floral aroma of wines (Sadoudi et al. 2012, 2017). *L. thermotolerans* is linked with the production of lactic acid, which is necessary for the correction of grape must acidity (Kapsopoulou et al. 2007).

The results obtained in the study of Câmara et al. (2019) showed that yeasts grown in nutrient-rich medium synthesised and accumulated a higher amount of glutathione and, as a result, had a rather high resistance to dehydration. In the case of the use of a nutrient-poor medium, large amounts of trehalose were accumulated. Nevertheless, this was not enough to increase the resistance of yeast to dehydration to the same level as was reached when glutathione accumulated in cells grown in a nutrient-rich medium (Câmara et al. 2019). Such data confirm the essential importance of glutathione, whose role as an intracellular protective compound was previously underestimated.

The goal of a number of studies was to search for ways to increase the quality (viability) of active dry preparations of various NCY (Roca-Domènech et al. 2016, 2018). This research direction is linked with our current knowledge that various non-*Saccharomyces* wine yeasts may be very interesting for the wine industry because of the production of secondary metabolites, such as glycerol, 2-phenylethyl acetate and isoamyl acetate (Garcia et al. 2002; Bely et al. 2008; Fleet 2008; Viana et al. 2011). *Schiz. pombe* is especially interesting for winemakers because of its high fermentative power, its capacity to reduce the gluconic acid content of must and its ability to efficiently metabolise L-malic acid into ethanol and carbon dioxide (Taillandier et al. 1991; Peinado et al. 2009; Benito et al. 2012). In the process of winemaking, *Schiz. pombe* yeast cells may be used for the partial or total consumption of L-malic acid, before being removed to prevent the production of an off-flavour (Silva et al. 2003). Therefore, the possibility of the production of *Schiz. pombe* strains of active dry yeast would supply winemakers with a new biotechnological tool for obtaining high-quality wines. In recent research, different techniques were used, with the goal of increasing the resistance of the cells of various strains of this yeast to the dehydration process; the best results were obtained when this yeast was dried in the presence of 10% trehalose. Moreover, for this dried yeast, the dehydration conditions were very important for their subsequent rehydration; optimal results can be obtained when 5 mM $MgSO_4$ is added to the rehydration solution. The combination of these dehydration-rehydration conditions increased the viability of some strains by up to 70% (Roca-Domènech et al. 2016). Similar results were obtained in another study in which the authors tried to find the optimal dehydration-rehydration conditions for *Starmerella bacillaris* yeast; this is necessary for winemakers to reduce the risk of sluggish fermentations of residual fructose, but it is extremely non-resistant to the drying process (Roca-Domènech et al. 2018). At reference conditions, the viability of the dehydrated cells of this yeast was not higher than 2%. Experiments showed that, for this yeast, the existence of 10% trehalose during the drying process led to an increase of *St. bacillaris* cell viability up to 18%. The optimal rehydration process of dehydrated *St. bacillaris* cells requires 1% raffinose or 0.5–1.0% galactose. The combination of optimal conditions for dehydration and rehydration had a synergistic effect and provided the possibility of increasing the viability of rehydrated dry yeast cells up to approximately 35% in the case of the addition of raffinose to the rehydration medium and up to 50% with galactose in the rehydration medium (Roca-Domènech et al. 2018). These results demonstrate that, besides intracellular trehalose, external trehalose also has protective effects to dehydration in various NCY and is one of the possible approaches for obtaining efficient preparations of various active dry yeasts.

Recently, a study was conducted that was devoted to the development of the optimal conditions for dehydration-rehydration of NCY used in the production of Korean persimmon wine and apple cider (Kim et al. 2019). Protective agents, composed primarily of 10% skim milk and six kinds of sugar (fructose, glucose, maltose, raffinose, sucrose and trehalose), were investigated for their ability to improve the survival of dried yeast cells of *Hanseniaspora uvarum, St. bacillaris, Pichia kluyveri, Meyerozyma caribbica* (synonym *Pichia caribbica*) and two strains of *Wickerhamomyces anomalus (synonym Pichia anomala).* Additionally, rehydration conditions for the recovery of dehydrated yeast preparations were also studied using four kinds of rehydration solutions. Each strain exhibited different optimal protectant and rehydration conditions. For *W. anomalus*, the addition of 10% trehalose and rehydration using 1% peptone water resulted in the highest survival rate. For *H. uvarum*, the best results were obtained with the addition of 10% sucrose and rehydration using 0.85% NaCl. For *P. kluyveri*, the optimal solutions were the addition of 10% fructose and rehydration using 0.85% NaCl. For *St. bacillaris*, the addition of 10% sucrose and rehydration using 0.85% NaCl was optimal, and for *M. caribbica* the addition of 10% trehalose and rehydration using phosphate buffer (Kim et al. 2019) was optimal. This research also showed the influence of the addition of different concentrations of two antioxidants (L-ascorbic acid or glutathione) on the viability of dehydrated yeast preparations. It appears that there were also differences for the various studied yeast strains. The optimal solutions were: 3–5 mM glutathione for *W. anomalus*, 5 mM glutathione for *H. uvarum* and *St. bacillaris*, and 5 mM L-ascorbic acid for *M. caribbica* (Kim et al. 2019).

This review chapter discussed the results of experiments with NCY dehydration-rehydration/reactivation, obtained mainly using a physiological approach. These studies provide the possibility to understand the role of various intracellular protective compounds, as well as to determine ways to "artificially" increase resistance to dehydration in initially non-resistant yeast species and strains. Some additional important information for the further development of this research area is presented in the next chapter, which is devoted to some biotechnological characteristics of NCY and the issues linked with their maintenance in dry active yeast preparations.

10.4 Biotechnology

In recent decades, microbiologists have paid more and more attention to the possible use of various species of NCY in biotechnology. First of all, it is linked to their applications in winemaking. Beyond the examples given in the previous chapter of this review, there are a number of other recent reports on the release of flavour and aroma compounds by NCY, which positively contribute to the organoleptic characteristics of wine (Andorrá et al. 2012; Whitener et al. 2015; Lachance 2016; Liu et al. 2016). Recently, it was also shown that *Hanseniaspora vineae* yeast strains may produce a significantly richer flavour compound diversity than traditional *Saccharomyces* wine yeast strains, including benzenoids, phenylpropanoids and acetate-derived compounds (Giorello et al. 2019). An interesting recent study was devoted to the formation of polymeric pigments in red wines through the sequential fermentation of flavanol-enriched musts with non-*Saccharomyces* yeasts (Escott et al. 2018). One of the strategies to obtain wines with reduced ethanol content is the sequential inoculation of non-*Saccharomyces* and *S. cerevisiae* yeasts; *Hanseniaspora uvarum* and *Candida membranaefaciens* yeasts were used for this goal in a recent study (Maturano et al. 2019). *Candida californica* and *Metschnikowia pulcherrima* were also used to decrease the amount of ethanol and sugar in wine (Aplin et al. 2019). *Torulaspora delbrueckii* was successfully used in pure and mixed secondary fermentations for sparkling wine. The sparkling wines obtained with *T. delbrueckii* showed different aromatic compositions and sensory profiles than those obtained with *S. cerevisiae* (Canonico et al. 2018). Various NCY species possess great flavour potential that can also be used in beer production; very interesting and promising results have been obtained with the use of all tested *Pichia* strains and especially with *P. kluyveri*, *Brettanomyces* species, *Wickerhamomyces anomalus* and *T. delbrueckii* (Basso et al. 2016; Canonico et al. 2017; Holt et al. 2018; Colomer et al. 2019).

These examples emphasise the underestimated significant importance of NCY in the classical biotechnological processes of winemaking and beer production. Correspondingly, the possibility of obtaining these yeast strains in a viable dry state becomes more and more likely. As already mentioned above, dry preparations of yeasts provide the possibility of the long-term storage of biotechnologically important strains, without losses of their viability, as well as storage at room temperatures. They have a big economic value because of the reduction of transportation and storage costs. Unfortunately, a lot of earlier studied strains of NCY were not resistant to the dehydration process (Pereira Ede et al. 2003; Matallana and Aranda 2016) and special procedures or pre-treatments had to be used to increase their resistance; some of these were described in the previous chapter of this review. It should be noted that antioxidant enzymes, such as catalase and superoxide dismutase, are good markers for studying the stress tolerance of non-*Saccharomyces* yeasts as their activity correlates with fermentative performance (Gamero-Sandemetrio et al. 2013; Matallana and Aranda 2016). At the same time, during the use of these dry yeast preparations for practical (industrial) goals, it is very important that the cells not only maintain a high viability rate in a dehydrated state but also that they do not lose their main biotechnological characteristics, which are important for the concrete process or the final product. Therefore, in some studies, this issue has been investigated. For example, in the work with dried preparations of *W. anomalus*, *H. uvarum*, *P. kluyveri*, *St. bacillaris* and *M. caribbica* yeasts, which might be important for the improvement of the quality of Korean persimmon wine and apple cider, the dehydration procedure did not change the characteristics of fermentation when these yeast strains were used as starter cultures (Kim et al. 2019). A study by Roca-Domènech et al. (2016) revealed that there were no significant changes in fermentation behaviour and the main volatile compounds that were detected in wines obtained when dehydrated preparations of *Schiz. pombe* were used, together with *S. cerevisiae* during laboratory-scale experiments; similar results were also obtained in the continuation of these studies (Roca-Domènech et al. 2018). The use of dry active preparations of another yeast strain, *St. bacillaris*, in co- and sequentially inoculated grape must fermentations, together with *S. cerevisiae* at the laboratory-scale production of wines, also did not show any changes in terms of the main volatile compounds. At the same time, they revealed that, in this case, there was also a positive effect of the use of dehydrated yeast preparations: an improvement in the fermentation performance behaviour (Roca-Domènech et al. 2018). Similar results of the more efficient use of dehydrated preparations of NCY, compared to their native preparations, were obtained in experiments with *Endomycopsis* (*Saccharomycopsis*) *fibuligera* (Rapoport et al. 1986). This yeast is the producer of amylolytic enzymes and is widely used as a microbial starter in various fermented foods (Hostinová 2002; Aidoo et al. 2006; Chi et al. 2009; Mi Lee et al. 2018). In studies performed with dehydrated yeast *E. fibuligera*, their use as the inoculum led to an essential increase in biomass yield, compared to when native yeast was used as the inoculum. Moreover, in these experiments, a higher production of both studied amylolytic enzymes, alpha-amylase and glucoamylase, was obtained. In anaerobic conditions, a higher intensity of ethanol production was revealed in experiments with dehydrated yeasts compared to native ones (Rapoport et al. 1986). To our knowledge, this study was the first in which the possibility of the higher efficiency of application of dehydrated yeasts, compared to native yeast culture, was shown. Later, such effects were also obtained for *S. cerevisiae* dehydrated cells (Rapoport et al. 1997; Borovikova et al. 2014), as well as for non-conventional wine yeasts, as shown above.

Another study direction of the possible applications of dried yeast for biotechnological goals is linked with their use for the protection and purification of the environment. The dry yeast *C. utilis* was shown to have high efficiency as a biofilter for heavy metal sorption from waste water. The rate of heavy metal sorption was essentially higher, especially in the beginning of the process, than when native yeast biomass was used for the same goals (Rapoport and Muter 1995; Muter et al. 2001a, 2001b, 2002; Muter and Rapoport 2013). These studies were also the first in which dehydrated yeast, used as a biofilter, was shown to be more efficient in the sorption of heavy metals than native yeasts. Later, similar results were obtained using other yeast species, including *S. cerevisiae.* An interesting application of the active dry wine yeasts, related to their higher sorption activity, was proposed for the additional improvement of wine quality. It is linked with the possibility of removing the unpleasant smell of volatile phenols from wine. Depending on the dehydration method, the sorption efficiency of dry yeasts for these compounds may be up to twice as high as that of the same native yeast (Pradelles et al. 2009). Together, these results show that, besides the well-known economic effects of using active dry yeasts to decrease transportation and storage expenses, there is also additional value in the application of these preparations for an essential increase in the efficiency of various biotechnological processes.

The rapid development of modern biotechnology is connected with the appearance of new biotechnological processes that use traditional and NCY. Correspondingly, this also increases the necessity of the application of dehydrated yeast preparations to new areas of industry, agriculture and medicine (Rapoport et al. 2016). Besides classical biotechnological processes (baking, ethanol, wine and beer production) and the processes mentioned above, they are used now or may be used in the near future as drug delivery systems, a source of beta-glucans, in animal feed, as new test-systems, in various biochemical analytical methods, in selective bio-transformations and for biocontrol (Rapoport et al. 2016); in the majority of these new approaches, technologies and processes, if not in all, NCY can be used. One of the examples of the possible application of dried preparations of NCY in animal feed is the use of dehydrated biomass of the red yeast *Rhodotorula glutinis.* Within its biomass, it synthesises and accumulates a rather essential amount of carotenoids, including beta-carotene (Buzzini et al. 2010; Mannazzu et al. 2015). Beta-carotene possesses many very valuable effects for cells and tissues. This compound is a precursor of vitamin A, an antioxidant, and has immunogenic and tumour-inhibiting activities (Amar et al. 2001; Bhosale et al. 2003; Frengova and Beshkova 2009; Anbazahan et al. 2014; Fernandez et al. 2015). It is used, or may be used, as an additive to the feeds for chicken, cattle and fish. At the same time, beta-carotene is not stable under storage conditions and can be easily denatured under the influence of environmental factors such as light, UV, heat or oxygen (Scita 1992; Modi et al. 1993). Dehydrated cells, which are in a state of anhydrobiosis, are very resistant to the influence of extreme environmental conditions. Regarding the dry cells of the red yeast *Rh. glutinis*, which accumulate high amounts of beta-carotene, their biomass can be stored in dark and cold conditions for long periods of time without beta-carotene losses. For comparison, storage of purified beta-carotene under the same conditions very quickly (over only 76 h) leads to its full loss (Bhosale et al. 2003). Another example of the modern application of viable dry preparations of NCY is their use as biocontrol agents. Many NCY species possess antimicrobial activities against various diseases in different agricultural plants, namely, *Candida sake*, *Cryptococcus laurentii*, *Cr. albidus*, *Cr. nodaensis*, *D. hansenii*, *Rh. glutinis*, *P. anomala* (now *Wickerhamomyces anomalus*), *P. guilliermondii* (now *Meyerozyma guilliermondii*) and *Metschnikowia pulcherrima* (Abadias et al. 2005; Druvefors and Schnurer 2005; Zhang et al. 2009; Andrade et al. 2014; Nunez et al. 2015, and others). Some yeast species can also be successfully used in the food industry for the bio-preservation of food products. It is clear that, for the distribution and storage of the active cultures of these yeasts, it would be much more efficient to have their dry active preparations. Some studies in this area were performed, for example, with *P. anomala* yeast, and the optimum conditions for the efficient dry biocontrol preparations of this yeast have been described (Melin et al. 2007, 2011).

10.5 Concluding Remarks

This review summarised the existing information on anhydrobiosis in NCY. On the one hand, experiments using these yeasts, which started about 50 years ago and initiated a new stage of more in-depth studies of anhydrobiosis in yeasts, have great theoretical value for basic microbiology and biology in general; they have made a very big contribution in the development of our understanding of this unique phenomenon of nature. On the other hand, at the moment, these studies are mainly connected with the practical needs of modern biotechnology. At the same time, this does not mean that they are only directed towards obtaining knowledge of the methodologies of the efficient production and use of new active dry microbial preparations for various biotechnological processes. In these studies, we also obtain additional knowledge on anhydrobiosis and the hidden potential of eukaryotic organisms. This information may be extrapolated to higher organisms and used, for example, in medicine too. It is clear that in the near future, we can expect further fast and essential progress in the biotechnological applications of NCY based on the current successes in the experimental research of their physiology, including those that are described in this book. Correspondingly, this will also lead to the necessity of additional studies directed towards the possibilities of obtaining new active dry preparations of biotechnologically important species of NCY; in its turn, this will lead to the accumulation of new information on the potential of eukaryotic organisms and the mechanisms of anhydrobiosis.

Acknowledgements This work was supported by TW–LV-LT Foundation for Scientific Cooperation grant 2016–2019 and grant of the European Regional Development Fund (ERDF) No. 1.1.1.1/16/A/113

References

Abadias M, Teixido N, Usall J, Solsona C, Vinasm I (2005) Survival of the postharvest biocontrol yeast *Candida sake* CPA-1 after dehydration by spray-drying. Biocontrol Sci Tech 15:835–846

Adler L, Gustafsson L (1980) Polyhydric alcohol production and intracellular amino acid pool in relation to halotolerance of the yeast *Debaryomyces hansenii*. Arch Microbiol 124:123–130

Aidoo KE, Nout MR, Sarkar PK (2006) Occurrence and function of yeasts in Asian indigenous fermented foods. FEMS Yeast Res 6:30–39

Amar EC, Kiron V, Satoh S, Watanabe T (2001) Influence of various dietary synthetic carotenoids on bio-defence mechanisms in rainbow trout, *Oncorhynchus mykiss* (Walbaum). Aquac Res 32:162–173

Anbazahan SM, Marim LSS, Yogeshwari G, Jagruthi C, Thirumurugan R, Arockiaraj J, Velanganni AAJ, Krishnamoorthy P, Balasundaram C, Harikrishnan R (2014) Immune response and disease resistance of carotenoids supplementation diet in *Cyprinus carpio* against *Aeromonas hydrophila*. Fish Shellfish Immunol 40:9–13

Andorrà I, Berradre M, Mas A, Esteve-Zarzoso B, Guillamon JM (2012) Effect of mixed culture fermentations on yeast populations and aroma profile. LWT – Food Sci Technol 49:8–13

Andrade MJ, Thorsen L, Rodrıguez A, Cordoba JJ, Jespersen L (2014) Inhibition of ochratoxigenic moulds by *Debaryomyces hansenii* strains for biopreservation of dry-cured meat products. Int J Food Microbiol 170:70–77

Aplin JJ, White KP, Edwards CG (2019) Growth and metabolism of non-Saccharomyces yeasts isolated from Washington state vineyards in media and high sugar grape musts. Food Microbiol 77:158–165

Barnett JA, Payne RW, Yarrow D (2000) Yeasts: characteristics and identification, 3rd edn. Cambridge University Press, Cambridge

Basso RF, Alcarde AR, Portugal CB (2016) Could non-Saccharomyces yeasts contribute on innovative brewing fermentations? Food Res Int 86:112–120

Beker MJ, Rapoport AI (1987) Conservation of yeasts by dehydration. Adv Biochem Engineer/Biotechnol 35:127–171

Bely M, Stoeckle P, Masneuf-Pomarède I, Dubourdieu D (2008) Impact of mixed *Torulaspora delbrueckii – Saccharomyces cerevisiae* culture on high-sugar fermentation. Int J Food Microbiol 122:312–320

Benito S, Palomero F, Morata A, Calderón F, Suárez-Lepe JA (2012) New applications for *Schizosaccharomyces pombe* in the alcoholic fermentation of red wines. Int J Food Sci Technol 47:2101–2108

Bhosale P, Jogdand VV, Gadre RV (2003) Stability of *b*-carotene in spray dried preparation of *Rhodotorula glutinis* mutant 32. J Appl Microbiol 95:584–590

Biryusova VI, Rapoport AI (1978) Cryofractographic investigation of the structure of yeast cells in an anabiotic state. Microbiol 47:245–251

Borovikova D, Scherbaka R, Patmalnieks A, Rapoport A (2014) Effects of yeast immobilisation on bioethanol production. Biotechnol Appl Biochem 61:33–39

Breuer U, Harms H (2006) Debaryomyces hansenii—an extremophilic yeast with biotechnological potential. Yeast 23:415–437

Buzzini P, Goretti M, Branda E, Turchetti B (2010) Basidiomycetous yeasts for production of carotenoids. In: Flickinger MC (ed) Encyclopedia of industrial biotechnology: bioprocess, bioseparation, and cell technology, vol 1. Wiley, New York, pp 469–481

Câmara AA Jr, Maréchal P-A, Tourdot-Maréchal R, Husson F (2019) Dehydration stress responses of yeasts *Torulaspora delbrueckii, Metschnikowia pulcherrima* and *Lachancea thermotolerans*: effects of glutathione and trehalose biosynthesis. Food Microbiol 79:137–146

Canonico L, Comitini F, Ciani M (2017) *Torulaspora delbrueckii* contribution in mixed brewing fermentations with different *Saccharomyces cerevisiae* strains. Int J Food Microbiol 259:7–13

Canonico L, Comitini F, Ciani M (2018) *Torulaspora delbrueckii* for secondary fermentation in sparkling wine production. Food Microbiol 74:100–106

Chen D, Liu S-Q (2016) Impact of simultaneous and sequential fermentation with Torulaspora *delbrueckii* and *Saccharomyces cerevisiae* on non-volatiles and volatiles of lychee wines. LWT – Food Sci Technol (Lebensmittel-Wissenschaft -Technol) 65:53–61

Chen K, Escott C, Loira I, del Fresno JM, Morata A, Tesfaye W, Calderon F, Suárez-Lepe JA, Han S, Benito S (2018) Use of non-*Saccharomyces* yeasts and oenological tannin in red winemaking: influence on colour, aroma and sensorial properties of young wines. Food Microbiol 69:51–63

Chi Z, Chi Z, Liu G, Wang F, Ju L, Zhang T (2009) Saccharomycopsis fibuligera and its applications in biotechnology. Biotechnol Adv 27:423–431

Colomer MS, Funch B, Forster J (2019) The raise of *Brettanomyces* yeast species for beer production. Cur Opinion Biotechnol 56:30–35

Crowe JH (1971) Anhydrobiosis: an unsolved problem. Am Nat 105:563–574

Crowe JH (2015) Anhydrobiosis: an unsolved problem with applications in human welfare. Subcell Biochem 71:263–280

De Francesco G, Turchetti B, Sileoni V, Marconi O, Perretti G (2015) Screening of new strains of *Saccharomycodes ludwigii* and *Zygosaccharomyces rouxii* to produce low-alcohol beer. J Inst Brew 121:113–121

Druvefors UA, Schnurer J (2005) Mold-inhibitory activity of different yeast species during airtight storage of wheat grain. FEMS Yeast Res 5:373–378

Dupont S, Rapoport A, Gervais P, Beney L (2014) The survival kit of *Saccharomyces cerevisiae* for anhydrobiosis. Appl Microbiol Biotechnol 98:8821–8834

Escott C, Del Fresno JM, Loira I, Morata A, Tesfaye W, González MC, Suárez-Lepe JA (2018) Formation of polymeric pigments in red wines through sequential fermentation of flavanol-enriched musts with non-*Saccharomyces* yeasts. Food Chem 239:975–983

Espindola AS, Gomes DS, Panek AD, Eleutherio EC (2003) The role of glutathione in yeast dehydration tolerance. Cryobiology 47:236–241

Fernandez I, Lopez-Joven C, Andree KB, Roque A, Gisbert E (2015) Vitamin A supplementation enhances Senegalese sole (*Solea senegalensis*) early juvenile's immunocompetence: new insights on potential underlying pathways. Fish Shellfish Immunol 46:703–709

Fleet GH (2008) Wine yeasts for the future. FEMS Yeast Res 8:979–995

Franca MB, Panek AD, Eleutherio ECA (2005) The role of cytoplasmic catalase in dehydration tolerance of *Saccharomyces cerevisiae*. Cell Stress Chaper 10:167–170

Frengova GI, Beshkova DM (2009) Carotenoids from *Rhodotorula* and *Phaffia*: yeasts of biotechnological importance. J Ind Microbiol Biotechnol 36:163–180

Gamero-Sandemetrio E, Gomez-Pastor R, Matallana E (2013) Zymogram profiling of superoxide dismutase and catalase activities allows *Saccharomyces* and non-*Saccharomyces* species differentiation and correlates to their fermentation performance. Appl Microbiol Biotechnol 97:4563–4576

Garcia A, Carcel C, Dalau L, Samson A, Aguera E, Agosin E, Gunata Z (2002) Influence of a mixed culture with *Debaryomyces vanriji* and *Saccharomyces cerevisiae* on the volatiles in a Muscat wine. J Food Sci 67:1138–1143

Garcia-Gonzalez A, Ochoa JL (1999) Anti-inflammatory activity of *Debaryomyces hansenii* Cu, Zn-SOD. Arch Med Res 30:69–73

Giorello F, Valera MJ, Martin V, Parada A, Salzman V, Camesasca L, Fariña L, Boido E, Medina K, Dellacassa E, Berna L, Aguilar PS, Mas A, Gaggero C, Carrau F (2019) Genomic and transcriptomic basis of *Hanseniaspora vineae's* impact on flavor diversity and wine quality. Appl Environ Microbiol 85:e01959-18. https://doi.org/10.1128/AEM.01959-18

Gschaedler A (2017) Contribution of non-conventional yeasts in alcoholic beverages. Curr Opinion Food Sci 13:73–77

Hernandez-Garcia A (2011) Anhydrobiosis in bacteria: from physiology to applications. J Biosci 36:939–950

Hernandez-Garcia A (2014) Induced Anhydrobiosis: powerful method for the preservation of microorganisms. In: Harzevili FD, Chen H (eds) Microbial biotechnology: progress and trends. CRC Press/Taylor & Francis Group, Boca Raton, pp 113–130

Holt S, Mukherjee V, Lievens B, Verstrepen KJ, Thevelein JM (2018) Bioflavoring by non-conventional yeasts in sequential beer fermentations. Food Microbiol 72 55–66

Hostinová E (2002) Amylolytic enzymes produced by the yeast *Saccharomycopsis fibuligera*. Biologia 57:247–252

Kapsopoulou K, Mourtzini A, Anthoulas M, Nerantzis E (2007) Biological acidification during grape must fermentation using mixed cultures of *Kluyveromyces thermotolerans* and *Saccharomyces cerevisiae*. World J Microbiol Biotechnol 23:735–739

Khroustalyova GM, Adler L, Rapoport AI (2001) Exponential growth phase cells of the osmotolerant yeast *Debaryomyces hansenii* are extremely resistant to dehydration stress. Process Biochem 36:1163–1166

Khroustalyova G, Giovannitti G, Severini D, Scherbaka R, Turchetti B, Buzzini P, Rapoport A (2019) Anhydrobiosis in yeasts: Psychrotolerant yeasts are highly resistant to dehydration. Yeast, 2019;1–5. https://doi.org/10.1002/yea.3382

Kim D-H, Lee S-B, Jeon J-Y, Park H-D (2019) Development of air-blast dried non-*Saccharomyces* yeast starter for improving quality of Korean persimmon wine and apple cider. Int J Food Microbiol 290:193–204

Krallish I, Jeppsson H, Rapoport A, Hanh-Hagerdal B (1994) The influence of the intracellular polyol content on the viability of xylose utilizing yeasts during dehydration-rehydration process. Cell Biol Intern 18:459

Krallish I, Jeppsson H, Rapoport A, Hahn-Hagerdal B (1997) Effect of xylitol and trehalose on dry resistance of yeasts. Appl Microbiol Biotechnol 47:447–451

Kulikova-Borovikova D, Khroustalyova G, Chang C-R, Daugelavicius R, Yurkiv M, Ruchala J, Sibirny A, Rapoport A (2018) Anhydrobiosis in yeast: glutathione overproduction improves resistance to dehydration of a recombinant *Ogataea (Hansenula) polymorpha* strain. Process Biochem 71:41–44

Kulkarni P, Loira I, Morata A, Tesfaye W, González MC, Suárez-Lepe JA (2015) Use of non-*Saccharomyces* yeast strains coupled with ultrasound treatment as a novel technique to accelerate ageing on lees of red wines and its repercussion in sensorial parameters. LWT-Food Sci Technol 64:1255–1262

Lachance M-A (2016) *Metschnikowia*: half tetrads, a regicide, and the fountain of youth. Yeast 33:563–557

Lachance MA, Gilbert GD, Starmer WT (1995) Yeast communities associated with *Drosophila* species and related flies in an eastern oak-pine forest: a comparison with western communities. J Ind Microbiol 14:484–494

Liu PT, Lu L, Duan CQ, Yan GL (2016) The contribution of indigenous non-*Saccharomyces* wine yeast to improved aromatic quality of Cabernet Sauvignon wines by spontaneous fermentation. LWT – Food Sci Technol 71:356–363

Mannazzu I, Landolfo S, Lopes da Silva T, Buzzini P (2015) Red yeasts and carotenoid production: outlining a future for nonconventional yeasts of biotechnological interest. World J Microbiol Biotechnol 31:1665–1673

Matallana E, Aranda A (2016) Biotechnological impact of stress response on wine yeast. Lett Appl Microbiol 64:103–110

Maturano YP, Mestre MV, Kuchen B, Toro ME, Mercado LA, Vazquez F, Combina M (2019) Optimization of fermentation-relevant factors: a strategy to reduce ethanol in red wine by sequential culture of native yeasts. Int J Food Microbiol 289:40–48

Melin P, Hakansson S, Schnurer J (2007) Optimization of liquid and dry formulations of the biocontrol yeast *Pichia anomala*. Appl Microbiol Biotechnol 73:1008–1016

Melin P, Schnurer J, Hakansson S (2011) Formulation and stabilisation of the biocontrol yeast *Pichia anomala*. Antonie Van Leeuwenhoek 99:107–112

Mi Lee S, Jung JH, Seo J-A, Kim Y-S (2018) Bioformation of volatile and nonvolatile metabolites by *Saccharomycopsis fibuligera* KJJ81 cultivated under different conditions—carbon sources and cultivation times. Molecules 23:2762
Miller MW, Phaff HJ (1998) *Saccharomycodes* E.C. Hansen. In: Kurtzman CP, Fell JW (eds) The yeasts. A taxonomic study, 4th edn. Elsevier, New York, pp 372–373
Modi RC, Walton JC, Burton GW, Hughes L, Ingold KH, Hudsay DA, Maffatt DJ (1993) Oxidative degradation of *b*-carotene and *b*-apo-8-carotenal. Tetrahedron 49:911–992
Muter O, Rapoport A (2013) Some physiological aspects of chromium sorption by yeasts. Lambert Academic Publishing, Saarbrücken
Muter O, Patmalnieks A, Rapoport A (2001a) Interrelations of the yeast *Candida utilis* and Cr (VI): metal reduction and its distribution in the cell and medium. Process Biochem 36:963–970
Muter O, Millers D, Grigorjeva L, Ventina E, Rapoport A (2001b) Cr(VI) sorption by intact and dehydrated *Candida utilis* cells: differences in mechanisms. Process Biochem 37:505–511
Muter O, Lubinya I, Millers D, Grigorjeva L, Ventinya E, Rapoport A (2002) Cr (VI) sorption by intact and dehydrated *Candida utilis* cells in the presence of other metals. Process Biochem 38:123–131
Norkrans B (1966) Studies on marine occurring yeasts: growth related to pH, NaCl concentration and temperature. Arch Microbiol 54:374–392
Nunez F, Lara MS, Peromingo B, Delgado J, Sanchez-Montero L, Andrade MJ (2015) Selection and evaluation of *Debaryomyces hansenii* isolates as potential bioprotective agents against toxigenic penicillia in dry-fermented sausages. Food Microbiol 46:114–120
Palomero F, Morata A, Benito S, Calderón F, Suárez-Lepe JA (2009) New genera of yeasts for over-lees aging of red wine. Food Chem 112:432–441
Peinado RA, Maestre O, Mauricio JC, Moreno J (2009) Use of a *Schizosaccharomyces pombe* mutant to reduce the content in gluconic acid of must obtained from rotten grapes. J Agric Food Chem 57:2368–2377
Pereira Ede J, Panek AD, Eleutherio EC (2003) Protection against oxidation during dehydration of yeast. Cell Stress Chaperon 8:120–124
Pradelles R, Vichi S, Alexandre H, Chassagne D (2009) Influence of the drying processes of yeasts on their volatile phenol sorption capacity in model wine. Int J Food Microbiol 135:152–157
Rapoport AI (1973) Rejection of areas of damaged cytoplasm by microorganisms in a state of anabiosis. Microbiol 42:317–318
Rapoport A (2017) Anhydrobiosis and dehydration of yeasts. In: Sibirny A (ed) Biotechnology of yeasts and filamentous Fungi. Springer, Cham, pp 87–116
Rapoport AI, Beker ME (1983) Effect of sucrose and lactose on resistance of the yeast *Saccharomyces cerevisiae* to dehydration. Microbiol 52:556–559
Rapoport AI, Kostrikina NA (1973) Cytological investigation of an anabiosis state of yeast organisms. Proc Acad Sci USSR Biology 5:770–773. (in Russian)
Rapoport A, Muter O (1995) Biosorption of hexavalent chromium by yeasts. Process Biochem 30:145–149
Rapoport AI, Biryuzova VI, Meisel MN (1973) Nuclear structure of yeast cells in the anabiotic state. Doklady Biol Sci 213:469–471
Rapoport AI, Pomoshchnikova NA, Fateeva MV, Nikitina TN, Meissel MN (1981) Nucleic acids and resistance of yeast cells to dehydration. Microbiol 50:163–165
Rapoport AI, Dreimane MA, Beker ME (1986) Influence of the dehydration process on *Endomycopsis fibuligera* yeasts. Microbiol 55:542–545
Rapoport AI, Puzyrevskaya OM, Saubenova MG (1988) Polyols and resistance of yeasts to dehydration. Microbiol 57:269–271
Rapoport AI, Khroustalyova GM, Adler L, Crowe JH (1995) Anhydrobiosis of yeast: main factors which determine maintenance of cell viability. Sociedade Brasileira de Bioquimica e Biologia Molecular (SBBq). Programa e Resumos da XXIV Reuniao Anual. Caxambu, MG, p 16
Rapoport AI, Khroustalyova GM, Kuklina EN (1997) Anhydrobiosis in yeast: activation effect. Brazil J Medic Biologic Res 30:9–13

Rapoport A, Rusakova A, Khroustalyova G, Walker G (2014) Thermotolerance in *Saccharomyces cerevisiae* is linked to resistance to anhydrobiosis. Process Biochem 49:1889–1892
Rapoport A, Turchetti B, Buzzini P (2016) Application of anhydrobiosis and dehydration of yeasts for non-conventional biotechnological goals. World J Microbiol Biotechnol 32:104
Rapoport A, Golovina EA, Gervais P, Dupont S, Beney L (2019) Anhydrobiosis: inside yeast cells. Biotechnol Adv 37:51–67
Rebello S, Abraham A, Madhavan A, Sindhu R, Binod P, Bahuleyan AK, Aneesh EM, Pandey A (2018) Non-conventional yeast cell factories for sustainable bioprocesses. FEMS Microbiol Lett 365:fny222
Roca-Domènech G, López-Martínez G, Tirado V, Borrull A, Candelas O, Rozès N, Cordero-Otero R (2016) Viability enhancement of *Schizosaccharomyces pombe* cells during desiccation stress. J Microbiol Res 6:82–91
Roca-Domènech G, López-Martínez G, Barrera E, Poblet M, Rozès N, Cordero-Otero R (2018) Enhancing the tolerance of the *Starmerella bacillaris* wine strain to dehydration stress. Annals Microbiol 68:667–676
Romano P, Marchese R, Laurita C, Saleano G, Turbanti L (1999) Biotechnological suitability of *Saccharomycodes ludwigii* for fermented beverages. World J Microb Biot 15:451–454
Romano P, Fiore C, Paraggio M, Caruso M, Capece A (2003) Function of yeast species and strains in wine flavour. Int J Food Microbiol 86:169–180
Sadoudi M, Tourdot-Maréchal R, Rousseaux S, Steyer D, Gallardo-Chacón J-J, Ballester J, Vichi S, Guérin-Schneider R, Caixach J, Alexandre H (2012) Yeast–yeast interactions revealed by aromatic profile analysis of Sauvignon Blanc wine fermented by single or co-culture of non-*Saccharomyces* and *Saccharomyces* yeasts. Food Microbiol 32:243–253
Sadoudi M, Rousseaux S, David V, Alexandre H, Tourdot-Maréchal R (2017) *Metschnikowia pulcherrima* influences the expression of genes involved in PDH bypass and glyceropyruvic fermentation in *Saccharomyces cerevisiae*. Front Microbiol 8:1137
Scita G (1992) Stability of b-carotene under different laboratory conditions. In: Packer L (ed) Methods in enzymology, vol 213. Academic, San-Diego, pp 175–185
Silva S, Ramon Portugal F, Andrade P, Texera M, Strehaino P (2003) Malic acid consumption by dry immobilized cells of *Schizosaccharomyces pombe*. Amer J Enol Viticult 54:50–55
Slininger PJ, Bolen PL, Kurtzman CP (1987) *Pachysolen tannophilus*: properties and process considerations for ethanol production from D-xylose. Enzym Microb Technol 9:5–15
Taillandier P, Riba JP, Strehaiano P (1991) Malic degradation by *Schizosaccharomyces* yeasts included in alginate beads. Bioproc Engineer 7:141–144
Tilbury RH (1980) In: Skinner FA, Passmore SM, Davenport RR (eds) In biology and activities of yeasts. Academic, London, pp 153–176
Vejarano R (2018) *Saccharomycodes ludwigii*, control and potential uses in winemaking processes. Fermentation 4:71
Viana F, Belloch C, Vallés S, Manzanares P (2011) Monitoring a mixed starter of *Hanseniaspora vineae-Saccharomyces cerevisiae* in natural must: impact on 2-phenylethyl acetate production. Int J Food Microbiol 151:235–240
Whitener MEB, Stanstrup J, Carlin S, Divol B, Du Toit M (2015) Effect of non-*Saccharomyces* yeasts on the volatile chemical profile of Shiraz wine. Aust J Grape Wine Res 33:179–192
Zhang HY, Wang L, Ma LC, Dong Y, Jiang S, Xu B, Zheng XD (2009) Biocontrol of major postharvest pathogens on apple using *Rhodotorula glutinis* and its effects on postharvest quality parameters. BioControl 48:79–83

Chapter 11
Non-conventional Yeasts for Producing Alternative Beers

Ciro Sannino, Ambra Mezzasoma, Pietro Buzzini, and Benedetta Turchetti

Abstract Since immemorial time beer is the product of wort fermentation catalyzed by the cells of two main yeast species, namely, *Saccharomyces cerevisiae* and *Saccharomyces pastorianus*, with hop addition. However, in recent years the beer market changed completely and novel types of beers, such as low-alcohol and low-calorie beers and spontaneously fermented or flavored beers, have become very popular. These different beers vary for the nature and amounts of water, cereals, hops, and other additives, techniques used for production, storage, and consumption, and also yeasts used as starter. Indeed, some non-conventional yeasts have been recently proposed as starter cultures for brewing.

This chapter provides a comprehensive review of the most recent papers describing yeast species diversity used in brewing industry or found in spontaneous fermented beers. Many species were described for their fermentative aptitude as a single or co-starter, either at lab or pilot scale and, rarely, at the industrial scale. Ascomycetous species are generally used but a member of Basidiomycota (*Mrakia* sp.) was also considered. A particular focus in the chapter is dedicated to flavor production and to spontaneous fermentation.

Keywords Non-conventional yeasts · Brewing · Low-alcohol beer · Spontaneous fermentation · Bioflavoring

11.1 Introduction

On the global scale, beer is the third most widely consumed drink (after tea and coffee) and is the most popular alcoholic beverage (Rodhouse and Carbonero 2017). In 2016, the global beer consumption was 186.89 million kiloliters: Asia was the continent with the highest global market (33.9% of total sales worldwide), followed by

C. Sannino · A. Mezzasoma · P. Buzzini · B. Turchetti (✉)
Department of Agriculture, Food and Environmental Science,
Industrial Yeasts Collection DBVPG, University of Perugia, Perugia, Italy
e-mail: benedetta.turchetti@unipg.it

A. Sibirny (ed.), *Non-conventional Yeasts: from Basic Research to Application*,
https://doi.org/10.1007/978-3-030-21110-3_11

Europe (26%), and Central and South America (17%). The Czech Republic exhibited the highest per capita consumption of beer (143 l) followed by Namibia (108 l). Considering the main beer-consuming countries, United States, Mexico, Vietnam, India, and Philippines showed the highest percentage of increase of beer consumption compared to 2015 (Kirin Beer University Report Global Beer Consumption by Country in 2016). A significant consolidation of this sector has also been observed in several European countries, where revenues sometimes exceeded 300–1000% from 2009 to 2014 (Basso et al. 2016).

It was established that beer-like beverages were produced since 5000 BC in the regions of Mesopotamia and Ancient Egypt, where the exposure of barley (or wheat) to air caused its germination and drying, and induced spontaneous "contamination" by a mixture of native microorganisms (Varela 2016; Wang et al. 2016). Brewing involves the conversion of barley starch contained into barley kernels (although other grains such as wheat, rice, oats, rye, and corn can also be used) into monosaccharides which are then fermented to ethanol and CO_2 by yeast metabolism. Baker's yeast of the genus *Saccharomyces*, in particular *Saccharomyces cerevisiae*, was found to be the dominant species involved in spontaneous fermentation. Since the second half of the twentieth century, the concept of controlled fermentations, realized by inoculating wort with specific yeast starters, was proposed. Therefore, a ponderous activity of isolation and selection of a pool of domesticated yeast strains has been carried out (Sicard and Legras 2011). The use of *Saccharomyces* strains in controlled fermentation over decades is basically due to some positive aspects: (i) efficient ethanol production; (ii) alcoholic fermentation, also under oxidative conditions, as preferential metabolic pathway due to Crabtree effect (repression of respiration by high contents of glucose); and (iii) high ethanol tolerance. Secondary characteristics are those related to the ability to produce a balance combination of flavored compounds that give original tastes to beer (Steensels and Verstrepen 2014).

Many innovative types of beer are nowadays commercialized following some new consumer trends. They vary for the nature and amounts of water, cereals, hops and other additives, technologies used for their production, storage and consumption, and also yeasts used as starters (Nelson 2014).

The primary criterion for clustering brewing technology is based on the type of fermentation: (i) top-fermenting brewing (ale beer), obtained using *S. cerevisiae* with a process occurring generally at 18–24 °C, and (ii) bottom-fermenting brewing (lager beer), produced using *Saccharomyces pastorianus* at a temperature of around 8–14 °C. Ales are characterized by fruity aromas while lager beers are more neutral as the contain lower amounts of fruity flavors; nonetheless *S. pastorianus* exhibits high resistance to different stresses, which makes it very useful for commercial producing beer (Dunn and Sherlock 2008; Gibson and Liti 2015; Müller-Auffermann et al. 2015). Strains used as starters, even if belonging to the same species, showed peculiar strain-related genetic and physiological characteristics as well as exclusive fermentative capacity (Berlowska et al. 2015). In particular, *S. cerevisiae* strains used in brewing show a higher genetic variability if compared with *S. pastorianus*. The genome sequences of a wide pool of *S. cerevisiae* strains of environmental and technological origin, were recently compared: overall ale-beer strains clustered separately from the other strains used as commercial starters, but showed a higher intraspecific variability if compared with strains use as wine starters, probably due to the hybrids or polyploid and aneuploid status of some of them (Gonzalez et al. 2008; Gonçalves et al. 2016). On the other hand, strains of *S. pastorianus*, whose origin as allopolyploid hybrid of *S. cerevisiae* and *Saccharomyces eubayanus* was recently hypothesized (Libkind et al. 2011), exhibited a lower intraspecific variability probably due to their geographically very circumscribed origin (Capece et al. 2018a). At present, the fermentation performance of *S. eubayanus* was also explored in order to select the most suitable starter for lager beer (Eizaguirre et al. 2018).

Due to the general uniformity of the main products commercialized by the big brewing companies, consumer's trends are nowadays increasing the demand of novel types of beer. Therefore, the growing commercial demand of low-alcohol, alcohol-free, low-calories, gluten free, and functional beers, as well as novel-flavored beers, have pushed the brewing market toward beer innovation (Yeo and Liu 2014; Michel et al. 2016a; Capece et al. 2018a). To fulfill consumer exigencies, a number of innovative small breweries were launched all over the world. This has increased the production of "old style" beers and new craft beers with distinctive and individual flavors given by the addition of some ingredients, like berries, herbs, and spices or special substrates of fermentation (e.g., germinated rice, wine must) which gave a special quality to the final product (Aquilani et al. 2015; Basso et al. 2016). The use of new distinctive flavored ingredients has been strictly coupled with the study and selection of new yeast strains showing unknown characteristics. In this contest, a number of studies focused their attention on some innovative brewing processes involving the use of novel strains of *S. cerevisiae* and *S. pastorianus* improved for some technological features. However this approach resulted very difficult. Therefore, the isolation and selection of some non-*Saccharomyces* and non-conventional yeasts have been recently studied.

11.2 Non-*Saccharomyces* and Non-conventional Yeasts: General Considerations

The investigation for new brewer yeasts was recently expanded also to yeast species never used in this contest and belonging to the so-called non-*Saccharomyces*, or even to the non-conventional yeasts (Basso et al. 2016). Although the group of "conventional" yeasts currently includes all *Saccharomyces* species (including those recently reassigned to other genera, namely, *Kazachstania exigua*, formerly *Saccharomyces exiguus*, Vaughan-Martini and Martini 2011), together with some species especially involved in food manufacturing, that is, *Schizosaccharomyces pombe* and *Kluyveromyces lactis*, there is still no generally accepted definition on non-conventional yeasts. However, although sometimes non-*Saccharomyces* and non-conventional yeasts are considered as synonyms, Sibirny and Scheffers (2002) suggested that the term non-conventional yeasts is gradually losing significance and usefulness. Beyond any definition, both non-*Saccharomyces* and non-conventional yeasts represent the vast majority of yeast species so far described (the current description of yeast taxonomy accounts for more than 1600 species, Kurtzman et al. 2011), which represents only a very small portion of total predictable yeast diversity (Boekhout 2005), which have recently revealed innumerable promising biotechnological abilities for both food and non-food industry (Sibirny and Scheffers, 2002; Wolf et al. 2003; Buzzini and Vaughan-Martini 2006). Indeed, an increasing body of the academic and industrial research has recently paid its attention to some non-conventional yeast genera, namely, *Pichia, Saccharomycodes, Zygosaccharomyces, Scheffersomyces, Hanseniaspora, Torulaspora,* and *Cyberlindnera* among Ascomycota, and *Mrakia* among Basidiomycota, for their possible exploitation as starter cultures for brewery technologies.

Demands for increased productivity from wider substrate range, production of novel compounds at the industrial level, as well as changing of consumer preferences can lead to a great interest in further enhancing the number of non-conventional yeasts being used by selecting useful strains with novel and attractive properties for beer production (Fleet 2006; Romano et al. 2006). They generally present low fermentation yields and are more sensible to ethanol but, in some cases, are able to improve other characteristics such as texture and integration of aroma for the creation of a final peculiar bouquet. In a few years, non-*Saccharomyces* and non-conventional yeasts changed their role: from potentially undesirable microorganisms associated to food (including beer) spoilage to a principal player in realizing peculiar products to gain a market position in this rising segment (Ciani and Comitini 2011; González et al. 2013; Johnson 2013; Gschaedler 2017).

11.3 Flavor-Active Compounds in Beer

The volatile portion of beer includes over 800 different compounds, but only several tens of these are considered flavor-active (Olaniran et al. 2011). Among them some classes of components determine the aroma structure and influence the beer quality when they are found in concentrations above their sensory thresholds (Table 11.1): (i) higher alcohols; (ii) esters between fatty acids and higher alcohols or ethanol; (iii) aldehydes; and (iv) vicinal diketones (Pires et al. 2014). However, some of these compounds may lead to a negative flavoring impact sometimes in dependence of their concentration. Strecker aldehydes (aged flavor), aldehydes of the Maillard reaction (furfural), aldehydes of fatty acid oxidation (trans-2-nonenal), and ketone diacetyl (2,3-butanedione) are considered off-flavors in beer (Ravasio et al. 2018). On the other hand, higher alcohols in small quantity have positive impacts that became negative if they exceeded 300 mg/L. Also, esters are pleasant in moderate quantities, but if overproduced they can negatively affect the beer aroma

Table 11.1 Main volatile compounds detected in beer and relative organoleptic threshold

Compounds	Flavor in beer	Organoleptic threshold (ppm)
Higher alcohols		
Propan-1-ol (n-propanol)	Alcohol	800
2-Methyl propanol (isobutyl alcohol)	Alcohol	200
2-Methyl butanol (active amyl alcohol)	Alcohol, banana, medicinal, solvent	65
3-Methyl butanol (isoamyl alcohol)	Alcohol	70
2-Phenyl ethanol	Roses, sweetish, perfumed	125
Esters		
Ethyl acetate	Solvent, fruity, sweetish	30
Isoamyl acetate	Banana, apple, solvent, estery	1.2
2-Phenylethyl acetate	Roses, honey, apple, sweetish	3.8
Ethyl caproate	Sour apple	0.21
Ethyl caprylate	Sour apple	0.9
Carbonyl compounds		
Acetaldehyde	Green leaves, fruity	25
2,3-Butanedione (diacetyl)	Butter-scotch	0.15

(Olaniran et al. 2017). The perception of volatile compounds depends on their volatility that is due to multiple factors such as pH, temperature, salt concentration, ethanol level, binding to fats/oils, proteins, starch, and phenolic compounds (matrix effect), as well as on synergistic or antagonistic interactions occurring between them (masking effect) (San-Juan et al. 2011; Lytra et al. 2013; Holt et al. 2018).

Barley malt and hops are responsible for some of the aroma compounds in beer, but the main players, that influence the final volatile composition, are the yeasts used as starters, whose metabolic versatility in producing a volatile profile has been extensively studied (Ramos-Jeunehomme et al. 1991; Rossouw et al. 2008; Adams and Taylor 2012; Pires et al. 2014; Mertens et al. 2015; Olaniran et al. 2017; Holt et al. 2018). Recently, Gamero et al. (2016) screened a selection of 143 yeast species for their fermentation performance and their potential production of aroma compounds such as fusel alcohols derived from the Ehrlich pathway, acetate esters, and ethyl esters. This study confirmed that many non-conventional yeast species showed high diversity of aroma profiles and, when used in beer wort, confirmed their ability to produce different flavor compounds. This is only the last paper concerning the use of non-conventional yeast for beer production. These species and their role in the brewing process are summarized in Table 11.2 and are reviewed in the present chapter as pure culture or co-culture starter or as players in spontaneous fermentation.

Table 11.2 Diversity and impact of non-conventional yeasts in different types of beer

Species	Beer								Starters		S.b.	G.r.	References
	na/la	la	al	lm	gue	fr	ch	tchp	cs	ci			
Blastobotrys mokoenaii									■	■			Holt et al. (2017)
Brettanomyces anomalus				■	■		■		■	■	■		Lentz et al. (2014), Spitaels et al. (2015), Michel et al. (2016a) and Holt et al. (2017)
Brettanomyces bruxellensis/ Dekkera bruxellensis		■	■	■	■		■		■	■	■	■	Daenen et al. (2008), Bokulich et al. (2012), Lentz et al. (2014), Spitaels et al. (2015), Crauwels et al. (2015), Steensels et al. (2015), Holt et al. (2017), Barbosa Piló et al. (2018), De Roos et al. (2018), Wang et al. (2016) and Serra Colomer et al. (2019)
Brettanomyces custersianus/ Dekkera custersiana				■							■		De Roos et al. (2018)
Brettanomyces naardenensis									■	■			Holt et al. (2017)
Candida stellata									■	■			Holt et al. (2017)
Candida californica							■				■		Barbosa Piló et al. (2018)
Candida ethanolica										■	■		Tokpohozin et al. (2016)
Candida friedrichii				■									Spitaels et al. (2015)
Candida humilis							■				■		Barbosa Piló et al. (2018)
Candida incommunis									■				Lentz et al. (2014)
Candida sake							■				■		Barbosa Piló et al. (2018)
Candida shehatae	■												Li et al. (2013)
Candida solani							■				■		Barbosa Piló et al. (2018)
Candida sorboxylosa							■				■		Barbosa Piló et al. (2018)
Candida tropicalis	■						■	■	■	■	■		'N Guessan et al. (2010), Attchelouwa et al. (2017) and Barbosa Piló et al. (2018)
Candida vinaria							■				■		Barbosa Piló et al. (2018)
Candida zeylanoides							■				■		Barbosa Piló et al. (2018)

Citeromyces matritensis													Holt et al. (2017)
Cyberlindnera fabianii													van Rijswijck et al. (2017)
Debaryomyces hansenii													Spitaels et al. (2015), Tokpohozin et al. (2016) and Holt et al. (2017)
Debaryomyces marama													Spitaels et al. (2015)
Debaryomyces occidentalis													Park et al. (2014)
Galactomyces candidum													Barbosa Piló et al. (2018)
Galactomyces geotrichum													Barbosa Piló et al. (2018)
Hanseniaspora opuntiae													Barbosa Piló et al. (2018)
Hanseniaspora uvarum													De Roos et al. (2018)
Kazachstania africana (former *Kluyveromyces africanus*)													Lyumugabe et al. (2012)
Kazachstania exigua													Barbosa Piló et al. (2018)
Kazachstania spp.													De Roos et al. (2018)
Kluyveromyces marxianus													Carlquist et al. (2015) and Attchelouwa et al. (2017)
Kodamaea ohmeri													N' Guessan et al. (2011) and Barbosa Piló et al. (2018)
Kregervanrija spp.													De Roos et al. (2018)
Lachancea thermotolerans													Holt et al. (2017) and Domizio et al. (2016)
Metschnikowia reukaufii													Holt et al. (2017)
Meyerozyma caribbica													N' Guessan et al. (2011)

(continucd)

Table 11.2 (continued)

Species	Beer								Starters		S.b.	G.r.	References
	na/ la	la	al	lm	gue	fr	ch	tchp	cs	ci			
Meyerozyma guilliermondii				■			■	■			■		Barbosa Piló et al. (2018), Attchelouwa et al. (2017) and Spitaels et al. (2015)
Mrakia gelida	■								■				De Francesco et al. (2018)
Ogataea spp.				■							■		De Roos et al. (2018)
Phaffomyces opuntiae (former *Pichia opuntiae*)											■		Bokulich et al. (2012)
Pichia fermentans				■			■				■		Bokulich et al. (2012), Spitaels et al. (2015) and Barbosa Piló et al. (2018)
Pichia kluyveri	■			■					■	■	■		N' Guessan et al. (2011), Bokulich et al. (2012), Saerens and Swiegers (2014), (2016) and Holt et al. (2017)
Pichia kudriavzevii	■						■	■	■	■	■		'N Guessan et al. (2010), Bokulich et al. (2012), Tokpohozin et al. (2016) Holt et al. (2017), van Rijswijck et al. (2017) and Barbosa Piló et al. (2018)
Pichia membranifaciens				■							■	■	Spitaels et al. (2015), Smukowski Heil et al. (2017) and De Roos et al. (2018)
Pichia terricola (former *Issatchenkia terricola*)									■				Lentz et al. (2014)
Priceomyces carsonii				■									Spitaels et al. (2015)
Rhodotorula mucilaginosa							■	■			■		Bokulich et al. (2012), Attchelouwa et al. (2017) and Barbosa Piló et al. (2018)
Saccharomyces bayanus		■										■	King and Dickinson, (2011)
Saccharomyces cerevisiae var. boulardii			■						■	■			Capece et al. (2018b)
Saccharomyces eubayanus		■										■	Libkind et al. (2011), Gibson and Liti (2015), Pires et al. (2014), Mertens et al. (2015), Krogerus et al. (2016) and Magalhães et al. (2017)
Saccharomyces kudriavzevii				■							■	■	Bizaj et al. (2012), De Roos et al. (2018)
Saccharomycodes ludwigii	■					■	■		■				Brányik et al. (2008), Liu et al. (2011) and De Francesco et al. (2015)
Saccharomycopsis fibuligera												■	Wang et al. (2016)

Saturnispora diversa (former *Candida diversa*)									■				Lentz et al. (2014)
Schizosaccharomyces pombe								■			■		Lyumugabe et al. (2012)
Starmerella bacillaris (former *Candida zemplinina*)									■	■			Holt et al. (2017) and Estela-Escalante et al. (2016)
Starmerella bombicola									■	■			Holt et al. (2017)
Torulaspora delbrueckii	■		■				■		■	■	■		Gibson and Liti (2015), Basso et al. (2016), Canonico et al. (2016), Michel et al. (2016b), Holt et al. (2017), Barbosa Piló et al. (2018) and Bellut et al. (2018)
Trichosporon asahii								■			■		Attchelouwa et al. (2017)
Trichosporon coremiiforme								■			■		Attchelouwa et al. (2017)
Wickerhamomyces anomalus				■					■	■	■		Spitaels et al. (2015), Basso et al. (2016), Holt et al. (2017) and De Roos et al. (2018)
Williopsis saturnus	■					■			■				Liu and Quek, (2016)
Yarrowia lipolytica				■							■		De Roos et al. (2018)
Zygoascus hellenicus							■				■		Barbosa Piló et al. (2018)
Zygosaccharomyces rouxii	■								■				Sohrabvandi et al. (2009) and De Francesco et al. (2015)
Zygotorulaspora florentina									■	■			Holt et al. (2017)

na/la no-alcohol or low alcohol, *la* lager, *al* ale, *lm* lambic, *ACA* American coolship ale, *gue* gueuze, *fr* fruity, *ch* chichi, *tchp* Tchapalo/Tchoukoutou, African sorghum beer, *sc* starter culture, *ci* co-inoculum, *sb* spontaneous brewing, *gr* genetic relevance

11.4 Non-conventional Yeast for Producing Alcohol-Free and Low-Alcohol Beer

Alcohol-free and low-alcohol beer production is becoming economically interesting for its constantly growing market and lower tax burden. The terminology "alcohol-free" and "low-alcohol" beer and the corresponding statutory regulations concerning alcohol limits showed many differences in each individual country. In some European countries like Germany, Switzerland, Austria, Finland, and Portugal, the term "alcohol-free" is referred to a maximum alcohol limit of 0.5% (v/v) ethanol. This limit decreases to <0.1% (v/v) in Denmark and in the Netherlands as well as in the UK with the limit set at <0.05% (v/v). In the USA, the limit of <0.5% (v/v) is defined for low-alcohol or reduced-alcohol beer while <0.05% is defined alcohol-free. Other countries, like Spain or France, accept the term "alcohol-free" for beer with limits of 1.0% and 1.2% (v/v), respectively (Caluwaerts 1995; Munroe 1995; Montanari et al. 2011).

Alcohol-free and low-alcohol beers can be obtained by technological approaches such as physical elimination of alcohol from regular beer or arrested fermentation, or by using non-conventional yeast starters unable to ferment maltose and/or maltotriose (the principal sugars in the wort) but able to produce higher amounts of volatile compounds, which are usually lost using the above technological approaches (De Francesco et al. 2018). Some non-conventional yeasts, which are so far known as beverage spoilage yeasts (Michel et al. 2016a), are also able to reduce wort aldehydes, and so the unpleasant "wort-like" flavors that can also be masked by the production of high amounts of esters (Saison et al. 2010; De Francesco et al. 2015). Several non-*Saccharomyces* species have been studied for alcohol-free and low-alcohol beer production (Table 11.2) at lab, pilot, and industrial scale (Michel et al. 2016a).

11.4.1 *Pichia kluyveri*

Considering its limited glucose fermenting ability, *P. kluyveri* has recently been revealed as a possible starter for producing alcohol-free and low-alcohol beers also in consideration of its fast fermentation ability and its significant aptitude in changing hop compounds into positive aroma (Anfang et al. 2009; Saerens and Swiegers 2014). Saerens and Swiegers (2014) patented the use of this species for the production of beer containing from <0.1% v/v to <0.2% v/v ethanol together with isoamyl acetate, isoamyl alcohol, ethyl butyrate, ethyl hexanoate, and ethyloctanoate. Sequential fermentations with *S. cerevisiae* have also been tested for obtaining low-alcohol beers with a more complex aroma profile compared to single fermentations with *S. cerevisiae* (Saerens and Swiegers 2016). *P. kluyveri* showed exceptional levels of glycerol, isoamyl acetate, and ethyl acetate production (Gutiérrez et al. 2018). However, an arrested fermentation could permit the isoamyl-acetate to be produced at a level covering the negative effect of ethyl acetate. Additionally, it was demonstrated that in sequential fermentations with *S. cerevisiae*, the ethyl acetate concentration was consistently reduced to a level similar as in *S. cerevisiae* single fermentations (Holt et al. 2017).

11.4.2 *Saccharomycodes ludwigii*

S. ludwigii has been frequently investigated for producing alcohol-free and low-alcohol beer and is already used in some German, Italian, and Czech breweries thanks to its aroma composition associated with positive sensory descriptors, such as the low concentration of esters and of off-flavors (Huige et al. 1990; Meier-Dörnberg et al. 2014; De Francesco et al. 2015; Petruzzi et al. 2016). *S. ludwigii* is often isolated from spoiled wine and usually shows a strain-specific high tolerance toward ethanol and a high production of ethyl acetate and glycerol (Romano et al. 1999; Gutiérrez et al. 2018). This species is unable to ferment maltose and maltotriose producing low concentrations of alcohol when grown in wort. Haehn and Glaubitz (1933) and successively Huige et al. (1990) patented the production of low-alcohol and non-alcoholic beer, respectively, using *S. ludwigii* strains. More recently, Mohammadi et al. (2011) immobilized the strain DSM 3447, recently tested by Sohrabvandi et al. (2010), as a way to induce a slight fermentation of maltose under different temperatures. A honey-like beer, showing moderate fermentation (ethanol 0.48% v/v) and low concentrations of ethyl acetate, isoamyl acetate, and 4-vinylguaiacol, but high quantities of amyl alcohols and higher alcohols compared to commercial German alcohol-free wheat beers, was recently obtained by using *S. ludwigii* strains (Meier-Dörnberg et al. 2014). In more recent times, an *S. ludwigii* strain was selected as the best starter for producing low-alcohol beer (0.51% v/v ethanol) starting from a special wort obtained by high-temperature mashing in order to maximize the non-fermentable sugars (De Francesco et al. 2015). The concentration of diacetyl and 3-methylpropionaldehyde was lower than the perception threshold, differently from a previous study (Narziss et al. 1992). Additionally, the high levels of desirable esters were able to mask the off-flavors typically produced by the partial fermentation (De Francesco et al. 2015).

11.4.3 *Zygosaccharomyces rouxii*

Z. rouxii is well known spoilage yeast exhibiting a high tolerance to osmotic pressure (Michel et al. 2016a) that implements its ability to ferment different carbon sources in high-gravity brewing wort (Gibson 2011). Studies of the volatile profile obtained by *Z. rouxii* in soy sauce revealed the production of a high quantity of ethyl acetate, amyl alcohols, isoamyl alcohols, and other esters and higher alcohol typical of beer fermentation (Cao et al. 2010; Lee et al. 2013). Therefore *Z. rouxii* has become a possible candidate as a starter culture for producing low-alcohol or alcohol-free beer (Michel et al. 2016a). Two strains of *Z. rouxii* were studied by Sohrabvandi et al. (2010). Strain DSM 2531, unable to use maltose and maltotriose, resulted the best strain for this purpose (Sohrabvandi et al. 2010). On the contrary, the same strain immobilized in brewer's spent grain and used in beer wort produced up to 4.1% v/v alcohol (Mohammadi et al. 2011). Mortazavian et al. (2014) used two strains of *Z. rouxii* for the fermentation of beer wort at different temperatures (from 4 to 24 °C) reaching no more than 0.4% of alcohol content, but scarce amounts of positive sensory descriptors. In another report, five different strains of the same species were studied for producing low-alcohol beer (from 0.9 to 3.32% v/v ethanol) using a wort characterized by low concentrations of simple carbohydrates. Volatile profile was strain-specific, but in most cases the quantity of esters and acetaldehyde remained below the perception threshold, while the level of diacetyl, above the perception threshold, conferred an unpleasant sweetish buttery flavor (De Francesco et al. 2015).

Sohrabvandi et al. (2009) investigated the use of two strains of *Z. rouxii* in a sequential fermentation with *S. cerevisiae* for producing non-alcoholic beer. In this contest *S. cerevisiae* promptly metabolized glucose and fructose leaving the substrate devoid of fermentable sugars. As a consequence *Z. rouxii,* unable to uptake maltose, was forced to use ethanol as residual C sources, thus reducing its final content (from 1.56% to 0.36%) and obtaining slightly lower overall flavor acceptability if compared with single *S. cerevisiae* fermentation (Sohrabvandi et al. 2009).

11.4.4 Scheffersomyces shehatae

A patent deposited by Li et al. (2013) proposed the use of a maltose negative strain of *S. shehatae* (former *Candida shehatae*) for producing non-alcoholic beer at 14 °C. A concentration of ethanol <0.5% v/v was obtained together with no smell, ordinary color, taste, and flavor. *S. shehatae* was studied prevalently for its ability to ferment D-xylose into ethanol (Kastner et al. 1992, 1996) and to bioreduce some hop monoterpenes, such as myrcene, into positive flavor-active compounds in beer (Mets and Verzele 1968; Goretti et al. 2013).

11.4.5 Cyberlindnera mrakii

Liu and Quek (2016) studied the use of *C. mrakii* (former *Williopsis saturnus* var. *mrakii*) for producing a fruity beer; its performance was also compared with a commercial *S. cerevisiae* strain (Safale US-05). Ethanol produced at the end of fermentation was 5 time higher for *S. cerevisiae* than for *C. mrakii* (1.7% v/v). Besides, beer obtained by *C. mrakii* exhibited high concentrations of acetate esters, in particular isoamyl, ethyl, and 2-phenylethyl acetate, low amount of methyl esters, and no trace of acetaldehyde, differently from the beer obtained with *S. cerevisiae*. Additionally *C. mrakii* retained more terpenes and terpenoids, suggesting that such beer could display more of the aromatic hint of hops (Liu and Quek 2016).

11.4.6 *Wickerhamomyces anomalus*

W. anomalus (former *Pichia anomala*) currently shows a broad spectrum of carbohydrate assimilation, including maltose (Kurtzman 2011). These species exhibit a fermentative ability in the absence of oxygen, although a small tolerance to ethanol has been found. The volatile profile includes ethyl propanoate phenyl ethanol, 2-phenylethyl acetate, and ethyl acetate (Passoth et al. 2006). Accordingly, *W. anomalus* was suggested as an interesting candidate for future production of low-alcohol beer (Basso et al. 2016).

11.4.7 *Mrakia gelida and Mrakia blollopis*

The significance of yeasts in many fermented foods and beverages is related almost exclusively to Ascomycota, while basidiomycetous yeasts able to express a fermenting metabolism are involved only exceptionally in fermented products (Hallsworth 1998; Tamang and Fleet 2009). In the beer market, the use of basidiomycetous yeasts was considered only by a few authors. The first approach was proposed by Thomas-Hall et al. (2010), who described three new species of the genus *Mrakia*, showing that one of them (*Mrakia blollopis*) was able to produce an artisanal beer using a home brewing kit. This early trial revealed the production of a beer with low-alcohol content (maximum 2.7% v/v) that increased (5–6% v/v) when a commercial strain of *S. cerevisiae* was sequentially added (Thomas-Hall et al. 2010). The use of different strains of *Mrakia* spp. was also tested by Tsuji et al. (2016). The authors tested both ethanol productivity and ethanol tolerance. While all strains produced ethanol from glucose at 10 °C, only a few strains of *M. blollopis* showed high ethanol tolerance. More recently, *Mrakia gelida* was used for producing low-alcohol beer at 10 °C with a 22-day-long fermentation (De Francesco et al. 2018). Its brewing performance was compared with that of a commercial strain of *S. ludwigii*. After the primary fermentation maltose was little fermented, generating a beer with only 1.16 % v/v of ethanol. After the subsequent secondary fermentation in a bottle (with the addition of primary sugar), ethanol content increased up to 1.4% (v/v). The production of positive and negative aroma compounds was at a concentration below the perception threshold but the sensory profile of *M. gelida* beer in comparison to that of S. *ludwigii* resulted to be fruitier (De Francesco et al. 2018).

11.4.8 Other Miscellaneous Non-conventional Species

Strains of *Hanseniaspora valbyensis*, *Hanseniaspora vineae*, *Torulaspora delbrueckii*, *Zygosaccharomyces bailii*, and *Zygosaccharomyces kombuchaensis* isolated from kombucha, an effervescent, slightly sweet, and slightly acidic tea obtained by the fermentation of bacteria and yeast was tested for its ability to produce alcohol-free and low-alcohol beer. All the strains showed the ability to ferment glucose and fructose but not maltose. In addition, none of them presented the production of phenolic off-flavor. All the strains exhibited very high viability rates and a final alcohol content ranging from 0.34% to 0.48 v/v. Higher alcohols and acetate esters were produced at a lower concentration if compared to the performance of commercial *S. ludwigii* and *S. cerevisiae* (Bellut et al. 2018).

11.5 Non-conventional Yeast for Producing Other Types of Beer

11.5.1 Torulaspora delbrueckii

Among non-conventional yeasts, *T. delbrueckii* has become one of the most commercially used non-*Saccharomyces* species in wine industry (Kurtzman et al. 2011; Tataridis et al. 2013; Jolly et al. 2014; Belda et al. 2017). This species showed the ability to ferment glucose and fructose and tolerate ethanol up to 9–11% v/v (Ciani and Maccarelli 1998; Tataridis et al. 2013; Benito 2018). Although it was considered for decades a common contaminant in the brewing environments for its ability to ferment maltose, during wort fermentation it exhibited the production of many positive fruity aromas, such as β- phenylethanol, n-propanol, iso-butanol, amyl alcohol, and ethyl acetate, and the ability to potentially convert hop monoterpene alcohols into linalool, that is one of the main compounds that defines hop aroma in the beer (King and Dickinson 2000; Daenen et al. 2008; Pires et al. 2014; Etschmann et al. 2015; Basso et al. 2016).

Beers produced by pure cultures of *T. delbrueckii* and by mixed fermentations of *S. cerevisiae*/*T. delbrueckii* were recently obtained (Canonico et al. 2016; Michel et al. 2016b). *T. delbrueckii* exhibited strain-related flavor-forming abilities, hop and ethanol resistance, amino acid catabolism and anabolism, and phenolic off-flavor production. The main produced volatiles were 2-phenylethanol, amyl alcohol, ethyl acetate, n-propanol, and isobutanol (Michel et al. 2016b). Pure fermentation with *T. delbrueckii* and mixed fermentations *T. delbrueckii*/*S. cerevisiae* were performed by Canonico et al. (2016, 2017). In pure culture, *T. delbrueckii* produced beers with low-alcohol content (2.66% v/v) providing pleasant and aromatic taste. On the contrary, an interesting increment of the esters phenyl ethyl acetate, ethyl hexanoate, and ethyl octaonate were exhibited in mixed fermentations. The possibility of using *T. delbrueckii* in sequential fermentations with *S. cerevisiae* to complete the fermentation was evaluated by Holt et al. (2017). High values of ethyl acetate were found in the beers obtained in the sequential fermentations with *S. cerevisiae*, leading to beers exhibiting a clove-like aroma, typical of some wheat and blond beers (Tataridis et al. 2013). The ability of *T. delbrueckii* to ferment more slowly than commercial *S. cerevisiae* strains but exhibiting the production of complex and intense beers rich in esters and poor of phenol attributes was also found (Tataridis et al. 2013).

11.5.2 *Candida tropicalis*

C. tropicalis is one of the mostly known clinical yeast species after *Candida albicans*, behaving as an opportunistic pathogen (Moran et al. 2002). *C. tropicalis* is supposed to ferment different carbon sources including maltose. N'Guessan et al. (2010) used *C. tropicalis* and *S. cerevisiae* isolated from tchapalo (traditional West African sorghum beer) as pure culture and co-culture for the fermentation of sorghum wort. At the end of fermentation, the beer showed 20.5 g/L of lactic acid and ethanol concentration of 0.2% v/v. Higher alcohols and acetaldehyde were poorly produced and the only volatile compound that was detected in higher concentrations, although lower than the threshold, was 2-butanone. A strain of *C. tropicalis* was also used by Allouse-Boraud et al. (2015) to ferment regular malt wort at 30 °C. The product exhibited a final alcohol content of 1.5% v/v, low amount of succinate and lactate, and reduced aroma-active compounds.

11.5.3 *Cyberlindnera fabianii and Pichia kudriavzevii*

C. fabianii and *P. kudriavzevii* were tested as selective starters for beer production as pure culture and in co-cultivation with *S. cerevisiae* brewing strain. When in pure culture *C. fabianii* and *P. kudriavzevii* showed a low fermentation rate but high esters production if compared with *S. cerevisiae*. In co-fermentation, both *C. fabianii* and *P. kudriavzevii* exhibited an alcohol concentration of 3.5% and 3.8% v/v, respectively, and volatile esters significantly higher if compared with what is found in the brewers' yeast mono-culture (van Rijswijck et al. 2017).

11.5.4 *Starmerella bacillaris*

S. bacillaris (former *Candida zemplinina*) was firstly isolated from Tokaj wine and shows high osmo- and alcohol tolerance. It was unable to assimilate and ferment maltose (Lachance et al. 2011). The performances of two strains of this species were studied in wort fermentation and compared with a commercial starter of *S. cerevisiae* (S-23). Different synthetic media composed of a laboratory malt wort added with a few adjuncts (glucose syrup and/or apple juice) were used. The two strains grew better than the control strain and maintained higher viable cell counts at the end of the fermentation. When only glucose syrup was added to the wort, the percentage of alcohol was lower but when apple juice was used the production of ethanol increased. Unfortunately, no aroma composition was reported. Therefore, this research can be considered only a preliminary study of the possible use of *S. bacillaris* for producing craft beer (Estela-Escalante et al. 2016).

11.5.5 Saccharomyces *sp.* "boulardii"

The recognition of *Saccharomyces* sp. "*boulardii*" (former *Saccharomyces cerevisiae* var. *boulardii*) as a taxonomical variety is still a controversial question. Some authors continue to mention it simply as *S. boulardii* while recently it was classified as *Saccharomyces* sp. "*boulardii*" waiting for further taxonomic clarification (Vaughan-Martini and Martini 2011).

Strains belonging to this species have been recently used as a biotherapeutic agent for the treatment of *Clostridium difficile*-associated diarrhea (Vaughan-Martini and Martini 2011), and in this form they are still commercialized in some countries. The effect in human health of beer produced by a co-fermentation of the probiotic *Saccharomyces* sp. "*boulardii*" and a *S. cerevisiae* starter was recently considered (Capece et al. 2018b). At the end of the fermentation, the probiotic yeast cells resulted dominant on *S. cerevisiae*, and the cells remained viable in beers. No significant changes have been found in the volatile profile, while antioxidant activity and polyphenols content increased when compared with beer produced using *S. cerevisiae* (Capece et al. 2018b).

11.5.6 Other Miscellaneous Non-conventional Species

Lentz et al. (2014) selected 10 "wild" yeasts isolated of the species *Candida incommunis*, *Pichia terricola* (former *Issatchenkia terricola*), and *Saturnispora diversa* (former *Candida diversa*) from the fruits of pindo palm, loquat, hackberry, and blackberry in Florida and Pennsylvania (USA) in order to test them as single pure cultures in brewing. *P. terricola* and *S. diversa* showed an efficient attenuation of the wort and alcohol tolerance to ≥8%, while *C. incommunis* exhibited the highest attenuation (90%) and curiously the lowest alcohol tolerance (6%). The performances of the isolates resulted generally strains dependent (Lentz et al. 2014). More recently, a set of strains of the non-conventional species *Galactomyces candidus* (former *Galactomyces geotrichum*), *Kazachstania zonata*, *Starmera caribaea*, and *Yarrowia lipolytica* were screened for their ability of producing pleasant fruity aromas during the fermentation of different substrates, including wort (Gutiérrez et al. 2018). *K. zonata* and *Y. lipolytica* showed a fermentation aptitude similar to the control strains. Additionally, *K. zonata* produced higher glycerol yields than the controls. *G. candidus* and *S. caribaea* exhibited the poorest fermentation activity. They were suggested to be used in mixed fermentations with *S. cerevisiae* for enhancing aroma complexity or as pure cultures for the development of non-alcoholic beverages (Gutiérrez et al. 2018).

11.6 Spontaneous Brewing

Spontaneous fermentations are processes in which the use of selected microorganisms is not foreseen. Spontaneous brewing is based on the sequential activities of wild microorganisms that are not deliberately introduced into raw materials and that pilot the fermentation (Petruzzi et al. 2016).

11.6.1 European Spontaneous Beers

In Europe, the most famous spontaneous fermentation beers are represented by lambic beers and their particular modifications called gueuze and kriek. These beers are spontaneously fermented with naturally occurring microorganisms specific of a little Belgian region known as Pajottenland, near Brussels. In these beers, several genera of yeasts and bacteria coexist and change their population over time, due to the long fermentation time (which can last even several years) and complex mutations of pH and nutrients (Crauwels et al. 2015).

There are only a small number of scientific papers about these kinds of beers, but now it is demonstrated that *Brettanomyces bruxellensis* (former *Dekkera bruxellensis*) is the essential microbial player that outlines the uniqueness of lambic beers. However, the ecology of lambic beers is complex and it was investigated since 1977 by Van Oevelen et al. through several selective media. Recently the taxonomy structure of the yeast community in such beers was analyzed using modern community profiling techniques based on high-quality sequencing, denaturing gradient gel electrophoresis (DGGE), and mixed techniques based on molecular biology and mass spectrometry (Bokulich et al. 2012; Spitaels et al. 2015).

Lastly, De Roos et al. (2018) performed a culture-dependent plating analysis combined with culture-independent amplicon sequencing. Despite different techniques and sampling methods adopted, almost all studies evidenced quite similar microbial composition and no significant differences on industrially produced lambic beers and traditional ones (Spitaels et al. 2015). Generally, the lambic beer production split up into four different phases: (i) initial growth of the *Enterobacteriaceae* (1 month), which produces low concentrations of organic acids and ethanol; (ii) main fermentation phase (3 months) dominated by the *Saccharomyces* spp.; (iii) acidification phase (6 months) dominated by lactic acid bacteria, such as *Pediococcus damnosus* and *Lactobacillus brevis* and in which the yeasts *Brettanomyces* spp. start to increase; (iv) maturation phase (after 10 months), where *Brettanomyces* spp., mainly *B. bruxellensis*, spread and contribute to acid acetic production, drawing the complex bouquet flavor of lambic beers (Spitaels et al. 2015).

The knowledge on lambic beers and the involved microorganisms have developed and opened to improvements, like the practice of acidification of wort to enhance the microbial stability, and the control of biogenic amines production by *Enterobacteriaceae*, ensuring healthy products on markets (De Roos et al. 2018). The non-conventional yeasts mostly found in lambic beers compose the microbial core that is maintained during consecutive fermentations; they can be summarized as below: *Brettanomyces anomalus*, *B. bruxellensis*, *Brettanomyces custersianus*, *Hanseniaspora uvarum*, *Kazachstania* spp., *Kregervanrija* spp., *Ogataea* spp., *Pichia membranifaciens*, *Pichia fermentans*, *W. anomalus*, *Y. lipolytica*, and *Saccharomyces kudriavzevii* (Spitaels et al. 2015). Furthermore other yeast species were occasionally isolated on selective media and they seem to belong to cask wood and the brewery environment: *Naumovozyma castellii*, *Debaryomyces hansenii*, *Debaryomyces maramus*, *Candida friedrichii*, *Meyerozyma guilliermondii*, and *Priceomyces carsonii* (Spitaels et al. 2015).

11.6.2 The Dominant Non-conventional Genus: **Brettanomyces** *spp.*

It was shown that *Brettanomyces* spp. are the main yeast characters in all the beers obtained by spontaneous fermentations (Steensels et al. 2015). They can affect many sensorial features of fermented products; a set of terms are used to describe its presence as "burnt plastic," "barnyard," "medical," "horse sweat," "leather," and "fruity" (Crauwels et al. 2015). In oenology, *Brettanomyces* spp. are better known for their role in the spoilage of wine (Crauwels et al. 2015), while in brewing industry, especially in lambic and gueuze beer, they have been studied for a long time for their specific physiological and metabolic features. *Brettanomyces* strains are able to metabolize a large amount of monosaccharides, disaccharides, trisaccharides, and dextrins (Blomqvist et al. 2010). Under aerobic condition, as open-air fermentation, *Brettanomyces* spp. produce hardly any glycerol, but high levels of acetic acid, moreover, under oxygen availability and nutrient shortage, can use ethanol and acetic acid as a sole carbon source (Schifferdecker et al. 2014; Smith and Divol 2018). Holt et al. (2017) investigated 17 non-conventional yeast species for the production of an appealing profile of flavor esters and phenolics in the first phase of alcoholic fermentation, followed by inoculation with *S. cerevisiae* to complete the fermentation. They observed that *Brettanomyces* strains succeed in a significant transformation of coumaric and ferulic acid into volatile phenols such as 4-ethylphenol and 4-ethylguaiacol, showing a desirable feature for beer spicy phenolic aroma (Holt et al. 2017).

Brettanomyces species manage to survive in wooden casks for a long time, due to their ability to breakdown cellobiose trough β-glucosidases and use it as carboun source. This enzymatic activity can result in a significant increase of several flavor active compounds that, on the contrary are odourless and non-volatile while bound to sugar molecules. (Serra Colomer et al. 2019).

During fermentation, Vervoort et al. (2016) verified the fruitiness of beer, without the characteristic off-flavors that are typically developed by *Brettanomyces* yeasts. It can be obtained isolating *Brettanomyces* β-D-glucosidase enzymes and adding them during the brewing process without the yeast. A range of a few hundred different yeast strains were investigated for enzyme activities and some *B. anomalus* and *B. bruxellensis* strains presented exceptionally high values during the screening; their genomes were sequenced in order to predict the genes involved. The β-D-glucosidase-encoding genes were codon-optimized and expressed in *E. coli*, purified, and were characterized for their technological and industrial applications (Vervoort et al. 2016). The glycoside hydrolase activity of *Brettanomyces* strains were also found by Daenen et al. (2008) on glycosides from sour cherry. The β-D-glucosidase activity expressed by *Brettanomyces* species is used in the growing market of low-calorie beers. Indeed, these glucosidases are able to degrade dextrins into simple sugars, which are the main residual sugars in beers produced by *S. cerevisiae* fermentation (Capece et al. 2018a).

11.6.3 African Spontaneous Beers

In Africa, wheat or barley malt is replaced by sorghum grains and sorghum opaque beer is an important beverage (Polycarpe Kayodé et al. 2007). It shows higher levels of calories, vitamins, especially the B-group, and amino acids than European beers (Chevassus et al. 1976). Denominations, production processes, and sorghum varieties vary according to geographical localization (Lyumugabe et al. 2012).

In Côte d'Ivoire, as in many other African countries, women are the brewers; they prepare a traditional sorghum beer named "tchapalo" using final wort with a portion of previous beer or dried yeast harvested from former production as alcoholic fermentation starters (Sefa-Dedeh et al. 1999). Tchapalo involves two main steps composed by (i) a spontaneous lactic fermentation and (ii) an uncontrolled alcoholic fermentation (Djè et al. 2009). N'guessan et al. (2011) investigated yeast species associated with the alcoholic fermentation of tchapalo. They found *S. cerevisiae*-like species, *C. tropicalis*, *P. kudriavzevii*, *P. kluyveri*, *Kodamaea ohmeri*, and *Meyerozyma caribbica*. Attchelouwa et al. (2017) observed that tchapalo was contaminated after two days at room temperature and after three days at a refrigerated temperature (4 °C) mainly by acetic acid and lactic acid bacteria, and by yeasts belonging to *S. cerevisiae*, *C. tropicalis*, *Rhodotorula mucilaginosa*, *Trichosporon asahii*, *Kluyveromyces marxianus*, *M. guilliermondii*, and *Trichosporon coremi-*

iforme. Despite acetic acid bacteria produce the over-oxidation of ethanol, and a resulting sensory spoilage, their presence, as well as that of lactic acid bacteria, can be considered a positive contribution to the acidification and consequently to the inhibition of undesirable microorganisms (Attchelouwa et al. 2017).

Researchers founded *P. kudriavzevii* in "Ikigage" from Rwanda, *T. delbrueckii* in "Tchoukoutou" from Benin, *K. marxianus*, *Naganishia albida* (former *Cryptococcus albidus*), and *D. hansenii* in "Bili bili" or "Amgba" from Cameroon, *C. tropicalis*, *H. uvarum* (former *Kloeckera apiculata*), *W. anomalus* (former *Hansenula anomala*), *T. delbrueckii*, *S. pombe* and *Kazachstania africana* (former *Kluyveromyces africanus*) in "Pito" from Ghana (Lyumugabe et al. 2012). More recently, Tokpohozin et al. (2016) collected 36 samples of Tchoukoutou starters in 12 localities of 2 zones with a high production of African sorghum beer in Benin. They aimed to create a biomarker set in order to describe the yeast ecology of the leaven used as starter. The study highlighted that the microbial ecology of Tchoukoutou is well described by MALDI-TOF analysis and bio-molecular ones used as different biomarkers. Yeast isolates belonged to four species, including *S. cerevisiae* (75.17%) as the dominant yeast, followed by *P. kudriavzevii* (17.24%), *Candida ethanolica* (4.14%), and *D. hansenii* complex (3.45%) (Tokpohozin et al. 2016).

Whereas African beers are consumed during the active fermentation state, low pH levels are not always reached. Therefore, their shelf life is quite short (24–72 h) and some opportunistic pathogens can be found, such as *C. tropicalis* (N'Guessan et al. 2010)

11.6.4 American Spontaneous Brews

In the American continent, two spontaneous fermentations take place, divided mainly in two different geographical areas: in North America, a European heritage, American coolship ale can be found, that employs production methods similar to traditional Belgian lambic, while in South America the "chicha," a beer-like beverage coming from pre-Hispanic times, is principally present.

Bokulich et al. (2012) investigated the brewhouse resident microbiota in American coolship ale, whose complete three-year-production process can be resumed as follows: (i) exposure of the wort to the atmosphere in the coolship (week 1) and (ii) 184 weeks of fermentation in oak casks. Even bottle-refermented samples were analyzed. Surprisingly, American coolship ale showed a similar yeast community of European lambic beers. Early dominant species was *S. cerevisiae* (60–80%) but *P. kudriavzevii* (former *Candida krusei*), *P. fermentans*, *Cryptococcus albidus var. kuetzingii*, and *R. mucilaginosa* were also present. *S. cerevisiae* maintained its dominance during the first 11 weeks, when *B. bruxellensis* appeared in a low percentage (average relative abundance 7%) and became dominant until the end of fermentation; after bottling, populations of *Phaffomyces opuntiae* (former *Pichia opuntiae*), *P. fermentans*, *P. kluyveri*, *C. albidus var. kuetzingii*, and *P. kudriavzevii* were found sporadically (Bokulich et al. 2012).

In South America, especially in the Andean regions, people from small villages used the name "chicha" to refer to different beverages, fermented or not, prepared from various materials, such as cassava, beans, and maize (such as rice, oats, and quinoa) and sometimes with added fruits (such as bananas) (Morris 1974). Vallejo et al. (2013) isolated *S. cerevisiae* as the single yeast species at the end of fermentation from 10 samples of *chicha de jora* collected from different traditional breweries in the Cusco region, Peru. Rodríguez et al. (2017) attributed to *Saccharomyces uvarum* the fermentation of apple chicha, manufactured by aboriginal communities of Andean Patagonia (Argentina and Chile). Mendoza et al. (2017) showed by high-throughput sequencing and culture-dependent approaches that *S. cerevisiae* was the dominant species in an Argentinian maize-based *chicha*. Finally Barbosa Barbosa Piló et al. (2018) identified 26 yeast species by restriction polymorphism mitochondrial profiles, sampling 42 chicha products in two regions of Ecuador. *S. cerevisiae* was again the most prevalent species occurring in 33 of the total 42 samples, but even non-*Saccharomyces* species were identified: *T. delbrueckii* emerged as the second most identified species, followed by *P. kudriavzevii*, *Candida sake*, *B. bruxellensis*, *P. fermentans*, and *S. ludwigii*.

Depending on the type of beer sampled and so the different manufacturing processes and variety of raw materials mixed, the non-conventional yeast population was very different: the most complex yeast ecology was found on samples of chicha de jora, made by yellow maize grain, which is malted and milled to flour (jora). In seven-grain chicha, *B. bruxellensis*, *P. kudriavzevii*, *Pichia manshurica* were detected; in chicha de yuca isolates of non-*Saccharomyces* species were represented by *C. tropicalis*, *Hanseniaspora opuntiae*, *K. ohmeri*, *P. kudriavzevii*, and *T. delbrueckii*; and only in chicha de morocho the complete absence of *S. cerevisiae* was replaced by *C. sake*, *P. fermentans*, *S. ludwigii*, and *T. delbrueckii* (Barbosa Piló et al. 2018).

11.7 Concluding Remarks

The beer market changed completely and new common requirements have become evident. An increasing demand of low-alcohol, free-alcohol, and low-calorie beers, in addition to spontaneous fermented beer or rich in alternative flavors, has encouraged international research toward innovative fermentations using different substrates (carbon sources and hops) but also novel yeast starters (Gibson et al. 2017). Non-conventional yeast starters usually showed a lower alcohol production if compared with "traditional" *S. cerevisiae* and *S. pastorianus* performance, but a higher quantity of positive flavored compounds was frequently described. The abundant number of recently published papers cited in this chapter demonstrate how this sector is active and the potential role that non-conventional yeasts can play in expanding the market of novel types of beers. Although many different species of non-conventional yeasts were described for their fermentative aptitude, alone or in co-fermentation with commercial starters, only rarely a given process scaled up at the industrial scale. The almost totality of the studies cited in this chapter, although very innovative, have been only realized at the lab, or at least, at the pilot scale.

Our current knowledge of non-conventional yeasts for food and beverage production (including beer) is rather incomplete, and the list of yeast species approved by the US Food and Drug Administration or the European Food Safety Authority for being utilized in the food and beverage industry is extremely limited. Additionally, a few of the described species are suspected to assume a potentially pathogenic aptitude (even if at an opportunistic level) or to produce some health dangerous compounds (such as biogenic amines, methanol, urea) (Petruzzi et al. 2016).

Therefore, although the mass of studies cited in this chapter can undoubtedly open a new era for brewing, additional specific studies on the physiological, metabolic, and technological aptitudes of some non-conventional yeasts will be necessary, before focusing on their industrial implementation.

References

Adams S, Taylor AJ (2012) Oral processing and flavour sensing mechanisms. In: Chen J, Engelen L (eds) Food oral processing: fundamentals of eating and sensory perception. Blackwell Publishing Ltd, Oxford, pp 177–202

Allouse-Boraud WAM, N'Guessan FK, Djeni TND et al (2015) Fermentation profile of *Saccharomyces cerevisiae* and *Candida tropicalis* as starter cultures on barley malt medium. J Food Sci Technol 8:5236–5242

Anfang N, Brajkovich M, Goddard M (2009) Co-fermentation with *Pichia kluyveri* increases varietal thiol concentrations in Sauvignon Blanc. Aust J Grape Wine Res 15:1–8

Aquilani B, Laureti T, Poponi S, Secondi L (2015) Beer choice and consumption determinants when craft beers are tasted: an exploratory study of consumer preferences. Food Qual Prefer 41:214–224

Attchelouwa KC, Aka-Gbézo S, N'guessan FK et al (2017) Biochemical and microbiological changes during the Ivorian sorghum beer deterioration at different storage temperatures. Beverages. https://doi.org/10.3390/beverages3030043

Barbosa Piló F, Carvajal-Barriga EJ, Guamán-Burneo MC et al (2018) *Saccharomyces cerevisiae* populations and other yeasts associated with indigenous beers (*chicha*) of Ecuador. Braz J Microbiol 49:808–815

Basso RF, Alcarde AR, Portugal CB (2016) Could non-*Saccharomyces* yeasts contribute on innovative brewing fermentations? Food Res Int 86:112–120

Belda I, Ruiz J, Beisert B et al (2017) Influence of *Torulaspora delbrueckii* in varietal thiol (3-SH and 4-MSP) release in wine sequential fermentations. Int J Food Microbiol 257:183–191

Bellut K, Michel M, Zarnkow M et al (2018) Application of non-*Saccharomyces* yeasts isolated from Kombucha in the production of alcohol-free beer. Fermentation 4(3):66

Benito S (2018) The impact of *Torulaspora delbrueckii* yeast in winemaking. Appl Microbiol Biotechnol 102:3081–3094

Berlowska J, Kregiel D, Rajkowska K (2015) Biodiversity of brewery yeast strains and their fermentative activities. Yeast 32(1):289–300

Bizaj E, Cordente AG, Bellon JR et al (2012) A breeding strategy to harness flavor diversity of *Saccharomyces* interspecific hybrids and minimize hydrogen sulfide production. FEMS Yeast Res 12:456–465

Blomqvist J, Eberhard T, Schnürer J et al (2010) Fermentation characteristics of *Dekkera bruxellensis* strains. Appl Microbiol Biotechnol 87:1487–1497
Boekhout T (2005) Gut feeling for yeasts. Nature 434(7032):449–451
Bokulich NA, Bamforth CW, Mills DA (2012) Brewhouse-resident microbiota are responsible for multi-stage fermentation of American Coolship Ale. PLoS One 7(4):e35507
Brányik T, Vicente AA, Dostálek P et al (2008) A review of flavour formation in continuous beer fermentation. J Inst Brew 114(1):3–13
Buzzini P, Vaughan-Martini A (2006) Yeast biodiversity and biotechnology. In: Rosa CA, Peter G (eds) Biodiversity and ecophysiology of yeasts. Springer, Berlin, pp 533–559
Caluwaerts HJJ (1995) Process for the manufacture of an alcohol free beer having the organoleptic properties of a lager type pale beer. US Pat 5(384):135
Canonico L, Agarbati A, Comitini F et al (2016) *Torulaspora delbrueckii* in the brewing process: a new approach to enhance bioflavour and to reduce ethanol content. Food Microbiol 56:45–51
Canonico L, Comitini F, Ciani M (2017) *Torulaspora delbrueckii* contribution in mixed brewing fermentations with different *Saccharomyces cerevisiae* strains. Int J Food Microbiol 259:7–13
Cao X, Hou L, Lu M et al (2010) Genome shuffling of *Zygosaccharomyces rouxii* to accelerate and enhance the flavor formation of soy sauce. J Sci Food Agric 90:281–285
Capece A, Romaniello R, Siesto G et al (2018a) Conventional and non-conventional yeasts in beer production. Fermentation. https://doi.org/10.3390/fermentation4020038
Capece A, Romaniello R, Pietrafesa A et al (2018b) Use of *Saccharomyces cerevisiae* var. *boulardii* in co-fermentations with *S. cerevisiae* for the production of craft beers with potential healthy value-added. Int J Food Microbiol 284:22–30
Carlquist M, Gibson B, Yuceer YK et al (2015) Process engineering for bioflavour production with metabolically active yeasts – a mini-review. Yeast 32(1):123–143
Chevassus-Agnès S, Favier JC, Joseph A (1976) Technologie traditionnelle et valeur nutritive des “bières” de sorgho du Cameroun Cahiers de. Nutrition et de Diététique 11(2):89–104
Ciani M, Comitini F (2011) Non-*Saccharomyces* wine yeasts have a promising role in biotechnological approaches to winemaking. Ann Microbiol 61(1):25–32
Ciani M, Maccarelli F (1998) Oenological properties of non-*Saccharomyces* yeasts associated with winemaking. World J Microbiol Biotechnol 14:199–203
Crauwels S, Steensels S, Aerts G et al (2015) *Brettanomyces bruxellensis*, essential contributor in spontaneous beer fermentations providing novel opportunities for the brewing industry. Brew Sci 68:110–121
Daenen L, Sterckx F, Delvaux FR et al (2008) Evaluation of the glycoside hydrolase activity of a *Brettanomyces* strain on glycosides from sour cherry (*Prunus Cerasus* L.) used in the production of special fruit beers. FEMS Yeast Res 8:1103–1114
De Francesco G, Turchetti B, Sileoni V et al (2015) Screening of new strains of *Saccharomycodes ludwigii* and *Zygosaccharomyces rouxii* to produce low-alcohol beer. J Inst Brew 121:113–121
De Francesco G, Sannino C, Sileoni V et al (2018) *Mrakia gelida* in brewing process: an innovative production of low alcohol beer using a psychrophilic yeast strain. Food Microbiol 76:354–362
De Roos J, Vandamme P, De Vuyst L (2018) Wort substrate consumption and metabolite production during lambic beer fermentation and maturation explain the successive growth of specific bacterial and yeast species. Front Microbiol 9:1–20
Djè KM, Aka S, Zinzendorf NY et al (2009) Predominant lactic acid bacteria involved in the spontaneous fermentation step of *tchapalo* process, a traditional sorghum beer of Côte d’Ivoire. Res J Biol Sci 4(7):789–795
Domizio P, House JF, Joseph CML et al (2016) *Lachancea thermotolerans* as an alternative yeast for the production of beer. J Inst Brew 122:599–604
Dunn B, Sherlock G (2008) Reconstruction of the genome origins and evolution of the hybrid lager yeast *Saccharomyces pastorianus*. Genome Res 18:1610–1623
Eizaguirre JI, Peris D, Rodríguez ME et al (2018) Phylogeography of the wild Lager-brewing ancestor (*Saccharomyces eubayanus*) in Patagonia. Environ Microbiol 20(10):3732–3743

Estela-Escalante WDS, Rosales-Mendoza M, Moscosa-Santillán JE et al (2016) Evaluation of the fermentative potential of *Candida zemplinina* yeasts for craft beer fermentation. J Inst Brew 122:530–535

Etschmann M, Huth I, Walisko R et al (2015) Improving 2-phenylethanol and 6-pentyl-α-pyrone production with fungi by microparticle-enhanced cultivation (MPEC). Yeast 32:145–157

Fleet GH (2006) The commercial and community significance of yeasts in food and beverage production. In: Querol A, Fleet GH (eds) Yeasts in food and beverages. Springer, Berlin, pp 1–12

Gamero A, Quintilla R, Groenewald M et al (2016) High-throughput screening of a large collection of non-conventional yeasts reveals their potential for aroma formation in food fermentation. Food Microbiol 60:147–159

Gibson BR (2011) 125th anniversary review: improvement of higher gravity brewery fermentation via wort enrichment and supplementation. J Inst Brew 117:268–284

Gibson B, Liti G (2015) *Saccharomyces pastorianus*: genomic insights inspiring innovation for industry. Yeast 32:17–27

Gibson B, Geertman A, Hittinger CT et al (2017) New yeasts—new brews: modern approaches to brewing yeast design and development FEMS yeast research, vol 17, p fox038

Gonçalves M, Pontes A, Almeida P et al (2016) Distinct domestication trajectories in top-fermenting beer yeasts and wine yeasts. Curr Biol 26:2750–2761

Gonzalez SS, Barrio E, Querol A (2008) Molecular characterization of new natural hybrids of Saccharomyces cerevisiae and S. kudriavzevii in brewing. Appl Environ Microbiol 74(8):2314–2320

González R, Quirós M, Morales P (2013) Yeast respiration of sugars by non-*Saccharomyces* yeast species: a promising and barely explored approach to lowering alcohol content of wines. Trends Food Sci Tech 29(1):55–61

Goretti M, Turchetti B, Cramarossa MR et al (2013) Production of flavours and fragrances via bio-reduction of (4R)-(−)-carvone and (1R)-(−)-myrtenal by non-conventional yeast whole-cells. Molecules 18:5736–5748

Gschaedler A (2017) Contribution of non-conventional yeasts in alcoholic beverages. Curr Opin Food Sci 13:73–77

Gutiérrez A, Boekhout T, Gojkovic Z, Katz M (2018) Evaluation of non- yeasts in the fermentation of wine, beer and cider for the development of new beverages. J Inst Brew 124(4):389–402

Haehn H, Glaubitz M (1933) Beer manufacture, patent US1898047 A

Hallsworth JE (1998) Ethanol-induced water stress in yeast. J Ferment Bioeng 85:125–137

Holt S, Mukherjee V, Lievens B et al (2017) Bioflavoring by non-conventional yeasts in sequential beer fermentations. Food Microbiol 72:55–66

Holt S, Mukherjee V, Lievens B, Verstrepen KJ, Thevelein JM (2018) Bioflavoring by non-conventional yeasts in sequential beer fermentations. Food Microbiol 72:55–66

Huige NG, Sanchez GW, Leidig AR (1990) Process for preparing a nonalcoholic (less than 0.5 volume percent alcohol) malt beverage, Patent US4970082 A

Johnson E (2013) Biotechnology of non-*Saccharomyces* yeasts—the ascomycetes. Appl Microbiol Biotechnol 97(2):503–517

Jolly N, Varela C, Pretorius I (2014) Not your ordinary yeast: non-*Saccharomyces* yeasts in wine production uncovered. FEMS Yeast Res 14:215–237

Kastner JR, Ahmad M, Jones WJ et al (1992) Viability of *Candida shehatae* in D-xylose fermentations with added ethanol. Biotechnol Bioeng 40:1282–1285

Kastner JR, Roberts RS, Jones WJ (1996) Effect of pH on cellviability and product yields in d-xylose fermentations by *Candida shehatae*. Appl Microbiol Biotechnol 45:224–228

King A, Richard Dickinson J (2000) Biotransformation of monoterpene alcohols by *Saccharomyces cerevisiae*, *Torulaspora delbrueckii* and *Kluyveromyces lactis*. Yeast 16:499–506

Krogerus K, Arvas M, De Chiara M et al (2016) Ploidy influences the functional attributes of de novo lager yeast hybrids. Appl Microbiol Biot 100:7203–7222

Kurtzman CP (2011) Wickerhamomyces. In: Kurtzman CP, Fell JW, Boekhout T (eds) The yeasts a taxonomic study. Elsevier, Oxford, pp 899–917

Kurtzman C, Fell J, Boekhout T (2011) The yeasts a taxonomy study. Elsevier, Amsterdam

Lachance MA, Boekhout T, Scorzetti G et al (2011) Candida. In: Kurtzman CP, Fell JW, Boekhout T (eds) The yeasts: a taxonomic study. Elsevier, Oxford, pp 987–1277

Lee KE, Lee SM, Choi YH et al (2013) Comparative volatile profiles in soy sauce according to inoculated microorganisms. Biosci Biotechnol Biochem 77:2192–2200

Lentz M, Putzke T, Hessler R et al (2014) Genetic and physiological characterization of yeast isolated from ripe fruit and analysis of fermentation and brewing potential. J Inst Brew 120:559–564

Li H, Liu Y, Zhang W (2013) Method for Manufacturing Alcohol-Free Beer through *Candida shehatae*. China Patent CN102220198

Libkind D, Hittinger CT, Valerio E et al (2011) Microbe domestication and the identification of the wild genetic stock of lager-brewing yeast. Proc Natl Acad Sci U S A 108:14539–14544

Liu SQ, Quek AY (2016) Evaluation of beer fermentation with a novel yeast *Williopsis saturnus*. Food Technol Biotechnol 54(4):403–412

Liu Y, Li H, Du JH (2011) Non-alcoholic beer production by *Saccharomycodes ludwigii*. Food Sci 15:186–190

Lytra G, Tempere S, Le Floch A et al (2013) Study of sensory interactions among red wine fruity esters in a model solution. J Agric Food Chem 61:8504–8513

Lyumugabe F, Gros J, Nzungize J et al (2012) Characteristics of African traditional beers brewed with Sorghum Malt: a review. Biotechnol Agron Soc Environ 16(4):509–530

Magalhães F, Krogerus K, Vidgren V et al (2017) Improved cider fermentation performance and quality with newly-generated *Saccharomyces cerevisiae* × *Saccharomyces eubayanus* hybrids. J Ind Microbiol Biot 44(8):1203–1213

Meier-Dörnberg T, Schneiderbanger H, Jacob F et al (2014) Alcohol-free wheat beer with maltose negative yeast strain *Saccharomycodes ludwigii*, in 3rd Young Scientists Symposium, Poster no. P.3.5. Available from: https://www.researchgate.net/publication/307992436_Alcohol-Free_Wheat_Beer_with_Maltose_Negative_Yeast_Strain_*Saccharomycodes_ludwigii*

Mendoza LM, Neef A, Vignolo G, Belloch C (2017) Yeast diversity during the fermentation of Andean chicha: a comparison of high-throughput sequencing and culture dependent approaches. Food Microbiol 67:1–10

Mertens S, Steensels J, Saels V et al (2015) A large set of newly created interspecific *Saccharomyces* hybrids increases aromatic diversity in lager beers. Appl Environ Microbiol 81:8202–8214

Mets MD, Verzele M (1968) The aroma of hops I. origin and identification of hop aroma substances. J Inst Brew 74:74–81

Michel M, Meier-Dörnberg T, Jacob F et al (2016a) Review: pure non-*Saccharomyces* starter cultures for beer fermentation with a focus on secondary metabolites and practical applications. J Inst Brew 122:569–587

Michel M, Kopecká J, Meier-Dörnberg T et al (2016b) Screening for new brewing yeasts in the non-*Saccharomyces* sector with *Torulaspora delbrueckii* as model. Yeast 33:129–144

Mohammadi A, Razavi SH, Mousavi SM et al (2011) A comparison between sugar consumption and ethanol production in wort by immobilized *Saccharomyces cerevisiae*, *Saccharomyces ludwigii* and *Saccharomyces rouxii* on brewer's spent grain. Braz J Microbiol 42:605–615

Montanari L, Marconi O, Mayer H et al (2011) Production of alcohol-free beer. In: Preedy VR (ed) Beer in health and disease prevention. Elsevier, New York, pp 61–75

Moran GP, Sullivan DJ, Coleman DC (2002) Emergence of non-*Candida albicans Candida* species as pathogens in: Calderone RA (Ed) *Candida* and Candidiasis ASM press, Washington, DC, pp 37–53

Morris C (1974) Reconstructing patterns of non-agricultural production in the Inca economy: archaeology and documents in institutional analysis. In: Moore CB (ed) Reconstructing complex societies. Supplement to American Schools of Oriental Research Bulletin No. 20, Cambridge, pp 49–60

Mortazavian A, Razavi S, Mousavi S et al (2014) The effect of *Saccharomyces* strain and fermentation conditions on quality parameters of non-alcoholic beer. J Paramed Sci 5:21–26
Müller-Auffermann K, Caldera A, Jacob F et al (2015) Characterization of different bottom fermenting *Saccharomyces pastorianus* brewing yeast strains. Brew Sci 68:46–57
Munroe JH (1995) Fermentation In: Dekker M, Hardwick W A (Eds) Handbook of Brewing CRC Press, New York, p 323–353
N'Guessan FK, N'Dri DY, Camara F et al (2010) *Saccharomyces cerevisiae* and *Candida tropicalis* as starter cultures for the alcoholic fermentation of *tchapalo*, a traditional sorghum beer. World J Microbiol Biotechnol 26:693–699
N'guessan KF, Brou K, Noe'mie J et al (2011) Identification of yeasts during alcoholic fermentation of *tchapalo*, a traditional sorghum beer of Côte d'Ivoire. A Van Leeuw J 99:855–864
Narziss L, Miedaner H, Kern E et al (1992) Technology and composition of non-alcoholic beers. Brauwelt Int 4:396–410
Nelson M (2014) The geography of beer. Springer, New York, p 9
Olaniran AO, Maharaj YR, Pillay B (2011) Effects of fermentation temperature on the composition of beer volatile compounds, organoleptic quality and spent yeast density. Electro J Biotech 14
Olaniran AO, Hiralal L, Mokoena MP, Pillay B (2017) Flavour-active volatile compounds in beer: production, regulation and control. J Inst Brew 123(1):13–23
Park JY, Lee JY, Choi SH, Ko HM, Kim IC, Lee HB, Bai S (2014) Construction of dextrin and isomaltose-assimilating brewer's yeasts for production of low-carbohydrate beer. Biotechnol Lett 36(8):1693–1699
Passoth V, Fredlund E, Druvefors UÄ et al (2006) Biotechnology, physiology and genetics of the yeast *Pichia anomala*. FEMS Yeast Res 6(1):3–13
Petruzzi L, Corbo MR, Mn S et al (2016) Brewer's yeast in controlled and uncontrolled fermentations, with a focus on novel, nonconventional, and superior strains. Food Rev Int 32(4):341–363
Pires EJ, Teixeira JA, Brányik T et al (2014) Yeast: the soul of beer's aroma and a review of flavour-active esters and higher alcohols produced by the brewing yeast. Appl Microbiol Biotechnol 98:1937–1949
Polycarpe Kayodé AP, Hounhouigan DJ, Nout MJR et al (2007) Household production of sorghum beer in Benin: technological and socio-economic aspects. Int J Consum Stud 31:258–264
Ramos-Jeunehomme C, Laub R, Masschelein CA (1991) Why is ester formation in brewery fermentations yeast strain dependent? In: 23rd European brewery convention congress. Oxford University, Lisbon, pp 257–264
Ravasio D, Carlin S, Boekhout T, Groenewald M et al (2018) Adding flavor to beverages with non-conventional yeasts. Fermentation 4:15. https://doi.org/10.3390/fermentation4010015
Rodhouse L, Carbonero F (2017) Overview of craft brewing specificities and potentially associated microbiota. Crit Rev Food Sci Nutr 14:1–12
Rodríguez ME, Pérez-Través L, Sangorrín MP et al (2017) Saccharomyces uvarum is responsible for the traditional fermentation of apple chicha in Patagonia. FEMS Yeast Res 17:1–11
Romano P, Marchese R, Laurita C et al (1999) Biotechnological suitability of *Saccharomycodes ludwigii* for fermented beverages. World J Microbiol Biotechnol 15:451–454
Romano P, Capece A, Jespersen L (2006) Taxonomic and ecological diversity of foods and beverage yeasts. In: Querol A, Fleet GH (eds) Yeasts in food and beverages. Springer, Berlin, pp 13–54
Rossouw D, Naes T, Bauer FF (2008) Linking gene regulation and the exo-metabolome: a comparative transcriptomics approach to identify genes that impact on the production of volatile aroma compounds in yeast. BMC Genomics 9:530
Saerens S, Swiegers JH (2014) Enhancement of beer flavor by a combination of *Pichia* yeast and different hop varieties, patent US20140234480 A1
Saerens S, Swiegers JH (2016) Production of low-alcohol or alcohol-free beer with *Pichia kluyveri* yeast strains. U.S. Patent US2016/0010042
Saison D, De Schutter DP, Vanbeneden N et al (2010) Decrease of aged beer aroma by the reducing activity of brewing yeast. J Agric Food Chem 58:3107–3115

San-Juan F, Ferreira V, Cacho J et al (2011) Quality and aromatic sensory descriptors (mainly fresh and dry fruit character) of Spanish red wines can be predicted from their aroma-active chemical composition. J Agric Food Chem 59:7916–7924
Schifferdecker AJ, Dashko S, Ishchuk OP et al (2014) The wine and beer yeast *Dekkera bruxellensis*. Yeast 31:323–332
Sefa-Dedeh S, Sanni AI, Tetteh G et al (1999) Yeasts in the traditional brewing of *Pito* in Ghana. Word J Microb Biot 15:593–597
Serra Colomer M, Funch B, Forster J (2019) The raise of *Brettanomyces* yeast species for beer production. Curr Opin Biotech 56:30–35
Sibirny AA, Scheffers L (2002) Thematic section 'biochemistry, genetics, biotechnology and ecology of non-conventional yeasts. FEMS Yeast Res 2:293
Sicard D, Legras JL (2011) Bread, beer and wine: yeast domestication in the *Saccharomyces* Sensu stricto complex. C R Biol 334(3):229–236
Smith BD, Divol B (2018) The carbon consumption pattern of the spoilage yeast *Brettanomyces bruxellensis* in synthetic wine-like medium. Food Microbiol 73:39–48
Smukowski Heil C, Burton JN, Liachko I, Friedrich A, Hanson NA, Morris CL, Schacherer J, Shendure J, Thomas JH, Dunham MJ (2017) Identification of a novel interspecific hybrid yeast from a metagenomic spontaneously inoculated beer sample using Hi-C. Yeast 35:71–84
Sohrabvandi S, Hadi Razavi S, Mousavi M et al (2009) Application of *Saccharomyces rouxii* for the production of non-alcoholic beer. Food Sci Biotechnol 18:1132–1137
Sohrabvandi S, Hadi Razavi S, Mousavi M et al (2010) Characteristics of different brewer's yeast strains used for nonalcoholic beverage fermentation in media containing different fermentable sugars. Int J Biotechnol 8:178–185
Spitaels F, Wieme AD, Janssens M et al (2015) The microbial diversity of an industrially produced lambic beer shares members of a traditionally produced one and reveals a core microbiota for lambic beer fermentation. Food Microbiol 49:23–32
Steensels J, Verstrepen KJ (2014) Taming wild yeast: potential of conventional and nonconventional yeasts in industrial fermentations. Annu Rev Microbiol 68(1):61–80
Steensels J, Daenen L, Malcorps P et al (2015) *Brettanomyces* yeasts- from spoilage organisms to valuable contributors to industrial fermentations. Int J Food Microbiol 206:24–38
Tamang JP, Fleet GH (2009) Yeasts diversity in fermented foods and beverages. In: Satyanarayana T, Kunze G (eds) Yeast biotechnology: diversity and applications. Springer, Dordrecht, pp 169–198
Tataridis P, Kanelis A, Logotetis S et al (2013) Use of non-*Saccharomyces Torulaspora delbrueckii* yeast strains in winemaking and brewing. J Nat Sci (124):415–426
Thomas-Hall SR, Turchetti B, Buzzini P et al (2010) Cold-adapted yeasts from Antarctica and the Italian Alps-description of three novel species: *Mrakia robertii* sp. nov., *Mrakia blollopis* sp. nov. and *Mrakiella niccombsii* sp. nov. Extremophiles 14:47–59
Tokpohozin SE, Lauterbach A, Fisher S et al (2016) Phenotypical and molecular characterization of yeast content in the starter of *"Tchoukoutou"*, a beninese African sorghum beer. Eur Food R Technol 242(12):2147–2160
Tsuji M, Kudoh S, Hoshino T (2016) Ethanol productivity of cryophilic basidiomycetous yeast *Mrakia* spp. correlates with ethanol tolerance. Mycoscience 57:42–50
Vallejo JM, Miranda P, Flores-Félix JD et al (2013) Atypical yeasts identified as *Saccharomyces cerevisiae* by MALDI-TOF MS and gene sequencing are the main responsible of fermentation of *chicha*, a traditional beverage from Peru. Syst Appl Microbiol 36:560–564
Van Oevelen D, Spaepen M, Timmermans P et al (1977) Microbiological aspect o spontaneous wort fermentation in the production of limbic and gueuze. J Inst Brew 83:356–360
van Rijswijck IMH, Wolkers-Rooijackers JCM, Abee T et al (2017) Performance of non-conventional yeasts in co-culture with brewers' yeast for steering ethanol and aroma production. Microb Biotechnol 10:1591–1602
Varela C (2016) The impact of non-*Saccharomyces* yeasts in the production of alcoholic beverages. Appl Microbiol Biotechnol 100:9861–9874

Vaughan-Martini A, Martini A (2011) *Saccharomyces*. In: Kurtzman CP, Fell JW, Boekhout T (eds) The yeasts: a taxonomic study Elsevier, Oxford, pp 733–746
Vervoort Y, Herrera-Malaver B, Mertens S et al (2016) Characterization of the recombinant *Brettanomyces anomalus* β-glucosidase and its potential for bioflavouring. J Appl Microbiol 121:721–733
Wang J, Liu L, Ball T et al (2016) Reavealing a 5,000-y-old beer recipe in China. Proc Natl Acad Sci U S A 113:6444–6448
Wolf K, Breunig K, Barth G (2003) Non conventional yeasts in genetics, biochemistry and biotechnology. Springer, Berlin
Yeo HQ, Liu SQ (2014) An overview of selected specialty beers: developments, challenges and prospects. Int J Food Sci Technol 49:1607–1618

Chapter 12
Yeasts for Bioconversion of Crude Glycerol to High-Value Chemicals

Marta Semkiv and Andriy Sibirny

Abstract Biodiesel production is a fast-growing industry. Biodiesel is obtained through transesterification of different kinds of oils with methanol. This process results in a formation of substantial amounts (up to 10% of a total product mass) of the by-product fraction that mainly contains glycerol but also some toxic contaminations (spent catalyst, salts after neutralization, residual methanol, methyl esters, and free fatty acids), and that is therefore called crude glycerol. Efficient utilization of this fraction is imperative to the sustainability of the biodiesel industry. This review describes different methods of valorization of the crude glycerol fraction with the focus on biotechnological processes conducted by yeasts. In particular, production of organic acids, polyols, ethanol, microbial oil, carotenoids, γ-decalactone, sophorolipids, heterologous proteins, and biomass is discussed.

Keywords Biodiesel · Crude glycerol · *Yarrowia lipolytica* · *Pichia pastoris* · Oleaginous yeasts · Citric acid · Erythritol · SCO · Ethanol · Carotenoids

12.1 Biodiesel as an Emerging Sustainable Biofuel

Global warming (caused by increased carbon dioxide emissions), environmental pollution, and the threat of exhaustion of world resources of fossil fuels are some of the many concerns that had arisen in twentieth century and followed humanity into the new millennium. The 2030 Agenda for Sustainable Development, adopted by all United Nations Member States in 2015, is a call for action to fight major global issues. At its heart are the 17 Sustainable Development Goals, among them goal #7

M. Semkiv (✉)
Department of Molecular Genetics and Biotechnology, Institute of Cell Biology NAS of Ukraine, Lviv, Ukraine

A. Sibirny
Department of Molecular Genetics and Biotechnology, Institute of Cell Biology, NAS of Ukraine, Lviv, Ukraine

Department of Biotechnology and Microbiology, University of Rzeszow, Rzeszow, Poland

A. Sibirny (ed.), *Non-conventional Yeasts: from Basic Research to Application*, https://doi.org/10.1007/978-3-030-21110-3_12

(ensure access to affordable, reliable, sustainable and modern energy for all) and goal #13 (take urgent action to combat climate change and its impacts) (https://sustainabledevelopment.un.org/?menu=1300).

Search for "green", renewable energy resources that can be used instead of fossil fuels brought to humankind attention among others such options as biogas, bioethanol, and biodiesel. These biofuels are produced from renewable biological material and after combustion generate an acceptable level of emissions gases (Quispe et al. 2013).

Biodiesel is a liquid biofuel obtained by transesterification (Fig. 12.1) of triglycerides from vegetable oils or animal fats with an alcohol (Fukuda et al. 2001). Biodiesel can be used in diesel engines alone or blended with diesel oil.

First vegetable oil-powered engine was produced by the French company Otto and presented at the Paris Exhibition in 1900. Rudolph Diesel also experimented with castor oil and animal oils as a possible fuel for locomotive engines (Radley 2016). But due to the success of petroleum and diesel oil as the cheap fuels, these studies did not gain further attention until early 1980 when the study with the sunflower oil was conducted in South Africa prompted by diesel oil embargo (Ma and Hanna 1999) and the National Program of Vegetable Oils for Energy Purposes (PRO-OIL) started to gain momentum in Brazil. But vegetable oils contain, besides triacylglycerols, some additional compounds (free fatty acids, phospholipids, sterols, etc.) that complicate its direct usage as a fuel for engines (Kegl 2008). These problems can be overcome with the chemical modifications of vegetable oil, such as cracking, esterification, and transesterification. The process of fatty acids transester-

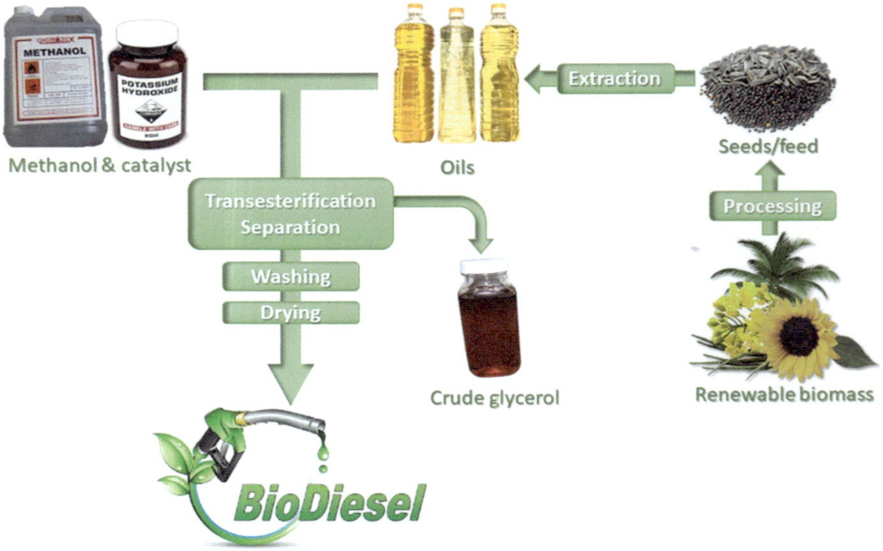

Fig. 12.1 Biodiesel production by transesterification of triglycerides from vegetable oils with methanol. (Adapted from Smirnov et al. 2018)

ification was developed to obtain a mixture of long-chain monoalkylic esters (biodiesel). In Europe, biodiesel is produced on an industrial scale since 1992 (Quispe et al. 2013). Stimulated by tax exemptions and other governmental incentives, world's biodiesel production increases every year and is expected to reach 110,000 ML in 2020 (Rodrigues et al. 2017).

Such drastic increase in production was supported by the fact that biodiesel fuel has a lot of advantages over the petroleum and diesel oil: it has a higher flash point and flammability point which ensures safer storage, loading/unloading, and handling of this material by the drivers and operators (Knothe et al. 2005); it degrades more rapidly than diesel fuel, minimizing the environmental consequences in case of accidental spills; after combustion, it produces less emissions of contaminants (carbon monoxide, particulate matter, polycyclic aromatic hydrocarbons, aldehydes, carcinogenic substances) and has no jeopardy due to emissions of sulfur dioxide (SO_2); and it has good lubricating properties (Romano and Sorichetti 2011). Unfortunately, there are several aspects in which biodiesel is inferior to petroleum diesel: it has the lower calorific value, which causes slightly higher fuel consumption; it has slightly higher emissions of nitrous oxide; it is less stable than diesel fuel and therefore it has short shelf life (up to 6 months); and in pure form, it may degrade equipment made from plastic or natural rubber; it may dissolve the deposits of sediments from diesel fuel in storage tanks and fuel lines and flushed them into the car engine (Romano and Sorichetti 2011). Also, biodiesel has a high freezing point – at low temperatures, it tends to quickly lose fluidity that may lead to clogging of filters and damage to the starting system of the engine (Munoz et al. 2012). Biodiesel exhibits more corrosive behavior than diesel oil due to the following factors: it is more hygroscopic and captured water can itself act on the corrosion or cause the hydrolysis of biodiesel or promote microbial growth; the presence of impurities like methanol, free glycerol, free fatty acid, and catalyst residues (Na and K) also promotes metallic corrosion; and biodiesel dissolves more metallic parts than diesel due to its good lubricity, and these trace metals in solution enhance biodiesel degradation and therefore corrosion (Haseeb et al. 2011; Singh et al. 2012). The intensity of biodiesel self-oxidation and its corrosive behavior depends on the amount of the double bonds in unsaturated fatty acids, so the oils with high concentrations of polyunsaturated fatty acids are undesirable for biodiesel production (Borsato et al. 2012). All mentioned disadvantages are significantly reduced when biodiesel is used in blends with diesel fuel (Romano and Sorichetti 2011). Blends with diesel fuel are designated as "B∗," where "∗" is the number reflecting the percentage of biodiesel in the blend. For example, "B2" indicates a blend with 2% biodiesel and 98% petroleum diesel (Romano and Sorichetti 2011). In Brazil, government initiates increase in biodiesel percentage in the blends almost every year. For example, in 2008, all diesel blends sold had to be at least B3; in 2009, B4; and in 2010, B5 (Mota et al. 2009). B20, a mixture of 20% biodiesel and 80% diesel, is scheduled for introduction in 2020 (Pousa et al. 2007).

Theoretically, biodiesel can be produced from any source containing oil, but not all sources produce biodiesel that complies with strict international standards for the quality fuel. The most commonly used vegetable oils are rapeseed (in European

Union), soybean (in United States of America, Brazil, Argentina), palm (in Asian and Central American countries), and sunflower oils (Romano and Sorichetti 2011) (Fig. 12.2). Latest advantages in bioengineering even allowed the obtaining of oil-seed crops with high content of specific fatty acids (e.g., high omega-7 monounsaturated fatty acid), which improved ignition quality and oxidative stability of produced biodiesel (Beaudoin et al. 2014). But there are some issues in using oilseed plants: for their cultivation, productive agricultural lands are used that can be otherwise exploited to obtain more food. That's why attempts have been made to produce biodiesel from nonedible feedstock (second-generation biodiesel). Nonedible, drought-tolerant oil plants (e.g., jojoba, Jatropha) that were grown on nonused land can be considered as such feedstock. Another good example can be waste cooking oil – a residue from cooking process of industries, restaurants, and bars. Although it was shown to be somewhat inferior in comparison with virgin vegetable oils due to the high content of free fatty acids that can be converted to soaps during transesterification process (Thompson and He 2006), after some adjustments (e.g., changing the catalysis from basic to acidic), it can be used for biodiesel production. Finally,

Fig. 12.2 Feedstock for production of the first-, second-, and third-generation biodiesel

very promising source of oils is microalgae which were included in the so-called third-generation biofuels group (Rodrigues et al. 2017). Being fast-growing, autotrophic microorganisms (Yang et al. 2011), they produce 200 more oil per unit of area than the best-performing oil plants (Demirbas 2009).

The most common process of biodiesel production is the transesterification (also called alcoholysis) of vegetable oil (or animal fat) with short-chain alcohols, including methanol, ethanol, butanol, and amylic alcohol. This reaction results in the formation of a mixture of esters of the fatty acids and glycerol (Meher et al. 2006). The glycerol layer is denser than the ester one; therefore, it deposits at the bottom of the reactor and can be removed by simple decantation. Methanol (CH_3OH) and ethanol (C_2H_5OH) are the most widely used alcohols for biodiesel production. Most of the production facilities prefer methanol as it offers easier ester phase separation from glycerol, better alcohol recovery, and higher reaction speed (Munoz et al. 2012). Some enterprises consider ethanol route as more ecologically friendly, though in ethanolysis the mixture of esters and glycerol is more stable, complicating the separation and purification of biodiesel fraction (Meher et al. 2006). Important parameter of transesterification reaction is alcohol/oil ration. The stoichiometry of the reaction requires three molecules of alcohol for each molecule of triacylglycerides; however, as the reaction is reversible, excess amount of one of the reagents is required to drive the process to completion. So the most biodiesel facilities use at least 6 to 1 M ratio of methanol to oil. In case of ethanol, molar ratio between 9:1 and 12:1 gave the best results (Sinha et al. 2008). The majority of the excess alcohol (up to 80%) ends up in the crude glycerol fraction after the reaction, so producers tend to recover the alcohol by distillation and reuse it (Miesiac 2003).

Different catalysts can be used for transesterification reaction: basic or acidic, homogeneous or heterogeneous, and chemical or biological. The most commonly used process of biodiesel production driven by basic catalysis is faster than in case of acidic catalysis, and the end product has less corrosive properties (Ma and Hanna 1999). Substances used for basic catalysis are the alkoxides (Schwab et al. 1987) and hydroxides (Aksoy et al. 1990) of sodium or potassium. The alkoxides of alkali cations such as potassium methoxide (CH_3ONa) are the most reactive catalysts, but the hydroxides (KOH and NaOH) are more accessible in price, so they are being used more (Munoz et al. 2012). KOH is more expensive than NaOH; however, there is less soap formation using KOH (Fukuda et al. 2001). The excessive amounts of basic catalyst can decrease the acidity index of biodiesel, but it can also cause the formation of soaps, hampering the separation of glycerol from esters. That's why the base catalysis is preferable when the oils do not contain a lot of water and free fatty acids. In case of high free fatty acids content, a pretreatment by saponification or the acid catalysis for previous esterification of these acids is recommended (Marchetti et al. 2007). Heterogeneous acid catalysts have many advantages: they simultaneously promote alcoholysis of triglycerides and esterification of free fatty acids, reduce the number of needed purification steps, and, as it forms the insoluble phase, give the possibility to recover and reuse the catalyst in a continuous process (Munoz et al. 2012). The biological or enzymatic catalysis is also a very perspective option as it is more specific, does not promote side reactions, and allows the simple recovery of glycerol, the total esterification of free fatty acids, and the use of mild conditions in the process. The main drawbacks of this technology are the high cost of enzyme production, extraction, and purification as well as their instability in solution (Singh and Singh 2010). These problems can be partially resolved by the immobilization of enzymes, which allows their reuse. Many processes of biodiesel production using immobilized lipases have been developed (Shieh et al. 2003).

Although transesterification is the most important step in biodiesel production, additional steps are required to obtain a quality product (Meher et al. 2006). As was already mentioned, the mix of two components (esters and glycerol) should be separated after the completion of the transesterification reaction, and esters fraction should be purified from the excess of catalyst, water, and alcohol.

There are some other technological processes that can be used for biodiesel production, for example, the esterification catalyzed by the sulfonic or sulfuric acids. This reaction is quite slow and requires temperatures above 100 °C for over 3 hours and a large alcohol excess, but the obtained esters yield is very high (99%) (Al-Widyan and Al-Shyoukh 2002; Fukuda et al. 2001).

12.2 Crude Glycerol as a by-Product of Biodiesel Industry

The by-products of biodiesel production are glycerol, biodiesel washing wastewaters, methanol, and solid residues (Varanda et al. 2011). As was already mentioned, glycerol is a major by-product of transesterification reaction that can and has to be removed from biodiesel as it can promote the formation of deposits and sediments, reducing the engine life (Munoz et al. 2012). The fraction which is being removed after the transesterification reaction contains not only glycerol but many other compounds. It is most commonly referred to as crude glycerol. Crude glycerol makes up to 10% of a total product mass, so 1 kg of crude glycerol is produced per 12.6 L of biodiesel (Dobroth et al. 2011). Due to rapid increase in biodiesel production, by 2020, global annual crude glycerol production is expected to reach of 4200 ML (Okoye and Hameed 2016). In fact, the global market is flooded with excessive crude glycerol, which led to a drastic decrease in its price from $400 per ton in 2001 to less than $100 per ton in 2011 (Quispe et al. 2013). Large-scale biodiesel producers refine obtained crude glycerol to a chemically pure substance and sell it to the food, pharmaceutical, or cosmetics industries. However, the process of crude glycerol purification is quite expensive and inaccessible for small- to moderate-scale biodiesel producers (Thompson and He 2006). As more and more crude glycerol is generated, its disposal starts to be a problem. Biodiesel producers together with researchers must seek alternative applications for crude glycerol. Some of them have already been developed – e.g., using crude glycerol as a cheap organic solvent, as a raw feedstock for the production of value-added compound, as building block to biomaterial synthesis, etc. (Yang et al. 2012).

Utilization of unpurified crude glycerol as a feedstock in industrial processes is hampered by the inconsistent nature of this fraction: its content varies strongly, depending on the kind of oil and the process employed for biodiesel synthesis. It can be in liquid or solid (Nanda et al. 2014) state and generally has high pH (above 10) and low density. The main contaminants in crude glycerol are water, ash (mainly originated from the KOH catalyst), calcium, magnesium, phosphorous, sulfur, methanol (or other alcohol that was used for biodiesel production), soaps (products of reaction between free fatty acids from oil and basic catalyst), free fatty acids (FFA), methyl esters of fatty acids (FAME), and glycerides. Glycerides, soaps, FFA, and FAME are sometimes generally named nonglycerol organic matter (NGOM). Methanol can be partially eliminated from crude glycerol fraction by biodiesel producers; that's why its content varies from 0.5% up to 20–30% when such process did not take place. Glycerol content in crude glycerol samples also differs significantly, ranging from 20 to 80 wt% (Hu et al. 2012a). Thompson and He analyzed 7 types of crude glycerol produced from different feedstock and found out that all of them have more or less similar chemical composition except of the crude glycerol produced from waste cooking oil which had much more soaps and dissolved unreacted glycerides and esters (Thompson and He 2006). In contrast to these findings, when Hansen et al. analyzed 11 crude glycerol samples from different biodiesel plants in Australia, the substantial differences between these samples were demonstrated (Hansen et al. 2009).

The impurities in crude glycerol may significantly affect its implementation in industrial processes. It was shown that soap and methanol inhibit the production of docosahexaenoic acid from crude glycerol by algae (Pyle et al. 2008) and the high content of Na or K can substantially suppress the microbial activity during the anaerobic digestion of crude glycerol for production of biogas (Santibanez et al. 2011). Interestingly, some studies showed that certain impurities in crude glycerol can actually benefit the production of certain compounds such as bio-oil (Xiu et al. 2010) and polyurethane foams (Hu et al. 2012b). It is necessary to analyze the composition of crude glycerol to develop the right way of its application.

12.3 Potential Applications of Crude Glycerol

12.3.1 Crude Glycerol Purification

As was already mentioned, successful utilization of crude glycerol adds to biodiesel industry profitability. The most obvious application of crude glycerol is to refine it to high-purity glycerol which can be used by food, cosmetic, and pharmaceutical industries. Three types of refined glycerol can be distinguished based on their purity: (1) "technical grade" glycerol that can be used as a reagent for chemical synthesis but not in food or pharmacy; (2) United States Pharmacopeia (USP), glycerol appropriate for food and pharmaceutical products; and (3) Food Chemicals Codex (FCC), glycerol appropriate for use in food (Quispe et al. 2013).

The choice of the refining process for crude glycerol should be based on its chemical composition and the desired level of purity of the final product. The typical refining process includes three stages: acidification/neutralization, a vacuum evaporation to remove methanol (or ethanol) and water, and the final refining to achieve high purity. On the first stage, a strong acid (e.g., phosphoric acid, hydrochloric acid, or sulfuric acid) is added followed by neutralization of solution with, for example, NaOH. Several reiterations of these two steps can be performed (Javani et al. 2012). Acid reacts with soaps and convert them into free fatty acids, which results in a formation of a cloudy solution. After settling for some time, this solution separates into three phases: top fatty acid phase, middle phase with glycerol and methanol (or ethanol), and bottom phase with the inorganic salts (Rodrigues et al. 2017). Nanda et al. compared the performance of hydrochloric acid, sulfuric acid, and phosphoric acid in the purification process and found out that phosphoric acid was superior to the others as it provided reduction of the time period required for phases separation, the formed precipitates were found to be easily separated by filtration, and the obtained phosphates could be directly used as a fertilizer and as buffer solution (Nanda et al. 2014).

On the second stage, residual alcohol (methanol or ethanol) is removed by a vacuum distillation that prevents glycerol decomposition caused by higher temperatures. Alcohol is removed in order to be used again in transesterification reaction. But sometimes biodiesel producers even do not perform the recovery of methanol as using new methanol is more cost-effective (Bohon et al. 2011). Obtained after the second stage of purification, glycerol can be used as "technical grade" raw material, whereas the attainment of a "food grade" glycerol requires the third stage of purification. Final purification can be achieved, for example, through ion exchange, activated carbon absorption, or membrane separation technology (MST) (Rodrigues et al. 2017).

The overall process of crude glycerol purification is cumbersome, energy-consuming, and expensive, thus prohibitive for small- and average-scale producers who must develop other feasible uses for this feedstock.

12.3.2 Production of Heat and Energy from Crude Glycerol

At present, more than 2000 uses for glycerol are known, but most of them require purified glycerol (Quispe et al. 2013). Potential applications for nonpurified crude glycerol are heat and energy production (e.g., by combustion, production of H2 or syngas through steam reforming, production of electricity with microbial fuel cells, production of biomethane by anaerobic digestion), implementation as an addition to a compost or an animal feed, and thermochemical or biological conversions for value-added products (Claude 1999).

Renewable energy can be produced from crude glycerol by means of thermochemical (e.g., pyrolysis and gasification) or biological processes (e.g., biological fuel cells, hydrogen generation, and anaerobic digestion) (Plácido and Capareda 2016).

Johnson and Taconi reported that the combustion of crude glycerol is a viable strategy for its disposal, though not for the large producers of biodiesel (Johnson and Taconi 2007). The heat of combustion of the crude glycerol obtained after trans-esterification of first-use vegetable oils is higher than that of pure glycerol (Lide 1999), and for crude glycerol obtained from waste cooking oil, this value is even higher (Thompson and He 2006). Co-combustion of the crude glycerol with other biomass may proffer good results (Thompson and He 2006). But still crude glycerol is considered to be cheap, low-quality fuel due to its drawbacks: relatively low heating value, high self-ignition temperature, and high emissions and salt content. Also, crude glycerol combustion causes the formation of highly toxic acrolein, though its emission can be brought to the acceptable levels (Rodrigues et al. 2017). Therefore, it is better to process crude glycerol in order to obtain more valued fuels – e.g., hydrogen or biomethane.

Hydrogen can be produced from glycerol through thermochemical or biological reactions. Examples of thermochemical reactions are pyrolysis and gasification, steam reforming (Sánchez et al. 2010), supercritical water reforming (Byrd et al. 2008), or aqueous phase reforming (Tuza et al. 2014).

Pyrolysis is a process of chemically decomposing organic materials at elevated temperatures (>300 °C) in the absence of oxygen and, typically, under pressure. This process results in a formation of a gas phase (syngas, the mixture of hydrogen and carbon monoxide), liquid phase (bio-oil), and a solid phase (biochar). In several studies, crude glycerol was used as an auxiliary compound to pyrolyze different types of feedstocks (swine manure, lignite, olive kernel, corn straw, etc.), and the addition of crude glycerol has been found to increase hydrogen and light hydrocarbons concentrations in syngas and the quality of bio-oil (Cheng et al. 2014; Delgado et al. 2013; Manara and Zabaniotou 2013; Skoulou et al. 2012). These observations reveal the possibility to use crude glycerol as a co-substrate for pyrolysis at the thermal conversion plant which can use other agricultural residues of biodiesel production (plants' stems, leafs, pressed seeds, etc.) as main substrate (Plácido and Capareda 2016).

Glycerol steam reforming (GSR) provides a possibility to produce high amounts of hydrogen using existing steam reforming units. But direct application of crude glycerol as a feedstock for steam reforming arises some issues connected with difficulties in purification of the formed hydrogen and the high deposition of carbon and coke during the process which eventually leads to catalyst inactivation (Rodrigues et al. 2017).

Biological processes that can be employed for hydrogen production from glycerol are dark fermentation and photofermentation (Ghosh et al. 2012b; Rossi et al. 2011). Dark fermentation is performed by anaerobic or facultative anaerobic microorganisms such as *Enterobacter aerogenes* (Sarma et al. 2013), *Escherichia coli* (Gonzalez et al. 2008), *Klebsiella* sp. (Chookaew et al. 2014), and *Clostridium pasteurianum* (Lo et al. 2013). The efficiency of this process can be improved by modifying the microorganisms using selection (Varrone et al. 2013) or genetic engineering

(Gonzalez et al. 2008), by determining the optimal composition of the culture media and the optimal reactor conditions (Ngo et al. 2011), and by reducing the inhibitor (e.g., methanol and saponified free fatty acids) concentrations (Sarma et al. 2014).

Photofermentation is the process of hydrogen production from organic feedstock (usually, organic acids) in the presence of light. Certain purple non-sulfur photosynthetic bacteria can directly transform glycerol into bio-hydrogen (Ghosh et al. 2012a; Sabourin-Provost and Hallenbeck 2009). For example, this process can be efficiently carried out by bacterium *Rhodopseudomonas palustris* (Ghosh et al. 2012a), which can use also crude glycerol although its growth is inhibited by saponified free fatty acids from this feedstock (Pott et al. 2013). Alleviation of this inhibition can be achieved by the pH adjustment and the precipitation of saponified free fatty acids with calcium salts (Pott et al. 2014).

Biogas (mixture of methane, carbon dioxide, and other gases) can be generated in the process called anaerobic digestion, during which microorganisms break down biodegradable material in the absence of oxygen. Crude glycerol could be employed as the main (Hutnan et al. 2013) or an additional carbon source (Siles Lopez et al. 2009) during anaerobic digestion. Crude glycerol has been co-digested with waste compounds such as sewage sludge, manure, and food wastes, which has improved the methane yield from these raw materials (Alvarez et al. 2010; Nartker et al. 2014). This implies that crude glycerol can be sold to anaerobic digestion plants or the biodiesel producers can equip their facilities with anaerobic digestion reactors to process crude glycerol together with the other organic wastes left from their main activity (Plácido and Capareda 2016).

Also, it was shown that crude glycerol can be used as substrate for microbial fuel cells, which produce electricity (Feng et al. 2011).

12.3.3 Using Crude Glycerol in Agriculture

It has been suggested that crude glycerol can be composted (Brown 2007) and used in fertilizers or as an animal feed supplement (Chung et al. 2007).

It was attempted to add crude glycerol to the ration of dairy cows (Chung et al. 2007; DeFrain et al. 2004), pigs (Kijora et al. 1995), broiler chickens (Cerrate et al. 2006), and laying hens (Lammers et al. 2008). Thompson and He showed that crude glycerol obtained from the first-use oil samples can be used as a source of carbohydrates whereas the crude glycerol from waste cooking oil can be used as a fat supplement in animal feed (Thompson and He 2006). At the same time, scientists express concerns about crude glycerol implementation as an animal food additive as there is no information about long-term impact of impurities present in crude glycerol (e.g., methanol).

12.3.4 Production of Chemical Compounds from Crude Glycerol

Glycerol can be used as a building block for production of chemical compounds through thermochemical or biological processes. Some of this processes can employ crude glycerol as well, although impurities in crude glycerol often cause undesirable side reactions, products of which as well as impurities themselves contaminate end product. Here, we will recount a few compounds that can be obtained from glycerol and then concentrate our attention on the possible implementation of crude glycerol as a feedstock for bioconversion using yeasts.

Crude glycerol can be thermochemically converted into propylene glycol (Chiu et al. 2006; Dasari et al. 2005), acetol (Chiu et al. 2006), biopolyols and polyurethane foams (Hu et al. 2012b; Luo et al. 2013), acrolein (Cheng et al. 2013), etc. Hydrothermal electrolysis of crude glycerol in alkaline biodiesel wastewater at high temperatures and pressures produces lactic acid (Yuksel et al. 201[illegible]). Crude glycerol can be used for glycerolysis of castor oil methyl esters to achieve monoglycerides and diglycerides which can be used in the plastic industries (Echeverri et al. 2013). Glycerol carbonate can be produced from crude glycerol through transesterification with dimethyl carbonate (Plácido and Capareda 2016). It is a valuable compound that can be used for production of polyesters, polycarbonates, polyurethanes, polyamides, surfactants, lubricating oils, cosmetics, and electrolytic carriers in lithium ion batteries (Ochoa-Gómez et al. 2009; Okoye and Hameed 2016) and as a substitute for ethylene or propylene carbonate. Also crude glycerol can be converted to mono-, di-, and triacetylesthers in the process of acetylation via esterification (Dosuna-Rodríguez and Gaigneaux 2012; Gonçalves et al. 2008). Triacetin (triacetylated glycerol esther) can be used in the cosmetic, pharmaceutical, tobacco, and food industries or as a fuel additive. Mono- and diacetin have uses as cryogenics and as feedstocks for the production of biodegradable polyesters (Rodrigues et al. 2017).

But not all processes that use glycerol as a building block can successfully employ crude glycerol. For example, when succinic acid-based polyesters were synthesized using crude glycerol, they didn't show rubbery behavior at room conditions due to incorporation of impurities from the glycerol source to the polymer backbone, whereas polymers synthesized from pure or "technical grade" glycerol possessed desired properties (Valerio et al. 2015).

Biological crude glycerol transformation has several advantages compared to chemical conversion. It is more specific in terms of produced products, has higher tolerance to impurity, and is more environmentally friendly (Clomburg and Gonzalez 2013). There are a number of microorganisms which can metabolize glycerol, such as representatives of the genera *Escherichia*, *Klebsiella*, *Enterobacter*, *Gluconobacter*, *Clostridium*, *Candida*, *Aspergillus*, etc. (Solomon et al. 1995). Though when impurities in crude glycerol reach high concentrations, they can inhibit growth of the microorganisms. For example, the salts, methanol, and fatty acids present in crude glycerol were reported to inhibit *C. pasteurianum* growth (Venkataramanan et al. 2012), and the fermentation behavior of the bacteria was only restored when fatty acids were removed by acid precipitation. Similarly, crude glycerol was found to significantly inhibit growth and production of 1,3-propanediol by *Citrobacter freundii* due to the high concentrations of free fatty acids and free methyl esters (Anand and Saxena 2012).

Conversion of crude glycerol by microorganisms can be used for the production of the wide range of chemicals: oxalic and docosahexaenoic acid (Ethier et al. 2011), polyhydroxyalkanoate (PHA) and polyhydroxybutyrate (PHB) (Dobroth et al. 2011), 1,3-propanediol (Casali et al. 2012), 2,3-butanediol (Biebl et al. 1998), dihydroxyacetone (Liu et al. 2013b), surfactants (Sousa et al. 2012), eicosahexanoic acid (Athalye et al. 2009), ethanol (Oh et al. 2011), glyceric acid (De Ley and Frateur 1970), n-butanol (Jensen et al. 2012), propanoic acid (Bertleff et al. 2005), trehalose (Ohtake and Wang 2011), single-cell oil (Chatzifragkou et al. 2011; Garlapati et al. 2016), amino acids (Meiswinkel et al. 2013), organic solvent-tolerant lipase (Volpato et al. 2008), lignoceric acid (Habe et al. 2008), and many others. Despite such impressive list of compounds that can be produced from crude glycerol, at present, an industrial process using crude glycerol does not exist, mainly due to the high cost of product recovery (Plácido and Capareda 2016). We can almost apply for this situation a paraphrased old industry proverb about lignin: "You can make anything out of crude glycerol, except money." Hopefully, with the improvement of technologies of crude glycerol conversion, they would be finally industrialized.

12.4 Crude Glycerol as a Feedstock for Biotechnological Processes Using Yeasts

12.4.1 Glycerol Intake and Conversion by Yeasts

12.4.1.1 Transport of Glycerol Through the Plasma Membrane

Since glycerol is fairly common in the environment, it is not surprising that many yeasts can use it as a source of carbon and energy. For that glycerol should be first transported inside the yeast cell. The molecular mechanisms of glycerol movement through the plasma membrane were best investigated in the yeast *Saccharomyces cerevisiae*, for other yeasts information on this subject is fragmentary. Previously, it was deemed that glycerol is able to penetrate *S. cerevisiae* membrane by diffusion (Gancedo et al. 1968; Heredia et al. 1968), passive transport through the channel formed by protein Fps1 (Luyten et al. 1995; Sutherland et al. 1997), and active transport through the channels formed by proteins Gup1 and Gup2 (Holst et al. 2000). It all turned out to be not the case as it was discovered that glycerol is imported through the glycerol/H^+-symporter Stl1 (Ferreira et al. 2005). The deletion of the *STL1* gene completely abolished the active transport of glycerol, and the corresponding deletion strain of *S. cerevisiae* was incapable of growing on glycerol as the sole source of carbon nutrition. Similar H^+ and Na^+-glycerol symporters have been described for the halotolerant yeast *Debaryomyces hansenii* (Lucas et al. 1990), *Pichia sorbitophila* (Lages and Lucas 1995), and *Zygosaccharomyces rouxii* (van Zyl et al. 1990).

In many non-*Saccharomyces* yeasts, glycerol import into the cells may actually be performed by facilitated diffusion through the membrane facilitator proteins – homologues of Fps1 protein. For example, it has been established (Liu et al. 2013a) that expression of the *FPS2* gene from yeast *Pachysolen tannophilus* in *S. cerevisiae* complements the deletion of the *STL1* gene, whereas the expression of its own *FPS1* gene does not provide growth restoration on glycerol. The same effect was achieved by expression of the homologues of the *FPS1* gene from different types of nonconventional yeasts (*Candida jadinii*, *Pichia pastoris*[sometimes referred to as *Komagataella pastoris or Komagataella phaffii*], and *Yarrowia lipolytica*) in *S. cerevisiae stl1*Δ mutant (Klein et al. 2016). This high level of growth persists even after deletion of the *STL1* gene (Klein et al., 2016). However, in order to finally confirm the role of these transporters in the glycerol import, scientists should perform their deletion and overexpression in the yeast species from which they originate.

12.4.1.2 Ways of Catabolism of Glycerol in Yeast

Catabolism of glycerol in the yeast cells is carried out through glycerol-3-phosphate (phosphorylation pathway) or dihydroxyacetone (oxidative pathway of glycerol utilization) (Fig. 12.3). Obtained in this process, dihydroxyacetone phosphate can either be included into the central metabolism through conversion to glyceraldehyde-3-phosphate under the action of a triose phosphate isomerase or may be a substrate for the synthesis of lipids. The conversion of glycerol through glycerol-3-phosphate is catalyzed by the enzymes glycerol kinase and FAD-dependent glycerol-3-phosphate dehydrogenase. In *S. cerevisiae*, glycerol kinase is encoded by the gene *GUT1* (GlycerolUpTake) (Pavlik et al. 1993; Sprague and Cronan 1977), and glycerol-3-phosphate dehydrogenase is encoded by the gene *GUT2* (Ronnow and Kielland-Brandt 1993). This pathway seems to be the only way of glycerol conversion in *S. cerevisiae* as *gut1*Δ and *gut2*Δ mutants are incapable to utilize glycerol (Sprague and Cronan 1977). This pathway also has been identified in many other yeasts, such as *D. hansenii* (Adler et al. 1985), *Z. rouxii* (Pribylova et al. 2007), and *Candida glycerinogenes* (Wang et al. 2000).

But many other yeast species convert glycerol through dihydroxyacetone. The first step in this pathway is the oxidation of glycerol to dihydroxyacetone with glycerol dehydrogenase, which is encoded by the gene *GCY1*. The second step is the phosphorylation of dihydroxyacetone to dihydroxyacetone phosphate with dihydroxyacetone kinase, which is encoded by genes *DAK1* and *DAK2*. The mentioned genes are present in the *S. cerevisiae* genome (Jung et al. 2012; Norbeck and Blomberg 1997), but the corresponding enzymes are rather involved in regulation of the concentration of glycerol during hyperosmotic stress (Blomberg 2000), than in glycerol utilization as a carbon source. At the same time, in other yeasts, this pathway is more important.

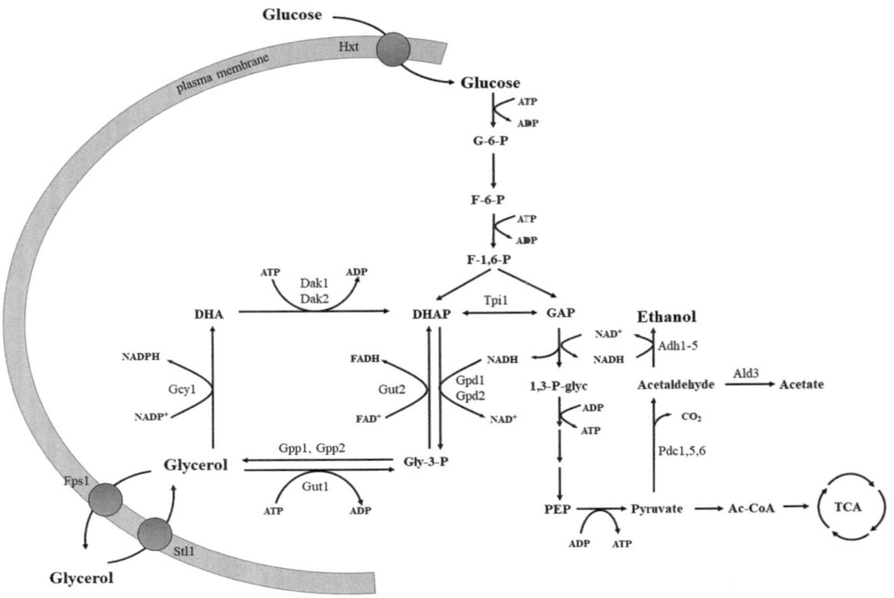

Fig. 12.3 Glycerol metabolism and ethanol production in *S. cerevisiae*. Abbreviations: *G-6-P* glucose-6-phosphate, *F-6-P* fructose-6-phosphate, *F-1,6-P* fructose-1,6-biphosphate, *DHAP* dihydroxyacetone phosphate, *DHA* dihydroxyacetone, *GAP* glyceraldehyde-3-phosphate, *Gly-3-P* glycerol-3-phosphate, *1,3-P-glyc* 1,3-phosphoglycerate, *PEP* phosphoenolpyruvate, *Ac-CoA* acetyl coenzyme A, *TCA* tricarboxylic acid cycle, *Tpi1* triose phosphate isomerase, *Adh1-5* alcohol dehydrogenases, *Pdc1,5,6* pyruvate decarboxylases, *Ald3* aldehyde dehydrogenase, *Gpd1, Gpd2* cytosolic glycerol-3-phosphate dehydrogenases, *Gpp1, Gpp2* glycerol-3-phosphate phosphatases, *Gut1* glycerol kinase, *Gut2* glycerol-3-phosphate dehydrogenase, *Gcy1* glycerol dehydrogenase, *Dak1, Dak2* dihydroxyacetone kinase, *Hxt, Stl1, Fps1* membrane transporters (Semkiv et al. 2017)

There is not much information available about the role of one or another pathway of glycerol utilization in particular yeast species. Tani and Yamada divided the studied yeast species into three groups: (1) yeast that converts glycerol through glycerol-3-phosphate (e.g., *Candida boidinii*), (2) yeast that converts glycerol through dihydroxyacetone (e.g., *Hansenula ofunaensis*), and (3) yeast that can use both pathways (e.g., *Candida valida*) (Tani and Yamada 1987). However, these observations were based on the presence of certain enzymes activities, and not on the analysis of the deletion mutants, and therefore cannot be considered as a final proof. Convincing evidence of the predominant function of the dihydroxyacetone pathway of glycerol utilization was obtained after analysis of *Schizosaccharomyces pombe* mutant with the deletion of the gene *GLD1* which encodes glycerol dehydrogenase (Matsuzawa et al. 2010).

It must be mentioned that most strains of the conventional yeast *S. cerevisiae* exhibit rather poor growth on the medium containing glycerol as a sole carbon source (Swinnen et al. 2013). Some other yeasts grow on glycerol much better. For example, when the ability of 42 different types of yeast to grow on glycerol was tested, *Pichia jadinii* and *Pichia anomala* revealed the highest growth rate – about three-fold higher than in *S. cerevisiae* (Lages et al. 1999). Also nonconventional yeasts *Y. lipolytica*, *P. pastoris*, and *P. tannophilus* have good ability to grow on glycerol, so pure and crude glycerol can be used as a feedstock for biotechnological processes which employ these yeasts (Klein et al. 2017).

12.4.2 Bioconversion of Crude Glycerol to Value-Added Compounds by Yeasts

12.4.2.1 Production of Organic Acids

Citric (CA) and Isocitric (ICA) Acids Crude glycerol can be used as a feedstock for the production of organic acids, in particular, citric acid (CA) and isocitric acid (ICA). Being tricarboxylic acid (TCA) cycle intermediates, both these acids play a central role in metabolism of aerobic organisms and can be produced in excessive amounts and excreted by certain fungi, bacteria, and yeasts in specific growth conditions (Fig. 12.4). CA is of interest for many industrial applications due to its nontoxic, acidulant, buffering, and chelating properties (Rzechonek et al. 2019). For example, CA is used as an acidity regulator, preservative, and flavor enhancer in the food and beverage industry, as an antioxidant or a buffering system for improving stability of pharmaceuticals, as a detergent component in dishwasher cleaners, and as a cross-linker in the production of biodegradable polymers in cosmetic, metallurgy, textile, and other industries (Karaffa and Kubicek 2003). Annual world CA production exceeded two million tons in 2015, growing at 3–5% per year (Ciriminna et al. 2017).

For the first time, CA was isolated from citrus fruits. Currently, it is commercially produced by fermentation of beet or cane molasses as well as glucose syrup by the mycelial fungus *Aspergillus niger* (Karaffa and Kubicek 2003). This method offers high product yield but has several disadvantages: molasses need to be treated with toxic ferrocyanides to remove the excess of trace elements which are harmful to *A. niger*, production is associated with the accumulation of significant amounts of heavy metal-contaminated wastewater and solid waste gypsum (about 16 tons waste/ton CA), possible substrate spectrum for *A. niger* is quite narrow, etc. (Kamzolova et al. 2015). That is why during the last decades, some yeast species have been considered as a substitute of *A. niger* for CA production. These yeasts have broader substrate range, lower sensitivity to heavy metals and oxygen limitations, and CA yield comparable to *A. niger* (Kamzolova et al. 201[illegible]). In particular, some yeast species can use pure and crude glycerol as a carbon substrate for CA production.

When forty yeast species were tested concerning their growth in crude and commercial glycerol, four yeast strains (*Lindnera saturnus* UFLA CES-Y677, *Y. lipolytica* UFLA CM-Y9.4, *Rhodotorula glutinis* NCYC 2439, and *Cryptococcus*

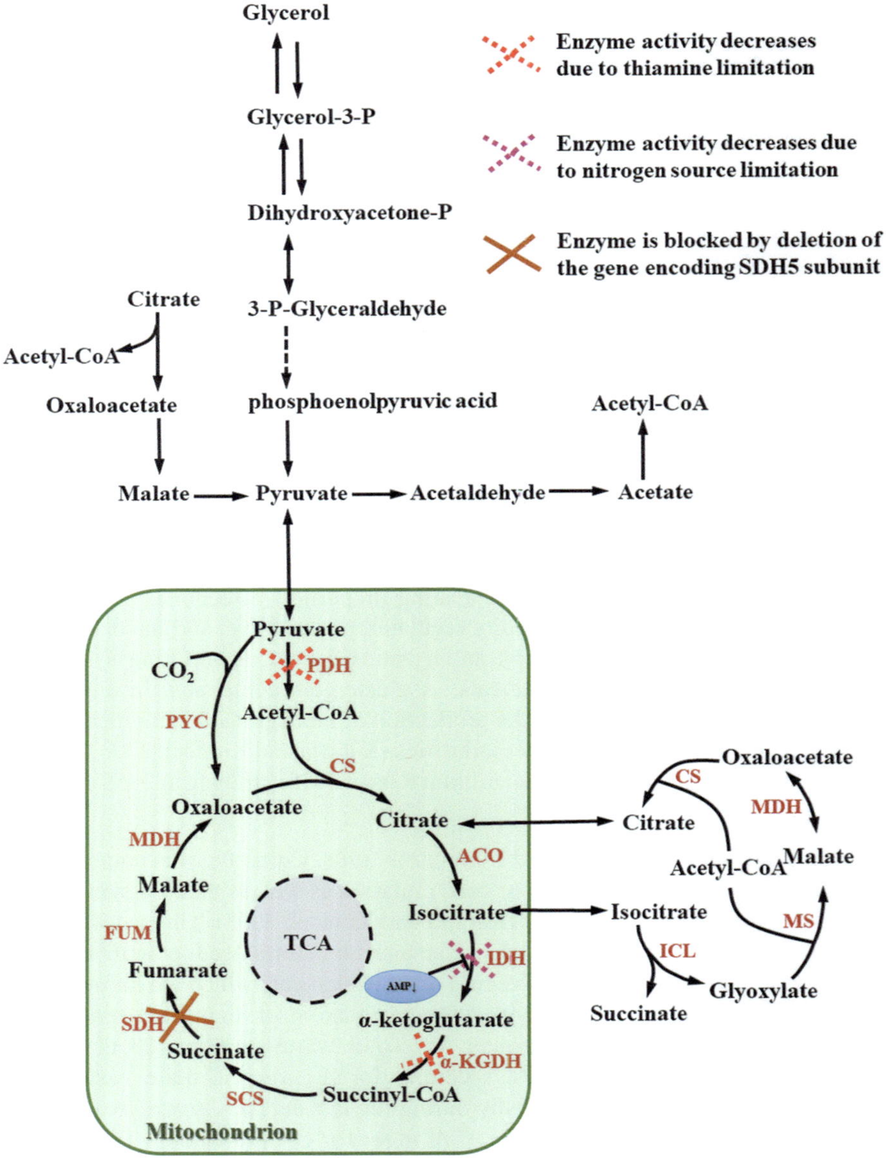

Fig. 12.4 CA, ICA, PA, KGA, and SA production in *Y. lipolytica*. Modified from (Gao et al. 2016a). *PDH* pyruvate dehydrogenase, *CS* citrate synthase, *ACO* aconitase, *IDH* isocitrate dehydrogenase, *α-KGDH* α-ketoglutarate dehydrogenase, *SCS* succinyl-CoA synthetase, *SDH* succinic dehydrogenase, *FUM* fumarase, *MDH* malate dehydrogenase, *PYC* pyruvate carboxylase, *ICL* isocitrate lyase, *MS* malate synthase

curvatus NCYC 476) were found to be able to grow in these conditions, and among them, *Y. lipolytica* had shown the highest level of glycerol intake (Souza et al. 2014).

When organic acids formation was tested for 66 yeast strains of different genera (*Candida, Pichia, Saccharomyces, Torulopsis*, and *Yarrowia*) in the medium containing pure glycerol as a carbon source, 41 strains belonging mainly to species *Y. lipolytica* produced acids in contrast to 25 strains of the genera *Debaryomyces, Candida, Pichia, Saccharomyces*, and *Torulopsis* (Kamzolova et al. 2011). All in all, *Y. lipolytica* seems to be one of the best candidates for crude or pure glycerol conversion to CA (and other valuable compounds; see next paragraphs).

Y. lipolytica is widely known for its ability to produce lipids (Groenewald et al. 2014), polyols, and organic acids (Kamzolova et al. 2015; Rymowicz et al. 2010) from the wide range of substrates, including unspecific carbon sources such as fatty acids, alkanes, plant oils, fats, waste cooking oil, or crude glycerol (Mironczuk et al. 2016). In particular, crude glycerol is easily utilized by the yeast *Y. lipolytica* despite its highly contaminated nature (Papanikolaou et al. 2002a).

Y. lipolytica is able to secrete high amounts of intermediate organic acids, such as pyruvic, citric, isocitric, or 2-oxoglutaric acids, into the medium under the conditions of excessive amounts of carbon source and limited amounts of some of the other growth factors (e.g., N-source, thiamine, or by the mineral salt components P, S, or Mg). Nitrogen limitation causes the secretion of CA and ICA mixture (Holz et al. 2009). Secretion of ICA was previously considered to be undesirable side process of CA production with *Y. lipolytica* due to the fact that ICA has an inferior buffer capacity and chelating ability compared to CA, and the crystallization of CA during the purification process is disturbed by ICA contaminations >5% (Forster et al. 2007). Therefore, previously, researchers mainly aimed to decrease ICA production (Finogenova et al. 2002; Forster et al. 2007). But recently, the development of the reliable method of CA/ICA separation via methyl esterification has risen the interest in ICA production (Heretsch et al. 2008). ICA is much more expensive compared to CA and can be used as a chiral building block for chemical synthesis of complex natural products; as a useful pharmaceutical, food, and beverage additive; and in cosmetics and detergents (Heretsch et al. 2008). Also, monopotassium salt of ICA has been used in several biochemical analyses (assays of aconitate hydratase, NAD-isocitrate dehydrogenase, NADP-isocitrate dehydrogenase, isocitrate lyase) (Kamzolova et al. 2011). Due to this emerging areas of ICA application, researchers started to consider the switch of the CA/ICA ratio toward the ICA production to be a favorable outcome (Holz et al. 2009; Rzechonek et al. 2019).

The CA/ICA ratio formed by *Y. lipolytica* mainly depends on the substrate, cultivation conditions (air saturation, intracellular iron and zinc content, etc.), and the strain used (Forster et al. 2007). Wild-type strains secrete mainly CA and about 8–16% ICA on carbohydrates or glycerol as sole carbon source and approximately 50–65% CA and 35–50% ICA on the gluconeogenetic substrates alkanes and the renewable triglycerides, ethanol or acetate (Finogenova et al. 2005). Interestingly, when crude glycerol was used for *Y. lipolytica* fermentation, CA/

ICA pattern was shifted more toward ICA in comparison with the fermentation on the medium with pure glycerol as carbon source (Kamzolova et al. 2011; Rzechonek et al. 2019). This may be caused by the contaminants present in crude glycerol, e.g., free fatty acids.

Y. lipolytica utilizes glycerol through glycerol-3-phosphate (phosphorylation pathway) (Makri et al. 2010), and activities of enzymes connected with oxidative pathway of glycerol utilization were not detected (Morgunov et al. 2013). Therefore, during assimilation of glycerol from the cultivation medium, *Y. lipolytica* cells possess high activities of glycerol kinase, NAD-dependent glycerol-3-phosphate dehydrogenase, and (in some *Y. lipolytica* strains) FAD-dependent glycerol-3-phosphate dehydrogenase (Morgunov et al. 2004). During assimilation of the crude glycerol, induction of glyoxylate cycle enzymes isocitrate lyase and malate synthase was also observed. This induction was probably associated with the active assimilation of fatty acids from glycerol-containing wastes (Morgunov et al. 2013). Glycerol kinase activity in *Y. lipolytica* cells was found to be affected by pH and salt presence – low pH or high amounts of salts in the cultivation medium inhibit glycerol kinase and decrease overall velocity of glycerol assimilation (Tomaszewska et al. 2014a).

As was already mentioned, CA production by *Y. lipolytica* requires specific cultivation conditions. Acid formation does not occur in the exponential growth phase but is very active in the stationary growth phase (Kamzolova et al. 2015). Yeast growth limitation has to be achieved through the restriction of mineral components in the medium, such as nitrogen, phosphorus, sulfur, or magnesium (Imandi et al. 2008). It was shown that when yeast growth was limited with phosphorus or sulfur, a significant amount of ICA was produced; therefore, nitrogen limitation is more preferable for CA production (Kamzolova et al. 2011). Optimal pH for CA production in *Y. lipolytica* is 4.5–6.5. Lower pH triggers polyols synthesis (Egermeier et al. 2017) although recombinant *Y. lipolytica* with overexpression of the genes *GUT1* (encoding glycerol kinase) and *GUT2* (encoding glycerol-3-phosphate dehydrogenase) were shown to be able to produce CA at pH 3 (Rzechonek et al. 2019). The optimal temperature for CA production is about 28 °C, and the optimal dissolved oxygen concentration (pO2) 50% (of air saturation) (Morgunov et al. 2013). Less active aeration during the process causes an undesirable decrease in the activity of some mitochondrial enzymes (citrate synthase, aconitase, malate dehydrogenase, and NADP-dependent isocitrate dehydrogenase) involved in the synthesis of CA (Kamzolova et al. 2011). Addition of exogenous CA in a moderate amount at early stage of CA production stimulates the biosynthesis of endogenous CA by the yeast cells (Kamzolova et al. 2015). Different studies report either increase (Morgunov et al. 2013) or decrease (Kamzolova et al. 2011) in CA production when crude glycerol was used as carbon source instead of a pure glycerol. Also, crude glycerol may enable either increase in yeast growth due to the presence of contaminants that can be used as nutritional elements by yeasts or (in higher concentrations) growth inhibition due to high concentration of toxic compounds (Souza et al. 2014). Besides optimization of the culture conditions, over the years, *Y. lipolytica* mutants with improved CA production or modified CA/ICA ratio were obtained by UV irradiation combined with negative selection on acetate-containing medium, selection on fluoroacetate, genetic engineering, etc. (Forster et al. 2007; Holz et al. 2009; Morgunov et al. 2013; Tomaszewska et al. 2014a).

Also, ability to produce high amounts of CA from crude glycerol under nitrogen-limiting conditions has been shown for some *Candida* species, such as strains *Candida parapsilosis* ATCC 7330 and *Candida guilliermondii* ATCC 9058 (West 2013). Unfortunately, despite all these developments, there is no large industrial process of CA production with yeasts at present, although there are indications for recent process developments using *Y. lipolytica* (Fickers et al. 2005).

Pyruvic Acid Besides CA and ICA, *Y. lipolytica* can excrete pyruvic acid (PA) and α-ketoglutaric acid (KGA) under the conditions of excessive amounts of carbon source and limited amounts of thiamine in the cultivation medium. *Y. lipolytica* is a thiamine-auxotrophic yeast due to the inability to synthesize the pyrimidine structure of the thiamine molecule (Yin et al. 2012). When thiamine in the cultivation medium is depleted, activities of thiamine-dependent enzymes (pyruvate dehydrogenase, transketolase, α-ketoglutarate dehydrogenase) start to decrease, blocking the conversion of the corresponding substrates (Fig. 12.4). Big disadvantage of this process is the simultaneous production of PA and KGA because these acids are similar in physical and chemical properties which make the procedure of their separation quite tricky. That is why if production is directed toward PA, researches try to decrease the accumulation of KGA by *Y. lipolytica* and vice versa.

PA is used as a diet supplementary (Stanko et al. 1992) for the production of L-tryptophan (Nakazawa et al. 1972), L-tyrosine, and 3,4-dihydroxyphenyl alanine (Yamada et al. 1972) and as a substrate for enzyme activities assays in biochemistry and medicine. PA can be produced by chemical process or with the help of certain bacteria, basidiomycetes, or yeasts (Morgunov et al. 2004). For example, extensively studied PA producer is yeast *Candida glabrata* (former *Torulopsis glabrata*), selected strain of which is able to produce 67.8 g/L of PA from glucose (Yonehara and Miyata 1994). *Y. lipolytica* was found to produce PA not only from glucose but also from glycerol. When 18 strains of the genera *Candida* and *Yarrowia* were tested for their ability to produce PA using glucose or glycerol as carbon sources, seven efficient PA producers were identified, and *Y. lipolytica* strain 374/4 was the best among them (Morgunov et al. 2004). It produced 1.6-fold higher amount of PA in glycerol-containing medium than in case of the medium with glucose. Also, amount of accumulated KGA was 4 times lower than that of PA. In glycerol-containing medium with 2 μg/L of thiamine strain *Y. lipolytica* 374/4 accumulated 61.3 g/L of PA by the 78th hour of cultivation. It was confirmed that activities of thiamine-dependent enzymes, such as transketolase, pyruvate dehydrogenase, and α-ketoglutarate dehydrogenase, decreased substantially after the transition of yeast cells from the exponential growth phase to growth retardation phase caused by the exhaustion of thiamine in the medium.

The overexpression of the genes encoding α-ketoglutarate dehydrogenase complex is another possible approach to further increase the production of PA and restrict the accumulation of KGA in *Y. lipolytica* (Holz et al. 2011).

However, mentioned experiments with PA production by *Y. lipolytica* were performed in the defined medium with the addition of pure glycerol. When yeast was grown in complex polypeptone-containing medium, no PA production was observed, probably due to high thiamine content in this medium (Yonehara and Miyata 1994). To our knowledge, no attempts have been made to produce PA from crude glycerol, but it was successfully used for the production of KGA (Otto et al. 2012).

α-Ketoglutaric Acid The α-ketoglutaric acid (KGA) is used as a dietary supplement, in the agrochemical and pharmaceutical industries, as a building block for the synthesis of heterocycles and elastomers, etc. (Sauer et al. 2008). Currently, KGA is synthesized chemically from diethyl succinate and diethyl oxalate or by transamination of glyoxylic acid with sodium glutamate and a copper catalyst (Otto et al. 2011), but these methods are ecologically harmful. KGA production via microbial fermentation has a potential to be more profitable and environmentally friendly. KGA can be produced by bacteria *Pseudomonas fluorescens*, *Serratia marcescens*, *Bacillus* ssp., *Corynebacterium glutamicum*, and *Arthrobacter paraffineus* or by yeasts *Candida* ssp., *Pichia* ssp., *C. glabrata*, and *Y. lipolytica* (Otto et al. 2011). Most studies on KGA production by yeasts concentrate on the last two species.

KGA production by these yeasts is triggered by the thiamine limitation and excess of carbon source and accompanied by the production of PA as a major by-product and fumarate, malate, and succinate as minor by-products (Otto et al. 2012). However, when n-alkanes, plant oils, fatty acids, or their derivatives are used as substrates for KGA production, PA is not accumulating due to the fact that substrate is degraded via acetyl-CoA omitting pyruvate (Finogenova et al. 2005).

KGA production from pure and crude glycerol is accompanied by PA accumulation, which needs to be minimized. In an attempt to decrease the by-product yield during KGA production from crude glycerol by *Y. lipolytica*, Otto et al. constructed recombinant *Y. lipolytica* strains with the overexpression of gene *FUM1* (encoding fumarase), *PYC1* (encoding pyruvate carboxylase), or both mentioned genes (Otto et al. 2012). Multicopy integration of the gene *FUM1* into *Y. lipolytica* genome caused a significant reduction of the production of PA, fumarate, and malate during crude glycerol conversion to KGA. In contrast, the overexpression of *PYC1* gene or both genes *FUM1* and *PYC1* caused an increased accumulation of the mentioned by-products (Otto et al. 2012). Besides that, the production of KGA decreased in the strain with the *PYC1* gene overexpression in comparison to the initial *Y. lipolytica* strain. Interestingly, in other study, overexpression of the heterologous pyruvate carboxylase genes *ScPYC1* from *S. cerevisiae* and *RoPYC2* from *Rhizopus oryzae* in *Y. lipolytica* strain WSH-Z06 caused the increase of KGA yields by 24.5 and 35.3% and the decrease of PA yields by 51.9 and 69.8%, respectively (Yin et al. 2012). In a 3-L fermenter, the recombinant strain with *RoPYC2* gene overexpression produced the highest amount of KGA – 62.5 g/L with a decrease in PA yield from 35.2 to 13.5 g/L (Yin et al. 2012).

The optimal pH for KGA production is around 3.5; therefore, a two-stage pH control strategy was developed for KGA synthesis from crude glycerol in the bioreactor. For the beginning of cultivation, pH was set on 5.0 and aeration rate at 50%

(growth phase). Then for the KGA production phase, pH was reduced to 3.8 and pO2 to 10% (Otto et al. 2012). At the beginning of production phase, large amounts of by-product PA were synthesized, but as the glycerol in the medium was exhausted, cells started to reuptake PA and converted it to KGA. PA reduction in the engineered yeasts with overexpression of heterologous gene-encoding pyruvate carboxylase (*ScPYC1*or *RoPYC2*) was much faster than those in the initial *Y. lipolytica* strain WSH-Z06 (Yin et al. 2012).

Similar effect – decreased PA and increased KGA concentrations – was achieved for KGA and PA producing yeast *C. glabrata* by stimulation of pyruvate carboxylase activity by the increased supply of its cofactor biotin (Zhang et al. 2009).

Also KGA/PA ratio was increased in the recombinant *C. glabrata* strain with the overexpression of heterologous gene *PDC1* (encoding pyruvate decarboxylase) from *S. cerevisiae* (Zhang et al. 2009). Further improvement in KGA production is possible by modifying the activities of NAD- and NADP-dependent isocitrate dehydrogenases (catalyzes the oxidation of isocitrate to KGA) and the mitochondrial α-ketoglutarate dehydrogenase complex (catalyzes the oxidative decarboxylation of KGA to succinyl-CoA) (Otto et al. 2012).

Succinic Acid Succinic acid (SA) can be used as a precursor of many important chemicals in surfactant, food, and pharmaceutical industries (Yan et al. 2014). The US Department of Energy has listed SA among the top twelve potential biogenerated chemical building blocks for the future (White et al. 2004).

Currently, SA is mainly produced by petroleum-based chemical process, but efforts are made to implement bio-based methods of SA production from sustainable feedstock (Mazière et al. 2017). Many SA producers were found among bacteria (Sanchez et al. 2005), e.g., *Actinobacillus succinogenes* and *A. succiniciproducens*. *Basfia succiniciproducens* has been shown to convert crude glycerol to SA (Scholten et al. 2009). Yeasts are also considered as SA producers owing to their tolerance to high acidity. Glycerol is considered as one of the most promising feedstock for commercially feasible SA production (Tan et al. 2014).

It comes as no surprise that *Y. lipolytica* was proposed as a potential SA producer from crude glycerol. But wild-type strains of *Y. lipolytica* almost did not accumulate SA during cultivation. Previously, production of KGA by *Y. lipolytica* was combined with subsequent decarboxylation of KGA by hydrogen peroxide to obtain SA (Kamzolova et al. 2009).

To obtain *Y. lipolytica* strain directly producing SA, Gao et al. decided to block (Fig. 12.4) further SA metabolism in TCA cycle (Gao et al. 2016a). The succinate dehydrogenase complex oxidizes SA to fumaric acid with the simultaneous reduction of the ubiquinone to ubiquinol (Cecchini 2003). This complex consists of five subunits (SDH1–5), among them SDH1 and SDH2 are catalytic subunits, SDH3 and SDH4 perform a role of molecular anchors, and SDH5 is required for SDH complex stability and activity (Oyedotun and Lemire 2004). In the study of Gao et al., the gene-encoding SDH5 subunit (*Ylsdh5*) was deleted in *Y. lipolytica* strain Po1f to obtain recombinant strain PGC01003. This strain demonstrated a weak growth in glucose-containing medium in contrast to the *sdh1* or *sdh2* deletion strains which have been shown to lose their ability to grow in glucose (Yuzbashev et al. 2010). PGC01003 indeed accumulated and secreted enhanced amounts of SA. It also accumulated substantial amounts of acetic acid (approximately 6 g/L) due to the emerged imbalance between glycolysis and TCA cycle. When the cultivation media and conditions were optimized, strain PGC01003 produced 43 g/of SA from crude glycerol during batch cultivation and 160 g/L of SA during fed-batch cultivation (Gao et al. 2016a).

Using in situ fibrous bed bioreactor (*is*FBB) under the optimal conditions (20 g sugarcane bagasse as immobilization material, 120 g/L crude glycerol as carbon source, and 4 L min^{-1} of aeration rate), SA production by PGC01003 was increased to 53.6 g/L during batch cultivation and 209.7 g/L during fed-batch cultivation. Methanol was shown to inhibit the cell growth and SA production under described conditions when its initial concentration was more than 5 g/L (Li et al. 2018a). Obtained *Y. lipolytica* strain and developed *is*FBB have a great potential for industrial implementation.

Lactic Acid Due to its chemical properties, lactic acid (LA) is of interest for many industrial applications, e.g., food, pharmaceutical, leather, textile, and chemical industries (Datta and Henry 2006). Also, L- and D-optical isomers of LA are the components used for the production of biodegradable polylactic acid that can be used in automobile, packaging, and cosmetic industries (Abdel-Rahman et al. 2013).

LA can be produced by homo- and heterofermentative lactic acid bacteria, fungi (e.g., genus *Rhizopus*), yeasts (*Saccharomyces* and *Kluyveromyces* genera), and microalgae (e.g., *Scenedesmus obliquus*) (Abdel-Rahman et al. 2013). Developed industrial processes use carbohydrates as a substrate for LA production. There have been reported bacteria and fungi with the efficient LA production from glycerol as a sole carbon source (Mazumdar et al. 2013; Vodnar et al. 2013). For example, the fungus *R. oryzae* was able to produce about 48 g/L of LA from 75 g/L of crude glycerol supplemented with lucerne green juice (Vodnar et al. 2013).

Yeasts are perspective microorganisms for LA production as they are robust, resistant to low pHs, and widely used for industrial bioprocesses (Sauer et al. 2010). The methylotrophic yeast *P. pastoris* can be used for crude glycerol conversion, as it efficiently utilizes glycerol and methanol as carbon sources and it is resistant to enhanced methanol concentrations (Looser et al. 2015). *P. pastoris* even accumulates more biomass when crude glycerol is used as a carbon source in comparison to the growth on the medium with pure glycerol (Anastacio et al. 2014).

When de Lima et al. had expressed the gene *LDH* encoding lactate dehydrogenase from *Bos taurus* under the control of the strong constitutive promoter *GAP1* in *P. pastoris*, obtained recombinant strains produced only 10% of the theoretically possible LA amount (de Lima et al. 2016). Therefore, they additionally overexpressed in this strain heterologous gene-encoding LA transporter Jen1p from *S. cerevisiae* or homologous gene-encoding putative LA transporter PAS which was identified by amino acid sequence similarity with Jen1p. Performance of the obtained recombinant strains was evaluated in fed-batch fermentation which was composed of two phases. First phase of cellular growth was carried out under sufficient oxygen supply followed by LA production stage accompanied with a single-pulse addition of 4% crude glycerol and hypoxia conditions. Under such conditions, the strain containing homologous PAS transporter showed the highest LA yield of approximately 0.7 g/g of glycerol. Simultaneously, acetic acid formation by this strain was reduced by half. In conclusion, recombinant *P. pastoris* strain with *LDH* and *PAS* overexpression can be used for efficient LA production from crude glycerol in two-phase fed-batch fermentation (de Lima et al. 2016).

12.4.2.2 Production of Erythritol and Other Polyols

Erythritol Polyols are noncyclic hydrogenated carbohydrates which can be used as low caloric sweeteners in food and beverages industry (Park et al. 2016). Such compounds as erythritol, glycerol, ribitol, arabitol, mannitol, xylitol, and sorbitol belong to this group. Chemical method of polyols production (catalytic hydrogenation of sugars under high temperature and pressure) is not very efficient, so production of polyols is chiefly carried out by microbial conversion of carbohydrates (Park et al. 2016). For the first time, production of polyols was observed by Röhr during citric acid fermentation by *A. niger* (Röhr et al. 1983). In yeasts, polyols chiefly play role of osmolytes, which protect yeast cells against osmotic stress, but they are also produced during yeasts growth under normal conditions (Kayingo et al. 2001).

Erythritol is one of the most important polyols, whose production by microorganisms is being extensively studied. Erythritol is a four-carbon polyol with no optical activity. It occurs naturally in alcoholic beverages, in mushrooms, and as a component of fruits such as pears, melons, and grapes (Bernt et al. 1996). Erythritol exhibits about 70% of the relative sweetness of sucrose, but its caloric value is very low and the majority of consumed compound is quickly excreted, so it is considered to be "zero-calorie sweetener" (Rzechonek et al. 2018). Also, erythritol is safe for diabetic patients, has no carcinogenic or teratogenic potential (Bernt et al. 1996), and does not contribute to tooth decay (Rzechonek et al. 2018). In fact, it was even suggested to prevent dental caries (Hashino et al. 2013) and improve endothelial function in patients with type II diabetes (Flint et al. 2014). However, as in the case of other polyols, the excessive consumption of erythritol may cause a laxative effect (Oku and Nakamura 2007). Due to its positive properties, erythritol ought to be widely used in food and pharmaceutics, but its application is somewhat restricted by its relatively high retail price.

Among the microorganisms capable of overproducing erythritol, there are osmophilic yeasts from the genera *Pichia*, *Zygopichia*, *Candida*, *Debaryomyces*, *Moniliella*, *Torula*, *Torulopsis*, *Trigonopsis*, *Trichosporon*, *Trichosporonoides*, *Pseudozyma*, and *Ustilago* (Jeya et al. 2009; Moon et al. 2010), some fungi and lactic acid bacteria (Veiga-Da-Cunha et al. 1992), etc. The mechanism of erythritol formation in osmophilic yeasts is not completely identified. It possibly proceeds through transketolase (TK) reaction between fructose-6-phosphate and glyceraldehyde-3-phosphate. Obtained in this reaction, erythrose-4-phosphate and xylulose-5-phosphate could be dephosphorylated and reduced to erythritol and arabitol as verified in *C. magnoliae* (Park et al. 2005). The gene-encoding enzyme responsible for erythrose-4 phosphate dephosphorylation has not yet been identified (Rzechonek et al. 2018). Erythrose is reduced to erythritol by NADPH-dependent enzyme erythrose reductase (ER) (Lee et al. 2010). ER has been identified and described in a few yeasts, such as *Moniliella megachiliensis* (Kobayashi et al. 2013), *Candida magnoliae* (Lee et al. 2010), and *Y. lipolytica* (Janek et al. 2017). TK and ER activities in some yeasts are strongly influenced by the presence of NaCl (Park et al. 2011; Sawada et al. 2009). On the whole, changes in osmotic pressure (caused by high-sugar concentration or salt addition) induce erythritol accumulation in yeast cells. Other factors that influence erythritol biosynthesis are pH and temperature of medium, type of substrate and its concentration, sources of nitrogen and phosphorus, and additional factors such as chloride, copper, and manganese ions (Jeya et al. 2009; Tomaszewska et al. 2014a).

Aureobasidium sp., *Pseudozyma tsukubaensis* (Tomaszewska et al. 2014a), *Moniliella pollinis*, *M. megachiliensis*, and recently also *Y. lipolytica* (Rzechonek et al. 2018) were implemented for industrial erythritol production. On the industrial scale, erythritol is predominantly obtained from glucose derived from wheat or cornstarch hydrolysates (Moon et al. 2010). Glycerol has been previously reported as not suitable for erythritol biosynthesis (Jeya et al. 2009), but soon it was found that yeast *Y. lipolytica* produced erythritol in the cultivation medium with glycerol as the sole carbon source at pH 3.0 even better than in the medium with glucose (Rymowicz et al. 2009). Even in the medium containing both glucose and glycerol, *Y. lipolytica* first utilizes glycerol (Papanikolaou et al. 2002b). As was already mentioned, *Y. lipolytica* produces mainly CA during glycerol fermentation at pH 4.5–6.5; however, at pH 3.0, CA concentration is very low as polyols production (Fig. 12.5) starts to prevail (Tomaszewska et al. 2014a). Low pH value during erythritol biosynthesis is an advantage, because it protects the culture against bacterial contamination and thereby supports the development of continuous fermentation procedures (Tomaszewska et al. 2012). Erythritol production from glycerol in *Y. lipolytica* is also stimulated by salt addition as it increases activities of TK and ER. Moreover, the presence of salt in the medium improved not only erythritol yield but also the selectivity of biosynthesis, decreasing the amount of by-product polyols. Crude glycerol contains salt contamination, which can additionally induce erythritol production. For example, notable increase in erythritol yield was observed for the *Y. lipolytica* strain Wratislavia K1 when crude glycerol was used instead of pure glycerol as carbon source for erythritol production (Tomaszewska et al. 2014a). Glycerol itself generates higher osmotic stress than glucose which is beneficial to erythritol production (Yang et al. 2016). Another advantage of glycerol as a substrate for erythritol production is the composition of by-products after fermentation. When sugars are used as a substrate, glycerol occurs as one of the main by-products of erythritol production, which is quite difficult to separate from erythritol (Rzechonek et al. 2018). When glycerol is used as carbon source, it can be completely depleted from the cultivation medium and the presence of other by-products

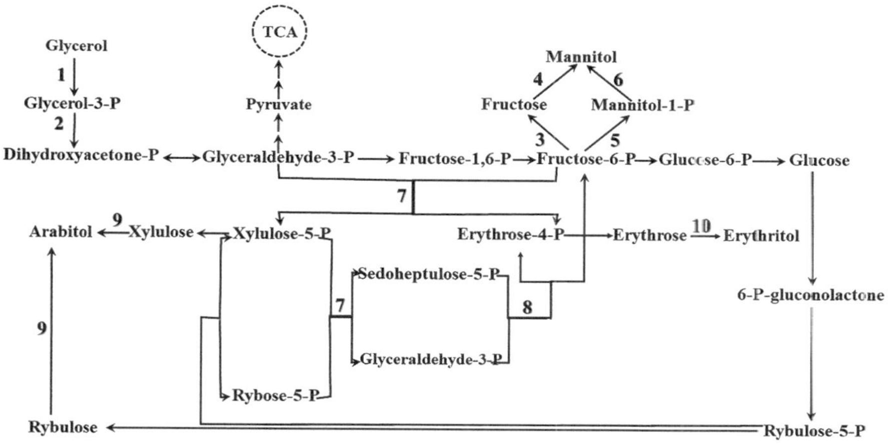

Fig. 12.5 Hypothetical pathways of glycerol conversion into polyols in *Y. lipolytica* (Tomaszewska et al. 2014a). *1* glycerol kinase, *2* glycerol-3-P dehydrogenase, *3* mannitol dehydrogenase, *4* hexokinase, *5* mannitol-1-P dehydrogenase, *6* mannitol-1-phosphatase, *7* transketolase, *8* transaldolase, *9* arabitol dehydrogenase, *10* erythrose reductase

may be reduced as well (Mironczuk et al. 2014; Mironczuk et al. 2015; Rymowicz et al. 2009). Therefore, using glycerol and especially cheap crude glycerol as a substrate could ensure cost-effective erythritol production by *Y. lipolytica*.

Erythritol synthesis from glycerol could be improved in several directions: (a) optimization of the cultural medium composition, (b) optimization of fermentation setup, and (c) modification of the *Y. lipolytica* strains used in the biosynthesis by mutagenesis or genetic engineering.

When artificial neural network model and genetic algorithm were used to predict the optimal medium composition to maximize erythritol production from crude glycerol, it was found out that medium with 232.39 g/L crude glycerol, 1.57 g/L urea, and 31.03 g/L NaCl led to predictive maximum erythritol concentration of 110.7 g/L (Yang et al. 2016). In real experiment in such conditions, *Y. lipolytica* strain A16 produced 109.2 g/L of erythritol, which is approximately 2 times higher than the usual values obtained in batch fermentations (40 to 60 g/L) (Rywinska et al. 2015; Yang et al. 2016). Also it was found out that erythritol production can be improved by medium supplementation with Mn^{2+} (25 mg/L), Cu^{2+} (2.5 mg/L) (Tomaszewska et al. 2014b), thiamine, yeast extract (Rywinska et al. 2015), and surfactant Span 20 (Rakicka et al. 2016b).

In respect of the fermentation setup, improved productivity is obtained in fed-batch systems where the amount of substrate is renewed at least once or, sometimes, a few times (Rymowicz et al. 2009; Rywinska et al. 2015; Tomaszewska et al. 2014b; Yang et al. 2014a).

Another modification is repeated fed-batch cultures (RBC). In this system, after consumption of the substrate, fermentation medium is separated into a concentrated fraction of microorganisms and another fraction enriched with the fermentation product. Later, fresh substrate is added to the microorganism fraction and another cycle of the production begins. This may be repeated one or several times (Mironczuk et al. 2014). In such system, after replacing 30% of the working volume several times, researchers were able to obtain 224 g/L of erythritol with a 0.78 g/g product yield from pure glycerol (Rzechonek et al. 2018). In the RBC with crude glycerol, the average erythritol concentration oscillated from 81.3 to 180.8 g/L during the course of fermentation (Mironczuk et al. 2014). In the continuous fermentation performed in a chemostat, erythritol production from pure glycerol was 103.4 g/L with a productivity of 1.12 g/L h and a yield of 0.52 g/g. When crude glycerol was used as a substrate, these values were lower – 81.9 g/L of erythritol with productivity of 0.9 g/L h and yield of 0.4 g/g (Rakicka et al. 2016a).

Several modified *Y. lipolytica* strains with improved erythritol production were obtained. For example, *Y. lipolytica* strain Wratislavia K1 was isolated by two-stage selection: first, acetate-negative mutant incapable of growth on acetate as the sole carbon and energy source was isolated after exposure to UV irradiation, and then the most productive strain was isolated in the course of continuous citric acid production from glucose in nitrogen-limited chemostat at a dilution rate of 0.016 h^{-1} (Tomaszewska et al. 2014a). Strain Wratislavia K1 produced the highest amounts of erythritol among other acetate-negative mutants, probably due to its inability to reutilize this polyol. Rymowicz et al. reported that using crude glycerol (300 g/L) in a fed-batch system with *Y. lipolytica* Wratislavia K1 resulted in a production of 170 g/L of erythritol with a yield of 0.56 g/g and productivity of 1.0 g/L/h (Rymowicz et al. 2009).

Later, the strain Wratislavia K1 was genetically modified by overexpression of the native gene *GUT1* and *S. cerevisiae* gene *SUC2* gene (encoding sucrose hydrolyzing enzyme invertase). The obtained recombinant strain possessed ability to utilize sucrose and to assimilate glycerol faster than the initial strain (Rakicka et al. 2017). This strain was tested in two-stage fermentation: first, abundant biomass was accumulated during growth in the medium containing industrial raw molasses as a sole carbon source; then, stage of polyols production started after the addition of 4% NaCl and 150 g/L of crude glycerol. In such conditions, recombinant strain produced 100.65 g/L of polyols, with productivity of 1.09 g/L/h and yield of 0.67 g/g. This way, efficient polyol production from inexpensive raw materials was achieved (Rakicka et al. 2017).

Genes encoding the two first enzymes of glycerol assimilation (*GUT1* and *GUT2*) have been overexpressed in *Y. lipolytica* strain A101. Overexpression of *GUT2* gene did not improve erythritol production, and overexpression of *GUT1* gene or co-overexpression of both genes caused a 24% or 35% increase in erythritol productivity, respectively (Mironczuk et al. 2016).

Also attempt has been made to hinder erythritol reutilization by *Y. lipolytica*. For that, the gene *EYK1* encoding erythrulose kinase was identified and deleted in *Y. lipolytica*. Obtained recombinant strain was unable to use erythritol as a carbon source and revealed 26% improvement in erythritol productivity (Carly et al. 2017).

Another osmotolerant yeast, *M. megachiliensis*, has been found to be able to utilize nonrefined glycerol waste derived from palm oil or beef tallow and convert it to erythritol. When 200 g/L of crude glycerol was used as a carbon source for *M. megachiliensis* fermentation, the yield of erythritol was approximately 60% (Kobayashi et al. 2015).

Mannitol Mannitol is a six-carbon polyol that can be used in the food, pharmaceutical, and medical industries. It has similar characteristics to other polyols: sweet taste; low caloric value; naturally found in fruits, vegetables, and mushrooms; etc. (Khan et al. 2009). Mannitol is industrially produced by catalytic hydrogenation of fructose and glucose mixture at high temperature and pressure, which requires a high purity of substrates. Microbial production of mannitol (Fig. 12.5) is an appealing alternative (Saha and Racine 2011). Mannitol can be produced by yeasts *C. magnoliae*, *C. zeylanoides*, the fungi *Aspergillus*, and bacteria, especially lactic acid bacteria (Saha and Racine 2011). Fructose- and glucose-containing media have been found to be the most suitable for the production of mannitol. Osmophilic yeast *C. magnoliae* has been shown to produce 209 g/L of mannitol from fructose/glucose mixture with an impressive yield of 83% (Song et al. 2002). *C. magnoliae* was reported to produce 213 g/L of mannitol from similar substrate in fed-batch fermentation (Lee et al. 2003). Fructose/glucose medium supplementation with Ca^{2+} and Cu^{2+} further increased the production of mannitol with *C. magnoliae* strain HH-01 reaching 223 g/L (Lee et al. 2007b).

There is not much information about mannitol production from glycerol. Mannitol is a major by-product during glycerol conversion to erythritol by *Y. lipolytica*. Some strains of *Y. lipolytica* produce mannitol as the only polyol in nitrogen-limited conditions. For example, *Y. lipolytica* LFMB strain 19 produced 19.4 g/L of mannitol from 90 g/L of glycerol (Chatzifragkou et al. 2011). When several *Y. lipolytica* were analyzed, strains A UV'1 and A-15 were found to be able to produce significant amounts (up to 27.6 g/L) of mannitol (Tomaszewska et al. 2012). NaCl salt addition to the medium improves erythritol biosynthesis and simultaneously inhibits mannitol formation by these strains. In the fed-batch process, the A UV'1 strain produced 91.6 g/L erythritol and 38.1 g/L mannitol from pure glycerol as a carbon source (Tomaszewska et al. 2012).

The resting cells of yeast *C. magnoliae* were found to be able to produce exclusively mannitol (up to 51 g/L) from pure glycerol. Mannitol yield in this process was as high as 50% (Khan et al. 2009).

Arabitol Arabitol is a five-carbon polyol that can be used in similar manner as erythritol and mannitol, that is, as a natural sweetener, a dental caries reducer, and a sugar substitute for diabetic patients (Gare 2002). Besides usual polyols' properties, arabitol can be transformed into several groups of chemicals like its enantiomer xylitol, arabonic/arabinoic acid, etc. (White et al. 2004).

Arabitol can be produced by osmophilic yeast species such as *Debaryomyces*, *Candida*, *Pichia*, *Wickerhamomyces* (*Hansenula*), and *Saccharomycopsis* (*Endomycopsis*) (Fig. 12.5) (Koganti et al. 2011). When 214 yeast strains, many osmotolerant, were analyzed in regard to their ability to produce arabitol from glycerol, the genera *Debaryomyces* and *Geotrichum* had the largest numbers of strains that produced noticeable amounts (≥5 g/L) of polyols from glycerol. *Debaryomyces* strains tended to produce predominantly arabitol, whereas *Geotrichum* strains produced arabitol and mannitol. Strain *D. hansenii* SBP-1 was chosen for further studies as it produced high amounts of arabitol and the minimum of non-arabitol polyols. The best conditions for arabitol production from glycerol with this strain were temperature 30 °C, initial glycerol concentration 150 g/L, high content of dissolved oxygen, and presence of glucose or xylose. Addition of sorbitol or NaCl salt, on the other hand, inhibited arabitol production (Koganti et al. 2011).

Another important natural sweetener, xylitol, can be produced by bioconversion of xylose. Glycerol was successfully used as a co-substrate during xylitol production from xylose or sugarcane bagasse hemicellulosic hydrolysate by yeasts *S. cerevisiae* (Kogje and Ghosalkar 2017), *C. guilliermondii* (Arruda et al. 2015), or *Candida tropicalis* (Ko et al. 2006).

12.4.2.3 Production of Ethanol

Bioethanol produced from sustainable carbohydrate feedstock can be used as a petrol's substitute or additive in order to alleviate environmental pollution. So-called "first-generation" bioethanol is produced from sugarcane, corn, or sugar beets. Considerable efforts have been made toward the development of profitable technology for "second-generation" ethanol production from lignocellulosic feedstock (Kurylenko et al. 2016). However, application of raw cellulosic material requires its complicated and costly physicochemical pretreatment and enzymatic hydrolysis (Li et al. 2018b). That is why crude glycerol is being considered as an inexpensive feedstock that may not need any pretreatment prior to the start of alcoholic fermentation. The cost of ethanol production from glycerol was estimated to be almost 40% lower compared with production from corn-derived sugars (Yazdani and Gonzalez 2007).

But, as in the most microorganisms glycerol utilization occurs through respiratory metabolism, there have been few reports on microbial conversion of glycerol to ethanol by use of wild-type strains (Yazdani and Gonzalez 2007). In particular, *Paenibacillus macerans* (Gupta et al. 2009) and *E. aerogenes* (Ito et al. 2005) were reported to produce ethanol under anaerobic conditions from pure or crude glycerol, respectively. An engineered *E. coli* strain with overexpression of genes involved in the fermentative pathway of glycerol utilization produced 21 g/L of ethanol from 60 g/L of pure glycerol under microaerobic conditions (Durnin et al. 2009). An engineered *Klebsiella pneumoniae* strain has been shown to achieve 25 g/L ethanol on crude glycerol (Oh et al. 2011).

As ethanol production is the redox-neutral process and biomass accumulation is accompanied with NAD^+ reduction to $NADH(H^+)$, in anaerobic conditions, cells need some way to consume excessive $NADH(H^+)$. That is why bacteria during glycerol fermentation under anaerobic conditions usually produce also by-products (mainly 1,2-propanediol) that can serve as an electron sink. But the newly isolated bacteria, identified as nonpathogenic *Kluyvera cryocrescens*, was able to convert biodiesel-derived crude glycerol to ethanol without 1,2-propanediol production (Choi et al. 2011). It was assumed that some impurities from crude glycerol played the role of external electron acceptor for *K. cryocrescens* in this process. When glycerol fermentation was performed in the presence of limited oxygen, *K. cryocrescens* revealed even higher ethanol productivity and yield than under strict anaerobic condition. It produced up to 27 g/L ethanol from crude glycerol under microaerobic batch fermentation (Choi et al. 2011).

However, ethanol-producing bacteria possess drawbacks: they are susceptible to phagolysis, to high ethanol concentrations, and to toxic compounds in crude glycerol; they may be pathogenic. Yeasts are more robust in terms of ethanol production in industrial setup.

As was already mentioned, catabolism of glycerol in the yeast cells is carried out through glycerol-3-phosphate or dihydroxyacetone. Some researchers considered these pathways to be respiratory and fermentative route of glycerol metabolism, respectively (though without solid justification) (Hong et al. 2010; Yu et al. 2010). That is why mainly genes coding for glycerol dehydrogenase (Gcy) and dihydroxyacetone kinase (Dak) were overexpressed with the aim to improve glycerol conversion to ethanol in yeasts.

Ethanol is produced from pyruvate via pyruvate decarboxylation to acetaldehyde, followed by the reduction of acetaldehyde to form ethanol (Fig. 12.3). These two reactions are catalyzed by the enzymes pyruvate decarboxylase (Pdc) and alcohol dehydrogenase (Adh), respectively. The *PDC* and *ADH* genes expression levels strongly affect ethanol yield during alcoholic fermentation (Nikel et al. 2010). Distinctive traits of Pdc and Adh enzymes in baker's yeast *S. cerevisiae* render them to be one of the best ethanol producer. *S. cerevisiae* prefers fermentative metabolism of glucose over respiratory metabolism almost in all conditions except when the yeast cultures grow under high oxygen supply and very low glucose concentration in the cultivation medium (van Hoek et al. 1998).

Unfortunately, *S. cerevisiae* poorly grows on glycerol if no growth-supporting supplements (such as amino acids and nucleic bases) are added (Swinnen et al. 2013). The growth of *S. cerevisiae* on glycerol may be improved by evolutionary adaptation (Ochoa-Estopier et al. 2011), expression of heterologous glycerol transporters (Klein et al. 2016), etc.

Several rounds of metabolic engineering were performed in order to improve ethanol production from glycerol by *S. cerevisiae* (Yu et al. 2012; Yu et al. 2010). First, by overexpression of the genes encoding glycerol dehydrogenase, dihydroxyacetone kinase, and presumable glycerol uptake protein Gup1 (whose role in glycerol transport was later disproved), the overall ethanol production was enhanced by 3.4-fold and reached 2.4 g/L (Yu et al. 2010). Later, genes *GPD2* (coding glycerol-3-phosphate dehydrogenase involved in glycerol synthesis de novo) and *FPS1* (coding for glycerol facilitator involved in glycerol export from *S. cerevisiae* cells) were deleted causing further increase of ethanol production to 4.4 g/L. Finally, overexpression of pyruvate decarboxylase and alcohol dehydrogenase genes allowed to obtain recombinant *S. cerevisiae* strain which accumulated 5.4 g/L of ethanol from glycerol (Yu et al. 2012).

Methylotrophic thermotolerant yeast *Ogataea (Hansenula) polymorpha* was suggested to be a better choice for crude glycerol conversion to ethanol as it is less susceptible to the toxic effects of methanol and heavy metals which may be present in crude glycerol. To improve production of ethanol from glycerol, *O. polymorpha* strain DL1 (currently reclassified as *Ogataea parapolymorpha* (Suh and Zhou 2010)) was engineered to express genes encoding pyruvate decarboxylase (*pdc*) and alcohol dehydrogenase II (*adhB*) from *Zymomonas mobilis* under the control of the glyceraldehyde-3-phosphate dehydrogenase (GAPDH) promoter. Corresponding strain produced 2.74 g/L of ethanol from glycerol which was 3.3-fold higher in comparison with the parental strain DL1. Further, genes encoding glycerol dehydrogenase (*dhaD*) and dihydroxyacetone kinase (*dhaKLM*) from *K. pneumoniae* were expressed in this strain, which resulted in production of 3.1 g/L ethanol in obtained recombinant strain (Hong et al. 2010).

Overexpression of homologous gene *ADH1* on the base of DL1 *adh1Δ* strain did not result in increase of ethanol production from glycerol (Suwannarangsee et al. 2010). However, simultaneous overexpression of the genes *PDC1* and *ADH1* in *O. polymorpha* strain NCYC495 led to the increase in ethanol production from glycerol. Obtained recombinant strain produced up to 5.0 g/L of ethanol from glycerol under the increased to 45 °C fermentation temperature (Kata et al. 2016). Further improvement of this strain was achieved by overexpression of genes involved either in oxidative or phosphorylative pathway of glycerol catabolism, as well as heterologous gene coding for glycerol transporter *FPS1* from *P. pastoris*. The resultant strains produced up to 10.7 g/L of ethanol from pure glycerol; however, during the fermentation of crude glycerol, ethanol production reached only up to 3.6 g/L of ethanol, probably due to inhibitory influence of impurities in crude glycerol (Semkiv et al., accepted).

Recently, the new species *P. tannophilus* has been isolated as a yeast organism capable of effective fermenting xylose to ethanol (Kurtzman 1983). It was reported that *P. tannophilus* could accumulate 4 g/L ethanol on glycerol under aerobic growth (Maleszka et al. 1982). *P. tannophilus* strain CBS4044 produced 17.5 g/L of ethanol from 5% (v/v) crude glycerol in bioreactor set to 450 rpm and 0.05 L/min airflow. During fed-batch fermentation, ethanol accumulation had reached 28.1 g/L (Liu et al. 2012). Also high fermentation efficiency (up to 90% yield of ethanol relative to the theoretical limit) has been shown for poly(vinyl alcohol) cryogel-immobilized cells of *P. tannophilus* strain Y-475 (Stepanov and Efremenko 2017).

The impurities (ash, methanol, salts, etc.) and variability of crude glycerol were not found to have negative effect on the viability and ethanol production of *P. tannophilus* (Liu et al. 2012). But it was shown that *P. tannophilus* ceased to grow when ethanol was added to the cultivation medium in concentration 40 g/L (Zhao et al. 2010). The ethanol tolerance of *P. tannophilus* needs to be improved, e.g., through adaptive evolution or UV mutagenesis and selection for more ethanol-tolerant strains (Watanabe et al. 2011). But overall, *P. tannophilus* is a robust microorganism which can be easily adapted for ethanol production from nonrefined crude glycerol feedstock.

To conclude, several attempts have been made to produce ethanol from glycerol using microbial fermentation, and in most cases, pure glycerol was used as substrate (Durnin et al. 2009; Gupta et al. 2009; Hong et al. 2010; Kata et al. 2016; Suwannarangsee et al. 2010; Yu et al. 2012; Yu et al. 2010).

12.4.2.4 Microbial Oil (Lipids or Triacylglycerol) and Lipase Production

Perhaps the most interesting and well-studied sphere of potential crude glycerol implementation is the production of microbial oil, which can be used as an alternative to vegetable oil in biodiesel industry. As was already mentioned, vegetable oils are the main raw materials for biodiesel production (Fig. 12.2). The cost of the oils accounts for 70–85% of the total biodiesel production cost (Miao and Wu 2006). Traditional oil-rich crops are cultivated using arable lands which triggers "food versus fuel" debate. Attempts are constantly being made to find new cheap, renewable, and nonedible feedstock for biodiesel production. A very promising potential alternative is represented by microbial lipids, also referred to as single-cell oils (SCO) (Khot et al. 2012). SCO can be used for the production of "third-generation" biodiesel and other oleochemicals (e.g., surfactants, lubricants) or as a source of valuable polyunsaturated fatty acids, which cannot be synthesized by mammals (Biermann et al. 2011; Fakas et al. 2006). SCO can be obtained from oleaginous microorganisms that accumulate substantial quantities of lipids in their cells (20–70% of dry cell weight) (Saenge et al. 2011). Oleaginous species were found among bacteria, yeast, filamentous fungi, and microalgae. Microalgae are the most intensively studied among them. Huge advantage of these organisms is the autotrophic nutrition type. They are able to utilize and sequester carbon dioxide from the atmosphere; therefore, they do not need any additional carbon source (Miao and Wu 2006). Lipid content in oleaginous marine algae cells varies from 20 to 50% (w/w) with fatty acids (FA) composition similar to vegetable oils, and they present higher yields per square meter of used lands than palm oil (Chisti 2007). However, in comparison to bacteria and yeasts, algae require larger acreages to cultivation and a continuous source of light to promote photosynthesis. Besides, they have longer cultivation period and lower biomass and lipids yield than, for example, oleaginous yeasts.

Of the known yeasts species, 3–10% belong to the oleaginous yeasts, including representatives of the genera *Yarrowia*, *Candida*, *Rhodotorula*, *Rhodosporidium*, *Cryptococcus*, *Sporidiobolus*, *Kodamaea*, *Pseudozyma*, *Trichosporon*, and *Lipomyces* (Ageitos et al. 2011). Many oleaginous yeast species were isolated from soil (Saenge et al. 2011) and plant surfaces (Clément-Mathieu et al. 2008). Yeasts accumulate lipids not only as constituents of the membrane but also in the form of triacylglycerols within intracellular lipid bodies (Fig. 12.6). They have similar FA content to that of many plant oils and exhibit shorter life cycle and higher growth rate and lipids production than microalgae (Li et al. 2008). Yeasts can be easily grown in bioreactor independently of location, climate, and season and harvested within a few days. Type and quantity of produced lipids can be customized by adjustment of cultivation medium composition or by metabolic engineering of yeast strains (Blazeck et al. 2014).

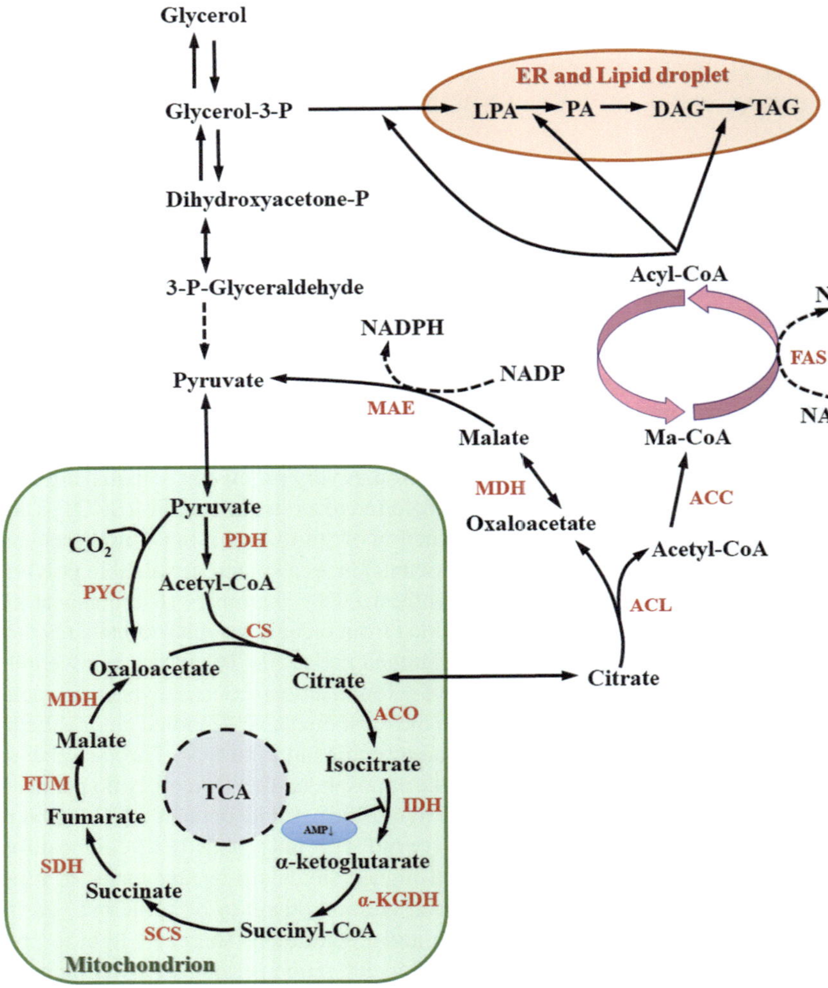

Fig. 12.6 Hypothetical pathway of glycerol conversion into lipids in oleaginous yeasts. Modified from (McNeil and Stuart 2018). Components of the TCA cycle designated as in Fig. 12.4. Key enzymes shown in red: *MAE* malic enzyme, *MDH* malate dehydrogenase, *ACL* ATP citrate lyase, *ACC* acetyl-CoA carboxylase, *FAS* fatty acid synthetase. Metabolites: *LPA* lysophosphatidic acid, *PA* phosphatidic acid, *DAG* diacylglycerol, *TAG* triacylglycerol, *Ma-CoA* malonyl-CoA

Despite all these advantages, yeast oil production is still limited to the lab scale due to its high cost (Koutinas et al. 2014). As heterotrophic organisms, yeasts need a carbon source for biomass and lipids production. The use of a low-cost carbon feedstock is an important step toward the feasible SCO production with yeasts. Many possible options of raw material were suggested over the years, such as cane molasses, fish meal wastewater, rice straw hydrolysate (Liu et al. 2016), spent yeast from brewery industry wastewater, sludge (Zhang et al. 2017), whey, wastewaters of animal fat treatment (Papanikolaou et al. 2002a), etc. Some of these applications not only provide a raw material with almost zero cost but also help to solve the problem of wastes disposal.

Many oleaginous yeasts are able to use glycerol as carbon source; hence, crude glycerol could also be used as a cheap feedstock for SCO production. Zhang et al. found that crude glycerol conversion to biodiesel is a process with a positive energy balance (Zhang et al. 2016). The maximum theoretical lipid yield is 0.3 g lipids/g glycerol since 32 moles of glycerol are used to produce 1 mole of triglyceride (Yang et al. 2014b); real yield is, of course, much lower and depends on many factors.

One of the most important factors is C/N ratio. Lipid accumulation in oleaginous yeasts is mainly triggered when carbon excess in the environment is associated with nitrogen limitation (André et al. 2009). A high C/N ratio enhances lipid accumulation, whereas biomass production is favored by a low C/N ratio (Saenge et al. 2011). For example, when *Lipomyces starkeyi* strain DSM 70295 was cultivated in the medium with C/N ratio of 150, obtained lipid content was 68% of dry cell weight (DCW), whereas with a C/N ratio of 60, a lipid content was only 40% (Angerbauer et al. 2008). This phenomenon is explained by the fact that nitrogen limitation activates the enzyme AMP-deaminase, which starts to degrade the mitochondrial AMP in order to release $NH4^+$ ions. The decrease of AMP inhibits the enzyme isocitrate dehydrogenase, blocking the TCA cycle and promoting the accumulation of acetyl-CoA which is subsequently used for the synthesis of fatty acids (Beopoulos et al. 2011).

It is also important to consider the type of used nitrogen source. Organic nitrogen sources, such as peptone or yeast extract, were shown to be more beneficial for cell growth and lipid accumulation by some yeast (*Trichosporon cutaneum*, *Trichosporon fermentans*, *Trichosporonoides spathulata*, *R. glutinis*), whereas for other yeast strains (*Cryptococcus albidus var. albidus* and *Y. lipolytica* QU21) inorganic nitrogen sources, such as NH_4Cl and $(NH_4)_2SO_4$, were preferable (Cheirsilp et al. 2011; Liu et al. 2016; Poli et al. 2014).

Other factors that can positively influence lipid accumulation were a controlled pH regime (Chen et al. 2018a; Manowattana et al. 2018; Saenge et al. 2011), lower dissolved oxygen level (Manowattana et al. 2018; Yen and Zhang 2011), and addition of inorganic salts such as sodium, calcium, potassium, or magnesium salt (Saenge et al. 2011) or organic acids such as acetic, succinic, or citric acid (Manowattana et al. 2018).

There are varied observations considering influence of impurities from crude glycerol on growth and lipids accumulation in oleaginous yeasts. Some studies reported that not all batches of crude glycerol are appropriate substrates for SCO production (Dobrowolski et al. 2016; Qiao et al. 2015). Others had found that impurities can positively affect cells growth and certain metabolite production (Chatzifragkou and Papanikolaou 2012; Signori et al. 2016). Gao et al. had found that addition of methyl oleate, sodium oleate, and NaCl impurities increased lipid production by oleaginous yeast *Rhodosporidium toruloides* 32489, whereas methanol had a negative effect on lipids accumulation, and the net effect of all studied compounds was positive (Gao et al. 2016b). The negative effect of methanol on the cell growth was also reported on *T. fermentans*, *T. cutaneum* (Liu et al. 2016), *L. starkeyi* (Liu et al. 2017), etc. However, Chen et al. used methanol in concentration 1.4% (w/v) to assist the lipid production with oleaginous yeast *Trichosporon oleaginosus* cultivated under non-sterilize conditions (Chen et al. 2018b). Overall, impurities from crude glycerol start to have a negative impact on yeast growth and lipids accumulation if their level exceeds certain threshold which is strain-specific. Yeast strain adaptation to stressful conditions can be one of the possible strategies for dealing with inhibitor problems (Shen et al. 2011). Other strategy is to develop proper cultivation methodology to avoid the growth inhibition by substrate (Signori et al. 2016). In this respect, fed-batch fermentation (when substrate is fed to the reactor through multiple steps) is proved to be superior to the batch fermentation. Fed-batch strategy allowed Koutinas et al. to obtain and achieve very high lipids accumulation: biomass concentration of 185 g/L with the 76% lipid content and lipid productivity of 1 g/L/h (Koutinas et al. 2014).

Yeast cells start to accumulate lipids in the growth phase, and this process intensifies after nitrogen depletion, so lipid content reaches its maximum value in the early stationary phase (Beopoulos et al. 2008). But later, lipids can be degraded into free FA, so well-timed biomass harvest is essential to maximize lipids content. It was suggested that lipid turnover can be repressed in double limited media (Papanikolaou et al. 2004), for example, in media with a very low concentration of nitrogen and magnesium (Bellou et al. 2016). Dynamics of lipids accumulation can also be modified by changing the initial yeast inoculum age (Kuttiraja et al. 2016).

FA content is the important characteristic of the SCO as a potential biodiesel feedstock. FA composition (i.e., the chain length and degree of unsaturation) of lipids influences the quality of biodiesel (Pinzi et al. 2009). Higher saturated fatty acid content promotes increase in viscosity, density, and melting point of the produced biodiesel. On the other hand, polyunsaturated fatty esters have low cetane number and reduced oxidative stability, which is also undesirable for a diesel fuel (Knothe 2008). Therefore, the best raw materials for biodiesel production are rich in monounsaturated fatty acids such as oleic acid. It was reported that yeast oil similarly to palm oil and Jatropha oil have the highest percentages of monounsaturated FA, which makes it perspective source for biodiesel production (Liang et al. 2010). The difference in fatty acid content among samples of yeast oils may occur due to the type of substrate that was used, culture conditions, and the age of the cells (Fakas et al. 2009).

The most common FA in oleaginous yeast are palmitic (C16:0), palmitoleic (C16:1), stearic (C18:0), oleic (C18:1), and linoleic acid (C18:2) that account for over 90% of the total FA content (Signori et al. 2016). Some studies report that using crude glycerol as a substrate for lipids production increases the content of the oleic acid (Ramirez-Castrillon et al. 2017; Signori et al. 2016) or linoleic acid (Spier et al. 2015). In particular, *R. glutinis* produces high amounts of linoleic acid when glycerol or, especially, crude glycerol is used as a carbon source (Easterling et al. 2009; Yen et al. 2012). As a polyunsaturated FA, linoleic acid is not very desirable for biodiesel production, but it is considered as a nutritionally essential fatty acid ω6. Other cultivation conditions can also influence the FA content: treatment of the oleaginous red yeast *Sporidiobolus pararoseus* KM281507 with intense light or supplementation of the medium with 1.5% olive oil increases the unsaturated FA content in this yeast (Chaiyaso and Manowattana 2018), *Y. lipolytica* goes through significant alteration of FA composition during the transition to the citric acid formation phase (Kamzolova et al. 2011), etc. Moreover, FA content is species- and even strain-dependent. The high monounsaturated FA content of crude lipids in oleaginous yeasts provides the possibility of the yeast oil use as a biodiesel feedstock in cold-climate countries.

The most extensively studied oleaginous yeasts are *Y. lipolytica* (Papanikolaou and Aggelis 2002), *R. toruloides* (Shen et al. 2013), and *R. glutinis* (Chi et al. 2011). Of these, *Y. lipolytica* is able to grow on various substrates, and it is regarded as the model organism to understand the lipid synthesis (Beopoulos et al. 2009). Initially, *Y. lipolytica* was considered to be able to accumulate significant lipid amounts only when fatty materials were employed as substrate (Papanikolaou et al. 2007). But later, it turned out that *Y. lipolytica* can use virtually any substrate for lipids production, including pure and crude glycerol. For example, *Y. lipolytica* strain UFLA CM-Y9.4 that was selected for its ability to grow in 30% crude glycerol accumulated 63.4% of lipids (w/w) in this conditions (Souza et al. 2014). Other *Y. lipolytica* strain, SKY7, under optimal cultivation conditions (initial glycerol concentration 112.5 g/L, C/N molar ratio of 100, and with 5% v/v inoculum supplementation) accumulated lipids to the concentration of 43.8% w/w with a biomass yield of 14.8 g/L (Kuttiraja et al. 2016). In some processes, employing *Y lipolytica* crude glycerol was used as co-substrate: strain JMY4086 accumulated 31% lipids of CDW (cell dry weight) using molasses and crude glycerol (Rakicka et al. 2015), and strain TISTR 5151 accumulated 68% lipids of CDW on decanter effluent from palm oil mill supplemented with crude glycerol (Louhasakul and Cheirsilp 2013).

The major fatty acid in the lipids produced with *Y lipolytica* is oleic acid (C18:1); palmitic (C16:0) and linoleic (C18:2) acids are also detected in high quantities (Makri et al. 2010; Poli et al. 2014). However, the level of SCO production in *Y. lipolytica* is moderate: wild-type strain usually accumulates no more than 30% CDW as neutral lipids (Munch et al. 2015). As was already mentioned, in stationary phase, *Y. lipolytica* starts to produce other important compounds (citric acid, acetic acid, mannitol) and, at the same time, partially degrades accumulated lipids (Papanikolaou et al. 2013). Therefore, it could be advantageous to use alternate oleaginous yeast species for the SCO production without side products.

Several studies aimed to find the strain with the best ability to convert crude or pure glycerol into SCO. When 12 different yeast strains were analyzed using crude glycerol as the main carbon source, *Lipomyces lipofer* NRRL Y-1155 stood out above the other strains, achieving 9.48 g/l biomass, 57.64% lipid content, and 5.46 g/l lipid production (Spier et al. 2015). When 33 yeasts strains belonging to 19 species were screened for the ability to grow and produce intracellular lipids in a pure glycerol-based medium, *Candida freyschussii* ATCC 18737 was selected. During cultivation with continuous feeding of crude glycerol at the rate of 5.5 g/L/h, it produced 28 g/L of lipids with volumetric productivity of 0.28 g/L/h (Raimondi et al. 2014). Two-step selection revealed 23 oleaginous yeasts among 387 yeast strains. These strains belonged to the species *Candida silvae, Kodamaea ohmeri, Meyerozyma caribbica, Pichia manshurica, Cryptococcus* cf. *podzolicus, Cryptococcus laurentii, Rhodosporidium fluviale, Rhodotorula taiwanensis* and *Sporidiobolus ruineniae*. *R. fluviale* DMKU-RK253 accumulated the highest quantity of lipids – 65.2% of CDW by shaking flask cultivation in crude glycerol (Polburee et al. 2015). When Nile red (a red phenoxazine dye, which selectively stains lipophilic substances) was used for a high-throughput screening, yeast *Meyerozyma (Pichia) guilliermondii* BI281A was selected. It was able to produce lipids with 74% monounsaturated FA content in the medium with crude glycerol from a biodiesel refinery (Ramirez-Castrillon et al. 2017). Strain TYC-2187 was isolated from wild grapes and identified as *Pseudozyma* sp. After 48 h of growth in the medium with 80 g/L crude glycerol and 10 g/L yeast extract as nitrogen source, it produced 15.7 g/L of lipids (Takakuwa et al. 2013). Probably the highest lipids content (up to 74% of CDW) was obtained for *R. toruloides* AS2.1389 in the medium with 50 g/L of crude glycerol (Xu et al. 2012). Other oleaginous yeast species with good ability to convert glycerol into SCO are *R. toruloides* DSM 4444, *L. starkeyi* DSM 70295 and *T. oleaginosus* (previously *C. curvatus*) DSM 70022 (Signori et al. 2016), *T. fermentans* CICC 1368 and *T. cutaneum* AS 2.0571 (Liu et al. 2016), *Wickerhamomyces anomalus* CCMA 0358 and *Cryptococcus humicola* CCMA 0346 (Souza et al. 2017), and *Rhodosporidium babjevae* and *Rhodosporidium diobovatum* (Munch et al. 2015). Yu et al. even made the attempt to produce triacylglycerol from glycerol with genetically modified non-oleaginous yeast *S. cerevisiae* with overexpression of genes encoding glycerol kinase (*GUT1*), diacylglycerol acyltransferase (*DGA1*), and phospholipid diacylglycerol acyltransferase (*LRO1*). Although lipids production in engineered strain increased in comparison with the WT strain, it still produced very low amounts of lipids – only 23.0 mg/L (Yu et al. 2013).

Crude glycerol was also used as co-substrate together with other substances that play role of the nitrogen source. For example, it was used in lipid production with such oleaginous yeasts and additional substrates: *R. glutinis* and waste solution collected from the brewing company (called thin stillage) (Yen et al. 2012), *T. oleaginosus* and corn steep liquor with recycled de-oiled yeast autolysate (Thiru et al. 2011), *R. glutinis* and rapeseed meal (Uckun Kiran et al. 2013), or sunflower meal (Leiva-Candia et al. 2015). The last process allows for simultaneous utilization of different waste materials obtained over the course of biodiesel production.

Glycerol could be used both as carbon source for the production of microbiological lipase or as a substrate for mono-, di-, and triacylglycerol production with this enzyme. For example, crude glycerol was used as a starting material for production of valuable mono-, di-, and triacylglycerol using commercial lipase from *Candida* sp. The optimum conditions for acylglycerol production were a glycerol to fatty acid molar ratio of 6:1, 100 mg of lipase with the reaction temperature and time of 40 °C and 24 h, respectively (Binhayeeding et al. 2017).

Lipases are now scarcely used as feed enzymes. Their wider use as feed additives could be beneficial for animal nutrition as hydrolysis of lipids increases their energy value and improves nutrient use efficiency (Magdouli et al. 2017). Wild-type strain of *Y. lipolytica* produces lipase, and its secretion is known to be induced by the presence of triacylglycerols and fatty acids in the cultivation medium (Benjamin and Pandey 1996). Lee et al. reported that *Y. lipolytica* strain NRRL Y-2178 is capable of alkaline lipase synthesis when glycerol is used as a carbon source (Lee et al. 2007a). Researchers had shown that microbiological lipase production from glycerol is only possible when one or more inducers of lipase synthesis (such as olive oil, waste cooking oil) are added to the medium (Fabiszewska et al. 2014; Goncalves et al. 2013).

When crude glycerol and crustacean waste and olive oil were used for the lipase production with *Y. lipolytica*, the obtained lipase activity was 38 U/mL (Magdouli et al. 2017).

To conclude, crude glycerol could be used together with other waste materials for the production of SCO and lipase with oleaginous yeasts.

12.4.2.5 Production of Other Compounds

Carotenoids Carotenoids are the naturally occurring pigments responsible for the bright red, yellow, or orange hues in many vegetables and fruits, some microorganisms, and animals. Carotenoids are precursor of vitamin A (Johnson and Schroeder 1996) and hormones (Vershinin 1999) and have photoprotective, antioxidant, and immunostimulating properties (Johnson and Schroeder 1996; Moline et al. 2009; Vershinin 1999). Carotenoids are widely used in medicine and in cosmetic, food, and feed industries. They can be synthesized chemically or by carotenoids accumulating microorganisms. Few of such microorganisms are already used for carotenoids production at the industrial scale, e.g., *Phaffia rhodozyma* (new designation *Xanthophyllomyces dendrorhous*) and *Haematococcus pluvialis* are used for astaxanthin production and *Blakeslea trispora* for β-carotene production (Dufosse 2006).

Already mentioned red oleaginous yeasts are one of the high potential natural carotenoids sources. In particular, carotenoids are produced by yeasts of the genera *Rhodosporidium*, *Rhodotorula*, *Sporobolomyces*, *Sporidiobolus*, and *Xanthophyllomyces* (Frengova and Beshkova 2009). Crude glycerol is appealing cheap substrate for carotenoids production with these yeasts.

When eighteen yeasts belonging to the species *Rhodotorula cresolica*, *R. glutinis*, *Rhodotorula mucilaginosa*, *Rhodosporidium paludigenum*, *R. toruloides*, *Sporobolomyces coprosmae*, *Sporobolomyces ruberrimus*, *Sporidiobolus salmonicolor*, and *Sporobolomyces oryzicola* were analyzed regarding their ability to grow in glycerol-containing medium, strains *R. glutinis* C2.5t1, *R. mucilaginosa* DBVPG 6094, *R. mucilaginosa* C71t0, *S. oryzicola* CBS 7228, and *R. paludigenum* CBS 6566 fully utilized glycerol within the first 48 h of fermentation (Cutzu et al. 2013). Surprisingly, carotenoids yields were higher in the medium with glycerol than in the medium with glucose as a sole carbon source, although glucose is a preferred carbon source for red yeasts (Cutzu et al. 2013; Taccari et al. 2012). Among selected strains, *R. glutinis* C2.5t1 showed the highest level of β-carotene production from glycerol; therefore, it was subjected to UV mutagenesis to further improve its production efficiency. Among the obtained mutants, one yellow-colored strain 400A15 produced 280% higher amounts of β-carotene at the expense of torulene and torularhodin production. After optimization of the medium composition, total carotenoids accumulation by the mutant 400A15 has reached 14.07 ± 1.45 mg/L (Cutzu et al. 2013).

Other research reports carotenoids production of 135.25 mg/L by *R. glutinis* strain TISTR 5159 in fed-batch fermentation in a stirred tank bioreactor under optimized conditions (pH controlled at 6.0, aeration rate at 2 vvm, glycerol concentration of 9.5%, and C/N ratio of 85) (Saenge et al. 2011). Also, glycerol was exploited as a carbon source for astaxanthin production with *P. rhodozyma* (Kusdiyantini et al. 1998) and torularhodin and β-carotene production by *S. ruberrimus* (Razani et al. 2007).

One of the highest carotenoids yields was reported for the strain *S. pararoseus* KM281507, which produced 109.75 ± 0.21 mg/L of β-carotene and 151.00 ± 2.71 mg/L of total carotenoids under optimized conditions in the airlift bioreactor (Manowattana et al. 2018). Researchers have found that β-carotene and total carotenoids production by this strain increased during cultivation under the uncontrolled pH regime (with pH gradually decreasing from 5.63 to 2.72) in contrast to controlled pH cultivation which augmented biomass and lipids production. Light can also improve carotenogenesis by oleaginous red yeast, as carotenoids protect yeast cells from light-induced cell damage (Mata-Gomez et al. 2014). Other factors that can positively influence carotenoids production were the high aeration rate and dissolved oxygen level (Manowattana et al. 2018), addition of the organic acids (e.g., acetic, succinic, and citric acid, which are the source of acetyl-CoA that is the precursor of carotenoids biosynthesis), and addition of olive oil or Tween 60 (Chaiyaso and Manowattana 2018). Impurities in crude glycerol, such as salts and methanol, can inhibit biomass accumulation and carotenoids production by red oleaginous yeasts. That's why very high initial concentration of crude glycerol in the medium should be avoided (Chaiyaso and Manowattana 2018).

γ-Decalactone The flavor compounds are commonly used in the food, cosmetic, and pharmaceutical industries. They can be produced by chemical synthesis or extracted from plants. Both these processes have disadvantages: chemical production is a complicated multistep process that can result in generation of undesirable racemic mixtures of compounds (Longo and Sanromán 2006), whereas plant extraction does not provide acceptable product yield and cost-efficiency. These factors increase the appeal of microbial fermentation as a cheap and ecological way of the flavor compounds' production (Romero-Guido et al. 2011). Among the important aroma producers are yeasts *Sporobolomyces, Pichia, Candida*, *Rhodotorula*, and *Yarrowia* (Braga and Belo 2016).

Lactones are the flavor chemicals with a characteristic "fruity" aroma. Among these compounds, γ-decalactone (with an aroma of peach) is the most widely produced (Pereira de Andrade et al. 2017). The microbial γ-decalactone production occurs mostly through peroxisomal β-oxidation of ricinoleic acid, which is the major constituent (80%) of castor oil. Crude glycerol is being considered as an alternative source for γ-decalactone production. Widely studied yeast *Y. lipolytica* as well as not so extensively studied yeast *L. saturnus* were tested regarding their ability to produce γ-decalactone from castor oil or crude glycerol (Pereira de Andrade et al. 2017; Soares et al. 2017). The yeast *L. saturnus* has been recently shown to produce isoamyl acetate (banana flavor) by isoamyl alcohol biotransformation in beet molasses (Yilmaztekin and Tay 2013). Although *Y. lipolytica* strain CCMA 0242 efficiently produced γ-decalactone from castor oil, it produced negligible amounts of γ-decalactone (2.5 mg/L) from crude glycerol. Therefore, crude glycerol showed no potential as a substrate for the production of aroma by *Y. lipolytica* CCMA 0242 (Pereira de Andrade et al. 2017). On the contrary, *L. saturnus* strain CCMA 0243 produced 5.8 g/L of γ-decalactone after 120 hr. of cultivation in the medium with 10% of crude glycerol, which was 2.7 times more γ-decalactone than it produced from 10% castor oil. Therefore, crude glycerol is a prospective substrate for γ-decalactone production by *L. saturnus* CCMA 0243 (Soares et al. 2017).

Sophorolipids Biosurfactants (BS) are surface-active compounds produced by microorganisms from sustainable feedstock, which are used as emulsifiers and detergents (Kitamoto et al. 2002). Sophorolipids (SL) are the glycolipid BSs which consist of a hydrophobic fatty acid tail and a hydrophilic carbohydrate head, sophorose. SL are produced from glucose and/or hydrophobic materials including vegetable oils, fatty acids, FAMEs, and alkanes by yeasts *Starmerella bombicola* (Ashby et al. 2006), *Candida apicola* (Hommel et al. 1994), *Wickerhamiella domercqiae* (Chen et al. 2006), *Candida batistae* (Konishi et al. 2008), etc. *S. (Candida) bombicola* is considered to be the conventional SL-producing yeast. It was shown that *S. bombicola* produces insignificant amounts of SL during fermentation on pure glycerol, but the use of the biodiesel coproduct steam (which comprised 40% glycerol, 34% hexane-soluble substrates, and 26% water) as feedstock increased the SL yield to 60 g/L (Ashby and Solaiman 2010).

New strain ZM1502 was isolated from withered leaves by its ability to produce BS from glycerol, and it was identified as *Candida floricola*. It produced only acid-form SLs, whereas *S. bombicola* produces mainly lactone-form SLs with small amounts of acid-form SLs (Konishi et al. 2017). The absence of the lactone forms in the metabolites of *C. floricola* is likely due to lack of a specific lactone esterase. *C. floricola* strains ZM1502 and CBS 7290 produced more than 3.5 g/L acid-form SLs from 20% glycerol. These results suggest that *C. floricola* could be used for selective production of acid-form SLs (which are in some aspects superior to lactone-form SLs) from crude glycerol (Konishi et al. 2017).

Heterologous Proteins The methylotrophic yeast *P. pastoris* has many traits which make it a remarkable host for heterologous protein production, among them being the following: (1) similarly to bacteria, *P. pastoris* grows fast and can be easily subjected to genetic manipulations; (2) it has the subcellular machinery that provides characteristic eukaryotic posttranslational modifications, such as proteolytic processing, glycosylation, folding, and disulfide bond formation; (3) it is more inclined to respiratory rather than fermentative metabolism, even at high-biomass level, which allows its growth to high cellular concentration on defined minimal medium and prevents the accumulation of unfavorable secondary metabolites as acetic acid and ethanol; (4) *P. pastoris* secretes very low levels of native proteins; therefore, secreted recombinant protein usually comprises the majority of the total protein in the medium, which greatly facilitates its purification; (5) *P. pastoris* is a GRAS (generally recognized as safe) organism; and (6) it has strong constitutive or inducible promoters which provide efficient expression of heterologous genes (Cereghino et al. 2002; Cereghino and Cregg 2000; Cregg et al. 2000). In particular, there are two efficient promoters which are commonly used for foreign gene expression in *P. pastoris*: strong constitutive promoter of glyceraldehyde-3-phosphate dehydrogenase gene (*GAP1*) (Jiang et al. 2012) and methanol-inducible promoter from the alcohol oxidase I gene (*AOX1*) that is strongly repressed in the presence of glucose and glycerol but induced over 1000-fold with methanol as the sole carbon source (Macauley-Patrick et al. 2005).

P. pastoris consumes glycerol very efficiently due to the presence of four genes coding for glycerol transporters which provide high, specific glycerol uptake rates (Mattanovich et al. 2009). That is why glycerol is often used as a carbon source in *P. pastoris* fermentations. Untreated crude glycerol can substitute analytical glycerol in this process as *P. pastoris* is able to tolerate and use methanol and its growth is not inhibited by free fatty acids (Anastacio et al. 2014).

Heterologous protein production with crude glycerol as a carbon source can consist of one or two stages. When p*GAP1* promoter is used for heterologous gene expression, protein is synthesized in a single step at the same time with biomass accumulation, in crude glycerol-containing medium. This is the faster way; however, some studies have indicated that constitutive expression of recombinant proteins could generate cytotoxic effects in *P. pastoris* (Cereghino and Cregg 2000; Macauley-Patrick et al. 2005). If the recombinant yeast strains had the heterologous gene under the control of the inducible p*AOX1* promoter, a two-stage process is employed: first, biomass is produced in the complex medium with crude glycerol, and then heterologous gene production is carried out in minimal medium with methanol addition (Anastacio et al. 2014). While the former medium containing rich nutritional supplements supported high growth of yeast before induction, the latter medium ensured easier purification of the secreted protein (Aoki et al. 2003).

Crude glycerol was used as a carbon source for one- or two-stage production of the following heterologous proteins: phytase (Tang et al. 2009), recombinant human erythropoietin (Çelik et al. 2008), bovine chymosin (Noseda et al. 2014; Noseda et al. 2016), cysteine proteinase (NsCys) of northern shrimp *Pandalus borealis* (Aoki et al. 2003), highly thermostable β-mannanase (ReTMan26) from a thermophilic *Bacillus subtilis* (TBS2) (Luo et al. 2018), α-amylase (Anastacio et al. 2014), etc.

In some of these cases, impurities from crude glycerol, such as 0.2% and 0.3% (w/v) soap (Luo et al. 2018) or NaCl, KCl, and K_2SO_4 salts (Anastacio et al. 2014), were found to inhibit *P. pastoris* growth and heterologous protein production, necessitating the corresponding adjustments of crude glycerol concentration in the medium. But overall, it can be concluded that crude glycerol without any purification steps may be directly used as carbon source for protein production in *P. pastoris* (Anastacio et al. 2014).

Amino Acids-Rich Biomass Yeast biomass itself is a valuable product as it can be used as a nutrient-rich additive for animal feeding. Important parameters that influence nutritional value of fodder yeasts are protein content (with recommended level of 40–52%), the content of essential amino acids (EAA), the amount of polyunsaturated fatty acids (PUFA), and the content of calcium, magnesium, copper, iron, zinc, etc. (Boze et al. 2008).

It has been shown that crude glycerol can be used for fodder yeast production by *Y. lipolytica* with good yield and productivity (Juszczyk and Rymowicz 2009). The European Feed Manufacturers' Federation authorized the sale of *Y. lipolytica* fodder yeast produced from crude glycerol.

Juszczyk et al. among 21 *Y. lipolytica* strains isolated from different environments selected strain S6 with the highest level of biomass accumulation on glycerol (Juszczyk et al. 2013). This strain was used for production of yeast biomass in bioreactor (pH 3.5) with pure glycerol or crude glycerol (in concentration 25 g/L) as a carbon source. Volumetric biomass production was 11.7 g/L from pure glycerol and 12.3 g/L from crude glycerol. Yeast biomass obtained on crude glycerol was characterized by higher content of proteins (42–45%), essential amino acids (45.4 g/100 g of protein), and ash (i.e., K, Na, Mg, Ca, Cu, Zn) in comparison with the biomass obtained on pure glycerol. All biomass samples were characterized by high content of unsaturated fatty acids. Sample from crude glycerol contained higher amounts of lysine, threonine, and phenylalanine/tyrosine than the FAO/WHO standard of whole egg. However, the amount of sulfuric amino acids (methionine and cysteine) in strain S6 biomass was much lower than in the whole egg, which decreased its nutritional value. Therefore, biomass of strain S6 might be suitable for fodder production when compiled with cereals, which are known to contain high levels of sulfuric amino acids but low amounts of lysine, isoleucine, and threonine. In conclusion, crude glycerol is superior to pure glycerol as a carbon source in respect of produced *Y. lipolytica* biomass amount and content (Juszczyk et al. 2013).

12.5 Conclusions

Biodiesel industry produces huge amounts of contaminated crude glycerol as a by-product. Nowadays, in some countries, crude glycerol is treated as industrial wastewater or simply incinerated, undermining the right of biodiesel to be called "green fuel." This review describes different possible applications of crude glycerol in yeast biotechnology (some of them are summarized in Table 12.1), among which crude glycerol conversion to the new portions of biodiesel is, perhaps, the most inventive and feasible. But none of these methods has been implemented at large scale yet, so this sphere needs further development.

Table 12.1 Some of the compounds that could be produced by yeasts from crude or pure glycerol. Product yield is expressed through g/g of accumulated biomass, denoted[b], or g/g of consumed glycerol, denoted[s]

Compound	Organism	Carbon source	Culture system	Total amount produced (g/L)	Product yield (g/g)	References
Citric acid	*Y. lipolytica* N15	Crude glycerol (100 g/L)	Batch	71	0.9[s]	Kamzolova et al. (2011)
	Y. lipolytica NG40/UV7	Crude glycerol (pulse added)	Fed-batch	112	0.9[s]	Morgunov et al. (2013)
	Y. lipolytica AJD pADUTGut1	Pure glycerol (150 g/L)	Batch	60.4	0.4[s]	Mironczuk et al. (2016)
	C. parapsilosis ATCC 7330	Crude glycerol (60 g/L)	Shake flask	11.3	0.22[s]	West (2013)
Isocitric acid	*Y. lipolytica* AJD pADUTGut1/2	Crude glycerol (150 g/L)	Batch	42.5	0.28[s]	Rzechonek et al. (2019)
Pyruvic acid	*Y. lipolytica* 374/4	Pure glycerol (30 + 30 + 30 g/L)	Fed-batch	61.3	0.71[s]	Morgunov et al. (2004)
α-Ketoglutaric acid	*Y. lipolytica* H355A(FUM1)	Crude glycerol (150 + 123 g/L)	Fed-batch	134.1	0.47[s]	Otto et al. (2012)
	Y. lipolytica-RoPYC2	Pure glycerol (100 g/L)	Two-stage batch	62.5	5.17[b]	Yin et al. (2012)
Succinic acid	*Y. lipolytica* PGC01003 (*Ylsdh5*Δ)	Crude glycerol (100 g/L + 6 additional feedings)	Fed-batch	160.2	0.4[s]	Gao et al. (2016a)
		Crude glycerol (120 g/L)	Batch (*is*FBB)	53.6	0.45[s]	Li et al. (2018a)
Lactic acid	*P. pastoris* GLS (*his4*⁻ + pGAP-LDH+ pPGK-PAS)	Pure glycerol (20 + 40 g/L)	Two-stage fed-batch	28.3	0.7[s]	de Lima et al. (2016)
Erythritol	*Y. lipolytica* Wratislavia K1	Crude glycerol (300 g/L)	Fed-batch	170	0.56[s]	Rymowicz et al. (2009)
	Y. lipolytica A16	Crude glycerol (232.39 g/L)	Batch	109.2	0.47[s]	Yang et al. (2016)

Mannitol	Resting cells of *C. magnoliae* (NCIM 3470)	Pure glycerol (100 g/L)	Shake flask	51	0.50[s]	Khan et al. (2009)
	Y. lipolytica A UV'1	Pure glycerol (150 g/L)	Batch	27.6	0.16[s]	Tomaszewska et al. (2012)
Arabitol	*D. hansenii* SBP-1	Crude glycerol (150 g/L)	Shake flask	~14	0.5[s]	Koganti et al. (2011)
Ethanol	*S. cerevisiae* YPH499 *fps1Δ gpd2Δ (pGcyaDak AdhPdc, pGupCas)*	Pure glycerol (20 g/L)	Shake flask	5.42	0.31[s]	Yu et al. (2012)
	O. polymorpha NCYC495/ADH1/PDC1/GCY1/DAK1	Pure glycerol (150 g/L)	Shake flask	10.7	0.13[s]	Semkiv et al. (accepted)
	P. tannophilus CBS4044	Crude glycerol 5% (v/v)	Batch	17.5	0.28[s]	Liu et al. (2012)
Lipids	*R. glutinis* TISTR 5159	Crude glycerol (95 g/L)	Batch	4.33	0.53[b]	Saenge et al. (2011)
	S. pararoseus KM281507	Crude glycerol (55 g/L)	Batch	6.58	0.50[b]	Manowattana et al. (2018)
	Y. lipolytica SKY7	Crude glycerol (112.5 g/L)	Batch	6.36	0.43[b] 0.192[s]	Kuttiraja et al. (2016)
	L. Lipofer NRRL Y-1155	Crude glycerol (60.9 g/L)	Shake flask	5.46	0.58[b]	Spier et al. (2015)
	C. freyschussii ATCC 18737	Crude glycerol (40 g/L + continuous feeding 5.5 g/l/h)	Continuous	28	0.34[b]	Raimondi et al. (2014)
	Pseudozyma sp. TYC-2187	Crude glycerol (80 g/L)	Shake flask	15.7	0.55[b]	Takakuwa et al. (2013)
	R. toruloides AS2.1389	Crude glycerol A (60 g/L)	Batch	18.5	0.70[b]	Xu et al. (2012)
		Crude glycerol B (60 g/L)		13.4	0.74[b]	

(continued)

Table 12.1 (continued)

Compound	Organism	Carbon source	Culture system	Total amount produced (g/L)	Product yield (g/g)	References
Carotenoids	*R. glutinis* TISTR 5159	Crude glycerol (95 g/L)	Batch	0.125	0.015[b]	Saenge et al. (2011)
	S. pararoseus KM281507	Crude glycerol (55 g/L)	Batch	0.151	0.009[b]	Manowattana et al. (2018)
γ-Decalactone	*L. saturnus* CCMA 0243	Crude glycerol (100 g/L)	Shake flask	5.8	n.a.	Soares et al. (2017)
Sophorolipids	*C. floricola* ZM1502	Pure glycerol (200 g/L)	Shake flask	3.5	n.a.	Konishi et al. (2017)
Biomass	*Y. lipolytica* S6	Crude glycerol (25 g/L)	Batch	12.3	0.52[s]	Juszczyk et al. (2013)

References

Abdel-Rahman MA, Tashiro Y, Sonomoto K (2013) Recent advances in lactic acid production by microbial fermentation processes. Biotechnol Adv 31:877–902. https://doi.org/10.1016/j.biotechadv.2013.04.002

Adler L, Blomberg A, Nilsson A (1985) Glycerol metabolism and osmoregulation in the salt-tolerant yeast Debaryomyces hansenii. J Bacteriol 162:300–306

Ageitos JM, Vallejo JA, Veiga-Crespo P, Villa TG (2011) Oily yeasts as oleaginous cell factories. Appl Microbiol Biotechnol 90:1219–1227. https://doi.org/10.1007/s00253-011-3200-z

Aksoy HA, Becerik I, Karaosmanoglu F, Yatmaz HC, Civelekoglu H (1990) Utilization prospects of Turkish raisin seed oil as an alternative engine fuel. Fuel 69:600–603

Alvarez JA, Otero L, Lema JM (2010) A methodology for optimising feed composition for anaerobic co-digestion of agro-industrial wastes. Bioresour Technol 101:1153–1158. https://doi.org/10.1016/j.biortech.2009.09.061

Al-Widyan MI, Al-Shyoukh AO (2002) Experimental evaluation of the transesterification of waste palm oil into biodiesel. Bioresour Technol 85:253–256

Anand P, Saxena RK (2012) A comparative study of solvent-assisted pretreatment of biodiesel derived crude glycerol on growth and 1,3-propanediol production from Citrobacter freundii. New Biotechnol 29:199–205. https://doi.org/10.1016/j.nbt.2011.05.010

Anastacio GS, Santos KO, Suarez PA, Torres FA, De Marco JL, Parachin NS (2014) Utilization of glycerin byproduct derived from soybean oil biodiesel as a carbon source for heterologous protein production in Pichia pastoris. Bioresour Technol 152:505–510. https://doi.org/10.1016/j.biortech.2013.11.042

André A et al (2009) Biotechnological conversions of bio-diesel-derived crude glycerol by Yarrowia lipolytica strains. Eng Life Sci 9:468–478. https://doi.org/10.1002/elsc.200900063

Angerbauer C, Siebenhofer M, Mittelbach M, Guebitz GM (2008) Conversion of sewage sludge into lipids by Lipomyces starkeyi for biodiesel production. Bioresour Technol 99:3051–3056. https://doi.org/10.1016/j.biortech.2007.06.045

Aoki H, Nazmul Ahsan M, Watabe S (2003) Heterologous expression in Pichia pastoris and single-step purification of a cysteine proteinase from northern shrimp. Protein Expr Purif 31:213–221

Arruda PV, Hernández AF, Felipe MGA (2015) Glycerol use as strategy to improve xylitol production from sugarcane bagasse hemicellulosic hydrolysate. Eur Intl J Sci Technol 4:93–104

Ashby RD, Solaiman DK (2010) The influence of increasing media methanol concentration on sophorolipid biosynthesis from glycerol-based feedstocks. Biotechnol Lett 32:1429–1437. https://doi.org/10.1007/s10529-010-0310-0

Ashby RD, Solaiman DK, Foglia TA (2006) The use of fatty acid esters to enhance free acid sophorolipid synthesis. Biotechnol Lett 28:253–260. https://doi.org/10.1007/s10529-005-5527-y

Athalye SK, Garcia RA, Wen Z (2009) Use of biodiesel-derived crude glycerol for producing eicosapentaenoic acid (EPA) by the fungus Pythium irregulare. J Agric Food Chem 57:2739–2744. https://doi.org/10.1021/jf803922w

Beaudoin F, Sayanova O, Haslam RP, Bancroft I, Napier JA (2014) Oleaginous crops as integrated production platforms for food, feed, fuel and renewable industrial feedstock. OCL 21:D606

Bellou S, Triantaphyllidou IE, Mizerakis P, Aggelis G (2016) High lipid accumulation in Yarrowia lipolytica cultivated under double limitation of nitrogen and magnesium. J Biotechnol 234:116–126. https://doi.org/10.1016/j.jbiotec.2016.08.001

Benjamin S, Pandey A (1996) Optimization of liquid media for lipase production by *Candida rugosa*. Bioresour Technol 55:167–170

Beopoulos A et al (2008) Control of lipid accumulation in the yeast Yarrowia lipolytica. Appl Environ Microbiol 74:7779–7789. https://doi.org/10.1128/AEM.01412-08

Beopoulos A, Chardot T, Nicaud JM (2009) Yarrowia lipolytica: a model and a tool to understand the mechanisms implicated in lipid accumulation. Biochimie 91:692–696. https://doi.org/10.1016/j.biochi.2009.02.004

Beopoulos A, Nicaud JM, Gaillardin C (2011) An overview of lipid metabolism in yeasts and its impact on biotechnological processes. Appl Microbiol Biotechnol 90:1193–1206. https://doi.org/10.1007/s00253-011-3212-8

Bernt WO, Borzelleca JF, Flamm G, Munro IC (1996) Erythritol: a review of biological and toxicological studies regulatory toxicology and pharmacology. RTP 24:S191–S197. https://doi.org/10.1006/rtph.1996.0098

Bertleff W, Roeper M, Sava X (2005) Carbonylation. In: Ullmann's encyclopedia of industrial chemistry. Wiley-VCH, Weinheim. doi:https://doi.org/10.1002/14356007.a05_217

Biebl H, Zeng AP, Menzel K, Deckwer WD (1998) Fermentation of glycerol to 1,3-propanediol and 2,3-butanediol by Klebsiella pneumoniae. Appl Microbiol Biotechnol 50:24–29

Biermann U, Bornscheuer U, Meier MAR, Metzger JO, Schäfer HJ (2011) Oils and fats as renewable raw materials in chemistry. Angew Chem Int Ed 50:3854–3871. https://doi.org/10.1002/anie.201002767

Binhayeeding N, Klomklao S, Sangkharak K (2017) Utilization of waste glycerol from biodiesel process as a substrate for mono-, Di-, and triacylglycerol production. Energy Procedia 138:895–900

Blazeck J et al (2014) Harnessing Yarrowia lipolytica lipogenesis to create a platform for lipid and biofuel production. Nat Commun 5:3131. https://doi.org/10.1038/ncomms4131

Blomberg A (2000) Metabolic surprises in Saccharomyces cerevisiae during adaptation to saline conditions: questions, some answers and a model. FEMS Microbiol Lett 182:1–8

Bohon MD, Metzger BA, Linak WP, King CJ, Roberts WL (2011) Glycerol combustion and emissions. Proc Combust Inst 33:2717–2724

Borsato D, Maia ECR, Dall'Antonia LH, da Silva HC, Pereira JL (2012) Kinetics of oxidation of biodiesel from soybean oil mixed with TBHQ: determination of storage time. Química Nova 35:733–737

Boze H, Moulin G, Galzy P (2008) Production of microbial biomass. In: Rehm HJ, Reed G (eds) Biotechnology: enzymes, biomass, food and feed, vol 9, 2nd edn. Wiley-VCH Verlag GmbH, Weinheim. https://doi.org/10.1002/9783527620920.ch5

Braga A, Belo I (2016) Biotechnological production of gamma-decalactone, a peach like aroma, by Yarrowia lipolytica. World J Microbiol Biotechnol 32:169. https://doi.org/10.1007/s11274-016-2116-2

Brown R (2007) Biodiesel co-product markets in Wyoming. International center for appropriate & sustainable technology, Lakewood

Byrd AJ, Pant KK, Gupta RB (2008) Hydrogen production from glycerol by reforming in supercritical water over Ru/Al2O3 catalyst. Fuel 87:2956–2960

Carly F, Gamboa-Melendez H, Vandermies M, Damblon C, Nicaud JM, Fickers P (2017) Identification and characterization of EYK1, a key gene for erythritol catabolism in Yarrowia lipolytica. Appl Microbiol Biotechnol 101:6587–6596. https://doi.org/10.1007/s00253-017-8361-y

Casali S, Gungormusler M, Bertin L, Fava F, Azbar N (2012) Development of a biofilm technology for the production of 1,3-propanediol (1,3-PDO) from crude glycerol. Biochem Eng J 64:84–90

Cecchini G (2003) Function and structure of complex II of the respiratory chain. Annu Rev Biochem 72:77–109. https://doi.org/10.1146/annurev.biochem.72.121801.161700

Çelik E, Ozbay N, Oktar N, Çalık P (2008) Use of biodiesel byproduct crude glycerol as the carbon source for fermentation processes by recombinant Pichia pastoris. Ind Eng Chem Res 47:2985–2990. https://doi.org/10.1021/ie071613o

Cereghino JL, Cregg JM (2000) Heterologous protein expression in the methylotrophic yeast Pichia pastoris. FEMS Microbiol Rev 24:45–66. https://doi.org/10.1111/j.1574-6976.2000.tb00532.x

Cereghino GP, Cereghino JL, Ilgen C, Cregg JM (2002) Production of recombinant proteins in fermenter cultures of the yeast Pichia pastoris. Curr Opin Biotechnol 13:329–332

Cerrate S, Yan F, Wang Z, Coto C, Sacakli P, Waldroupand PW (2006) Evaluation of glycerine from biodiesel production as a feed ingredient for broilers. Int J Poult Sci 5:1001–1007

Chaiyaso T, Manowattana A (2018) Enhancement of carotenoids and lipids production by oleaginous red yeast Sporidiobolus pararoseus KM281507. Prep Biochem Biotechnol 48:13–23. https://doi.org/10.1080/10826068.2017.1381620

Chatzifragkou A, Papanikolaou S (2012) Effect of impurities in biodiesel-derived waste glycerol on the performance and feasibility of biotechnological processes. Appl Microbiol Biotechnol 95:13–27. https://doi.org/10.1007/s00253-012-4111-3

Chatzifragkou A et al (2011) Biotechnological conversions of biodiesel derived waste glycerol by yeast and fungal species. Energy 36:1097–1108

Cheirsilp B, Suwannarat W, Niyomdecha R (2011) Mixed culture of oleaginous yeast Rhodotorula glutinis and microalga Chlorella vulgaris for lipid production from industrial wastes and its use as biodiesel feedstock. New Biotechnol 28:362–368. https://doi.org/10.1016/j.nbt.2011.01.004

Chen J, Song X, Zhang H, Qu Y (2006) Production, sturucture elucidation and anticancer properties of sophololipid from *Wickerhamiella domercqiae*. Enzym Microb Technol 39:501–506

Chen J, Zhang X, Drogui P, Tyagi RD (2018a) The pH-based fed-batch for lipid production from Trichosporon oleaginosus with crude glycerol. Bioresour Technol 259:237–243. https://doi.org/10.1016/j.biortech.2018.03.045

Chen J, Zhang X, Tyagi RD, Drogui P (2018b) Utilization of methanol in crude glycerol to assist lipid production in non-sterilized fermentation from Trichosporon oleaginosus. Bioresour Technol 253:8–15. https://doi.org/10.1016/j.biortech.2018.01.008

Cheng L, Liu L, Ye XP (2013) Acrolein production from crude glycerol in sub- and super-critical water. J Am Oil Chem Soc 90:601–610. https://doi.org/10.1007/s11746-012-2189-5

Cheng D, Wang L, Shahbazi A, Xiu S, Zhang B (2014) Characterization of the physical and chemical properties of the distillate fractions of crude bio-oil produced by the glycerol-assisted liquefaction of swine manure. Fuel 130:251–256

Chi Z, Zheng Y, Jiang A, Chen S (2011) Lipid production by culturing oleaginous yeast and algae with food waste and municipal wastewater in an integrated process. Appl Biochem Biotechnol 165:442–453. https://doi.org/10.1007/s12010-011-9263-6

Chisti Y (2007) Biodiesel from microalgae. Biotechnol Adv 25:294–306. https://doi.org/10.1016/j.biotechadv.2007.02.001

Chiu CW, Dasari MA, Sutterlin WR, Suppes GJ (2006) Removal of residual catalyst from simulated biodiesel's crude glycerol for glycerol hydrogenolysis to propylene glycol. Ind Eng Chem Res 45:791–795

Choi WJ, Hartono MR, Chan WH, Yeo SS (2011) Ethanol production from biodiesel-derived crude glycerol by newly isolated Kluyvera cryocrescens. Appl Microbiol Biotechnol 89:1255–1264. https://doi.org/10.1007/s00253-010-3076-3

Chookaew T, Prasertsan P, Ren ZJ (2014) Two-stage conversion of crude glycerol to energy using dark fermentation linked with microbial fuel cell or microbial electrolysis cell. New Biotechnol 31:179–184. https://doi.org/10.1016/j.nbt.2013.12.004

Chung YH, Rico DE, Martinez CM, Cassidy TW, Noirot V, Ames A, Varga GA (2007) Effects of feeding dry glycerin to early postpartum Holstein dairy cows on lactational performance and metabolic profiles. J Dairy Sci 90:5682–5691. https://doi.org/10.3168/jds.2007-0426

Ciriminna R, Meneguzzo F, Delisi R, Pagliaro M (2017) Citric acid: emerging applications of key biotechnology industrial product. Chem Cent J 11:22. https://doi.org/10.1186/s13065-017-0251-y

Claude S (1999) Research of new outlets for glycerol—recent developments in France. Lipid Sci Technol 101:101–104

Clément-Mathieu G, Chain F, Marchand G, Bélanger R (2008) Leaf and powdery mildew colonization by glycolipid-producing Pseudozyma species. Fungal Ecol 1:69–77

Clomburg JM, Gonzalez R (2013) Anaerobic fermentation of glycerol: a platform for renewable fuels and chemicals. Trends Biotechnol 31:20–28. https://doi.org/10.1016/j.tibtech.2012.10.006

Cregg JM, Cereghino JL, Shi J, Higgins DR (2000) Recombinant protein expression in Pichia pastoris. Mol Biotechnol 16:23–52. https://doi.org/10.1385/MB:16:1:23
Cutzu R et al (2013) From crude glycerol to carotenoids by using a *Rhodotorula glutinis* mutant. World J Microbiol Biotechnol 29:1009–1017. https://doi.org/10.1007/s11274-013-1264-x
Dasari MA, Kiatsimkul PP, Sutterlin WR, Suppes GJ (2005) Lowpressure hydrogenolysis of glycerol to propylene glycol. Appl Catal a-General 281:225–231
Datta R, Henry M (2006) Lactic acid: recent advances in products, processes and technologies – a review. J Chem Technol Biotechnol 81:1119–1129. https://doi.org/10.1002/jctb.1486
De Ley J, Frateur J (1970) The status of the generic name Gluconobacter. Int J Syst Bacteriol 20:83–95
de Lima PB et al (2016) Novel homologous lactate transporter improves L-lactic acid production from glycerol in recombinant strains of Pichia pastoris. Microb Cell Factories 15:158. https://doi.org/10.1186/s12934-016-0557-9
DeFrain JM, Hippen AR, Kalscheur KF, Jardon PW (2004) Feeding glycerol to transition dairy cows: effects on blood metabolites and lactation performance. J Dairy Sci 87:4195–4206. https://doi.org/10.3168/jds.S0022-0302(04)73564-X
Delgado R, Rosas JG, Gómez N, Martínez O, Sanchez ME, Cara J (2013) Energy valorisation of crude glycerol and corn straw by means of slow co-pyrolysis: production and characterisation of gas, char and bio-oil. Fuel 112:31–37. https://doi.org/10.1016/j.fuel.2013.05.005
Demirbas A (2009) Fuels from biomass. In: Biohydrogen: For future engine fuel demands. Springer, London, pp 43–59. https://doi.org/10.1007/978-1-84882-511-6_2
Dobroth ZT, Hu S, Coats ER, McDonald AG (2011) Polyhydroxybutyrate synthesis on biodiesel wastewater using mixed microbial consortia. Bioresour Technol 102:3352–3359. https://doi.org/10.1016/j.biortech.2010.11.053
Dobrowolski A, Mitula P, Rymowicz W, Mironczuk AM (2016) Efficient conversion of crude glycerol from various industrial wastes into single cell oil by yeast Yarrowia lipolytica. Bioresour Technol 207:237–243. https://doi.org/10.1016/j.biortech.2016.02.039
Dosuna-Rodríguez I, Gaigneaux EM (2012) Glycerol acetylation catalysed by ion exchange resins. Catal Today 195:14–21
Dufosse L (2006) Microbial production of food grade pigments. Food Technol Biotechnol 44:313–321
Durnin G, Clomburg J, Yeates Z, Alvarez PJ, Zygourakis K, Campbell P, Gonzalez R (2009) Understanding and harnessing the microaerobic metabolism of glycerol in Escherichia coli. Biotechnol Bioeng 103:148–161. https://doi.org/10.1002/bit.22246
Easterling ER, French WT, Hernandez R, Licha M (2009) The effect of glycerol as a sole and secondary substrate on the growth and fatty acid composition of *Rhodotorula glutinis*. Bioresour Technol 100:356–361. https://doi.org/10.1016/j.biortech.2008.05.030
Echeverri DA, Perez WA, Rios LA (2013) Synthesis of maleated-castor oil glycerides from biodiesel-derived crude glycerol. Ind Crop Prod 49:299–303
Egermeier M, Russmayer H, Sauer M, Marx H (2017) Metabolic flexibility of *Yarrowia lipolytica* growing on glycerol. Front Microbiol 8:49. https://doi.org/10.3389/fmicb.2017.00049
Ethier S, Woisard K, Vaughan D, Wen Z (2011) Continuous culture of the microalgae Schizochytrium limacinum on biodiesel-derived crude glycerol for producing docosahexaenoic acid. Bioresour Technol 102:88–93. https://doi.org/10.1016/j.biortech.2010.05.021
Fabiszewska AU, Stolarzewicz I, Zamojska W, Białecka-Florjańczyk E (2014) Carbon sources influencing profile of lipases produced by yeast Yarrowia lipolytica. Appl Biochem Microbiol 50:404–441
Fakas S, Papanikolaou S, Galiotou-Panayotou M, Komaitis M, Aggelis G (2006) Lipids of Cunninghamella echinulata with emphasis to gamma-linolenic acid distribution among lipid classes. Appl Microbiol Biotechnol 73:676–683. https://doi.org/10.1007/s00253-006-0506-3
Fakas S, Papanikolaou S, Batsos A, Galiotou-Panayotou M, Mallouchos A, Aggelis G (2009) Evaluating renewable carbon sources as substrates for single cell oil production by

Cunninghamella echinulata and Mortierella isabellina. Biomass Bioenergy 33(4):573–580. https://doi.org/10.1016/j.biombioe.2008.09.006

Feng Y, Yang Q, Wang X, Liu Y, Lee H, Ren N (2011) Treatment of biodiesel production wastes with simultaneous electricity generation using a single-chamber microbial fuel cell. Bioresour Technol 102:411–415. https://doi.org/10.1016/j.biortech.2010.05.059

Ferreira C et al (2005) A member of the sugar transporter family, Stl1p is the glycerol/H+ symporter in *Saccharomyces cerevisiae*. Mol Biol Cell 16:2068–2076. https://doi.org/10.1091/mbc.E04-10-0884

Fickers P, Benetti PH, Wache Y, Marty A, Mauersberger S, Smit MS, Nicaud JM (2005) Hydrophobic substrate utilisation by the yeast Yarrowia lipolytica, and its potential applications. FEMS Yeast Res 5:527–543. https://doi.org/10.1016/j.femsyr.2004.09.004

Finogenova TV et al (2002) Biosynthesis of citric and isocitric acids from ethanol by mutant Yarrowia lipolytica N 1 under continuous cultivation. Appl Microbiol Biotechnol 59:493–500. https://doi.org/10.1007/s00253-002-1022-8

Finogenova TV, Morgunov IG, Kamzolova SV, Cherniavskaia OG (2005) Organic acid production by the yeast Yarrowia lipolytica (a review). Prikl Biokhim Mikrobiol 41:478–486

Flint N et al (2014) Effects of erythritol on endothelial function in patients with type 2 diabetes mellitus: a pilot study. Acta Diabetol 51:513–516. https://doi.org/10.1007/s00592-013-0534-2

Forster A, Jacobs K, Juretzek T, Mauersberger S, Barth G (2007) Overexpression of the ICL1 gene changes the product ratio of citric acid production by Yarrowia lipolytica. Appl Microbiol Biotechnol 77:861–869. https://doi.org/10.1007/s00253-007-1205-4

Frengova GI, Beshkova DM (2009) Carotenoids from Rhodotorula and Phaffia: yeasts of biotechnological importance. J Ind Microbiol Biotechnol 36:163–180. https://doi.org/10.1007/s10295-008-0492-9

Fukuda H, Kondo A, Noda H (2001) Biodiesel fuel production by transesterification of oils. J Biosci Bioeng 92:405–416

Gancedo C, Gancedo JM, Sols A (1968) Glycerol metabolism in yeasts. Pathways of utilization and production. Eur J Biochem 5:165–172

Gao C et al (2016a) Robust succinic acid production from crude glycerol using engineered Yarrowia lipolytica. Biotechnol Biofuels 9:179. https://doi.org/10.1186/s13068-016-0597-8

Gao Z, Ma Y, Wang Q, Zhang M, Wang J, Liu Y (2016b) Effect of crude glycerol impurities on lipid preparation by Rhodosporidium toruloides yeast 32489. Bioresour Technol 218:373–379. https://doi.org/10.1016/j.biortech.2016.06.088

Gare F (2002) The sweet miracle of xylitol. Basic Health Publications, North Bergen

Garlapati VK, Shankar U, Budhiraja A (2016) Bioconversion technologies of crude glycerol to value added industrial products. Biotechnol Rep 9:9–14. https://doi.org/10.1016/j.btre.2015.11.002

Ghosh D, Sobro IF, Hallenbeck PC (2012a) Stoichiometric conversion of biodiesel derived crude glycerol to hydrogen: response surface methodology study of the effects of light intensity and crude glycerol and glutamate concentration. Bioresour Technol 106:154–160. https://doi.org/10.1016/j.biortech.2011.12.021

Ghosh D, Tourigny A, Hallenbeck PC (2012b) Near stoichiometric reforming of biodiesel derived crude glycerol to hydrogen by photofermentation. Int J Hydrog Energy 37:2273–2277

Gonçalves VLC, Pinto BP, Silva JC, Mota CJA (2008) Acetylation of glycerol catalyzed by different solid acids. Catal Today 133-135:673–677

Goncalves FAG, Colen G, Takahashi JA (2013) Optimization of cultivation conditions for extracellular lipase production by *Yarrowia lipolytica* using response surface method. Afr J Biotechnol 12:2270–2278

Gonzalez R, Murarka A, Dharmadi Y, Yazdani SS (2008) A new model for the anaerobic fermentation of glycerol in enteric bacteria: trunk and auxiliary pathways in Escherichia coli. Metab Eng 10:234–245. https://doi.org/10.1016/j.ymben.2008.05.001

Groenewald M, Boekhout T, Neuveglise C, Gaillardin C, van Dijck PW, Wyss M (2014) Yarrowia lipolytica: safety assessment of an oleaginous yeast with a great industrial potential. Crit Rev Microbiol 40:187–206. https://doi.org/10.3109/1040841X.2013.770386

Gupta A, Murarka A, Campbell P, Gonzalez R (2009) Anaerobic fermentation of glycerol in Paenibacillus macerans: metabolic pathways and environmental determinants. Appl Environ Microbiol 75:5871–5883. https://doi.org/10.1128/AEM.01246-09
Habe H, Morita T, Sakaki K, Kitamoto D (2008) Isolation and characterization of thermotolerant fungi producing lignoceric acid from glycerol. J Oleo Sci 57:251–255
Hansen CF, Hernandez A, Mullan BP, Moore K, Trezona-Murray M, King RH, Pluske JR (2009) A chemical analysis of samples of crude glycerol from the production of biodiesel in Australia, and the effects of feeding crude glycerol to growing-finishing pigs on performance, plasma metabolites and meat quality at slaughter. Anim Product Sci 49:154–161
Haseeb ASMA, Fazal MA, Jahirul MI, Masjuki HH (2011) Compatibility of automotive materials in biodiesel: a review. Fuel 90:922–931
Hashino E, Kuboniwa M, Alghamdi SA, Yamaguchi M, Yamamoto R, Cho H, Amano A (2013) Erythritol alters microstructure and metabolomic profiles of biofilm composed of Streptococcus gordonii and Porphyromonas gingivalis. Mol Oral Microbiol 28:435–451. https://doi.org/10.1111/omi.12037
Heredia CF, Sols A, DelaFuente G (1968) Specificity of the constitutive hexose transport in yeast. Eur J Biochem 5:321–329
Heretsch P, Thomas F, Aurich A, Krautscheid H, Sicker D, Giannis A (2008) Syntheses with a chiral building block from the citric acid cycle: (2R,3S)-isocitric acid by fermentation of sunflower oil. Angew Chem 47:1958–1960. https://doi.org/10.1002/anie.200705000
Holst B, Lunde C, Lages F, Oliveira R, Lucas C, Kielland-Brandt MC (2000) GUP1 and its close homologue GUP2, encoding multimembrane-spanning proteins involved in active glycerol uptake in *Saccharomyces cerevisiae*. Mol Microbiol 37:108–124
Holz M, Forster A, Mauersberger S, Barth G (2009) Aconitase overexpression changes the product ratio of citric acid production by Yarrowia lipolytica. Appl Microbiol Biotechnol 81:1087–1096. https://doi.org/10.1007/s00253-008-1725-6
Holz M et al (2011) Overexpression of alpha-ketoglutarate dehydrogenase in Yarrowia lipolytica and its effect on production of organic acids. Appl Microbiol Biotechnol 89:1519–1526. https://doi.org/10.1007/s00253-010-2957-9
Hommel RK, Weber L, Weiss A, Himmelireich U, Rilke O, Kleber H-P (1994) Production of sophorose lipid by *Candida (Torulopsis) apicola* grown on glucose. J Biotechnol 33:147–155
Hong WK, Kim CH, Heo SY, Luo LH, Oh BR, Seo JW (2010) Enhanced production of ethanol from glycerol by engineered Hansenula polymorpha expressing pyruvate decarboxylase and aldehyde dehydrogenase genes from Zymomonas mobilis. Biotechnol Lett 32:1077–1082. https://doi.org/10.1007/s10529-010-0259-z
Hu S, Luo X, Wan C, Li Y (2012a) Characterization of crude glycerol from biodiesel plants. J Agric Food Chem 60:5915–5921. https://doi.org/10.1021/jf3008629
Hu S, Wan C, Li Y (2012b) Production and characterization of biopolyols and polyurethane foams from crude glycerol based liquefaction of soybean straw. Bioresour Technol 103:227–233. https://doi.org/10.1016/j.biortech.2011.09.125
Hutnan M, Kolesarova N, Bodik I, Czolderova M (2013) Long-term monodigestion of crude glycerol in a UASB reactor. Bioresour Technol 130:88–96. https://doi.org/10.1016/j.biortech.2012.12.003
Imandi SB, Bandaru VV, Somalanka SR, Bandaru SR, Garapati HR (2008) Application of statistical experimental designs for the optimization of medium constituents for the production of citric acid from pineapple waste. Bioresour Technol 99:4445–4450. https://doi.org/10.1016/j.biortech.2007.08.071
Ito T, Nakashimada Y, Senba K, Matsui T, Nishio N (2005) Hydrogen and ethanol production from glycerol-containing wastes discharged after biodiesel manufacturing process. J Biosci Bioeng 100:260–265. https://doi.org/10.1263/jbb.100.260
Janek T, Dobrowolski A, Biegalska A, Mironczuk AM (2017) Characterization of erythrose reductase from Yarrowia lipolytica and its influence on erythritol synthesis. Microb Cell Factories 16:118. https://doi.org/10.1186/s12934-017-0733-6

Javani A, Hasheminejad M, Tahvildari K, Tabatabaei M (2012) High quality potassium phosphate production through step-by-step glycerol purification: a strategy to economize biodiesel production. Bioresour Technol 104:788–790. https://doi.org/10.1016/j.biortech.2011.09.134
Jensen TO, Kvist T, Mikkelsen MJ, Christensen PV, Westermann P (2012) Fermentation of crude glycerol from biodiesel production by Clostridium pasteurianum. J Ind Microbiol Biotechnol 39:709–717. https://doi.org/10.1007/s10295-011-1077-6
Jeya M et al (2009) Isolation of a novel high erythritol-producing Pseudozyma tsukubaensis and scale-up of erythritol fermentation to industrial level. Appl Microbiol Biotechnol 83:225–231. https://doi.org/10.1007/s00253-009-1871-5
Jiang XP, Yin ML, Chen P, Yang Q (2012) Constitutive expression, purification and characterization of bovine prochymosin in Pichia pastoris GS115. World J Microbiol Biotechnol 28:2087–2093. https://doi.org/10.1007/s11274-012-1012-7
Johnson EA, Schroeder WA (1996) Microbial carotenoids. Adv Biochem Eng Biotechnol 53:119–178
Johnson DT, Taconi KA (2007) The glycerin glut: options for the value-added conversion of crude glycerol resulting from biodiesel production. Environ Prog 26:338–348
Jung JY, Kim TY, Ng CY, Oh MK (2012) Characterization of GCY1 in *Saccharomyces cerevisiae* by metabolic profiling. J Appl Microbiol 113:1468–1478. https://doi.org/10.1111/jam.12013
Juszczyk P, Rymowicz W (2009) Characterization of microbial biomass production from glycerin waste by various yeast strains. In: Aggelis G (ed) Microbial conversions of raw glycerol. Nova Science Publishers, New York, pp 125–135
Juszczyk P, Tomaszewska L, Kita A, Rymowicz W (2013) Biomass production by novel strains of *Yarrowia lipolytica* using raw glycerol, derived from biodiesel production. Bioresour Technol 137:124–131. https://doi.org/10.1016/j.biortech.2013.03.010
Kamzolova SV, Yusupova AI, Vinokurova NG, Fedotcheva NI, Kondrashova MN, Finogenova TV, Morgunov IG (2009) Chemically assisted microbial production of succinic acid by the yeast Yarrowia lipolytica grown on ethanol. Appl Microbiol Biotechnol 83:1027–1034. https://doi.org/10.1007/s00253-009-1948-1
Kamzolova SV, Fatykhova AR, Dedyukhina EG, Anastassiadis SG, Golovchenko NP, Morgunov IG (2011) Citric acid production by yeast grown on glycerol-containing waste from Biodiesel Industry. Food Technol Biotechnol 49:65–74
Kamzolova SV, Vinokurova NG, Lunina JN, Zelenkova NF, Morgunov IG (2015) Production of technical-grade sodium citrate from glycerol-containing biodiesel waste by Yarrowia lipolytica. Bioresour Technol 193:250–255. https://doi.org/10.1016/j.biortech.2015.06.092
Karaffa L, Kubicek CP (2003) Aspergillus niger citric acid accumulation: do we understand this well working black box? Appl Microbiol Biotechnol 61:189–196. https://doi.org/10.1007/s00253-002-1201-7
Kata I, Semkiv MV, Ruchala J, Dmytruk KV, Sibirny AA (2016) Overexpression of the genes PDC1 and ADH1 activates glycerol conversion to ethanol in the thermotolerant yeast Ogataea (Hansenula) polymorpha. Yeast 33:471–478. https://doi.org/10.1002/yea.3175
Kayingo G, Kilian SG, Prior BA (2001) Conservation and release of osmolytes by yeasts during hypo-osmotic stress. Arch Microbiol 177:29–35. https://doi.org/10.1007/s00203-001-0358-2
Kegl B (2008) Effects of biodiesel on emissions of a bus diesel engine. Bioresour Technol 99:863–873. https://doi.org/10.1016/j.biortech.2007.01.021
Khan A, Bhide A, Gadre R (2009) Mannitol production from glycerol by resting cells of Candida magnoliae. Bioresour Technol 100:4911–4913. https://doi.org/10.1016/j.biortech.2009.04.048
Khot M, Kamat S, Zinjarde S, Pant A, Chopade B, Ravikumar A (2012) Single cell oil of oleaginous fungi from the tropical mangrove wetlands as a potential feedstock for biodiesel. Microb Cell Factories 11:71. https://doi.org/10.1186/1475-2859-11-71
Kijora C, Bergner H, Kupsch RD, Hagemann L (1995) Glycerol as a feed component in fattening pigs. Archiv fur Tierernahrung 47:345–360
Kitamoto D, Isoda H, Nakahara T (2002) Functions and potential applications of glycolipid biosurfactants--from energy-saving materials to gene delivery carriers. J Biosci Bioeng 94:187–201

Klein M, Islam Z-u, Knudsen PB, Carrillo M, Swinnen S, Workman M, Nevoigt E (2016) The expression of glycerol facilitators from various yeast species improves growth on glycerol of *Saccharomyces cerevisiae*. Metabolic Eng Commun 3:252–257. https://doi.org/10.1016/j.meteno.2016.09.001

Klein M, Swinnen S, Thevelein JM, Nevoigt E (2017) Glycerol metabolism and transport in yeast and fungi: established knowledge and ambiguities. Environ Microbiol 19:878–893. https://doi.org/10.1111/1462-2920.13617

Knothe G (2008) "Designer" Biodiesel: optimizing fatty ester composition to improve fuel properties. Energy Fuels 22:1358–1364 doi:https://doi.org/10.1021/ef700639e

Knothe G, van Gerpen J, Krahl J (2005) The biodiesel handbook. AOCS Press, Champaign

Ko BS, Kim J, Kim JH (2006) Production of xylitol from D-xylose by a xylitol dehydrogenase gene-disrupted mutant of Candida tropicalis. Appl Environ Microbiol 72:4207–4213. https://doi.org/10.1128/AEM.02699-05

Kobayashi Y, Yoshida J, Iwata H, Koyama Y, Kato J, Ogihara J, Kasumi T (2013) Gene expression and function involved in polyol biosynthesis of Trichosporonoides megachiliensis under hyper-osmotic stress. J Biosci Bioeng 115:645–650. https://doi.org/10.1016/j.jbiosc.2012.12.004

Kobayashi Y, Iwata H, Mizushima D, Ogihara J, Kasumi T (2015) Erythritol production by Moniliella megachiliensis using nonrefined glycerol waste as carbon source. Lett Appl Microbiol 60:475–480. https://doi.org/10.1111/lam.12391

Koganti S, Kuo TM, Kurtzman CP, Smith N, Ju LK (2011) Production of arabitol from glycerol: strain screening and study of factors affecting production yield. Appl Microbiol Biotechnol 90:257–267. https://doi.org/10.1007/s00253-010-3015-3

Kogje AB, Ghosalkar A (2017) Xylitol production by genetically modified industrial strain of Saccharomyces cerevisiae using glycerol as co-substrate. J Ind Microbiol Biotechnol 44:961–971. https://doi.org/10.1007/s10295-017-1914-3

Konishi M, Fukuoka T, Morita T, Imura T, Kitamoto D (2008) Production of new types of sophorolipids by Candida batistae. J Oleo Sci 57:359–369

Konishi M, Morita T, Fukuoka T, Imura T, Uemura S, Iwabuchi H, Kitamoto D (2017) Selective production of acid-form Sophorolipids from glycerol by Candida floricola. J Oleo Sci 66:1365–1373. https://doi.org/10.5650/jos.ess17116

Koutinas AA, Chatzifragkou A, Kopsahelis N, Papanikolaou S, Kookos I (2014) Design and techno-economic evaluation of microbial oil production as a renewable resource for biodiesel and oleochemical production. Fuel 116:566–577

Kurtzman CP (1983) Biology and physiology of the D-xylose fermenting yeast Pachysolen tannophilus. Pentoses and Lignin 27:73–83

Kurylenko O et al (2016) New approaches for improving the production of the 1st and 2nd generation ethanol by yeast. Acta Biochim Pol 63:31–38. https://doi.org/10.18388/abp.2015_1156

Kusdiyantini E, Gaudin P, Goma G, Blanc PJ (1998) Growth kinetics and astaxanthin production of Phaffia rhodozyma on glycerol as a carbon source during batch fermentation. Biotechnol Lett 20:929–934. https://doi.org/10.1023/a:1005445808265

Kuttiraja M, Douha A, Valero JR, Tyagi RD (2016) Elucidating the effect of glycerol concentration and C/N ratio on lipid production using Yarrowia lipolytica SKY7. Appl Biochem Biotechnol 180:1586–1600. https://doi.org/10.1007/s12010-016-2189-2

Lages F, Lucas C (1995) Characterization of a glycerol/H+ symport in the halotolerant yeast Pichia sorbitophila. Yeast 11:111–119. https://doi.org/10.1002/yea.320110203

Lages F, Silva-Graca M, Lucas C (1999) Active glycerol uptake is a mechanism underlying halotolerance in yeasts: a study of 42 species. Microbiology 145(Pt 9):2577–2585. https://doi.org/10.1099/00221287-145-9-2577

Lammers PJ et al (2008) Nitrogen-corrected apparent metabolizable energy value of crude glycerol for laying hens. Poult Sci 87:104–107. https://doi.org/10.3382/ps.2007-00255

Lee JK, Koo BS, Kim SY, Hyun HH (2003) Purification and characterization of a novel mannitol dehydrogenase from a newly isolated strain of Candida magnoliae. Appl Environ Microbiol 69:4438–4447

Lee GH, Bae JH, Suh MJ, Kim IH, Hou CT, Kim HR (2007a) New finding and optimal production of a novel extracellular alkaline lipase from Yarrowia lipolytica NRRL Y-2178. J Microbiol Biotechnol 17:1054–1057

Lee JK, Oh DK, Song HY, Kim IW (2007b) Ca2+ and Cu2+ supplementation increases mannitol production by Candida magnoliae. Biotechnol Lett 29:291–294. https://doi.org/10.1007/s10529-006-9230-4

Lee DH, Lee YJ, Ryu YW, Seo JH (2010) Molecular cloning and biochemical characterization of a novel erythrose reductase from Candida magnoliae JH110. Microb Cell Factories 9:43. https://doi.org/10.1186/1475-2859-9-43

Leiva-Candia DE, Tsakona S, Kopsahelis N, Garcia IL, Papanikolaou S, Dorado MP, Koutinas AA (2015) Biorefining of by-product streams from sunflower-based biodiesel production plants for integrated synthesis of microbial oil and value-added co-products. Bioresour Technol 190:57–65. https://doi.org/10.1016/j.biortech.2015.03.114

Li Q, Du W, Liu D (2008) Perspectives of microbial oils for biodiesel production. Appl Microbiol Biotechnol 80:749–756. https://doi.org/10.1007/s00253-008-1625-9

Li C, Gao S, Yang X, Lin CSK (2018a) Green and sustainable succinic acid production from crude glycerol by engineered Yarrowia lipolytica via agricultural residue based in situ fibrous bed bioreactor. Bioresour Technol 249:612–619. https://doi.org/10.1016/j.biortech.2017.10.011

Li J, Lu M, Guo X, Zhang H, Li Y, Han L (2018b) Insights into the improvement of alkaline hydrogen peroxide (AHP) pretreatment on the enzymatic hydrolysis of corn Stover: chemical and microstructural analyses. Bioresour Technol 265:1–7. https://doi.org/10.1016/j.biortech.2018.05.082

Liang Y, Cui Y, Trushenski J, Blackburn JW (2010) Converting crude glycerol derived from yellow grease to lipids through yeast fermentation. Bioresour Technol 101:7581–7586. https://doi.org/10.1016/j.biortech.2010.04.061

Lide DR (1999) Handbook of chemistry and physics. 80th edn. CRC Press, Boca Raton

Liu X, Jensen PR, Workman M (2012) Bioconversion of crude glycerol feedstocks into ethanol by Pachysolen tannophilus. Bioresour Technol 104:579–586. https://doi.org/10.1016/j.biortech.2011.10.065

Liu X, Mortensen UH, Workman M (2013a) Expression and functional studies of genes involved in transport and metabolism of glycerol in Pachysolen tannophilus. Microb Cell Factories 12:27. https://doi.org/10.1186/1475-2859-12-27

Liu YP, Sun Y, Tan C, Li H, Zheng XJ, Jin KQ, Wang G (2013b) Efficient production of dihydroxyacetone from biodiesel-derived crude glycerol by newly isolated Gluconobacter frateurii. Bioresour Technol 142:384–389. https://doi.org/10.1016/j.biortech.2013.05.055

Liu LP, Hu Y, Lou WY, Li N, Wu H, Zong MH (2016) Use of crude glycerol as sole carbon source for microbial lipid production by oleaginous yeasts. Appl Biochem Biotechnol. https://doi.org/10.1007/s12010-016-2340-0

Liu L, Zong M, Hu Y, Li N, Lou W, Wu H (2017) Efficient microbial oil production on crude glycerol by Lipomyces starkeyi AS 2.1560 and its kinetics. Process Biochem 58:230–238

Lo Y, Chen X, Huang C (2013) Dark fermentative hydrogen production with crude glycerol from biodiesel industry using indigenous hydrogen-producing bacteria. Int J Hydrog Energy 38:15815–15822

Longo MA, Sanromán MA (2006) Production of food aroma compounds: microbial and enzymatic methodologies. Food Technol Biotechnol 44:335–353

Looser V et al (2015) Cultivation strategies to enhance productivity of Pichia pastoris: a review. Biotechnol Adv 33:1177–1193. https://doi.org/10.1016/j.biotechadv.2015.05.008

Louhasakul Y, Cheirsilp B (2013) Industrial waste utilization for low-cost production of raw material oil through microbial fermentation. Appl Biochem Biotechnol 169:110–122. https://doi.org/10.1007/s12010-012-9965-4

Lucas C, Da Costa M, van Uden N (1990) Osmoregulatory active sodium-glycerol cotransport in the Halotolerant yeast Debaryomyces hansenii. Yeast 6:187–191

Luo X, Hu S, Zhang X, Li Y (2013) Thermochemical conversion of crude glycerol to biopolyols for the production of polyurethane foams. Bioresour Technol 139:323–329. https://doi.org/10.1016/j.biortech.2013.04.011

Luo Z, Miao J, Luo W, Li G, Du Y, Yu X (2018) Crude glycerol from biodiesel as a carbon source for production of a recombinant highly thermostable beta-mannanase by Pichia pastoris. Biotechnol Lett 40:135–141. https://doi.org/10.1007/s10529-017-2451-x

Luyten K, Albertyn J, Skibbe WF, Prior BA, Ramos J, Thevelein JM, Hohmann S (1995) Fps1, a yeast member of the MIP family of channel proteins, is a facilitator for glycerol uptake and efflux and is inactive under osmotic stress. EMBO J 14:1360–1371

Ma F, Hanna MA (1999) Biodiesel production: a review. Bioresour Technol 70:1–15

Macauley-Patrick S, Fazenda ML, McNeil B, Harvey LM (2005) Heterologous protein production using the Pichia pastoris expression system. Yeast 22:249–270. https://doi.org/10.1002/yea.1208

Magdouli S, Guedri T, Tarek R, Brar SK, Blais JF (2017) Valorization of raw glycerol and crustacean waste into value added products by Yarrowia lipolytica. Bioresour Technol 243:57–68. https://doi.org/10.1016/j.biortech.2017.06.074

Makri A, Fakas S, Aggelis G (2010) Metabolic activities of biotechnological interest in Yarrowia lipolytica grown on glycerol in repeated batch cultures. Bioresour Technol 101:2351–2358. https://doi.org/10.1016/j.biortech.2009.11.024

Maleszka R, Wang P, Schneider H (1982) Ethanol production from D-galactose and glycerol by Pachysolen tannophilus. Enzym Microb Technol 4:349–352

Manara P, Zabaniotou A (2013) Co-pyrolysis of biodiesel-derived glycerol with Greek lignite: a laboratory study. J Anal Appl Pyrolysis 100:166–172

Manowattana A, Techapun C, Watanabe M, Chaiyaso T (2018) Bioconversion of biodiesel-derived crude glycerol into lipids and carotenoids by an oleaginous red yeast Sporidiobolus pararoseus KM281507 in an airlift bioreactor. J Biosci Bioeng 125:59–66. https://doi.org/10.1016/j.jbiosc.2017.07.014

Marchetti JM, Miguel VU, Errazu AF (2007) Possible methods for biodiesel production. Renew Sust Energ Rev 11:1300–1311

Mata-Gomez LC, Montanez JC, Mendez-Zavala A, Aguilar CN (2014) Biotechnological production of carotenoids by yeasts: an overview. Microb Cell Factories 13:12. https://doi.org/10.1186/1475-2859-13-12

Matsuzawa T, Ohashi T, Hosomi A, Tanaka N, Tohda H, Takegawa K (2010) The gld1+ gene encoding glycerol dehydrogenase is required for glycerol metabolism in Schizosaccharomyces pombe. Appl Microbiol Biotechnol 87:715–727. https://doi.org/10.1007/s00253-010-2586-3

Mattanovich D et al (2009) Genome, secretome and glucose transport highlight unique features of the protein production host Pichia pastoris. Microb Cell Factories 8:29. https://doi.org/10.1186/1475-2859-8-29

Mazière A, Prinsen P, García A, Luque R, Len C (2017) A review of progress in (bio)catalytic routes from/to renewable succinic acid. Biofuels Bioprod Biorefin 11:908–931. https://doi.org/10.1002/bbb.1785

Mazumdar S, Blankschien MD, Clomburg JM, Gonzalez R (2013) Efficient synthesis of L-lactic acid from glycerol by metabolically engineered *Escherichia coli*. Microb Cell Factories 12:7. https://doi.org/10.1186/1475-2859-12-7

McNeil BA, Stuart DT (2018) Lipomyces starkeyi: an emerging cell factory for production of lipids, oleochemicals and biotechnology applications. World J Microbiol Biotechnol 34:147. https://doi.org/10.1007/s11274-018-2532-6

Meher LC, Vidya Sagar D, Naik SN (2006) Technical aspects of biodiesel production by transesterification – a review. Renew Sust Energ Rev 10:248–268

Meiswinkel TM, Rittmann D, Lindner SN, Wendisch VF (2013) Crude glycerol-based production of amino acids and putrescine by Corynebacterium glutamicum. Bioresour Technol 145:254–258. https://doi.org/10.1016/j.biortech.2013.02.053

Miao X, Wu Q (2006) Biodiesel production from heterotrophic microalgal oil. Bioresour Technol 97:841–846. https://doi.org/10.1016/j.biortech.2005.04.008

Miesiac I (2003) Methods for utilization of the glycerine fraction from methanolysis of rape oil. Przemysl Chemiczny 42:1045–1047

Mironczuk AM, Furgala J, Rakicka M, Rymowicz W (2014) Enhanced production of erythritol by Yarrowia lipolytica on glycerol in repeated batch cultures. J Ind Microbiol Biotechnol 41:57–64. https://doi.org/10.1007/s10295-013-1380-5

Mironczuk AM, Rakicka M, Biegalska A, Rymowicz W, Dobrowolski A (2015) A two-stage fermentation process of erythritol production by yeast Y. lipolytica from molasses and glycerol. Bioresour Technol 198:445–455. https://doi.org/10.1016/j.biortech.2015.09.008

Mironczuk AM, Rzechonek DA, Biegalska A, Rakicka M, Dobrowolski A (2016) A novel strain of Yarrowia lipolytica as a platform for value-added product synthesis from glycerol. Biotechnol Biofuels 9:180. https://doi.org/10.1186/s13068-016-0593-z

Moline M, Libkind D, Dieguez Mdel C, van Broock M (2009) Photoprotective role of carotenoids in yeasts: response to UV-B of pigmented and naturally-occurring albino strains. J Photochem Photobiol B 95:156–161. https://doi.org/10.1016/j.jphotobiol.2009.02.006

Moon HJ, Jeya M, Kim IW, Lee JK (2010) Biotechnological production of erythritol and its applications. Appl Microbiol Biotechnol 86:1017–1025. https://doi.org/10.1007/s00253-010-2496-4

Morgunov IG, Kamzolova SV, Perevoznikova OA, Shishkanova NV, Finogenova TV (2004) Pyruvic acid production by a thiamine auxotroph of Yarrowia lipolytica. Process Biochem 39:1469–1474. https://doi.org/10.1016/S0032-9592(03)00259-0

Morgunov IG, Kamzolova SV, Lunina JN (2013) The citric acid production from raw glycerol by Yarrowia lipolytica yeast and its regulation. Appl Microbiol Biotechnol 97:7387–7397. https://doi.org/10.1007/s00253-013-5054-z

Mota CJA, Pinto BP, Lyra JT (2009) Obtenção de ésteres de glicerol para uso como aditivos a combustível. Paper presented at the III Congresso da rede brasileira de biodiesel, Brasília, DF,

Munch G, Sestric R, Sparling R, Levin DB, Cicek N (2015) Lipid production in the under-characterized oleaginous yeasts, Rhodosporidium babjevae and Rhodosporidium diobovatum, from biodiesel-derived waste glycerol. Bioresour Technol 185:49–55. https://doi.org/10.1016/j.biortech.2015.02.051

Munoz RAA, Fernandes DM, Santos DQ, Barbosa TGG, Sousa RMF (2012) Biodiesel: production, characterization, metallic corrosion and analytical methods for contaminants. In: Fang Z (ed) Biodiesel. IntechOpen doi:https://doi.org/10.5772/53655

Nakazawa H, Enei H, Okumura S, Yamada H (1972) Synthesis of L-tryptophan from pyruvate. Agric Biol Chem 32:2528–2532

Nanda MR, Yuan Z, Qin W, Poirier MA, Chunbao X (2014) Purification of crude glycerol using acidification: effects of acid types and product. Character Austin J Chem Eng 1:1004

Nartker S, Ammerman M, Aurandt J, Stogsdil M, Hayden O, Antle C (2014) Increasing biogas production from sewage sludge anaerobic co-digestion process by adding crude glycerol from biodiesel industry. Waste Manag 34:2567–2571. https://doi.org/10.1016/j.wasman.2014.08.017

Ngo TA, Kim M, Sim SJ (2011) High-yield biohydrogen production from biodiesel manufacturing waste by Thermotoga neapolitana. Int J Hydrog Energy 36:5836–5842

Nikel PI, Ramirez MC, Pettinari MJ, Mendez BS, Galvagno MA (2010) Ethanol synthesis from glycerol by *Escherichia coli* redox mutants expressing adhE from Leuconostoc mesenteroides. J Appl Microbiol 109:492–504. https://doi.org/10.1111/j.1365-2672.2010.04668.x

Norbeck J, Blomberg A (1997) Metabolic and regulatory changes associated with growth of *Saccharomyces cerevisiae* in 1.4 M NaCl. evidence for osmotic induction of glycerol dissimilation via the dihydroxyacetone pathway. J Biol Chem 272:5544–5554

Noseda DG, Blasco M, Recupero M, Galvagno MA (2014) Bioprocess and downstream optimization of recombinant bovine chymosin B in Pichia (Komagataella) pastoris under methanol-inducible AOXI promoter. Protein Expr Purif 104:85–91. https://doi.org/10.1016/j.pep.2014.09.014

Noseda DG, Recupero M, Blasco M, Bozzo J, Galvagno MA (2016) Production in stirred-tank bioreactor of recombinant bovine chymosin B by a high-level expression transformant clone of Pichia pastoris. Protein Expr Purif 123:112–121. https://doi.org/10.1016/j.pep.2016.03.008

Ochoa-Estopier A, Lesage J, Gorret N, Guillouet SE (2011) Kinetic analysis of a Saccharomyces cerevisiae strain adapted for improved growth on glycerol: implications for the development of yeast bioprocesses on glycerol. Bioresour Technol 102:1521–1527. https://doi.org/10.1016/j.biortech.2010.08.003

Ochoa-Gómez JR et al (2009) Synthesis of glycerol carbonate from glycerol and dimethyl carbonate by transesterification: catalyst screening and reaction optimization. Appl Catal A Gen 366:315–324. https://doi.org/10.1016/j.apcata.2009.07.020

Oh BR et al (2011) Efficient production of ethanol from crude glycerol by a *Klebsiella pneumoniae* mutant strain. Bioresour Technol 102:3918–3922. https://doi.org/10.1016/j.biortech.2010.12.007

Ohtake S, Wang YJ (2011) Trehalose: current use and future applications. J Pharm Sci 100:2020–2053. https://doi.org/10.1002/jps.22458

Okoye PU, Hameed BH (2016) Review on recent progress in catalytic carboxylation and acetylation of glycerol as a byproduct of biodiesel production. Renew Sust Energ Rev 53:558–574

Oku T, Nakamura S (2007) Threshold for transitory diarrhea induced by ingestion of xylitol and lactitol in young male and female adults. J Nutr Sci Vitaminol 53:13–20

Otto C, Yovkova V, Barth G (2011) Overproduction and secretion of alpha-ketoglutaric acid by microorganisms. Appl Microbiol Biotechnol 92:689–695. https://doi.org/10.1007/s00253-011-3597-4

Otto C, Yovkova V, Aurich A, Mauersberger S, Barth G (2012) Variation of the by-product spectrum during alpha-ketoglutaric acid production from raw glycerol by overexpression of fumarase and pyruvate carboxylase genes in Yarrowia lipolytica. Appl Microbiol Biotechnol 95:905–917. https://doi.org/10.1007/s00253-012-4085-1

Oyedotun KS, Lemire BD (2004) The quaternary structure of the *Saccharomyces cerevisiae* succinate dehydrogenase. Homology modeling, cofactor docking, and molecular dynamics simulation studies. J Biol Chem 279:9424–9431. https://doi.org/10.1074/jbc.M311876200

Papanikolaou S, Aggelis G (2002) Lipid production by Yarrowia lipolytica growing on industrial glycerol in a single-stage continuous culture. Bioresour Technol 82:43–49

Papanikolaou S, Chevalot I, Komaitis M, Marc I, Aggelis G (2002a) Single cell oil production by Yarrowia lipolytica growing on an industrial derivative of animal fat in batch cultures. Appl Microbiol Biotechnol 58:308–312. https://doi.org/10.1007/s00253-001-0897-0

Papanikolaou S, Muniglia L, Chevalot I, Aggelis G, Marc I (2002b) Yarrowia lipolytica as a potential producer of citric acid from raw glycerol. J Appl Microbiol 92:737–744

Papanikolaou S, Sarantou S, Komaitis M, Aggelis G (2004) Repression of reserve lipid turnover in Cunninghamella echinulata and Mortierella isabellina cultivated in multiple-limited media. J Appl Microbiol 97:867–875. https://doi.org/10.1111/j.1365-2672.2004.02376.x

Papanikolaou S, Chevalot I, Galiotou-Panayotou M, Komaitis M, Marc I, Aggelis G (2007) Industrial derivative of tallow: a promising renewable substrate for microbial lipid, single-cell protein and lipase production by Yarrowia lipolytica. Electron J Biotechnol 10:425–435

Papanikolaou S, Beopoulos A, Koletti A, Thevenieau F, Koutinas AA, Nicaud JM, Aggelis G (2013) Importance of the methyl-citrate cycle on glycerol metabolism in the yeast Yarrowia lipolytica. J Biotechnol 168:303–314

Park YC, Lee DY, Lee DH, Kim HJ, Ryu YW, Seo JH (2005) Proteomics and physiology of erythritol-producing strains. J Chromatogr B, Analyt Technol Biomed Life Sci 815:251–260 doi:https://doi.org/10.1016/j.jchromb.2004.10.065

Park EH, Lee HY, Ryu YW, Seo JH, Kim MD (2011) Role of osmotic and salt stress in the expression of erythrose reductase in Candida magnoliae. J Microbiol Biotechnol 21:1064–1068

Park YC, Oh EJ, Jo JH, Jin YS, Seo JH (2016) Recent advances in biological production of sugar alcohols. Curr Opin Biotechnol 37:105–113. https://doi.org/10.1016/j.copbio.2015.11.006

Pavlik P, Simon M, Schuster T, Ruis H (1993) The glycerol kinase (GUT1) gene of *Saccharomyces cerevisiae*: cloning and characterization. Curr Genet 24:21–25

Pereira de Andrade D, Carvalho BF, Schwan RF, Dias DR (2017) Production of gamma-Decalactone by yeast strains under different conditions. Food Technol Biotechnol 55:225–230. https://doi.org/10.17113/ftb.55.02.17.5009

Pinzi S, Garcia IL, Lopez-Gimenez FJ, Luque de Castro MD, Dorado G, Dorado MP (2009) The ideal vegetable oil-based biodiesel composition: a review of social, economical and technical implications. Energy Fuel 23:2325–2341. https://doi.org/10.1021/ef801098a

Plácido J, Capareda S (2016) Conversion of residues and by-products from the biodiesel industry into value-added products. Bioresour Bioproc 3:23. https://doi.org/10.1186/s40643-016-0100-1

Polburee P, Yongmanitchai W, Lertwattanasakul N, Ohashi T, Fujiyama K, Limtong S (2015) Characterization of oleaginous yeasts accumulating high levels of lipid when cultivated in glycerol and their potential for lipid production from biodiesel-derived crude glycerol. Fungal Biol 119:1194–1204. https://doi.org/10.1016/j.funbio.2015.09.002

Poli JS, da Silva MA, Siqueira EP, Pasa VM, Rosa CA, Valente P (2014) Microbial lipid produced by Yarrowia lipolytica QU21 using industrial waste: a potential feedstock for biodiesel production. Bioresour Technol 161:320–326. https://doi.org/10.1016/j.biortech.2014.03.083

Pott RW, Howe CJ, Dennis JS (2013) Photofermentation of crude glycerol from biodiesel using Rhodopseudomonas palustris: comparison with organic acids and the identification of inhibitory compounds. Bioresour Technol 130:725–730. https://doi.org/10.1016/j.biortech.2012.11.126

Pott RW, Howe CJ, Dennis JS (2014) The purification of crude glycerol derived from biodiesel manufacture and its use as a substrate by Rhodopseudomonas palustris to produce hydrogen. Bioresour Technol 152:464–470. https://doi.org/10.1016/j.biortech.2013.10.094

Pousa GPAG, Santos ALF, Suarez PAZ (2007) History and policy of biodiesel in Brazil. Energy Policy 35:5393–5398

Pribylova L, de Montigny J, Sychrova H (2007) Osmoresistant yeast Zygosaccharomyces rouxii: the two most studied wild-type strains (ATCC 2623 and ATCC 42981) differ in osmotolerance and glycerol metabolism. Yeast 24:171–180. https://doi.org/10.1002/yea.1470

Pyle DJ, Garcia RA, Wen Z (2008) Producing docosahexaenoic acid (DHA)-rich algae from biodiesel-derived crude glycerol: effects of impurities on DHA production and algal biomass composition. J Agric Food Chem 56:3933–3939. https://doi.org/10.1021/jf800602s

Qiao K et al (2015) Engineering lipid overproduction in the oleaginous yeast Yarrowia lipolytica. Metab Eng 29:56–65. https://doi.org/10.1016/j.ymben.2015.02.005

Quispe CAG, Coronado CJR, Carvalho JA Jr (2013) Glycerol: production, consumption, prices, characterization and new trends in combustion. Renew Sust Energ Rev 27:475–493. https://doi.org/10.1016/j.rser.2013.06.017

Radley S (2016) Biodiesel - Feedstocks, production and applications. Scitus Academics LLC

Raimondi S, Rossi M, Leonardi A, Bianchi MM, Rinaldi T, Amaretti A (2014) Getting lipids from glycerol: new perspectives on biotechnological exploitation of *Candida freyschussii*. Microb Cell Factories 13:83. https://doi.org/10.1186/1475-2859-13-83

Rakicka M, Lazar Z, Dulermo T, Fickers P, Nicaud JM (2015) Lipid production by the oleaginous yeast Yarrowia lipolytica using industrial by-products under different culture conditions. Biotechnol Biofuels 8:104. https://doi.org/10.1186/s13068-015-0286-z

Rakicka M, Rukowicz B, Rywinska A, Lazar Z, Rymowicz W (2016a) Technology of efficient continuous erythritol production from glycerol. J Clean Prod 139:905–913

Rakicka M, Rywinska A, Cybulski K, Rymowicz W (2016b) Enhanced production of erythritol and mannitol by Yarrowia lipolytica in media containing surfactants. Brazilian J Microbiol 47:417–423. https://doi.org/10.1016/j.bjm.2016.01.011

Rakicka M, Biegalska A, Rymowicz W, Dobrowolski A, Mironczuk AM (2017) Polyol production from waste materials by genetically modified Yarrowia lipolytica. Bioresour Technol 243:393–399. https://doi.org/10.1016/j.biortech.2017.06.137

Ramirez-Castrillon M et al (2017) The oleaginous yeast Meyerozyma guilliermondii BI281A as a new potential biodiesel feedstock: selection and lipid production optimization. Front Microbiol 8:1776. https://doi.org/10.3389/fmicb.2017.01776

Razani SH, Mousavi SM, Yeganeh HM, Marc I (2007) Fatty acid and carotenoid production by Sporobolomyces ruberrimus when using technical glycerol and ammonium sulfate. J Microbiol Biotechnol 17:1591–1597

Rodrigues A, Bordado JC, RGd S (2017) Upgrading the glycerol from biodiesel production as a source of energy carriers and chemicals—a technological review for three chemical pathways. Energies 10:1817

Röhr M, Kubicek CP, Zehentgruber O, Orthofer R (1983) A balance of oxygen and carbon conversion rates during pilot plant citric acid fermentation by Aspergillus Niger. Intl J Microbiol 1:19–25

Romano SD, Sorichetti PA (2011) Introduction to biodiesel production. In: Romano SD, Sorichetti PA (eds) Dielectric spectroscopy in biodiesel production and characterization. Springer, London

Romero-Guido C et al (2011) Biochemistry of lactone formation in yeast and fungi and its utilisation for the production of flavour and fragrance compounds. Appl Microbiol Biotechnol 89:535–547. https://doi.org/10.1007/s00253-010-2945-0

Ronnow B, Kielland-Brandt MC (1993) GUT2, a gene for mitochondrial glycerol 3-phosphate dehydrogenase of Saccharomyces cerevisiae. Yeast 9:1121–1130. https://doi.org/10.1002/yea.320091013

Rossi DM, da Costa JB, de Souza EA, Peralba MCR, Samios D, Ayub MAZ (2011) Comparison of different pretreatment methods for hydrogen production using environmental microbial consortia on residual glycerol from biodiesel. Int J Hydrog Energy 36:4814–4819

Rymowicz W, Rywinska A, Marcinkiewicz M (2009) High-yield production of erythritol from raw glycerol in fed-batch cultures of Yarrowia lipolytica. Biotechnol Lett 31:377–380. https://doi.org/10.1007/s10529-008-9884-1

Rymowicz W, Fatykhova AR, Kamzolova SV, Rywinska A, Morgunov IG (2010) Citric acid production from glycerol-containing waste of biodiesel industry by Yarrowia lipolytica in batch, repeated batch, and cell recycle regimes. Appl Microbiol Biotechnol 87:971–979. https://doi.org/10.1007/s00253-010-2561-z

Rywinska A, Marcinkiewicz M, Cibis E, Rymowicz W (2015) Optimization of medium composition for erythritol production from glycerol by Yarrowia lipolytica using response surface methodology. Prep Biochem Biotechnol 45:515–529. https://doi.org/10.1080/10826068.2014.940966

Rzechonek DA, Dobrowolski A, Rymowicz W, Mironczuk AM (2018) Recent advances in biological production of erythritol. Crit Rev Biotechnol 38:620–633. https://doi.org/10.1080/07388551.2017.1380598

Rzechonek DA, Dobrowolski A, Rymowicz W, Mironczuk AM (2019) Aseptic production of citric and isocitric acid from crude glycerol by genetically modified Yarrowia lipolytica. Bioresour Technol 271:340–344. https://doi.org/10.1016/j.biortech.2018.09.118

Sabourin-Provost G, Hallenbeck PC (2009) High yield conversion of a crude glycerol fraction from biodiesel production to hydrogen by photofermentation. Bioresour Technol 100:3513–3517. https://doi.org/10.1016/j.biortech.2009.03.027

Saenge C, Cheirsilp B, Suksaroge TT, Bourtoom T (2011) Potential use of oleaginous red yeast Rhodotorula glutinis for the bioconversion of crude glycerol from biodiesel plant to lipids and carotenoids. Process Biochem 46:210–218

Saha BC, Racine FM (2011) Biotechnological production of mannitol and its applications. Appl Microbiol Biotechnol 89:879–891. https://doi.org/10.1007/s00253-010-2979-3

Sanchez AM, Bennett GN, San KY (2005) Novel pathway engineering design of the anaerobic central metabolic pathway in Escherichia coli to increase succinate yield and productivity. Metab Eng 7:229–239. https://doi.org/10.1016/j.ymben.2005.03.001

Sánchez EA, D'Angelo MA, Comelli RA (2010) Hydrogen production from glycerol on Ni/Al2O3 catalyst. Int J Hydr Energy 35:5902–5907
Santibanez C, Teresa Varnero M, Bustamante M (2011) Residual glycerol from biodiesel manufacturing, waste or potential source of bioenergy: a review. Chilean J Agricult Res 71:469–475
Sarma SJ et al (2013) Evaluation of different supplementary nutrients for enhanced biohydrogen production by Enterobacter aerogenes NRRL B 407 using waste derived crude glycerol. Int J Hydrog Energy 38:15815–15822
Sarma SJ, Brar SK, Le Bihan Y, Buelna G, Soccol CR (2014) Mitigation of the inhibitory effect of soap by magnesium salt treatment of crude glycerol – a novel approach for enhanced biohydrogen production from the biodiesel industry waste. Bioresour Technol 151:49–53. https://doi.org/10.1016/j.biortech.2013.10.042
Sauer M, Porro D, Mattanovich D, Branduardi P (2008) Microbial production of organic acids: expanding the markets. Trends Biotechnol 26:100–108. https://doi.org/10.1016/j.tibtech.2007.11.006
Sauer M, Porro D, Mattanovich D, Branduardi P (2010) 16 years research on lactic acid production with yeast – ready for the market? Biotechnol Genet Eng Rev 27:229–256
Sawada K, Taki A, Yamakawa T, Seki M (2009) Key role for transketolase activity in erythritol production by Trichosporonoides megachiliensis SN-G42. J Biosci Bioeng 108:385–390. https://doi.org/10.1016/j.jbiosc.2009.05.008
Scholten E, Renz T, Thomas J (2009) Continuous cultivation approach for fermentative succinic acid production from crude glycerol by Basfia succiniciproducens DD1. Biotechnol Lett 31:1947–1951. https://doi.org/10.1007/s10529-009-0104-4
Schwab AW, Bagby MO, Freedman B (1987) Preparation and properties of diesel fuels from vegetable oils. Fuel 66:1372–1378
Semkiv MV, Dmytruk KV, Abbas CA, Sibirny AA (2017) Biotechnology of Glycerol Production and Conversion in Yeasts. In: Sibirny AA (ed) Biotechnology of yeasts and filamentous fungi. Springer, Cham, pp 117–148
Shen T, Junyong Z, Xiushan Y (2011) Evaluation of an adapted inhibitor-tolerant yeast strain for ethanol production from combined hydrolysate of softwood. Appl Energy 88:1792–1796
Shen H, Gong Z, Yang X, Jin G, Bai F, Zhao ZK (2013) Kinetics of continuous cultivation of the oleaginous yeast Rhodosporidium toruloides. J Biotechnol 168:85–89. https://doi.org/10.1016/j.jbiotec.2013.08.010
Shieh CJ, Liao HF, Lee CC (2003) Optimization of lipase-catalyzed biodiesel by response surface methodology. Bioresour Technol 88:103–106
Signori L, Ami D, Posteri R, Giuzzi A, Mereghetti P, Porro D, Branduardi P (2016) Assessing an effective feeding strategy to optimize crude glycerol utilization as sustainable carbon source for lipid accumulation in oleaginous yeasts. Microb Cell Factories 15:75. https://doi.org/10.1186/s12934-016-0467-x
Siles Lopez JA, Martin Santos Mde L, Chica Perez AF, Martin Martin A (2009) Anaerobic digestion of glycerol derived from biodiesel manufacturing. Bioresour Technol 100:5609–5615. https://doi.org/10.1016/j.biortech.2009.06.017
Singh SP, Singh D (2010) Biodiesel production through the use of different sources and characterization of oils and their esters as the substitute of diesel: a review. Renew Sust Energ Rev 14:200–216
Singh B, Korstad J, Sharma YC (2012) A critical review on corrosion of compression ignition (CI) engine parts by biodiesel and biodiesel blends and its inhibition. Renew Sust Energ Rev 16:3401–3408
Sinha S, Agarwal AK, Garg S (2008) Biodiesel development from rice bran oil: Transesterification process optimization and fuel characterization. Energy Convers Manag 49:1248–1257. https://doi.org/10.1016/jenconman200708010
Skoulou V, Manara P, Zabaniotou A (2012) H_2 enriched fuels from co-pyrolysis of crude glycerol with biomass. J Anal Appl Pyrolysis 97:198–204

Smirnov AA, Selishcheva SA, Yakovlev VA (2018) Acetalization catalysts for synthesis of valuable oxygenated fuel additives from glycerol. Catalysts 8:595

Soares GPA, Souza KST, Vilela LF, Schwan RF, Dias DR (2017) Gamma-decalactone production by Yarrowia lipolytica and Lindnera saturnus in crude glycerol. Prep Biochem Biotechnol 47:633–637. https://doi.org/10.1080/10826068.2017.1286601

Solomon BO, Zeng AP, Biebl H, Schlieker H, Posten C, Deckwer WD (1995) Comparison of the energetic efficiencies of hydrogen and oxychemicals formation in Klebsiella pneumoniae and Clostridium butyricum during anaerobic growth on glycerol. J Biotechnol 39:107–117

Song KH, Lee JK, Song JY, Hong SG, Baek H, Kim SY, Hyun HH (2002) Production of mannitol by a novel strain of Candida magnoliae. Biotechnol Lett 24(1):9–12

Sousa M, Melo VM, Rodrigues S, Sant'ana HB, Goncalves LR (2012) Screening of biosurfactant-producing Bacillus strains using glycerol from the biodiesel synthesis as main carbon source. Bioprocess Biosyst Eng 35:897–906. https://doi.org/10.1007/s00449-011-0674-0

Souza KS, Schwan RF, Dias DR (2014) Lipid and citric acid production by wild yeasts grown in glycerol. J Microbiol Biotechnol 24:497–506

Souza KS, Ramos CL, Schwan RF, Dias DR (2017) Lipid production by yeasts grown on crude glycerol from biodiesel industry. Prep Biochem Biotechnol 47:357–363. https://doi.org/10.1080/10826068.2016.1244689

Spier F, Buffon JG, Burkert CA (2015) Bioconversion of raw glycerol generated from the synthesis of biodiesel by different oleaginous yeasts: lipid content and fatty acid profile of biomass. Indian J Microbiol 55:415–422. https://doi.org/10.1007/s12088-015-0533-9

Sprague GF, Cronan JE (1977) Isolation and characterization of *Saccharomyces cerevisiae* mutants defective in glycerol catabolism. J Bacteriol 129:1335–1342

Stanko RT, Tietze DL, Arch JE (1992) Body composition, energy utilization, and nitrogen metabolism with a severely restricted diet supplemented with dihydroxyacetone and pyruvate. Am J Clin Nutr 55:771–776. https://doi.org/10.1093/ajcn/55.4.771

Stepanov N, Efremenko E (2017) Immobilised cells of Pachysolen tannophilus yeast for ethanol production from crude glycerol. New Biotechnol 34:54–58. https://doi.org/10.1016/j.nbt.2016.05.002

Suh SO, Zhou JJ (2010) Methylotrophic yeasts near Ogataea (Hansenula) polymorpha: a proposal of Ogataea angusta comb. nov. and Candida parapolymorpha sp. nov. FEMS Yeast Res 10:631–638. https://doi.org/10.1111/j.1567-1364.2010.00634.x

Sutherland FC et al (1997) Characteristics of Fps1-dependent and -independent glycerol transport in Saccharomyces cerevisiae. J Bacteriol 179:7790–7795

Suwannarangsee S et al (2010) Characterization of alcohol dehydrogenase 1 of the thermotolerant methylotrophic yeast Hansenula polymorpha. Appl Microbiol Biotechnol 88:497–507. https://doi.org/10.1007/s00253-010-2752-7

Swinnen S, Klein M, Carrillo M, McInnes J, Nguyen HT, Nevoigt E (2013) Re-evaluation of glycerol utilization in Saccharomyces cerevisiae: characterization of an isolate that grows on glycerol without supporting supplements. Biotechnol Biofuels 6:157. https://doi.org/10.1186/1754-6834-6-157

Taccari M, Canonico L, Comitini F, Mannazzu I, Ciani M (2012) Screening of yeasts for growth on crude glycerol and optimization of biomass production. Bioresour Technol 110:488–495. https://doi.org/10.1016/j.biortech.2012.01.109

Takakuwa N, Nagahama S, Matsumura H, Kinoshita M, Ohnishi M (2013) Efficient conversion of crude glycerol into triacylglycerol by the yeast Pseudozyma sp. TYC-2187 for biodiesel production. J Oleo Sci 62:605–612

Tan JP, Md Jahim J, Wu TY, Harun S, Kim BH, Mohammad AW (2014) Insight into biomass as a renewable carbon source for the production of succinic acid and the factors affecting the metabolic flux toward higher succinate yield. Indust Eng Chem Res 53:16123–16134. https://doi.org/10.1021/ie502178j

Tang S, Boehme L, Lam H, Zhang Z (2009) Pichia pastoris fermentation for phytase production using crude glycerol from biodiesel production as the sole carbon source. Biochem Eng J 43:157–162

Tani Y, Yamada K (1987) Glycerol metabolism in Methylotrophic yeasts. Agric Biol Chem 51:1927–1933. https://doi.org/10.1080/00021369.1987.10868322

Thiru M, Sankh S, Rangaswamy V (2011) Process for biodiesel production from Cryptococcus curvatus. Bioresour Technol 102:10436–10440. https://doi.org/10.1016/j.biortech.2011.08.102

Thompson JC, He BB (2006) Characterization of crude glycerol from biodiesel production from multiple feedstocks. Appl Eng Agric 22:261–265

Tomaszewska L, Rywinska A, Gladkowski W (2012) Production of erythritol and mannitol by Yarrowia lipolytica yeast in media containing glycerol. J Ind Microbiol Biotechnol 39:1333–1343. https://doi.org/10.1007/s10295-012-1145-6

Tomaszewska L, Rakicka M, Rymowicz W, Rywinska A (2014a) A comparative study on glycerol metabolism to erythritol and citric acid in Yarrowia lipolytica yeast cells. FEMS Yeast Res 14:966–976. https://doi.org/10.1111/1567-1364.12184

Tomaszewska L, Rymowicz W, Rywinska A (2014b) Mineral supplementation increases erythrose reductase activity in erythritol biosynthesis from glycerol by Yarrowia lipolytica. Appl Biochem Biotechnol 172:3069–3078. https://doi.org/10.1007/s12010-014-0745-1

Tuza P, Manfro RL, Ribeiro NFP, Souza MMVM (2014) Production of renewable hydrogen by aqueous-phase reforming of glycerol over Ni-Cu catalysts derived from hydrotalcite precursors. Prog Sustain Energy Technol Gener Renew Energy 1:413–426

Uckun Kiran E, Trzcinski A, Webb C (2013) Microbial oil produced from biodiesel by-products could enhance overall production. Bioresour Technol 129:650–654. https://doi.org/10.1016/j.biortech.2012.11.152

Valerio O, Horvath T, Pond C, Misra M, Mohanty A (2015) Improved utilization of crude glycerol from biodiesel industries: synthesis and characterization of sustainable biobased polyesters. Ind Crop Prod 78:141–147

van Hoek P, Flikweert MT, van der Aart QJ, Steensma HY, van Dijken JP, Pronk JT (1998) Effects of pyruvate decarboxylase overproduction on flux distribution at the pyruvate branch point in *Saccharomyces cerevisiae*. Appl Environ Microbiol 64:2133–2140

van Zyl PJ, Kilian SG, Prior BA (1990) The role of an active transport mechanism in glycerol accumulation during osmoregulation by Zygosaccharomyces rouxii. Appl Microbiol Biotechnol 34:231–235

Varanda MG, Pinto G, Martins F (2011) Life cycle analysis of biodiesel production. Fuel Process Technol 92:1087–1094

Varrone C et al (2013) Enrichment of activated sludge for enhanced hydrogen production from crude glycerol. Int J Hydrog Energy 38:1319–1331

Veiga-Da-Cunha M, Firme P, Romao MV, Santos H (1992) Application of C nuclear magnetic resonance to elucidate the unexpected biosynthesis of erythritol by Leuconostoc oenos. Appl Environ Microbiol 58:2271–2279

Venkataramanan KP, Boatman JJ, Kurniawan Y, Taconi KA, Bothun GD, Scholz C (2012) Impact of impurities in biodiesel-derived crude glycerol on the fermentation by Clostridium pasteurianum ATCC 6013. Appl Microbiol Biotechnol 93:1325–1335. https://doi.org/10.1007/s00253-011-3766-5

Vershinin A (1999) Biological functions of carotenoids--diversity and evolution. Biofactors 10:99–104

Vodnar DC, Dulf FV, Pop OL, Socaciu C (2013) L (+)-lactic acid production by pellet-form Rhizopus oryzae NRRL 395 on biodiesel crude glycerol. Microb Cell Factories 12:92. https://doi.org/10.1186/1475-2859-12-92

Volpato G, Rodrigues RC, Heck JX, Ayub MAZ (2008) Production of organic solvent tolerant lipase by Staphylococcus caseolyticus EX17 using raw glycerol as substrate. J Chem Technol Biotechnol 83:821–828. https://doi.org/10.1002/jctb.1875

Wang Z, Zhuge J, Cao Y, Chen J, Fang H (2000) The key enzymes of metabolisms of glycerol in Candida glycerolgenesis. Wei sheng wu xue bao = Acta microbiologica Sinica 40:180–187

Watanabe T, Watanabe I, Yamamoto M, Ando A, Nakamura T (2011) A UV-induced mutant of Pichia stipitis with increased ethanol production from xylose and selection of a spontaneous mutant with increased ethanol tolerance. Bioresour Technol 102:1844–1848. https://doi.org/10.1016/j.biortech.2010.09.087

West TP (2013) Citric acid production by Candida species grown on a soy-based crude glycerol. Prep Biochem Biotechnol 43:601–611. https://doi.org/10.1080/10826068.2012.762929

White J et al. (2004) Top value added chemicals from biomass: results of screening for potential candidates from sugars and synthesis gas, vol 1. U.S. Department of Energy [Office of] Energy Efficiency and Renewable Energy, Washington, DC

Xiu S, Shahbazi A, Shirley V, Mims MR, Wallace CW (2010) Effectiveness and mechanisms of crude glycerol on the biofuel production from swine manure through hydrothermal pyrolysis. J Anal Appl Pyrolysis 87:194–198

Xu J, Zhao X, Wang W, Du W, Liu D (2012) Microbial conversion of biodiesel byproduct glycerol to triacylglycerols by oleaginous yeast Rhodosporidium toruloides and the individual effect of some impurities on lipid production. Biochem Eng J 65:30–36

Yamada H, Kumagai H, Kashima N, Torii H, Enei H, Okumura S (1972) Synthesis of L-tyrosine from pyruvate, ammonia and phenol by crystalline tyrosine phenol lyase. Biochem Biophys Res Commun 46:370–374

Yan Q, Zheng P, Dong JJ, Sun ZH (2014) A fibrous bed bioreactor to improve the productivity of succinic acid by Actinobacillus succinogenes. J Chem Technol Biotechnol 89(11):1760–1766. https://doi.org/10.1002/jctb.4257

Yang J, Xu M, Zhang X, Hu Q, Sommerfeld M, Chen Y (2011) Life-cycle analysis on biodiesel production from microalgae: water footprint and nutrients balance. Bioresour Technol 102:159–165. https://doi.org/10.1016/j.biortech.2010.07.017

Yang F, Hanna MA, Sun R (2012) Value-added uses for crude glycerol – a byproduct of biodiesel production. Biotechnol Biofuels 5:13. https://doi.org/10.1186/1754-6834-5-13

Yang LB, Zhan XB, Zheng ZY, Wu JR, Gao MJ, Lin CC (2014a) A novel osmotic pressure control fed-batch fermentation strategy for improvement of erythritol production by Yarrowia lipolytica from glycerol. Bioresour Technol 151:120–127. https://doi.org/10.1016/j.biortech.2013.10.031

Yang X, Jin G, Gong Z, Shen H, Bai F, Zhao ZK (2014b) Recycling biodiesel-derived glycerol by the oleaginous yeast *Rhodosporidium toruloides* Y4 through the two-stage lipid production process. Biochem Eng J 91:86–91

Yang LB, Zhan XB, Zhu L, Gao MJ, Lin CC (2016) Optimization of a low-cost hyperosmotic medium and establishing the fermentation kinetics of erythritol production by Yarrowia lipolytica from crude glycerol. Prep Biochem Biotechnol 46:376–383. https://doi.org/10.1080/10826068.2015.1045604

Yazdani SS, Gonzalez R (2007) Anaerobic fermentation of glycerol: a path to economic viability for the biofuels industry. Curr Opin Biotechnol 18:213–219. https://doi.org/10.1016/j.copbio.2007.05.002

Yen HW, Zhang Z (2011) Effects of dissolved oxygen level on cell growth and total lipid accumulation in the cultivation of Rhodotorula glutinis. J Biosci Bioeng 112:71–74. https://doi.org/10.1016/j.jbiosc.2011.03.013

Yen HW, Yang YC, Yu YH (2012) Using crude glycerol and thin stillage for the production of microbial lipids through the cultivation of *Rhodotorula glutinis*. J Biosci Bioeng 114:453–456. https://doi.org/10.1016/j.jbiosc.2012.04.022

Yilmaztekin M, Tay S (2013) Enhanced production of isoamyl acetate via biotransformation with *Lindnera saturnus* by in situ product removal with macroporous adsorption resins. Paper presented at the 4th European yeast flavour workshop

Yin X, Madzak C, Du G, Zhou J, Chen J (2012) Enhanced alpha-ketoglutaric acid production in Yarrowia lipolytica WSH-Z06 by regulation of the pyruvate carboxylation pathway. Appl Microbiol Biotechnol 96:1527–1537. https://doi.org/10.1007/s00253-012-4192-z

Yonehara T, Miyata R (1994) Fermentative production of pyruvate from glucose by Torulopsis glabrata. J Ferment Bioeng 78:155–159

Yu KO, Kim SW, Han SO (2010) Engineering of glycerol utilization pathway for ethanol production by Saccharomyces cerevisiae. Bioresour Technol 101:4157–4161. https://doi.org/10.1016/j.biortech.2010.01.066

Yu KO, Jung J, Ramzi AB, Kim SW, Park C, Han SO (2012) Improvement of ethanol yield from glycerol via conversion of pyruvate to ethanol in metabolically engineered Saccharomyces cerevisiae. Appl Biochem Biotechnol 166:856–865. https://doi.org/10.1007/s12010-011-9475-9

Yu KO, Jung J, Ramzi AB, Choe SH, Kim SW, Park C, Han SO (2013) Development of a *Saccharomyces cerevisiae* strain for increasing the accumulation of triacylglycerol as a microbial oil feedstock for biodiesel production using glycerol as a substrate. Biotechnol Bioeng 110:343–347. https://doi.org/10.1002/bit.24623

Yuksel A, Sasaki M, Goto M (2011) A new green technology: hydrothermal electrolysis for the treatment of biodiesel wastewater. Res Chem Intermed 37:131–143. https://doi.org/10.1007/s11164-011-0260-8

Yuzbashev TV et al (2010) Production of succinic acid at low pH by a recombinant strain of the aerobic yeast Yarrowia lipolytica. Biotechnol Bioeng 107:673–682. https://doi.org/10.1002/bit.22859

Zhang D, Liang N, Shi Z, Liu L, Chen J, Du G (2009) Enhancement of α-ketoglutarate production in Torulopsis glabrata: redistribution of carbon flux from pyruvate to α-ketoglutarate. Biotechnol Bioprocess Eng 14:134–139. https://doi.org/10.1007/s12257-008-0169-2

Zhang X, Yan S, Tyagi RD, Surampalli RY, Valero JR (2016) Energy balance of biofuel production from biological conversion of crude glycerol. J Environ Manag 170: 169–176. https://doi.org/10.1016/j.jenvman.2015.09.031

Zhang X, Chen J, Yan S, Tyagi RD, Surampalli RY, Li J (2017) Lipid production for biodiesel from sludge and crude glycerol. Water Environ Res 89:424–439 doi:https://doi.org/10.2175/106143017X14839994523424

Zhao L, Yu J, Zhang X, Tan T (2010) The ethanol tolerance of Pachysolen tannophilus in fermentation on xylose. Appl Biochem Biotechnol 160:378–385. https://doi.org/10.1007/s12010-008-8308-y

Chapter 13
Frontier in Antifungal Treatments Against Major Human Fungal Opportunistic Pathogen *Candida* Species and Medically Important Fungi

Nitnipa Soontorngun, Pichayada Somboon, and Kwanruthai Watchaputi

Abstract The extensive uses of antifungal agents to treat fungal infections have created a global public health issue of drug resistance. *Candida* species and other pathogenic mycoses are the leading causes of invasive fungal infections and high mortality rate in human population. A limited number of antifungal agents show fungicidal effect; repetitive uses of fungistatic drugs lead to the development of drug resistance. New antifungal agents with a broader spectrum of activity and novel mechanisms of action have been recently developed to fight against resistant fungal strains and clinical isolates. This would create a possibility to investigate antifungal combinations in vitro, in yeast and in animal models. If the drugs have different mechanisms of action, increased drug potency and efficacy with reduced toxicity may be rewarded. Recently, there are several investigational antifungal agents on repurposing drugs and natural products from different sources—plants, microbial, and marines—which may be considered when designing antifungal drug combinations. Imminently, more classes of antifungals from natural products may be added to the current antifungal armamentarium. Therefore, the new frontier of combination therapy and natural drug discovery should continue to be pursued with anticipation while excessive and repetitive antifungal usage shall be refrained.

Keywords Antifungal · *Candida* · Combination therapy · Drug resistance · Natural product · Yeast model

---

N. Soontorngun (✉) · P. Somboon · K. Watchaputi
Division of Biochemical Technology, School of Bioresources and Technology, King Mongkut's University of Technology Thonburi, Bangkok, Thailand
e-mail: nitnipa.soo@kmutt.ac.th

A. Sibirny (ed.), *Non-conventional Yeasts: from Basic Research to Application*,
https://doi.org/10.1007/978-3-030-21110-3_13

13.1 Model Organisms for Studying Yeast and Fungal Pathogens

Candida infections are associated with high risk of mortality rate that could be as high as 90%, depending on associated risk factors (De Rosa et al. 2016; Campoy and Adrio 2017a). The predominant causing pathogen of all forms of candidiasis is *Candida albicans* (Kołaczkowska and Kołaczkowski 2016). Approximately 1.2 billion individuals are affected worldwide with at least 1.5 million deaths each year from fungal-associated infections and diseases (Campoy and Adrio 2017a; Chang et al. 2017). *C. albicans* is a diploid fungal opportunistic human pathogen. This yeast is a commensal organism found in the gastrointestinal tracts of healthy human hosts, but it may become a severe and persistent pathogen in people with compromised immunity. In fact, *C. albicans* is responsible for a large proportion of human fungal-associated diseases, ranging from superficial infections such as oral thrush to life-threatening systemic candidiasis. Invasive fungal infections especially in immunocompromised patients have significantly increased risks. Fungal species are widely distributed in environments and other organic substrates. There are approximately 611,000 species of fungi, (Mora et al. 2011), although only about 600 species are human pathogens (Brown et al. 2012a). Major risk factors associated to invasive fungal infections include neutropenia, haematological malignancies, bone marrow transplantation, chemotherapy, HIV infection, invasive medical procedures, and uses of immune suppressive agents (Baddley 2011a). Other risk factors are malnutrition, solid organ transplantation, severe burns or prolonged stays in intensive care, and major surgery (Baddley 2011b). Infection can be transmitted by inhalation of spores (aspergillosis, cryptococcosis, histoplasmosis), percutaneous inoculation in cutaneous and subcutaneous infections (dermatophytosis, madura foot), penetration through the mucosa by commensal organisms such as *C. albicans*, and the ingestion of a toxin in contaminated food or drink (gastrointestinal disease). Infections may be mild and only superficial (dermatophytosis and *Tinea versicolor*) or may cause life-threatening systemic infections (candidiasis, aspergillosis and mucormycosis). However, host immunity and physiological condition are important factors that confer the symptoms of the fungal disease caused by a given fungal agent. For example, *Candida* spp. may infect a local site (mucocutaneous or cutaneous candidiasis, onychomycosis) or cause systemic infections (renal, liver abscess, lung, and central nervous system). Recent reviews describe several different approaches for combating invasive mycoses, including the application of repurposing drugs such as inhibitors of HIV aspartic peptidase and immunotherapy for the development of therapeutic vaccines (Taborda and Nosanchuk 2017).

In addition, non-*albicans Candida* species are also reported as causing agents and correlated with increased virulence and death (De Rosa et al. 2016; Bassetti et al. 2006). Some of these emerging species exhibit resistance to antifungal agent triazoles and/or amphotericin B which are indicated as *Candida glabrata*, *Candida parapsilosis,* and *Candida tropicalis* (De Rosa et al. 2016; Richter et al. 2005). To counteract with the fungicidal and fungistatic effects against antifungal drugs,

Candida develops mechanisms to resist current clinical drugs. The mechanisms of antifungal resistance have been reported at the molecular level which can be divided into three main mechanisms. The first is reducing the drug concentration inside the fungal cells; second is decreasing the affinity of the drug for the drug's target; and the last is compensating the drug effect by modification of metabolism (Vandeputte et al. 2011a). One of the important drug-resistant mechanisms in *Candida* species is reducing intracellular accumulation by upregulation of drug efflux pumps. At least two families of multidrug transporters which include the ABC (ATP-binding cassette) transporter family and the major facilitator superfamily (MFS) have been demonstrated to confer resistance to one of the most prevalent antifungal drug used in clinical settings, azole class, in several *Candida* species (Jiang et al. 2012). *CDR1* and *CDR2* are the major drug efflux transporters which are associated with azole resistance in *C. albicans*. Moreover, the other ABC transporters involved in azole resistance have been reported in *C. albicans* and other *Candida* species, including *C. glabrata* (*CgCDR1*, *CgCDR2*, *CgSNQ2*) and *AFR1* in *C. neoformans (*Campoy and Adrio 2017b*)*. Furthermore, upregulated expression of the MFS transporter gene Mdr1 causes reduction of fluconazole resistance in the pathogenic yeast *C. albicans* and *C. dubliniensis* (Campoy and Adrio 2017b; Mogavero et al. 2011).

Antifungal drugs/agents are shown to attack different targets, including fungus-specific components of the cell wall or cell membrane, or cellular processes such as metabolism, DNA synthesis, mitochondrial function, or the stress response, while some display specificity to fungus-specific transporters (Perfect 2017). Despite the wealth of information on processes responsible for drug resistance in *C. albicans* and other pathogenic fungi (Scorzoni et al. 2017), less is known about the transcription factors involved in this phenomenon. In the budding yeast *Saccharomyces cerevisiae*, zinc cluster proteins form a major class of transcriptional regulators, with some members playing a prominent role in controlling the expression of multidrug resistance genes (MacPherson et al. 2006; Moye-Rowley 2003a). Zinc cluster proteins are unique to fungi and may represent potential targets for antifungal action. They belong to a subfamily of zinc finger proteins and are characterized by the well-conserved zinc cluster motif, $CysX_2CysX_6CysX_{5\text{-}12}CysX_2CysX_{6\text{-}8}Cys$. The six conserved cysteines bind to two zinc atoms, coordinating the proper folding of the zinc finger which is essential for DNA recognition (reviewed in (MacPherson et al. 2006)). In *S. cerevisiae*, zinc cluster proteins Pdrl and Pdr3 positively control the expression of *PDR5*, *SNQ2,* and *YOR1*, encoding ABC transporters that confer resistance to various cytotoxic compounds (Kolaczkowska and Goffeau 1999; Georgiev 2000). Other zinc cluster proteins (such as Yrrl, Rdrl, and Stb5) are also part of the PDR network and modulate resistance to some toxic compounds (reviewed in (MacPherson et al. 2006; Moye-Rowley 2003b)).

In *C. albicans*, four well-characterized zinc cluster proteins (Fcrl, Tacl, Upc2, and Mrrl) have been implicated in modulating drug resistance. Expression of the *FCR1* (fluconazole resistance 1) gene in a *S. cerevisiae* strain lacking *PDR1* and *PDR3* increases resistance to fluconazole (Talibi and Raymond 1999). Surprisingly, deletion of *FCR1* in *C. albicans* results in increased resistance to fluconazole (Talibi and Raymond 1999). Tacl (transcriptional activator of CDR genes) activates the expression of *CDR1* and *CDR2* encoding multidrug transporters (5). Interestingly, a *TAC1–2* mutant recovered from an azole-resistant strain is responsible for the constitutive overexpression of the *CDR1* and *CDR2* genes (Coste et al. 2004). Independent work from two laboratories showed that Upc2 regulates transcription of many ergosterol biosynthetic (*ERG*) genes, including *ERG11* (MacPherson et al. 2006; MacPherson et al. 2005; Silver et al. 2004). The Δ*upc2* strain is sensitive to various drugs, including azoles (MacPherson et al. 2005; Silver et al. 2004). This phenotype can be explained by the fact that removal of *UPC2* prevents induction of *ERG* gene expression upon azole treatment (without affecting their basal mRNA levels). A study has identified a new zinc cluster regulator responsible for the overexpression of a multidrug transporter gene *MDR1*, called Mrrl (Morschhäuser et al. 2007). Deletion of MRR1 abolishes the overexpression of *MDR1* in clinical azole-resistant isolates as well as the induction of this gene upon treatment with substrates of Mdrl such as benomyl or hydrogen peroxide. In addition, this study also identified point mutations responsible for mediating azole resistance on the *MRR1* allele. Previously, it is shown that deletion of *RDS2*, encoding another zinc cluster protein, sensitizes cells to ketoconazole in *S. cerevisiae* (Akache and Turcotte 2002). Rds2 is implicated in transcriptional regulation of genes involved in gluconeogenesis (Soontorngun et al. 2007). A clear ortholog of *RDS2* is found in *C. albicans* and is called *CWT1* (for cell wall transcription factor) (Moreno et al. 2003). A strain lacking *CWT1* is susceptible to membrane perturbing agents and has altered cell wall composition (Moreno et al. 2003). Overall, these transcription factors play a role in mediating antifungal drug resistance. Recently, small-molecule targeting of *C. glabrata* Pdr1 transcription factor-binding site in Gal11A KIX domain of mediator is reported as a novel therapeutic approach in fungal infection. The lead compound (iKIX1) is shown to impede Pdr1-dependent gene induction and resensitize azole-resistant *C. glabrata* in vitro and in animal models (Nishikawa et al. 2016).

Undeniably, the three most prevalent fungal pathogens belong to *Candida, Aspergillus*, and *Cryptococcus* genera. Their infections are normally found in invasive fungal infections (IFIs) and are a major cause of morbidity and mortality among immunocompromised patients such as cancer, organ transplant recipients, or AIDS (Pappas et al. 2010). *Aspergillus* species are the second most infectious fungi after *Candida* species (Pappas et al. 2010). *Aspergillus* is the leading cause of invasive mold infections especially in immunocompromised patients, with an estimated 300,000 cases worldwide every year (Chowdhary et al. 2014; Brown et al. 2012b). *Aspergillus* spp. include *A. fumigatus, A. flayus,* and *A. niger* that are members of the family *Trichocomaceae* of the *Ascomycetes.* They are haploid filamentous fungi that grow as a mycelium of branched or multinucleate hyphae (Kwon-Chung and Sugui 2013). Infection with these species often begins with inhalation of conidia (asexual spores) which germinate in the lungs and begin to grow hyphally. According to an abundant asexual reproduction cycle, this fungus produces billions of airborne conidia and has the ability of surviving in very different environments, such as those with temperatures up to 60 °C (Kwon-Chung and Sugui 2013).

The main antifungal drugs used in the treatment for aspergillosis are triazoles, including itraconazole (ITC), voriconazole (VRC), and posaconazole (POS).

Furthermore, new triazole, isavuconazole, has been proven for activity against *Aspergillus* (Miceli and Kauffman 2015). Each triazole is suitable for different symptoms. For instance, although VRC is recommended as a first-line therapy for IA (Maschmeyer et al. 2007; Walsh et al. 2008a), ITC is still commonly used for chronic and allergic noninvasive forms of aspergillosis (Walsh et al. 2008a; Howard et al. 2010), and POS was shown to reduce the number of invasive fungal infections in neutropenic patients (Verweij et al. 2016). Importantly, triazoles are the only orally available anti-*Aspergillus* agents. Therefore, they are essential for long-term therapy.

Although the most common antifungal resistance occurs in *Candida* species, resistance in other types of less common fungi is also a problem. Emerging resistance to the triazoles which are first-line treatment is commonly found, and it threatens the effectiveness of life-saving medications (Susan et al. 2009). Drug resistance in *Aspergillus* occurs through both with and without having treatment but from agricultural products treated with antifungal drugs. Unfortunately, up to 12% of *Aspergillus* infections are predicted to be resistant to treatment (Rivero-Menendez et al. 2016). More specifically, patients with stem cell and organ transplants in the United States were identified in the antifungal resistance of *Aspergillus* specimens up to 7% (Brown et al. 2012b; Arendrup 2014). Additionally, there are some alternative therapies to triazoles that can function as rescue treatments, such as echinocandins or amphotericin B (Walsh et al. 2008a).

C. neoformans is the only major pathogen of humans that is a member of the phylum Basidiomycetes. It is usually found as haploid yeast-forming cells, and although it can grow in a filamentous form, hyphae are rarely seen in infected patients, suggesting that morphological transitions do not play a role in virulence. *C. neoformans* is almost exclusively an opportunistic pathogen, entering the body through the lungs and, in immunocompromised patients, disseminating to the central nervous system. A significant proportion of AIDS patients contract it, and those who do need lifelong antifungal therapy. Unlike aspergillosis, cryptococcal infections are rare among patients with immune-mediated inflammatory diseases (IMIDs) treated with biologics. IMIDs are a group of diseases including rheumatoid arthritis (RA), inflammatory bowel disease (IBD), psoriasis, psoriatic arthritis (PsA), and ankylosing spondylitis (AS) (Vallabhaneni and Chiller 2016). Cryptococcal infection treatment normally uses azoles or amphotericin; however, TNF-α inhibitors could enhance the response to treatment of patient and avoid death (Tsiodras et al. 2008).

13.2 Major Classes of Antifungal Drugs

There are only a limited number of available drugs to fight against invasive fungal infections. Only some molecular classes that target distinct fungal metabolic pathways are currently used in clinical practice to treat fungal infections (Vandeputte et al. 2011b). Treatment of fungal infection is difficult due to developed resistance, limited numbers of available antifungal, and side effects due to toxicity of drugs (Alborzi et al. 2012). Numerous lines of evidence suggest that extensive and repetitive use of antifungal drugs during ongoing treatment of fungal infections permit this opportunistic yeast to acquire new mechanisms of drug resistance (Sanglard 2002). Currently, improvement of three classes of antifungals available—polyenes, azoles, and echinocandins—has been done for the development of future generation to enhance efficacy or reduce toxicity (Perfect 2017).

13.2.1 Polyenes

Polyenes are cyclic amphiphilic organic molecules known as macrolides. Polyene drugs are believed to target ergosterol, the main sterol component of fungal membranes (Fig. 13.1). Amphotericin B (AmB) is one of polyenes that was one of the gold standards for the treatment of systemic fungal infections. Indeed, AmB is active against most yeasts and filamentous fungi. It is recommended for the treatment of infections caused by *Candida*, *Aspergillus*, *Fusarium*, *Mucor*, *Rhizopus*, *Scedosporium*, *Trichosporon*, and *Cryptococcus*. The initial formulation was amphotericin B deoxycholate (D-AmB), which was developed in the 1950s. For many decades, D-AmB was the only antifungal agent available for the treatment of invasive fungal diseases. However, D-AmB had significant dose-limiting toxicity. Then, new less toxic formulations of AmB were developed. Liposomal

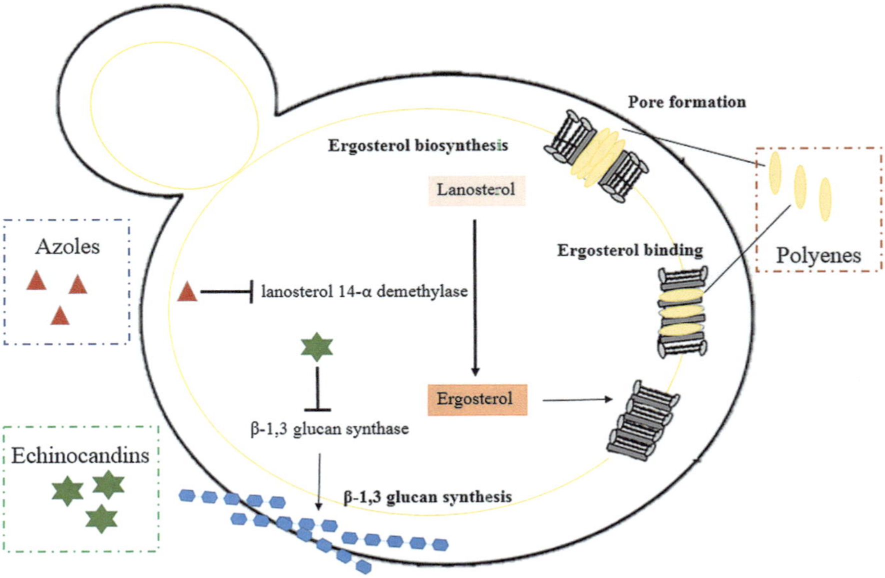

Fig. 13.1 Mechanisms of action of current antifungal drugs

amphotericin B (AmBisome®; L-AmB) is a unique lipid formulation of amphotericin B that has been used for nearly 20 years to treat a broad range of fungal infections. L-AmB had significantly reduced toxicity, while antifungal activity is retained (Stone et al. 2016).

13.2.2 *Azoles*

Azoles are the most widely used antifungals in clinical trials. Azoles target at lanosterol demethylase which is an enzyme in the ergosterol biosynthesis pathway (Fig. 13.1). Lanosterol demethylase is encoded by *ERG11* in *C. albicans* and *C. neoformans*, and by *CYP51A* and *CYP51B* in *A. fumigatus*. The azoles cause accumulation of 14α-methylergosta-8, 24(28)-dien-6β, 3α-diol which is a toxic sterol generated by Erg3. Moreover, azoles cause reduction of ergosterol and disruption of fungal cell membrane integrity. Azoles can be classified into two groups, based on the molecular structure, containing imidazoles or triazoles. The first developed azoles are imidazoles, such as clotrimazole, miconazole, and ketoconazole. The imidazoles have high toxicity and severe side effects. Thus, the triazoles are generated. Itraconazole and fluconazole are the first generation of triazoles which show a broad antifungal activity and improved safety profiles (Zheng et al. 2018). Fluconazole is a member which exhibits high and reliable bioavailability (Chang et al. 2017). Voriconazole and posaconazole are the second generation of triazoles that are approved by the US Food and Drug Administration (FDA). They have a broad spectrum of activity and exhibit effective antifungal activity against *Candida* species (Lei et al. 2018).

13.2.3 *Echinocandins*

Echinocandins are a new class of antifungals which are cyclic lipopeptide molecules. The echinocandins become active by inhibiting fungal 1,3-β-glucan synthase (Fig. 13.1) which is a major enzyme complex in cell wall synthesis (Chang et al. 2017). There are three classes of echinocandins which were approved by the European Medicines Agency (EMEA), containing caspofungin, anidulafungin, and micafungin (Kofla and Ruhnke 2011). Echinocandins exhibit a fungicidal effect against *Candida* spp. and fungistatic effect against *Aspergillus* spp. However, echinocandins did not exhibit antifungal activity against *Cryptococcus* spp. (Perlin 2011). These three echinocandins have reached the market in many countries, leaving some others in early clinical trials.

13.3 Current Recommendations for Antifungal Therapy

A promising strategy against antifungal drug-resistant fungi is to extend the life span and efficacy of the currently clinically used drugs by using combination therapy. Combining drugs has the potential to enhance the effectiveness and specificity compared to individual drug treatments and can slow the evolution of resistance (Spitzer et al. 2017). Combination of different antifungal drugs has been shown to promote drug synergy or some additive effect against many fungi although antagonism has also been observed (Spitzer et al. 2017; Vazquez 2007; Robbins et al. 2016).

13.3.1 Use of Drug Combination

13.3.1.1 Candidiasis

Candidiasis is one of the opportunistic fungal infections that causes high morbidity and mortality among immunocompromised patients (Ahmad et al. 2017). There are at least 90% of *Candida* species that cause invasive fungal infection, including *C. albicans*, *C. glabrata*, *C. tropicalis*, *C. parapsilosis*, and *C. krusei* (Pappas et al. 2016). Topical therapy is recommended for primary treatment of patients with an initial episode of mild disease (Pappas et al. 2009). The recommended treatment for moderate-to-severe disease is oral fluconazole 100–200 mg (3 mg/kg) once daily for 7–14 days. The recommended treatment for systemic infection is oral fluconazole 200–400 mg (3–6 mg/kg) daily for 14–21 days (Armstrong et al. 2016). Treatment of candidiasis is challenging due to the resistance that can develop during therapy and the limited number of available antifungal compounds (Cruz et al. 2018). Most *Candida* species except for *C. krusei* and *C. glabrata* respond to oral fluconazole. However, treatment of vulvovaginal candidiasis caused by *C. glabrata* is a problem. Azole treatment such as voriconazole is frequently unsuccessful against *C. glabrata*. From the latest guidelines published by the Infectious Diseases Society of America, combination treatment is not recommended for invasive candidiasis. However, some combination therapy is recommended for treatment, for example, in complicated vulvovaginal candidiasis that requires intravaginal administration of topical agents for 5–7 days or oral administration of fluconazole 150 mg every 72 hours for three doses (Pappas et al. 2016). Moreover, some reports showed that four of five patients received echinocandin in combination with fluconazole or L-AMB are alive and with good graft function at 1-year follow-up (Bartoletti et al. 2013).

13.3.1.2 Aspergillosis

Invasive aspergillosis is associated with high morbidity and mortality in immunocompromised patients, especially in cancer patients or who had to undergo hematopoietic stem cell transplantation (HSCT) (Pappas et al. 2016). Invasive *Aspergillus* infections are a serious problem worldwide, especially in the Middle East region which is associated with a burden on healthcare services. Antifungal agents including D-AMB, ABLC, L-AMB, ABCD, itraconazole, voriconazole, posaconazole, and caspofungin were approved by FDA for the treatment of invasive aspergillosis. Caspofungin is indicated for treatment of patients who are refractory to or intolerant of other approved therapies. The currently recommended dosage of caspofungin in adults consists of a single 70-mg loading dose on day 1, followed by 50 mg/day thereafter, administered by slow IV infusion of ~1 h (Walsh et al. 2008b). Some combination study was investigated against invasive aspergillosis. In a study, voriconazole and anidulafungin or placebo are randomly assigned to treatment with 454 patients with hematological malignancies (HMs) or hematopoietic cell transplantation (HCT) and suspected or documented invasive aspergillosis. Primary analysis is done in the modified intention-to-treat population of 277 patients in whom invasive aspergillosis was confirmed. Mortality rates at 6 weeks of patients who have invasive aspergillosis are lower in combination therapy than monotherapy (Marr et al. 2015).

13.3.1.3 Cryptococcosis

Cryptococcosis is a global invasive fungal infection associated with significant morbidity and mortality. Since the last Infectious Diseases Society of America (IDSA) guidelines in 2000, the extended-spectrum azoles, posaconazole and voriconazole, and the echinocandins have become available as new antifungal drugs which exhibit anti-cryptococcosis. The former have been studied clinically in salvage situations. Moreover, the lipid polyene formulations and drug combination studies have been considered for treatment with cryptococcosis. Some combination therapies are recommended for cryptococcosis treatment. For primary therapy in patients at low risk for therapeutic failure, the treatment is considered induction therapy with a combination of D-AmB and flucytosine for only 2 weeks, following consolidation with fluconazole (800 mg [12 mg/kg] per day orally) for 8 weeks. In some *C. gattii* infection, for very large and multiple cryptococcomas, a combination of D-AmB and flucytosine therapy for 4–6 weeks is considered, followed by fluconazole for 6–18 months, depending on whether surgery was performed (Perfect et al. 2010).

13.3.1.4 Mucormycosis

Mucormycosis is an invasive fungal infection that causes morbidity and mortality in patients especially in hematologic malignancies. Mucormycosis is primarily caused by *Mucorales*, a filamentous fungus of the *Mucormycetes* class. Mucormycosis exhibits a rapid progression in patients with diabetes mellitus or immunocompromised including hematologic malignancies and stem cell and solid organ transplants (Farmakiotis and Kontoyiannis 2016). Evidence-based recommendations guideline for the diagnosis and treatment of mucormycosis was developed by using the evidence criteria set forth by the American Infectious Diseases Society. Diagnosis of mucormycosis depends on the histology and/or identification of the isolate at the species level in case of absence of validated biomarkers. For primary treatment of mucormycosis, D-AmB is an antifungal agent that has been approved by the US Food and Drug Administration. However, D-AmB was replaced by lipid formulations of AmB, including L-AmB, AmB lipid complex (ABLC), and AmB colloidal dispersion (ABCD), because of its toxicity (Skiada et al. 2013). Posaconazole exhibits an effective antifungal activity against mucormycosis when compared with itraconazole and posaconazole (Skiada et al. 2013). Posaconazole monotherapy is not recommended to be a primary treatment for mucormycosis. However, the European Conference on Infections in Leukemia (ECIL) recommended posaconazole as an option for mucormycosis patients intolerant of AmB or who need to maintain the therapy (Greenberg et al. 2006). However, combination of posaconazole and L-AmB is the option for second-line treatment (Skiada et al. 2013).

In a study, combination of AmB and posaconazole is demonstrated against the hyphae of Zygomycetes. In vitro studies have demonstrated the synergistic effect of combination between posaconazole and amphotericin B against hyphae (Perkhofer et al. 2008). In animal model study, L-AmB combined with caspofungin improves the survival of diabetic ketoacidotic mice infected with *Rhizopus oryzae* (Spellberg et al. 2005). Combination of L-AmB with anidulafungin or micafungin can improve survival in infected mice with *Rhizopus oryzae* compared to placebo or monotherapy (Ibrahim et al. 2008).

13.3.1.5 Coccidioidomycosis

Coccidioidomycosis is a systemic fungal infection caused by the dimorphic fungi *Coccidioides immitis* and *Coccidioides posadasii*. The extent of infection and the immune system of the host are associated with clinical manifestations. Treatment of *Coccidioides* infection mostly used fluconazole for mild-to-moderate infection and AmB formulations for severe infection. Combination study was determined against coccidioidomycosis in nine children who had *Coccidioides* infections. The children were treated with a combination of voriconazole and caspofungin after failing conventional therapy consisting of a triazole, AmB, or a combination of both. The result found that eight of nine patients currently get better (Levy et al. 2013).

13.3.2 Use of Natural Antifungal Products

Undeniably, new antifungal agents are still in need due to drug resistance (Beardsley et al. 2018). This ability also allowed fungi to easily resist other agents within the same class because of the existing signature backbone and the increasing difficulty to create new drugs by modifying existing chemical classes. It is a challenge for scientists to search new classes of antifungal agents. Natural products (NPs) are getting more attention as powerful sources to discover antimicrobial agents, including antibacterial and antifungal agents (Marinelli 2009). NPs are chemicals normally called "secondary metabolites" with diverse structures and biological activities that originate from bacterial, fungal, plant, and marine animal sources (Savoia 2012; Radulovic et al. 2013; Bhatnagar and Kim 2012). In general, secondary metabolites are not essential for the growth, development, or reproduction of an organism but possibly help in adapting surrounding environment or defense mechanism against predators which assist in the survival of the organism (Dias et al. 2012; Croteau et al. 2000). According to biological properties, it has been revealed that NPs are important and valuable agents in pharmaceuticals, herbicides, and insecticides.

After the discovery of penicillin from *Penicillium notatum* by Alexander Fleming in 1929, thousands of bioactive secondary metabolites have been characterized and called this period of time "Golden Age" during the 1940s and 1950s. Unfortunately, the rate of discovery has then been slowed down (Marinelli 2009). Since 2000, 22 new antibiotics have been launched for treating infections in humans, but only 5 of these represented new compound classes. Among them, three classes originated from NPs, including the lipopeptide daptomycin, the pleuromutilin retapamulin, and the tiacumicin fidaxomicin (Harvey et al. 2015a). Unlike antibacterial agents, during 30 years, only echinocandin has been introduced as an antifungal agent originated from NP (Roemer and Krysan 2014). NPs used in the treatment of infections by bacteria, fungi, parasite, and virus are reviewed in (Richter et al. 2005). In the past, discovery of antimicrobial agents from NPs has been limited by the difficulties in identification and small library of identified compounds. Nowadays, microbial genomics and metabolomics technology have definitively provided such powerful tools to exploit and bring attention back to NPs' discovery (Harvey et al. 2015b).

13.3.2.1 Antifungal Compounds of Plant Origin

Plants have been used in traditional medicine for a long time. It is estimated that 14–28% of species of higher plants are used in therapy. Their compositions such as extracts and essential oils are important in the treatment of infectious diseases. Plants are the largest source of active compounds that could be used in medicine. The most famous and well-known example to date would be the synthesis of the anti-inflammatory agent, acetylsalicyclic acid or aspirin derived from the natural product, salicin isolated from the bark of the willow tree *Salix alba* L. (Facts et al. 2004). Morphine is a commercially important drug, first isolated in 1803 from

Papaver somniferum L. (opium poppy) (Dias et al. 2012). However, the most well-known NP from the plant *Taxus brevifolia* (Pacific Yew) is paclitaxel or taxol which is the most widely used breast cancer drug (Cragg 1998). Plant NPs used for antifungal treatment are normally identified from herbal plants (Martin and Ernst 2004), for example, berberine (Fig. 13.2), which has various biological activities against bacteria, protozoa, and fungi (Tan et al. 2011). Berberine is a protoberberine isoquinoline alkaloid compound found in a variety of plants, such as *Berberis aquifolium, Berberis vulgaris, Berberis aristata, Hydrastis canadensis, Phellodendron amurense, Coptis chinensis*, and *Tinospora cordifolia* (Mišík et al. 1995; Bezakova et al. 1996; Chen et al. 2008). Importantly, many studies have shown that berberine has a significant antifungal effect against *Candida albicans* and *Candida tropicalis* strains and biofilms (Iwazaki et al. 2010; Li et al. 2013; Dhamgaye et al. 2014; Shao et al. 2016). Plant oil is also a source of antifungal agents, for example, castor oil, the oil from the seeds of *Ricinus communis*. Castor oil contains ricinoleic acid which is a substrate in the preparation of synthesis of undecylenic acid. Undecylenic acid (Fig. 13.2) is mainly used in the treatment of superficial mycoses and *Candida albicans (Shi* et al. 2016*)*.

In addition to their aroma, plants' essential oils are rich in bioactive compounds with various biological activities, for example, antioxidant, anti-inflammatory, and antimicrobial. Thymol (Fig. 13.1) is the main monoterpene phenol found in essential oils isolated from various plants of *Lamiaceae* family such as *Thymus, Ocimum, Origanum, Monarda* genera, and other plants in the *Verbenaceae, Scrophulariaceae, Ranuncolaceae*, and *Apiaceae* families (Marchese et al. 2016). Thymol exhibits antifungal activity against various fungi, for instance, *C. albicans*, *Cryptococcus neoformans*, and *Aspergillus fumigatus* through targeting membrane integrity and reducing ergosterol content and inhibition of plasma membrane H + ATPase,

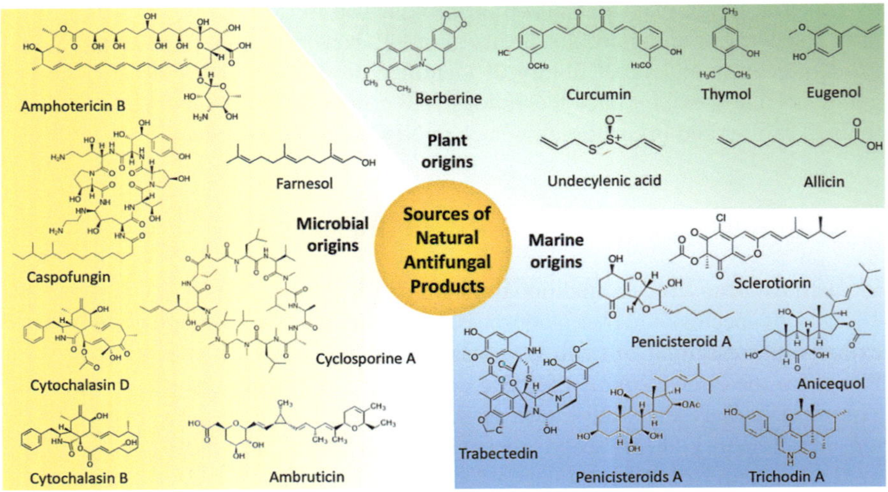

Fig. 13.2 Sources of natural antifungal products and examples

resulting in cellular acidification (Guo et al. 2009; Faria et al. 2011; Ahmad et al. 2010a). Additionally, thymol is synergistic with amphotericin B, fluconazole, and itraconazole and additive in combination with amphotericin B, fluconazole, and ketoconazole (Kim et al. 2008).

Eugenol (Fig. 13.2) and methyleugenol, which are present in high levels in the extracts of basil and clove, have also been reported for their antifungal properties. They disrupt ergosterol biosynthesis and provide synergistic effect with fluconazole against fluconazole-susceptible and resistant isolates of *C. albicans (Ahmad* et al. 2010b*)*. Moreover, similar to thymol, eugenol has been shown to inhibit the plasma membrane H + ATPase of fungal cells (Ahmad et al. 2010a). Enhancing antifungal activity of drug especially azoles has been reported for various other phenolic compounds found in plant extracts, such as berberine, curcumin (Fig. 13.2), and dihydrobenzaldehydes. They synergized fluconazole activity against *C. albicans* and *C. neoformans* through inhibition of the oxidative stress response pathway (Iwazaki et al. 2010; Faria et al. 2011; Sharma et al. 2010). Allicin (Fig. 13.2) is a compound isolated from garlic extract. Its mechanism of action reveals that the disruption of the vacuolar function requires the presence of ergosterol in the membrane (Borjihan et al. 2009). It also synergizes with amphotericin B and fluconazole against *C. albicans (Borjihan* et al. 2009*)*.

13.3.2.2 Antifungal Compounds of Microbial Origin

Microorganisms are excellent producers of useful NPs which many of them have been applied as antibiotics, antifungal agents, anticancer agents, immunosuppressants, and bioherbicides, as described in (Negri et al. 2014). Most of NPs with antimicrobial activities are produced by the bacteria *Streptomyces* species as described in (Golinska et al. 2014; Watve et al. 2001). *Streptomyces* species produced many important compounds in several antimicrobial classes: macrolides (tylosin, spiramycin); aminoglycosides (neomycin, kanamycin,); β-lactams (cephamycin, carbapenems); tetracyclines (tetracycline, chlortetracycline, oxytetracycline); polyenes (candicidin, amphotericin B, nystatin); peptides (actinomycin); and chloramphenicol (Watve et al. 2001). Echinocandins including caspofungin (Fig. 13.2), micafungin, and anidulafungin are the major antifungal drugs currently used as first-line therapy for candidemia/invasive candidiasis and as second-line drugs for aspergillosis because of their high activity against fluconazole-resistant *Candida* spp. and filamentous fungi (Negri et al. 2014). Echinocandins, such as caspofungin, inhibit the synthesis of glucan synthase enzyme which is involved in the synthesis of β(1,3)-d-glucan of the fungal cell wall (Sucher et al. 2009).

Echinocandins were initially obtained by fermentation of filamentous fungi including *Aspergillus* sp. (Barrett 2002a). NP drugs are safe and have minimal drug interactions and favorable pharmacokinetics (Barrett 2002b; Denning 2003; Cappelletty and Eiselstein-McKitrick 2007). However, they are normally modified from natural structures with the goal of increasing activity, for example, azole and echinocandin (Negri et al. 2014). Amphotericin B (Fig. 13.2) was originally isolated from *Streptomyces nodosus* in 1955 and still considered as the most effective antifungal drug (Dutcher 1968). Amphotericin B binds to ergosterol, creating holes in the fungal membrane (Mesa-Arango et al. 2012). Currently, amphotericin B is formulated with lipids, aiming to deduce the dose-dependent toxicity such as renal impairment and hypokalemia (Laniado-Laborin and Cabrales-Vargas 2009).

Besides, fungal diversity represents an abundant source for the discovery of novel compounds with antimicrobial activities. They are often viewed as food as exemplified by mushrooms, in the preparation of alcoholic beverages (yeasts), and for medication purpose as traditional medicine. Nowadays, their applications have extended to enzymes, biological control, antimicrobial agents, and other biological products (Marinelli 2009). The most well-known natural product derived from a fungus is penicillin from *Penicillium notatum* discovered by Fleming in 1929 (Mann 2000). Another NP derived from *Penicillium* sp. is griseofulvin, which was the antifungal drug available since the early time of antifungal agent discovery (Campoy and Adrio 2017b; Papich 2016). Griseofulvin functions by interfering microtubule assembly and inhibiting mitosis through binding to tubulin (Papich 2016). Sordarin was first isolated in 1969 from the fungus *Sordaria araneosa (Hauser and Sigg 1971).* Currently, they are available as semisynthetic natural products that are structurally modified from their natural structure (Odds 2001). Sordarins demonstrate potent and selective inhibitory activity of fungal protein synthesis by stabilizing the EF2/ribosome complex (Domínguez et al. 1998). There are many studies reporting excellent antifungal activity of sordarins against diverse model organisms, for example, *Saccharomyces cerevisiae* and *Candida albicans*, but are not yet available for clinical use (Justice et al. 1998; Shastry et al. 2001; Domínguez and Martín 1998; Dominguez et al. 1999).

Farnesol (Fig. 13.2) is a sesquiterpene alcohol acting as a quorum-sensing molecule which inhibits the yeast-to-hypha transition and compromises *C. albicans* biofilm formation (Derengowski et al. 2009). Farnesol has also been shown to inhibit the formation of *C. dubliniensis*, *C. tropicalis*, and *C. parapsilosis* biofilms (Weber et al. 2010). Furthermore, there were synergistic effects between farnesol and fluconazole/5-flurocytosine, while there were antagonistic effects between farnesol and terbinafine/itraconazole, respectively, on the biofilms formed by the resistant strains (Xia et al. 2017). On the other hand, cyclosporine A (Fig. 13.2) is a calcineurin inhibitor derived from the fungus *Tolypocladium inflatum* in 1971 (Tedesco and Haragsim 2012). It presents both antifungal activity and immunological properties along with immunosuppression activity useful for solid organ transplant patients (Tedesco and Haragsim 2012). For antifungal activity, cyclosporine A was shown to exhibit synergism with conventional antifungals (amphotericin, fluconazole, voriconazole, caspofungin) against *C. albicans, C. parapsilosis*, *C. sensus-tricto*, *C. metapsilosis*, and *C. orthopsilosis* (de Aguiar Cordeiro et al. 2014; Marchetti et al. 2000).

Alkaloid and polyketide metabolites are the major antimicrobial compounds produced by fungi. Polyketides are a class of secondary metabolites produced by almost all living organisms including fungi (O'Hagan 1991). Due to diversity in the family of polyketides, they are the most abundant medicinal sources that have been shown to display a wide range of potentially useful therapeutic values like antibiotic, anticancer, antifungal, hypolipidemic, and immunosuppressive properties (Cragg and Newman 2013). Normally, polyketides are biologically active organic compounds due to their complex structure. Many pharmaceuticals are derived from or inspired by polyketides. For example, waikialoid A and waikialide A, which are the complexes of alkaloid and polyketide metabolites produced by *Aspergillus* sp. derived from Hawaiian soil, exhibit antifungal activity and demonstrate inhibition activity on biofilm formation against *Candida albicans (Wang* et al. 2012*)*. Two other alkaloid metabolites, named shearinines D and E and produced by *Penicillium* sp., exhibit antibiofilm activity by blocking the outgrowth of hyphae at a relatively late stage of biofilm development of *C. albicans (You* et al. 2013*)*. Interestingly shearinines D and E exhibited synergistic activities with amphotericin B against different species of *Candida* species including *C. albicans, C. glabrata, C. parapsilosis, C. tropicalis,* and *C. kefyr* (You et al. 2013).

Polyketide glycosides from *Bionectria ochroleuca*, a fungus cultured on Cheerios breakfast cereal exhibited potent biofilm inhibition against *C. albicans* (Wang et al. 2014). Khafrefungin is a polyketide isolated from an endophytic fungus that shows a broad antifungal spectrum against *C. albicans*, *C. neoformans*, and *S. cerevisiae* (Mandala et al. 1997a; Wakabayashi et al. 2001). Antifungal activity against *S. cerevisiae* and pathogenic fungi is exerted through inducing of the ceramide accumulation and the IPC synthase inhibition (Mandala et al. 1997b). Ambruticins (Fig. 13.2) are polyketides of an unusual structure produced by the strains of *Sorangium (Polyangium) cellulosum* (Ringel et al. 1977). The proposed mode of action of ambruticin confers to Hik1 binding, and that results in an inappropriate cellular response, including intracellular accumulation of glycerol and, ultimately, cell death. On the other hand, ambruticin action would result in the emergence of dephosphorylated Ssk1 (active form) and overstimulation of the HOG pathway and thus result in cell death (Vetcher et al. 2007).

Xylaria is the most important member of *Xylariaceae* in the order of *Xylariales*. *Xylaria* is a fungal species normally found as saprophytes and endophytes. Endophytes synthesize various bioactive secondary metabolites that are normally found as bioactive compounds in host plants. Most of the *Xylaria* species that are found as endophytes are reported for the production of diverse bioactive compounds (Song et al. 2014). *Xylaria* fungi are fast-growing and generally found in tropical forests such as the forests in Thailand. *Xylaria sp.* extract is a promising source of novel antifungals that mediate induction of intracellular reactive oxygen species levels and plasma membrane rupture in some oxidant-sensitive strains (Somboon et al. 2017). The diversity of bioactive compounds produced by *Xylaria* species are, for example, terpenoids, cytochalasins, mullein, alkaloids, polyketides, and aromatic compounds as well as volatile organic compounds (Song et al. 2014). Some compounds exert both antimicrobial and cytotoxic activities, such as cytoshalasin D and cytochalasin Q. Most of the compounds have complex structure such as zorfimarins, helvolic acid, and cytochalasins. Cytochalasins are mycotoxins produced by most of the fungi, including fungi in the genera of *Penicillium, Aspergillus, Zygosporium, Phoma, Metarhizium, Chaetomium, Rosellinia, Ascochyta,*

Hypoxylon, Xylaria, and *Phomopsis (Scherlach* et al. 2010*)*. Cytochalasins are named by Carter which refer to words in Greek cytos (a cell) and chalasis (relaxation) regarding the effect of this compound on fibroblasts (Carter 1967). Many cytochalasins have been reported from several *Xylaria* species, for example, cytochalasin B, cytochalasin D, and cytochalasin Q as well as their derivatives (Tarman et al. 2011; Pongcharoen et al. 2007; Wei et al. 2015).

Cytochalasins are fungal metabolites that affect cellular morphology, inhibit cell division, and cause apoptosis (Haidle and Myers 2004). Cytochalasins are involved in fungal virulence and food spoilage, and there is a possibility to be involved in the maintenance of symbiotic fungus–plant relationships (Scherlach et al. 2010). Cytochalasin binds to the fast-growing plus end of the microfilament called "barbed" end and monomeric actin resulting in blocking both the polymerization and depolymerization of actin which caps the end of the new actin filament (Scherlach et al. 2010). Cytochalasin does not affect nuclear division, thus cells contain more nuclease over time which is called enucleate (Carter 1967). In addition to actin inhibition, cytochalasin can also inhibit other mechanisms of cell, such as glucose uptake inhibition and antiangiogenesis. Cytochalasin B, firstly named phomin, is the first cytochalasin isolated (Rothweiler and Tamm 1966). It has been used as a tool for the study of cell division and cell motility that are based on the active formation and degradation of actin microfilaments (Pinkofsky et al. 2000). Moreover, cytochalasin B (Fig. 13.2) inhibits glucose uptake by binding to Trp^{388} and Trp^{412}, which are important amino acids located at the export site of human GLUT1 (Pinkofsky et al. 2000). Cytochalasin H regulates plant growth, cytochalasin D (Fig. 13.2) inhibits protein synthesis, and cytochalasin E prevents angiogenesis (Scherlach et al. 2010).

13.3.2.3 Antifungal Compounds of Marine Origin

Because of the unique biodiversity of ocean, marine organisms may be a possible source for potential drug candidates (Haefner 2003). Nowadays, there are 13 natural products or derivatives originated from marine organisms in different phases of the clinical pipeline, and many of them are in the preclinical pipeline (Mayer et al. 2010). However, most of the natural products isolated from marine organisms exhibit anticancer activity (Dias et al. 2012). For example, Trabectedin (Fig. 13.2) or Ecteinascidin 743 or ET-743 became the first marine anticancer drug to be approved in the European Union in October 2007 (Haefner 2003). Trabectedin has been approved by the European Agency for the Evaluation of Medicinal Products (EMEA) and is completing key phase III studies in the United States for approval (Haefner 2003; Mayer et al. 2010). Spisulosine, isolated from the marine clam *Spisula polynyma*, exhibited substantial selective activity against tumor cells compared to normal cells (Alvarez-Miranda et al. 2003). It advances to phase I clinical trials against solid tumors but was withdrawn in late 2006 (Cuadros et al. 2000; Salcedo et al. 2003). Halichondrin B has been isolated from several sponges including *Halichondria okadai*, *Axinella* sp., and *Phakellia carteri*, and its analogs including Halichondrin E-7389 have been selected for further development and is currently in phase III clinical trials for the treatment of breast carcinoma (Uemura et al. 1985; Aicher et al. 1992; Pettit et al. 1991; Pettit et al. 1993; Litaudon et al. 1994; Chin et al. 2006). However, some compounds exhibit side effects, resulting in the withdrawal from clinical trials such as cryptophycin (Trimurtulu et al. 1994). Furthermore, there are many compounds isolated from marine organisms that show anticancer activity. For instance, a group of diterpenes including 4-acetoxydictylolactone, dictyolides A and B, and nordictyolide derived from the brown alga, *Dictyota dichotoma*, present antitumor activities (Ishitsuka et al. 1988; Faulkner 2001). Pyridine and pyridinone compounds obtained from marine fungi exhibit antifungal activity, for example, trichodin A (Fig. 13.2) isolated from *Trichoderma* sp. MF106 shows antifungal activity against *C. albicans* (Wu et al. 2014). Steroids isolated from *P. chrysogenum* QEN-24S called penicisteroid A (Fig. 13.1) exhibit antifungal activity against *A. niger* and *Alternaria brassicae (Gao* et al. 2011*)*. Trichodermaketone A (Fig. 13.2) derived from *T. koningii* belongs to quinone compounds and exhibits synergistic antifungal activity with ketoconazole against *C. albicans* (Song et al. 2010). Another quinone, helicascolide C, was isolated from *Daldinia eschscholzii* KT32 and showed antifungal activity against phytopathogenic fungus *Cladosporium cucumerinu* (Tarman et al. 2012). Moreover, there are other known antifungal compounds derived from marine organisms, for example, sclerotiorin and anicequol (Fig. 13.2) (Gao et al. 2011; Bao et al. 2010).

13.4 Conclusion

In past years, we have gained deeper insights into the molecular mechanisms for fungal drug resistance, virulence, and host–pathogen interaction. Through much efforts, several new potential antifungal agents and targets have been identified via different approaches, including chemical genomics-based screenings, high-throughput screenings of natural products and repurposing drugs, and in silico structure-based drug-binding assays. Despite the urgent need for effective antifungal agents, only a few promising molecules are in the pipeline for early clinical development stages, and still limited classes of antifungals are available in the market. This situation is in contrast to the increasing incidence of human mortality and morbidity due to fungal infection, especially in immunocompromised patients. To accelerate antifungal drug discovery from new sources such as natural products, regulation of biosynthetic genes required for ample production of natural antifungal agents and fungal stress responses to antifungal treatments are additional important domains of investigation. Ultimately, collaboration among researchers, clinicians, and pharmaceutical companies in conjunction with the government policy to reduce unnecessary antifungal uses will enable successful antifungal drug discovery and management of fungal infections against human fungal pathogens

Acknowledgments This work is supported by Thailand Research Fund, National Research Council of Thailand, and King Mongkut's University of Technology Thonburi through the KMUTT 55th Anniversary commemorative fund.

References

Ahmad A et al (2010a) Proton translocating ATPase mediated fungicidal activity of eugenol and thymol. Fitoterapia 81(8):1157–1162

Ahmad A et al (2010b) In vitro synergy of eugenol and methyleugenol with fluconazole against clinical Candida isolates. J Med Microbiol 59(Pt 10):1178–1184

Ahmad A et al (2017) Synergistic antifungal effect of cyclized chalcone derivatives and fluconazole against *Candida albicans*. Med Chem Comm 8(12):2195–2207

Aicher TD et al (1992) Total synthesis of halichondrin B and norhalichondrin B. J Am Chem Soc 114(8):3162–3164

Akache B, Turcotte B (2002) New regulators of drug sensitivity in the family of yeast zinc cluster proteins. J Biol Chem 277(24):21254–21260

Alborzi A, Moeini M, Haddadi P (2012) Antifungal susceptibility of the Aspergillus species by Etest and CLSI reference methods. Arch Iran Med 15(7):429

Alvarez-Miranda M et al 2003 Characterization of the mechanism of action of ES-285, a novel antitumor drug from Mactromeris poynyma. In Clinical cancer research. American Association for Cancer Research, Philadelphia

Arendrup MC (2014) Update on antifungal resistance in Aspergillus and Candida. Clin Microbiol Infect 20:42–48

Armstrong AW, Bukhalo M, Blauvelt A (2016) A clinician's guide to the diagnosis and treatment of Candidiasis in patients with psoriasis. Am J Clin Dermatol 17(4):329–336

Baddley JW (2011a) Clinical risk factors for invasive aspergillosis. Med Mycol 49(Suppl 1):S7–s12

Baddley JW (2011b) Clinical risk factors for invasive aspergillosis. Med Mycol 49(Suppl 1):S7–S12

Bao L et al (2010) (-)-Sclerotiorin from an unidentified marine fungus as an anti-meiotic and antifungal agent. Nat Prod Commun 5(11):1789–1792

Barrett D (2002a) From natural products to clinically useful antifungals. Biochim Biophys Acta 1587(2–3):224–233

Barrett D (2002b) From natural products to clinically useful antifungals. Biochim Biophys Acta (BBA) Mol Basis Dis 1587(2):224–233

Bartoletti M et al (2013) Incidence and outcome of early Candida peritonitis after liver and pancreas transplantation. Mycoses 56(2):162–167

Bassetti M et al (2006) Epidemiological trends in nosocomial candidemia in intensive care. BMC Infect Dis 6:21–21

Beardsley J et al (2018) Responding to the emergence of antifungal drug resistance: perspectives from the bench and the bedside. Future Microbiol 13:1175–1191

Bezakova L et al (1996) Lipoxygenase inhibition and antioxidant properties of bisbenzylisoqunoline alkaloids isolated from *Mahonia aquifolium*. Pharmazie 51(10):758–761

Bhatnagar I, Kim SK (2012) Pharmacologically prospective antibiotic agents and their sources: a marine microbial perspective. Environ Toxicol Pharmacol 34(3):631–643

Borjihan H et al (2009) The vacuole-targeting fungicidal activity of amphotericin B against the pathogenic fungus Candida albicans and its enhancement by allicin. J Antibiot (Tokyo) 62(12):691–697

Brown GD, Denning DW, Levitz SM (2012a) Tackling human fungal infections. Science 336(6082):647–647

Brown GD et al (2012b) Hidden Killers: Human Fungal Infections. Sci Transl Med 4(165):165rv13

Campoy S, Adrio JL (2017a) Antifungals. Biochem Pharmacol 133:86–96
Campoy S, Adrio JL (2017b) Antifungals. Biochem Pharmacol 133:86–96
Cappelletty D, Eiselstein-McKitrick K (2007) The Echinocandins. Pharmacotherapy 27(3):369–388
Carter SB (1967) Effects of cytochalasins on mammalian cells. Nature 213(5073):261–264
Chang YL et al (2017) New facets of antifungal therapy. Virulence 8(2):222–236
Chen J et al (2008) Analysis of major alkaloids in Rhizoma coptidis by capillary electrophoresis-electrospray-time of flight mass spectrometry with different background electrolytes. Electrophoresis 29(10):2135–2147
Chin Y-W et al (2006) Drug discovery from natural sources. AAPS J 8(2):E239–E253
Chowdhary A et al (2014) Exploring azole antifungal drug resistance in Aspergillus fumigatus with special reference to resistance mechanisms. Future Microbiol 9(5):697–711
Coste AT et al (2004) TAC1, transcriptional activator of CDR genes, is a new transcription factor involved in the regulation of Candida albicans ABC transporters CDR1 and CDR2. Eukaryot Cell 3(6):1639–1652
Cragg GM (1998) Paclitaxel (Taxol®): a success story with valuable lessons for natural product drug discovery and development. Med Res Rev 18(5):315–331
Cragg GM, Newman DJ (2013) Natural products: A continuing source of novel drug leads. Biochim Biophys Acta Gen Subj 1830(6):3670–3695
Croteau R, Kutchan TM, Lewis NG (2000) Natural products (secondary metabolites). In: Biochemistry and molecular biology of plants, vol 24. West Sussex, Chichester, pp 1250–1319
Cruz LIB et al (2018) Anti-Candida albicans activity of thiazolylhydrazone derivatives in invertebrate and murine models. J Fungi (Basel) 4(4):1–14
Cuadros R et al (2000) The marine compound spisulosine, an inhibitor of cell proliferation, promotes the disassembly of actin stress fibers. Cancer Lett 152(1):23–29
de Aguiar Cordeiro R et al (2014) The calcineurin inhibitor cyclosporin A exhibits synergism with antifungals against Candida parapsilosis species complex. J Med Microbiol 63(7):936–944
De Rosa FG et al (2016) Appropriate treatment of invasive candidiasis in ICU: timing, colonization index, Candida Score & Biomarkers, Towards de-Escalation? Turkish J Anaesthesiol Reanimat 44(6):279–282
Denning DW (2003) Echinocandin antifungal drugs. Lancet 362(9390):1142–1151
Derengowski LS et al (2009) Antimicrobial effect of farnesol, a Candida albicans quorum sensing molecule, on Paracoccidioides brasiliensis growth and morphogenesis. Ann Clin Microbiol Antimicrob 8:13–13
Dhamgaye S et al (2014) Molecular mechanisms of action of herbal antifungal alkaloid berberine, in Candida albicans. PLoS One 9(8):e104554
Dias DA, Urban S, Roessner U (2012) A historical overview of natural products in drug discovery. Meta 2(2):303–336
Domínguez JM, Martín JJ (1998) Identification of elongation factor 2 as the essential protein targeted by sordarins in Candida albicans. Antimicrob Agents Chemother 42(9):2279–2283
Domínguez JM et al (1998) Sordarins: a new class of antifungals with selective inhibition of the protein synthesis elongation cycle in yeasts. Antimicrob Agents Chemother 42(9) 2274–2278
Dominguez JM, Gomez-Lorenzo MG, Martin JJ (1999) Sordarin inhibits fungal protein synthesis by blocking translocation differently to fusidic acid. J Biol Chem 274(32):22423–22427
Dutcher JD (1968) The discovery and development of amphotericin B. Chest 54:296–298
Facts et al (2004) The review of natural products: formerly lawrence review of natural products published by facts and comparisons. Lippincott Williams & Wilkins, London
Faria NC et al (2011) Enhanced activity of antifungal drugs using natural phenolics against yeast strains of Candida and Cryptococcus. Lett Appl Microbiol 52(5):506–513
Farmakiotis D, Kontoyiannis DP (2016) Mucormycoses. Infect Dis Clin N Am 30(1):143–163
Faulkner DJ (2001) Marine natural products. Nat Prod Rep 18(1):1R–49R
Gao S-S et al (2011) Penicisteroids A and B, antifungal and cytotoxic polyoxygenated steroids from the marine alga-derived endophytic fungus Penicillium chrysogenum QEN-24S. Bioorg Med Chem Lett 21(10):2894–2897

Georgiev VS (2000) Membrane transporters and antifungal drug resistance. Curr Drug Targets 1(3):261–284
Golinska P et al (2014) Biogenic synthesis of metal nanoparticles from actinomycetes: biomedical applications and cytotoxicity. Appl Microbiol Biotechnol 98(19):8083–8097
Greenberg R et al (2006) Posaconazole as salvage therapy for zygomycosis. Antimicrob Agents Chemother 50(1):126–133
Guo N et al (2009) Antifungal activity of thymol against clinical isolates of fluconazole-sensitive and -resistant Candida albicans. J Med Microbiol 58.(Pt 8:1074–1079
Haefner B (2003) Drugs from the deep: marine natural products as drug candidates. Drug Discov Today 8(12):536–544
Haidle AM, Myers AG (2004) An enantioselective, modular, and general route to the cytochalasins: Synthesis of L-696,474 and cytochalasin B. Proc Natl Acad Sci U S A 101(33):12048–12053
Harvey AL, Edrada-Ebel R, Quinn RJ (2015a) The re-emergence of natural products for drug discovery in the genomics era. Nat Rev Drug Discov 14(2):111–129
Harvey AL, Edrada-Ebel R, Quinn RJ (2015b) The re-emergence of natural products for drug discovery in the genomics era. Nat Rev Drug Discov 14(2):111–129
Hauser D, Sigg HP (1971) Isolation and decomposition of sordarin. Helv Chim Acta 54(4):1178–1190
Howard SJ, Pasqualotto AC, Denning DW (2010) Azole resistance in allergic bronchopulmonary aspergillosis and Aspergillus bronchitis. Clin Microbiol Infect 16(6):683–688
Ibrahim AS et al (2008) Combination echinocandin-polyene treatment of murine mucormycosis. Antimicrob Agents Chemother 52(4):1556–1558
Ishitsuka MO, Kusumi T, Kakisawa H (1988) Antitumor xenicane and norxenicane lactones from the brown alga Dictyota dichotoma. J Org Chem 53(21):5010–5013
Iwazaki RS et al (2010) In vitro antifungal activity of the berberine and its synergism with fluconazole. Antonie Van Leeuwenhoek 97(2):201
Jiang C et al (2012) Mechanisms of azole resistance in 52 clinical isolates of *Candida tropicalis* in China. J Antimicrob Chemother 68(4):778–785
Justice MC et al (1998) Elongation factor 2 as a novel target for selective inhibition of fungal protein synthesis. J Biol Chem 273(6):3148–3151
Kim J et al (2008) Chemosensitization prevents tolerance of Aspergillus fumigatus to antimycotic drugs. Biochem Biophys Res Commun 372(1):266–271
Kofla G, Ruhnke M (2011) Pharmacology and metabolism of anidulafungin, caspofungin and micafungin in the treatment of invasive candidosis-review of the literature. Eur J Med Res 16(4):159
Kolaczkowska A, Goffeau A (1999) Regulation of pleiotropic drug resistance in yeast. Drug Resist Updat 2(6):403–414
Kołaczkowska A, Kołaczkowski M (2016) Drug resistance mechanisms and their regulation in non-albicans Candida species. J Antimicrob Chemother 71(6):1438–1450
Kwon-Chung KJ, Sugui JA (2013) Aspergillus fumigatus – what makes the species a ubiquitous human fungal pathogen? PLoS Pathog 9(12):e1003743
Laniado-Laborin R, Cabrales-Vargas MN (2009) Amphotericin B: side effects and toxicity. Rev Iberoam Micol 26(4):223–227
Lei J, Xu J, Wang T (2018) In vitro susceptibility of Candida spp. to fluconazole, itraconazole and voriconazole and the correlation between triazoles susceptibility: results from a five-year study. Journal de mycologie medicale 28(2):310–313
Levy ER et al (2013) Treatment of pediatric refractory coccidioidomycosis with combination voriconazole and caspofungin: a retrospective case series. Clin Infect Dis 56(11):1573–1578
Li D-D et al (2013) Fluconazole assists berberine to kill fluconazole-resistant *Candida albicans*. Antimicrob Agents Chemother 57:6016–6027
Litaudon M et al (1994) Isohomohalichondrin B, a new antitumour polyether macrolide from the New Zealand deep-water sponge Lissodendoryx sp. Tetrahedron Lett 35(50):9435–9438

MacPherson S et al (2005) Candida albicans zinc cluster protein Upc2p confers resistance to antifungal drugs and is an activator of ergosterol biosynthetic genes. Antimicrob Agents Chemother 49(5):1745–1752
MacPherson S, Larochelle M, Turcotte B (2006) A fungal family of transcriptional regulators: the zinc cluster proteins. Microbiol Mol Biol Rev 70(3):583–604
Mandala SM et al (1997a) Khafrefungin, a novel inhibitor of sphingolipid synthesis. J Biol Chem 272(51):32709–32714
Mandala SM et al (1997b) Khafrefungin, a novel inhibitor of sphingolipid synthesis. J Biol Chem 272(51):32709–32714
Mann J (2000) Murder, magic, and medicine. Oxford University Press, New York
Marchese A et al (2016) Antibacterial and antifungal activities of thymol: A brief review of the literature. Food Chem 210:402–414
Marchetti O et al (2000) Potent synergism of the combination of fluconazole and cyclosporine in Candida albicans. Antimicrob Agents Chemother 44(9):2373–2381
Marinelli F (2009) Chapter 2. from microbial products to novel drugs that target a multitude of disease indications. 2009/04/21 ed. Methods Enzymol 458:29–58
Marr KA et al (2015) Combination antifungal therapy for invasive aspergillosis: a randomized trial. Ann Intern Med 162(2):81–89
Martin KW, Ernst E (2004) Herbal medicines for treatment of fungal infections: a systematic review of controlled clinical trials. Mycoses 47(3–4):87–92
Maschmeyer G, Haas A, Cornely OA (2007) Invasive aspergillosis: epidemiology, diagnosis and management in immunocompromised patients. Drugs 67(11):1567–1601
Mayer AM et al (2010) The odyssey of marine pharmaceuticals: a current pipeline perspective. Trends Pharmacol Sci 31(6):255–265
Mesa-Arango AC, Scorzoni L, Zaragoza O (2012) It only takes one to do many jobs: Amphotericin B as antifungal and immunomodulatory drug. Front Microbiol 3:286
Miceli MH, Kauffman CA (2015) Isavuconazole: A New Broad-Spectrum Triazole Antifungal Agent. Clin Infect Dis 61(10):1558–1565
Mišík V et al (1995) Lipoxygenase inhibition and antioxidant properties of protoberberine and aporphine alkaloids isolated from *Mahonia aquifolium*. Planta Med 61(04):372–373
Mogavero S et al (2011) Differential requirement of the transcription factor Mcm1 for activation of the *Candida albicans* multidrug efflux pump MDR1 by its regulators Mrr1 and Cap1. Antimicrob Agents Chemother 55:2061
Mora C et al (2011) How many species are there on earth and in the ocean? PLoS Biol 9(8):e1001127
Moreno I et al (2003) Characterization of a *Candida albicans* gene encoding a putative transcriptional factor required for cell wall integrity. FEMS Microbiol Lett 226(1):159–167
Morschhäuser J et al (2007) The transcription factor Mrr1p controls expression of the MDR1 efflux pump and mediates multidrug resistance in Candida albicans. PLoS Pathog 3(11):e164
Moye-Rowley WS (2003a) Regulation of the transcriptional response to oxidative stress in fungi: similarities and differences. Eukaryot Cell 2(3):381–389
Moye-Rowley WS (2003b) Transcriptional control of multidrug resistance in the yeast Saccharomyces. Prog Nucleic Acid Res Mol Biol 73:251–279
Negri M et al (2014) Early state research on antifungal natural products. Molecules 19(3):2925–2956
Nishikawa JL et al (2016) Inhibiting fungal multidrug resistance by disrupting an activator-mediator interaction. Nature 530(7591):485–489
O'Hagan D (1991) The polyketide metabolites. Ellis Horwood Ltd, Chichester
Odds FC (2001) Sordarin antifungal agents. Expert Opin Ther Pat 11(2):283–294
Papich MG (2016) Griseofulvin. In Papich MG (ed) Saunders handbook of veterinary drugs, 4th edn. W.B. Saunders, St. Louis, pp 367–368
Pappas PG et al (2009) Clinical practice guidelines for the management of candidiasis: 2009 update by the Infectious Diseases Society of America. Clin Infect Dis 48(5):503–535

Pappas PG et al (2010) Invasive fungal infections among organ transplant recipients: results of the Transplant-Associated Infection Surveillance Network (TRANSNET). Clin Infect Dis 50(8):1101–1111
Pappas PG et al (2016) Clinical practice guideline for the management of Candidiasis: 2016 update by the infectious diseases Society of America. Clin Infect Dis 62(4):e1–e50
Perfect JR (2017) The antifungal pipeline: a reality check. Nat Rev Drug Discov 16(9):603–616
Perfect JR et al (2010) Clinical practice guidelines for the management of cryptococcal disease: 2010 update by the infectious diseases society of america. Clin Infect Dis 50(3):291–322
Perkhofer S et al (2008) Posaconazole enhances the activity of amphotericin B against hyphae of zygomycetes in vitro. Antimicrob Agents Chemother 52(7):2636–2638
Perlin DS (2011) Current perspectives on echinocandin class drugs. Future Microbiol 6(4):441–457
Pettit GR et al (1991) Antineoplastic agents. 219. Isolation and structure of the cell growth inhibitory constituents from the western Pacific marine sponge Axinella sp. J Med Chem 34(11):3339–3340
Pettit GR et al (1993) Antineoplastic agents. 251. Isolation and structure of stylostatin 1 from the Papua New Guinea marine sponge Stylotella sp. [Erratum to document cited in CA118 (5): 36157a]. J Org Chem 58(11):3222–3222
Pinkofsky HB, Dwyer DS, Bradley RJ (2000) The inhibition of GLUT1 glucose transport and cytochalasin B binding activity by tricyclic antidepressants. Life Sci 66(3):271–278
Pongcharoen W et al (2007) Cytotoxic metabolites from the wood-decayed fungus Xylaria sp. BCC 9653. Chem Pharm Bull 55(11):1647–1648
Radulovic NS et al (2013) Antimicrobial plant metabolites: structural diversity and mechanism of action. Curr Med Chem 20(7):932–952
Richter SS et al (2005) Antifungal susceptibilities of Candida species causing vulvovaginitis and epidemiology of recurrent cases. J Clin Microbiol 43(5):2155–2162
Ringel S et al (1977) Ambruticin (W7783), a new antifungal antibiotic. J Antibiot 30(5):371–375
Rivero-Menendez O et al (2016) Triazole resistance in Aspergillus spp.: a worldwide problem? J Fungi (Basel, Switzerland) 2(3):21
Robbins N, Wright GD, Cowen LE (2016) Antifungal drugs: the current armamentarium and development of new agents. Microbiol Spectr 4(5):903–922
Roemer T, Krysan DJ (2014) Antifungal drug development: challenges, unmet clinical needs, and new approaches. Cold Spring Harb Perspect Med 4(5):a019703
Rothweiler W, Tamm C (1966) Isolation and structure of Phomin. Experientia 22(11):750–752
Salcedo M et al (2003) The marine antitumor compound ES 285 activates EGD receptors. In: Clinical cancer research. American Association for Cancer Research, Philadelphia
Sanglard D (2002) Resistance of human fungal pathogens to antifungal drugs. Curr Opin Microbiol 5(4):379–385
Savoia D (2012) Plant-derived antimicrobial compounds: alternatives to antibiotics. Future Microbiol 7(8):979–990
Scherlach K et al (2010) The chemistry and biology of cytochalasans. Nat Prod Rep 27(6):869–886
Scorzoni L et al (2017) Antifungal therapy: new advances in the understanding and treatment of mycosis. Front Microbiol 8:36
Shao J et al (2016) Antiproliferation of berberine in combination with fluconazole from the perspectives of reactive oxygen species, ergosterol and drug efflux in a fluconazole-resistant Candida tropicalis isolate. Front Microbiol (7):1516
Sharma M et al (2010) Synergistic anticandidal activity of pure polyphenol curcumin I in combination with azoles and polyenes generates reactive oxygen species leading to apoptosis. FEMS Yeast Res 10(5):570–578
Shastry M et al (2001) Species-specific inhibition of fungal protein synthesis by sordarin: identification of a sordarin-specificity region in eukaryotic elongation factor 2. Microbiology 147(Pt 2):383–390
Shi D et al (2016) Antifungal effects of undecylenic acid on the biofilm formation of Candida albicans. Int J Clin Pharmacol Ther 54(5):343–353

Silver PM, Oliver BG, White TC (2004) Role of *Candida albicans* transcription factor Upc2p in drug resistance and sterol metabolism. Eukaryot Cell 3(6):1391–1397

Skiada A et al (2013) Diagnosis and treatment of mucormycosis in patients with hematological malignancies: guidelines from the 3rd European Conference on Infections in Leukemia (ECIL 3). Haematologica 98(4):492–504

Somboon P et al (2017) Fungicide Xylaria sp. BCC 1067 extract induces reactive oxygen species and activates multidrug resistance system in Saccharomyces cerevisiae. Future Microbiol 12:417–440

Song F et al (2010) Trichodermaketones A– D and 7-O-methylkoninginin D from the marine fungus Trichoderma koningii. J Nat Prod 73(5):806–810

Song F et al (2014) Secondary metabolites from the genus Xylaria and their bioactivities. Chem Biodivers 11(5):673–694

Soontorngun N et al (2007) Regulation of gluconeogenesis in *Saccharomyces cerevisiae* is mediated by activator and repressor functions of Rds2. Mol Cell Biol 27(22):7895–7905

Spellberg B et al (2005) Combination therapy with amphotericin B lipid complex and caspofungin acetate of disseminated zygomycosis in diabetic ketoacidotic mice. Antimicrob Agents Chemother 49(2):830–832

Spitzer M, Robbins N, Wright GD (2017) Combinatorial strategies for combating invasive fungal infections. Virulence 8(2):169–185

Stone NR et al (2016) Liposomal amphotericin B (AmBisome®): a review of the pharmacokinetics, pharmacodynamics, clinical experience and future directions. Drugs 76(4):485–500

Sucher AJ, Chahine EB, Balcer HE (2009) Echinocandins: the newest class of antifungals. Ann Pharmacother 43(10):1647–1657

Susan JH et al (2009) Frequency and evolution of azole resistance in *Aspergillus fumigatus* associated with treatment failure. Emerg Infect Dis J 15(7):1068

Taborda CP, Nosanchuk JD (2017) Editorial: vaccines, immunotherapy and new antifungal therapy against fungi: updates in the New Frontier. Front Microbiol 8:1743

Talibi D, Raymond M (1999) Isolation of a Putative *Candida albicans* Transcriptional Regulator Involved in Pleiotropic Drug Resistance by Functional Complementation of a pdr1 pdr3 Mutation in *Saccharomyces cerevisiae*. J Bacteriol 181(1):231–240

Tan W et al (2011) Berberine hydrochloride: anticancer activity and nanoparticulate delivery system. Int J Nanomedicine 6:1773–1777

Tarman K et al (2011) Biological and chemical study of two Indonesian marine endophytic fungi. Planta Med 77(12):SL71

Tarman K et al (2012) Helicascolide C, a new lactone from an Indonesian marine algicolous strain of Daldinia eschscholzii (Xylariaceae, Ascomycota). Phytochem Lett 5(1):83–86

Tedesco D, Haragsim L (2012) Cyclosporine: a review. J Transplant 2012:230386–230386

Trimurtulu G et al (1994) Total structures of cryptophycins, potent antitumor depsipeptides from the blue-green alga Nostoc sp. strain GSV 224. J Am Chem Soc 116(11):4729–4737

Tsiodras S et al (2008) Fungal infections complicating tumor necrosis factor alpha blockade therapy. Mayo Clin Proc 83(2):181–194

Uemura D et al (1985) Norhalichondrin A: an antitumor polyether macrolide from a marine sponge. J Am Chem Soc 107(16):4796–4798

Vallabhaneni S, Chiller TM (2016) Fungal infections and new biologic therapies. Curr Rheumatol Rep 18(5):29

Vandeputte P et al (2011a) Molecular mechanisms of resistance to 5-fluorocytosine in laboratory mutants of *Candida glabrata*. Mycopathologia 171(1):11–21

Vandeputte P, Ferrari S, Coste AT (2011b) Antifungal resistance and new strategies to control fungal infections. Intl J Microbiol 2012

Vazquez JA (2007) Combination antifungal therapy: the new frontier. Future Microbiol 2(2):115–139

Verweij PE et al (2016) Azole resistance in *Aspergillus fumigatus*: Can we retain the clinical use of mold-active antifungal azoles? Clin Infect Dis 62(3):362–368

Vetcher L et al (2007) The antifungal polyketide ambruticin targets the HOG pathway. Antimicrob Agents Chemother 51(10):3734–3736
Wakabayashi T, Mori K, Kobayashi S (2001) Total synthesis and structural elucidation of khafrefungin. J Am Chem Soc 123(7):1372–1375
Walsh TJ et al (2008a) Treatment of aspergillosis: clinical practice guidelines of the Infectious Diseases Society of America. Clin Infect Dis 46(3):327–360
Walsh TJ et al (2008b) Treatment of aspergillosis: clinical practice guidelines of the Infectious Diseases Society of America. Clin Infect Dis 46(3):327–360
Wang X et al (2012) Waikialoid A Suppresses Hyphal Morphogenesis and Inhibits Biofilm Development in Pathogenic Candida albicans. J Nat Prod 75(4):707–715
Wang B et al (2014) Polyketide glycosides from Bionectria ochroleuca inhibit Candida albicans biofilm formation. J Nat Prod 77(10):2273–2279
Watve MG et al (2001) How many antibiotics are produced by the genus Streptomyces? Arch Microbiol 176(5):386–390
Weber K, Schulz B, Ruhnke M (2010) The quorum-sensing molecule E,E-farnesol—its variable secretion and its impact on the growth and metabolism of Candida species. Yeast 27(9):727–739
Wei H et al (2015) Sesquiterpenes and other constituents of Xylaria sp. NC1214, a fungal endophyte of the moss Hypnum sp. Phytochemistry 118:102–108
Wu B et al (2014) Two new antibiotic pyridones produced by a marine fungus, Trichoderma sp. strain MF106. Mar Drugs 12(3):1208–1219
Xia J et al (2017) In vitro inhibitory effects of farnesol and interactions between farnesol and antifungals against biofilms of Candida albicans resistant strains. Biofouling 33(4):283–293
You J et al (2013) Small-molecule suppressors of Candida albicans biofilm formation synergistically enhance the antifungal activity of amphotericin B against clinical Candida isolates. ACS Chem Biol 8(4):840–848
Zheng Y-H et al (2018) An insight into new strategies to combat antifungal drug resistance. Drug Des Devel Ther 12:3807

Chapter 14
Glucose Sensing and Regulation in Yeasts

Olena G. Stasyk and Oleh V. Stasyk

Abstract Glucose as a favorite carbon source exerts dominant regulatory effects on yeast cell metabolism via several signaling pathways. They act in concert to fine-tune glucose transport, metabolism, and transcriptional machineries in response to altering exogenous glucose concentrations. In this chapter, we review the Snf3/Rgt2-mediated "sensor/receptor-repressor" pathway, the Mig1-Hxk2 glucose repression pathway, and the cAMP/PKA pathway, and how they are coordinated. The information on how these pathways operate in the yeast cell was gathered primarily in bakers' yeast *S. cerevisiae* ecologically adapted to aerobic glucose fermentation. The comparative analysis of glucose-sensing mechanisms in other so-called "nonconventional" yeasts and examples of biotechnological applications of the mutants with altered glucose regulation are also provided.

Keywords Glucose sensing · Signal transduction · Catabolite repression · Transcriptional regulation · Nonconventional yeasts

14.1 Mechanisms of Glucose Regulation in *S. cerevisiae* and Their Physiological Significance

Among various carbon sources which yeast cells can utilize, glucose is a preferable one. Glucose exerts dominant regulatory effects on cell metabolism via several coordinated signaling pathways that act at transcriptional, post-transcriptional, and post-translational levels. These pathways were studied primarily on the model of bakers' yeast *S. cerevisiae* which is ecologically adapted to aerobic glucose fermentation. Hereby we will review the gathered up-to-date information on

O. G. Stasyk
Department of Biochemistry, Ivan Franko National University of Lviv, Lviv, Ukraine

O. V. Stasyk (✉)
Department of Cell Signaling, Institute of Cell Biology, NAS of Ukraine, Lviv, Ukraine
e-mail: stasyk@cellbiol.lviv.ua

A. Sibirny (ed.), *Non-conventional Yeasts: from Basic Research to Application*,
https://doi.org/10.1007/978-3-030-21110-3_14

glucose-dependent regulatory pathways in *S. cerevisiae* as well as in other so-called "nonconventional" yeasts, where the data are available.

In the presence of glucose, metabolism of *S. cerevisiae* cells is switched to a fermentative mode even when they are exposed to oxygen (Crabtree effect, for review: Diaz-Ruiz et al. 2011; Pfeiffer and Morley 2014), and glucose transport and metabolism are upregulated. Simultaneously, catabolic activities involved in utilization of alternative carbon sources, including gluconeogenesis and respiration, are suppressed (reviewed in Rolland et al. 2002). This metabolic switch results from a crosstalk between several glucose-signaling pathways: (1) the Rgt2/Snf3 glucose induction pathway (also called the "sensor/receptor-repressor (SRR)" pathway) that regulates glucose uptake into the cells (reviewed in Kaniak et al. 2004; Johnston and Kim 2005); (2) the Snf1/Mig1 main glucose repression pathway that negatively regulates expression of the genes involved in the utilization of alternative carbon sources (reviewed in Hedbacker and Carlson 2008; Kayikci and Nielsen 2015; Coccetti et al. 2018); and (3) cAMP-dependent protein kinase A (PKA) pathway that senses cell's energy status and also known to affect expression of the *HXT* genes (reviewed in Thevelein et al. 2005; Conrad et al. 2014). These pathways and relevant aspects of their functional coordination are briefly characterized below.

It has to be mentioned that complete exhaustion of glucose from the growth medium leads to a drop in ATP generation and decrease in cell viability. Also, under conditions of glucose deprivation, the general translational machinery is inhibited (Ashe et al. 2000; Boender et al. 2011).

Understanding of the molecular mechanisms of glucose regulation, and of glucose-dependent transcriptional repression in particular, is of great importance for the development of biotechnological processes where it is desirable to regulate carbon fluxes and redirect them toward products of interests (Chen and Nielsen 2013; Dai et al. 2015).

14.1.1 Nontransporting Glucose Receptors Snf3/Rgt2-Mediated Pathway and Regulation of the HXT *Genes in* S. cerevisiae

Yeast cells efficiently sense glucose available for growth both extra- and intracellularly. In *S. cerevisiae,* the Snf3/Rgt2 signaling pathway represents a sensory cascade for detecting extracellular glucose levels and regulates primarily the expression of the seven main hexose transporter (*HXT*) genes in this yeast (Ozcan et al. 1996; Ozcan and Johnston 1999; Kaniak et al. 2004; Horák 2013). Since energy generation by fermentation is rather inefficient, yeast cells can pump large amounts of glucose through glycolysis by strongly elevating its uptake – the rate-limiting step of glucose metabolism (Boles and Hollenberg 1997). This signal further regulates the glucose uptake and is also translated into glucose repression (see below).

S. cerevisiae Snf3/Rgt2 receptors belong to the family of the *HXT* (HeXose transporter) proteins. However, contrary to the multiple *S. cerevisiae* functional hexose carriers (Hxt1–17 and Gal2), they lost the ability to transport glucose and became specialized membrane sensors that detect extracellular glucose levels (Boles and Hollenberg 1997; Ozcan et al. 1998; Wieczorke et al. 1999). Snf3 has been demonstrated to sense low glucose levels, while Rgt2, high levels of glucose (Ozcan et al. 1996; Ozcan and Johnston 1999). The Snf3/Rgt2-mediated signal is translated to promoters of the glucose-sensitive genes, in particular the *HXT* genes encoding functional hexose transporters–facilitators (Fig. 14.1).

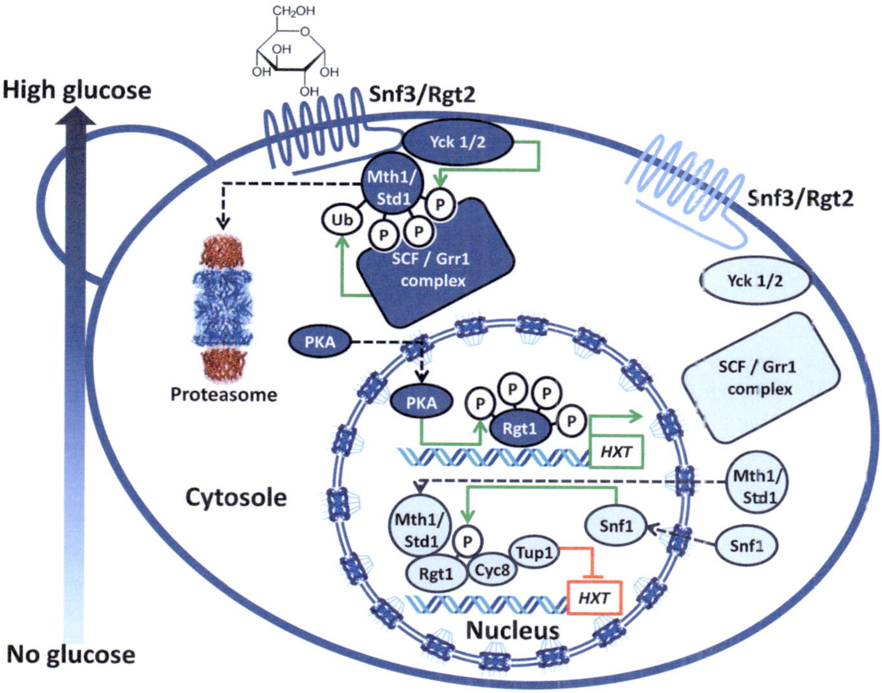

Fig. 14.1 Glucose-sensing "sensor/receptor–repressor (SRR)" pathway mediated by nontransporting glucose receptors Snf3 and Rgt2. Glucose (structure is depicted) is sensed by Snf3/Rtg2 leading to Yck1/2 activation and thereby phosphorylation of the regulatory proteins Mth1 and Std1. Activated Mth1 and Std1 are recognized by the SCFGrr1ubiquitin ligase and are degraded in the proteasome. During glucose starvation, Mth1/Std1 complex is not degraded and localizes to the nucleus, where it interacts with transcriptional repressor Rgt1. This allows Rgt1 binding and recruitment of the general repressor Cyc8/Tup1 to the promoters of the repressible *HXT* genes. Binding of Rgt1 to DNA takes place upon its Snf1-mediated phosphorylation, whereas dissociation is triggered by PKA-mediated hyperphosphorylation. PKA phosphorylation of Rgt1 occurs only when Mth/Std1 is degraded. Dark blue boxes indicate proteins' state at high glucose, while light blue boxes at low or no glucose conditions. Green arrows represent activation, red lines or boxes represent inhibition, and dotted arrows indicate protein relocation events

It has to be mentioned that nontransporting hexose sensors appear to be yeast-specific and have not been identified in higher eukaryotes. At the same time, they seem to be present in majority if not all yeast species, as exemplified by the *Kluyveromyces lactis* Rag4 (Betina et al., 2001), *Candida albicans* Hgt4 (Brown et al. 2006), *Hansenula polymorpha* Hxs1 (Stasyk et al. 2008), and *Pichia pastoris* Gss1 (Polupanov et al. 2012). At the same time, such transporters as exemplified by *U. maydis* Hxt1 (Schuler et al. 2015), *Neurospora crassa* Hgt1 and Hgt2 (Wang et al. 2017), and *Hansenula polymorpha* Gcr1 (Stasyk et al. 2004, 2018) may represent a novel class of yeast and fungal hexose transporting receptors or transceptors that participate in sugar signaling for catabolite repression and are interconnected with the PKA pathways (see below). No such hexose transporter homolog is present in *S. cerevisiae* genome and in the genomes of the related yeasts.

As a rule, low-affinity transporters, represented by *S. cerevisiae* Hxt1, are up-induced when glucose is abundant (sensed by Rgt2), whereas expression of high-affinity transporters, represented by Hxt7, is upregulated only at low extracellular glucose (sensed by Snf3) (Özcan 2002). At the same time, transcription of the high-affinity transporters is repressed by high glucose concentrations (see below for the mechanism involved). Thus, Snf3/Rgt2-mediated regulation of transcriptional activity of specific hexose transporters allows fast adaptation of transport capacity to glucose availability. Such regulation of the *HXT* genes is crucial in the context of glucose uptake being the critical regulatory step of glucose metabolism (Galazzo and Bailey 1990). Interestingly, to maintain intracellular glucose homeostasis, sensors Snf3 and Rgt2 were also proposed to sense an internal-to-external ratio of glucose concentrations and to modulate glucose uptake accordingly. The apparent affinity for glucose was found to reversely correlate with the intracellular glucose concentrations (Karhumaa et al. 2010). Thus, this data conforms to an alternating access model for transporter-like sensors, where intracellular ligand inhibits binding of extracellular ligand. Summarizing, intracellular glucose concentration in *S. cerevisiae* is maintained via tight regulation of hexose transporters ensuring that glucose uptake is sufficient almost regardless of its external concentrations (Broach 2012).

External glucose signal in the SRR pathway is transduced from Snf3 and Rgt2 to the membrane-associated type I casein kinases Yck1 andYck2. It was proposed that these kinases are required for degradation of two paralogous regulatory proteins Mth1 and Std1 (Fig. 14.1) (Schmidt et al. 1999; Moriya and Johnston 2004).

However, it was also reported that Mth1 is most probably not directly phosphorylated by Yck1/2 and is degraded in the nucleus independent of them (Pasula et al. 2010). Therefore, the Snf3/Rgt2 glucose sensors most probably transmit glucose signal to a yet unidentified protein(s) acting yet upstream of Mth1.

Of note, although Ycks were postulated to function downstream of nontransporting sensors Snf3 and Rgt2, it was recently found that overexpressed Yck1 failed to restore glucose signaling in a glucose sensor mutant. And, conversely, the signaling defect of a *yck* mutant could be suppressed by the overexpression of a glucose sensor Rgt2. In addition, it was found that phosphorylation of Yck consensus sites in Rgt2 C-terminal tail is required for corepressor binding (Snowdon and Johnston 2016). These combined data tentatively suggest that Snf3/Rgt2 pathway may be eventually revised, so Ycks act upstream or at the level of the glucose sensors. Additional studies are required to validate this hypothesis.

Thus, according to the currently accepted model (Fig. 14.1), under high glucose conditions, Mth1 and Std1 are recruited to the plasma membrane and, upon phosphorylation, undergo ubiquitination by SCF (Grr1) ubiquitin protein ligase and are targeted for degradation through the ubiquitin–proteasome pathway (Flick 2003; Moriya and Johnston 2004; Spielewoy et al. 2004; Kim and Johnston 2006). In addition, under excess glucose, transcriptional repressor Mig1 (see below) represses expression of the *MTH1* gene, which in turn is translated into repression of (some of) the *HXT* genes (Kim and Johnston 2006).

Mth1 and Std1 are activators of Rgt1, the transcriptional repressor acting on a number of glucose-induced genes, including the *HXT* and hexokinase genes (Palomino et al. 2006). Upon degradation of Mth1/Std1, Rgt1 is subject to degradation as well, which in turn relieves expression of its gene targets. For disassociation of Rgt1 from the respective promoters, its hyperphosphorylation mediated by protein kinase A, which is translocated to the nucleus in response to glucose, is required (Roy et al. 2014). Conversely, in the absence of glucose, Mth1 and Std1 act upon Rgt1, concealing PKA phosphorylation sites on Rgt1 which in turn leads to Rgt1 binding to promoters of the repressible genes (Fig. 14.1) (Flick 2003; Polish et al. 2005).

Finally, phosphorylation of Rgt1 by a catalytic α-subunit of the AMP-activated kinase (AMPK), Snf1, triggers its promoter-binding capability and repressor activity (Palomino et al. 2006). Unlike PKA, Snf1 locates to the nucleus during glucose deprivation (Vincent et al. 2001). Under low glucose conditions, DNA-bound Rgt1 in complex with Mth1/Std1 recruits the general transcriptional repressor complex composed of Cyc8 and Tup1 to inhibit the expression of the *HXT* genes (Lakshmanan et al. 2003; Polish et al. 2005; Horák 2013) (Fig. 14.1). Of note, the interaction between Rgt1 and Snf1 kinase is important for the general glucose repression mechanism by providing a "gradual" derepression of the *HXT* genes.

A new mechanism by which Snf1 activity is regulated by the Std1 protein and its regulator Sip5 was recently proposed (Simpson-Lavy et al. 2017; Simpson-Lavy and Kupiec 2018). It was elucidated that protein kinase Vhs1 phosphorylates Sip5 in response to glucose availability, disengaging it from Std1 in the nucleus and promoting the sequestering of Std1 into cytoplasmic aggregates, which have the properties of liquid drops and reside at the nucleus–vacuole junction. The process is reversible and requires protein chaperones, i.e., these Std1-containing puncta dissolve when glucose becomes scarce again (Simpson-Lavy and Kupiec 2018).

14.1.2 *Mig1-Hxk2 Glucose Repression Pathway and Snf1 Kinase*

It is well established that in the presence of glucose, a number of *S. cerevisiae* genes involved in the utilization of alternative carbon sources, gluconeogenesis, and mitochondrial functions are repressed (for review, see Gancedo 1998; Conrad et al. 2014; Kayikci and Nielsen 2015). It has also been shown that glycolytic enzyme hexokinase II (Hxk2) is required for this transcriptional repression effect (Moreno et al. 2005). Furthermore, it was established that yeast Hxk2 is in fact an intracellular sensor of glucose availability with a dual cytosolic and nuclear localization (Herrero et al. 1998; Randez-Gil et al. 1998a, b; Rodríguez et al. 2001; Moreno et al. 2005) (Fig. 14.2).

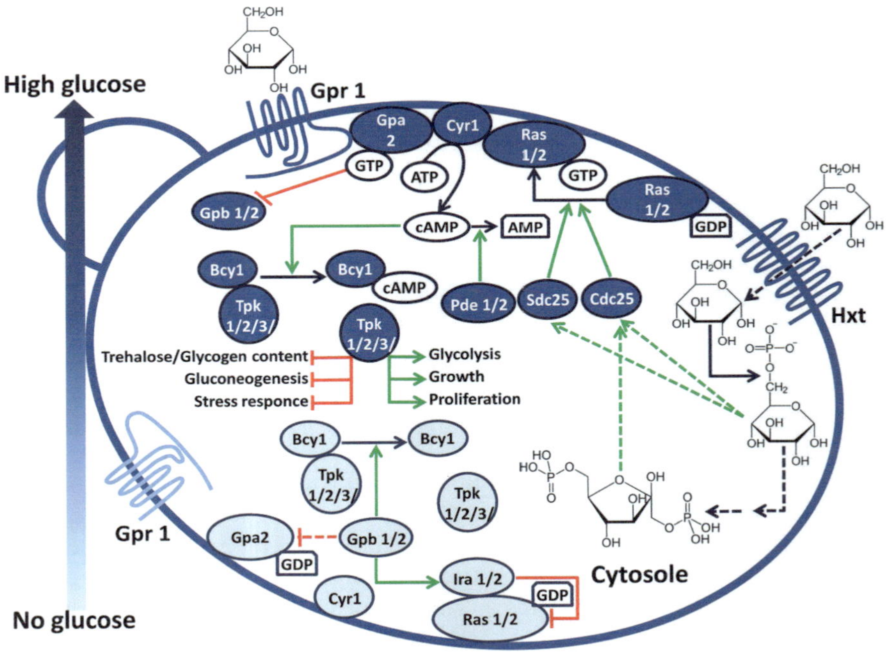

Fig. 14.2 Snf1 and Mig1-Hxk2 pathway of glucose repression in ***S. cerevisiae*.** High glucose concentrations render Snf1 inactive which leaves the transcription factor Mig1 and Hxk2 dephosphorylated by Reg1-Glc7 and, therefore, Mig1 being present in the nucleus where it exerts, together with the Ssn6/ Tup1 complex and Med8, repression effect on the genes involved in the utilization of alternative carbon sources. Mechanism responsible for the reduction of Snf1 activity is not fully understood, but dephosphorylation by Reg1-Glc7 is assumed to be important. When glucose becomes limited, Snf1 is active and phosphorylates both Mig1 and Hxk2, allowing their translocation out of the nucleus and thereby release of glucose repression and Med8, Adr1 and Cat8-stimulated expression of the glucose-repressible genes. Snf1 activity under low glucose conditions is stimulated by ADP, which protects Snf1 from Reg1-Glc7-mediated dephosphorylation. Dark blue boxes indicate proteins' state at high glucose concentrations, while light blue boxes at low or no glucose conditions. Green arrows represent activation, red lines or boxes represent inhibition, and dotted arrows indicate protein relocation events

In the nucleus, Hxk2 was implicated in the inhibition of the expression of a number of genes including a well-known target of glucose repression in *S. cerevisiae* – the invertase-encoding *SUC2*. Hxk2 action was suggested to be mediated by the transcription repressor Mig1 (Herrero et al. 1998; Ahuatzi et al. 2004, 2007; Moreno et al. 2005). It was also demonstrated that Hxk2 interacts with and regulates negatively Med8 – a subunit of the RNA polymerase II mediator complex (De La Cera et al. 2002).

This is interesting because although Med8 itself represses the *HXK2* gene expression, it also simultaneously induces expression of *SUC2* (Fig. 14.2) (Chaves et al. 1999).

A key transcriptional repressor of the glucose repression mechanism in *S. cerevisiae*, Mig1, controls expression of the genes encoding proteins functioning in metabolism and transport of such alternative carbon sources as maltose, galactose, and sucrose (Nehlin and Ronne 1990; Treitel et al. 1998).

As shown in Fig. 14.2, in glucose-rich conditions, Mig1 is located in the nucleus in a dephosphorylated form (Gancedo 1998; Ahuatzi et al. 2007). It binds to target gene promoters in a complex with the general Ssn6-Tup1 corepressor (Treitel and Carlson 1995; Wu and Trumbly 1998). Importantly, although Mig1 has also been shown to directly interact with Hxk2, alternate distribution of Hxk2 between the nucleus and cytoplasm appeared to depend on Mig1 through a yet unknown mechanism (Ahuatzi et al. 2004; Moreno et al. 2005). Nuclear export of Mig1 is triggered by the Snf1-mediated phosphorylation of its specific Ser311 residue. However, interaction with Hxk2 protects this residue and counteracts Mig1 translocation to the nucleus (Ahuatzi et al. 2007). Hxk2-Mig1, besides interacting with Med8, also binds to Cyc8 and Tup1 complex, thus forming the repressor complex sensitive to extracellular glucose levels (Fig. 14.2) (Papamichos-Chronakis et al. 2004). It was shown recently that different functional domains of Tup1 general transcriptional repressor, namely, Ssn6 interaction, N-terminal, and C-terminal repression domains, play differential roles in glucose repression and maltose metabolism (Lin et al. 2017).

Importantly, the mentioned Ser311 residue on Mig1 is required for regulation of the *SUC2* gene, and its phosphorylation by Snf1 causes Mig1 dissociation from Cyc8-Tup1 (Papamichos-Chronakis et al. 2004; Ahuatzi et al. 2007). Besides Mig1, Hxk2 is also phosphorylated by Snf1 at Ser14 under low glucose conditions (Kriegel et al. 1994; Fernańdez-García et al. 2012). Similar to Mig1, it is phosphorylated form of Hxk2 that is exported from the nucleus (Peláez et al. 2009), whereas both Hxk2 and Mig1, in order to relocalize to the nucleus, have to be dephosphorylated (Ahuatzi et al. 2007) (Fig. 14.2).

It was also found that Hxk2 supports the integrity of the repressor complex responsible for transcriptional repression of the *SUC2* gene, in which the DNA-binding Mig1 and Mig2 repressors and the regulatory proteins Snf1 and Reg1 participate (Vega et al. 2016). At low glucose conditions, Hxk2 is in its open configuration and leaves the repressor complex, which induces its dissociation from the promoter. Conversely, a closed conformation which Hxk2 adopts at high glucose conditions promotes its binding to Mig1 protein and the reassembly of the active repressor complex. Thus, it was proposed that Hxk2 protein functions as an intracellular sensor of cytoplasmic glucose levels that operates by changing its conformation and transduces the signal directly to the repressible promoters (Vega et al. 2016).

Dephosphorylation of both Mig1 and Hxk2 is mediated by Glc7, the catalytic subunit of protein phosphatase 1 (PP1) (Fig. 14.2) (Randez-Gil et al. 1998a, b; Rubenstein et al. 2008; Fernańdez-García et al. 2012). Since Glc7 is known to have a broad target specificity, its function in glucose repression is provided by its regulatory subunit, Reg1 (Tu and Carlson 1995; Rubenstein et al. 2008). It was reported that Mig1 is dephosphorylated, besides the glucose-dependent Glc7-Reg1 mechanism, also by a mechanism independent of glucose and Glc7-Reg1 (Shashkova et al. 2017). Therefore, Mig1 dephosphorylation appears to be controlled in a very complex manner apparently required for a rapid and sensitive regulation in response to altering exogenous glucose concentrations.

Notably, Snf1 protein kinase is another important target of the Reg1-Glc7 phosphatase complex (Sanz et al. 2000; Rubenstein et al. 2008). It was observed that deficiency in the regulatory protein Ssb, a chaperone homologous to the human Hsp70, leads to the defects in glucose repression in *S. cerevisiae* cells, due to the ability of Ssb protein to bridge between the Snf1 and Glc7 complexes (Hubscher et al. 2016). Ssb was demonstrated to perform its function via C-terminal binding to the 14–3-3 protein Bmh. Overexpression of either Ssb or Bmh enabled Glc7 to dephosphorylate Snf1 even in *Δreg1* cells, thus efficiently suppressing transcriptional deregulation. It was proposed that Ssb and Bmh may comprise a new chaperone module involved in the fine-tuning of Snf1 phosphorylation-dependent switch between fermentation and respiration in *S. cerevisiae* (Hubscher et al. 2016).

The Snf1 protein kinase, the yeast homolog of a highly conserved in eukaryotic cells AMP-activated protein kinase – AMPK, is a central element integrating signaling for glucose repression and signaling that balances the cellular energy levels. Together with one of the three β-subunits, Sip1, Sip2, or Gal83, and regulatory γ -subunit, Snf4, the catalytic α-subunit, Snf1, forms a heterotrimeric complex that regulates various cellular processes through a repertoire of transcription factors and pathways. Due to their functional versatility, understanding of the regulation of AMPKs in different organisms is of general interest, and much knowledge about this enzyme has been gathered by exploiting *S. cerevisiae* as a model organism (Hedbacker and Carlson 2008; Petranovic et al. 2010; Coccetti et al. 2018).

The Snf1 complex is involved, besides the regulation of *HXT*s and glucose-repressed genes, in a plethora of cellular processes important for cell proliferation, aging, energy homeostasis, autophagy, and stress response (Mitchelhill et al. 1994; Ashrafi et al. 2000; Wang et al. 2001; Sanz et al. 2016; Zhang and Cao 2017). Although the main role of the Snf1 complex is believed to be regulation of adaptation to glucose availability (Celenza and Carlson 1986), it also regulates metabolic enzymes involved in fatty acid metabolism and carbohydrate storage and is involved in a GCN4-dependent translation and amino acid biosynthesis (Hedbacker and Carlson 2008; Usaite et al. 2009). Snf1 has also been shown to directly interact with the general transcriptional apparatus and implicated in chromatin modification

(Kuchin et al. 2000). In addition to response to glucose limitation and other stress factors, new unconventional roles of Snf1 complex have recently emerged, even at physiological glucose-repressing conditions (Coccetti et al. 2018).

Snf1 kinase complex has a well-recognized dual role in glucose repression, both as an activator and as a repressor. Activation of Snf1 allows cells to switch from fermentation to respiration. It is assumed that Snf1 is activated via phosphorylation at Thr 210 by three redundant upstream kinases, Sak1, Elm1, and Tos3 and inactivated via dephosphorylation mediated by Reg1-Glc7 (Fig. 14.2) (McCartney and Schmidt 2001; Hong et al. 2003; Rubenstein et al. 2008). Intriguingly, activities of Reg1-Glc7 or the three upstream kinases are in fact independent of glucose concentrations, although phosphorylation of Snf1 was shown to be elevated under low glucose conditions (McCartney and Schmidt 2001; Rubenstein et al. 2008). Thus, the accessibility of Snf1 Thr210 for the action of Reg1-Glc7 has been proposed as a mechanism for the regulation of Snf1 activity (Rubenstein et al. 2008). Such regulation is based on a competition between the two alternative direct interactions of Reg1 with Glc7 or Snf1 (Ludin et al. 1998; Elbing et al. 2006). In other words, if Thr210 of Snf1 is masked by the interaction between Reg1 and Snf1, this therefore prevents against Snf1 dephosphorylation by Glc7 (Fig. 14.2) (Rubenstein et al. 2008; Tabba et al. 2010). It is assumed that ADP binding to the regulatory subunit protects Thr210 from dephosphorylation (Mayer et al. 2011). Recent data suggest that under glucose-rich conditions, Snf1 dephosphorylation is possibly mediated, besides Glc7-Reg1/2, also by the PP2A-like phosphatase, Sit4, and the PP2C phosphatase, Ptc1 (Castermans et al. 2012; Ruiz et al. 2012, 2013). Since ADP and not AMP was demonstrated to protect against dephosphorylation, thereby activating both yeast Snf1 and mammalian AMPK, there may be implicated an evolutionary conserved regulatory mechanism (Mayer et al. 2011; Xiao et al. 2011). Derepression of a number of glucose-repressible genes involved in gluconeogenesis and beta-oxidation requires transcriptional regulators Cat8 and Adr1 which depend on the Snf1-mediated induction (Rahner et al. 1999; Young et al. 2003, Tachibana et al. 2007) (Fig. 14.2).

Still, the mechanism by which the Snf1-Mig1 pathway is regulated on the molecular level is not entirely elucidated. In the recent study that applied a "single-cell" analytical approach and mathematical modeling, "a close link between the glucose uptake rate, which determines the glycolytic rate, and the activity of the Snf1-Mig1 system" has been proposed (Welkenhuysen et al. 2017). It was shown that differences in the amount of hexose transporter molecules (e.g., Hxt7) in the cell could cause cell-to-cell variability in the activation of Snf1-Mig1 pathway. This model suggested a formerly unrecognized regulatory step of the Snf1-Mig1 pathway at the level of Mig1 dephosphorylation and pointed at the transport of Mig1 in and out of the nucleus as a major source of variability between individual cells (Welkenhuysen et al. 2017).

Notably, Snf1 was shown to be required for the extension of chronological life span (CLS) by caloric restriction in chronologically aging yeast cells. Spontaneous mutations in transcriptional corepressor Cyc8 that specifically mediates repression through the transcriptional repressor Mig1, and in the *MIG1* gene itself, were found to suppress the short chronological life span in *S. cerevisiae* delta *snf1* mutant (Maqani et al. 2018). Deleting transcription factor gene *CAT8*, which is responsible for activating the genes of the glyoxylate and gluconeogenesis pathways, completely blocked CLS extension by the point cyc8 mutations, identifying these pathways as the key Snf1-regulated CLS determinants (Maqani et al. 2018).

S. cerevisiae Snf1 has also been implicated in the process of rapid decay of transcripts encoding proteins required for aerobic metabolism, such as alcohol dehydrogenase Adh2, upon the switch from respiration to fermentation. An intriguing funding has been reported that altering the *ADH2* gene promoter so it becomes Snf1-independent prevented glucose-induced mRNA decay expressed from such a modified promoter. It was proposed that Snf1 may influence mRNA stability by altering the transcription factor Adr1 recruitment activity. On the other hand, inhibiting the protein kinase A pathway (see below) did not affect glucose-induced decay of Adh2 mRNA (Braun et al. 2016).

14.1.3 cAMP/PKA Pathway of S. cerevisiae *and Functions of Grp1 and Ras1/2*

Yeast cells, of *S. cerevisiae* in particular, proliferating in a complete nutrient medium with a fermentable (so-called favorite) carbon source display phenotypes indicative of the high activity of the PKA (protein kinase A) pathway, whereas cells incubated with a nonfermentable carbon source or those at stationary phase of growth display the alternative phenotypes, characteristic for the low PKA activity. The latter include the low growth rate, upregulated stress–response pathways, low expression of ribosomal proteins, and higher accumulation of reserve carbohydrates, such as trehalose and glycogen, to name several of the most important ones (Thevelein and De Winde 1999). It is also well established that within minutes of exposure of *S. cerevisiae* cells to glucose, a spike of the cAMP level occurs, which is followed by a PKA-mediated activation of the enzyme trehalase, which, in turn, evokes mobilization of disaccharide trehalose and its hydrolysis into glucose monomers (Thevelein 1984; Wang et al. 2004; Santangelo 2006). Since for its sustained activation not only glucose but also a complete growth medium is required, the PKA pathway is apparently responsible for the complex monitoring, whether all conditions are appropriate for cell proliferation (Thevelein 2004; Conrad et al. 2014).

The mechanisms by which glucose triggers activation of the cAMP-PKA pathway and alters downstream sensitive cellular metabolism have been elucidated in detail (Conrad et al. 2014; Steyfkens et al. 2018). It was established that the regulation of the PKA pathway in response to alterations in glucose concentrations involves transmembrane GPCR (G-protein-coupled receptor) protein Gpr1 (Fig. 14.3) (Yun et al. 1997; Xue et al. 1998). Besides glucose, sucrose also acts as a ligand for Gpr1 (Kraakman et al. 1999; Lemaire et al. 2004). This protein is coupled to and interacts with a GTP-bound G-protein alpha-subunit, Gpa2, which inter-

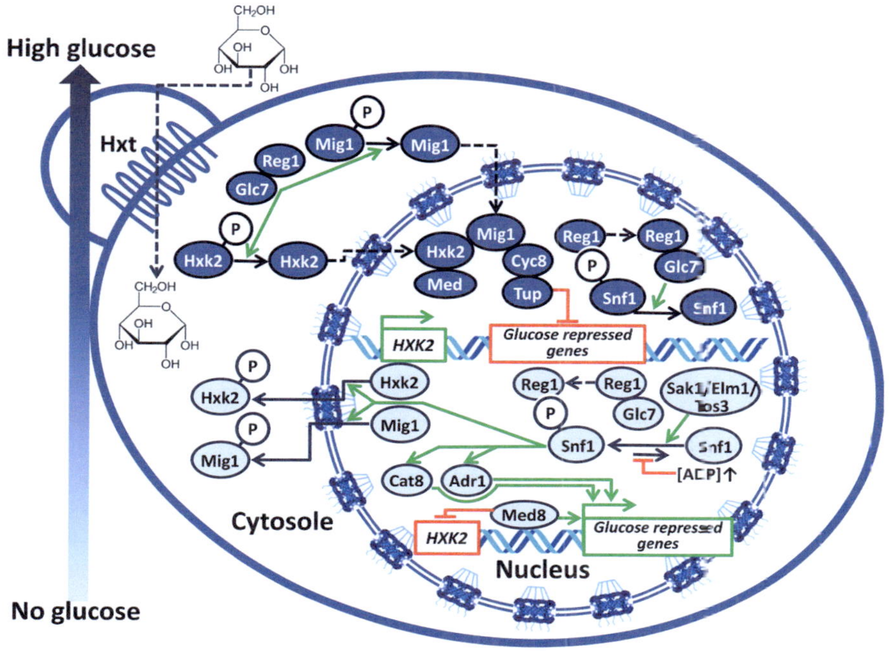

Fig. 14.3 cAMP/PKA glucose-signaling pathway. The plasma membrane protein Gpr1 senses the presence of extracellular glucose, what triggers the exchange from GDP- to GTP-bound Gpa2. GTP-bound Gpa2 activates Cyr1 which results in conversion of ATP into cAMP. Next, cAMP activates Tpk1/2/3, the catalytic subunits of PKA, dissociating it from a regulatory subunit Bcy1. Cyr1 is also stimulated by the GTP-bound Ras1/2, activated in turn by Cdc25 and Sdc25. It is proposed that glucose-6-phosphate mediates stimulation of Cdc25 and Sdc25 by glucose. During glucose starvation, cAMP/PKA signaling is attenuated by the action of Gpb1/2. Gpb1/2 stimulates Ira1/2, which catalyzes the hydrolysis of GTP-bound Ras1/2. Gpb1/2 also augments the association of Bcy1 and Tpk1/2/3. GTP-bound Gpa2 inhibits Gpb1/2, whereas Gpb1/2 may also possess some negative effect upon Gpa2. Additionally, two phosphodiesterases, Pde1 and Pde2, counteract PKA signaling in *S. cerevisiae*. Dark blue boxes indicate proteins' state at high glucose, while light blue boxes at low or no glucose conditions. Green arrows represent activation and red lines or boxes represent inhibition

acts further downstream with the adenylate cyclase Cyr1, the latter converting ATP into cAMP (Matsumoto 1982; Yun et al. 1997; Peeters et al. 2006; Steyfkens et al. 2018). It has been demonstrated that, in response to glucose, intracellular cAMP increases in a Gpr1-dependent manner, although a direct interaction of Gpr1 with glucose has not been proved yet (Yun et al. 1998).

Based on the abovementioned observations, the cAMP/PKA pathway operates in the following way: (1) glucose induces Gpr1 and thereby promotes the exchange of GDP-bound Gpa2 for a GTP-bound; (2) the GTP-bound Gpa2 translates the signal by stimulating Cyr1 and thus elevates the cytoplasmic concentration of cAMP; and

(3) elevated cAMP activates PKA (Fig. 14.3) (Toda et al. 1987; Kraakman et al. 1999).

In *S. cerevisiae*, PKA is a heterotetramer that consists of two catalytic subunits encoded by three different genes *TPK1–3* and a dimeric regulatory subunit, encoded by the *BCY1* gene (Toda et al. 1987). *TPK1–3* are known to have many diverse downstream targets involved in regulating stress response, gluconeogenesis, and glycolysis, as well as cell proliferation (Toda et al. 1987; Thevelein and De Winde 1999; Broach 2012). In addition to Gpa2, adenylate cyclase activity and consequently increased PKA activity can be stimulated by the two alternative GTP-binding proteins, Ras1and Ras2, in response to the signal derived from glycolysis (Fig. 14.3) (Toda et al. 1985; Rolland et al. 2000; Colombo et al. 2004). Ras1/2 activity is also modulated: the guanine nucleotide exchange factors Cdc25 and Sdc25 activate it (Boy-Marcotte et al. 1996), and the two GTPase-activating proteins Ira1 and Ira2 inhibit Ras1/2 (Tanaka et al. 1990). It was found that for Ras1/2 to be stimulated by glucose, its uptake and conversion into glucose 6-phosphate is required. Therefore, it is assumed, but not yet confirmed, that glucose 6-phosphate probably activates Ras1/2 through Cdc25 and/or Sdc25 (Beullens et al. 1988; Colombo et al. 2004). Ras1/2 inhibitor, Ira1/2, is functionally stabilized by the two kelch-repeat proteins, Gpb1 and Gpb2 (Harashima et al. 2006). Gpb1/2 interferes with the Gpr1–Gpa2 interaction (Harashima and Heitman 2005) and with the interaction between Tpk1/2/3 and the PKA-regulatory subunit, Bcy1, thus attenuating the PKA activity (Peeters et al. 2006). It was also reported that Gpb1/2 activity can be inhibited by a GTP-bound Gpa2 (Peeters et al. 2006). In addition, phosphodiesterases Pde1 and Pde2 have been identified as lowering cAMP levels by converting cAMP to AMP and, thus, counteracting the cAMP/PKA-signaling pathway (Fig. 14.3) (Ma et al. 1999). It has been also suggested that Ras activation might be dependent on vacuolar ATPase activity which is regulated by ATP level and interdependent cytosolic pH (Dechant et al. 2010).

It was shown just recently that the GPCR (Gpr1)-based extracellular glucose-sensing mechanism is unable to activate Cyr1 and stimulate cAMP synthesis as long as adenylate cyclase is not simultaneously activated by the Ras proteins. Importantly, the latter stimulation requires glucose transport into the cells, as well as its glycolytic conversion to fructose-1,6-bisphosphate, which activates Ras by directly stimulating its guanine nucleotide exchange factor, Cdc25 (Peeters et al. 2017; Steyfkens et al. 2018).

14.1.4 Crosstalk and Interactions of Yeast Glucose-Signaling Pathways

Yeast cells coordinate the signaling pathways into a crosstalking network to modulate their integrated response to glucose availability as schematically depicted in Fig. 14.4 (reviewed in Kim et al. 2013; Shashkova et al. 2015). For instance,

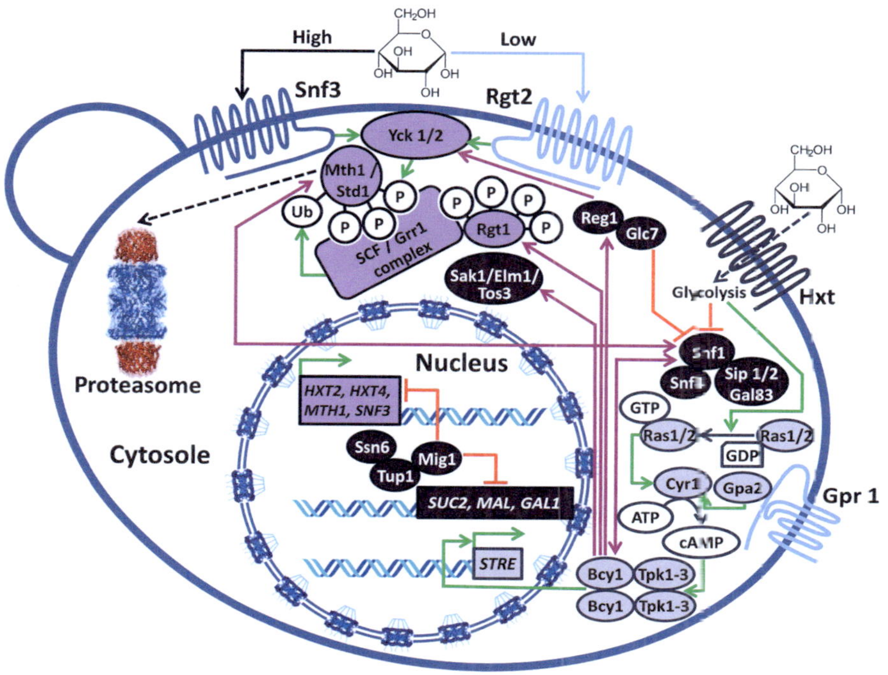

Fig. 14.4 Simplified scheme of the interactions between Snf1/Mig1 glucose repression pathway, cAMP–PKA, and Rgt2/Snf3 pathways in response to varying glucose levels. In the presence of glucose, Mig1 represses expression of the components of Rgt2/Snf3 pathway, Mth1 and Snf3, and some Hxt proteins; at the same time, Snf1 phosphatase, Glc7-Reg1, participates in activation of Yck1/2 kinases of Rgt2/Snf3 pathway; in the presence of glucose and preferred nutrients, the PKA pathway affects Snf1 activity and repression pathway. See more explanatory details on the cross-talk of the pathways in the main text. Components of the SRR pathway are depicted in light violet boxes; Snf1/Mig1 pathway, in dark violet boxes; and cAMP–PKA pathway – in light blue boxes. Green arrows represent activation, red lines or boxes represent inhibition, pink arrows represent crosstalk effects between the pathways, and dotted arrows indicate protein inter-compartment relocation events

generation of the glucose repression signal that inhibits the activity of the yeast AMPK kinase Snf1 requires glucose uptake and metabolism, and these signals must be coordinated. This circuit provides cells with a flexible mechanism that integrates different stimuli of individual pathways and fine-tunes the general regulatory response (Shashkova et al. 2015).

First of all, the Snf1 and Rgt2/Snf3 pathways interconnect at the level of regulation of gene expression. Thus, Mig1 in cooperation with Mig2, a Mig1 paralog which is not regulated by Snf1, are involved in repression of hexose transporter genes *HXT2* and *HXT4* and the genes encoding components of Rgt2/Snf3 pathway, *MTH1* and *SNF3*, under high glucose conditions (Fig. 14.4) (Özcan et al. 1996; Kaniak et al. 2004; Kim et al. 2006; Westholm et al. 2008).

This suggests that expression of the *HXT* genes in response to glucose is controlled by Snf1 by bifaced mechanism mediated by Rgt1 and Mig1 or interplay between induction and repression. A mathematical kinetic model describing this regulatory network predicted that interconnectivity between the Snf1 and Rgt1/Snf3 pathways ensures that resulting transport affinity correctly complies with actual external glucose concentrations (KuttyKrishnan et al. 2011; Bisson et al. 2016).

It was also shown that Snf1 regulates transcription of *HXT* genes by physically interacting with the key components of the Rgt2/Snf3 pathway (Fig. 14.4) (Tomas-Cobos and Sanz 2002). Accordingly, Std1 and Mth1 were not ubiquitinated and degraded even under high glucose conditions if Snf1 was artificially activated (Pasula et al. 2007). The observed effect was suggested to be due to inhibition of the nuclear export of Std1 and Mth1, and direct interaction of Snf1 with these two proteins was proposed (Hubbard et al. 1994; Tomas-Cobos and Sanz 2002; Kuchin et al. 2003). Also, for glucose-induced Mth1 degradation, Snf1 SUMOylation is apparently required (Simpson-Lavy and Johnston 2013). Other components of the glucose repression pathway are involved in the Rgt2/Snf3 signaling as well. For instance, the Snf1 phosphatase Glc7-Reg1 is an upstream activator of the Yck1/2 kinases, acting on Std1 and Mth1repressors (Fig. 14.4) (Moriya and Johnston 2004; Gadura et al. 2006).

There is also a reciprocal regulation in *S. cerevisiae* cells, i.e., the Rgt2/Snf3 pathway also interferes with the Snf1 pathway. It was shown that loss of Snf1 activity, e.g., due to a *SNF4* deletion, may be compensated by multicopy expression of Std1, suggesting that Std1 might affect glucose repression through its interaction with Snf1 (Hubbard et al. 1994). It has also been reported that overexpression of Std1 stabilizes active Snf1 complex (Kuchin et al. 2003). Transcriptional factors Mig1 and Mig2 have partially overlapping target promoters and are functionally partially redundant (Lutfiyya et al. 1998; Westholm et al. 2008). However, their expression is controlled by different mechanisms: Mig2 is under transcriptional control of the Rgt2/Snf3 pathway, and it also affects the repression pathway by being a transcriptional repressor of Mig1 (Kaniak et al. 2004). Thus, there are two different mechanisms by which Rgt2/Snf3 pathway fine-tunes glucose repression: first, by stabilizing the Snf1 complex and protecting the Snf1 functionally important Thr210 from dephosphorylation, and second, by affecting glucose repression through Mig2 (Fig. 14.4).

On the other hand, activity of the PKA pathway is needed to establish glucose repression. Thus, reduced PKA activity resulted in impaired glucose repression (Mbonyi et al. 1990), whereas overexpression of this pathway component led to the suppression of growth defects in delta *snf4* mutant (Hubbard et al. 1994). It was also reported that regulatory subunit Bcy1 and Ras GAP proteins Ira1/2 are required for activation of the Snf1 pathway in response to glucose limitation (Barrett et al. 2012). Notably, Bcy1 is also phosphorylated in an Snf1-dependent manner (Braun et al. 2014) (Fig, 4).

Although, as mentioned above, it was assumed that upstream kinases phosphorylate Snf1 constitutively and independently of exogenous glucose concentration (Rubenstein et al. 2008), it appeared that mutations in the putative PKA phosphorylation sites in kinase Sak1 (Hedbacker et al. 2004) result in a modest increase in Snf1 phosphorylation when glucose is present (Barrett et al. 2012). It was also demonstrated that Snf1 activity is upregulated independently of Sak1 status if cAMP-PKA pathway is downregulated, suggesting that other components of Snf1-dependent cascade are targets of PKA (Barrett et al. 2012). For example, such a target could be Reg1-Glc7 (PP1 phosphatase) complex, since its activation by glucose involves Php1-dependent post-translational modification (Castermans et al. 2012).

Furthermore, it was proposed that the cAMP-PKA pathway affects the spatial localization and activation of the Snf1 complex. For instance, under glucose limitation, the complex containing the Sip1 β-subunit localizes in proximity to the vacuole, while under high glucose it has a cytosolic localization (Vincent et al. 2001; Hedbacker et al. 2004). It was observed that in strains lacking the Bcy1 regulatory subunit, Sip1 is constitutively adjacent to the vacuole (Hedbacker et al. 2004). Thus, it seems that both Rgt2/Snf3pathway and the PKA pathway fine-tune the Snf1 pathway in response to alterations in glucose levels (Shashkova et al. 2015).

In addition, Rgt2/Snf3 and cAMP-PKA pathways are interacting too. As mentioned above, adenylate cyclase activation requires simultaneous action of the activated Gpr1 and Ras proteins (Peeters et al. 2017). Glucose activation of cAMP synthesis in yeasts thus requires the interplay of an extracellular nontransporting receptor-based sensing system with an intracellular sensing system of its catabolite (fructose-1,6-bisphosphate). It was the first discovered example of such a combined sensing system. However, recent data also suggest that, similarly, hexokinase product glucose-6-phosphate may be involved in signaling for glucose catabolite repression (Suppi et al. 2013; Vega et al. 2016). It also remains unclear whether the high-affinity glucose transporters, Hxt6 and Hxt7, which are expressed in the medium devoid of glucose, also play a role in glucose-induced activation of the cAMP-PKA pathway. Alternatively, the kelch-repeat protein that bypasses adenylate cyclase could also be responsible for this mode of glucose-triggered activation of PKA, without involvement of cAMP signaling (Peeters et al. 2006, 2007).

14.2 Glucose Regulation in "Nonconventional" Yeasts

As mentioned above, *S. cerevisiae*, which has been the main model organism to decipher how glucose-sensing and glucose-signaling mechanisms in the yeast cell are organized and interact, is ecologically adapted to aerobic glucose fermentation (Pfeiffer and Morley 2014). This lifestyle might have specifically influenced glucose signaling in this yeast (Dashko et al. 2014). Therefore, majority of other, so-called "nonconventional" and mostly Crabtree-negative yeast species very possibly exhibit a differently organized glucose-sensing machinery. This notion gets its support as more and more experimental and genomic data have been gathered on various yeast species. Below, we briefly review and summarize the findings on glucose signaling in diverse yeast species with emphasis on the very recent reports and representative yeasts not covered in the older review articles.

14.2.1 Schizosaccharomyces pombe

In the fission yeast *Schizosaccharomyces pombe*, which is, similar to *S. cerevisiae*, a Crabtree-positive (able to ferment glucose aerobically) yeast species, mutations have been indentified that protected cells from a 2-deoxyglucose (2-DG) toxicity. Mutants in the putative odr1 gene, encoding an uncharacterized hydrolase, led to a strong 2-DG resistance. It was speculated that Odr1 effect is similar to the effect of *S. cerevisiae* Dog1 and Dog2 phosphatases, which dephosphorylate toxic catabolite 2-DG-6-phosphate synthesized by hexokinase. The screen of a haploid deletion library with 2-DG-resistant mutants identified the genes *S. pombe* snf5, ypa1, pas1, and pho7. Notably, counterparts of the mentioned *S. pombe* genes in *S. cerevisiae* are not involved in mechanisms related to 2-DG resistance, pointing at possible principal differences between the budding and fission yeasts in this respect (Vishwanatha et al. 2016). Nevertheless, both yeasts currently serve as unicellular models for cancer cells' Warburg effect, including studies on the mode of action and resistance to 2-DG (Vishwanatha and D'Souza 2017).

S. pombe mutants resistant to glucose repression (ird5, ird13, and ird14) were characterized by the increased expression level of the fructose bisphosphatase fbp1 gene and by high tolerance to the oxidative stress. Their analysis also suggested that *S. pombe*, alike *S. cerevisiae*, possesses multiple mechanisms of glucose detection and signaling (Palabiyik et al. 2013), although PKA pathway has been proposed as a primary one (Hoffman 2005a).

In addition, two genes encoding proteins highly similar to Snf1-like protein kinase in *S. pombe*, ssp2(+) and ppk9(+), have been identified. Deletion of ssp2(+), but not ppk9(+), caused a partial defect in glucose derepression of *S. pombe* invertase, fructose bisphosphatase, and glycerol dehydrogenase, as well as defects in assimilation of sucrose and glycerol. Scr1, a putative transcription factor responsible for glucose repression, localized to the nucleus under glucose-rich conditions and to the cytoplasm during glucose starvation in the wild-type cells, but it had an invariably nuclear localization in the *ssp2Δ* mutant. It was also demonstrated that Ssp2 regulates phosphorylation and subcellular localization of Scr1 in response to glucose. It was proposed that *S. pombe* cells harbor a signaling pathway that operates similar to its Snf1-dependent analog from *S. cerevisiae* (Matsuzawa et al. 2012).

Of note, fission yeast *S. pombe* and budding yeast *S. cerevisiae* exhibit glucose metabolism in some aspects similar to that of cancer cells that display increased glycolytic rate and mitochondrial dysfunctions. Both yeasts have been used as models for such malignant human cells, including studies on 2-DG metabolism and signaling (for review on the subject, see Ref. Vishwanatha and D'Souza 2017). For a more detailed review of glucose signaling, in particular on *S. pombe* PKA pathway, please see previous reviews (Hoffman 2005a, b).

14.2.2 Kluyveromyces lactis

To investigate the evolution of the Cat8-Sip4-controlled transcriptional network, the comparative response to carbon limitation of Crabtree-positive *S. cerevisiae* to that of Crabtree-negative *Kluyveromyces lactis* has been studied (Mehlgarten et al. 2015). It was found that in *K. lactis*, but not in *S. cerevisiae*, the Sip4 protein plays an essential role in C2-carbon sources assimilation, including induction of the glyoxylate cycle. Both *Kl*Cat8 and *Kl*Sip4 were also involved in the regulation of lactose metabolism in *K. lactis*. The evidence was presented that *Kl*Sip4 counteracts *Kl*Cat8-mediated transcription activation of certain genes by competing for binding to some but not all CSREs (carbon source responsive elements) in their promoters. Also, the hierarchical relationship of Cat8 and Sip4 transcription factors and the set of target genes differs between bakers' yeast and *K. lactis*. Therefore, a role of Sip4 in controlling anabolic metabolism has diverged between the species and has been largely lost in the *Saccharomyces* lineage. These data contribute to the present knowledge on the genetic differences that support the particular metabolic lifestyle of the individual yeast species (Mehlgarten et al. 2015).

However, the function of the Snf1 complex in cell wall biosynthesis, established in *S. cerevisiae*, was found to be conserved in *K. lactis*. Sensitivities of *K. lactis* mutants in the components of this complex (*KlSNF1, KlGAL83, KlSNF4*) could be epistatically supressed by auxiliary mutations in glycolytic genes (*KlPFK1*, *KlPFK2*, *KlPGI1*) or in the *KlMIG1* gene. It was also revealed that, contrary to *S. cerevisiae* counterpart, a substantial amount of KlMig1 protein remained phosphorylated and in cytosol even under high extracellular glucose concentrations (Rippert et al. 2017).

In order to elucidate which functions of yeast hexokinase are evolutionarily conserved, the *K. lactis* global cellular response to the *hxk1* null mutation has been analyzed. In the *hxk1* mutant proteome, 45 proteins were identified as exhibiting at least threefold enhanced (most of the proteins) or reduced levels as compared to the wild-type strain. They were primarily involved in the carbohydrate and fatty acid metabolism, protein turnover, general stress response, and chromatin remodeling. Therefore, *Kl*Hxk1 was proposed to be a conserved multifunctional enzyme that has a repressive effect on a multitude of metabolic pathways also in Crabtree-negative yeasts (Mates et al. 2014).

As mentioned, *Sc*Hxk2 has a dual function combining the catalytic role in glycolysis with transcriptional regulation of glucose-repressible genes. Different studies proposed different upstream serine/threonine protein kinases for accomplishing *Sc*Hxk2 phosphorylation at serine 15 that leads to its translocation from the nucleaus into the cytosol. Experimental evidence has been presented that it is Tda1 kinase which is essential for such a modification, especially at limited external glucose availability, whereas Snf1, Mck1, PKA, and Sch9 are rather dispensable. Notably, a putative Tda1 homolog from *K. lactis* was shown to functionally complement *S. cerevisiae tda1* mutant by phosphorylating evolutionarily conserved serine 15 amino acid residue of hexokinase (Kaps et al. 2015).

In *K. lactis*, expression of the main glucose transporter gene *RAG1* is controlled by extracellular glucose through a signaling cascade similar to *S. cerevisiae* Snf3/Rgt2/Rgt1 "sensor/receptor–repressor (SRR)" signaling pathway and by the proficiency of glycolysis. It was demonstrated that *Kl*Rgt1 repressor mediates the effect of intracellular metabolism on *Kl*Rag1 expression by targeting the localization and stability of the Rag4, a single *K. lactis* nontransporting sensor. It was proposed that such a retrocontrol, conserved between *K. lactis* and *S. cerevisiae*, might efficiently prevent yeast cells from unnecessary glucose transport and its intracellular accumulation (Cairey-Remonnay et al. 2015).

For a further review on physiology of *K. lactis*, its pathways of glucose regulation in particular, the readers are redirected to older in-depth reviews on the subject (Rubio-Texeira 2005; Horák 2013; Rodicio and Heinisch 2013).

14.2.3 Kluyveromyces marxianus

2-DG-resistant mutants of the thermotolerant ethanologenic yeast *Kluyveromyces marxianus* were isolated and classified into three complementation groups. One of the mutants was characterized by the enhanced xylose utilization ability in the presence of glucose. This ability was shown to be due to mutation in the RAG5 gene encoding *K. marxianus* hexokinase. The corresponding mutation led to a single amino acid substitution (G270S) that impaired hexokinase enzymatic activity. Thus, similar to *S. cerevisiae*, hexokinase apparently participates in glucose repression in a not too distantly related *Kluyveromyces* species (Suprayogi et al. 2015).

Recently, *K. marxianus* mutants disrupted in the orthologs of *S. cerevisiae* Mig1 and Hxk2 have been described. The mutants were demonstrated to exhibit similar functional alteration with respect to glucose repression mechanism as observed in *S. cerevisiae*, suggesting a functional conservation of the main glucose repression pathway between these two yeasts (Nurcholis et al. 2019).

14.2.4 Candida albicans

Although *Candida albicans* is the leading cause of human fungal infections, it is also a member of the physiological human microbiome that consists of thousands of microbial species. Among the most abundant microbiota-derived metabolites found on human mucosal surfaces that control *C. albicans* proliferation at physiological concentrations and pH values are weak organic acids. Several genes, among them *CaMIG1*, have been identified which are required for the resistance of *C. albicans* to one or more organic acids. Consistent with glucose being an upstream activator of Mig1, glucose presence was required for organic acid resistance in the wild-type cells. It was proposed that *C. albicans* Mig1 plays a central role in orchestrating a transcriptional yeast cell response program to resist the fungistatic effect of metabolites produced by the gastrointestinal microbiota (Cottier et al. 2015). Also, at least certain individual hexose transporters have been implicated into pathogenicity and stress response in *C. albicans*. In RNA-sequencing experiments, it was also found that sensitivity of *C. albicans* to weak organic acids, e.g., acetic acid, is dependent on glucose sensing mediated by one of its putative glucose transporters, namely, Hgt16 (Cottier et al. 2015).

It was shown that glucose control in *S. cerevisiae* involves the antioxidant enzyme Cu/Zn superoxide dismutase (SOD1) and that generated by SOD1 H_2O_2 stabilizes the casein kinase Yck1 upon glucose sensing (Reddi and Culotta 2013). It was established that the two cytosolic *C. albicans* SOD enzymes, Cu/Zn SOD1 and Mn containing SOD3, both stabilized *S. cerevisiae* Yck1 and functionally complemented *S. cerevisiae sod1Δ* mutant. However, functions of SODs in glucose control in the two yeasts were found to be diverged, since, contrary to *S. cerevisiae*, Yck/Rgt1 pathway in *C. albicans* was not under control of SOD enzymes. Instead, in their native host, both *C. albicans* SODs functioned to repress glucose transporter genes in response to glucose in *Ca*Mig1-mediated repression pathway. Thus, in *C. albicans*, the SODs negatively regulate glucose uptake, while in *S. cerevisiae*, SOD1 induces glucose uptake (Broxton et al. 2018). Therefore, it was proposed that the role of SOD enzymes in glucose control has been rewired in Crabtree-negative opportunistic pathogen *C. albicans* to accommodate its adaptation to low glucose levels. Notably, *C. albicans* and *S. cerevisiae* also exhibit highly divergent glucose uptake modes (Sexton et al. 2007).

The above reports present a good example of why many of the established findings on glucose-signaling networks in *S. cerevisiae* model have to be generalized for Crabtree-negative yeasts with a justified cation. Although many aspects of glucose regulation are similar between *C. albicans* and bakers' yeast, e.g., the involvement of Rgt1 and Mig1 repressors, specific regulatory pathways in *C. albicans* have been apparently fine-tuned to meet the ecological adaptations of an opportunistic human pathogen. For further information and detailed reviews on *C. albicans* glucose-sensing network, the readers are referred to previous publications on this subject (Zaragoza et al. 2000; Brown et al. 2006, 2009; Sexton et al. 2007; Sabina and Brown 2009; Horák 2013).

14.2.5 Yarrowia lypolitica

In the alkane-utilizing oleaginous yeast *Yarrowia lipolytica*, production of the extracellular lipase encoded by the *LIP2* gene in the presence of glucose was demonstrated to be negatively regulated by hexokinase Hxk1 (Fickers et al. 2005). This fact points at a conserved role of hexokinase in catabolite repression in this evolutionarily distant to *S. cerevisiae* yeast. In addition, establishment of this fact and further studies on the molecular mechanisms governing glucose regulation may eventually lead to a new metabolic engineering and strain improvement in *Y. lipolytica*, the organism recently proposed as an effective biodiesel-producing microbial cell factory (Yan et al. 2017). In addition, a better understanding of *Y. lypolitica* carbon catabolite repression machinery may add in constructing superior strain-producers of citric acid (Sabra et al. 2017). It was also observed that glucose has a relatively mild repressive effect on xylose and cellobiose utilization in *Y. lipolytica*. Manipulating with these mechanisms may help in constructing superior *Y. lipolytica* strains for enhanced conversion of biomass-derived fermentable sugars to chemicals and fuels (Ryu et al. 2016).

14.2.6 Dekkera bruxellensis

Dekkera bruxellensis, while being a phylogenetically distant relative of *S. cerevisiae* and a closer relative of methylotrophic yeast, shares with the former the ability to grow without oxygen and the ability to produce ethanol under aerobic conditions. *D. bruxellensis* diverged from the *Saccharomyces* lineage approximately 200 million years ago, before the onset of the whole genome duplication. *D. bruxellensis* is known as a competitor of *S. cerevisiae* in the fermentation environment and is characterized by a remarkable adaptability. In spite of its relatively slower growth, *D. bruxellensis* outcompetes *S. cerevisiae* in glucose-limited cultures, indicating a more efficient energy metabolism and/or higher affinity for glucose (Blomqvist and Passoth 2015).

It is assumed that partial increase in ploidy and rewiring of certain promoters have enabled evolution of the fermentative lifestyle in *D. bruxellensis*. In comparison to *S. cerevisiae*, *D. bruxellensis* also exhibits a broader spectrum of the utilized carbon and nitrogen sources. For instance, it is able to utilize nitrates. However, little data are available on the mechanism of glucose regulation in this yeast.

It was revealed nevertheless that in nonoptimal growth conditions, *D. bruxellensis* tends to switch to the respiratory metabolism and also converts less glucose to ethanol than *S. cerevisiae* does when nitrogen supply is limited (De Barros Pita et al. 2013).

Also, in contrast to *S. cerevisiae*, galactose is a nonfermentable carbon source in this yeast. But, similar to bakers' yeast, *D. bruxellensis* exhibits the "short-term Crabtree effect," and the expression of genes involved in respiratory metabolism and galactose utilization is repressed by glucose (Moktaduzzaman et al. 2015).

It can be predicted that further study of *D. bruxellensis* glucose-signaling networks will add to our understanding of what steps evolutionary adaptation took to adapt a particular species to a fermenative lifestyle.

14.2.7 Spathaspora passalidarum

For being an efficient ethanologenic yeast, it is required to possess abilities of utilization of various sugars including xylose and arabinose from lignocellulosic biomass. Interestingly, new strain isolate of the thermotolerant yeast *Spathaspora passalidarum* demonstrated no significant glucose repression in experiments with mixed sugars, the observation consistent with its strong resistance to 2-DG. In addition, inhibitor of oxidative respiration, antimycin A, showed no effect on the growth of this *S. passalidarum* strain in xylose-supplemented medium. Thus, due to its peculiar glucose repression signaling, which has to be studied in more detail, this yeast may present an unusual alternative to bakers' yeast as a fermenter of sugar mixtures (Rodrussamee et al. 2018).

14.2.8 Xanthophyllomyces dendrorhous

Recent studies have demonstrated that glucose repression negatively regulates carotenogenesis in the yeast *Xanthophyllomyces dendrorhous*. This regulation was mediated by Mig1 and corepressor complex Cyc8-Tup1, components of a classical glucose repression mechanism in *S. cerevisiae*. Of note, *X. dendrorhous* Mig1, Cyc8, and Tup1 were functionally active when heterologously expressed in *S. cerevisiae* (Alcaíno et al. 2016; Córdova et al. 2016).

14.2.9 Methylotrophic Yeasts: Pichia (Komagataella) pastoris

The methylotrophic, Crabtree-negative yeasts, such as *Pichia pastoris*, *Hansenula polymorpha*, and *Candida boidinii*, are widely used as heterologous protein production hosts (Stasyk 2017). Strong inducible promoters derived from methanol utilization genes or constitutive promoters of glycolytic genes are typically used to drive foreign gene expression. Notably, genes involved in methanol utilization are not only repressed in the presence of glucose but also disaccharides in case of *H. polymorpha*, or by glycerol in case of *P. pastoris* (for a comprehensive review on the regulation of yeast methanol utilization pathways, see Hartner et al. (2006)).

Interestingly, contrary to the previous assumptions, it was recently proposed that the entire methanol assimilation pathway in the methylotrophic yeast *P. pastoris* localizes to peroxisomes rather than relying on the part of the cytosolic pentose phosphate pathway for xylulose-5-phosphate regeneration. For this purpose, *P. pastoris* (and presumably also other methylotrophic yeasts) have evolved a duplicated set of methanol-inducible enzymes of formaldehyde assimilation targeted to peroxisomes (Rußmayer et al. 2015).

It was also established in global microarray analysis that *P. pastoris* regulates its gene-specific response to different carbon sources at the transcriptional, rather than translational, level (Prielhofer et al. 2015). Transcript-specific translational responses were found to be minimal, while extensive transcriptional regulation was observed for cells grown on alternative carbon sources. Interestingly, despite their lower growth rates, global translation was most active in methanol-grown *P. pastoris* cells, as compared to cells grown under excess of glycerol or glucose. Peroxisomal and methanol utilization genes were confirmed to be subject to carbon substrate (glucose or glycerol) repression, but were also found to be, in addition to induction by methanol, strongly derepressed in glucose-limiting conditions (as often applied in fed-batch cultivation modes) (Prielhofer et al. 2015).

Methanol utilization in yeast including *P. pastoris* depends on the induction of the expression of genes whose products are required for methanol metabolism and the key enzyme products reside in peroxisomes. Mxr1 (methanol expression regulator 1) has been identified as an essential factor for induction of peroxisomal alcohol oxidase (*AOX1*) and normal peroxisomal biogenesis in response to methanol. Mxr1 is a large protein with a zinc finger DNA-binding domain near its N terminus that exhibits certain similarity to *S. cerevisiae* Adr1. Mxr1 has been shown to localize to the nucleus in the cells grown on methanol or other gluconeogenic substrates, however, it was cytosolic in glucose-containing medium. It was proposed that *P. pastoris* Mxr1 is a deviant homolog of *S. cerevisiae* Adr1 that has gained new functions adapted to methylotrophic metabolism (Lin-Cereghino et al. 2006).

PpNrg1, a Cys2His2 zinc finger transcriptional repressor, was found to be localized to the nucleus and to participate in the repression of *P. pastoris* P_{AOX1} and of the number of other genes involved in methanol utilization and peroxisome biogenesis in glucose- and glycerol-grown cells. Some binding sites of PpNrg1 overlapped with sites of PpMxr1, a putative P_{AOX1} inducer (Wang et al. 2016).

In a kinase screening assay in *P. pastoris*, 27 target kinases have been identified as involved in cell growth or P_{AOX1} regulation. Out of them, five kinases were involved in P_{AOX1} activation and five in P_{AOX1} repression, most of which appeared to be related to Snf1 and Hog1 pathways. Notably, one of these kinases has been identified as *P. pastoris* Snf1 subunit *Pp*Gal83 which is involved in the positive regulation of P_{AOX1} (Chen and Nielsen 2013).

It was reported that *P. pastoris* mutants in the genes encoding putative homologs of transcriptional repressors *MIG1* and *MIG2* were deficient in catabolite repression of peroxisomal alcohol oxidase in glycerol, but not in glucose medium. Comparative genome-wide RNA-sequencing analysis revealed that methanol catabolism and peroxisome biogenesis pathways were remarkably upregulated (the peroxisomal alcohol oxidase genes *AOX1* and *AOX2* higher than 30-fold) in *P. pastoris Δmig1Δmig2* mutant cells grown in glycerol. The binding of an essential transcription activator, Mit1, to the AOX1 promoter in the *mig* double mutant was enhanced as well (Shi et al. 2018).

Two putative hexose transporter homologs, *Pp*Hxt1 and *Pp*Hxt2, have been identified and characterized in *P. pastoris* (Zhang et al. 2010). When expressed in a *S. cerevisiae* hxt-null mutant strain that is unable to transport and grow on the monosaccharides, either protein restored growth on glucose or fructose. Both *P. pastoris HXT* genes were transcriptionally regulated by glucose: transcript levels of *PpHXT1* were up-induced by high glucose levels, whereas levels of *PpHXT2* transcript were relatively less abundant and fully induced only at low levels of glucose. While *Pp*Hxt2 was apparently important for fermentative growth, because it was essential for growth on glucose or fructose when respiration was inhibited, deletion of *PpHXT1*, but not *PpHXT2*, led to derepression (impaired catabolite repression) of the alcohol oxidase *AOX1* gene in the presence of hexoses. The authors proposed that sugar-inducible P_{AOX1} in a delta *hxt1* strain may be a promising tool for methanol-free expression of heterologous proteins (Zhang et al. 2010).

Also, a single homolog of nontransporting yeast glucose sensors, termed *Pp*Gss1 (**g**lucose **s**ensor), has been identified in *P. pastoris*. Deletion of the corresponding gene strongly negatively affected growth on glucose and concomitantly glucose repression and glucose-triggered autophagic peroxisome degradation in the mutant cells (Polupanov et al. 2012). Interestingly, C-terminal extension of the nontransporting sensor Gss1 appeared to be not compulsory for signaling for glucose catabolite repression. In addition, substitution of one conserved amino acid R180K in Gss1 produced no apparent phenotype, contrary to analogous mutation in other similar glucose sensors from *S. cerevisiae*, *C. albicans*, or *H. polymorpha* that led to constitutively signaling sensors' mutant forms. It was not conclusively elucidated, however, whether such an effect was due to effector exclusion (deficient glucose uptake) or because of the specific sensing functions carried out by this novel sensor (Polupanov and Sibirny 2014).

As mentioned, expression of the *P. pastoris* methanol-inducible P_{AOX1} used widely for the production of recombinant proteins is regulated by the zinc finger protein Mxr1 (Lin-Cereghino et al. 2006). Also, the putative biotin starvation and methanol-inducible zinc finger protein named Rop (repressor of phosphoenolpyruvate carboxykinase (PEPCK)) has been identified in *P. pastoris* (Kumar and Rangarajan 2011). It was found that Rop repressor negatively regulates P_{AOX1} in a nutrient-rich medium. Deletion of the *ROP* gene resulted in an enhanced expression of *AOXI* and faster growth, whereas Rop overexpression led to repression of *AOXI* and growth retardation of *P. pastoris* cells in YPM medium. Intriguingly, deletion or overexpression of Rop had no effect on the *AOXI* gene expression and methylotrophic growth in a standard "minimal" medium with yeast nitrogen base and methanol (YNM). In vitro DNA binding studies identified the 5′ CYCCNY 3′ motif in the P_{AOX1} as a binding site for both Mxr1 and Rop. The latter was also found to translocate from cytosol to nucleus in *P. pastoris* cells cultured in YPM but not YNM. It was proposed that Mxr1 and Rop exhibit the same DNA binding specificity but regulate methanol metabolism antagonistically in *P. pastoris* (Kumar and Rangarajan 2011).

14.2.10 Hansenula (Ogataea) polymorpha

Yet in the early works on *H. polymorpha*, it was shown that regulation of peroxisomal alcohol (or methanol) oxidase (AOX), i.e., its repression by glucose and depression as well as induction by methanol, is controlled at the level of transcription (Roggenkamp et al. 1984).

In *H. polymorpha* mutants that exhibited insensitivity to glucose repression of methanol assimilation and peroxisome formation, it was shown that upon prolonged glucose metabolism this phenotype is masked by catabolite inactivation and degradation of the methylotrophic enzymes. Thus, glucose repression and catabolite inactivation (which involves pexophagy) appeared to be controlled by the distinct sets of genes (Roggenkamp 1988). On one of the isolated *H. polymorpha* mutant which exhibited constitutive peroxisome biogenesis in glucose medium, it was demonstrated that transferring its methanol-grown cells to glucose triggers profound pexophagy. Thus, again, it was proposed that the two glucose-triggered regulatory processes are controlled by genetic elements which are, at least in part, distinct (Parpinello et al. 1998).

Ethanol continued to repress methanol metabolism in such mutants. These data suggested that, similar to another methylotrophic species *Pichia methanolica*, glucose and ethanol have distinct repression pathways in *H. polymorpha* (Roggenkamp 1988). Thus, the existence of several autonomous mechanisms of the catabolite repression by carbon sources of alcohol oxidase in the methylotrophic yeasts has been recognized early on (Sibirny et al. 1987, 1988).

It has been also recognized that mutants of the methylotrophic yeast deficient in glucose repression may be perturbed in glucose transport system (Alamäe and Simisker 1994; Stasyk et al. 1994).

Interestingly, it was demonstrated that *H. polymorpha* alcohol (methanol) oxidase gene promoter P_{MOX} confers a glucose-repressible expression of lacZ reporter gene in *S. cerevisiae*. A 200-bps long region has been identified in P_{MOX}, termed MOX-B, responsible for glucose regulation, which also harbored a putative consensus binding site for Adr1, a transcription factor involved in the derepression of *S. cerevisiae* peroxisomal proteins. This data suggested that Adr1 may be a conserved regulator for genes encoding peroxisomal proteins in various *Saccharomycetaceae* yeast (Pereira and Hollenberg 1996).

Recently, *H. polymorpha* P_{MOX} has been studied with respect to glucose catabolite repression at the single-cell level. The authors reported that decoupling single cells from population activity revealed a hitherto underrated ultrasensitivity of P_{MOX} to glucose repression. While previous (population-based) studies suggested full P_{MOX} derepression at extracellular glucose concentrations of ~1 g/L(−1), glucose repression on a single-cell level occurred already at concentrations as low as 5 × 10(−4) g/L(−1) (Dusny and Schmid 2016).

H. polymorpha maltase gene (*HpMAL1*) is essential for the growth of this yeast on maltose and sucrose (Alamäe et al. 2003). Interestingly, expression of the *S. cerevisiae* maltase gene *MAL62* from its own promoter was induced by maltose and sucrose, and repressed by glucose if expressed in *H. polymorpha* delta *mal1* mutant, and vice versa, *HpMAL1* promoter was also recognized and correctly regulated by carbon sources in a *S. cerevisiae* maltase-negative mutant. Therefore, it was proposed that transcriptional regulators of *S. cerevisiae MAL* genes (including Mig1 repressor) can affect the expression of the *H. polymorpha* maltase gene and that homologs of these proteins may exist in *H. polymorpha*. It was also proposed that, due to its strength and strict regulation, *HpMAL1* gene promoter may be a suitable platform for certain biotechnological applications, for instance, maltose-inducible and glucose-repressible production of recombinant proteins (Alamäe et al. 2003). Analysis of 2-DG-resistant mutants suggested that *H. polymorpha* most probably possesses a repressor protein that in the presence of glucose can downregulate expression of both maltase and enzymes of methanol oxidation (Alamäe and Liiv 1998).

Regulation of synthesis of the sugar-repressed alcohol oxidase, catalase and maltase in the isolated hexose kinase-negative, glucokinase-negative and double kinase-negative mutants of *H. polymorpha* was studied. It was found that in mutants possessing either one of the glucose-phosphorylating enzymes, glucose repression of the mentioned enzymes remained intact, whereas in a double kinase-negative mutant, it was abolished. It was concluded that glucose repression in *H. polymorpha* requires a glucose-phosphorylating enzyme. In line with this hypothesis, the presence of fructose-phosphorylating hexokinase was specifically required for fructose repression of alcohol oxidase, catalase, and maltase. Hence, glucose-6-phosphate seems to be a so-called "corepressor" molecule that triggers hexose repression signaling in this yeast (Kramarenko et al. 2000).

It was also reported that *H. polymorpha* glucokinase gene, *HpGLK1*, was able to functionally complement glucose, but not fructose, repression deficiency in double kinase-negative mutant. Accordingly, the enzyme was shown to be able to phosphorylate glucose, mannose, and 2-DG, but not fructose. When transformed into *S. cerevisiae* triple kinase-negative mutant, *HpGLK1* could not functionally complement glucose repression defect with regard to the synthesis of invertase and maltase. Thus, it was confirmed that hexose phosphorylation is required for functional repression signaling in *H. polymorpha*. Moreover, hexokinase as a protein is not strictly essential for this process and, contrary to the situation in *S. cerevisiae*, it can be functionally substituted by glucokinase I in repression signaling (Laht et al. 2002). The *H. polymorpha* hexokinase gene *HpHXK1* was cloned by functional complementation of the mentioned double kinase mutant. It was verified that indeed hexokinase is specifically needed for the establishment of fructose repression in *H. polymorpha* (Karp et al. 2003).

In subsequent studies, expression of recombinant β-glucuronidase under alternative *H. polymorpha MOX*, *FMD*, *MPP1*, and *MAL1* gene promoters in response to mono- and disaccharides was monitored in mutants deficient in (1) hexokinase, (2) hexokinase and glucokinase, (3) maltose permease, or (iv) maltase (Suppi et al. 2013). It was confirmed that glucose and fructose repress the mentioned promoters only if transported into and phosphorylated in the cell. Therefore, the hypothesis that glucose-6-phosphate is a sugar repression signaling metabolite for *H. polymorpha* got further support. However, it was also observed that glucose and fructose strongly activated expression from *MOX*, *FMD*, *MPP1*, and *MAL1* promoters in double hexokinase and glucokinase deletion strain, intriguingly indicating that unphosphorylated monosaccharides themselves have an apparent promoter-derepressing effect. It was also shown that disaccharides maltose and sucrose must be internalized and split into monosaccharides to exert their repression on P_{MOX}. In other words, it is monosaccharide metabolites that trigger repression signaling. Since derepression of methanol-induced promoters *MOX* and *FMD* in glucose medium readily occurs in hexokinase- or double kinase-negative mutants, these strains were proposed as hosts for methanol-free production of foreign proteins under control of one of the mentioned gene promoters (Suppi et al. 2013).

A hexose transporter homolog has been identified in *H. polymorpha* by functional complementation of 2-DG resistant mutant, designated *HpGcr1* (*G*lucose *C*atabolite *R*epression) (Stasyk et al. 2004). Deficiency in the *GCR1* gene led to a pleiotropic phenotype that included partially impaired glucose transport, and, correspondingly, impaired growth on glucose, and constitutive presence of peroxisomes and peroxisomal enzymes in glucose-grown cells. In addition to glucose, mannose, trehalose, and fructose failed to repress peroxisomal alcohol oxidase in *gcr1* delta mutant cells. Catabolite repression of the peroxisomal enzymes and peroxisome biogenesis elicited by ethanol, sucrose, and maltose was found to be intact in *gcr1* mutant cells. Certain phenotypic differences between the constructed *gcr1* delta mutant and UV-induced *gcr1–2* mutant (Stasyk et al. 1997) that carries single amino acid substitution S85F suggested that Gcr1 participates in glucose repression mechanism not as a regular hexose transporter but also or rather as a specific sensor protein (Stasyk et al. 2004).

It was reported recently that potential *Hp*Gcr1 orthologs are found only in the genomes of a few phylogenetically closely related to *H. polymorpha* yeasts but are absent in all other yeasts whose genomes are publicly available (Stasyk et al. 2018). On the other hand, other closest *Hp*Gcr1 homologs were found to be high-affinity glucose symporters or putative transceptors from filamentous fungi. The latter fact suggested a possible *Hp*Gcr1 origin due to a specific archaic gene retention or via horizontal gene transfer from Eurotiales fungi to *H. polymorpha*'s clade. However, phylogenetic analysis failed to resolve between these two options. Of note, putative *Hp*Gcr1 homolog is absent in the genome of *P.pastoris* but is present in *C. boidinii*. It was also found that, similar to the yeast nontransporting glucose sensors, the substitution of the conserved arginine residue converted *Hp*Gcr1^{R165K} into a constitutively signaling form. Synthesis of the mutant *Hp*Gcr1^{R165K} in delta *gcr1Δ* strain did not restore glucose transport or repression, but instead profoundly impaired growth independent of the carbon source used (Stasyk et al. 2018).

Intriguingly, *gcr1Δ* was also impaired in transcriptional induction of repressible peroxisomal alcohol oxidase and, correspondingly, in growth on methanol. This data suggested that *Hp*Gcr1 has signaling function also in the absence of glucose. Moreover, overexpression of the functional transporter *Hp*Hxt1 in *gcr1Δ* partially restored growth on glucose and glucose repression but did not rescue impaired growth on methanol. Heterologous expression of *Hp*Gcr1 in a *S. cerevisiae* hxt-null strain did not restore glucose uptake due to this protein mislocalization. However, *Hp*Gcr1 overexpression in *H. polymorpha* led to an increased sensitivity to extracellular 2-deoxyglucose, strongly suggesting *Hp*Gcr1 is a functional glucose carrier. The combined data suggested that *Hp*Gcr1 may be the first representative of a novel type of yeast glucose transceptor (Stasyk et al. 2018).

It is important to emphasize that transporting receptors or transceptors for a number of signaling molecules are well known in yeast and are mainly involved in transducing the signals of nutrient availability via the PKA pathway (for review see Conrad et al. 2014; Steyfkens et al. 2018). However, no such transceptors were previously known for glucose or other sugars, although some candidates have been recently suggested (Schuler et al. 2015).

In addition to *Hp*Gcr1, two other hexose transporter homologs were identified in the *H. polymorpha* genome and functionally characterized (Stasyk et al. 2008). One protein, *Hp*Hxs1 (hexose sensor), exhibited the highest degree of primary sequence similarity to yeast nontransporting glucose sensors, alike *S. cerevisiae* Snf3 and Rgt2. Another, *Hp*Hxt1 (hexose transporter), was the most similar to functional yeast glucose carriers-facilitators. When heterologously overexpressed in *S. cerevisiae*, hexose transporterless mutant, *Hp*Hxs1, did not restore growth on glucose or fructose, suggesting it is nonfunctional as a carrier. To the contrary, *Hp*Hxt1 was functional in this heterologous host. In *H. polymorpha*, Hxs1 was found to be expressed at a moderately low level and was required for glucose induction of the expression of functional transporter Hxt1, a low-affinity carrier. Importantly, similar to other yeast nontransporting sensors, C-terminal region of Hxs1 was essential for its sensor function, and a conserved amino acid substitution R203K in the Hxs1 sequence converted the protein into a constitutively signaling form. Importantly, Hxs1 was not required for glucose repression or glucose-induced pexophagy. However, hxs1 mutation led to a significantly but transiently impaired AOX repression in response to fructose, probably due to the stronger defect in transport of this hexose in the mutant strain. It was concluded that in the Crabtree-negative yeast *H. polymorpha*, the single transporter-like sensor Hxs1 mediates signaling in the hexose induction pathway, whereas the rate of hexose uptake affects the strength of hexose catabolite repression (Stasyk et al. 2008).

The peculiar phenotype of the *H. polymorpha* mutants affected in a putative transceptor Gcr1 was utilized to design a modified expression platform that relies on the strong and strictly regulated by carbon sources promoter of peroxisomal alcohol oxidase P_{MOX} (Krasovska et al. 2007). As mentioned above, expression from P_{MOX} in the wild-type *H. polymorpha* host strain is induced by methanol and is partially derepressed in glycerol or xylose media, whereas in the presence of hexoses, disaccharides, or ethanol, it is fully repressed. It is known that P_{MOX}-induced methanol requirement for maximal induction of the gene expression in large-scale fermentations is a significant drawback, as methanol is highly toxic and flammable, supports relatively slow growth rate and methylotrophic growth, and also requires energy-consuming and expensive aeration.

H. polymorpha mutants deficient in glucose repression of P_{MOX} due to *gcr1* mutation and unidentified secondary mutations were utilized as expression hosts for a number of foreign proteins. The mutants exhibited pronounced repression defects only by hexoses and xylose, but not by disaccharides or ethanol. The developed modified two-carbon source mode expression platform thus utilized convenient sugar substrates for growth (e.g., sucrose or maltose) and induction of recombinant protein expression (e.g., glucose or xylose) (Krasovska et al. 2007).

Efficient and strictly regulated by sugar substrates expression of three recombinant proteins, namely, a secreted glucose oxidase from the fungus *Aspergillus niger*, a secreted human mini pro-insulin, and an intracellular hepatitis B virus surface antigen (HBsAg), in these mutant hosts was demonstrated. The proposed modified expression platform thus preserved the favorable regulatable nature of P_{MOX} without need for methanol, making a convenient alternative to the traditional *H. polymorpha* expression hosts (Krasovska et al. 2007). The productivity of this *H. polymorpha* mutant system has been later demonstrated on the example of highly efficient HBsAg producer (Krasovska et al. 2013).

To elucidate whether Snf1-dependent regulatory pathway is conserved in *H. polymorpha*, *HpTUP1* and *HpSNF1* were identified and disrupted. Deletion of *HpTUP1* did not affect glucose repression of the major methylotrophic metabolism genes (e.g., *MOX, DAS*), suggesting that function of Tup1 in repression signaling established in *S. cerevisiae* is not conserved in *H. polymorpha*. In contrast, the deletion of *HpSNF1* impaired activation of the methylotrophic genes in the absence of glucose, suggesting its conserved function in transcriptional induction of glucose-repressible genes. The previously identified Adr1-like factor (Pereira and Hollenberg 1996) has been proposed to interact with *Hp*Snf1 in such a regulation (Oliveira et al. 2003).

Finally, it was demonstrated that deficiency in the putative *H. polymorpha* homologs of transcriptional repressors Mig1 (*Hp*Mig1 and *Hp*Mig2), as well as *Hp*Tup1 homolog, only partially and differentially affects repression of peroxisomal alcohol oxidase by sugars and ethanol (Stasyk et al. 2007). Specifically, glucose repression defect was more prominent in the double *mig1mig2* mutant relative to each of the single mutants. However, this defect was much weaker in comparison with that observed in *H. polymorpha gcr1* mutant, suggesting other transcriptional repressors are involved in the process.

It was reported earlier that deficiency in *Hp*Tup1 leads to impairment of glucose- or ethanol-induced macropexophagy (Leão-Helder et al. 2004). In *H. polymorpha mig1mig2* double-deletion cells, glucose-triggered macropexophagy was also substantially impaired. These findings suggested that homologs of the elements of the *S. cerevisiae* main repression pathway have pleiotropic but specialized functions in catabolite regulation in *H. polymorpha* (Stasyk et al. 2007).

14.2.11 Candida boidinii

Collection of *C. boidinii* mutants exhibiting peroxisomal alcohol oxidase activity when grown on glucose and methanol mixture was isolated using 2-DG-resistance selection scheme. Analysis of their representatives showed that although glucose repression of peroxisomal enzyme was impaired, glucose-induced inactivation was not. In addition, ethanol-triggered repression was intact in the mutants. Thus, these early works suggested that, similar to *H. polymorpha* and other methylotrophic yeasts, processes of glucose and ethanol catabolite repression, as well as catabolite inactivation, are controlled by separate sets of genes (Sakai et al. 1998).

Expression profile of several *C. boidinii* promoters of the genes involved in methanol metabolism and peroxisome functioning has been evaluated. It was revealed that although expression of the promoters of alcohol oxidase and dihydroxyacetone synthase genes (P_{AOD1} and P_{DAS1}, respectively) was completely repressed by glucose, formate-induced expression of formate dehydrogenase gene P_{FDH1} was not (Yurimoto et al. 2000).

Similarly, D-alanine-induced activity of D-amino acid oxidase and catalase activity were not completely repressed by glucose in *C. boidinii*, as monitored on mRNA and protein levels (Sakai et al. 1998).

The putative *C. boidinii* homolog of the Mig1 repressor involved in glucose repression in *S. cerevisiae* has been identified. Disruption of the *CbMIG1* gene had no apparent negative effect on utilization of various carbon sources. In the *mig1* deletion mutant, no AOD1 expression has been detected, tentatively suggesting that Mig1 is dispensible for glucose repression in this yeast. However, AOD1 expression was increased in the early stage of adaptation to methanol upon transfer of the delta *mig1* mutant cells from glucose medium to methanol. Furthermore, localization of *Cb*Mig1 in glucose-grown cells was primarily nuclear, whereas in methanol-grown cells, it was cytosolic. These data suggested that, similar to *H. polymorpha*, *Cb*Mig1 negatively regulates methanol-inducible gene expression in response to glucose, and that activity and localization of AOD1-inducing transcriptional factors interplay with CbMig1 (Zhai et al. 2012).

14.3 Future Perspectives

Several very recent findings suggest possible future directions of the research in the field of glucose signaling in yeasts and some novel perspective biotechnological applications based on this knowledge. For instance, one such direction may concern the phenomenon of epigenetic transcriptional memory of the promoters of the repressible genes, e.g., *S. cerevisiae GAL* genes or genes of respirative metabolism whose transcription is mediated by Tup1 (Sood et al. 2017).

A very interesting observation that warrants further study was recently reported that the increased heme synthesis in *S. cerevisiae*, even under glucose-repressive conditions, induced transcription of the genes required for the tricarboxylic acid (TCA) cycle, electron transport chain, and oxidative phosphorylation, essentially leading to a switch from fermentation to respiration. These results tentatively indicated that glucose-mediated repression signaling in *S. cerevisiae* may involve, or may be due, at least in part, to the low intracellular heme level (Zhang et al. 2017).

Recent report suggests that low RNA polymerase III activity results in upregulation of *S. cerevisiae* hexose transporter Hxt2 independently of major glucose-signaling pathways, recognized as on/off switch of either positive or negative *HXT* gene regulation. Furthermore, Rgt1/Ssn6-Tup1 complex, which has a dual function in gene transcription as a repressor-activator complex, apparently contributes to such *HXT2* transcriptional activation (Adamczyk and Szatkowska 2017). If confirmed, it has to be further elucidated what is the physiological significance of such a mechanism.

An intricate link between (the regulation of) autophagic vacuolar degradation of cell constituents, cell energy status, and glucose repression was recently reported. That is, nonselective autophagy was found to be induced in response to abrupt carbon starvation when *S. cerevisiae* cells are grown with glycerol but not glucose as the carbon source. Autophagy was shown to play an important role in cell viability during a prolonged carbon starvation. Also, the extent of activation of such carbon starvation-induced autophagy positively correlated with cells' oxygen consumption rate and intracellular ATP levels, pointing to a crosstalk between autophagy induction and respiratory metabolism (Adachi et al. 2017). Overall, these findings suggested that carbon starvation-induced autophagy is negatively regulated by carbon catabolite repression – the link that has to be further experimentally validated.

A *S. cerevisiae* strain able to utilize aminosugar glucosamine as a sole carbon and nitrogen source due to expression of the heterologous glucosamine-6-phosphate deaminase from *Y. lypolitica* has been described. It was found that glucosamine was able to exert a repression effect on a number of glucose-repressible enzymes, although the effect was weaker than that exerted by glucose. It was proposed that the availability of such a recombinant strain may provide a new tool in addressing a number of specific questions about glucose-signaling pathways in *S. cerevisiae* (Flores and Gancedo 2018).

By examining *S. cerevisiae tps1* mutant, deficient in trehalose-6-phosphate synthase, complemented by a range of heterologous *TPS1* genes from different organisms, it was established that glucose repression of gluconeogenic genes is dependent on trehalose-6-phosphate (T6P), a product of (Tps1). It was suggested that the TPS protein is indispensable for growth on fermentable sugars, and T6P, therefore, has a critical role as a sensing molecule that promotes both sugar fermentation and glucose repression in bakers' yeast. It remains to be experimentally addressed whether the same similar regulation also takes place in "nonconventional" yeasts (Vicente et al. 2018).

A deeper knowledge of glucose-dependent signaling mechanism in various yeasts is necessary to address several biotechnological challenges. It is known, for instance, that efficient utilization of different sugars from their mixtures upon various biotechnological applications requires modulation of glucose repression mechanisms. Accordingly, bioethanol production processes with *S. cerevisiae* using lignocellulosic biomass as a feedstock are challenged by the efficiency of simultaneous utilization of pentose and hexose sugars from biomass hydrolysates. Not only the pentose uptake into the cell represents a crucial role for the efficiency of the process, but glucose repression of the broad subset of genes poses a problem that has to be solved (Oehling et al. 2018). Various genetic perturbations in *S. cerevisiae* are known to alleviate the effects of glucose repression on consumption of alternative sugars or their mixtures. It was recently reported that mutations in the glucose-phosphorylating enzymes (Hxk1, Hxk2, Glk1) that conferred resistance to 2-DG were sufficient to provide simultaneous glucose and xylose utilization due to a reduction in the rate of glucose phosphorylation and impaired glucose repression (Lane et al. 2018). It was also found that varying expression levels of hexokinase expressed under inducible promoter also led to the simultaneous utilization of glucose and xylose. Therefore, it was proposed that intracellular rate of glucose phosphorylation is a key regulatory factor for efficient utilization of sugar mixtures in *S. cerevisiae* (Lane et al. 2018).

Also, glucose suppresses maltose metabolism in bakers' yeast cells incubated on lean dough, which negatively affects its fermentation. It was revealed that overexpression of a regulatory subunit of Snf1 kinase, Snf4, and deletions of the regulatory subunits of the type I protein phosphatase Reg1 and Reg2, especially when combined, significantly alleviated glucose repression of maltose metabolism, thereby enhancing maltose metabolism and leavening ability to varying degrees. Such mutant strains were proposed as superior fermenters for applications that involve leavening ability (Lin et al. 2018).

To address the question of why most of the natural isolates of *S. cerevisiae* poorly utilize xylose, a unique natural isolate (YB-2625) which has superior xylose metabolism capability in the presence of mixed sugars was analyzed. This strain appeared to exhibit the decreased transcriptional levels of the genes regulating glucose repression (*MIG1*, *MIG2*) as well as *HXK2*. In addition, a number of genes encoding antioxidant enzymes (e.g., *CTT1*, *CTA1*, *SOD2*, and *PRX1*) were upregulated in this strain. It was proposed that enhanced oxidative stress tolerance and relief of glucose repression are the two major mechanisms that potentially provide superior xylose utilization in *S. cerevisiae* (Cheng et al. 2018).

Alleviation of glucose repression by deleting *MIG1* or *HXK2* also improved productivity of *S. cerevisiae* producer of lactic acid. The positive effect was due to a more efficient generation of ATP and NAD+ via respiration, because ATP depletion through extensive export of lactate and protons was one of the major reasons for the impaired growth of the original strain producer. It was proposed that mutations affecting glucose repression could be applied to improve *S. cerevisiae* strains for various bioprocesses where increasing biomass yield or respiratory flux is desirable.

For further reading on the subject of manipulations with catabolite repression and glucose-signaling pathways in *S. cerevisiae* and other yeasts to construct better fermenters of alternative sugars, such as xylose or arabinoase, see recent reviews of Radecka et al. (2015) and Hou et al. (2017).

Acknowledgments OGS and OVS were supported in part from the grant of Ministry of Education and Science of Ukraine #0117 U001226 and OVS by grant No. 0115 U004200 of the program "Molecular and cellular biotechnologies for medicine, industry and agriculture" of National Academy of Sciences of Ukraine.

References

Adachi A, Koizumi M, Ohsumi Y (2017) Autophagy induction under carbon starvation conditions is negatively regulated by carbon catabolite repression. J Biol Chem. https://doi.org/10.1074/jbc.M117.817510

Adamczyk M, Szatkowska R (2017) Low RNA polymerase III activity results in up regulation of HXT2 glucose transporter independently of glucose signaling and despite changing environment. PLoS One. https://doi.org/10.1371/journal.pone.0185516

Ahuatzi D, Herrero P, De La Cera T, Moreno F (2004) The glucose-regulated nuclear localization of hexokinase 2 in *Saccharomyces cerevisiae* is Mig1-dependent. J Biol Chem. https://doi.org/10.1074/jbc.M313431200

Ahuatzi D, Riera A, Peláez R et al (2007) Hxk2 regulates the phosphorylation state of Mig1 and therefore its nucleocytoplasmic distribution. J Biol Chem. https://doi.org/10.1074/jbc.M606854200

Alamäe T, Liiv L (1998) Glucose repression of maltase and methanol-oxidizing enzymes in the methylotrophic yeast *Hansenula polymorpha*: isolation and study of regulatory mutants. Folia Microbiol (Praha) 43:443–452

Alamäe T, Simisker J (1994) Isolation and preliminary characterization of Pichia pinus mutants insensitive to glucose repression. Yeast 10:1459–1466. https://doi.org/10.1002/yea.320101109

Alamäe T, Pärn P, Viigand K, Karp H (2003) Regulation of the *Hansenula polymorpha* maltase gene promoter in *H. polymorpha* and *Saccharomyces cerevisiae*. FEMS Yeast Res 4:165–173. https://doi.org/10.1016/S1567-1356(03)00142-9

Alcaíno J, Bravo N, Córdova P et al (2016) The involvement of Mig1 from xanthophyllomyces dendrorhous in catabolic repression: an active mechanism contributing to the regulation of carotenoid production. PLoS One. https://doi.org/10.1371/journal.pone.0162838

Ashe MP, De Long SK, Sachs AB (2000) Glucose depletion rapidly inhibits translation initiation in yeast. Mol Biol Cell 11:833–848. https://doi.org/10.1091/mbc.11.3.833

Ashrafi K, Lin SS, Manchester JK, Gordon JI (2000) Sip2p and its partner Snf1p kinase affect aging in S. cerevisiae. Genes Dev 14:1872–1885. https://doi.org/10.1101/gad.14.15.1872

Barrett L, Orlova M, Maziarz M, Kuchin S (2012) Protein kinase a contributes to the negative control of SNF1 protein kinase in *Saccharomyces cerevisiae*. Eukaryot Cell. https://doi.org/10.1128/EC.05061-11

Beullens M, Mbonyi K, Geerts L et al (1988) Studies on the mechanism of the glucose-induced cAMP signal in glycolysis and glucose repression mutants of the yeast *Saccharomyces cerevisiae*. Eur J Biochem 172:227–231

Bisson LF, Fan Q, Walker GA (2016) Sugar and glycerol transport in *Saccharomyces cerevisiae*. Adv Exp Med Biol 892:125–168

Blomqvist J, Passoth V (2015) *Dekkera bruxellensis*-spoilage yeast with biotechnological potential, and a model for yeast evolution, physiology and competitiveness. FEMS Yeast Res 15(4):fov021

Boender LGM, Almering MJH, Dijk M et al (2011) Extreme calorie restriction and energy source starvation in *Saccharomyces cerevisiae* represent distinct physiological states. Biochim Biophys Acta, Mol Cell Res. https://doi.org/10.1016/j.bbamcr.2011.07.008

Boles E, Hollenberg CP (1997) The molecular genetics of hexose transport in yeasts. FEMS Microbiol Rev 21:85–111

Boy-Marcotte E, Ikonomi P, Jacquet M (1996) SDC25, a dispensable Ras guanine nucleotide exchange factor of *Saccharomyces cerevisiae* differs from CDC25 by its regulation. Mol Biol Cell. https://doi.org/10.1091/mbc.7.4.529

Braun KA, Vaga S, Dombek KM et al (2014) Phosphoproteomic analysis identifies proteins involved in transcription-coupled mRNA decay as targets of Snf1 signaling. Sci Signal. https://doi.org/10.1126/scisignal.2005000

Braun KA, Dombek KM, Young ET (2016) Snf1-dependent transcription confers glucose-induced decay upon the mRNA product. Mol Cell Biol. https://doi.org/10.1128/MCB.00436-15

Broach JR (2012) Nutritional control of growth and development in yeast. Genetics 192:73–105
Brown V, Sexton JA, Johnston M (2006) A glucose sensor in *Candida albicans*. Eukaryot Cell 5:1726–1737. https://doi.org/10.1128/EC.00186-06
Brown V, Sabina J, Johnston M (2009) Specialized sugar sensing in diverse fungi. Curr Biol 19:436–441. https://doi.org/10.1016/j.cub.2009.01.056
Broxton CN, He B, Bruno VM, Culotta VC (2018) A role for *Candida albicans* superoxide dismutase enzymes in glucose signaling. Biochem Biophys Res Commun. https://doi.org/10.1016/j.bbrc.2017.11.084
Cairey-Remonnay A, Deffaud J, Wésolowski-Louvel M et al (2015) Glycolysis controls plasma membrane glucose sensors to promote glucose signaling in yeasts. Mol Cell Biol 35:747–757. https://doi.org/10.1128/MCB.00515-14
Castermans D, Somers I, Kriel J et al (2012) Glucose-induced posttranslational activation of protein phosphatases PP2A and PP1 in yeast. Cell Res. https://doi.org/10.1038/cr.2012.20
Celenza JL, Carlson M (1986) A yeast gene that is essential for release from glucose repression encodes a protein kinase. Science (80-). https://doi.org/10.1126/science.3526554
Chaves RS, Herrero P, Moreno F (1999) Med8, a subunit of the mediator CTD complex of RNA polymerase II, directly binds to regulatory elements of SUC2 and HXK2 genes. Biochem Biophys Res Commun. https://doi.org/10.1006/bbrc.1998.9954
Chen Y, Nielsen J (2013) Advances in metabolic pathway and strain engineering paving the way for sustainable production of chemical building blocks. Curr Opin Biotechnol
Cheng C, Tang RQ, Xiong L et al (2018) Association of improved oxidative stress tolerance and alleviation of glucose repression with superior xylose-utilization capability by a natural isolate of *Saccharomyces cerevisiae*. Biotechnol Biofuels. https://doi.org/10.1186/s13068-018-1018-y
Coccetti P, Nicastro R, Tripodi F (2018) Conventional and emerging roles of the energy sensor Snf1/AMPK in *Saccharomyces cerevisiae*. Microb Cell 5:482–494. https://doi.org/10.15698/mic2018.11.655
Colombo S, Ronchetti D, Thevelein JM et al (2004) Activation state of the Ras2 protein and glucose-induced signaling in *Saccharomyces cerevisiae*. J Biol Chem 279:46715–46722. https://doi.org/10.1074/jbc.M405136200
Conrad M, Schothorst J, Kankipati HN et al (2014) Nutrient sensing and signaling in the yeast *Saccharomyces cerevisiae*. FEMS Microbiol Rev 38:254–299. https://doi.org/10.1111/1574-6976.12065
Córdova P, Alcaíno J, Bravo N et al (2016) Regulation of carotenogenesis in the red yeast Xanthophyllomyces dendrorhous: the role of the transcriptional co-repressor complex Cyc8-Tup1 involved in catabolic repression. Microb Cell Factories. https://doi.org/10.1186/s12934-016-0597-1
Cottier F, Tan ASM, Xu X et al (2015) MIG1 regulates resistance of *Candida albicans* against the fungistatic effect of weak organic acids. Eukaryot Cell. https://doi.org/10.1128/EC.00129-15
Dai Z, Liu Y, Guo J, et al (2015) Yeast synthetic biology for high-value metabolites. FEMS Yeast Res 15(1):1–11
Dashko S, Zhou N, Compagno C, Piskur J (2014) Why, when, and how did yeast evolve alcoholic fermentation? FEMS Yeast Res:14
De Barros Pita W, Silva DC, Simões DA, et al (2013) Physiology and gene expression profiles of *Dekkera bruxellensis* in response to carbon and nitrogen availability. Antonie van Leeuwenhoek, Int J Gen Mol Microbiol. https://doi.org/10.1007/s10482-013-9998-x
De La Cera T, Herrero P, Moreno-Herrero F et al (2002) Mediator factor Med8p interacts with the hexokinase 2: implication in the glucose signalling pathway of *Saccharomyces cerevisiae*. J Mol Biol. https://doi.org/10.1016/S0022-2836(02)00377-7
Dechant R, Binda M, Lee SS et al (2010) Cytosolic pH is a second messenger for glucose and regulates the PKA pathway through V-ATPase. EMBO J. https://doi.org/10.1038/emboj.2010.138
Diaz-Ruiz R, Rigoulet M, Devin A (2011) The Warburg and Crabtree effects: on the origin of cancer cell energy metabolism and of yeast glucose repression. Biochim Biophys Acta Bioenerg 1807:568–576

Dusny C, Schmid A (2016) The *MOX* promoter in *Hansenula polymorpha* is ultrasensitive to glucose-mediated carbon catabolite repression. FEMS Yeast Res. https://doi.org/10.1093/femsyr/fow067

Elbing K, McCartney RR, Schmidt MC (2006) Purification and characterization of the three Snf1-activating kinases of *Saccharomyces cerevisiae*. Biochem J. https://doi.org/10.1042/BJ20051213

Fernańdez-García P, Peláez R, Herrero P, Moreno F (2012) Phosphorylation of yeast hexokinase 2 regulates its nucleocytoplasmic shuttling. J Biol Chem. https://doi.org/10.1074/jbc.M112.401679

Fickers P, Nicaud JM, Destain J, Thonart P (2005) Involvement of hexokinase Hxk1 in glucose catabolite repression of LIP2 encoding extracellular lipase in the yeast *Yarrowia lipolytica*. Curr Microbiol. https://doi.org/10.1007/s00284-004-4401-9

Flick KM (2003) Grr1-dependent inactivation of Mth1 mediates glucose-induced dissociation of Rgt1 from *HXT* gene promoters. Mol Biol Cell. https://doi.org/10.1091/mbc.E03-03-0135

Flores C-L, Gancedo C (2018) Construction and characterization of a *Saccharomyces cerevisiae* strain able to grow on glucosamine as sole carbon and nitrogen source. Sci Rep 8:16949. https://doi.org/10.1038/s41598-018-35045-8

Gadura N, Robinson LC, Michels CA (2006) Glc7-Reg1 phosphatase signals to Yck1,2 casein kinase 1 to regulate transport activity and glucose-induced inactivation of saccharomyces maltose permease. Genetics. https://doi.org/10.1534/genetics.105.051698

Galazzo JL, Bailey JE (1990) Fermentation pathway kinetics and metabolic flux control in suspended and immobilized *Saccharomyces cerevisiae*. Enzym Microb Technol 12:162–172. https://doi.org/10.1016/0141-0229(90)90033-M

Gancedo JM (1998) Yeast carbon catabolite repression. Microbiol Mol Biol Rev 62:334–361

Harashima T, Heitman J (2005) Galpha subunit Gpa2 recruits kelch repeat subunits that inhibit receptor-G protein coupling during cAMP-induced dimorphic transitions in *Saccharomyces cerevisiae*. Mol Biol Cell. https://doi.org/10.1091/mbc.E05-05-0403

Harashima T, Anderson S, Yates JR, Heitman J (2006) The kelch proteins Gpb1 and Gpb2 inhibit Ras activity via association with the yeast RasGAP neurofibromin homologs Ira1 and Ira2. Mol Cell. https://doi.org/10.1016/j.molcel.2006.05.011

Hartner FS, Glieder A, Wolf K et al (2006) Regulation of methanol utilisation pathway genes in yeasts. Microb Cell Factories 5:39. https://doi.org/10.1186/1475-2859-5-39

Hedbacker K, Carlson M (2008) SNF1/AMPK pathways in yeast. Front Biosci 13:2408–2420

Hedbacker K, Townley R, Carlson M (2004) Cyclic AMP-dependent protein kinase regulates the subcellular localization of Snf1-Sip1 protein kinase. Society. https://doi.org/10.1128/MCB.24.5.1836

Herrero P, Martínez-Campa C, Moreno F (1998) The hexokinase 2 protein participates in regulatory DNA-protein complexes necessary for glucose repression of the SUC2 gene in *Saccharomyces cerevisiae*. FEBS Lett. https://doi.org/10.1016/S0014-5793(98)00872-2

Hoffman CS (2005a) Glucose sensing via the protein kinase a pathway in *Schizosaccharomyces pombe*. Biochem Soc Trans 33:257–260. https://doi.org/10.1042/BST0330257

Hoffman CS (2005b) Except in every detail: comparing and contrasting G-protein signaling in *Saccharomyces cerevisiae* and *Schizosaccharomyces pombe*. Eukaryot Cell 4(3):495–503

Hong S-P, Leiper FC, Woods A et al (2003) Activation of yeast Snf1 and mammalian AMP-activated protein kinase by upstream kinases. Proc Natl Acad Sci. https://doi.org/10.1073/pnas.1533136100

Horák J (2013) Regulations of sugar transporters: insights from yeast. Curr Genet 1–31

Hou J, Qiu C, Shen Y et al (2017) Engineering of *Saccharomyces cerevisiae* for the efficient co-utilization of glucose and xylose. FEMS Yeast Res 17. https://doi.org/10.1093/femsyr/fox034

Hubbard EJ, Jiang R, Carlson M (1994) Dosage-dependent modulation of glucose repression by MSN3 (STD1) in *Saccharomyces cerevisiae*. Mol Cell Biol. https://doi.org/10.1128/MCB.14.3.1972

Hubscher V, Mudholkar K, Chiabudini M et al (2016) The Hsp70 homolog Ssb and the 14-3-3 protein Bmh1 jointly regulate transcription of glucose repressed genes in *Saccharomyces cerevisiae*. Nucleic Acids Res. https://doi.org/10.1093/nar/gkw168

Johnston M, Kim J-H (2005) Nutrient sensing through the plasma membrane of eukaryotic cells. Biochem Soc:247–252. https://doi.org/10.1042/BST0330247

Kaniak A, Xue Z, Macool D et al (2004) Regulatory network connecting two glucose signal transduction pathways in *Saccharomyces cerevisiae*. Eukaryot Cell. https://doi.org/10.1128/EC.3.1.221-231.2004

Kaps S, Kettner K, Migotti R et al (2015) Protein kinase Ymr291w/Tda1 is essential for glucose signaling in *Saccharomyces cerevisiae* on the level of hexokinase isoenzyme *Sc*Hxk2 phosphorylation. J Biol Chem 290:6243–6255. https://doi.org/10.1074/jbc.M114.595074

Karhumaa K, Wu B, Kielland-Brandt MC (2010) Conditions with high intracellular glucose inhibit sensing through glucose sensor SnF3 in *Saccharomyces cerevisiae*. J Cell Biochem 110:920–925. https://doi.org/10.1002/jcb.22605

Karp H, Järviste A, Kriegel TM, Alamäe T (2003) Cloning and biochemical characterization of hexokinase from the methylotrophic yeast *Hansenula polymorpha*. Curr Genet 44:268–276. https://doi.org/10.1007/s00294-003-0448-6

Kayikci Ö, Nielsen J (2015) Glucose repression in *Saccharomyces cerevisiae*. FEMS Yeast Res 15

Kim J-H, Johnston M (2006) Two glucose-sensing pathways converge on Rgt1 to regulate expression of glucose transporter genes in *Saccharomyces cerevisiae*. J Biol Chem 281:26144–26149. https://doi.org/10.1074/jbc.M603636200

Kim JH, Brachet V, Moriya H, Johnston M (2006) Integration of transcriptional and posttranslational regulation in a glucose signal transduction pathway in *Saccharomyces cerevisiae*. Eukaryot Cell 5:167–173. https://doi.org/10.1128/EC.5.1.167-173.2006

Kim JH, Roy A, Jouandot D, Cho KH (2013) The glucose signaling network in yeast. Biochim Biophys Acta Gen Subj 1830(11):5204–5210

Kraakman L, Lemaire K, Ma P et al (1999) A *Saccharomyces cerevisiae* G-protein coupled receptor, Gpr1, is specifically required for glucose activation of the cAMP pathway during the transition to growth on glucose. Mol Microbiol 32:1002–1012

Kramarenko T, Karp H, Järviste A, Alamäe T (2000) Sugar repression in the methylotrophic yeast *Hansenula polymorpha* studied by using hexokinase-negative, glucokinase-negative and double kinase-negative mutants. Folia Microbiol (Praha) 45:521–529

Krasovska OS, Stasyk OG, Nahorny VO et al (2007) Glucose-induced production of recombinant proteins in *Hansenula polymorpha* mutants deficient in catabolite repression. Biotechnol Bioeng 97:858–870. https://doi.org/10.1002/bit.21284

Krasovska OS, Stasyk OV, Sibirny AA (2013) Stable overproducer of hepatitis B surface antigen in the methylotrophic yeast *Hansenula polymorpha* due to multiple integration of heterologous auxotrophic selective markers and defect in peroxisome biogenesis. Appl Microbiol Biotechnol. https://doi.org/10.1007/s00253-013-5223-0

Kriegel TM, Vojtek AB, Clifton D et al (1994) *In vivo* phosphorylation site of hexokinase 2 in *Saccharomyces cerevisiae*. Biochemistry. https://doi.org/10.1021/bi00167a019

Kuchin S, Treich I, Carlson M (2000) A regulatory shortcut between the Snf1 protein kinase and RNA polymerase II holoenzyme. Proc Natl Acad Sci U S A. https://doi.org/10.1073/pnas.140109897

Kuchin S, Vyas VK, Kanter E et al (2003) Std1p (Msn3p) positively regulates the Snf1 kinase in *Saccharomyces cerevisiae*. Genetics 163(2):507–14

Kumar NV, Rangarajan PN (2011) Catabolite repression of phosphoenolpyruvate carboxykinase by a zinc finger protein under biotinand pyruvate carboxylase-deficient conditions in *Pichia pastoris*. Microbiology. https://doi.org/10.1099/mic.0.053488-0

KuttyKrishnan S, Sabina J, Langton L, et al (2011) A quantitative model of glucose signaling in yeast reveals an incoherent feed forward loop leading to a specific, transient pulse of transcription. In: lecture notes in computer science (including subseries lecture notes in artificial intelligence and lecture notes in bioinformatics). Springer, Berlin

Laht S, Karp H, Kotka P et al (2002) Cloning and characterization of glucokinase from a methylotrophic yeast *Hansenula polymorpha*: different effects on glucose repression in *H. polymorpha* and *Saccharomyces cerevisiae*. Gene 296:195–203. https://doi.org/10.1016/S0378-1119(02)00859-4

Lakshmanan J, Mosley AL, Özcan S (2003) Repression of transcription by Rgt1 in the absence of glucose requires Std1 and Mth1. Curr Genet. https://doi.org/10.1007/s00294-003-0423-2

Lane S, Xu H, Oh EJ et al (2018) Glucose repression can be alleviated by reducing glucose phosphorylation rate in *Saccharomyces cerevisiae*. Sci Rep. https://doi.org/10.1038/s41598-018-20804-4

Leão-Helder AN, Krikken AM, Lunenborg MGJ et al (2004) *Hansenula polymorpha* Tup1p is important for peroxisome degradation. FEMS Yeast Res. https://doi.org/10.1016/j.femsyr.2004.04.006

Lemaire K, Van De Velde S, Van Dijck P, Thevelein JM (2004) Glucose and sucrose act as agonist and mannose as antagonist ligands of the G protein-coupled receptor Gpr1 in the yeast *Saccharomyces cerevisiae*. Mol Cell. https://doi.org/10.1016/j.molcel.2004.10.004

Lin X, Yu AQ, Zhang CY et al (2017) Functional analysis of the global repressor Tup1 for maltose metabolism in *Saccharomyces cerevisiae*: different roles of the functional domains. Microb Cell Factories. https://doi.org/10.1186/s12934-017-0806-6

Lin X, Zhang CY, Meng L et al (2018) Overexpression of *SNF4* and deletions of REG1- and REG2-enhanced maltose metabolism and leavening ability of baker's yeast in lean dough. J Ind Microbiol Biotechnol. https://doi.org/10.1007/s10295-018-2058-9

Lin-Cereghino GP, Godfrey L, de la Cruz BJ et al (2006) Mxr1p, a key regulator of the methanol utilization pathway and peroxisomal genes in *Pichia pastoris*. Mol Cell Biol. https://doi.org/10.1128/MCB.26.3.883-897.2006

Ludin K, Jiang R, Carlson M (1998) Glucose-regulated interaction of a regulatory subunit of protein phosphatase 1 with the Snf1 protein kinase in *Saccharomyces cerevisiae*. Proc Natl Acad Sci. https://doi.org/10.1073/pnas.95.11.6245

Lutfiyya LL, Iyer VR, DeRisi J et al (1998) Characterization of three related glucose repressors and genes they regulate in *Saccharomyces cerevisiae*. Genetics 150(4):1377–1391

Ma P, Wera S, Van Dijck P, Thevelein JM (1999) The *PDE1*-encoded low-affinity phosphodiesterase in the yeast *Saccharomyces cerevisiae* has a specific function in controlling agonist-induced cAMP signaling. Mol Biol Cell. https://doi.org/10.1091/mbc.10.1.91

Maqani N, Fine RD, Shahid M, et al (2018) Spontaneous mutations in *CYC8* and *MIG1* suppress the short chronological lifespan of budding yeast lacking SNF1/AMPK. Microb cell. Doi: https://doi.org/10.15698/mic2018.05.630

Mates N, Kettner K, Heidenreich F et al (2014) Proteomic and functional consequences of hexokinase deficiency in glucose-repressible *Kluyveromyces lactis*. Mol Cell Proteom. https://doi.org/10.1074/mcp.M113.032714

Matsumoto K (1982) Isolation and characterization of yeast mutants deficient in adenylate cyclase and cAMP-dependent protein kinase. Proc Natl Acad Sci. https://doi.org/10.1073/pnas.79.7.2355

Matsuzawa T, Fujita Y, Tohda H, Takegawa K (2012) Snf1-like protein kinase ssp2 regulates glucose derepression in *Schizosaccharomyces pombe*. Eukaryot Cell. https://doi.org/10.1128/EC.05268-11

Mayer FV, Heath R, Underwood E et al (2011) ADP regulates SNF1, the *Saccharomyces cerevisiae* homolog of AMP-activated protein kinase. Cell Metab. https://doi.org/10.1016/j.cmet.2011.09.009

Mbonyi K, Van Aelst L, Arguelles JC et al (1990) Glucose-induced Hyperaccumulation of cyclic AMP and defective glucose repression in yeast strains with reduced activity of cyclic AMP-dependent protein kinase. Mol Cell Biol. https://doi.org/10.1128/MCB.10.9.4518.Updated

McCartney RR, Schmidt MC (2001) Regulation of Snf1 kinase. Activation requires phosphorylation of threonine 210 by an upstream kinase as well as a distinct step mediated by the Snf4 subunit. J biol Chem. https://doi.org/10.1074/jbc.M104418200

Mehlgarten C, Krijger JJ, Lemnian I et al (2015) Divergent evolution of the transcriptional network controlled by Snf1-interacting protein Sip4 in budding yeasts. PLoS One. https://doi.org/10.1371/journal.pone.0139464

Mitchelhill KI, Stapleton D, Gao G et al (1994) Mammalian AMP-activated protein kinase shares structural and functional homology with the catalytic domain of yeast Snf1 protein kinase. J Biol Chem. https://doi.org/10.1074/jbc.271.2.611

Moktaduzzaman M, Galafassi S, Capusoni C et al (2015) Galactose utilization sheds new light on sugar metabolism in the sequenced strain *Dekkera bruxellensis* CBS 2499. FEMS Yeast Res. https://doi.org/10.1093/femsyr/fou009

Moreno F, Ahuatzi D, Riera A et al (2005) Glucose sensing through the Hxk2-dependent signalling pathway. Biochem Soc Trans 33:265–268. https://doi.org/10.1042/BST0330265

Moriya H, Johnston M (2004) Glucose sensing and signaling in *Saccharomyces cerevisiae* through the Rgt2 glucose sensor and casein kinase I. Proc Natl Acad Sci 101:1572–1577. https://doi.org/10.1073/pnas.0305901101

Nehlin JO, Ronne H (1990) Yeast *MIG1* repressor is related to the mammalian early growth response and Wilms' tumour finger proteins. EMBO J. https://doi.org/10.1016/0008-8846(83)90137-0

Nurcholis M, Nitiyon S, Suprayogi et al (2019) Functional analysis of Mig1 and Rag5 as expressional regulators in thermotolerant yeast *Kluyveromyces marxianus*. Appl Microbiol Biotechnol 103:395–410. https://doi.org/10.1007/s00253-018-9462-y

Oehling V, Klaassen P, Frick O et al (2018) l-arabinose triggers its own uptake via induction of the arabinose-specific Gal2p transporter in an industrial *Saccharomyces cerevisiae* strain. Biotechnol Biofuels. https://doi.org/10.1186/s13068-018-1231-8

Oliveira MA, Genu V, Salmazo APT et al (2003) The transcription factor Snf1p is involved in a Tup1p-independent manner in the glucose regulation of the major methanol metabolism genes of *Hansenula polymorpha*. Genet Mol Biol 26:521–528. https://doi.org/10.1590/S1415-47572003000400017

Özcan S (2002) Two different signals regulate repression and induction of gene expression by glucose. J Biol Chem. https://doi.org/10.1074/jbc.M208726200

Ozcan S, Johnston M (1999) Function and regulation of yeast hexose transporters. Microbiol Mol Biol Rev 63:554–569

Ozcan S, Dover J, Rosenwald AG et al (1996) Two glucose transporters in *Saccharomyces cerevisiae* are glucose sensors that generate a signal for induction of gene expression. Proc Natl Acad Sci 93:12428–12432. https://doi.org/10.1073/pnas.93.22.12428

Özcan S, Dover J, Rosenwald AG et al (1996) Two glucose transporters in *Saccharomyces cerevisiae* are glucose sensors that generate a signal for induction of gene expression. Proc Natl Acad Sci U S A 93:12428–12432

Ozcan S, Dover J, Johnston M (1998) Glucose sensing and signaling by two glucose sensor in the yeast Saccharomyces cerevisae. EMBO J 17:2573–2655

Palabiyik B, Jafari Ghods F, Onay Ucar E (2013) Effects of glucose sensing/signaling on oxidative stress response in glucose repression mutants of *Schizosaccharomyces pombe*. Genet Mol Res 12:5046–5056. https://doi.org/10.4238/2013.October.25.3

Palomino A, Herrero P, Moreno F (2006) Tpk3 and Snf1 protein kinases regulate Rgt1 association with *Saccharomyces cerevisiae HXK2* promoter. Nucleic Acids Res. https://doi.org/10.1093/nar/gkl028

Papamichos-Chronakis M, Gligoris T, Tzamarias D (2004) The Snf1 kinase controls glucose repression in yeast by modulating interactions between the Mig1 repressor and the Cyc8-Tup1 co-repressor. EMBO Rep. https://doi.org/10.1038/sj.embor.7400120

Parpinello G, Berardi E, Strabbioli R (1998) A regulatory mutant of *Hansenula polymorpha* exhibiting methanol utilization metabolism and peroxisome proliferation in glucose. J Bacteriol

Pasula S, Jouandot D, Kim JH (2007) Biochemical evidence for glucose-independent induction of HXT expression in *Saccharomyces cerevisiae*. FEBS Lett. https://doi.org/10.1016/j.febslet.2007.06.013

Pasula S, Chakraborty S, Choi JH, Kim JH (2010) Role of casein kinase 1 in the glucose sensor-mediated signaling pathway in yeast. BMC Cell Biol. https://doi.org/10.1088/1468-6996/11/4/044302

Peeters T, Louwet W, Gelade R et al (2006) Kelch-repeat proteins interacting with the G protein Gpa2 bypass adenylate cyclase for direct regulation of protein kinase A in yeast. Proc Natl Acad Sci 103:13034–13039. https://doi.org/10.1073/pnas.0509644103

Peeters T, Versele M, Thevelein JM (2007) Directly from Gα to protein kinase A: the kelch repeat protein bypass of adenylate cyclase. Trends Biochem Sci 32:547–554

Peeters K, Van Leemputte F, Fischer B et al (2017) Fructose-1,6-bisphosphate couples glycolytic flux to activation of Ras. Nat Commun 8. https://doi.org/10.1038/s41467-017-01019-z

Peláez R, Herrero P, Moreno F (2009) Nuclear export of the yeast hexokinase 2 protein requires the Xpo1 (Crm1)-dependent pathway. J Biol Chem 284:20548–20555. https://doi.org/10.1074/jbc.M109.013730

Pereira GG, Hollenberg CP (1996) Conserved regulation of the *Hansenula polymorpha MOX* promoter in *Saccharomyces cerevisiae* reveals insights in the transcriptional activation by Adr1p. Eur J Biochem. https://doi.org/10.1111/j.1432-1033.1996.0181q.x

Petranovic D, Tyo K, Vemuri GN, Nielsen J (2010) Prospects of yeast systems biology for human health: integrating lipid, protein and energy metabolism. FEMS Yeast Res 10(8):1046–1059

Pfeiffer T, Morley A (2014) An evolutionary perspective on the Crabtree effect. Front Mol Biosci. https://doi.org/10.3389/fmolb.2014.00017

Polish JA, Kim JH, Johnston M (2005) How the Rgt1 transcription factor of *Saccharomyces cerevisiae* is regulated by glucose. Genetics. https://doi.org/10.1534/genetics.104.034512

Polupanov AS, Sibirny AA (2014) Cytoplasmic extension peptide of *Pichia pastoris* glucose sensor Gss1 is not compulsory for glucose signalling. Cell Biol Int. https://doi.org/10.1002/cbin.10189

Polupanov AS, Nazarko VY, Sibirny AA (2012) Gss1 protein of the methylotrophic yeast *Pichia pastoris* is involved in glucose sensing, pexophagy and catabolite repression. Int J Biochem Cell Biol 44:1906–1918. https://doi.org/10.1016/j.biocel.2012.07.017

Prielhofer R, Cartwright SP, Graf AB et al (2015) *Pichia pastoris* regulates its gene-specific response to different carbon sources at the transcriptional, rather than the translational, level. BMC Genomics. https://doi.org/10.1186/s12864-015-1393-8

Radecka D, Mukherjee V, Mateo RQ et al (2015) Looking beyond Saccharomyces: the potential of non-conventional yeast species for desirable traits in bioethanol fermentation. FEMS Yeast Res

Rahner A, Hiesinger M, Schüller HJ (1999) Deregulation of gluconeogenic structural genes by variants of the transcriptional activator Cat8p of the yeast *Saccharomyces cerevisiae*. Mol Microbiol. https://doi.org/10.1046/j.1365-2958.1999.01588.x

Randez-Gil F, Herrero P, Sanz P et al (1998a) Hexokinase PII has a double cytosolic-nuclear localisation in *Saccharomyces cerevisiae*. FEBS Lett. https://doi.org/10.1016/S0014-5793(98)00289-0

Randez-Gil F, Sanz P, Entian K-D, Prieto JA (1998b) Carbon source-dependent phosphorylation of hexokinase PII and its role in the glucose-signaling response in yeast. Mol Cell Biol 18:2940–2948. https://doi.org/10.1128/MCB.18.5.2940

Reddi AR, Culotta VC (2013) SOD1 integrates signals from oxygen and glucose to repress respiration. Cell. https://doi.org/10.1016/j.cell.2012.11.046

Rippert D, Backhaus K, Rodicio R, Heinisch JJ (2017) Cell wall synthesis and central carbohydrate metabolism are interconnected by the SNF1/Mig1 pathway in *Kluyveromyces lactis*. Eur J Cell Biol 96:70–81. https://doi.org/10.1016/j.ejcb.2016.12.004

Rodicio R, Heinisch JJ (2013) Yeast on the milky way: genetics, physiology and biotechnology of *Kluyveromyces lactis*. Yeast. https://doi.org/10.1002/yea.2954

Rodríguez A, De La Cera T, Herrero P, Moreno F (2001) The hexokinase 2 protein regulates the expression of the GLK1, HXK1 and HXK2 genes of *Saccharomyces cerevisiae*. Biochem J 355:625–631

Rodrussamee N, Sattayawat P, Yamada M (2018) Highly efficient conversion of xylose to ethanol without glucose repression by newly isolated thermotolerant *Spathaspora passalidarum* CMUWF1-2. BMC Microbiol. https://doi.org/10.1186/s12866-018-1218-4

Roggenkamp R (1988) Constitutive appearance of peroxisomes in a regulatory mutant of the methylotrophic yeast *Hansenula polymorpha*. Mol Gen Genet 213:535–540

Roggenkamp R, Janowicz Z, Stanikowski B, Hollenberg CP (1984) Biosynthesis and regulation of the peroxisomal methanol oxidase from the methylotrophic yeast *Hansenula polymorpha*. Mol Gen Genet. https://doi.org/10.1007/bf00425563

Rolland F, De Winde JH, Lemaire K et al (2000) Glucose-induced cAMP signalling in yeast requires both a G-protein coupled receptor system for extracellular glucose detection and a separable hexose kinase-dependent sensing process. Mol Microbiol 38:348–358. https://doi.org/10.1046/j.1365-2958.2000.02125.x

Rolland F, Winderickx J, Thevelein JM (2002) Glucose-sensing and -signalling mechanisms in yeast. In: FEMS yeast research. Oxford University Press, Oxford, pp 183–201

Roy A, Kim YB, Cho KH, Kim JH (2014) Glucose starvation-induced turnover of the yeast glucose transporter Hxt1. Biochim Biophys Acta Gen Subj 1840:2878–2885. https://doi.org/10.1016/j.bbagen.2014.05.004

Rubenstein EM, McCartney RR, Zhang C et al (2008) Access denied: Snf1 activation loop phosphorylation is controlled by availability of the phosphorylated threonine 210 to the PP1 phosphatase. J Biol Chem. https://doi.org/10.1074/jbc.M707957200

Rubio-Texeira M (2005) A comparative analysis of the GAL genetic switch between not-so-distant cousins: *Saccharomyces cerevisiae* versus *Kluyveromyces lactis*. FEMS Yeast Res

Ruiz A, Liu Y, Xu X, Carlson M (2012) Heterotrimer-independent regulation of activation-loop phosphorylation of Snf1 protein kinase involves two protein phosphatases. Proc Natl Acad Sci. https://doi.org/10.1073/pnas.1206280109

Ruiz A, Xu X, Carlson M (2013) Ptc1 protein phosphatase 2C contributes to glucose regulation of SNF1/AMP-activated protein kinase (AMPK) in *Saccharomyces cerevisiae*. J Biol Chem. https://doi.org/10.1074/jbc.M113.503763

Rußmayer H, Buchetics M, Gruber C et al (2015) Systems-level organization of yeast methylotrophic lifestyle. BMC Biol. https://doi.org/10.1016/j.proeng.2015.11.007

Ryu S, Hipp J, Trinha CT (2016) Activating and elucidating metabolism of complex sugars in *Yarrowia lipolytica*. Appl Environ Microbiol. https://doi.org/10.1128/AEM.03582-15

Sabina J, Brown V (2009) Glucose sensing network in *Candida albicans*: a sweet spot for fungal morphogenesis. Eukaryot Cell 8:1314–1320

Sabra W, Bommareddy RR, Maheshwari G et al (2017) Substrates and oxygen dependent citric acid production by *Yarrowia lipolytica*: insights through transcriptome and fluxome analyses. Microb Cell Factories. https://doi.org/10.1186/s12934-017-0690-0

Sakai Y, Yurimoto H, Matsuo H, Kato N (1998) Regulation of peroxisomal proteins and organelle proliferation by multiple carbon sources in the methylotrophic yeast, *Candida boidinii*. Yeast. https://doi.org/10.1002/(SICI)1097-0061(19980930)14:13<1175::AID-YEA319>3.0.CO;2-7

Santangelo GM (2006) Glucose signaling in *Saccharomyces cerevisiae*. Microbiol Mol Biol Rev 70:253–282. https://doi.org/10.1128/MMBR.70.1.253-282.2006

Sanz P, Alms GR, Haystead TA, Carlson M (2000) Regulatory interactions between the Reg1-Glc7 protein phosphatase and the Snf1 protein kinase. Mol Cell Biol. https://doi.org/10.1128/MCB.20.4.1321-1328.2000

Sanz P, Viana R, Garcia-Gimeno MA (2016) AMPK in yeast: the SNF1 (sucrose non-fermenting 1) protein kinase complex. Experientia Suppl 2012:353–374

Schmidt MC, McCartney RR, Zhang X et al (1999) Std1 and Mth1 proteins interact with the glucose sensors to control glucose-regulated gene expression in *Saccharomyces cerevisiae*. Mol Cell Biol. https://doi.org/10.1128/MCB.19.7.4561

Schuler D, Wahl R, Wippel K et al (2015) Hxt1, a monosaccharide transporter and sensor required for virulence of the maize pathogen *Ustilago maydis*. New Phytol 206:1086–1100. https://doi.org/10.1111/nph.13314

Sexton JA, Brown V, Johnston M (2007) Regulation of sugar transport and metabolism by the *Candida albicans* Rgt1 transcriptional repressor. Yeast. https://doi.org/10.1002/yea.1514

Shashkova S, Welkenhuysen N, Hohmann S (2015) Molecular communication: crosstalk between the Snf1 and other signaling pathways. FEMS Yeast Res:15

Shashkova S, Wollman AJM, Leake MC, Hohmann S (2017) The yeast Mig1 transcriptional repressor is dephosphorylated by glucose-dependent and -independent mechanisms. FEMS Microbiol Lett:364. https://doi.org/10.1093/femsle/fnx133

Shi L, Wang X, Wang J et al (2018) Transcriptome analysis of *Δmig1Δmig2* mutant reveals their roles in methanol catabolism, peroxisome biogenesis and autophagy in methylotrophic yeast *Pichia pastoris*. Genes and Genomics. https://doi.org/10.1007/s13258-017-0641-5

Sibirny AA, Titorenko VI, Efremov BD, Tolstorukov II (1987) Multiplicity of mechanisms of carbon catabolite repression involved in the synthesis of alcohol oxidase in the methylotrophic yeast Pichia pinus. Yeast. https://doi.org/10.1002/yea.320030404

Sibirny AA, Titorenko VI, Gonchar M V., et al (1988) Genetic control of methanol utilization in yeasts. J Basic Microbiol 29:462

Simpson-Lavy KJ, Johnston M (2013) SUMOylation regulates the *SNF1* protein kinase. Proc Natl Acad Sci U S A. https://doi.org/10.1073/pnas.1304839110/-/DCSupplemental.www.pnas.org/cgi/doi/10.1073/pnas.1304839110

Simpson-Lavy K, Kupiec M (2018) A reversible liquid drop aggregation controls glucose response in yeast. Curr Genet 64:785–788

Simpson-Lavy K, Xu T, Johnston M, Kupiec M (2017) The Std1 activator of the Snf1/AMPK kinase controls glucose response in yeast by a regulated protein aggregation. Mol Cell. https://doi.org/10.1016/j.molcel.2017.11.016

Snowdon C, Johnston M (2016) A novel role for yeast casein kinases in glucose sensing and signaling. Mol Biol Cell 27:3369–3375. https://doi.org/10.1091/mbc.E16-05-0342

Sood V, Cajigas I, D'Urso A et al (2017) Epigenetic transcriptional memory of GAL genes depends on growth in glucose and the tup1 transcription factor in *Saccharomyces cerevisiae*. Genetics. https://doi.org/10.1534/genetics.117.201632

Spielewoy N, Flick K, Kalashnikova TI et al (2004) Regulation and recognition of SCF-Grr1 targets in the glucose and amino acid signaling pathways. Mol Cell Biol. https://doi.org/10.1128/MCB.24.20.8994-9005.2004

Stasyk O (2017) Methylotrophic yeasts as producers of recombinant proteins. In: Biotechnology of yeasts and filamentous fungi. Springer, Cham

Stasyk OV, Petryshyn AV, Sybirny AA (1994) Impairment of glucose uptake as a possible cause of catabolite repression deficiency in mutants of the methylotrophic yeast *Hansenula polymorpha*. Folia Microbiol. https://doi.org/10.1007/BF02814101

Stasyk OV, Ksheminskaya GP, Kulachkovskii AR, Sibirnyi AA (1997) Mutants of the methylotrophic yeast *Hansenula polymorpha* with impaired catabolite repression. Microbiology 66:631–636

Stasyk OV, Stasyk OG, Komduur J et al (2004) A hexose transporter homologue controls glucose repression in the methylotrophic yeast *Hansenula polymorpha*. J Biol Chem 279. https://doi.org/10.1074/jbc.M310960200

Stasyk OG, Van Zutphen T, Ah Kang H et al (2007) The role of *Hansenula polymorpha* *MIG1* homologues in catabolite repression and pexophagy. FEMS Yeast Res 7. https://doi.org/10.1111/j.1567-1364.2007.00286.x

Stasyk OG, Maidan MM, Stasyk OV et al (2008) Identification of hexose transporter-like sensor HXS1 and functional hexose transporter *HXT1* in the methylotrophic yeast *Hansenula polymorpha*. Eukaryot Cell 7. https://doi.org/10.1128/EC.00028-08

Stasyk OG, Denega IO, Padhorny D et al (2018) Glucose regulation in the methylotrophic yeast Hansenula (Ogataea) polymorpha is mediated by a putative transceptor Gcr1. Int J Biochem Cell Biol. https://doi.org/10.1016/j.biocel.2018.08.002

Steyfkens F, Zhang Z, Van Zeebroeck G, Thevelein JM (2018) Multiple transceptors for macro- and micro-nutrients control diverse cellular properties through the PKA pathway in yeast: a paradigm for the rapidly expanding world of eukaryotic nutrient transceptors up to those in human cells. Front. Pharmacol 9:191

Suppi S, Michelson T, Viigand K, Alamäe T (2013) Repression vs. activation of *MOX*, *FMD*, *MPP1* and MAL1 promoters by sugars in *Hansenula polymorpha*: the outcome depends on cell's ability to phosphorylate sugar. FEMS Yeast Res 13:219–232. https://doi.org/10.1111/1567-1364.12023

Suprayogi NMT, Lertwattanasakul N et al (2015) A *Kluyveromyces marxianus* 2-deoxyglucose-resistant mutant with enhanced activity of xylose utilization. Int Microbiol. https://doi.org/10.2436/20.1501.01.255

Tabba S, Mangat S, McCartney R, Schmidt MC (2010) PP1 phosphatase-binding motif in Reg1 protein of *Saccharomyces cerevisiae* is required for interaction with both the PP1 phosphatase Glc7 and the Snf1 protein kinase. Cell Signal. https://doi.org/10.1016/j.cellsig.2010.02.003

Tachibana C, Biddick R, Law GL, Young ET (2007) A poised initiation complex is activated by SNF1. J Biol Chem. https://doi.org/10.1074/jbc.M707363200

Tanaka K, Nakafuku M, Satoh T et al (1990) S. cerevisiae genes IRA1 and IRA2 encode proteins that may be functionally equivalent to mammalian ras GTPase activating protein. Cell. https://doi.org/10.1016/0092-8674(90)90094-U

Thevelein JM (1984) Cyclic-AMP content and trehalase activation in vegetative cells and ascospores of yeast. Arch Microbiol. https://doi.org/10.1007/BF00425409

Thevelein J (2004) Nutrient sensing through the plasma membrane of eukaryotic cells, Cirencester, UK, 25–29 September, 2004. FEMS Yeast Res, In, pp 303–305

Thevelein JM, De Winde JH (1999) Novel sensing mechanisms and targets for the cAMP-protein kinase a pathway in the yeast *Saccharomyces cerevisiae*. Mol Microbiol 33:904–918

Thevelein JM, Geladé R, Holsbeeks I et al (2005) Nutrient sensing systems for rapid activation of the protein kinase a pathway in yeast. Biochem Soc Trans. https://doi.org/10.1042/BST0330253

Toda T, Uno I, Ishikawa T, et al (1985) In yeast, RAS proteins are controlling elements of adenylate cyclase. Cell. https://doi.org/10.1016/0092-8674(85)90305-8

Toda T, Cameron S, Sass Pet al (1987) Three different genes in *S. cerevisiae* encode the catalytic subunits of the cAMP-dependent protein kinase. Cell. https://doi.org/10.1016/0092-8674(87)90223-6

Tomás-Cobos L, Sanz P (2002) Active Snf1 protein kinase inhibits expression of the *Saccharomyces cerevisiae* HXT1 glucose transporter gene. Biochem J 368:657–663. https://doi.org/10.1042/BJ20020984

Treitel MA, Carlson M (1995) Repression by SSN6-TUP1 is directed by MIG1, a repressor/activator protein. Proc Natl Acad Sci. https://doi.org/10.1073/pnas.92.8.3132

Treitel MA, Kuchin S, Carlson M (1998) Snf1 protein kinase regulates phosphorylation of the Mig1 repressor in *Saccharomyces cerevisiae*. Mol Cell Biol. https://doi.org/10.1128/MCB.18.11.6273

Tu J, Carlson M (1995) *REG1* binds to protein phosphatase type 1 and regulates glucose repression in *Saccharomyces cerevisiae*. EMBO J. https://doi.org/10.1016/j.ymeth.2014.10.001

Usaite R, Jewett MC, Oliveira AP et al (2009) Reconstruction of the yeast Snf1 kinase regulatory network reveals its role as a global energy regulator. Mol Syst Biol. https://doi.org/10.1038/msb.2009.67

Vega M, Riera A, Fernandez-Cid A et al (2016) Hexokinase 2 is an intracellular glucose sensor of yeast cells that maintains the structure and activity of mig1 protein repressor complex. J Biol Chem 291. https://doi.org/10.1074/jbc.M115.711408

Vicente RL, Spina L, Gómez JPL et al (2018) Trehalose-6-phosphate promotes fermentation and glucose repression in *Saccharomyces cerevisiae*. Microb Cell (Graz, Austria) 5:444–459. https://doi.org/10.15698/mic2018.10.651

Vincent O, Townley R, Kuchin S, Carlson M (2001) Subcellular localization of the Snf1 kinase is regulated by specific β subunits and a novel glucose signaling mechanism. Genes Dev. https://doi.org/10.1101/gad.879301

Vishwanatha A, D'Souza CJM (2017) Multifaceted effects of antimetabolite and anticancer drug, 2-deoxyglucose on eukaryotic cancer models budding and fission yeast. IUBMB Life 69:137–147

Vishwanatha A, Rallis C, Bevkal Subramanyaswamy S et al (2016) Identification of nuclear genes affecting 2-Deoxyglucose resistance in *Schizosaccharomyces pombe*. FEMS Yeast Res. https://doi.org/10.1093/femsyr/fow061

Wang Z, Wilson WA, Fujino MA, Roach PJ (2001) Antagonistic controls of autophagy and glycogen accumulation by Snf1p, the yeast homolog of AMP-activated protein kinase, and the cyclin-dependent kinase Pho85p. Mol Cell Biol. https://doi.org/10.1128/MCB.21.17.5742-5752.2001

Wang Y, Pierce M, Schneper L et al (2004) Ras and Gpa2 mediate one branch of a redundant glucose signaling pathway in yeast. PLoS Biol 2. https://doi.org/10.1371/journal.pbio.0020128

Wang X, Cai M, Shi L et al (2016) PpNrg1 is a transcriptional repressor for glucose and glycerol repression of *AOX1* promoter in methylotrophic yeast *Pichia pastoris*. Biotechnol Lett. https://doi.org/10.1007/s10529-015-1972-4

Welkenhuysen N, Borgqvist J, Backman M et al (2017) Single-cell study links metabolism with nutrient signaling and reveals sources of variability. BMC Syst Biol. https://doi.org/10.1186/s12918-017-0435-z

Westholm JO, Nordberg N, Murén E et al (2008) Combinatorial control of gene expression by the three yeast repressors Mig1, Mig2 and Mig3. BMC Genomics. https://doi.org/10.1186/1471-2164-9-601

Wieczorke R, Krampe S, Weierstall T et al (1999) Concurrent knock-out of at least 20 transporter genes is required to block uptake of hexoses in *Saccharomyces cerevisiae*. FEBS Lett 464:123–128. https://doi.org/10.1016/S0014-5793(99)01698-1

Wu J, Trumbly RJ (1998) Multiple regulatory proteins mediate repression and activation by interaction with the yeast Mig1 binding site. Yeast. https://doi.org/10.1002/(SICI)1097-0061(199808)14:11<985::AID-YEA294>3.0.CO;2-C

Xiao B, Sanders MJ, Underwood E et al (2011) Structure of mammalian AMPK and its regulation by ADP. Nature. https://doi.org/10.1038/nature09932

Xue Y, Batlle M, Hirsch JP (1998) GPR1 encodes a putative G protein-coupled receptor that associates with the Gpa2p G(α) subunit and functions in a Ras-independent pathway. EMBO J. https://doi.org/10.1093/emboj/17.7.1996

Yan J, Yan Y, Madzak C, Han B (2017) Harnessing biodiesel-producing microbes: from genetic engineering of lipase to metabolic engineering of fatty acid biosynthetic pathway. Crit Rev Biotechnol

Young ET, Dombek KM, Tachibana C, Ideker T (2003) Multiple pathways are co-regulated by the protein kinase Snf1 and the transcription factors Adr1 and Cat8. J Biol Chem. https://doi.org/10.1074/jbc.M301981200

Yun CW, Tamaki H, Nakayama R et al (1997) G-protein coupled receptor from yeast *Saccharomyces cerevisiae*. Biochem Biophys Res Commun. https://doi.org/10.1006/bbrc.1997.7649

Yun CW, Tamaki H, Nakayama R et al (1998) Gpr1p, a putative G-protein coupled receptor, regulates glucose-dependent cellular cAMP level in yeast *Saccharomyces cerevisiae*. Biochem Biophys Res Commun. https://doi.org/10.1006/bbrc.1998.9600

Yurimoto H, Komeda T, Lim CR et al (2000) Regulation and evaluation of five methanol-inducible promoters in the methylotrophic yeast *Candida boidinii*. Biochim Biophys Acta Gene Struct Expr. https://doi.org/10.1016/S0167-4781(00)00157-3

Zaragoza O, Rodríguez C, Gancedo C (2000) Isolation of the *MIG1* gene from *Candida albicans* and effects of its disruption on catabolite repression. J Bacteriol. https://doi.org/10.1128/JB.182.2.320-326.2000

Zhai Z, Yurimoto H, Sakai Y (2012) Molecular characterization of *Candida boidinii MIG1* and its role in the regulation of methanol-inducible gene expression. Yeast. https://doi.org/10.1002/yea.2909

Zhang N, Cao L (2017) Starvation signals in yeast are integrated to coordinate metabolic reprogramming and stress response to ensure longevity. Curr Genet 63:839–843

Zhang P, Zhang W, Zhou X et al (2010) Catabolite repression of Aox in *Pichia pastoris* is dependent on hexose transporter *Pp*Hxt1 and pexophagy. Appl Environ Microbiol. https://doi.org/10.1128/AEM.00607-10

Zhang T, Bu P, Zeng J, Vancura A (2017) Increased heme synthesis in yeast induces a metabolic switch from fermentation to respiration even under conditions of glucose repression. J Biol Chem. https://doi.org/10.1074/jbc.M117.790923

Chapter 15
L-Cysteine Metabolism Found in *Saccharomyces cerevisiae* and *Ogataea parapolymorpha*

Hiroshi Takagi

Abstract Sulfur's cellular requirements can be met by the cell's uptake of sulfur-containing amino acids. The requirements can also be fulfilled by the cell's assimilation of inorganic sulfur into organic compounds, such as L-homocysteine (Hcy) and L-cysteine (Cys), which are used for the biosynthesis of L-methionine (Met) and L-glutathione (GSH), respectively. Cys can be synthesized via the sulfur assimilation pathway in microorganisms and plants, but not the corresponding pathway in animals. *Saccharomyces cerevisiae*, which is the conventional yeast, synthesizes Cys from Hcy via a reverse trans-sulfuration pathway. It has been concluded that Cys is synthesized exclusively by L-cystathionine β-synthase and L-cystathionine γ-lyase. A promising host strain for high-level production of GSH is the thermotolerant methylotrophic yeast *Ogataea parapolymorpha* (formerly *Hansenula polymorpha*). Domain analyses of the serine *O*-acetyltransferase (SAT) in the non-conventional yeast *Ogataea parapolymorpha* (OpSat1) and those of other fungal SATs have demonstrated that these proteins have a mitochondrial targeting sequence (MTS) at the N-terminus that differs markedly from the classical bacterial and plant SATs. OpSat1 is functionally interchangeable with the *E. coli* SAT, i.e., CysE, even though compared to CysE, OpSat1 has far lower enzymatic activity, with marginal feedback inhibition by Cys. In light of the key role of OpSat1 in the regulation of the pathway of Cys biosynthesis in *O. parapolymorpha*, and its crucial role in sulfur metabolism, it is apparent that OpSat1 could be a target for the metabolic engineering used to generate yeast strains that produce sulfur-containing metabolites such as GSH.

Keywords L-Cysteine · Sulfur · Sulfate · Thiosulfate · *Saccharomyces cerevisiae* · *Ogataea parapolymorpha* · *O*-Acetyl-L-serine · L-Serine *O*-Acetyltransferase · Feedback inhibition · Mitochondria

H. Takagi (✉)
Division of Biological Science, Graduate School of Science and Technology, Nara Institute of Science and Technology, Nara, Japan
e-mail: hiro@bs.naist.jp

A. Sibirny (ed.), *Non-conventional Yeasts: from Basic Research to Application*,
https://doi.org/10.1007/978-3-030-21110-3_15

15.1 Introduction

Sulfur (S) is required for a number of cellular processes, including the following: (1) the synthesis of the organic compounds L-methionine (Met), L-cysteine (Cys), and Cys-containing peptide, L-glutathione (GSH), (2) the stabilization of protein conformation through disulfide-bonds, enzyme reactions (with an iron-sulfur cluster as the prosthetic group), (3) the metabolism of secondary products (e.g., glucosinolates, sulfated compounds), (4) the response against environmental stresses (e.g., involving GSH or phytochelatins), and (5) the redox cycle (involving glutaredoxins and thioredoxins). Most plants and microorganisms (but not animals) assimilate environmental inorganic forms of sulfur such as sulfate (SO_4^{2-}) and thiosulfate ($S_2O_3^{2-}$) into the organic sulfur-containing amino acids Met and Cys. Many enzymes are needed for the de novo biosynthesis of Cys and Met in the yeast *Saccharomyces cerevisiae*, as well as for recycling organic sulfur metabolites (Thomas and Surdin-Kerjan 1997). Cys, Met, and GSH are important compounds used in the pharmaceutical, food, and cosmetic industries, comprising a world market of over 7 billion USD per year (Takagi and Ohtsu 2017).

As shown in Fig. 15.1, Cys is crucial for protein folding, assembly, and stability through the formation of disulfide-bonds. Cys is also needed for iron/sulfur (Fe/S) clusters of the catalytic domain of some enzymes (e.g., aconitase), cytochromes, and degradative L-serine (Ser) dehydratase of the bacterium *Escherichia coli* (Cicchillo et al. 2004). Cys also functions as a master antioxidant molecule. Thioredoxin and glutaredoxin, which use GSH, are involved in the protection of cells under oxidative stress conditions. Periplasmic Cys was recently proposed to protect *E. coli* cells from hydrogen peroxide in the environment produced by phagocytes (Ohtsu et al. 2010, 2015). Cys is used in the food industry as a flavor enhancer, dietary supplement, and dough conditioner; in the pharmaceutical industry as a skin discoloration preventive and an expectorant agent; and in the cosmetic industry as a whitening agent and a perm assistant. Approximately 5000 tons of Cys is used globally each year, and are now used to supply Cys, e.g., the hydrolysis of keratin, enzymatic synthesis, and fermentation (Takagi and Ohtsu 2017). Extraction methods

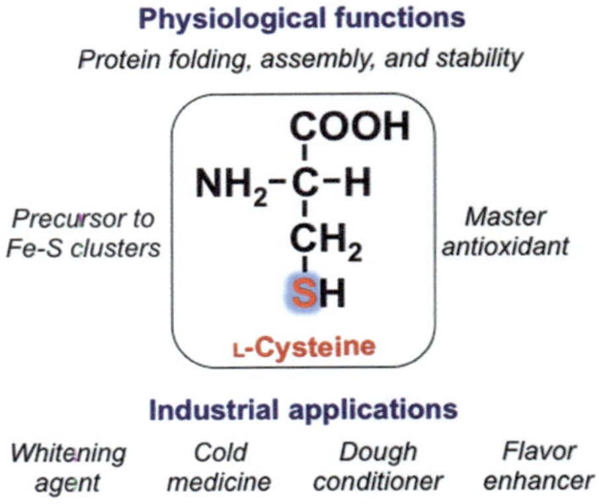

Fig. 15.1 The physiological functions and industrial applications of Cys

result in insufficient Cys yields, waste treatment problems, and unpleasant odors (Hunt 1985). Although Cys is considered "generally recognized as safe (GRAS)" substance by the US Food and Drug Administration, synthetic or biotechnological products are preferable for consumer use. This is due to the severe problem of bovine spongiform encephalopathy (mad cow disease) in products of animal origin (Wada and Takagi 2006).

Figure 15.2 illustrates how *E. coli* and *S. cerevisiae* metabolize sulfur. In microorganisms and plants, Cys can be synthesized via the sulfur assimilation pathway. In enteric bacteria such as *E. coli* and *Salmonella typhimurium*, Cys synthesized from sulfide proceeds by a two-step reaction. First, L-serine (Ser) *O*-acetyltransferase (SAT) (EC 2.3.1.30) catalyzes the synthesis of *O*-acetyl-L-serine (OAS) from Ser and acetyl-CoA. Then, OAS is sequentially condensed with sulfide by OAS sulfhydrylase (OASS) (EC 4.2.99.8) to form Cys (the OAS pathway) (Fig. 15.3a). In contrast, some microorganisms (e.g., *S. cerevisiae*, *Lactococcus lactis* (Sperandio et al. 2005), and *Pseudomonas putida* (Vermeji and Kertesz 1999)) synthesizes Cys from L-homocysteine (Hcy) via L-cystathionine (CTT) comprised of L-cystathionine

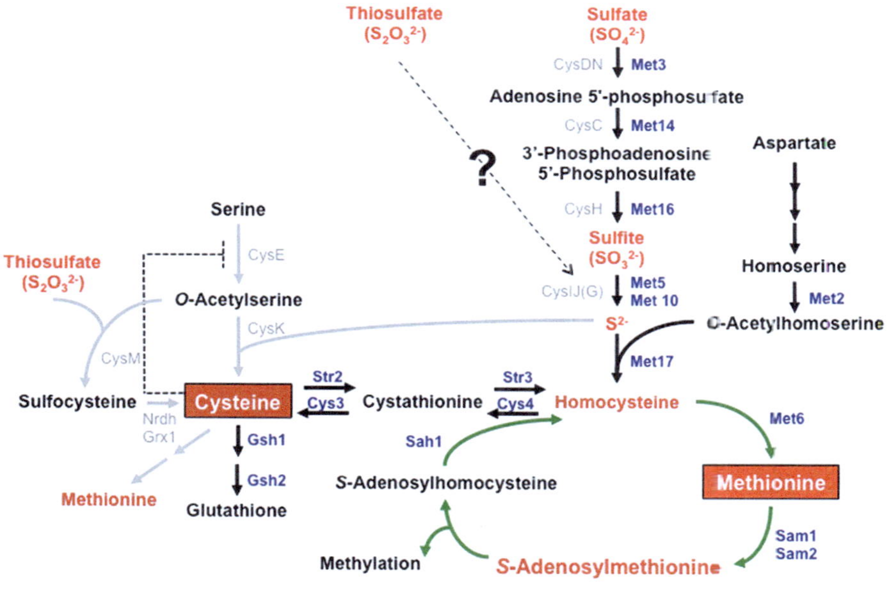

Fig. 15.2 Metabolic pathways of Cys and Met in *E. coli* and *S. cerevisiae*. Protein names of *S. cerevisiae*: Met3, ATP sulfurylase; Met14, adenylylsulfate kinase; Met16, 3′-phosphoadenylsulfate reductase; Met5, sulfite reductase β-subunit; Met10, sulfite reductase α-subunit; Met17, *O*-acetyl-L-homoserine (OAH)/*O*-acetyl-L-serine (OAS) sulfhydrylase; Met2, L-homoserine-*O*-acetyltransferase; Met6, cobalamin-independent Met synthase; Sam1, Sam2, *S*-adenosylmethionine (SAM) synthetase; Sah1, SAM hydrolase; Cys4, L-cystathionine (CTT) β-synthase; Cys3, CTT γ-lyase; Str2, CTT γ-synthase; Str3, Peroxisomal CTT β-lyase; Gsh1, γ-glutamylcysteine synthetase; and Gsh2, glutathione (GSH) synthetase. Activity of *E. coli* CysE (L-serine *O*-acetyltransferase) is subject to feedback inhibition by Cys

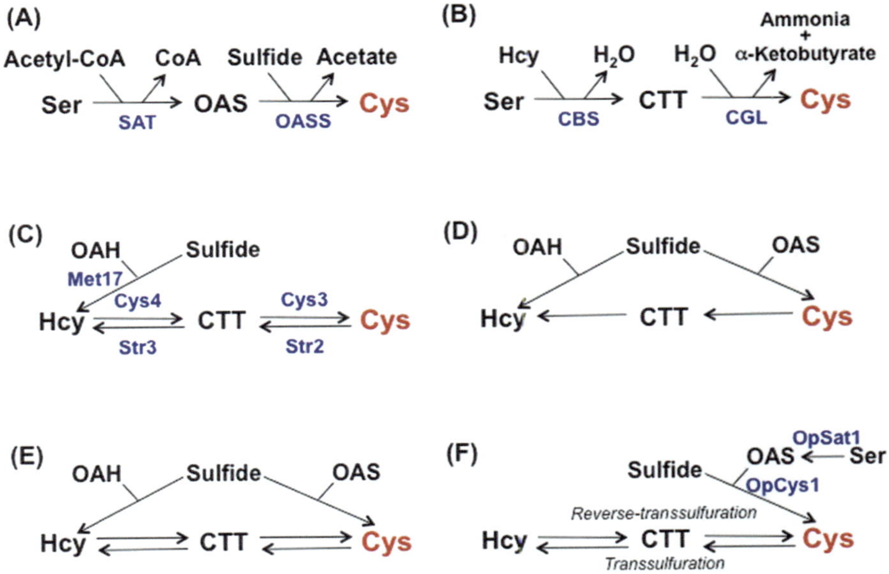

Fig. 15.3 Various pathways of Cys biosynthesis in *E. coli* and *Salmonella typhimurium* (the OAS pathway) (**a**), *S. cerevisiae*, *Lactococcus lactis*, and *Pseudomonas putida* (the CTT pathway) (**b**), *S. cerevisiae* (**c**), *Schizosaccharomyces pombe* (**d**), *Aspergillus nidulans* (**e**), and *O. parapolymorpha* (**f**). Protein names: SAT, L-serine *O*-acetyltransferase; OASS, OAS sulfhydrylase; CBS, CTT β-synthase; CGL, CTT γ-lyase; OpSat1, *O. parapolymorpha* SAT, OpCys1, *O. parapolymorpha* OASS

β-synthase (CBS) (EC 4.2.1.22) and L-cystathionine γ-lyase (CGL) (EC 4.4.1.1) through reverse trans-sulfuration (i.e., the CTT pathway) (Fig. 15.3b). Fungi such as *Aspergillus nidulans* (Paszewski and Grabski 1974) and *Neurospora crassa* (Jacobson and Metzenberg 1977) are known to produce Cys by both the OAS and CTT pathways.

The metabolic pathway of Cys in *E. coli* has been well-studied, but there have been few investigations of the metabolic pathway of Cys in other microorganisms. This chapter focuses on the recent findings regarding Cys and sulfur metabolism in a conventional yeast (*S. cerevisiae*) and non-conventional yeast species (*Ogataea parapolymorpha*).

15.2 Cys Metabolism in *Escherichia coli*

The biosynthesis of Cys from Ser in *E. coli* proceeds in a two-step pathway. Ser undergoes a substitution of its β-hydroxyl with a thiol in two steps. SAT (the *cysE* gene product) catalyzes the acetylation by acetyl-CoA of the β-hydroxyl of Ser to give OAS. The second step, the α,β-elimination of acetate from OAS and the addition of H_2S to give Cys, is then catalyzed by OASS-A (the *cysK* gene product). The first reaction catalyzed by SAT is the rate-limiting step of Cys biosynthesis, because SAT activity is sensitive to feedback inhibition by Cys. The OAS pathway was thought to be the sole Cys biosynthetic pathway. *E. coli* does not have a reverse trans-sulfuration pathway, which converts CTT (the product of Met metabolism) into Cys (Figs. 15.2 and 15.3a). The *E. coli cysE* mutants have exhibited Cys auxotrophy (Kredich and Tomkins 1966).

The biosynthesis of Cys in *E. coli* is regulated at both the level of enzymatic activity (feedback inhibition) and the level of gene expression. The Cys regulon (known as "cys-regulon") comprises the gene for the biosynthesis of Cys and the genes for the uptake and reduction in oxidized sulfur sources, such as sulfate and thiosulfate. The expression of these genes (except for *cysE* and *cysG*) is under the control of the transcriptional activator CysB. CysB requires the inducer *N*-acetyl Ser (which is formed from OAS by a chemical reaction) for its activity and for sulfur limitation. Sulfide and thiosulfate act as anti-inducers for CysB. A detailed discussion of the mechanisms of the cys-regulon can be found elsewhere (Kredich 1996).

In the bacterium *E. coli*, the degradation of Cys is catalyzed mainly by Cys desulfhydrase (CD) (Soda 1987). Enzymes with CD activity in *E. coli* have been extensively investigated. Awano et al. (2003) reported that tryptophanase (TNA; EC 4.1.99.1, the *tnaA* gene product) and CTT β-lyase (CBL; EC 4.4.1.8, the *metC* gene product) catalyzed the CD reaction and acted on Cys degradation in *E. coli*. With the use of the plasmid gene library of *E. coli*, three additional proteins with CD activity were then identified in *E. coli* as two OASSs (the *cysK* gene product OASS-A and the *cysM* gene product OASS-B) and the protein MalY (Awano et al. 2005). The growth phenotype and the results of transcriptional analyses indicate that the CD activity of TNA primarily contributes to Cys degradation in vivo, but unidentified proteins with CD activity that may be induced by Cys could also present in *E. coli* (Awano et al. 2005).

Export systems for Cys in *E. coli* have also been well-studied. Two transporter proteins, YdeD and YfiK, have been identified as Cys exporters (Daßler et al. 2000; Franke et al. 2003). YdeD and YfiK are membrane-integral proteins and they belong to different protein families. In addition, Cys is exported from the cytoplasm to the periplasm by an ATP-binding cassette-type transporter CydDC required for cytochrome assembly in *E. coli* (Pittman et al. 2002). In addition to these transporters, there are 37 open reading frames (ORFs) that are suspected drug transporter genes in *E. coli* (Nishino and Yamaguchi 2001). Yamada et al. (2006) reported that among the 33 putative drug transporter genes tested, the multidrug transporter Bcr (a member of the major facilitator family) is involved in Cys export. Amino acid transport assays showed that Bcr overexpression conferring resistance to both bicyclomycin and tetracycline specifically promotes the export of Cys driven by energy derived from the proton gradient. With the use of a systematic and comprehensive collection of gene-disrupted *E. coli* K-12 mutants, the *tolC* gene (encoding the outer membrane channel TolC) was recently identified as a novel gene involved in Cys export

(Wiriyathanawudhiwong et al. 2009). Gene expression analyses revealed that the *tolC* gene is essential for Cys tolerance in *E. coli* cells.

15.3 Cys Metabolism in the Conventional Yeast *Saccharomyces cerevisiae*

Certain fungal species have also been reported to have the OAS pathway (Paszewski and Grabski 1974; Jacobson and Metzenberg 1977), but in both yeast and fungal species, Cys biosynthesis can also be achieved independently of the OAS pathway. Sulfide is condensed with *O*-acetyl-L-homoserine (OAH) to generate Hcy, which is converted to CTT and then to Cys through reverse trans-sulfuration (the CTT pathway) (Figs. 15.2 and 15.3b). *S. cerevisiae* is of evolutional interest in terms of its pathway and its regulation; some *S. cerevisiae* strains were shown to have sufficient SAT activity, but *S. cerevisiae* strains with the S288C background had deficient SAT (Ono et al. 1999). The genome sequence data of *S. cerevisiae* also showed that the ORFs corresponding to the *CYS1* and *CYS2* genes, mutations of which cause SAT deficiency (Ono et al. 1984, 1988), were missing in *S. cerevisiae* strain S288C. It should be noted that *S. cerevisiae* has an enzyme (the *MET17/MET25* gene product (D'Andrea et al. 1987)) that acts as both designated *O*-acetyl-L-homoserine/*O*-acetyl-L-serine sulfhydrylase (OAHS/OASS) in vitro assays (Yamagata et al. 1974). However, with regard to *S. cerevisiae*, previous studies the authors of several studies concluded that detectable SAT and OAHS/OASS do not constitute a Cys biosynthetic pathway and that Cys is synthesized exclusively through the CTT pathway constituted by CBS (the *CYS4* gene product) and CGL (the *CYS3* gene product) (Cherest and Surdin-Kerjan 1992; Ono et al. 1999). Therefore, for a better understanding of the role of the OAS pathway in *S. cerevisiae*, a detailed comparison of the Cys biosynthetic pathway in *S. cerevisiae* with that of *E. coli* is necessary.

Interestingly, two *S. cerevisiae* strains (IS66-4C and DKD-5C) have detectable activities of SAT and OASS, but they synthesize Cys exclusively via CTT by CBS and CGL (Ono et al. 1988, 1999). To untangle this peculiar feature, the *E. coli* genes, *cysE* and *cysK* (encoding SAT and OASS-A, respectively) were introduced into *S. cerevisiae* Cys auxotrophs (Takagi et al. 2003). Although the cells expressing the *E. coli* SAT grew on medium lacking Cys, the cells expressing the *E. coli* OASS-A did not grow at all, and the cells expressing both enzymes grew very well without Cys. The results indicated that *S. cerevisiae* SAT cannot support the biosynthesis of Cys, but *S. cerevisiae* OASS produces Cys if a sufficient amount of OAS is provided by the *E. coli* SAT. In addition, the *E. coli* OASS-A was expressed in the *cys3* and *cys4* disruptant, both of which had endogenous SAT activity but are Cys-dependent due to blockade of the CTT pathway. However, all of the above-mentioned transformants showed no growth at all on minimal medium, indicating that Cys was not fully synthesized via the chimeric OAS pathway. It was concluded that detectable SAT and OASS do not constitute the Cys synthetic pathway and that Cys is synthesized exclusively via the CTT pathway. It appears that *S. cerevisiae* SAT does not have a metabolic role in vivo because of its very low activity or localization, thus controlling the OAS level required for sulfate assimilation, but playing no role in the direct synthesis of Cys.

It is not yet known why some *S. cerevisiae* strains have detectable SAT activity that is not involved in Cys synthesis in vivo. We speculate that *S. cerevisiae* SAT is a novel type of enzyme that is transported into the nucleus (after being synthesized on ribosomes) in order to directly control the level of OAS. Cys is also likely to regulate SAT both transcriptionally and post-translationally. No data are available regarding the protein sequence of SAT; this is in part because of instability of SAT during purification. The *A. nidulans cysA* gene that encodes SAT has been cloned and characterized (Grynberg et al. 2000). Its protein sequence showed high similarity to that of L-homoserine *O*-acetyltransferase (HAT) from yeasts and fungi, and it showed no homology with known SATs from plants or bacteria. The nucleotide sequences of the *MET2* gene, which encodes HAT, in *S. cerevisiae* strains S288C and IS66-4C were thus compared; no nucleotide substitutions or frame-shift were detected. The isolation of the *CYS1* and/or *CYS2* genes encoding SAT is necessary to elucidate the role of OAS in Cys biosynthesis in *S. cerevisiae*,

Regarding sulfur assimilation, *E. coli* has two pathways for Cys synthesis: a sulfate pathway and a thiosulfate pathway (Fig. 15.2). The sulfate pathway consumes two ATP molecules and four NADPH molecules to make one Cys molecule. The thiosulfate pathway spends only one NADPH molecule (Nakatani et al. 2012). Both pathways use OAS as a common carbon skeleton to incorporate sulfur. *S. cerevisiae* uses thiosulfate as a sole sulfur source (Marzluf 1997). It was recently indicated that thiosulfate is converted mainly into sulfite and then into sulfide prior to its use by the sulfate pathway (unpublished data), which is consistent with the results of a previous report (Thomas et al. 1992) (Fig. 15.2). We cultured *S. cerevisiae* cells in minimal (SD) medium in the presence of sulfate or thiosulfate as a sole sulfur source. We observed that compared to sulfate, thiosulfate was more effective as a source for improving the cell growth. Notably, a tenfold less concentration of thiosulfate than sulfate was still effective for cell growth. The thiosulfate-grown cells also rapidly consumed glucose in the medium and accelerated the production of ethanol in the log phase. However, the final productivity in the stationary phase was almost the same as that shown by the sulfate-source cells (Funahashi et al. 2015). In a further investigation of cellular metabolism, analyses of the carbon metabolic flux after labeling by ^{13}C-glucose and the NADPH level were conducted, and the results revealed that compared to sulfate thiosulfate maintained high levels of NADPH, with a decreased need for NADPH production by suppressed pentose phosphate pathway flux. In thiosulfate-grown cells, the glycolysis flux from glucose increased, but that of TCA decreased, leading to an increased flux toward ethanol from pyruvate. Based on the finding that thiosulfate is better sulfur source than sulfate for limiting consumption of ATP and NADPH, it was proposed that compared to sulfate, thiosulfate is an effective sulfur source for the biotechnological production of useful compounds (other than ethanol) in *S. cerevisiae* (Funahashi et al. 2015). Because thiosulfate possesses more reduced sulfur atom than sulfate, sulfur assimilation into Met and/or Cys should be feasible, saving consumption of cellular fundamental energy metabolites (ATP and NADPH). Therefore, intracellular sulfur metabolites were measured in *S. cerevisiae* cultured in the modified SD medium with 1 mM Na_2SO_4 (sulfate) or $Na_2S_2O_3$ (thiosulfate) as a sole sulfur source (Fig. 15.4). In the presence of thiosulfate, cellular level of organic sulfur compounds, such as Cys, Hcy, and GSH, was significantly increased compared with that in the presence of sulfate. These results suggest that thiosulfate as well as sulfate is metabolized to Hcy and the thiosulfate pathway is in part overlapped to that of sulfate in *S. cerevisiae* cells (unpublished data).

Based on these data and general knowledge, thiosulfate utilization can be summarized as follows: (1) Thiosulfate utilization results in a reduction of NADPH consumption in sulfur assimilation (Cys and Met biosynthesis), because thiosulfate has a more reduced form of sulfur atom than sulfate. (2) This results in a reduced need for NADPH production by the pentose phosphate pathway, and the actual reduction of flux to the pathway. (3) The reduced equivalent of carbon flux

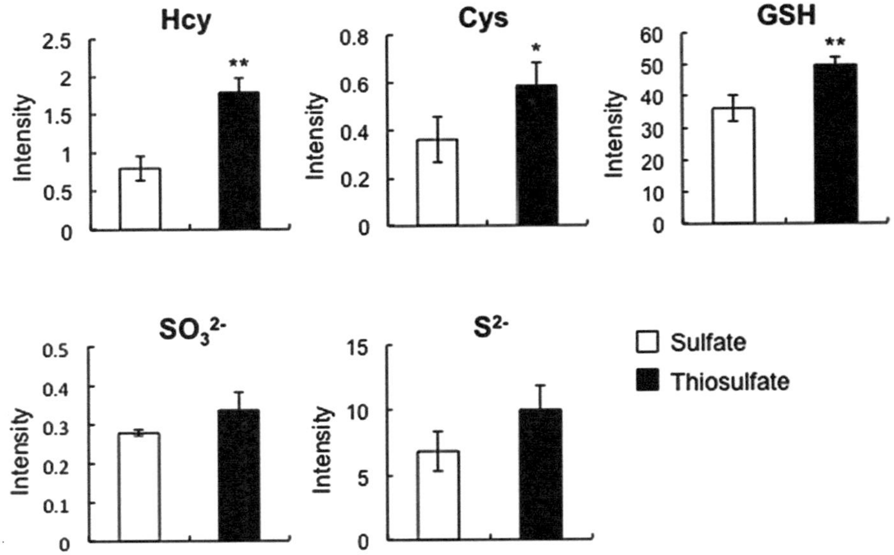

Fig. 15.4 The metabolome analysis of sulfur compounds in *S. cerevisiae*. Cells of *S. cerevisiae* BY4741 were precultured in the modified SD medium containing 1 mM sulfate as a sole sulfur source at 30 °C for overnight, and the precultures were inoculated into the modified SD medium containing 1 mM sulfate (open bar) or thiosulfate (closed bar) to set into OD_{600} of 0.05 in each culture. The cultures were cultivated to the exponential growth phase (OD_{600} of 0.4) at 30 °C with shaking. The collected cells were subjected to the metabolome analysis. The contents of sulfur compounds were expressed as "intensity" by the respective metabolite-to-D-camphor-10-sulfonic acid sodium salt (internal standard) ratio of peak area in the mass chromatography monitoring the m/z transition of the derivatives. The values are the means and standard deviations of three independent experiments. Statistically significant differences were determined by Student's *t* test ($*p < 0.05$, $**p < 0.01$, sulfate vs. thiosulfate)

alternatively turns toward the glycolytic pathway, leading to increased ATP production. (4) The enhanced ATP pool suppresses both another ATP production system of oxidative phosphorylation (respiration) and the TCA cycle activity (i.e., the Crabtree effect). (5) The accelerated glycolytic pathway and the suppressed TCA cycle cause an increase in the flux toward ethanol production from pyruvate. (6) NADH needed for the enhanced production of ethanol is essentially provided by the simultaneously accelerated glycolytic pathway probably in a manner that does not alter its cellular concentration in order to maintain the redox balance. It has thus been proposed, as noted earlier, that in *S. cerevisiae* thiosulfate is a more effective sulfur source than sulfate for the biotechnological production of useful materials other than ethanol (Takagi and Ohtsu 2017). Based on this concept, other products can be obtained by using thiosulfate as the sulfur source, and also molecular basis of the assimilation pathway and the responsible enzymes could be further clarified.

15.4 Cysteine Metabolism in the Non-conventional Yeast *Ogataea parapolymorpha*

In yeast and filamentous fungi, sulfide can be condensed with either OAS (to directly generate Cys) or with OAH (to generate Hcy, the precursor of Met). The resulting Cys and Hcy can be interconverted through a trans-sulfuration pathway. One of the non-conventional yeast strains, the thermotolerant methylotrophic yeast *Ogataea parapolymorpha* (formerly called *Hansenula polymorpha*) (Suh and Zhou 2010; Kurtzman 2011), is characterized by high tolerance to various stresses induced by xenobiotics (drugs), heavy metals, and environmental pollutants (Kaszycki et al. 2013). *O. parapolymorpha* has been widely studied for the overproduction of GHS and various heterologous proteins (Kang et al. 2001). Although this yeast has a trans-sulfuration pathway, Sohn et al. (2014) found that Cys biosynthesis from sulfide mediated by SAT activity was the only pathway in *O. parapolymorpha* for the synthesis of Cys from sulfur (Fig. 15.5). The sulfur metabolic pathway of *O. parapolymorpha* was systematically analyzed. The demonstration of the precise sulfur metabolic pathway of *O. parapolymorpha*, which was reconstructed based on genome sequences analyses and confirmed by systematic gene deletion experiments, revealed the absence of a de novo synthesis of Hcy from sulfur in this non-conventional yeast species. The direct biosynthesis of Cys from sulfide is therefore the only pathway for the synthesis of sulfur amino acids from inorganic sulfur in *O. parapolymorpha*, even though both directions of the trans-sulfuration pathway was also observed that Cys (and no other sulfur amino acid) could repress the expression of a subset of sulfur genes. This result suggested a central and exclusive role of Cys in the control of the metabolism of sulfur in *O. parapolymorpha*. Other research has shown that ^{35}S-Cys is incorporated into intracellular sulfur compounds such as GSH more efficiently than ^{35}S-Met in *O. parapolymorpha*; this also supports the concept of a Cys-centered sulfur pathway. These findings comprise the

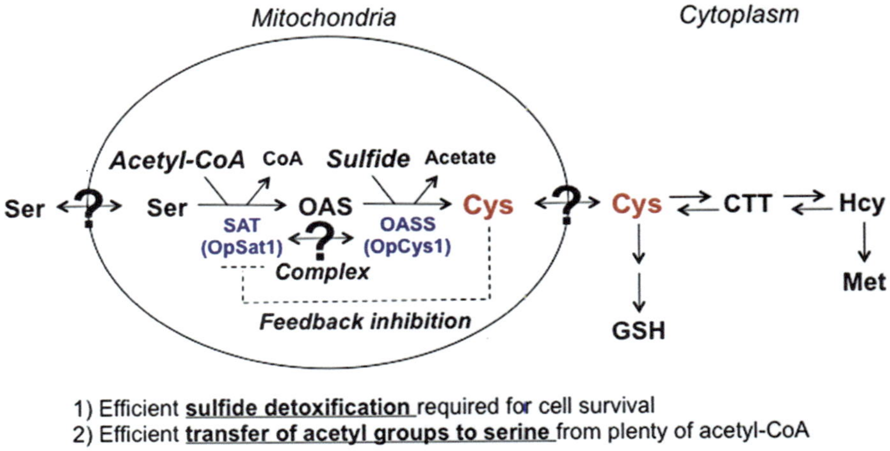

Fig. 15.5 Metabolic pathways of Cys in *O. parapolymorpha*. The physiological roles of OpSat1 localized in the mitochondria are proposed

first evidence of novel features of the *O. parapolymorpha* sulfur metabolic pathway. The features of this pathway are distinct from those observed in other yeast species and filamentous fungi (Sohn et al. 2014).

A comparative genomic study of the sulfur metabolism of several hemiascomycetous yeasts (Hébert et al. 2011) revealed that in *S. cerevisiae* and most hemiascomycetous yeasts Hcy is the base for the biosynthesis of sulfur amino acids (Fig. 15.3c). For example, in *S. cerevisiae*, sulfide is first incorporated into a three-carbon chain through the formation of Hcy, and this is followed by the synthesis of Cys and Met from Hcy via the methyl cycle (leading to the syntheses of Met and *S*-adenosyl-L-methionine) or via the reverse trans-sulfuration pathway (leading to the syntheses of GSH and Cys) (Penninckx 2002). In contrast, the fission yeast *Schizosaccharomyces pombe* and the filamentous fungus *A. nidulans* and have an additional mechanism for incorporating sulphide into carbon chains to provide Cys through the OAS pathway (Fig. 15.3d). In filamentous fungi, the interconversion of sulfur amino acids (Met-Hcy-Cys) occurs through both forward and reverse trans-sulfuration pathways (Fig. 15.3e). *S. pombe* has only one pathway with forward trans-sulfuration. The wild-type *S. pombe* strain was suggested to be able to use Met as a sulfur source only after Met is degraded, giving rise to sulfate (Brzywczy et al. 2002). Similar to the findings in *S. cerevisiae* and filamentous fungi, these fission yeast data imply that *O. parapolymorpha* also has both forward and reverse trans-sulfuration pathways (Fig. 15.3f). Like *S. cerevisiae*, *O. parapolymorpha* can thus grow in the presence of Met or Cys as the sole sulfur source. *O. parapolymorpha* has a novel feature, however, the lack of a direct incorporation of sulfide into Hcy, due to the absence of a *MET17*/*MET25* homolog. The biosynthesis of Met in *O. parapolymorpha* is therefore possible only through Cys and CTT, that is, the forward trans-sulfuration reaction. Intriguingly, the sulfur metabolism of

O. parapolymorpha is centered on Cys instead of Met, unlike other hemiascomycetes such as *S. cerevisiae* and *Kluyveromyces lactis* (Hébert et al. 2011).

Systematic analyses have revealed the structural characteristics and physiological function of SAT present in *O. parapolymorpha* (OpSat1) (Yeon et al. 2018). OpSat1 is functionally interchangeable with SAT in *E. coli* SAT, although their structural organization differs and their sequence identity is low. OpSat1 must be localized at the mitochondria to be fully active. These characteristics have shown that OpSat1 is a new mitochondrial SAT with a crucial role in the sulfur assimilatory metabolism in *O. parapolymorpha*. Sohn et al. (2014) reported that the results of their analysis of the sulfur metabolic pathway of *O. parapolymorpha*. In this yeast, despite the presence of trans-sulfuration pathway, it was indicated that the biosynthesis of Cys from sulfide mediated by SAT activity is the only pathway for synthesizing Cys from inorganic sulfur. The deletion of the OpSat1 gene resulted in the generation of the Cys auxotrophic phenotype. Sohn et al. (2014) also constructed the phylogenetic tree of SATs from various organisms (e.g., bacteria, yeast, fungi, and plants) in order to elucidate both the evolutionary origin and the structural features of OpSat1. All of the SATs in yeasts and fungi were predicted to have a mitochondrial targeting sequence (MTS) and a highly conserved α/β hydrolase 1 domain. The phylogenetic trees demonstrated that the SATs of yeasts and fungi had unique domain structures that differ markedly from those of classical SATs.

Due to the lack of OAH sulfhydrylase, *O. parapolymorpha* cannot incorporate sulfur into Hcy. This suggests that that the biosynthesis of Cys from sulfide and OAS is the only pathway. The *E. coli* SAT (CysE), in contrast, has both the essential function of Cys biosynthesis and an additional assimilation pathway from thiosulfate that uses the compound *S*-sulfocysteine. It was reported that both OpSat1 and CysE proteins were heterologously expressed in *O. parapolymorpha Opsat1*Δ and *E. coli cysE*Δ strains (Fig. 15.6a). The Cys auxotrophy of the *Opsat1*Δ strain was rescued by the expression of CysE. Similarly, OpSat1 expression fully restored the Cys auxotrophy of the *E. coli cysE*Δ strain. These results suggest that despite their different structures and evolutionary distance, CysE and OpSat1 could be functionally exchangeable as active SATs. In another experiment, OpSat1 and CysE were purified as recombinant proteins fused with GST and his-tag. The SAT activity was then measured. Compared to that of CysE, the relative activity of OpSat1 was approximately 300-fold lower. As previously reported (Nakamori et al. 1998; Takagi et al. 1999a), CysE activity is subjected to feedback inhibition by Cys. OpSat1 activity seemed to be relatively insensitive to the presence of Cys, suggesting that there is much less feedback inhibition by Cys (compared to CysE). In the presence of 1 μM Cys, the CysE activity was dramatically decreased to 40%, whereas the OpSat1 activity was 30% even in the presence of 1 mM Cys.

A high-performance liquid chromatography (HPLC) analysis of the metabolites was performed to examine the effects of OpSat1 deletion and overexpression on the intracellular levels of sulfur compounds. The GSH level in wild-type strain was markedly high compared to that of *S. cerevisiae*. The deletion of OpSat1 resulted in decreases in the intracellular levels of Cys and GSH, but not Hcy, which is likely due to the blockage of the synthesis of OAS. However, the overexpression of OpSat1

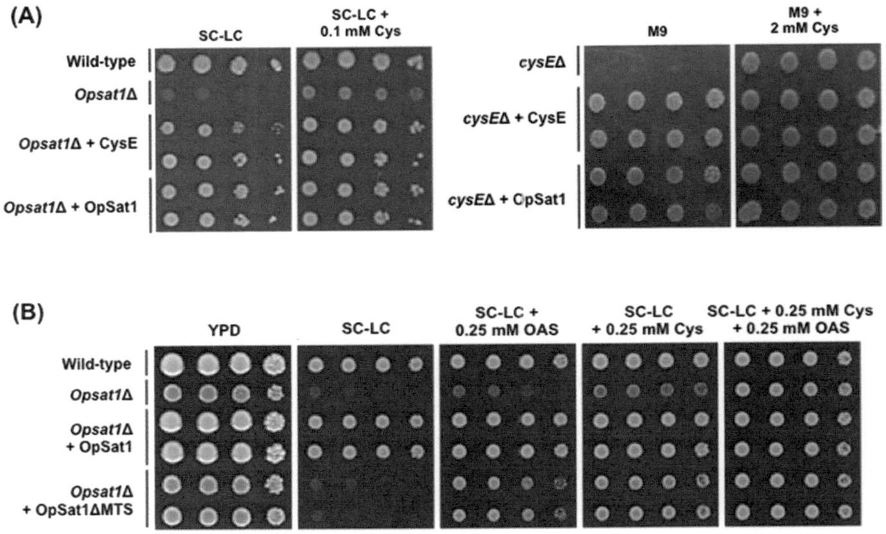

Fig. 15.6 Functional analysis of OpSat1. (**a**) Functional exchange of *E. coli* and *O. parapolymorpha* SATs (CysE and OpSat1, respectively). Cells of *O. parapolymorpha* harboring an empty vector (wild-type), *Opsat1*Δ harboring an empty vector (*Opsat1*Δ), the CysE expression plasmid (*Opsat1*Δ + CysE), and the OpSat1 expression plasmid (*Opsat1*Δ + OpSat1) were spotted onto SC-LC plates supplemented with and without 0.25 mM OAS or Cys. Cells of *E. coli cysE*Δ harboring an empty vector (*cysE*Δ), the CysE expression plasmid (*cysE*Δ + CysE), and the OpSat1 expression plasmid (*cysE*Δ + OpSat1) were serially diluted and spotted onto M9 minimal medium plates with and without 2 mM OAS or Cys. All plates were incubated at 37 °C for 2 days. (**b**) Functional analysis of mitochondrial targeting sequence (MTS)-deleted OpSat1. Cells of *O. parapolymorpha* harboring an empty vector (wild-type), *Opsat1*Δ harboring an empty vector (*Opsat1*Δ), the OpSat1 expression plasmid (*Opsat1*Δ + OpSat1), and the OpSat1-**Δ**MTS expression plasmid (*Opsat1*Δ + OpSat1ΔMTS) were serially diluted and spotted on YPD and SC-LC plates supplemented with 0.25 mM OAS, Cys, or both. All plates were incubated at 37 °C for 2 days. Yeast cells corresponding to an OD_{600} of 1, 0.1, 0.01, and 0.001 were spotted. (This figure is modified from Yeon et al. 2018)

and that of CysE under the control of a strong promoter did not increase the Cys or GSH levels. This suggests that the overexpression of wild-type SAT alone is not sufficient to increase the Cys and GSH levels. For the determination of the subcellular localization of OpSat1 with an N-terminal MTS, a yEGFP-fused construct (at the *Opsat1* promoter locus) was integrated on the chromosome. Fluorescence signals observed by confocal microscopy were detected mainly at the mitochondria. For a further validation of the function of MTS, yEGFP constructs with and without MTS were expressed under the control of the *Opsat1* promoter. The cells expressing the MTS-fused yEGFP evidently showed strong fluorescent signals, again mainly at the mitochondria, whereas the cells expressing the yEGFP construct without MTS had fluorescent signals that were diffused over the cytoplasm, indicating that OpSat1 is probably located in the mitochondria. The question of whether the mitochondrial localization is required for SAT activity was addressed by expressing the

MTS-deleted OpSat1 in *Opsat1*Δ cells (Fig. 15.6b). In the OpSat1 deletion mutant, the growth defect was rescued by providing an exogenous supply of either OAS or Cys, and the co-supplementation of both OAS and Cys rescued the growth defect even more efficiently. It was observed that the MTS-deleted OpSat1 could restore the OAS auxotrophy but not the Cys auxotrophy of *Opsat1*Δ cells. This interestingly result suggested that OpSat1 functions in the cytoplasm as SAT for the production of OAS, but in order for the full activity of OpSat1 to be expressed in Cys biosynthesis, its mitochondrial localization is required.

The reason why OpSat1 is localized in the mitochondria in *O. parapolymorpha* is not yet known (Fig. 15.5). Three cDNA clones encoding SAT have been isolated in the plant *Arabidopsis thaliana*: SAT-c, SAT-m, and SAT-p. The deduced amino acid sequence of the *A. thaliana* SAT isozymes is approximately 50% identical to that of the *E. coli* CysE. Noji et al. (1998) showed that SAT-m and SAT-p are feedback-insensitive isozymes up to 100 μM Cys, and SAT-c was inhibited by 10 μM Cys. The concentrations necessary for 50% inhibition (IC_{50}) of SAT-c under standard assay conditions was 1.8 μM, which is the same order of magnitude as that of the *E. coli* wild-type SAT (6 μM). The expression of the cDNA encoding *A. thaliana* feedback inhibition-insensitive SAT (SAT-m, SAT-p) improved the production of Cys in *E. coli* cells (Takagi et al. 1999b). These results suggested that the three SAT isoforms, localized in different organelles, are subjected to differing types of feedback regulation, presumably in order to play differing roles in the production of OAS and Cys in *A. thaliana*. It appears that SAT-c is responsible for strict regulation (by Cys feedback inhibition) of the concentration of OAS in the cytosol. Under the Cys limiting condition, SAT-c may be de-repressed and provide OAS for the biosynthesis of Cys and perhaps for inducing the gene expression of Cys biosynthetic enzymes. Most of the GSH and Met is synthesized in chloroplasts, and thus a sufficiently high amount of Cys is necessary in chloroplasts for the full biosyntheses of GSH and Met (Noji et al. 1998). The production of a large amount of OAS may therefore be required even under a high Cys concentration without the feedback inhibition by Cys in chloroplasts. In plants, mitochondria are important for synthesizing OAS, which is the Cys precursor (Haas et al. 2008). Another feedback-sensitive SAT is localized in the cell cytosol in watermelon. Taking all of the above findings into account, it can be speculated that in plant cells, cytosolic SAT is feedback-sensitive, and the isoform of SAT that is localized in organelles (i.e., chloroplasts and mitochondria) is feedback-insensitive.

Similarly to plants, yeasts generate sulfide as an intermediate of the sulfur assimilation. Sulfide is a potent inhibitor of cytochrome c oxidase in mitochondria. Efficient sulfide detoxification is thus mandatory for the survival of plant cells and yeast cells. One of the proposed physiological roles of OAS is thus the catalysis of sulfide detoxification in the mitochondria, in addition to Cys synthesis. The mitochondrial localization might provide another benefit, linked to the intracellular acetyl-CoA level, which is the substrate of SATs. In *S. cerevisiae*, the mitochondrial acetyl-CoA concentration is much higher compared to the concentrations in the other cellular compartments. In light of the low SAT activity, the mitochondrial localization could provide a way to transfer acetyl groups to serine from a high amount of acetyl-CoA more efficiently than in the cell cytoplasm.

15.5 Conclusions and Future Perspective

Recent advances in biochemical studies, genome sequencing, structural biology, and metabolome analyses of bacteria have enabled us the use of several new approaches to achieve the direct fermentation of Cys from glucose. With *E. coli*, a combination of two factors is needed to improve Cys fermentation as follows: (1) The enhancement of biosynthesis—the overexpression of the altered *cysE* gene encoding feedback inhibition-insensitive SAT is necessary (Nakamori et al. 1998; Takagi et al. 1999a; Kai et al. 2006). (2) Weakening the degradation, by the knockout of the genes encoding Cys desulfhydrases CDs (Awano et al., 2003, 2005). In addition, thiosulfate was reported to be a more effective sulfur source than the commonly used sulfate for producing Cys in *E. coli* (Nakatani et al. 2012), because thiosulfate is advantageous for saving the consumption of NADPH and related energy molecules. However, high intracellular Cys levels in *E. coli* were shown to be inhibitory or even toxic (Harris 1981; Park and Imlay 2003; Sørensen and Pederson 1991). Further improvements in Cys production are thus expected to be based on the amino acid efflux system, with overexpression of the genes involved in Cys transport (Yamada et al. 2006; Wiriyathanawudhiwong et al. 2009).

The wild-type *O. parapolymorpha* yeast strain accumulated more intracellular GSH (56 nM/OD_{600}) than *S. cerevisiae* (2 nM/OD_{600}), whereas the strain's Cys and Hcy levels are similar to those in *S. cerevisiae* (Kitajima et al. 2012). This nonconventional yeast has thus attracted significant attention as an industrial yeast strain for various biotechnological applications (e.g., Cys and GSH overproduction). In some plants, SAT and Cys synthase (CS), which is also called OASS, form a protein complex (Bogdanova and Hell 1997). In the plant *A. thaliana*, the rate-limiting enzyme in the strict regulation of OAS (a substrate of CS) is SAT-c. In most transgenic plants, the simultaneous overexpression of CS and SAT leads to higher total GSH levels than the overexpression of a single enzyme (Liszewska et al. 2001).

The co-expression of *E. coli* SAT and OASS-A used an additional Cys synthetic pathway in *S. cerevisiae*, leading to an increased production of GSH. As noted above, the overexpression of SATs from *O. parapolymorpha* (OpSat1) and *E. coli* (CysE) did not increase the intracellular level of GSH or that of Cys. Thus, a simple overexpression of wild-type SAT appears to be insufficient to increase these levels. Although OpSat1 seems to function in the cell cytoplasm as a SAT for producing OAS, its localization in the mitochondria is necessary for the full activity of OpSat1 in Cys biosynthesis. A plausible target for metabolic engineering for the design of Cys feedback-insensitive variants to produce high-value sulfur-containing metabolites (such as Cys and GSH) is thus OpSat1 (Wirtz and Hell 2003).

Acknowledgments We are greatly indebted to our co-researchers Dr. Hyun Ah Kang, Ji Yoon Yeon, and Su Jin Yoo (Chung-Ang University, Seoul, Korea) and Dr. Bun-ichiro Ono (Ritsumeikan University, Japan). I am also grateful to Dr. Shigeru Nakamori (Fukui Prefectural University, Fukui, Japan) and Dr. Iwao Ohtsu (Nara Institute Science and Technology, Nara, Japan). This review includes the work supported by a grant from Ajinomoto, Co., Inc., to H.T.

References

Awano N, Wada M, Kohdoh A, Oikawa T, Takagi H, Nakamori S (2003) Effect of cysteine desulfhydrase gene disruption on L-cysteine overproduction in *Escherichia coli*. Appl Microbiol Biotechnol 62:239–243

Awano N, Wada M, Mori H, Nakamori S, Takagi H (2005) Identification and functional analysis of *Escherichia coli* cysteine desulfhydrases. Appl Environ Microbiol 71:4149–4152

Bogdanova N, Hell R (1997) Cysteine synthesis in plants: protein-protein interactions of serine acetyltransferase from *Arabidopsis thaliana*. Plant J 11:251–262

Brzywczy J, Sienko M, Kucharska A, Paszewski A (2002) Sulphur amino acid synthesis in *Schizosaccharomyces pombe* represents a specific variant of Sulphur metabolism in fungi. Yeast 19:29–35

Cherest H, Surdin-Kerjan Y (1992) Genetic analysis of a new mutation conferring cysteine auxotrophy in *Saccharomyces cerevisiae*: updating of the sulfur metabolism pathway. Genetics 130:51–58

Cicchillo RM, Baker MA, Schnitzer EJ, Newman EB, Krebs C, Booker SJ (2004) *Escherichia coli* L -serine deaminase requires a [4Fe-4S] cluster in catalysis. J Biol Chem 279:32418–32425

D'Andrea R, Surdin-Kerjan Y, Pure G, Cherest H (1987) Molecular genetics of *met17* and *met25* mutants of *Saccharomyces cerevisiae*: intragenic complementation between mutations of a single structural gene. Mol Gen Genet 207:165–170

Daßler T, Maier T, Winterhalter C, Böck A (2000) Identification of a major facilitator protein from *Escherichia coli* involved in efflux of metabolites of the cysteine pathway. Mol Microbiol 36:1101–1112

Franke I, Resch A, Daßler T, Maier T, Böck A (2003) YfiK from *Escherichia coli* promotes export of *O*-acetylserine and cysteine. J Bacteriol 185:1161–1166

Funahashi E, Saiki K, Honda K, Sugiura Y, Kawano Y, Ohtsu I, Watanabe D, Wakabayashi Y, Abe T, Nakanishi T, Suematsu M, Takagi H (2015) A finding of thiosulfate pathway for synthesis of organic sulfur compounds in *Saccharomyces cerevisiae* and an improvement of ethanol production. J Biosci Bioeng 120:666–669

Grynberg M, Topczewski J, Godzik A, Paszewski A (2000) The *Aspergillus nidulans cysA* gene encodes a novel type of serine *O*-acetyltransferase which is homologous to homoserine *O*-acetyltransferases. Microbiology 146:2695–2703

Haas FH, Heeg C, Queiroz R, Bauer A, Wirtz M, Hell R (2008) Mitochondrial serine acetyltransferase functions as a pacemaker of cysteine synthesis in plant cells. Plant Physiol 148:1055–1067

Harris CL (1981) Cystine and growth inhibition of *Escherichia coli*: threonine deaminase as the target enzyme. J Bacteriol 145:1031–1035

Hébert A, Forquin-Gomez MP, Roux A, Aubert J, Junot C, Loux V, Heilier JF, Bonnarme P, Beckerich JM, Landaud S (2011) Exploration of sulfur metabolism in the yeast *Kluyveromyces lactis*. Appl Microbiol Biotechnol 91:1409–1423

Hunt S (1985) Degradation of amino acids accompanying in vitro protein hydrolysis. In: Barrett GC (ed) Chemistry and biology of the amino acids. Chapman & Hall, London, pp 3763–3798

Jacobson ES, Metzenberg RL (1977) Control of aryl sulfatase in a serine auxotroph of Neurospora. J Bacteriol 130:1397–1398

Kai Y, Kashiwagi T, Ishikawa K, Ziyatdinov MK, Redkina EI, Kiriukhin MY, Gusyatiner MM, Kobayashi S, Takagi H, Suzuki E (2006) Engineering of *Escherichia coli* L-serine *O*-acetyltransferase on the basis of crystal structure: desensitization to feedback inhibition by L-cysteine. Protein Eng Des Sel 19:163–167

Kang HA, Kang W, Hong WK, Kim MW, Kim JY, Sohn JH, Choi ES, Choe KB, Rhee SK (2001) Development of expression systems for the production of recombinant human serum albumin using the MOX promoter in *Hansenula polymorpha* DL-1. Biotechnol Bioeng 76:175–185

Kaszycki P, Walski T, Hachicho N, Heipieper HJ (2013) Biostimulation by methanol enables the methylotrophic yeasts *Hansenula polymorpha* and *Trichosporon* sp. to reveal high formaldehyde biodegradation potential as well as to adapt to this toxic pollutant. Appl Microbiol Biotechnol 97:5555–5564

Kitajima T, Jigami Y, Chiba Y (2012) Cytotoxic mechanism of selenomethionine in yeast. J Biol Chem 287:10032–10038

Kredich NM (1996) Biosynthesis of cysteine. In: Neidhardt FC, Curtiss RIII, Ingraham JL, Lin ECC, Low KB, Magasanik B, Reznicoff WS, Riley M, Schaechter M, Umbarger JE (eds) *Escherichia coli* and *Salmonella typhimurium*: cellular and molecular biology, 2nd edn. ASM, Washington, DC, pp 514–527

Kredich NM, Tomkins GM (1966) The enzymatic synthesis of L-cysteine in *Escherichia coli* and *Salmonella typhimurium*. J Biol Chem 241:4955–4965

Kurtzman CPA (2011) New methanol assimilating yeast, *Ogataea parapolymorpha*, the ascosporic state of *Candida parapolymorpha*. Antonie Van Leeuwenhoek 100:455–462

Liszewska F, Blaszczyk A, Sirko A (2001) Modification of non-protein thiols contents in transgenic tobacco plants producing bacterial enzymes of cysteine biosynthesis pathway. Acta Biochim Pol 48:647–656

Marzluf GA (1997) Molecular genetics of sulfur assimilation in filamentous fungi and yeast. Annu Rev Microbiol 51:73–96

Nakamori S, Kobayashi S, Kobayashi C, Takagi H (1998) Overproduction of L-cysteine and L-cystine by *Escherichia coli* strains with a genetically altered serine acetyltransferase. Appl Environ Microbiol 64:1607–1611

Nakatani T, Ohtsu I, Nonaka G, Wiriyathanawudhiwong N, Morigasaki S, Takagi H (2012) Enhancement of thioredoxin/glutaredoxin-mediated L-cysteine synthesis from *S*-sulfocysteine increases L-cysteine production in *Escherichia coli*. Microb Cell Factories 11:62. https://doi.org/10.1186/1475-2859-11-62

Nishino K, Yamaguchi A (2001) Analysis of a complete library of putative drug transporter genes in *Escherichia coli*. J Bacteriol 183:5803–5812

Noji M, Inoue K, Kimura N, Gouda A, Saito K (1998) Isoform-dependent differences in feedback regulation and subcellular localization of serine acetyltransferase involved in cysteine biosynthesis from *Arabidopsis thaliana*. J Biol Chem 273:32739–32745

Ohtsu I, Wiriyathanawudhiwong N, Morigasaki S, Nakatani T, Kadokura H, Takagi, H (2010) The L-cysteine/L-cystine shuttle system provides reducing equivalents to the periplasm in *Escherichia coli*. J Biol Chem 285: 17479–17487

Ohtsu I, Kawano Y, Suzuki M, Morigasaki S, Saiki K, Yamazaki S, Nonaka G, Takagi H (2015) Uptake of L-cystine via an ABC transporter contributes defense of oxidative stress in the L-cysteine export-dependent manner in *Escherichia coli*. PLoS One 10(4):e0120619

Ono B, Suruga T, Yamamoto M, Yamamoto S, Murata K, Kimura A, Shinoda S, Ohmori S (1984) Cystathionine accumulation in *Saccharomyces cerevisiae*. J Bacteriol 158:860–865

Ono B, Shirahige Y, Nanjoh A, Andoh N, Ohue H, Ishino-Arao Y (1988) Cysteine biosynthesis in *Saccharomyces cerevisiae*: mutation that confers cystathionine β-synthase deficiency. J Bacteriol 170:5883–5889

Ono B, Hazu T, Yoshida S, Kawato T, Shinoda S, Brzvwczy J, Paszewski A (1999) Cysteine biosynthesis in *Saccharomyces cerevisiae*: a new outlook on pathway and regulation. Yeast 15:1365–1375

Park S, Imlay JA (2003) High levels of intracellular cysteine promote oxidative DNA damage by driving the Fenton reaction. J Bacteriol 185:1942–1950

Paszewski A, Grabski J (1974) Regulation of S-amino acids biosynthesis in *Aspergillus nidulans*. Mol Gen Genet 132:307–320

Penninckx MJ (2002) An overview on glutathione in *Saccharomyces* versus nonconventional yeasts. FEMS Yeast Res 2:295–305

Pittman MS, Corker H, Wu G, Binet MB, Moir AJG, Poole RK (2002) Cysteine is exported from the *Escherichia coli* cytoplasm by CydDC, an ATP-binding cassette-type transporter required for cytochrome assembly. J Biol Chem 277:49841–49849

Soda K (1987) Microbial sulfur amino acids: an overview. In: Jakoby WB, Griffith OW (eds) Methods in enzymology, vol 143. Academic Press, Orlando, pp 453–459

Sohn MJ, Yoo SJ, Oh DB, Kwon O, Lee SY, Sibirny AA, Kang HA (2014) Novel cysteine-centered sulfur metabolic pathway in the thermotolerant methylotrophic yeast *Hansenula polymorpha*. PLoS One 9:e100725

Sørensen MA, Pederson S (1991) Cysteine even in low concentrations, induces transient amino acid starvation in *Escherichia coli*. J Bacteriol 173:5244–5246

Sperandio B, Polard P, Ehrlich DS, Renault P, Guédon E (2005) Sulfur amino acid metabolism and its control in *Lactococcus lactis* IL1403. J Bacteriol 187:3762–3778

Suh SO, Zhou JJ (2010) Methylotrophic yeasts near *Ogataea* (*Hansenula*) *polymorpha*: a proposal of *Ogataea angusta* comb. nov. and *Candida parapolymorpha* sp. nov. FEMS Yeast Res 10:631–638

Takagi H, Ohtsu (2017) l-Cysteine metabolism and fermentation in microorganisms. In: Yokota A, Ikeda M (eds) Amino acid fermentation. Springer, New York, pp 129–152

Takagi H, Kobayashi C, Kobayashi S, Nakamori S (1999a) PCR random mutagenesis into *Escherichia coli* serine acetyltransferase: isolation of the mutant enzymes that cause overproduction of l-cysteine and l-cystine due to the desensitization to feedback inhibition. FEBS Lett 452:323–327

Takagi H, Awano N, Kobayashi S, Noji M, Saito K, Nakamori S (1999b) Overproduction of l-cysteine and l-cystine by expression of genes for feedback inhibition-insensitive serine acetyltransferase from *Arabidopsis thaliana* in *Escherichia coli*. FEMS Microbiol Lett 179:453–459

Takagi H, Yoshioka K, Awano N, Nakamori S, Ono B (2003) Role of *Saccharomyces cerevisiae* serine *O*-acetyltransferase in cysteine biosynthesis. FEMS Microbiol Lett 218:291–297

Thomas D, Surdin-Kerjan Y (1997) Metabolism of sulfur amino acids in *Saccharomyces cerevisiae*. Microbiol Mol Biol Rev 61:503–532

Thomas D, Barbey R, Henry D, Surdin-Kerjan Y (1992) Physiological analysis of mutants of *Saccharomyces cerevisiae* impaired in sulphate assimilation. J Gen Microbiol 138:2021–2028

Vermeji P, Kertesz MA (1999) Pathways of assimilative sulfur metabolism in *Pseudomonas putida*. J Bacteriol 181:5833–5837

Wada M, Takagi H (2006) Metabolic pathways and biotechnological production of l-cysteine. Appl Microbiol Biotechnol 73:48–54

Wiriyathanawudhiwong N, Ohtsu I, Li ZD, Mori H, Takagi H (2009) The outer membrane TolC is involved in cysteine tolerance and overproduction in *Escherichia coli*. Appl Microbiol Biotechnol 81:903–913

Wirtz M, Hell R (2003) Production of cysteine for bacterial and plant biotechnology: application of cysteine feedback-insensitive isoforms of serine acetyltransferase. Amino Acids 24:195–203

Yamada S, Awano N, Inubushi K, Maeda E, Nakamori S, Nishino K, Yamaguchi A, Takagi H (2006) Effect of drug transporter genes on cysteine export and overproduction in *Escherichia coli*. Appl Environ Microbiol 72:4735–4742

Yamagata S, Takeshima K, Naiki N (1974) Evidence for the identity of *O*-acetylserine sulfhydrylase with *O*-acetylhomoserine sulfhydrylase in yeast. J Biochem 75:1221–1229

Yeon JY, Yoo SJ, Takagi H, Kahn HA (2018) A novel mitochondrial serine *O*-acetyltransferase, OpSAT1, plays a critical role in sulfur metabolism in the thermotolerant methylotrophic yeast *Ogataea parapolymorpha*. Sci Rep 8:2377

Chapter 16
Strategies at Bioreactor Scale for the Production of Recombinant Proteins in *Yarrowia lipolytica*

Marie Vandermies and Patrick Fickers

Abstract Recombinant protein production represents a multibillion-dollar market. Therefore, it constitutes an important research field both in academia and industry. The use of yeast as cell factory presents several advantages such as ease of genetic manipulation, growth at high cell density, and possibility of posttranslational modifications. *Yarrowia lipolytica* is considered as one of the most attractive hosts due to its ability to metabolize raw substrate, to express genes at high level, and to secrete protein in large amounts. In the recent years, several reviews were dedicated to genetic tools developed for this purpose. Although the construction of efficient cell factory for recombinant protein synthesis is important, the development of an efficient process for protein production constitutes an equally vital aspect. Indeed, a sports car could not drive fast on a gravel road. The aim of this review is to provide a comprehensive snapshot of process tools to consider for recombinant protein production in bioreactor using *Y. lipolytica* as a cell factory, in order to facilitate the decision-making for future strain and process engineering.

Keywords *Yarrowia lipolytica* · Recombinant protein · Bioreactor · Metabolic load · Molecular strategies · Process strategies

16.1 Introduction

Microbial recombinant protein production has a wide array applications and represents a multibillion-dollar market (Mattanovich et al. 2012; Vieira Gomes et al. 2018). As unicellular eukaryotes, yeasts offer a convenient host system combining simplicity of growth, ability of genetic engineering, and the posttranslational modifications. Among yeast cell factories, *Yarrowia lipolytica* appears as one of the most attractive. It has been recognized by the American Food and Drug Administration

---

M. Vandermies · P. Fickers (✉)
TERRA Teaching and Research Centre, Microbial Processes and Interactions, University of Liège-Gembloux AgroBio Tech, Gembloux, Belgium
e-mail: marie.vandermies@doct.uliege.be; pfickers@uliege.be

A. Sibirny (ed.), *Non-conventional Yeasts: from Basic Research to Application*, https://doi.org/10.1007/978-3-030-21110-3_16

(FDA) as an organism Generally Regarded As Safe (GRAS status), making it suitable for food and pharmaceutical applications (Groenewald et al. 2014). Compared to *S. cerevisiae*, it shows reduced hyperglycosylation issues (Vieira Gomes et al. 2018). One of the main features of *Y. lipolytica* is its extraordinary capacity to secrete proteins (for a recent review, see Celińska and Nicaud (2018)). *Y. lipolytica* strains have been isolated from diverse environments such as seawater, cheese, and waste sludge, explaining the adaptability of this yeast to a wide range of substrates and culture conditions (Nicaud 2012). So far, more than 150 recombinant proteins have been obtained using *Y. lipolytica*, but only 25 of them have been produced at bioreactor scale. In industry, *Y. lipolytica*-based processes are still scarce, despite recombinant protein productivity similar to that of other yeast cell factories, such as the methylotrophic yeast *Pichia pastoris*. While significant progress has been made to produce recombinant protein at reactor scale, process efficiency is still limited by issues such as metabolic burden, unmastered dimorphism, and oxygen requirements. Here, we provide a comprehensive overview of process tools to consider for recombinant protein production in bioreactor using *Y. lipolytica* as a cell factory, in order to facilitate the decision-making for future strain and process engineering. These tools are summarized in Table 16.1 for ease of reading different sections.

Table 16.1 Molecular and process strategies available for bioprocess optimization

Strategy	Options
Molecular strategies	
Mandatory	
Plasmid	Episomal or integrative; mono- or multi-copy
Promoter	Constitutive or inducible; strength
Optional	
Strain engineering	Auxotrophies; deletion of *AEP/AXP* and of specific genes, docking platforms
Protein engineering	Codon optimization; secretion
Dimorphism	Arrest in a defined morphotype; immobilized-cell technology
Process strategies	
Mandatory	
Culture mode	Batch, fed-batch, or continuous
Culture medium	Complex or defined; alternative carbon sources
Dissolved oxygen	Airflow and agitation; oxygen spikes; Vhb expression; ionic liquids; pressurization
pH	Tight control or *AEP/AXP* deletion
Inoculum and growth rate	High cell density
Working volume	Miniaturization, lab scale, or pilot scale
Optional	
Analytical and predictive models	Feed modeling; calculation of carbon and nitrogen fluxes
Co-production	Metabolite(s), single-cell protein, single-cell oil

16.2 Protein Production

Proteins of various origins can be produced in a recombinant manner. Most of the time, the resulting additional anabolic load leads to some metabolic perturbation. For efficient process development, these perturbations must be minimized.

16.2.1 Recombinant Proteins

In *Y. lipolytica*, recombinant protein production in bioreactor mainly focused on proteins from fungal origin (for a review, see (Madzak and Beckerich 2013)). However, examples of viral (Hamsa and Chattoo 1994), bacterial (Bhave and Chattoo 2003; Huang et al. 2011; Kar et al. 2012; Liu et al. 2015; Fickers et al. 2005a), vegetal (Chang et al. 1997, 1998a, b; Kim et al. 2000), or mammalian (Gasmi et al. 2011a, b, 2012; Madzak et al. 2000; Tiels et al. 2012) proteins have also been reported. These recombinant proteins are mainly enzymes with applications in green chemistry, food and feed, cosmetics, detergents, textiles and paper industries, biofuels, and waste stream depollution. Another field of application is the sector of medicine and pharmacy where these enzymes, such as lipases, are used for drugs synthesis (Botes et al. 2008; Maharajh et al. 2008a, b). Phenomenon of hyperglycosylation, commonly observed in *S. cerevisiae*, is not critical in *Y. lipolytica* (Jolivet et al. 2007; Kopečný et al. 2005), rendering this host more adapted for the production of therapeutic proteins such as virus antigen (Hamsa and Chattoo 1994), glucocerebrosidase (De Pourcq et al. 2012a), and lysosomal enzyme (Tiels et al. 2012). In addition, the construction of strains with humanized glycosylation pathway allowed glycosylation patterns similar to that of human glycoproteins (Callewaert et al. 2009). Finally, recombinant proteins may be also synthetized for the sake of process improvement such as the single-chain hemoglobin Vhb from *Vitreoscilla stercoraria* (Bhave and Chattoo 2003) that promotes cell growth in oxygen-limited environment, the invertase Suc2p from *S. cerevisiae* (Förster et al. 2007; Lazar et al. 2013), or the exo-inulinase Inu1p from *Kluyveromyces marxianus* (Cui et al. 2011), converting, respectively, sucrose and inulin into glucose and fructose that can further be metabolized by the cells. Reporter proteins, such as β-galactosidase (Kar et al. 2012; Fickers et al. 2005a) or yellow fluorescent protein (YFP, (Trassaert et al. 2017)), were also used to decipher promoter regulation during bioreactor operations. A non-exhaustive list of recent examples of recombinant proteins produced by *Y. lipolytica* in bioreactor can be found in Table 16.2.

Table 16.2 Recent examples of recombinant protein production in bioreactor using *Y. lipolytica*

Protein	Molecular strategies	Process strategies	Results	
Organism protein (size, sequence characteristics) → applications	Strain, copy number, *promoter*, signal sequence or intracellular	Cultivation mode, medium type, reactor volume, HCD if applicable	Enzyme activity /L Enzyme productivity /L	References
Bacteria				
Streptomyces hygroscopicus pro-transglutaminase (43.6 kDa, deglycosylated variant) → food industry	Po1h, multi-copy (/), *hp4d*, *XPR2* pre	FB, complex medium, 3 L, HCD	35,300 U.L^{-1} 300 U.L^{-1}.h^{-1}	Liu et al. (2015)
Bacillus subtilis β-mannanase (40.9 kDa, glycosylation, codon optimization) → paper, detergents, food, and feed industry	Po1g, multi-copy (6), *hp4d*, *XPR2* pre	B, defined medium (+ YE), 20 L	602,000 U.L^{-1} 26,000 U.L^{-1}.h^{-1}	YaPing et al. (2017)
	– –	FB, defined medium (+ YE), 20 L, HCD	1,024,000 U.L^{-1} 14,000 U.L^{-1}.h^{-1}	– –
Fungi				
Aspergillus niger β-Mannanase (41.3 kDa, glycosylation, codon optimization) → paper, detergents, food, and feed industry	Po1g, multi-copy (6), *hp4d*, *XPR2* pre	B, defined medium (+ YE), 20 L	1,375,000 U.L^{-1} 28,000 U.L^{-1}.h^{-1}	YaPing et al. (2017)
	– –	FB, defined medium (+ YE), 20 L, HCD	3,024,000 U.L^{-1} 42,000 U.L^{-1}.h^{-1}	– –
Rhodotorula araucariae epoxide hydrolase (46 kDa) → pharmaceuticals	Po1h, multi-copy (13–15), *hp4d*, intracellular	FB, defined medium (+ YE), /, HCD	11,000–12,000 U.L^{-1} /	Botes et al. (2008)
	– –	B, defined medium (+ YE), 15 L, HCD	22,750 U.L^{-1} /	Maharajh et al. (2008a)
	– –	FB, defined medium (+ YE), 15 L, HCD	194,700 U.L^{-1}	– –

(continued)

Table 16.2 (continued)

Protein	Molecular strategies	Process strategies	Results	
Organism protein (size, sequence characteristics) → applications	Strain, copy number, *promoter*, signal sequence or intracellular	Cultivation mode, medium type, reactor volume, HCD if applicable	Enzyme activity /L Enzyme productivity /L	References
	– –	Improved FB, defined medium (+ YE), 15 L, HCD	3440 U.L^{-1}.h^{-1} 232,257 U.L^{-1} 5813 U.L^{-1}.h^{-1} N.B.: ovoid cells	Maharajh et al. (2008b)
Aspergillus aculeatus endo-β-1,4-mannanase (42 kDa) → paper, detergents, food, and feed industry	Po1h, mono-copy, *hp4d*, *LIP2* prepro	B, defined medium, 2 L, HCD	685,000 nkat. L^{-1} 14,000 nkat. L^{-1}.h^{-1}	Roth et al. (2009)
	Po1h, multi-copy (8), *hp4d*, native signal sequence	– –	5,776,000 nkat. L^{-1} 120,000 nkat. L^{-1}.h^{-1}	– –
	Po1h, multi-copy (9), *hp4d*, native signal sequence	– –	6,719,000 nkat. L^{-1} 140,000 nkat. L^{-1}.h^{-1}	– –
	– –	FB, defined medium, 15 L, HCD	26,139,000 nkat.L^{-1} 373,000 nkat. L^{-1}.h^{-1}	– –
	– –	Improved FB, complex medium, 2 L, HCD	40,835,000 nkat.L^{-1} 913,000 nkat. L^{-1}.h^{-1}	van Zyl (2011)
Candida antarctica lipase B (34 kDa) → biofuels, fragrances, food industry, pharmaceuticals	JMY1212, mono-copy, *pPOX2*, *LIP2* prepro	FB, defined medium, 3 L	5090 U.L^{-1} (190 mg.L^{-1}) /	Emond et al. (2010)
Trametes versicolor laccase IIIb (55.6 kDa) → waste detoxification, textile, biofuels, organic synthesis, food industry	Po1g, mono-copy, *hp4d*, native signal sequence	B, complex medium, 5 L N.B.: Beet molasses as carbon source	4500 U.L^{-1} 47 U.L^{-1}.h^{-1}	Darvishi et al. (2017)

(continued)

Table 16.2 (continued)

Protein	Molecular strategies	Process strategies	Results	
Organism protein (size, sequence characteristics) → applications	Strain, copy number, *promoter*, signal sequence or intracellular	Cultivation mode, medium type, reactor volume, HCD if applicable	Enzyme activity /L Enzyme productivity /L	References
	– –	B, complex medium, 5 L N.B.: Sucrose as carbon source	6760 $U.L^{-1}$ 70 $U.L^{-1}.h^{-1}$	Darvishi et al. (2018)
Insect				
Sitophilus oryzae α-amylase (53.9 kDa, codon optimization) → paper, textile, food industry	Po1g, multi-copy (4), *hp4d*, *XPR2* pre + native signal sequence	B, complex medium, 5 L	81 $U.L^{-1}$ 2.8 $U.L^{-1}.h^{-1}$ N.B.: Part of the enzyme retained inside the cells	Celińska et al. (2015)
	– –	– –	142.8 $U.L^{-1}$ /	Celińska et al. (2016)
	A-101 (wt), multi-copy (4), *pTEF*, native signal sequence	– –	11.9 $U.L^{-1}$ / N.B.: Proteolysis	
	Po1g, multi-copy (4), *hp4d*, *XPR2* pre + native signal sequence	FB, complex medium, 5 L, HCD N.B.: Analysis of carbon and nitrogen fluxes	22,980 $U.L^{-1}$ 121.6 $U.L^{-1}.h^{-1}$ N.B.: Metabolite co-production	Celińska et al. (2017)
Human				
α2b-interferon (19 kDa, codon optimization) → pharmaceuticals	Po1d, mono-copy, *pPOX2*, *LIP2* pre + dipeptides	B, defined medium, 5 L	50 $mg.L^{-1}$ N.B.: Ovoid cells, proteolysis	Gasmi et al. (2011a)
	– –	FB, defined medium, 5 L, HCD	425 $mg.L^{-1}$ 16.9 $mg.L^{-1}.h^{-1}$ N.B.: Proteolysis	(Gasmi et al. 2011b)
Human α-glucosidase (110 kDa, glycosylation, codon optimization) → pharmaceuticals	Po1d, multi-copy (3), *pPOX2*, *LIP2* pre + 2 X-ala	FB, complex medium, 10 L	200 mg purified protein	(Tiels et al. 2012)

Abbreviations and symbols: B = batch mode; FB = fed-batch mode; HCD = high cell density; "/" = absence of information; "– –" = identical information

16.2.2 Metabolic Burden

The overexpression of recombinant genes in a host cell and the subsequent protein synthesis mobilizes cellular resources like energy, transcription and translation factors, and amino acids. This leads to a physiological disorder known as metabolic burden (Glick 1995; Görgens et al. 2001). It has been demonstrated that metabolic burden increases with the increase the of recombinant gene size (Görgens et al. 2001; Corchero and Villaverde 1998) and copy number (see below), and is further amplified by the expression level (Glick 1995). Process conditions such as limitation in nutrients or oxygen availability enhance the difficulty for cells to cope with metabolic burden (Glick 1995; van Rensburg et al. 2012). The main outcome of this metabolic burden is the reduction of the growth capacity and biomass yield (Glick 1995; Görgens et al. 2001; Roth et al. 2009; YaPing et al. 2017). Limitations in specific amino acids and energy may result in translational errors, impacting recombinant protein activity, stability, and possible immunogenicity (Glick 1995). In addition, this results in protein quality heterogeneity, which highly complexifies the purification scheme. When secreted, recombinant proteins compete with endogenous proteins for exportation, and may trigger the unfolded protein response, initiating additional activities of protein folding, vesicular transport, and protein degradation (Babour et al. 2008; Oh et al. 2010) that increase intracellular disorder (de Ruijter et al. 2018). Cell fitness may be further impaired by the activity of certain heterologous proteins, interfering with host cell functioning (Glick 1995; Hanquier et al. 2003).

To avoid or at least reduce the metabolic burden, molecular and process strategies must be considered. Integration of expression cassette in yeast chromosome must be favored over episomal plasmids, whose maintenance in the cells requires a substantial amount of energy. Synthetic gene with a codon-optimized sequence must be considered when codon bias of heterologous gene differs significantly to that of the recipient cells (Gasmi et al. 2011a; 2012; Tiels et al. 2012; YaPing et al. 2017; Celińska et al. 2015). This codon optimization is important, not only for protein translation but also for its correct folding (Angov 2011). Improvement of secretion efficiency (for details, see (Celińska et al. 2018)) and synthesis of proproteins (Liu et al. 2015; Madzak et al. 2000) will further contribute to reduce the metabolic burden. Determination of an optimal growth rate—not always corresponding to the maximum growth rate μ_{max}—during recombinant protein production can also promote a balance between cell growth and protein synthesis, especially when constitutive promoters are considered for recombinant gene expression (Maharajh et al. 2008b). In bioreactor-based processes, inducible promoters exploited in a fed-batch mode allow biomass to grow at a growth rate close to μ_{max} prior to switch to the protein production phase by addition of a specific inducer. Moreover, regulation of the expression level via dose-dependent induction is a promising exploratory path to modulate the metabolic burden. Medium optimization, fed-batch, and improved dissolved oxygen management help ensure sufficient nutrient and oxygen supply to recombinant cells facing metabolic burden.

16.3 Molecular Strategies

In order to maximize the recombinant protein productivity, the construction of the most suited cell factory is one of the key steps. Several parameters should be considered such as the host strain and the expression system. These points are discussed here below.

16.3.1 Host Strains

The most popular strains used for recombinant protein production are summarized in Table 16.3. The construction of the Po1 strain series constitutes a milestone for recombinant protein production in *Y. lipolytica*. These strains are unable to synthetize AEP and/or AXP extracellular proteases, which is of importance when producing secreted recombinant proteins. They can use sucrose as a carbon source due to the synthesis of recombinant invertase encoded by *SUC2* gene for *S. cerevisiae* (for details, see Madzak et al. (2000); Le Dall et al. (1994)). Po1g was fitted with an integrated docking platform for site-directed gene cloning. Other strains have been additionally deleted for lipase-encoding genes: *LIP2* in strain MTLY50 (Fickers et al. 2003) and *LIP2*, *LIP7*, and *LIP8* in strain JMY1212 (Bordes et al. 2007; Fickers et al. 2005b). Strains with decreased hyperglycosylation and glycosylation patterns similar to mammalian ones have been obtained by disruption (De Pourcq et al. 2012a, b; Jaafar et al. 2003; Park et al. 2008; Song et al. 2007) or overexpression (De Pourcq et al. 2012b) of specific genes involved in these pathways. These

Table 16.3 Main strains used for recombinant protein production

Name	Genotype	References
Po1d	*MatA, leu2-270, ura3-302::SUC2, xpr2-322*	Le Dall et al. (1994)
Po1e	*MatA, leu2-270, ura3-302::SUC2-URA3-pBR*[a]*, xpr2-322,*	Madzak et al. (2000)
Po1f	*MatA, leu2-270, ura3-302::SUC2, xpr2-322, axp12*	Madzak et al. (2000)
Po1g	*MatA, leu2-270, ura3-302::SUC2-URA3-pBR*[a]*, xpr2-322, axp1-2,*	Madzak et al. (2000)
Po1h	*MatA, ura3-302, xpr2-322, axp1-2*	Madzak (2003)
MTLY50	*MatA, leu2-270, ura3-302, xpr2-322, Δlip2*	Fickers et al. (2003)
JMY1212	*MatA, ura3-302, leu2-270-LEU2-zeta*[b]*, xpr2-322, Δlip2, Δlip7, Δlip8*	Bordes et al. (2007)
JMY2394	*MatA, leu2-270, ura3-302::SUC2, xpr2-322, Δku70*	Verbeke et al. (2013)
RIY176	*MatA, leu2-270, ura3-302::SUC2, xpr2-322, Δeyk1*	Trassaert et al. (2017)

[a]pBR322 docking platform
[b]zeta docking platform

genetic modifications have been gathered in the glyco-engineered strains patented by Universiteit Gent (Belgium), Vlaams Instituut voor Biotechnologie (Belgium), and Oxyrane UK Ltd. (Manchester, United Kingdom) in 2011 (Callewaert et al. 2009). Some of the above host strains have been made commercially available in an expression kit from Yeastern Biotech (Taipei, Taiwan) (for details, see Madzak et al. (2000); Nicaud et al. (2002)).

16.3.2 Expression Plasmids

Both episomal and integrative plasmids have been developed for gene expression. In *Y. lipolytica*, episomal vectors are avoided for industrial applications. They do not yield higher protein production titer in bioreactor as compared to integrative counterpart (Chang et al. 1998b). Moreover, they request a selection pressure (antibiotic or auxotrophy) for maintenance in cell that quickly becomes cumbersome at large-scale and/or during prolonged cultivations (Chang et al. 1998b, Celińska et al. 2015). For integrative plasmids, several integration options are available: random zeta integration, site-directed integration at pBR322, or zeta docking platform. Moreover, markerless integration methods based on the CRISPR-Cas9 technology have recently been developed (Borsenberger et al. 2018; Gao et al. 2016), Δ*ku70* strain for improved homologous recombination has been constructed (Verbeke et al. 2013), and intergenic sites with high gene expression levels for efficient targeted integration have been identified (Holkenbrink et al. 2018).

Expression vectors could be integrated as mono- or multi-copies. Multi-copy integration is based on the utilization of the defective allele *ura3d4* of *URA3* gene encoding leucine aminopeptidase (Le Dall et al. 1994). In the reported reactor-scale productions, mono-copy integration remains predominant. Multi-copy integration leads to genetic instability that globally increases with the number of copies of integrated expression cassette. This instability is further influenced by the culture conditions, recipient strain, integration locus, and individual cell behavior (Le Dall et al. 1994; Juretzek et al. 2001). However, multi-copy integration has been demonstrated as a rewarding strategy to improve process productivity. Most of the time, the gain in protein production is balanced by lower cell growth ability, reflecting the metabolic burden discussed earlier.

This pattern was later confirmed for fed-batch endo-1,4-β-mannanase production under the control of *pLIP2* promoter (see below). Multi-copy strains (eight and nine copies) presented a lower biomass yield, higher volumetric, and specific glucose uptake rates and an eight- to tenfold increased enzyme production titer (Roth et al. 2009). YaPing and colleagues constructed mono- and six-copy strains to produce in batch bioreactor two mannanases, namely, manA from *Aspergillus niger* and manB from *Bacillus subtilis*. Compared to mono-copy integration, multi-copy strategy resulted in 3.5-fold and 4.5-fold increased enzyme yield and productivity for manA and manB, respectively. By contrast, biomass yield was reduced by a 2.2- and 1.9-fold, respectively (YaPing et al. 2017). The only report on a positive effect of multi-copy integration is for *SUC2* gene in sucrose-based batch. It resulted in a higher growth rate (0.132 versus 0.096 h^{-1}) along with an improved enzyme yield (1991 versus 597 $U.L^{-1}$) and productivity (27.65 versus 8.29 $U.L^{-1}.h^{-1}$). This trend is most probably related to the increase of sucrose uptake rate observed with the increase of copy number of *SUC2* expression cassette (Lazar et al. 2013). Further developments of multi-copy strategies at large scale require overcoming copy number instability and define the clearest proportional relationship possible between copy number and expression intensification.

16.3.3 Promoters

Promoter of *AEP* gene encoding alkaline extracellular protease, namely, *pXPR2*, was developed in the first attempts to control recombinant gene expression in *Y. lipolytica* (Hamsa and Chattoo 1994; Heslot et al. 1990; Nicaud et al. 1989; Tharaud et al. 1992) and has been patented by Pfizer Inc. in 1993 (James and Strick 1998). *pXPR2* is strongly induced by high peptone concentrations at a pH above 6 (Ogrydziak and Scharf 1982). These specific conditions for induction led to the design of pioneer fed-batch strategies, detailed hereafter. They also fostered the development of more polyvalent promoters, not limited to peptone-containing media (Juretzek et al. 2001). p*hp4d* is a synthetic promoter combining four copies of the upstream activating sequences (UAS) of *XPR2* gene promoter ($UAS1_{XPR2}$) linked to the core element *LEU2* promoter (Madzak et al. 2000). Although induction levels obtained with this promoter were high, *php4d*-driven expression was found growth phase-dependent, a feature less compatible with bioreactor-based process (Nicaud et al. 2002). By contrast, promoter induction of *TEF* gene encoding translation elongation factor-1α (*pTEF*) was found uncoupled from cell growth and yielded induction level comparable to that of p*hp4d* (Müller and Dalbøge 2001). Tunable promoters were also developed based on *LEU2* or *TEF* core element and $UAS1_{XPR2}$ element. For these promoters, it has been demonstrated that increasing the copy number of UAS element (up to 32 copies) increases the strength of induction (Blazeck et al. 2011). This synthetic promoter design was then extended to UAS and core elements of other promoters (Blazeck et al. 2011; Blazeck et al. 2013; Dulermo et al. 2017; Shabbir Hussain et al. 2016). However, the utilization of strong constitutive promoter frequently exhausts the host cells (Glick 1995). As a consequence, regulated (i.e., inducible) promoters are often preferred for recombinant protein production in bioreactor process (Vieira Gomes et al. 2018) since they permit to dissociate the growth and protein production phases and thereby alleviate the metabolic burden generated by recombinant gene expression.

So far, well-established regulated promoters are those responding to hydrophobic substrates such as fatty acids, triglycerides, or alkanes (Nicaud et al. 2002; Juretzek et al. 2000). Applications based on these promoters, namely, *pPOX2* (promoter of acyl-CoA oxidase 2) and *pLIP2* (promoter of lipase Lip2p), reached the reactor scale. *pPOX2* induced by oleic acid was used to express fungal lipase genes

(Aloulou et al. 2007; Emond et al. 2010) and human proteins (Gasmi et al. 2011a, b, 2012; Tiels et al. 2012), in both complex and defined media. *pLIP2* has been limited so far to the production of recombinant lipase Lip2p (Fickers et al. 2005c), and its expression was found more sensitive to environmental conditions than *pPOX2*. It is influenced by nature and concentration of nitrogen nutrients and pH of the culture medium. Higher protein yields were obtained at pH 6 over pH 3 in defined medium (Sassi et al. 2017), or in the presence of tryptone N1 in complex medium (Fickers et al. 2004, 2005c). Host strains deleted for *Y. lipolytica* main lipase genes, namely, *LIP2*, *LIP7,* and *LIP8* (Fickers et al. 2003, 2011; Bordes et al. 2007), allow to clearly separate the growth phase from the protein production phase. Besides, disruption of lipase-encoding genes prevents host cell diversion from recombinant expression and avoids the presence of undesired proteins (i.e., lipases) in the culture supernatant (Gasmi et al. 2012; Emond et al. 2010). Notwithstanding suboptimal strain background, both *pPOX2* and *pLIP2* delivered high expression levels in reported reactor studies. *pLIP2* appeared more efficient than *pPOX2* according to flask experiments (Fickers et al. 2005c; Sassi et al. 2016). The main drawback of the utilization of hydrophobic inducers is their insolubility in aqueous culture medium. They require the formation of an emulsion that becomes difficult to maintain in large-scale bioreactor due to the lower mixing efficiency of the latter. In this regard, co-feeding with a hydrophilic carbon source (e.g., glucose, glycerol) could contribute to solve this issue (Sassi et al. 2016), but this strategy has not been implemented in reactor cultures yet.

Promoters responding to hydrophilic inducers offer an interesting alternative to *pPOX2* and *pLIP2,* especially in large-scale bioreactor. In this perspective, the characterization and functional dissection of upstream regions from genes *EYK1* and *EYD1*, encoding, respectively, an erythrulose kinase and an erythritol dehydrogenase, involved in erythritol catabolism in *Y. lipolytica*, led to the release of a series of engineered promoters with tunable strength (depending on the number of UAS). These promoters are inducible by erythritol and erythrulose and repressed by glycerol and glucose (Trassaert et al. 2017; Park et al. 2019). With these patented promoters (Nicaud et al. 2018), erythritol and erythrulose can be used as an inducer and a carbon source in wild-type strain, or as a free inducer in *Δeyk1* or *Δeyd1* strains. During the characterization of a *pEYK1*-derived promoter in chemostat bioreactor, the promoter was found to respond to erythritol and erythrulose in a dose-dependent manner (Trassaert et al. 2017), paving the way to a fine regulation of metabolic burden by gene titration.

16.3.4 Dimorphism

Y. lipolytica is a dimorphic yeast that can grow under ovoid, yeast-like to pseudohyphae to septate hyphae morphologies. Effectors of the dimorphic transition include pH, temperature, hydrodynamic stress, osmotic pressure, carbon and nitrogen sources, and nutrient or oxygen starvation (for a review, see Timoumi et al. (2018)).

Their influence has been shown to depend on the considered strain (Pérez-Campo and Domínguez 2001). Dimorphic transition is the reflection of important physiological modifications that may lower process outcomes, such as carbon source uptake (Palande et al. 2014; Zinjarde et al. 1998) or protein productivity (Gasmi et al. 2011a; Fickers et al. 2009; Madzak et al. 2005). Recombinant protein productivity was found dependent on the strain morphotype. Lipase Lip2p was produced at higher rate with hyphae (Fickers et al. 2009), while ovoid morphotype yielded to higher interferon and laccase production (Gasmi et al. 2011a; Madzak et al. 2005).

Currently, *Y. lipolytica* morphological state is not efficiently mastered in bioreactor. Individual cell behavior divergence, naturally occurring within a cell population, results in phenotypic heterogeneity, i.e., a broad distribution of cell sizes and shapes. This heterogeneity impacts the rheological properties of the culture broth, influencing heat, and mass transfer phenomena (Coelho et al. 2010; Kraiem et al. 2013). Therefore, monitoring and control of cell morphology are key parameters for process optimization and reproducibility. With the emergence of flow cytometry, *Y. lipolytica* cell shape distribution can be rapidly evaluated (Sassi et al. 2017; Timoumi et al. 2017a; Timoumi et al. 2017b) and even monitored online (Bouchedja et al. 2017). From there, automated actuation could be implemented to control the morphological heterogeneity of a cell population (Delvigne et al. 2014). A cost-effective solution to master *Y. lipolytica* morphology in bioreactor would consist in the utilization of host strains blocked in a defined morphotype, as recently suggested for improvement of fatty acid production (Xie 2017). Besides environmental effectors, several molecular mechanisms have been found to be involved in the dimorphic transition (Cervantes-Chávez et al. 2009; Cervantes-Chávez and Ruiz-Herrera 2006; Chang et al. 2007; Hurtado and Rachubinski 1999, 2002a, b; Hurtado et al. 2000; Li et al. 2014; Morales-Vargas et al. 2012; Torres-Guzman and Domínguez 1997). Strains growing constitutively in the yeast-like form are available, namely, *Δhoy1* (Torres-Guzman and Domínguez 1997), *Δmhy1* (Hurtado and Rachubinski 1999), *Δcla4* (Szabo 2001), *Δbem1* (Hurtado and Rachubinski 2002a), *Δylbmh1* (Hurtado and Rachubinski 2002b), and *Δste11* (Cervantes-Chávez and Ruiz-Herrera 2006). Filamentous morphology confers to the cells an increased retention ability to a support inside the bioreactor, opening the way to immobilized-cell technology (Vandermies et al. 2018). This technology has already proved its strength for succinic acid production (Li et al. 2018a, b), and could advantageously be applied to secreted recombinant proteins (Vandermies et al. 2018). With immobilized-cell bioreactors, downstream processing is considerably simplified since cells are retained in the reactor, greatly reducing bioprocess overall cost. Currently identified filamentous mutants, namely, *Δtpk1* (Cervantes-Chávez et al. 2009) and *Δylhls1 (Vandermies et al. 2018)*, can contribute to controlled cell immobilization inside bioreactors.

16.4 Process Strategies

The process strategy is, with the cell factory development step, a milestone in process optimization. How to grow production cells in bioreactor in a way that maximizes productivity and yield?

16.4.1 The Predominance of Fed-Batch

In most cases, bioreactor process optimization concerns feeding strategies, either for carbon source during cell growth or inducer during the protein production phase. Neither starving nor overfed cells are desirable, nor is a high nutrient concentration that could alter physical properties of the culture medium and impair cell growth or recombinant protein synthesis. Exploratory continuous processes in bioreactor (i.e., chemostat) have been reported, aiming to assess the optimal conditions for promoter induction (Chang et al. 1997; Trassaert et al. 2017). Semi-continuous cultures have been proposed to increase process productivity through time saving, for instance, the two-stage cyclic fed-batch process designed by Chang and colleagues for rice α-amylase production under the control of *pXPR2* (Chang et al. 1998a, b). In this process, parts of culture broth were regularly transferred from growth to production medium, while remaining cells in the growth phase were further fed with fresh growth medium. At industrial scale, however, the risks associated with continuous process—i.e., contaminations, strain mutations, and product instability—are too important to be taken, and this type of process is not considered in practice. The only example of *Y. lipolytica*-based continuous process at large scale was for single-cell oil production from petroleum in the 1970s (Barth and Gaillardin 1996).

By contrast to shake-flask, culture in bioreactor offers improved oxygenation conditions and medium homogeneity (Gasmi et al. 2012). For culture scale-up from shake-flask to pilot-scale bioreactors, an increase of three- to fivefold in enzyme productivity has been reported (Maharajh et al. 2008a; YaPing et al. 2017; Nicaud et al. 2002; Darvishi et al. 2017, 2018). For complex nonenzymatic therapeutic proteins, the increase in productivity can be even higher, with a 416-fold increase of human interferon concentration obtained in 5-L batch bioreactor as compared to 250 mL shake-flask culture (Gasmi et al. 2011a). The transition from batch to fed-batch mode drives the process further. In fed-batch processes, cell concentration in the range of 50–100 g of dry cell weight (DCW).L^{-1} is commonly obtained (Chang et al. 1997; Kim et al. 2000; Botes et al. 2008; Maharajh et al. 2008a, b; Roth et al. 2009; YaPing et al. 2017; Nicaud et al. 2002; Celińska et al. 2017), while protein productivity increases by two–threefold (YaPing et al. 2017; van Zyl 2011) to tenfold (Maharajh et al. 2008a; Nicaud et al. 2002).

Most fed-batch processes start with carbon source feeding (Huang et al. 2011; Liu et al. 2015; Kim et al. 2000; Gasmi et al. 2011b; Maharajh et al. 2008a, b; Roth et al. 2009; Nicaud et al. 2002) or complete culture medium feeding (Chang et al. 1997, 1998b; van Zyl 2011; Park et al. 2000). Most of the time, feeding of inducer starts at the end of a batch phase, i.e., when the main carbon source has been depleted (Emond et al. 2010). However, extending nutrient feeding during the recombinant protein production phase leads to higher biomass and protein yields, since the cells are supported in their growth-to-production phase transition (Gasmi et al. 2011b). Besides empirical feeding strategy, exponential feeding models could be used. They allow to fulfill the exact nutritional needs of exponentially growing cells in a bioreactor (Chang et al. 1997, 1998a; b; Gasmi et al. 2011b; Maharajh et al. 2008b; van Zyl 2011). These models aim to control growth rate, to prevent cell overfeeding at the beginning of the culture and nutrient starvation during exponential growth phase, and to avoid negative influence on recombinant gene expression (Maharajh et al. 2008b). Currently, feeding strategy based on one-step feeding (Huang et al. 2011; Kim et al. 2000), constant feed rates (Botes et al. 2008; Nicaud et al. 2002), and feed pulses at fixed times (Förster et al. 2007; Emond et al. 2010; Celińska et al. 2017) are preferred, mainly due to their convenience of implementation, particularly at large scale and even if they deliver suboptimal recombinant protein productivity. Modeling approaches to control process productivity, reviewed recently for *P. pastoris* (Theron et al. 2018), could be applied in the future for *Y. lipolytica* since that genome-scale metabolic models are available for this yeast (Huang et al. 2018; Loira et al. 2012; Mishra et al. 2018; Wei et al. 2017).

In fed-batch mode, inducible promoters are used to dissociate the growth and production phases, in order to alleviate the metabolic burden that recombinant protein synthesis brings on the host cells (Gasmi et al. 2012). In the first instance, biomass is allowed to form a concentration that sustains subsequent protein production (Chang et al. 1997; Gasmi et al. 2011a, b; Park et al. 2000). Then, the inducer is fed to the culture medium and cell growth consequently slows down to the profit of protein synthesis. Intracellular energy fluxes are diverted from biomass formation to recombinant protein synthesis. This strategy was successfully implemented for synthesis of human interferon (Gasmi et al. 2011a, b, 2012; Tiels et al. 2012).

Chang and colleagues used chemostat culture to determine optimal culture conditions that maximize *pXPR2* induction (Chang et al. 1997). Medium containing proteose peptone, and, to a lesser extent, neopeptone positively affected promoter induction, while pH of 6.8 was found optimal. They also defined the optimal ratio of carbon source to inducer to be used. These findings were then consistently applied in the following studies (Chang et al. 1998a, b). Besides, *Y. lipolytica* is characterized as an osmotolerant yeast, meaning that cells could withstand high osmotic pressure. Based on this, a one-step feeding process of proteose peptone (for *pXPR2* induction) and concentrated glycerol solution was used as a simpler feeding strategy (Kim et al. 2000). This strategy led to a 30% increased α-amylase yields. Promoter *pXPR2* has slowly fallen into disuse, mainly because most of the bioreactor processes are based on defined medium. Currently, expression of recombinant gene is based on *pPOX2* and *pLIP2* systems, even at industrial scale (Nicaud et al. 2002). To induce *pPOX2*, several oleic acid feeding strategies have been tested, including one-step addition (Gasmi et al. 2011a, 2012), periodic addition (Gasmi et al. 2011b), continuous feed rate (Gasmi et al. 2011b), or gradually increasing feeding rate (Emond et al. 2010). The amount of oleic acid available per unit of biomass, the modalities of oleic acid feeding, and the residual oleic acid concentration were found to be the main parameters influencing *pPOX2* induction, as high oleic acid levels can be toxic to the cells (Gasmi et al. 2011b). Similar relation between inducer concentration and induction levels has been reported for *pLIP2* induction in the presence of preferred inducers, namely, oleic acid and methyl oleate (Fickers et al. 2004, 2005a). However, more elaborated investigations on feeding strategies remain to be conducted to highlight the optimal conditions of *pLIP2* induction. For both *pPOX2* and *pLIP2*, hydrophobic inducer dispersion in the culture medium is of importance. However, the quality of the resulting emulsion is a technical constraint hampering inducible bioprocesses, particularly at large scale. A strategy to solve this issue consists of setting up a co-feeding strategy to replace part of the hydrophobic substrate by an alternative energy source, without affecting recombinant protein yield. Sassi and colleagues obtained in shake-flask higher induction levels of *pLIP2*-driven gene expression by combining oleic acid with glucose as a supplementary carbon source (Sassi et al. 2016). Finally, the induction of *pEYK1*-derived promoters has so far been only tested through pulse additions of erythritol and erythrulose during chemostat cultures (Trassaert et al. 2017). Implementation of other induction strategies in fed-batch shall unveil the potential of these promoters inducible by hydrophilic substrates regarding recombinant protein production. These are under investigation in our laboratory using the CalB lipase from *Candida antarctica* as a model protein (Park et al., unpublished results).

16.4.2 Culture Medium

Most of the defined media used for recombinant protein production in *Y. lipolytica* contain traces of yeast extract (Botes et al. 2008; YaPing et al. 2017), peptone (Chang et al. 1997, 1998a, b; Förster et al. 2007), or both (Kim et al. 2000). This increases protein productivity (Celińska et al. 2017). The complex media used consist mainly of original or modified versions of Yeast Peptone Dextrose (YPD) and Protein Production Broth (PPB), with the notable exception of a most basic mixture of yeast extract and glucose tested by van Zyl (van Zyl 2011). These media are simple to handle and may lead to higher biomass and protein yields (Celińska et al. 2015), but they suffer from non-defined composition and batch-to-batch variability, explaining why defined media are preferred for controlled applications such as therapeutic protein production (Gasmi et al. 2011a). Since protein-based media can trigger protease gene expression, they must be reserved to host strains devoid of extracellular protease activity (i.e., strain deleted for *AEP* and *AXP* genes).

Medium optimization focuses on selecting the adequate medium formula (Madzak et al. 2000), discriminating the best combination of carbon and nitrogen sources (Fickers et al. 2005c), or adjusting the nature and concentration of a higher number of nutrients to cell needs via design of experiment-based methods. Prior to bioreactor production, Gasmi and colleagues investigated the effect of minimum medium composition and nature of nitrogen source and inducer on cell growth and interferon production (Gasmi et al. 2011a). The concentration of PTM1 salt solution, used in *P. pastoris*-based bioprocesses, was assessed in batch reactor. The effects of each PTM1 component and other vitamins were tested in a Box-Behnken statistical experimental design (for details see (Gasmi et al. 2011a)). Similarly, Darvishi and colleagues employed the Taguchi's experimental design method to determine the optimal concentration of four major medium components (carbon source, nitrogen source, yeast extract, and thiamine) in bioreactor (Darvishi et al. 2017, 2018). Adapting the initial concentration of carbon source may further help shortening the lag phase inherent to the start of a culture (Kim et al. 2000).

Y. lipolytica is able to metabolize a wide variety of carbon sources including hydrophobic (e.g., fatty acids and alkanes) or hydrophilic (e.g., glucose, fructose, organic acids, and alcohols) substrates. These diverse carbon sources can thus be used in bioprocesses, either pure or obtained from industrial co-products. Comprehensive reviews on metabolic engineering for alternative substrate utilization have recently been published (Ledesma-Amaro et al. 2016; Spagnuolo et al. 2018). Taking advantage of low-cost substrates, the production of low-added value compounds is susceptible to reach a break-even point (Fickers et al. 2005a). Higher production yields are generally obtained with the pure compound, in contrast to industrial co-product that contains impurities in most of the cases (Cui et al. 2011; Darvishi et al. 2017, 2018). However, in substrate valorization, upper performance levels are not always targeted, provided that the alternative carbon source contributes to render the process cost-effective. The best-known example of alternative carbon source valorization is raw glycerol, a by-product of the biodiesel industry, cheaper than glucose. It has been used for recombinant protein production in several studies (Chang et al. 1997, 1998a, b; Kim et al. 2000; Trassaert et al. 2017; Celińska et al. 2015; Celińska et al. 2016, 2017; Park et al. 2000). For *pLIP2*-driven expression, industrial residue methyl oleate (Kar et al. 2012; Fickers et al. 2005a) and olive oil (Fickers et al. 2005c) have been used as both carbon source and inducer in bioreactor processes for lipase production. Other potential hydrophobic substrates for recombinant protein production include olive mill wastewater (Papanikolaou et al. 2008), waste cooking oil (Katre et al. 2017), and industrial derivative of tallow (Papanikolaou et al. 2007). All of these have been assessed so far for bioconversion, bioaccumulation, or metabolite production.

Utilization of a specific carbon source may require the overexpression of genes encoding enzymes that perform missing or deficient steps of substrate catabolism. Strain Po1g, a strain optimized for heterologous protein production (Madzak et al. 2000), secretes the invertase Suc2p from *S. cerevisiae*. This strain has been observed to catabolize sucrose from beet molasses for heterologous laccase production at bioreactor scale (Darvishi et al. 2017). In the previous reactor studies, part of the synthesized invertase was found to be associated with the cells (Förster et al. 2007; Lazar et al. 2013), and the phenomenon was amplified with the use of *XPR2* pre-sequence over the native signal peptide (Lazar et al. 2013). Cell-bound enzymatic activity seems to be more profitable for the cells than fully secreted enzymes released in the culture medium. *SUC2* gene expression in Po1g and in the study of Förster and colleagues was controlled by the inducible promoter *pXPR2*. According to the comparative study of Lazar and colleagues, similar biomass and enzyme yields can be obtained under control of the constitutive promoter *pTEF*. More recently, Guo and colleagues used *pTEF* for constitutive intracellular expression of six cellulolytic enzymes. Although resulting yields were lower than in glucose-based medium, pretreated cellulose sustained the production of a significant amount of recombinant lipase in shake-flasks (Guo et al. 2018). Keeping these observations in mind, exploiting engineered strains that have been validated for bioconversion, bioaccumulation, or metabolite production would contribute to reduce the costs of bioreactor operation. Among the possibilities to explore, galactose (Lazar and Gamboa-Meléndez 2015), lactoserum (Fickers et al., unpublished results), raw starch (Ledesma-Amaro et al. 2015), inulin (Cui et al. 2011), and xylose (Li and Alper 2016) appear to be promising options.

16.4.3 Oxygen Supply

Since *Y. lipolytica* is an obligate aerobe, ensuring sufficient oxygen intake is mandatory to minimize hyphal morphology (Zinjarde et al. 1998; Bellou et al. 2014; Ruiz-Herrera and Sentandreu 2002) and to achieve correct process rates and yields (as shown by Kar and colleagues for β-galactosidase production (Kar et al. 2008, 2012)). Fixed agitation (i.e., impeller rotation speed) and aeration (i.e., airflow) represent the simplest operating design for processes that are not highly oxygen demanding. They are easy to set up, especially at large scale, provided that process oxygen requirements are already known and that no unexpected cellular behavior does occur. Otherwise, cells may encounter a severe lack of oxygen (Huang et al. 2011; Darvishi et al. 2017), resulting in the aforementioned consequences. For optimal oxygen transfer rate (OTR) to the culture broth, airflow, and more frequently agitation, is manually or automatedly adjusted. For the latter, system accuracy regarding reaction time and response design is expected, to guarantee a compliant reaction to cell demand. Conversely, in the case of intensively growing cultures, the system could be subject to very rapid stirring speed (Celińska et al. 2016), that both are unrealistic at large scale and generates hydrodynamic stress to the cells (Kar et al. 2011). Oxygen-enriched air (Kar et al. 2012; Chang et al. 1997, 1998a, b; Kim et al. 2000; Park et al. 2000) and pure oxygen spikes (Gasmi et al. 2011b; Tiels et al. 2012) may be implemented to prevent or palliate to agitation and aeration limitations during lab experiments, but these technologies seem barely viable at large scale. Growth rate can be balanced by the choice of the culture medium composition, to comply with the oxygenation limitations of a system (the more culture medium is concentrated in nutrients, the more metabolic activity, and consequently oxygen demand, will be elevated) (Vandermies et al., unpublished results). One may also think to ponder the use of hydrophobic carbon sources in the culture medium, since hydrophobic substances like olive oil seem to limit oxygen transfer (Amaral et al. 2008).

Strategies to overcome oxygen limitations at larger scale rely on process engineering—improvement of oxygen solubility in the culture medium—and genetic engineering, improvement of cellular capacity to mobilize oxygen. To enhance *Y. lipolytica* oxygen uptake rates, Bhave and Chattoo expressed a hemoglobin-encoding gene from *Vitreoscilla stercoraria* (Vhb) and monitored the effects on cell growth in oxygen-limiting and non-limiting conditions (Bhave and Chattoo 2003). *VHB*-expressing cells performed significantly better than wild-type cells under oxygen limitation. In most reported studies on *Y. lipolytica*, the lower accepted DO value is systematically fixed at 20% (Gasmi et al. 2011b; Botes et al. 2008; Maharajh et al. 2008a, b; Roth et al. 2009; YaPing et al. 2017; Celińska et al. 2015; Nicaud et al. 2002; Emond et al. 2010; Darvishi et al. 2018; Celińska et al. 2016, 2017; van Zyl 2011). Oxygen availability exceeding cell requirements is less damageable since *Y. lipolytica* has proficient coping mechanisms (Biryukova et al. 2006; Lopes et al. 2013), opening the way for oxygen solubility forcing in pressurized bioreactors (Lopes et al. 2009, 2013). OTR could also be increased by using biphasic reactors, containing ionic liquids such as perfluorodecalin in addition to the aqueous culture medium (Amaral et al. 2008). This approach resulted in k_La (oxygen transfer coefficient) values increased up to 230% in YPD medium and is conceivable at large scale, especially in strains engineered to tolerate high ionic liquid concentrations (Walker et al. 2018). However, the utilization of these ionic liquids may raise environmental concerns given that perfluorodecalin belongs to the perfluorocarbons, a category of compounds decried by the Kyoto Protocol (United Nations Framework Convention on Climate Change 2008).

16.4.4 pH

Y. lipolytica is able to grow decently under a broad range of pH conditions (Egermeier et al. 2017; Kebabci and Cihangir 2012). However, pH values of culture medium affect recombinant protein production in several ways. Firstly, the synthesis of *Y. lipolytica* extracellular proteases production is pH-dependent. It is triggered at pH 2 to 6 for AXP and pH 6 to 9 for AEP (Young et al. 1996). If the host strain is not deleted for both extracellular proteases, process pH must be regulated around 6 to avoid any damages to secreted recombinant proteins, as reported for human interferon (Gasmi et al. 2011b). pH also contributes to the induction of some promoters. It is now established that full induction of *pXPR2*, the promoter of the gene coding for AEP, requires a pH of 6.8, suboptimal pH values leading to suboptimal recombinant protein yields (Förster et al. 2007). In fed-batch process based on *pXPR2*, pH can be switched to 6.8 to initiate protein production phase, as demonstrated by Kim and colleagues (Kim et al. 2000). p*LIP2* also seems to present a pH-dependent induction pattern in defined medium, with higher induction levels obtained at higher pH (Sassi et al. 2017). By contrary, *pPOX2* does not seem to suffer from pH dependency. It has been used with success between pH 5 (Gasmi et al. 2011a, b, 2012) and 6.8 (Tiels et al. 2012) in bioreactors. This characteristic is still to be assessed for the *pEYK1* and *pEYD1* series of promoters, limited for the moment to pH 6.8 (Trassaert et al. 2017; Park et al. 2019). pH is also known to be an effector of dimorphic transition. Most studies point out a tendency to adopt a yeast-like morphotype at acidic pH, and a hyphal morphotype at pH close to neutrality. Under defined culture conditions, however, cells may adopt another morphological behavior. Indeed, Timoumi and colleagues observed a predominance of filamentous cells at pH 4.5 and 7 versus ovoid cells at pH 5.6 during batch cultures (Timoumi et al. 2017a). Aside from morphological consequences, opting for a low pH value will allow *Y. lipolytica* cells to grow while limiting the risk of bacterial contamination, which is of great industrial interest. Finally, all these considerations are to ponder with pH influence on the recombinant protein, since inadequate pH value may be unfavorable to recombinant protein activity or even detrimental to its stability (Celińska et al. 2015, 2016).

16.4.5 Inoculum and Growth Rate

Surprisingly little attention has been paid to bioreactor inoculum size, and even less on seed cell physiology, despite the fact that these parameters directly influence the cell growth kinetics. Only a few examples were reported in the literature (Gasmi et al. 2011a, 2012; Madzak et al. 2005; Darvishi et al. 2017, 2018), and most case studies simply repeat the same preculture conditions for all experimental replicates without more precision. The same minor attention has been paid to biomass concentration at the beginning of the protein production phase. Most processes just end up the growth phase when the concentration reaches 20–30 g $DCW.L^{-1}$ (Chang et al. 1997, 1998a; Gasmi et al. 2011a; Madzak et al. 2000; Park et al. 2000), which is considered a biomass sufficient to support recombinant protein production as abovementioned. The only reported example of higher biomass is that of hGM-CSF (Gasmi et al. 2012). During the growth phase, glucose feeding was used to increase the final biomass up to 90 g $DCW.L^{-1}$ before starting the protein production phase. This strategy was developed to compensate the deleterious consequences of human protein synthesis on cell fitness. The effect of biomass concentration at the beginning of induction phase was assessed for human interferon production. Higher recombinant protein and biomass yields were obtained when growth phase was ended at 73 g $DCW.L^{-1}$ in comparison with 105 g $DCW.L^{-1}$ (Gasmi et al. 2011b).

The growth rate also affects the process performances. According to reported studies, protein production may be positively (Maharajh et al. 2008a) or negatively (Botes et al. 2008; Madzak et al. 2005; van Zyl 2011) affected by high growth rate.

In the first case, a higher biomass concentration produced higher enzyme overall yield, even if the specific protein production was decreased. In the second case, a lower growth rate resulted in increased volumetric and specific protein production. Madzak and colleagues explained this correlation by possible impaired folding and subsequent degradation of complex proteins at high growth rates (Madzak et al. 2005). An optimal growth rate of 0.1 h^{-1} was determined in chemostat at the beginning of the series of studies on rice α-amylase (Chang et al. 1997), and used consequently for enzyme production (Chang et al. 1998a; Chang et al. 1998b). Ever since, several studies have adopted a value of 0.1 h^{-1}, higher than the lowest reported growth rates (Botes et al. 2008; van Zyl 2011) but lower than the average maximal growth rate (Gasmi et al. 2012; van Zyl 2011; Celińska et al. 2016), however, without further investigation (Gasmi et al. 2011b, 2012; Park et al. 2000).

Arising from inoculum size and growth rate, high cell density cultures are a preferred strategy to optimize recombinant protein volumetric productivity, which is a key parameter of bioprocess cost-effectiveness (Vieira Gomes et al. 2018; Kim et al. 2000; Celińska et al. 2017). High cell density is favored by sufficient nutrient intake and is thereby typically associated with fed-batch setup (Liu et al. 2015; Chang et al. 1997, 1998a, b; Kim et al. 2000; Botes et al. 2008; Maharajh et al. 2008a, b; Roth et al. 2009; YaPing et al. 2017; Celińska et al. 2017; van Zyl 2011). Yet, in specific cases, highest cell density does not result in best protein yields (Madzak et al. 2000); therefore, a rapid assessment of the most adequate option may significantly contribute to process optimization.

16.4.6 Working Volume

For *Y. lipolytica*-based bioprocesses reported so far, the working volume usually ranged from 600 mL to 10 L. These values are both large enough to limit the number of possible replicate experiments and low enough to hinder the exploitability of results for scale-up. To palliate to such issues, alternative bioreactor designs are available. The DASbox® mini-bioreactor platform (Eppendorf, Hamburg, Germany), suited for 60–250 mL working volume, was used for lipid accumulation (Bouchedja et al. 2017) and erythrulose production (Carly et al. 2018). This platform was also run in chemostat mode to characterize the regulation of *pEYK1* promoter (Trassaert et al. 2017). The bioreactor system from 2Mag (Munich, Germany) is suited for 8–15 mL working volume and has been used in our laboratory for culture medium optimization (Vandermies et al., unpublished results). On the other hand, switching to industrial volumes implies modifications in the physical behavior of the culture broth, resulting in heterogeneous micro-environments inside the bioreactor that can impede the bioprocess if not correctly managed (Kar et al. 2011). So far, the only reported recombinant expression at pilot scale concerns the of β-galactosidase used as a reporter gene in a 500-L reactor (Kar et al. 2012).

16.4.7 Analytical and Predictive Bioprocess Models

Despite being scarcely used, bioprocess modeling offers insightful information to decrypt and optimize recombinant protein production process. For synthesis of epoxide hydrolase (EH) by fed-batch process, Maharajh and colleagues tested different glucose feeding rates with an analytical model (Maharajh et al. 2008a). The definition of a second-order polynomial equation allowed to describe biomass evolution and the relationship between specific EH activity and specific glucose feeding rate. In a second study, the effect of exponential glucose feeding rates on EH production was investigated (Maharajh et al. 2008b). Changing from a constant to an exponential feed rate modified the relationships between the different parameters. For both biomass productivity and yield, a direct relationship was established with exponential glucose feed rate. An inverse relationship was detected between specific EH activity and exponential feed rate, and between specific EH activity and biomass yield. The relationships between exponential feed rate and volumetric EH activity and exponential feed rate and volumetric EH productivity corresponded to second-order polynomials. These responses were introduced in Design-Expert 6 software (Stat-Ease, Inc., Minneapolis, MN, USA) to develop an integrated surface response model predicting the optimal exponential glucose feed leading to maximal EH production. Experimental validation of the model confirmed its accuracy (standard deviation was less than 10% between modeled and actual data, except for biomass productivity). Such a methodology could easily be exploited to optimize other fed-batch processes. In another study, Celińska and colleagues sought to unveil the distribution of carbon and nitrogen fluxes during insect α-amylase (SoAMY) fed-batch production (Celińska et al. 2017). As expected, most of the carbon and nitrogen sources were dedicated to biomass formation and SoAMY production, though part of the resources were diverted to metabolite production (namely, erythritol, mannitol, citric acid, and potentially other metabolites not analyzed during the study). Only 45% of carbon and 36% of nitrogen sources were recovered through this simplified analysis. The remaining carbon load was most probably released as CO_2 or incorporated into non-analyzed metabolites, while a high nitrogen load was still present in the culture medium at the end of the process. This kind of calculation shall help optimizing carbon and nitrogen loads in culture medium for recombinant protein production.

16.4.8 Coupling Recombinant Protein Production with Another Bioprocess

Usually, optimization effort is directed to the production of the recombinant protein. It is commonly accepted that, in order to obtain the highest protein yields, metabolite secretion or their accumulation inside the cells must be avoided. However, this challenge could be considered otherwise, by taking advantage of *Y. lipolytica* multipotency to increase the process rentability, rather than minimizing the formation of by-products. Co-production could include metabolite synthesis, or bioaccumulation in the form of single-cell protein or single-cell oil. This lead has been suggested by Celińska and colleagues (Celińska et al. 2017) for SoAMY production. As seen above, SoAMY secretion was accompanied by sizable titers of erythritol, mannitol, and citric acid that could, according to the authors, "significantly contribute to the process economy." This co-production principle is seldom explored but could be applied to recombinant enzymes secreted for alternative substrate valorization. For example, sucrose metabolization for citric acid production (Förster et al. 2007; Lazar et al. 2013) could be coupled with invertase recovery, and inulin metabolization for the formation of single-cell protein (Cui et al. 2011) could be coupled with inulinase recovery.

16.5 Concluding Remarks

As stated in the present review, *Y. lipolytica* can be considered as an efficient cell factory for the production of recombinant proteins. This nonconventional yeast delivers yields comparable to well-established system such as *S. cerevisiae* and *P. pastoris*. By contrast to *S. cerevisiae*, *Y. lipolytica* processes are not hampered by high hyperglycosylation issues or metabolite diversion toward ethanol production in aerobic conditions. Moreover, *Y. lipolytica* expression systems do not require the utilization of harmful and flammable inducers such as methanol in *P. pastoris*.

Recombinant processes using *Y. lipolytica* rely on strong advantages. Numerous molecular tools are available to optimize recombinant gene expression, including a diversity of promoters; high cell density is easily achievable when coupled with fed-batch mode; and other culture conditions (e.g., complex or defined medium, classical or alternative carbon source, pH value) can be adjusted to considered recombinant protein production, thanks to *Y. lipolytica* versatility.

However, industrial implementation of *Y. lipolytica*-based processes still suffers from technical issues such as high oxygen requirements, possible by-product formation, and uncontrolled dimorphism. Concerning oxygen supply, innovative approaches have been proposed to increase dissolved oxygen levels (i.e., ionic liquids, pressurized reactors) and *Y. lipolytica* oxygen uptake rate (i.e., *Vitreoscilla stercoraria* hemoglobin expression). In further developments, *Y. lipolytica* weaknesses regarding morphology and by-product formation shall be used as strengths. Rather than minimizing by-product formation, coupling recombinant protein together with metabolite synthesis or bioaccumulation could participate in bioprocess break-even. Recently, dimorphism has been reported as an asset to develop original reactor setup based on immobilized-cell technology. Currently restricted to succinic acid production, this technology could advantageously be extended to secreted recombinant protein production.

Acknowledgments M. Vandermies is recipient of a fellowship (n°12,688) from the Fonds pour la Formation à la Recherche dans l'Industrie et l'Agriculture (FRIA, Belgium).

References

Aloulou A, Rodriguez JA, Puccinelli D et al (2007) Purification and biochemical characterization of the LIP2 lipase from *Yarrowia lipolytica*. Biochim Biophys Acta BBA - Mol Cell Biol Lipids 1771:228–237. https://doi.org/10.1016/j.bbalip.2006.12.006

Amaral PFF, Freire MG, Rocha-Leão MHM et al (2008) Optimization of oxygen mass transfer in a multiphase bioreactor with perfluorodecalin as a second liquid phase. Biotechnol Bioeng 99:588–598. https://doi.org/10.1002/bit.21640

Angov E (2011) Codon usage: Nature's roadmap to expression and folding of proteins. Biotechnol J 6:650–659. https://doi.org/10.1002/biot.201000332

Babour A, Kabani M, Boisramé A, Beckerich J-M (2008) Characterization of Ire1 in the yeast *Yarrowia lipolytica* reveals an important role for the Sls1 nucleotide exchange factor in unfolded protein response regulation. Curr Genet 53:337. https://doi.org/10.1007/s00294-008-0190-1

Barth G, Gaillardin C (1996) *Yarrowia lipolytica*. In: Nonconventional Yeasts in Biotechnology. Springer, Berlin/Heidelberg, pp 313–388

Bellou S, Makri A, Triantaphyllidou I-E et al (2014) Morphological and metabolic shifts of *Yarrowia lipolytica* induced by alteration of the dissolved oxygen concentration in the growth environment. Microbiology 160:807–817. https://doi.org/10.1099/mic.0.074302-0

Bhave SL, Chattoo BB (2003) Expression of *Vitreoscilla* hemoglobin improves growth and levels of extracellular enzyme in *Yarrowia lipolytica*. Biotechnol Bioeng 84:658–666. https://doi.org/10.1002/bit.10817

Biryukova EN, Medentsev AG, Arinbasarova AY, Akimenko VK (2006) Tolerance of the yeast *Yarrowia lipolytica* to oxidative stress. Microbiology 75:243–247. https://doi.org/10.1134/S0026261706030015

Blazeck J, Liu L, Redden H, Alper H (2011) Tuning gene expression in *Yarrowia lipolytica* by a hybrid promoter approach. Appl Environ Microbiol 77:7905–7914. https://doi.org/10.1128/AEM.05763-11

Blazeck J, Reed B, Garg R et al (2013) Generalizing a hybrid synthetic promoter approach in *Yarrowia lipolytica*. Appl Microbiol Biotechnol 97:3037–3052. https://doi.org/10.1007/s00253-012-4421-5

Bordes F, Fudalej F, Dossat V et al (2007) A new recombinant protein expression system for high-throughput screening in the yeast *Yarrowia lipolytica*. J Microbiol Methods 70:493–502. https://doi.org/10.1016/j.mimet.2007.06.008

Borsenberger V, Onésime D, Lestrade D et al (2018) Multiple parameters drive the efficiency of CRISPR/Cas9-induced gene modifications in *Yarrowia lipolytica*. J Mol Biol 430:4293–4306. https://doi.org/10.1016/j.jmb.2018.08.024

Botes AL, Labuschagne M, Roth R, et al (2008) Recombinant Yeasts for Synthesizing Epoxide Hydrolases. U.S. patent, US 20080171359 A1

Bouchedja DN, Danthine S, Kar T et al (2017) Online flow cytometry, an interesting investigation process for monitoring lipid accumulation, dimorphism, and cells' growth in the oleaginous yeast *Yarrowia lipolytica* JMY 775. Bioresour Bioprocess 4:3. https://doi.org/10.1186/s40643-016-0132-6

Callewaert NLM, Vervecken W, Pourcq KJMD, et al (2009) *Yarrowia lipolytica* and *Pichia pastoris HAC1* nucleic acids U.S. Patent, US20090069232A1.

Carly F, Steels S, Telek S et al (2018) Identification and characterization of *EYD1*, encoding an erythritol dehydrogenase in *Yarrowia lipolytica* and its application to bioconvert erythritol into erythrulose. Bioresour Technol 247:963–969. https://doi.org/10.1016/j.biortech.2017.09.168

Celińska E, Nicaud J-M (2018) Filamentous fungi-like secretory pathway strayed in a yeast system: peculiarities of *Yarrowia lipolytica* secretory pathway underlying its extraordinary performance. Appl Microbiol Biotechnol 103(1):39–52. https://doi.org/10.1007/s00253-018-9450-2

Celińska E, Białas W, Borkowska M, Grajek W (2015) Cloning, expression, and purification of insect (*Sitophilus oryzae*) alpha-amylase, able to digest granular starch, in *Yarrowia lipolytica* host. Appl Microbiol Biotechnol 99:2727–2739. https://doi.org/10.1007/s00253-014-6314-2

Celińska E, Borkowska M, Białas W (2016) Evaluation of heterologous α-amylase production in two expression platforms dedicated for *Yarrowia lipolytica*: commercial Po1g–pYLSC (*php4d*) and custom-made A18–pYLTEF (*pTEF*). Yeast 33:165–181. https://doi.org/10.1002/yea.3149

Celińska E, Borkowska M, Białas W (2017) Enhanced production of insect raw-starch-digesting alpha-amylase accompanied by high erythritol synthesis in recombinant *Yarrowia lipolytica* fed-batch cultures at high-cell-densities. Process Biochem 52:78–85. https://doi.org/10.1016/j.procbio.2016.10.022

Celińska E, Borkowska M, Białas W et al (2018) Robust signal peptides for protein secretion in *Yarrowia lipolytica*: identification and characterization of novel secretory tags. Appl Microbiol Biotechnol 102:5221–5233. https://doi.org/10.1007/s00253-018-8966-9

Cervantes-Chávez JA, Ruiz-Herrera J (2006) *STE11* disruption reveals the central role of a *MAPK* pathway in dimorphism and mating in *Yarrowia lipolytica*. FEMS Yeast Res 6:801–815. https://doi.org/10.1111/j.1567-1364.2006.00084.x

Cervantes-Chávez JA, Kronberg F, Passeron S, Ruiz-Herrera J (2009) Regulatory role of the *PKA* pathway in dimorphism and mating in *Yarrowia lipolytica*. Fungal Genet Biol 46:390–399. https://doi.org/10.1016/j.fgb.2009.02.005

Chang CC, Ryu DDY, Park CS, Kim J-Y (1997) Enhancement of rice α-amylase production in recombinant *Yarrowia lipolytica*. J Ferment Bioeng 84:421–427. https://doi.org/10.1016/S0922-338X(97)82002-8

Chang CC, Ryu DDY, Park CS, Kim J-Y (1998a) Improvement of heterologous protein productivity using recombinant *Yarrowia lipolytica* and cyclic fed-batch process strategy. Biotechnol Bioeng 59:379–385. https://doi.org/10.1002/(SICI)1097-0290(19980805)59:3<379::AID-BIT15>3.0.CO;2-9

Chang CC, Ryu DDY, Park CS et al (1998b) Recombinant bioprocess optimization for heterologous protein production using two-stage, cyclic fed-batch culture. Appl Microbiol Biotechnol 49:531–537. https://doi.org/10.1007/s002530051209

Chang J, Fagarasanu A, Rachubinski RA (2007) Peroxisomal Peripheral Membrane Protein YlInp1p Is Required for Peroxisome Inheritance and Influences the Dimorphic Transition in the Yeast *Yarrowia lipolytica*. Eukaryot Cell 6:1528–1537. https://doi.org/10.1128/EC.00185-07

Coelho MAZ, Amaral PFF, Belo I (2010) *Yarrowia lipolytica*: an industrial workhorse. In: Méndez-Vilas A (ed) Current research, technology and education topics in applied microbiology and microbial biotechnology advances. Formatex Research Center, pp 930–940

Corchero JL, Villaverde A (1998) Plasmid maintenance in *Escherichia coli* recombinant cultures is dramatically, steadily, and specifically influenced by features of the encoded proteins. Biotechnol Bioeng 58:625–632. https://doi.org/10.1002/(SICI)1097-0290(19980620)58:6<625::AID-BIT8>3.0.CO;2-K

Cui W, Wang Q, Zhang F et al (2011) Direct conversion of inulin into single cell protein by the engineered *Yarrowia lipolytica* carrying inulinase gene. Process Biochem 46:1442–1448. https://doi.org/10.1016/j.procbio.2011.03.017

Darvishi F, Moradi M, Madzak C, Jolivalt C (2017) Production of Laccase by Recombinant *Yarrowia lipolytica* from Molasses: Bioprocess Development Using Statistical Modeling and Increase Productivity in Shake-Flask and Bioreactor Cultures. Appl Biochem Biotechnol 181:1228–1239. https://doi.org/10.1007/s12010-016-2280-8

Darvishi F, Moradi M, Jolivalt C, Madzak C (2018) Laccase production from sucrose by recombinant *Yarrowia lipolytica* and its application to decolorization of environmental pollutant dyes. Ecotoxicol Environ Saf 165:278–283. https://doi.org/10.1016/j.ecoenv.2018.09.026

De Pourcq K, Vervecken W, Dewerte I et al (2012a) Engineering the yeast *Yarrowia lipolytica* for the production of therapeutic proteins homogeneously glycosylated with Man8GlcNAc2 and Man5GlcNAc2. Microb Cell Factories 11:53. https://doi.org/10.1186/1475-2859-11-53

De Pourcq K, Tiels P, Hecke AV et al (2012b) Engineering *Yarrowia lipolytica* to Produce Glycoproteins Homogeneously Modified with the Universal Man3GlcNAc2 N-Glycan Core. PLOS ONE 7:e39976. https://doi.org/10.1371/journal.pone.0039976

de Ruijter JC, Koskela EV, Nandania J et al (2018) Understanding the metabolic burden of recombinant antibody production in *Saccharomyces cerevisiae* using a quantitative metabolomics approach. Yeast 35:331–341. https://doi.org/10.1002/yea.3298

Delvigne F, Zune Q, Lara AR et al (2014) Metabolic variability in bioprocessing: implications of microbial phenotypic heterogeneity. Trends Biotechnol 32:608–616. https://doi.org/10.1016/j.tibtech.2014.10.002

Dulermo R, Brunel F, Dulermo T et al (2017) Using a vector pool containing variable-strength promoters to optimize protein production in *Yarrowia lipolytica*. Microb Cell Factories 16(1):31. https://doi.org/10.1186/s12934-017-0647-3

Egermeier M, Russmayer H, Sauer M, Marx H (2017) Metabolic flexibility of *Yarrowia lipolytica* growing on glycerol. Front Microbiol 8:49. https://doi.org/10.3389/fmicb.2017.00049

Emond S, Montanier C, Nicaud J-M et al (2010) New efficient recombinant expression system to engineer *Candida antarctica* lipase B. Appl Environ Microbiol 76:2684–2687. https://doi.org/10.1128/AEM.03057-09

Fickers P, Le Dall M, Gaillardin C et al (2003) New disruption cassettes for rapid gene disruption and marker rescue in the yeast *Yarrowia lipolytica*. J Microbiol Methods 55:727–737. https://doi.org/10.1016/j.mimet.2003.07.003

Fickers P, Nicaud JM, Gaillardin C et al (2004) Carbon and nitrogen sources modulate lipase production in the yeast *Yarrowia lipolytica*. J Appl Microbiol 96:742–749. https://doi.org/10.1111/j.1365-2672.2004.02190.x

Fickers P, Destain J, Thonart P (2005a) Methyl oleate modulates *LIP2* expression in the lipolytic yeast *Yarrowia lipolytica*. Biotechnol Lett 27:1751–1754. https://doi.org/10.1007/s10529-005-3547-2

Fickers P, Fudalej F, Dall MTL et al (2005b) Identification and characterisation of *LIP7* and *LIP8* genes encoding two extracellular triacylglycerol lipases in the yeast *Yarrowia lipolytica*. Fungal Genet Biol 42:264–274. https://doi.org/10.1016/j.fgb.2004.12.003

Fickers P, Fudalej F, Nicaud J-M et al (2005c) Selection of new over-producing derivatives for the improvement of extracellular lipase production by the non-conventional yeast *Yarrowia lipolytica*. J Biotechnol 115:379–386. https://doi.org/10.1016/j.jbiotec.2004.09.014

Fickers P, Destain J, Thonart P (2009) Improvement of *Yarrowia lipolytica* lipase production by fed-batch fermentation. J Basic Microbiol 49:212–215. https://doi.org/10.1002/jobm.200800186

Fickers P, Marty A, Nicaud JM (2011) The lipases from *Yarrowia lipolytica*: genetics, production, regulation, biochemical characterization and biotechnological applications. Biotechnol Adv 29:632–644. https://doi.org/10.1016/j.biotechadv.2011.04.005

Förster A, Aurich A, Mauersberger S, Barth G (2007) Citric acid production from sucrose using a recombinant strain of the yeast *Yarrowia lipolytica*. Appl Microbiol Biotechnol 75:1409–1417. https://doi.org/10.1007/s00253-007-0958-0

Gao S, Tong Y, Wen Z et al (2016) Multiplex gene editing of the *Yarrowia lipolytica* genome using the CRISPR-Cas9 system. J Ind Microbiol Biotechnol 43:1085–1093. https://doi.org/10.1007/s10295-016-1789-8

Gasmi N, Ayed A, Nicaud J-M, Kallel H (2011a) Design of an efficient medium for heterologous protein production in *Yarrowia lipolytica*: case of human interferon alpha 2b. Microb Cell Factories 10:38

Gasmi N, Ayed A, Ammar BBH et al (2011b) Development of a cultivation process for the enhancement of human interferon alpha 2b production in the oleaginous yeast, *Yarrowia lipolytica*. Microb Cell Factories 10:90. https://doi.org/10.1186/1475-2859-10-90

Gasmi N, Lassoued R, Ayed A et al (2012) Production and characterization of human granulocyte–macrophage colony-stimulating factor (hGM-CSF) expressed in the oleaginous yeast *Yarrowia lipolytica*. Appl Microbiol Biotechnol 96:89–101. https://doi.org/10.1007/s00253-012-4141-x
Glick BR (1995) Metabolic load and heterologous gene expression. Biotechnol Adv 13:247–261. https://doi.org/10.1016/0734-9750(95)00004-A
Görgens JF, van Zyl WH, Knoetze JH, Hahn-Hägerdal B (2001) The metabolic burden of the *PGK1* and *ADH2* promoter systems for heterologous xylanase production by *Saccharomyces cerevisiae* in defined medium. Biotechnol Bioeng 73:238–245. https://doi.org/10.1002/bit.1056
Groenewald M, Boekhout T, Neuvéglise C et al (2014) *Yarrowia lipolytica*: safety assessment of an oleaginous yeast with a great industrial potential. Crit Rev Microbiol 40:187–206. https://doi.org/10.3109/1040841X.2013.770386
Guo Z, Robin J, Duquesne S et al (2018) Developing cellulolytic *Yarrowia lipolytica* as a platform for the production of valuable products in consolidated bioprocessing of cellulose. Biotechnol Biofuels 11:141. https://doi.org/10.1186/s13068-018-1144-6
Hamsa PV, Chattoo BB (1994) Cloning and growth-regulated expression of the gene encoding the hepatitis B virus middle surface antigen in *Yarrowia lipolytica*. Gene 143:165–170. https://doi.org/10.1016/0378-1119(94)90092-2
Hanquier J, Sorlet Y, Desplancq D et al (2003) A single mutation in the activation site of bovine trypsinogen enhances its accumulation in the fermentation broth of the yeast *Pichia pastoris*. Appl Environ Microbiol 69:1108–1113. https://doi.org/10.1128/AEM.69.2.1108-1113.2003
Heslot H, Nicaud JM, Fabre E et al (1990) Cloning of the alkaline extracellular protease gene of *Yarrowia lipolytica* and its use to express foreign genes. In: Nga BH, Lee YK (eds) Microbiological application in food biotechnology. Elsevier, Amsterdam, pp 27–45
Holkenbrink C, Dam MI, Kildegaard KR et al (2018) EasyCloneYALI: CRISPR/Cas9-based synthetic toolbox for engineering of the yeast *Yarrowia lipolytica*. Biotechnol J 13:1700543. https://doi.org/10.1002/biot.201700543
Huang Y-C, Chen Y-F, Chen C-Y et al (2011) Production of ferulic acid from lignocellulolytic agricultural biomass by *Thermobifida fusca* thermostable esterase produced in *Yarrowia lipolytica* transformant. Bioresour Technol 102:8117–8122. https://doi.org/10.1016/j.biortech.2011.05.062
Huang Y-Y, Jian X-X, Lv Y-B et al (2018) Enhanced squalene biosynthesis in *Yarrowia lipolytica* based on metabolically engineered acetyl-CoA metabolism. J Biotechnol 281:106–114. https://doi.org/10.1016/j.jbiotec.2018.07.001
Hurtado CA, Rachubinski RA (1999) *MHY1* encodes a C2H2-type zinc finger protein that promotes dimorphic transition in the yeast *Yarrowia lipolytica*. J Bacteriol 181:3051–3057
Hurtado CAR, Rachubinski RA (2002a) Isolation and Characterization of *YlBEM1*, a Gene Required for Cell Polarization and Differentiation in the Dimorphic Yeast *Yarrowia lipolytica*. Eukaryot Cell 1:526–537. https://doi.org/10.1128/EC.1.4.526-537.2002
Hurtado CAR, Rachubinski RA (2002b) *YlBMH1* encodes a 14-3-3 protein that promotes filamentous growth in the dimorphic yeast *Yarrowia lipolytica*. Microbiology 148:3725–3735. https://doi.org/10.1099/00221287-148-11-3725
Hurtado CAR, Beckerich J-M, Gaillardin C, Rachubinski RA (2000) A Rac Homolog Is Required for Induction of Hyphal Growth in the Dimorphic Yeast *Yarrowia lipolytica*. J Bacteriol 182:2376–2386. https://doi.org/10.1128/JB.182.9.2376-2386.2000
Jaafar L, León M, Zueco J (2003) Isolation of the *MNN9* gene of *Yarrowia lipolytica* (YlMNN9) and phenotype analysis of a mutant *ylmnn9Δ* strain. Yeast 20:633–644. https://doi.org/10.1002/yea.990
James LC, Strick CA (1998) Multiple integrative vectors and *Yarrowia lipolytica* transformants. U.S. patent, US5786212A
Jolivet P, Bordes F, Fudalej F et al (2007) Analysis of *Yarrowia lipolytica* extracellular lipase Lip2p glycosylation. FEMS Yeast Res 7:1317–1327. https://doi.org/10.1111/j.1567-1364.2007.00293.x

Juretzek T, Wang H-J, Nicaud J-M et al (2000) Comparison of promoters suitable for regulated overexpression of β-galactosidase in the alkane-utilizing yeast*Yarrowia lipolytica*. Biotechnol Bioprocess Eng 5:320–326. https://doi.org/10.1007/BF02942206

Juretzek T, Le Dall M-T, Mauersberger S et al (2001) Vectors for gene expression and amplification in the yeast *Yarrowia lipolytica*. Yeast 18:97–113. https://doi.org/10.1002/1097-0061(20010130)18:2<97::AID-YEA652>3.0.CO;2-U

Kar T, Delvigne F, Masson M et al (2008) Investigation of the effect of different extracellular factors on the lipase production by *Yarrowia lipolytica* on the basis of a scale-down approach. J Ind Microbiol Biotechnol 35:1053–1059. https://doi.org/10.1007/s10295-008-0382-1

Kar T, Delvigne F, Destain J, Thonart P (2011) Bioreactor scale-up and design on the basis of physiologically relevant parameters: application to the production of lipase by *Yarrowia lipolytica*. Biotechnol Agron Soc Env 11

Kar T, Destain J, Thonart P, Delvigne F (2012) Scale-down assessment of the sensitivity of *Yarrowia lipolytica* to oxygen transfer and foam management in bioreactors: investigation of the underlying physiological mechanisms. J Ind Microbiol Biotechnol 39:337–346. https://doi.org/10.1007/s10295-011-1030-8

Katre G, Ajmera N, Zinjarde S, RaviKumar A (2017) Mutants of *Yarrowia lipolytica* NCIM 3589 grown on waste cooking oil as a biofactory for biodiesel production. Microb Cell Factories 16. https://doi.org/10.1186/s12934-017-0790-x

Kebabci Ö, Cihangir N (2012) Comparison of three *Yarrowia lipolytica* strains for lipase production: NBRC 1658, IFO 1195, and a local strain. Turk J Biol 36:15–24

Kim J-W, Park TJ, Ryu DDY, Kim J-Y (2000) High cell density culture of *Yarrowia lipolytica* using a one-step feeding process. Biotechnol Prog 16:657–660. https://doi.org/10.1021/bp000037n

Kopečný D, Pethe C, Šebela M et al (2005) High-level expression and characterization of *Zea mays* cytokinin oxidase/dehydrogenase in *Yarrowia lipolytica*. Biochimie 87:1011–1022. https://doi.org/10.1016/j.biochi.2005.04.006

Kraiem H, Manon Y, Anne-Archard D, Fillaudeau L (2013) In-situ and ex-situ rheometry of high density *Yarrowia lipolytica* broth: determination of critical concentration and impact of yeast mycelial transition. In Proceedings of the 48ème congrès annuel du Groupe Français de Rhéologie-GFR 2013, Nantes, France, 28 October—30 October 2013.

Lazar Z, Gamboa-Meléndez H, Le Coq A-MC-, et al (2015) Awakening the endogenous Leloir pathway for efficient galactose utilization by *Yarrowia lipolytica*. Biotechnol Biofuels 8:185. https://doi.org/10.1186/s13068-015-0370-4

Lazar Z, Rossignol T, Verbeke J et al (2013) Optimized invertase expression and secretion cassette for improving *Yarrowia lipolytica* growth on sucrose for industrial applications. J Ind Microbiol Biotechnol 40:1273–1283. https://doi.org/10.1007/s10295-013-1323-1

Le Dall M-T, Nicaud J-M, Gaillardin C (1994) Multiple-copy integration in the yeast *Yarrowia lipolytica*. Curr Genet 26:38–44. https://doi.org/10.1007/BF00326302

Ledesma-Amaro R, Dulermo T, Nicaud JM (2015) Engineering *Yarrowia lipolytica* to produce biodiesel from raw starch. Biotechnol Biofuels 8:1

Ledesma-Amaro R, Lazar Z, Rakicka M et al (2016) Metabolic engineering of *Yarrowia lipolytica* to produce chemicals and fuels from xylose. Metab Eng 38:115–124. https://doi.org/10.1016/j.ymben.2016.07.001

Li H, Alper HS (2016) Enabling xylose utilization in *Yarrowia lipolytica* for lipid production. Biotechnol J 11:1230–1240. https://doi.org/10.1002/biot.201600210

Li M, Li Y-Q, Zhao X-F, Gao X-D (2014) Roles of the three Ras proteins in the regulation of dimorphic transition in the yeast *Yarrowia lipolytica*. FEMS Yeast Res 14:451–463. https://doi.org/10.1111/1567-1364.12129

Li C, Gao S, Yang X, Lin CSK (2018a) Green and sustainable succinic acid production from crude glycerol by engineered *Yarrowia lipolytica* via agricultural residue based in situ fibrous bed bioreactor. Bioresour Technol 249:612–619. https://doi.org/10.1016/j.biortech.2017.10.011

Li C, Gao S, Li X et al (2018b) Efficient metabolic evolution of engineered *Yarrowia lipolytica* for succinic acid production using a glucose-based medium in an in situ fibrous bioreactor under low-pH condition. Biotechnol Biofuels 11:236. https://doi.org/10.1186/s13068-018-1233-6
Liu S, Wan D, Wang M et al (2015) Overproduction of pro-transglutaminase from *Streptomyces hygroscopicus* in *Yarrowia lipolytica* and its biochemical characterization. BMC Biotechnol 15. https://doi.org/10.1186/s12896-015-0193-1
Loira N, Dulermo T, Nicaud J-M, Sherman DJ (2012) A genome-scale metabolic model of the lipid-accumulating yeast *Yarrowia lipolytica*. BMC Syst Biol 6:35. https://doi.org/10.1186/1752-0509-6-35
Lopes M, Gomes N, Mota M, Belo I (2009) *Yarrowia lipolytica* growth under increased air pressure: influence on enzyme production. Appl Biochem Biotechnol 159:46–53. https://doi.org/10.1007/s12010-008-8359-0
Lopes M, Mota M, Belo I (2013) Comparison of *Yarrowia lipolytica* and *Pichia pastoris* Cellular Response to Different Agents of Oxidative Stress. Appl Biochem Biotechnol 170:448–458. https://doi.org/10.1007/s12010-013-0205-3
Madzak C (2003) New tools for heterologous protein production in the yeast *Yarrowia lipolytica*. In: Pandalai SG (ed) Recent research developments in microbiology. Research Signpost, Trivandrum, pp 453–479
Madzak C, Beckerich J-M (2013) Heterologous protein expression and secretion in *Yarrowia lipolytica*. In: Barth G (ed) *Yarrowia lipolytica*. Springer, Berlin Heidelberg, pp 1–76
Madzak C, Tréton B, Blanchin-Roland S (2000) Strong hybrid promoters and integrative expression/secretion vectors for quasi-constitutive expression of heterologous proteins in the yeast *Yarrowia lipolytica*. J Mol Microbiol Biotechnol 2:207–216
Madzak C, Otterbein L, Chamkha M et al (2005) Heterologous production of a laccase from the basidiomycete *Pycnoporus cinnabarinus* in the dimorphic yeast *Yarrowia lipolytica*. FEMS Yeast Res 5:635–646. https://doi.org/10.1016/j.femsyr.2004.10.009
Maharajh D, Roth R, Lalloo R et al (2008a) Multi-copy expression and fed-batch production of *Rhodotorula araucariae* epoxide hydrolase in *Yarrowia lipolytica*. Appl Microbiol Biotechnol 79:235–244. https://doi.org/10.1007/s00253-008-1420-7
Maharajh D, Lalloo R, Görgens J (2008b) Effect of an exponential feeding regime on the production of *Rhodotorula araucariae* epoxide hydrolase in *Yarrowia lipolytica*. Lett Appl Microbiol 47:520–525. https://doi.org/10.1111/j.1472-765X.2008.02425.x
Mattanovich D, Branduardi P, Dato L et al (2012) Recombinant protein production in yeasts. In: Lorence A (ed) Recombinant gene expression. Humana Press, Totowa, pp 329–358
Mishra P, Lee N-R, Lakshmanan M et al (2018) Genome-scale model-driven strain design for dicarboxylic acid production in *Yarrowia lipolytica*. BMC Syst Biol 12(Suppl 2):12. https://doi.org/10.1186/s12918-018-0542-5
Morales-Vargas AT, Domínguez A, Ruiz-Herrera J (2012) Identification of dimorphism-involved genes of *Yarrowia lipolytica* by means of microarray analysis. Res Microbiol 163:378–387. https://doi.org/10.1016/j.resmic.2012.03.002
Müller S, Dalbøge H (2001) Yeast promoters suitable for expression cloning in yeast and heterologous expression of proteins in yeast. U.S. patent, US6265185.
Nicaud J-M (2012) *Yarrowia lipolytica*. Yeast 29:409–418. https://doi.org/10.1002/yea.2921
Nicaud J-M, Fabre E, Gaillardin C (1989) Expression of invertase activity in *Yarrowia lipolytica* and its use as a selective marker. Curr Genet 16:253–260. https://doi.org/10.1007/BF00422111
Nicaud J, Madzak C, Vandenbroek P et al (2002) Protein expression and secretion in the yeast *Yarrowia lipolytica*. FEMS Yeast Res 2:371–379. https://doi.org/10.1016/S1567-1356(02)00082-X
Nicaud J-M, Trassaert M, Thomas S, et al (2018) Inducible promoter for gene expression and synthetic biology. World patent application, WO2018210867 A1
Ogrydziak DM, Scharf SJ (1982) Alkaline extracellular protease produced by *Saccharomycopsis lipolytica* CX161-1B. J Gen Microbiol 128:1225–1234. https://doi.org/10.1099/00221287-128-6-1225

Oh MH, Cheon SA, Kang HA, Kim J-Y (2010) Functional characterization of the unconventional splicing of *Yarrowia lipolytica HAC1* mRNA induced by unfolded protein response. Yeast 27:443–452. https://doi.org/10.1002/yea.1762

Palande AS, Kulkarni SV, León-Ramirez C et al (2014) Dimorphism and hydrocarbon metabolism in *Yarrowia lipolytica* var. indica. Arch Microbiol 196:545–556. https://doi.org/10.1007/s00203-014-0990-2

Papanikolaou S, Chevalot I, Galiotou-Panayotou M, et al (2007) Industrial derivative of tallow: a promising renewable substrate for microbial lipid, single-cell protein and lipase production by *Yarrowia lipolytica*. Electron J Biotechnol 10:0–0. https://doi.org/10.2225/vol10-issue3-fulltext-8

Papanikolaou S, Galiotou-Panayotou M, Fakas S et al (2008) Citric acid production by *Yarrowia lipolytica* cultivated on olive-mill wastewater-based media. Bioresour Technol 99:2419–2428. https://doi.org/10.1016/j.biortech.2007.05.005

Park CS, Chang CC, Ryu DDY (2000) Expression and high-level secretion of *Trichoderma reesei* Endoglucanase I in *Yarrowia lipolytica*. Appl Biochem Biotechnol 87:1–16. https://doi.org/10.1385/ABAB:87:1:1

Park J-N, Song Y, Kim J-Y, et al (2008) A novel *YlMPO1* gene derived from *Yarrowia lipolytica* and a process for preparing a glycoprotein not being mannosylphosphorylated by using a mutated *Yarrowia lipolytica* in which *YlMPO1* gene is disrupted. World patent application, WO2008136564 A1

Park Y-K, Korpys P, Kubiak M, et al (2019) Engineering the architecture of erythritol-inducible promoters for regulated and enhanced gene expression in *Yarrowia lipolytica*. FEMS Yeast Res 19(1). https://doi.org/10.1093/femsyr/foy105

Pérez-Campo FM, Domínguez A (2001) Factors affecting the morphogenetic switch in *Yarrowia lipolytica*. Curr Microbiol 43:429–433. https://doi.org/10.1007/s002840010333

Roth R, Moodley V, van Zyl P (2009) Heterologous expression and optimized production of an *Aspergillus aculeatus* Endo-1,4-β-mannanase in *Yarrowia lipolytica*. Mol Biotechnol 43:112–120. https://doi.org/10.1007/s12033-009-9187-3

Ruiz-Herrera J, Sentandreu R (2002) Different effectors of dimorphism in *Yarrowia lipolytica*. Arch Microbiol 178:477–483. https://doi.org/10.1007/s00203-002-0478-3

Sassi H, Delvigne F, Kar T et al (2016) Deciphering how LIP2 and POX2 promoters can optimally regulate recombinant protein production in the yeast *Yarrowia lipolytica*. Microb Cell Factories 15(1):159. https://doi.org/10.1186/s12934-016-0558-8

Sassi H, Delvigne F, Kallel H, Fickers P (2017) pH and not cell morphology modulate pLIP2 induction in the dimorphic yeast *Yarrowia lipolytica*. Curr Microbiol 74:413–417. https://doi.org/10.1007/s00284-017-1207-0

Shabbir Hussain M, Gambill L, Smith S, Blenner MA (2016) Engineering promoter architecture in oleaginous yeast *Yarrowia lipolytica*. ACS Synth Biol 5:213–223. https://doi.org/10.1021/acssynbio.5b00100

Song Y, Choi MH, Park J-N et al (2007) Engineering of the yeast *Yarrowia lipolytica* for the production of glycoproteins lacking the outer-chain mannose residues of N-glycans. Appl Environ Microbiol 73:4446–4454. https://doi.org/10.1128/AEM.02058-06

Spagnuolo M, Shabbir Hussain M, Gambill L, Blenner M (2018) Alternative Substrate Metabolism in *Yarrowia lipolytica*. Front Microbiol 9:1077. https://doi.org/10.3389/fmicb.2018.01077

Szabo R (2001) Cla4 protein kinase is essential for filament formation and invasive growth of *Yarrowia lipolytica*. Mol Genet Genomics 265:172–179. https://doi.org/10.1007/s004380000405

Tharaud C, Ribet A-M, Costes C, Gaillardin C (1992) Secretion of human blood coagulation factor XIIIa by the yeast *Yarrowia lipolytica*. Gene 121:111–119. https://doi.org/10.1016/0378-1119(92)90168-O

Theron CW, Berrios J, Delvigne F, Fickers P (2018) Integrating metabolic modeling and population heterogeneity analysis into optimizing recombinant protein production by *Komagataella (Pichia) pastoris*. Appl Microbiol Biotechnol 102:63–80. https://doi.org/10.1007/s00253-017-8612-y

Tiels P, Baranova E, Piens K et al (2012) A bacterial glycosidase enables mannose-6-phosphate modification and improved cellular uptake of yeast-produced recombinant human lysosomal enzymes. Nat Biotechnol 30:1225–1231. https://doi.org/10.1038/nbt.2427

Timoumi A, Cléret M, Bideaux C et al (2017a) Dynamic behavior of *Yarrowia lipolytica* in response to pH perturbations: dependence of the stress response on the culture mode. Appl Microbiol Biotechnol 101:351–366. https://doi.org/10.1007/s00253-016-7856-2

Timoumi A, Bideaux C, Guillouet SE et al (2017b) Influence of oxygen availability on the metabolism and morphology of *Yarrowia lipolytica*: insights into the impact of glucose levels on dimorphism. Appl Microbiol Biotechnol 101:7317–7333. https://doi.org/10.1007/s00253-017-8446-7

Timoumi A, Guillouet SE, Molina-Jouve C et al (2018) Impacts of environmental conditions on product formation and morphology of *Yarrowia lipolytica*. Appl Microbiol Biotechnol 102:3831–3848. https://doi.org/10.1007/s00253-018-8870-3

Torres-Guzman JC, Domínguez A (1997) HOY1, a homeo gene required for hyphal formation in *Yarrowia lipolytica*. Mol Cell Biol 17:6283–6293

Trassaert M, Vandermies M, Carly F et al (2017) New inducible promoter for gene expression and synthetic biology in *Yarrowia lipolytica*. Microb Cell Factories 16(1):141. https://doi.org/10.1186/s12934-017-0755-0

United Nations Framework Convention on Climate Change (2008) Kyoto protocol reference manual on accounting of emissions and assigned amounts. https://unfccc.int/resource/docs/publications/08_unfccc_kp_ref_manual.pdf. Accessed on 2018-12-12 at 08:54

van Rensburg E, den Haan R, Smith J et al (2012) The metabolic burden of cellulase expression by recombinant *Saccharomyces cerevisiae* Y294 in aerobic batch culture. Appl Microbiol Biotechnol 96:197–209. https://doi.org/10.1007/s00253-012-4037-9

van Zyl PJ (2011) A method for producing a polypeptide in *Yarrowia lipolytica*. World patent application, WO2011148339 A1

Vandermies M, Kar T, Carly F et al (2018) *Yarrowia lipolytica* morphological mutant enables lasting in situ immobilization in bioreactor. Appl Microbiol Biotechnol. https://doi.org/10.1007/s00253-018-9006-5

Verbeke J, Beopoulos A, Nicaud J-M (2013) Efficient homologous recombination with short length flanking fragments in *Ku70* deficient *Yarrowia lipolytica* strains. Biotechnol Lett 35:571–576. https://doi.org/10.1007/s10529-012-1107-0

Vieira Gomes AM, Souza Carmo T, Silva Carvalho L et al (2018) Comparison of yeasts as hosts for recombinant protein production. Microorganisms 6:38. https://doi.org/10.3390/microorganisms6020038

Walker C, Ryu S, Trinh CT (2018) Exceptional solvent tolerance in *Yarrowia lipolytica* Is enhanced by sterols. Metab Eng 54:83-95. https://doi.org/10.1101/324681

Wei S, Jian X, Chen J et al (2017) Reconstruction of genome-scale metabolic model of *Yarrowia lipolytica* and its application in overproduction of triacylglycerol. Bioresour Bioprocess 4:51. https://doi.org/10.1186/s40643-017-0180-6

Xie D (2017) Integrating Cellular and Bioprocess Engineering in the Non-Conventional Yeast *Yarrowia lipolytica* for Biodiesel Production: A Review. Front Bioeng Biotechnol 5:65. https://doi.org/10.3389/fbioe.2017.00065

YaPing W, Ben R, Ling Z, Lixin M (2017) High-level expression of two thermophilic β-mannanases in *Yarrowia lipolytica*. Protein Expr Purif 133:1–7. https://doi.org/10.1016/j.pep.2017.02.008

Young TW, Wadeson A, Glover DJ et al (1996) The extracellular acid protease gene of *Yarrowia lipolytica*: sequence and pH-regulated transcription. Microbiology 142:2913–2921. https://doi.org/10.1099/13500872-142-10-2913

Zinjarde SS, Pant A, Deshpande MV (1998) Dimorphic transition in *Yarrowia lipolytica* isolated from oil-polluted sea water. Mycol Res 102:553–558. https://doi.org/10.1017/S0953756297005418

MIX
Papier aus verantwortungsvollen Quellen
Paper from responsible sources
FSC® C105338

If you have any concerns about our products,
you can contact us on
ProductSafety@springernature.com

In case Publisher is established outside the EU,
the EU authorized representative is:
Springer Nature Customer Service Center GmbH
Europaplatz 3, 69115 Heidelberg, Germany

Printed by Libri Plureos GmbH
in Hamburg, Germany